Polarization Effects in Semiconductors

From Ab InitioTheory to Device Applications

Colin Wood · Debdeep Jena
Editors

Polarization Effects in Semiconductors

From Ab InitioTheory to Device Applications

Colin Wood
US Office of Naval Research
Arlington, VA
USA

Debdeep Jena
University of Notre Dame
Notre Dame, IN
USA

Library of Congress Control Number: 2007932418

ISBN 978-0-387-36831-3 e-ISBN 978-0-387- 68319-5

Printed on acid-free paper.

9 8 7 6 5 4 3 2 1

springer.com

Preface

In the last two decades basic semiconductor research has increasingly focussed attention away from cubic crystal III-V and II-VI compounds toward the "wide-bandgap semiconductors" Ga, Al, In binary and ternary nitrides, and silicon carbide. These smaller anion compounds pack more densely in hexagonal crystals, with the consequence of very high spontaneous and deformation induced electrostatic polarization.

The importance and potential functionality of was rapidly recognized in the late 1990s. In 2000 the Multi-disciplinary University Research Initiative (MURI) of "Polarization Effects in Wide Bandgap Semiconductors" was initiated by the Office of Naval Research to accelerate and consolidate understanding, engineering and device application of the extra electro-physical parameter space. The winning program, "Polaris" by teams centered at UC San Diego, and Cornell University, was one of the most productive in the history of the MURI program, and is testament to the value of sponsored collaborative research so ardently defended by the Director of Defense Research and Engineering (DDR&E) Office of the Secretary of Defense (OSD).

Electronic polarization has profound consequences on the electrostatics and electrodynamics of epitaxial films and heterostructures. Polaris team members developed a comprehensive scientific understanding and made many conceptual advances in polarization-related semiconductor physics. As a result, many electronic and optical devices have been significantly improved and novel devices conceived and realized.

This book is an attempt to ensure that the pioneering advances of the Polaris investigators are collected and expanded to allow efficient recognition and understanding of the many new scientific and engineering principles, considerations and applications developed in the 5 year program.

Each chapter addresses aspects of polarization effects from a different perspective, and for different purposes. There is some overlap in the introductory content in several chapters, albeit each with its own unique flavor. The editors decided to retain this format as it serves to make each chapter self-contained, so that readers have the option of perusing them independently without loss of continuity, and to guide readers more effectively to the subtle differences in perspectives.

Arlington, VA *Colin Wood*
Notre Dame, IN *Debdeep Jena*

Contributors

P. Alpay
Materials Science and Engineering Program & Institute of Materials Sciences, University of Connecticut, Storrs, CT 06269 USA, e-mail: p.alpay@ims.uconn.edu

O. Ambacher
Institute of Micro- and Nanotechnologies, Technical University Ilmenau, D-98693 Ilmenau, Germany, e-mail: oliver.ambacher@tu-ilmenau.de

P. M. Asbeck
Department of Electrical and Computer Engineering, University of California, San Diego, CA 92093 USA, e-mail: asbeck@ece.ucsd.edu

J. Bernholc
Center for High Performance Simulation & Department of Physics, North Carolina State University, Raleigh, NC 27695, USA, e-mail: bernholc@ncsu.edu

P. Boguslawski
Institue of Physics, Polish Academy of Sciences, Al. Lotnikow 32/46, 02-668 Warsaw, Poland, e-mail: bogus@ifpan.edu.pl

R. Butté
Laboratory of Advanced Semiconductors for Photonics and Electronics (LASPE), EPFL, Station 3, CH-1015, Lausanne, Switzerland, e-mail: raphael.butte@epfl.ch

M. V. S. Chandrashekhar
Electrical and Computer Engineering, Cornell University, Ithaca, NY 14853, USA, e-mail: mc296@cornell.edu

V. Cimalla
Institute of Micro- and Nanotechnologies, Technical University Ilmenau, D-98693 Ilmenau, Germany, e-mail: volker.cimalla@tu-ilmenau.de

N. Grandjean
Laboratory of Advanced Semiconductors for Photonics and Electronics (LASPE),

EPFL, Station 3, CH-1015, Lausanne, Switzerland,
e-mail: nicolas.grandjean@epfl.ch

D. Jena
Department of Electrical Engineering, University of Notre Dame, Notre Dame, IN 46556 USA, e-mail: djena@nd.edu

G. Koley
Department of Electrical Engineering, University of South Carolina, Columbia, SC 29208, USA, e-mail: koley@engr.sc.edu

J. Leach
Department of Electrical and Computer Engineering, Virginia Commonwealth University, Richmond, VA 23284, USA, e-mail: s2jleach@vcu.edu

J. Mantese
United Technologies Research Center, 411 Silver Lane, MS 129-45, East Hartford, CT 06108, USA, e-mail: mantesjv@utrc.utc.com

H. Morkoç
Department of Electrical and Computer Engineering, Virginia Commonwealth University, Richmond, VA 23284, USA, e-mail: hmorkoc@vcu.edu

J. Singh
Electrical Engineering and Computer Science Dept. University of Michigan, Ann Arbor, MI 48109, USA, e-mail: singh@engin.umich.edu

M. Singh
Electrical Engineering and Computer Science Dept. University of Michigan, Ann Arbor, MI 48109, USA, e-mail: msingh@eecs.umich.edu

M. G. Spencer
Electrical and Computer Engineering Dept. Cornell University, Ithaca, NY 14853, USA, e-mail: spencer@ece.cornell.edu

C. I. Thomas
Electrical and Computer Engineering Dept. Cornell University, Ithaca, NY 14853, USA, e-mail: cit5@cornell.edu

Y.-R. Wu
Assistant Professor, National Taiwan University, Graduate Institute of Photonics and Optoelectronics and Department of Electrical Engineering, MD 617, No 1. Roosevelt Road Sec. 4, Taipei 10617, Taiwan, e-mail: yrwu@cc.ee.ntu.edu.tw

E. T. Yu
Electrical and Computer Engineering Dept. University of California, San Diego, CA 92093 USA e-mail: ety@ece.ucsd.edu

Contents

Theoretical Approach to Polarization Effects in Semiconductors

Piotr Boguslawski and J. Bernholc

1 Introduction

As a rule, investigations of physical effects in solids are motivated by the need of understanding at a fundamental level, which facilitates their effective application in the fabrication of devices. The problem of electrical polarization of piezoelectric, ferroelectric, and pyroelectric solids is no exception. In the last 15 years we have witnessed very intensive investigations of the theory of spontaneous polarization, as well as of the dielectric response of crystals to external perturbations. Our current understanding stems from the development of electronic structure calculations based on first principles, and subsequently from evolution of appropriate theoretical approaches allowing for both a proper definition of polarization and accurate calculations. From the experimental side, much of the impetus came from experimental work devoted to, e.g., GaN-like group-III nitrides, in which internal electric fields of both pyro- and piezoelectric origin are large, determining the properties of quantum structures and devices [1]. Spectacular progress in this area has led to innovative devices described in several chapters of this book.

There are two important issues clarified by first principles calculations in the last two decades. The first achievement was to provide a link between *microscopic* distribution of electrons determined by first-principles calculations and the description based on classical *macroscopic* electrostatics. In particular, calculations have shown how actual charge densities and dipole layers at surfaces and/or interfaces in semiconductor heterostructures look like at the atomic scale, what are their localization and origin, etc. These basic concepts and ingredients of classical electrodynamics developed during the last two centuries are now visualized by first principles theory. Typical results are described in Sections 6 and 7.

The second success is a demonstration of the fact that spontaneous polarization and piezoelectric effects are bulk properties of solids, and thus may be studied by calculations performed for infinite crystals [2, 3]. This subject was first treated in the paper by Martin [4], who showed that the piezoelectric tensor of an insulator is a bulk quantity. In fact, the development of elegant theoretical approaches

[2, 3] has enabled efficient calculations, and the results are in good agreement with experiment. Furthermore, a wealth of new information has been obtained. In this chapter, we briefly summarize the theoretical approaches and illustrate them with appropriate examples.

As an example of the role played by the electric field, we discuss field-induced electromigration of hydrogen in GaN/AlN heterojunctions rather than the impact of electric field on the electronic structure, which was discussed at length in many papers [1] and other chapters of this book.

2 Basic Electrostatics

The distribution of free carriers and the magnitude of electric currents in a semiconductor are determined by electric fields. In a "classical" semiconductor or semiconductor structure, such as a Si-based junction, the field is of external origin. The electric field inside the structure is determined by, e.g., the applied voltage, which is screened by electrons and nuclei of the solid. The situation is more complex in ferroelectrics, pyroelectrics, and piezoelectrics, where there may exist an internal field even in the absence of the external one, due to the presence of non-vanishing electric polarization. Independent of the origin of the fields, electrostatics of macroscopic media introduce basic concepts and relationships briefly recalled below.

In a bulk homogeneous insulator, the electric field $\boldsymbol{E}$, polarization $\boldsymbol{P}$, and the electric displacement $\boldsymbol{D}$ are related by:

$$\boldsymbol{D} = \varepsilon \boldsymbol{E} = \boldsymbol{E} + 4\pi \boldsymbol{P}, \tag{1}$$

see the classical textbook by Jackson [5]. Here ε is the static dielectric constant. According to Poisson's equation, $\boldsymbol{D}$ is determined by the external electric charge density qn:

$$\mathrm{div}\boldsymbol{D} = 4\pi\, qn\,(\mathbf{r}), \tag{2}$$

where q is the elementary (positive) charge. To solve this equation one needs boundary conditions. For a finite solid, boundary conditions at surfaces or at interfaces between layers of a heterostructure play an important role, since they determine discontinuities at the boundaries and thus the magnitudes of fields in the whole system. The continuity condition at the surface reads [5]

$$\Delta \boldsymbol{D} \bullet \mathbf{i} = 4\pi\, \sigma. \tag{3}$$

Here, $\Delta \boldsymbol{D} = \boldsymbol{D}_2 - \boldsymbol{D}_1$, where $\boldsymbol{D}_2$ is outside and $\boldsymbol{D}_1$ inside the solid, and $\mathbf{i}$ is the unit vector normal to the surface. Thus, the change in the normal component of the displacement is determined by the surface charge density σ. It should be stressed that σ is of *external* origin, as it follows from the Poisson's equation, Eq. 2, and does not include polarization charges. For example, σ can be the macroscopic charge density on the plates of a charged capacitor, which generates the electric field. Discontinuity of polarization at the surface or, in general, at an interface between two dielectrics, is related to the polarization surface charge density σ_{pol} by

$$-\Delta \boldsymbol{P} \bullet \mathbf{i} = \sigma_{\text{pol}} \tag{4}$$

Eqs. 3–4 indicate that the field in the *bulk* of an insulator is defined by the *surface* charge. This is the basis of operation of sensors, where the adsorption of a molecule changes the electric characteristics of a device. Moreover, Eqs. 3–4 show that in the absence of external charges at interfaces of, e.g., a superlattice, discontinuities of the field and the polarization are related by $\Delta \boldsymbol{E} \bullet \mathbf{i} = -4\pi\, \Delta \boldsymbol{P} \bullet \mathbf{i}$.

Similarly, the discontinuity of the electrostatic potential ϕ is determined by the dipole layer at the surface (or interface). In analogy with Eq. 3, the discontinuity of the electrostatic potential is given by

$$\Delta\, \phi = 4\pi\, \delta, \tag{5}$$

where δ is the surface dipole density. Clearly, as already indicated, the continuity equations apply both to free surfaces and to interfaces in heterostructures and superlattices. Both are of interest in this review.

To highlight the difference between the situations relevant for the discussions that follow, we show in Fig. 1 three cases of an insulator with and without surface charges. In the first case, two opposite surfaces are charged. This situation holds for a solid placed in a charged capacitor. The surface charge densities are of external origin (capacitor plates), and they induce a non-vanishing displacement field $\boldsymbol{D}$ in the solid in which polarization $\boldsymbol{P}$ and electric field $\boldsymbol{E}$ are parallel. The polarization induced by the external electric field consists of a relative displacement of positive and negative charges inside a solid, which screen (weaken) these fields.

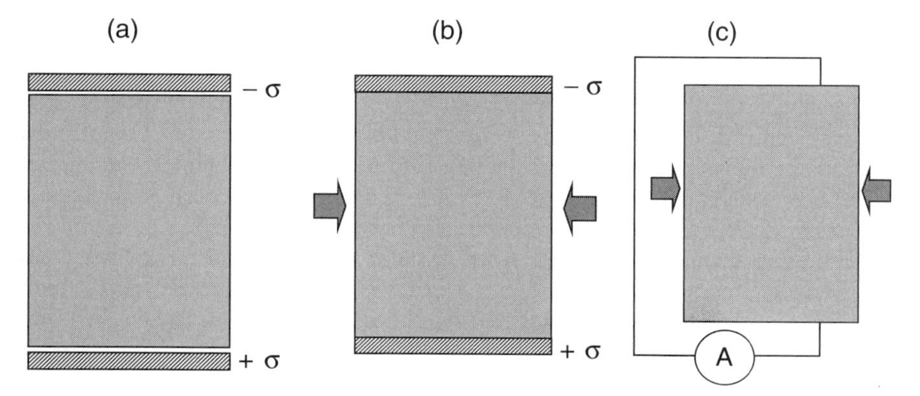

Fig. 1 Surface charges of opposite sign, shown as gray layers, produced by (a) external reservoirs, (b) the piezoelectric effect, with the gray arrows indicating the external pressure, and (c) a strained solid shown in (b) but with short-circuited surfaces, which led to a flow of a current. The configurations shown in (b) and (c) corresponds to, e.g., a zinc-blende crystal strained uniaxially in the [110] direction, which induces polarization oriented in the [001] direction.

The second case looks similar, but now the surface charges are obtained when a piezoelectric is appropriately strained. In this case there is no external charge, thus $\boldsymbol{D} = 0, \boldsymbol{E} = -4\pi \boldsymbol{P}$, i.e., $\boldsymbol{P}$ and $\boldsymbol{E}$ are antiparallel. The surface charge density owns its origin to bulk piezoelectric polarization. By applying stress that produces polarization, an internal electric current is induced, leading to the separation of charge and formation of charged surface layers. The fact that polarization is related to current flow is exploited in the modern approach to theory of polarization, briefly summarized in Sec. 5.

Finally, in the case of a strained piezoelectric, the electric field may be eliminated by short-circuiting the two charged surfaces (i.e., $\boldsymbol{E} = 0$, but $\boldsymbol{P} \neq 0$ and $\boldsymbol{D} = 4\pi \boldsymbol{P}$). This case is of interest for the calculations of polarization presented below, which explicitly assume a vanishing electric field. The strain-induced electric field explains the difference between longitudinal and transverse optical phonons in a polar semiconductor like GaAs: in the former case phonon-induced atomic displacements induce not only the polarization, but also an electric field of opposite sign, which provides an additional restoring force that increases the frequency of the longitudinal phonon relative to the transverse branch.

3 Polarization

In general, polarization of a solid in a finite electric field may be expressed as a sum

$$\boldsymbol{P} = \boldsymbol{P}_0 + \boldsymbol{P}_{el} + \boldsymbol{P}_{lat}, \tag{6}$$

where $\boldsymbol{P}_0$ is the spontaneous polarization of the lattice at equilibrium and zero electric field, while $\boldsymbol{P}_{el}$ and $\boldsymbol{P}_{lat}$ correspond to the electron and lattice contributions. The low symmetry of several types of lattices allows the presence of non-vanishing bulk polarization at zero field. In particular, a crystal with the Wurtzite structure is pyroelectric, and its polarization is oriented parallel to the [0001] direction, i.e., only along the c-axis. It is always present and has the same sign independently of the field. In contrast, in ferroelectrics such as the Perovskites, the orientation of polarization can be reversed by an external electric field. In real samples, the intrinsic polarization is in general screened by free carriers that originate from ionized native defects present in the bulk of the sample, and/or by surface charges.

In the regime of linear response, the two latter components of Eq. (6), which are induced by either electric field or stress, may be expressed as follows. $\boldsymbol{P}_{el}$ is the electronic screening obtained under the condition that the atoms are not displaced from their equilibrium sites,

$$\boldsymbol{P}_{el} = (1/4\pi)(\varepsilon_\infty - 1)\boldsymbol{E}, \tag{7}$$

where ε_∞ is the electron component of the dielectric tensor, i.e., the high-frequency dielectric constant describing the response of the electron gas at frequencies higher than the phonon frequencies, but lower than electronic excitations. The full dielectric response to an external electric field, given by the static dielectric constant introduced in Eq. 1, includes also the lattice contribution, i.e., possible displacements of the nuclei from their equilibrium positions.

$\boldsymbol{P}_{lat}$ is induced by the lattice response and describes the piezoelectric effect, i.e., lattice polarization in vanishing electric field. The stress-induced polarization consists of two terms:

$$\boldsymbol{P}_{lat} = (q/\Omega)Z^{*}\boldsymbol{u} + \kappa\,\mathbf{e}. \tag{8}$$

The first is given by the relative displacements of sublattices in ionic crystals. The ionic displacements $\boldsymbol{u}_i$ relative to the cell center satisfy $\Sigma\boldsymbol{u}_i = 0$. For example, in the zinc blende structure the relative displacement $\boldsymbol{u}$ of cation and anion sublattices change the dipole moment of the unit cell by $Z^{*}\boldsymbol{u}$, where Z^{*} is the Born effective charge and Ω is the unit cell volume. The second term describes polarization induced by strain when the sublattices are not displaced relative to each other, i.e., the atomic coordinates are rescaled by the macroscopic stress, but the electrons are allowed to relax due to the stress-induced changes in distances between atoms. κ is the so-called clamped-ion component of the piezoelectric tensor, and $\mathbf{e}$ is the strain field tensor.

Finally, it is important to observe that the quantity accessible experimentally is not the polarization itself, but rather its changes induced by temperature changes (pyroelectricity), by external pressure (piezoelectricity), or electric field [3]. Similarly, as will be described below, in theoretical calculations polarization of a system is defined and calculated with respect to a non-polar phase.

Before presenting the recently developed advanced theory we illustrate in the next section the basic concepts underlying the macroscopic Eqs. (1–5) with examples of results obtained by microscopic *ab initio* calculations.

4 *Ab Initio* Calculations of the Electronic Structure

From the theoretical point of view, electrostatics of solids poses a question of how to compute the response of a dielectric to two most important external perturbations, i.e., the electric field and pressure. The electric field in a material is determined by the distribution of atoms and their electrons. To support an electric field the material must be an insulator, otherwise the field would be screened by electrons. Calculations of the vibronic, dielectric, piezoelectric, and ferroelectric properties based on density functional perturbation theory are described in Ref. [6]. An alternative approach involving analysis of electron redistribution and the phases of electron wavefunctions leads to a relatively simple, intuitive description of polar materials. It is presented briefly in the next Section. In both cases, the starting point is provided by *ab initio* calculations of the electronic structure based on the density functional theory (DFT).

For a given configuration of atoms, DFT [7] accurately predicts the distribution of electrons through the solution of Schrödinger-like Kohn-Sham equations for the one-electron wave functions ψ_i (in Rydbergs):

$$\left\{-\nabla^2 + V_N(\vec{r}) + \int \frac{2\rho(\vec{r}')d\vec{r}'}{|\vec{r}-\vec{r}'|} + \mu_{xc}[\rho(\vec{r})]\right\} \psi_i(\vec{r}) = \varepsilon_i \psi_i(\vec{r}), \quad i = 1, \ldots M. \tag{9}$$

Here, the first term represents the kinetic energy, the second is the potential due to nuclei (or atomic cores in the pseudopotential approach), the third is the classical electron-electron repulsion potential, and the fourth, the so-called exchange and correlation potential, accounts for the Pauli principle and spin effects. Summing over squares of the M occupied wave functions gives the microscopic electron density ρ.

Semiconductors of interest for this chapter contain either 2 atoms in the unit cell (GaAs or GaN in the zinc-blende (*zb*) phase) or 4 (GaN and AlN in the Wurtzite (*w*) structure). More complex materials, including those containing defects or surfaces, can be modeled with a large, periodically-repeated unit cell, containing tens or hundreds of atoms. This is also the case for superlattices. The methods for solving DFT equations for perfect crystals or large "supercells" are by now quite advanced and several techniques exist for obtaining the wave functions and optimizing the structural parameters of a material by minimizing the total energy of the system [8]. For example, the methods that we use employ real-space grids and multi-grid acceleration techniques [9]. The use of grids leads to effective parallelization over hundreds of processors through domain decomposition, thereby enabling accurate calculations for rather large supercells.

5 Modern Theory of Polarization

The polarization of a solid can be expressed as a sum of ionic and electronic contributions

$$\vec{P} = \vec{P}_{ion} + \vec{P}_{el} = \frac{q}{V_{sample}} \left[\sum_l Z_l \vec{b}_l - \int_{sample} \vec{r}\rho(\vec{r})d\vec{r} \right] \tag{10}$$

where Z_l and $\vec{b}_l$ are the ionic charge and position of the l-th atom in the solid, $\rho(\vec{r})$ is the electron density, and V_{sample} is the volume of the sample. In the pseudopotential approach, Z_l is the charge of the ionic core and only valence electrons are taken into account. However, calculations directly based on this definition were never performed for at least two reasons. First, a straightforward summation over charges would need to include the entire sample and its surfaces, which is prohibitively expensive in terms of computational time. The second and more important reason that motivated an impressive theoretical effort in the last two decades is the following. It is empirically known that polarization (or rather polarization changes, which are actually measured, see Sec. 3) is a bulk property, and therefore independent of the state of surface of the sample, its reconstruction, etc. The bulk nature of polarization of a piezoelectric insulator was first demonstrated by Martin [4]. However, the calculations for bulk systems use periodic unit cells; there are no surfaces and the field must be zero in order to satisfy the periodic boundary conditions. Furthermore, different unit cells can be used to describe a periodic structure and different cells cut through different bonds. A calculation of a dipole moment thus turns out to give different values, depending on what unit cell is used. Only for the case of well-localized electrons, *e.g.*, for negative ions, is the dipole moment per unit cell uniquely defined. This is the well-known Clausius-Mossotti limit, used to relate the dielectric constant to atomic polarizabilities in ionic solids.

It was also recognized that polarization of an infinite crystal is not a uniquely-defined quantity [10] and is in fact multi-valued [11]. Consequently, the physical quantity that is to be evaluated is the *difference* between macroscopic polarizations of a solid in two different states. Typically, one of these states is a non-polar reference state. This approach enables calculations for infinite crystals, where surface effects are absent by construction. Furthermore, comparisons between the polar and non-polar states often provide deep insight into the physics of polarization in a very elegant way. The first example of an approach along this line is presented in Sec. 7 for BeO, where the spontaneous polarization of Wurtzite *w*-BeO was evaluated using the zinc blende *zb*-BeO phase as a reference system with zero polarization.

An important point is that there must exist an adiabatic transformation of the system under consideration from the non-polar to the polar configuration, which keeps the solid insulating. For example, in the case discussed in Sec. 7, one can imagine the construction of a polar *w*-BeO phase from the infinite cubic non-polar *zb*-BeO crystal by generating a suitable sequence of stacking faults. A reference non-polar system may be also an artificial one: in the case of BN polar nanotubes, the non-polar nanotube consisted of artificial atoms that were 50% B and 50% N, and the built-up of polarization was followed by smoothly changing the properties of atoms (computing polarization many times along the way) until the nanotube become a real system with 100% B and N atoms [12].

The details of the theoretical approach are found in the papers by King-Smith and Vanderbilt [2] and by Resta [3]. Below, we outline only the main points. In quantum mechanics, all the information required for an unequivocal computation of polarization is contained in the system's wave functions. However, unlike the electron density $\rho(\vec{r})$, which is the squared *modulus* of the wave function, the polarization is fundamentally related to the *phase* of the wave function [2, 3]. Let's assume that the system is adiabatically transformed from a non-polar "reference" configuration to a polar one. During the transformation, an adiabatic current $\vec{j}$ must flow through the unit cell to account for the charge redistribution and the resulting polarization. The origin of this current, or more generally the intimate relationship of polarization with electric current, may intuitively be understood based on Fig. 1, which shows the piezoelectric effect. Initially, a crystal is unstrained and unpolarized. Application of strain induces both the distortion of the crystal and its polarization, which in turn is displayed by the presence of surface charge densities σ, due to electric currents that have flown through the crystal. This current may be measured when the crystal's surfaces are short-circuited, as shown in Fig. 1c. In a similar way, polarization of w-BeO layers in a w-BeO/zb-BeO superlattice discussed in Sec. 7 and shown in Fig. 3 is related to the 2-dimensional charge densities accumulated at consecutive interfaces. These charge densities screen the polarization field present in the wurtzite part of the superlattice and can be viewed as the result of current flow created during an imaginary transformation of this part from a non-polar zinc-blende structure to the polar wurtzite structure.

Writing the complex wave functions as $|\psi(\vec{r})|\exp\{i\varphi(\vec{r})\}$, the current

$$\vec{j} = i(\psi\nabla\psi^* - \psi^*\nabla\psi) \tag{11}$$

will depend on the phases $\varphi(\vec{r})$ of system's wave functions. The geometric phase (or Berry phase) technique provides an elegant solution for the problem of computing electronic polarization in a periodic system by linking the phase evolution of the system's wave functions to the current flowing during the transformation and thus to the polarization change:

$$\Delta\vec{P}_{el} = \int_0^{\Delta t} \vec{j}(t)dt. \tag{12}$$

The ionic part of polarization $\vec{P}_{ion}$ is computed with a trivial summation over point charges of nuclei (or atomic cores).

More formally, let us assume that the system is adiabatically transformed from a non-polar reference configuration to a polar one. The transition may be parameterized with a variable λ, which changes respectively from 0 to 1. Resta [3] has proposed that the corresponding change in polarization is calculated using

$$\Delta\vec{P} = \int_0^1 (\partial\vec{P}/\partial\lambda)d\lambda \tag{13}$$

where within the LDA

$$\partial P_\alpha/\partial\lambda =$$
$$\frac{ifq\hbar}{N\Omega m_e}\sum_k\sum_{n=1}^{M}\sum_{m=M+1}^{\infty}\frac{<\psi_{kn}^{(\lambda)}\left|\hat{p}_\alpha\right|\psi_{kn}^{(\lambda)}><\psi_{km}^{(\lambda)}\left|\partial V_{KS}^{(\lambda)}/\partial\lambda\right|\psi_{km}^{(\lambda)}>}{(\varepsilon_{kn}^{(\lambda)}-\varepsilon_{km}^{(\lambda)})^2}+c.c. \tag{14}$$

where α is the Cartesian direction, V_{KS} is the self-consistent Kohn-Sham potential, m_e is the electron mass, N is the number of the unit cells in the crystal, M is the number of the occupied bands, ε_{kn} is the eigenenergy of the state k in the n-th band, and f is the occupation number of states in the valence band ($f = 2$ when spin degeneracy in taken into account).

After algebraic transformations described in detail in Ref. [2] one may obtain the electronic polarization per unit cell calculated from the periodic parts of the occupied wave functions of the solid, $\psi_{nk}(\vec{r}) = u_{nk}(\vec{r})e^{i\vec{k}\vec{r}}$, as

$$\vec{P}_{el} = -\frac{2iq}{(2\pi)^3} \sum_{nocc} \int_{BZ} d\vec{k} \langle u_{nk}| \vec{\nabla}_k |u_{nk}\rangle . \tag{15}$$

If we denote by $\vec{G}_i$ the reciprocal lattice vectors defined by $\vec{G}_i \cdot \vec{R}_j = 2\pi m$, where $\vec{R}_j$ belongs to the real-space lattice of all allowable translations of the unit cell, we can introduce the electronic (or Berry) phase for the system as

$$\varphi_{el} = \Omega \vec{G}_i \cdot \vec{P}_{el}/q. \tag{16}$$

By following the change in the Berry phase during the transformation (which must keep the system insulating) we can obtain the difference in polarization. However, because the Berry phase (as a true angular variable) is calculated modulo 2π, the polarization can only be obtained modulo $2q\vec{R}/\Omega$. The difference in total polarization, $\Delta\vec{P}$, will then be well defined if $|\Delta\vec{P}| << |2q\vec{R}/\Omega|$ or, if the change in the Berry phase is accurately monitored along a continuous path, as was done for the case of BN nanotubes, the multiple of $2q\vec{R}/\Omega$ in the Berry phase is also determined. This is important to ensure appropriate cancellation between the electronic and ionic terms, which can separately undergo large changes in their Berry phases [12].

Alternatively, one can transform the periodic wave functions into "bonding" functions, which tend to be localized around individual atoms or bonds in the unit cell. These are the so-called Wannier functions, which form an orthonormal basis and are defined as

$$W_n(\vec{r}) = V_{cell}/(2\pi)^3 \int \psi_{nk}(\vec{r})d\vec{k}, \tag{17}$$

where the integral is over the Brillouin Zone. The electron density is easily expressed in terms of Wannier functions

$$\rho(\vec{r}) = \frac{1}{(2\pi)^3} \sum_{nocc} \int |\psi_{nk}(\vec{r})|^2 d\vec{k} = \sum_{nocc} |W_n(\vec{r})|^2. \tag{18}$$

The Wannier functions lead to an amazingly simple expression for polarization, namely

$$\vec{P}_{el} = -q \int \vec{r}\rho(\vec{r})d\vec{r} = -2q \sum_{nocc} \langle W_n|\vec{r}|W_n\rangle = -2q \sum_{nocc} \vec{r}_n, \tag{19}$$

which involves only the centers of Wannier functions $\vec{r}_n = \langle W_n|\vec{r}|W_n\rangle$. Thus, the difference in polarization, $\Delta\vec{P}$, is obtained by simply treating these centers as point charges that move during the transformation. No ambiguity arises, because the centers of the Wannier functions are uniquely assigned to each unit cell of the crystal. We may thus think of the electronic charge as being localized into point charges $-q$ located at the Wannier centers associated with the occupied states in each unit cell. In this picture, a quantum mechanical system of crystal electrons is mapped onto a

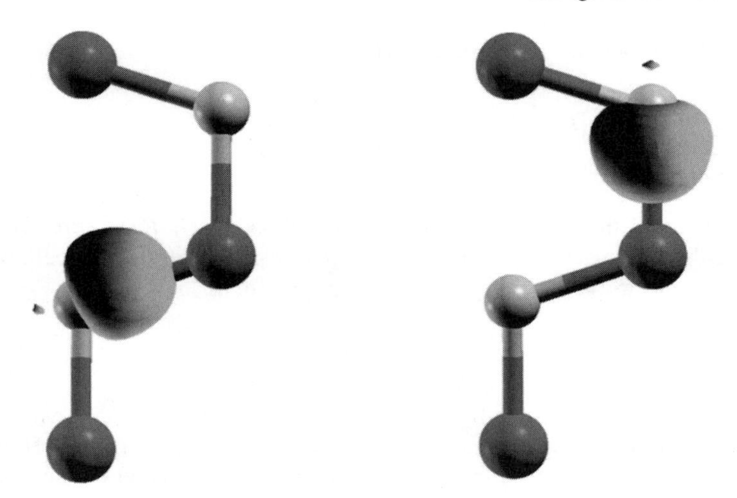

Fig. 2 Examples of "sp^3" Wannier functions associated with the N atoms in AlN. Light and dark spheres represent N and Al atoms, respectively.

classical system of point charges, in analogy with the Clausius-Mossotti limit. As an example, in Fig. 2 we show two of the eight Wannier functions corresponding to the doubly occupied orbitals in the unit cell of *w*-GaN. Several algorithms have been developed to efficiently calculate Wannier functions [13, 14].

6 Polarization at Interfaces: Interface Dipoles

6.1 Averaging Microscopic Charges and Field

The microscopic electric fields in solids, which are local and varying at length scales shorter than the lattice constant, are related to the microscopic charge density of nuclei and electrons by Poisson's equation. The electron density can be calculated from Schrödinger's equation or, when using density functional theory, from the self-consistent Kohn-Sham equations (9). However, in heterojunctions and other systems in which the effective mass description of the free carriers is valid, one is interested in macroscopic fields, which are slowly varying on the scale of a lattice constant.

To use Eqs. 1–5 on the macroscopic scale, it is convenient to average the electron density, the electrostatic potential, and the fields. In practice, the smallest possible and meaningful volume is that of the unit cell. In a heterojunction, the most convenient approach is to first average in the (x, y) plane of the interface, and then average over an appropriate lattice period in the perpendicular direction. One should notice that the average electron density n in Eqs. 1–5 is a macroscopic average of the microscopic charge density of valence electrons $\rho(\boldsymbol{r})$.

6.2 AlAs/GaAs Superlattice

To illustrate the conceptual aspects of linking macroscopic electrostatics with quantum-mechanical calculations, we summarize below the results of a study of electronic structure of interfaces in AlAs/GaAs superlattices [15], a simple but relevant example for the remainder of this chapter. The system studied was a (001)-oriented $(GaAs)_3(AlAs)_3$ superlattice consisting of three layers of each component. In the *zb* structure, the consecutive (001) planes contain only one kind of atoms. Their positions along the [001] axis are shown in the upper part of Fig. 3, and are denoted by Ga, Al, and As, respectively. The authors have used first-principles pseudopotential calculations and Fig. 3 also displays the average valence electron density n and the electrostatic potential energy V for this system.

We first analyze the electron density and begin with the obvious statement that the average electron density in bulk of a III-V compound is 8 electrons per unit cell. As is evident from the figure, in spite of the very short period of the superlattice, the calculated average electron density is 8 in the middle of each layer.

Next, conventional ideal interface planes of this system are placed on As atoms; one of them is in the middle of the figure, at $z = 0$, while the two others are at its edges. In the vicinity of interfaces the electron density differs from 8 by about 0.5 %. More precisely, across each interface the density has a typical dipolar shape and the sign of the dipoles alternates at consecutive interfaces. These results indicate that both the bulk region of each layer and interfaces between them are well defined. This, in turn, is necessary to correctly delineate the interfaces, and properly distinguish between bulk and interface effects. The calculated interface dipole layers, which have zero thickness in classical textbooks, turn out to extend over about 4 atomic monolayers, i.e., about 5 Å, in this particular case. This is indeed negligible in macroscopic samples. The dipoles are formed due to charge transfer between the nearest Ga-As and Al-As bonds only at the interface, which explains their strong spatial localization. Very similar results are found for (110)- and (111)-oriented superlattices [15].

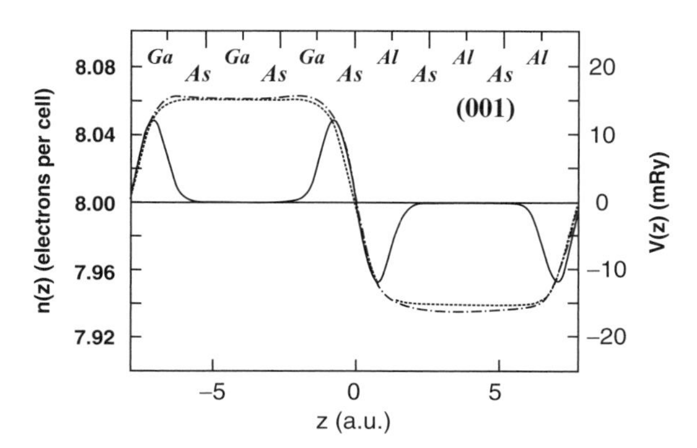

Fig. 3 Macroscopic averages of electron density n (full line) and potential energy V (dashed-dotted line) of (001)-oriented $(GaAs)_3(AlAs)_3$ superlattice. See Ref. [15] for details. Figure reprinted with permission from Ref. [15]. Copyright (1988) by the American Physical Society.

In agreement with Eq. 4, the interface dipoles induce a difference in electrostatic energy, which is 0.41 eV. This energy is constant inside the GaAs and AlAs layers, which indicates that the electric field vanishes, as expected. It is the electrostatic contribution to the band offset in a GaAs/AlAs heterojunction.

7 Spontaneous Polarization in the Wurtzite Structure: BeO

The first calculations of spontaneous polarization in a crystal with the wurtzite symmetry, BeO, were performed by Posternak *et al.* [16]. They have used an infinite superlattice (SL) consisting of alternating layers of BeO in two phases, namely *w* and *zb*, shown in the upper panel of Fig. 4. The essential idea underlying this approach is that the *zb* phase has no intrinsic polarization due to its high symmetry, and thus the polarization in this system is entirely due to the spontaneous polarization of the wurtzite phase. Thus, the *zb* phase plays here the role of the non-polarized reference system, as mentioned earlier.

The results are shown in Fig. 4. The lower panel displays the averaged valence electron density and the potential energy. The average electron density n is practically constant in the middle of each layer and significant deviations from the average value only occur close to interfaces. Therefore, similar to the case of

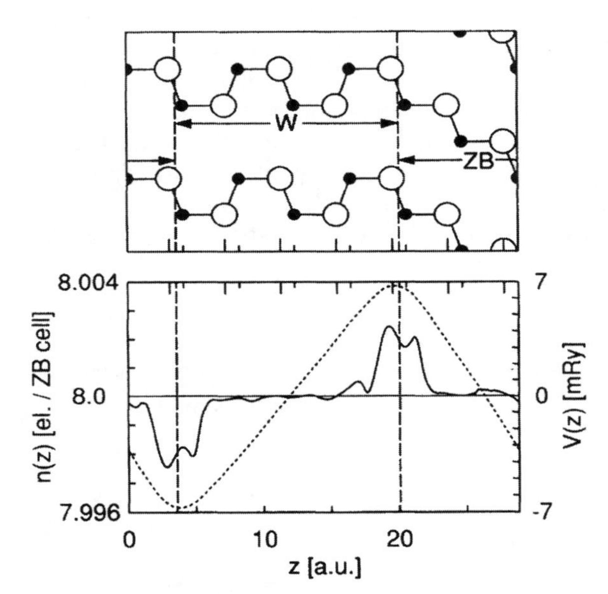

Fig. 4 Upper panel: configuration of atoms in the large unit cell (i.e., one period of the SL). Solid and open circles represent Be and O atoms, respectively. With the usual conventions for the *w* structure, the positive *c*-axis direction points towards the right. Vertical dashed lines indicate the conventional interfaces between wurtzite and zinc blende layers. Lower panel: macroscopic averages of the electron density n (solid line) and of the electrostatic potential energy V (dotted line) in the case of ideal wurtzite geometry. Figure reprinted with permission from Ref. [16]. Copyright (1990) by the American Physical Society.

interface dipoles in GaAs/AsAs SL, interfacial monopoles are strongly localized and limited to about 2 monolayers, i.e., 4 Å. As expected, charge accumulation and depletion regions alternate at consecutive interfaces, generating an electrostatic potential via the Poisson's equation. Since the crystal is neutral, the formation of interface monopoles is due to charge transfer between interfaces and they have equal magnitudes and alternating signs.

Except for the regions at interfaces, the potential varies linearly in space, which corresponds to a constant electric field. Since polarization in the *zb* phase vanishes by symmetry, the polarization present in the system is entirely due to the layers of the wurtzite phase and has been calculated from the slope shown in Fig. 4. Finally, one should note the absence of interface dipoles, since at both sides of each interface there are the same Be-O bonds, and atoms in both *zb* and *w* structures have the same coordination of nearest neighbors.

The monopole contribution, i.e., the interface charge density, is related to the difference between bulk polarizations by the continuity equation, Eq. 4. In a general case, where *zb*-BeO and *w*-BeO layers are of different widths, Eq. 4 may be written in the form

$$\sigma_{\text{int}} = (P_A - P_B)(\iota_\text{A} + \iota_\text{B})/(\iota_\text{A}\,\varepsilon_\text{B} + \iota_\text{B}\,\varepsilon_\text{A}), \tag{20}$$

where ι is the width of a given layer, and ε its static dielectric constant. In particular, when $\iota_\text{A} \neq \iota_\text{B}$ electric fields also differ.

As was mentioned earlier, the quantity of interest is the average electric field defined in Sec. 6 rather than the polarization. Electric fields are determined not only by the local charge distribution but also by the boundary conditions. A standard approach to compute the electronic structure of a semiconductor heterostructure is to use an appropriate superlattice. In this case, the use of periodic boundary conditions requires that the electrostatic potential is periodic, which results in electric fields in the two different parts of the superlattice having opposite signs, with a vanishing average over the full superlattice period. Moreover, the vanishing average electric field corresponds to short-circuiting of the two free surfaces in a finite sample.

We end this Section by mentioning that in the case of ideal *zb* and *w* layers of BeO shown in Fig. 4, the calculated electric field is $0.59 \times 10^9\,\text{V/m}$. It is also possible to determine the piezoelectric constant of *w*-BeO, by varying the geometry of atoms in the supercell from the ideal one to the appropriately distorted.

8 GaN/AlN Superlattice: Spontaneous Polarization and Piezoelectricity

The same approach has been used to study spontaneous polarization and piezoelectric effects in a GaN/AlN superlattice, in which both compounds have the wurtzite structure [17]. In this system, the two effects discussed above, i.e., the spontaneous polarization and the formation of interface dipoles, are simultaneously present. Moreover, in addition to the spontaneous polarization of GaN and AlN, there is a piezoelectric contribution to the polarization induced by the difference in lattice constants of AlN and GaN, and the imposed assumption of pseudomorphic growth, i.e., the condition that the in-plane lattice constants of GaN and AlN are equal. The choice of the substrate (GaN, AlN, or an intermediate case) determines in which layer(s) the piezoelectric effect is present.

Figure 5 displays the atomic structure of the superlattice, the total average charge density (equal to zero in the bulk regions of every layer), and the macroscopic electrostatic potential. Qualitatively, these results can be viewed as a superposition of those from Figs. 3 and 4. First, the potential varies linearly within each layer, which demonstrates the presence of electric fields, and second, there are large 'discontinuities' at the interfaces that originate from formation of dipole layers via charge transfer between the nearest Al-N and Ga-N bonds at interfaces. It should be stressed that the electrostatic potential is continuous at the microscopic scale, and by a discontinuity ΔV we denote the difference between its averaged values at each side of the interface, as schematically shown in Fig. 5b.

These effects may be analyzed by decomposing the total charge density into a monopole and dipole components discussed in the previous examples. The results

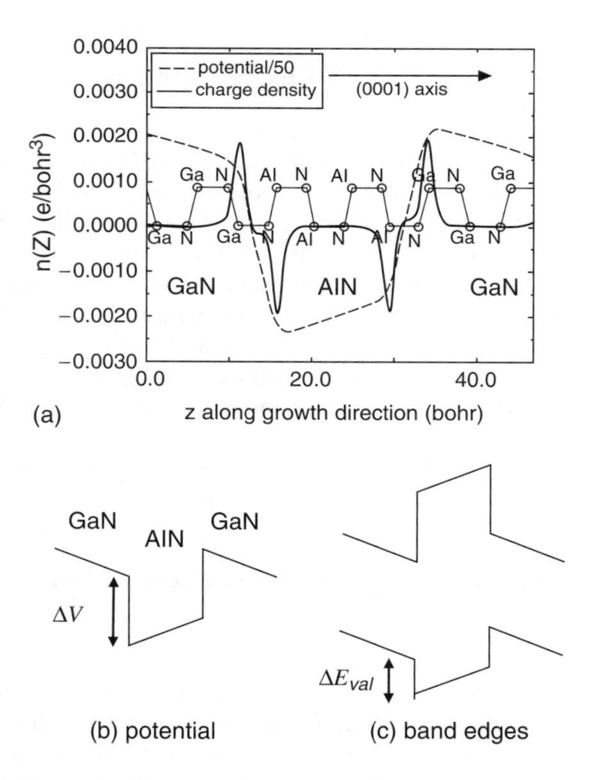

Fig. 5 (a) Total (electronic plus ionic) density, and the electrostatic potential (in Hartrees), for AlN/GaN superlattice lattice-matched to GaN. The magnitude of the electric field is about 10^9 V/cm. Note that the interfacial region is limited to about 5 Å. (Figure reprinted with permission from Ref. [17]. Copyright (1997) by the American Physical Society.) (b-c) Schematic of the potential energy shown in (a), and of the band offsets of GaN/AlN superlattice.

are shown in Fig. 6. While an exact definition of the monopole and dipole contributions is not possible in the low-symmetry wurtzite lattice, which in particular lacks an inversion plane perpendicular to the *c*-axis, the procedure used by the authors is satisfactorily accurate, see [17] for details.

The monopole component gives rise to macroscopic electric fields and is due to both the spontaneous polarization and the piezoelectric effects. In principle, it can be calculated *a la* Posternak *et al.* [16] separately for GaN and AlN using appropriate superlattices. However, it is the difference of polarizations between GaN and AlN, and between the generated electrostatic fields, which is of practical importance. The calculated fields are about 10^9 V/m, similarly to those in BeO.

For obtaining the full spectrum of polar properties, it is easier and more accurate to use the Berry phase approach. The values of the spontaneous polarization P_{eq}, effective charge Z, and the two components of the piezoelectric tensor calculated using this method are given in Table 1. Similar values were obtained in Ref. [18]. For comparison, the results for ZnO are also presented.

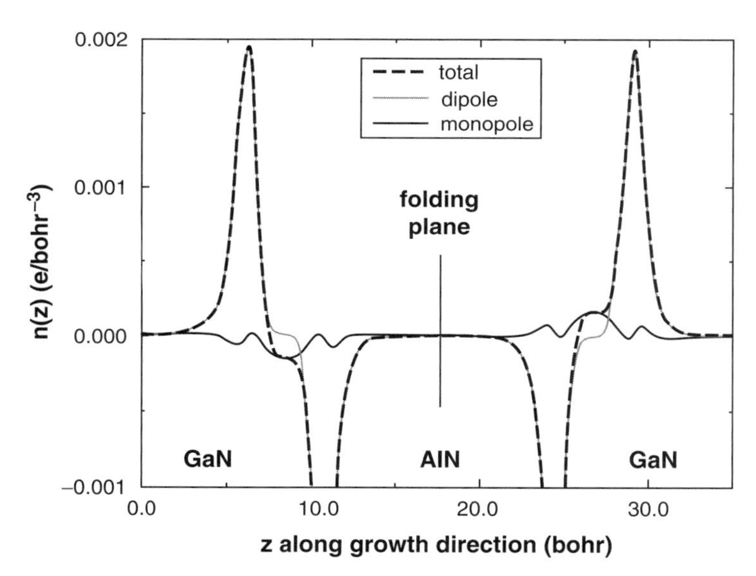

Fig. 6 Total electron density (dash-dotted), and its monopole (solid) and dipole (dashed) components for the superlattice shown in Fig. 5. Figure reprinted with permission from Ref. [17]. Copyright (1997) by the American Physical Society.

Table 1 Spontaneous polarization P_{eq}, (in C/m^2) effective charge Z^*, and the two components of the piezoelectric tensor (in C/m^2) calculated in Ref. [19].

	P_{eq}	Z^*	e_{33}	e_{31}
AlN	−0.081	−2.70	1.46	−0.60
GaN	−0.029	−2.72	0.73	−0.49
InN	−0.032	−3.02	0.97	−0.57
ZnO	−0.057	−2.11	0.89	−0.51

Given the similarity of the [0001] direction in wurtzite with the [111] direction in zinc-blende crystals, it is possible and meaningful to compare the respective components of the piezoelectric tensor of the nitrides with those of other *zb*-III-V and *zb*-II-VI semiconductors. The detailed results of Ref. [19] show that:

(i) The absolute values of the piezoelectric constants are an order of magnitude higher in group-III nitride crystals than in other III-V and II-VI crystals.
(ii) Similarly, the spontaneous polarization is very large and only about 4 times smaller than in typical perovskite ferroelectric crystals.
(iii) The sign of the piezoelectric constants is the same as in II-VI compounds but opposite to the III-V ones. This effect is due to the high ionicity of III-N nitrides, as manifested by the high values of the effective charges Z, which are close to those of ZnO (see Table 1), and higher than those of phosphides or arsenides. Consequently, the internal-strain ionic contribution, which is proportional to Z, dominates the clamped-ion term in Eq. 8.
(iv) Among the nitrides, the spontaneous polarization of AlN is the strongest because of its largest deviation from the ideal wurtzite structural parameters ($c/a = \sqrt{8/3}$, and $u = 0.375$).

Finally, for application purposes, alloys with tailored chemical compositions are often of greater interest than pure binary compounds. For this reason, the important physical properties of III-N alloys, like GaInN, have been investigated as a function of chemical composition. The obtained results show that lattice parameters follow Vegard's law, i.e., they change linearly with composition. On the other hand, polarization in III-N alloys turns out to be a non-linear function of the chemical composition of the alloy [18, 20, 21]. This holds for both the spontaneous polarization and the piezoelectric constants. The main reason for these non-linear dependences is the lattice mismatch, i.e., the difference between the lattice constants of the nitrides, which results in internal distortions in alloys.

As in the case of GaAs/AlAs superlattices, potential discontinuities contribute to band energy discontinuities or band energy offsets at heterointerfaces, as shown in Fig. 5c. The offsets are found to depend on the substrate: for AlN on GaN $\Delta E_{val} = 0.20\,\text{eV}$, and for GaN on AlN $\Delta E_{val} = 0.85\,\text{eV}$ [17]. The non-equivalence of the two cases is strain-induced, caused by strong piezoelectricity of the nitrides. Similar results were obtained by Buongiorno Nardelli *et al.* [22] for zinc-blende GaN/AlN superlattices.

9 Electric Field-Driven Diffusion and Segregation of Dopants in Superlattices

9.1 Introduction

Polarization in bulk III-N nitrides results in the presence of electric fields discussed in the previous Sections. The impact of these fields on the electronic properties has been briefly mentioned above and is discussed in detail in several chapters of this book. The aim of this Section is to point out that electric fields, if strong enough, alter not only the electronic structure, but also the distribution of dopants and defects in nitride-based quantum structures. They may also result in electrodiffusion of charged dopants.

Hydrogen in GaN/AlN superlattices is chosen as the key example; it is the dominant impurity that passivates dopants in structures grown by metalorganic chemical vapor deposition. As will be shown below, the distribution of H in a GaN/AlN superlattice is determined not only by its solubility, but also by the presence of electric fields. We show that these fields give rise to a field-driven electrodiffusion of charged impurities, which can lead to, e.g., an accumulation of hydrogen close to the appropriate interfaces.

In general, properties of semiconductor structures rely entirely on accurate control of doping and stoichiometry. These, in turn, are determined by solubility of impurities and the presence of native defects. For this reason, the issue of equilibrium concentrations of dopants and defects in bulk semiconductors has been extensively studied in the last decade. The first principles theory of doping efficiency [23, 24] allows one to relate the equilibrium concentrations of dopants with the conditions of crystal growth. Moreover, in the wide-gap III-N nitrides self-compensation, i.e., non-intentional formation of native defects, plays an important role [25]. However, since semiconductor systems of current interest are not *bulk* materials but rather *epitaxial heterostructures*, such as GaN/AlN, a question arises about the equilibrium distribution of dopants and defects in heterostructures. Experimentally, thermally activated segregation of impurities through semiconductor heterointerfaces has been investigated for Si/SiGe heterostructures, where the equilibrium concentrations of dopants in Si and SiGe layers differ by a factor of two to three [26–29]. The segregation is stronger in III-V heterostructures [28, 30, 31], where concentration differences may reach two orders of magnitude and affect the luminescence efficiency of light-emitting diodes [32]. A theory of interfacial segregation has been recently formulated in modern terms [33], allowing studies of this effect by accurate *ab initio* methods.

To illustrate the general features that determine segregation, typical dopants and defects in GaN/AlN structures are discussed. The results show that the distribution of dopants in a heterojunction cannot in general be obtained from calculations performed for isolated bulk crystals. The large differences in segregation properties between cation- and anion-substituting dopants are discussed and explained, as are those between shallow and deep ones, and the impact of the defect's charge state (the Fermi level position) on segregation.

9.2 Interfacial Segregation

Consider first the basic concepts describing the doping of a homogeneous solid: The equilibrium concentration of dopants is defined for given conditions of growth under the assumption that equilibrium is maintained between the growing sample and the reservoirs of relevant atomic species characterized by their chemical potentials μ [23, 24]. At these conditions, insertion of an impurity atom into a solid induces a change in the total energy of the system, which is the formation energy E_{form}. For example, E_{form} of a neutral interstitial H in GaN is

$$E_{form} = E_{tot}(\text{GaN : H})\text{-}E_{tot}(\text{GaN})\text{-}\mu(\text{H}), \tag{21}$$

where E_{tot} is the total energy of the crystal with or without the impurity. Neglecting small changes in the vibrational entropy, one finds that the equilibrium concentration of dopants is equal to

$$N_{imp} = N_0 \exp(\text{-}E_{form}/k_\text{B}T), \tag{22}$$

where N_0 is the density of lattice sites that can be occupied by a given dopant.

Turning to interfacial segregation in an A/B heterostructure, the equilibrium concentrations of dopants at the A and B components, N_{imp}(A) and N_{imp}(B) respectively, are determined by the condition that the chemical potential of the dopant is the same in the whole system, and in particular at both sides of the interface. Using this fact one finds that the segregation coefficient, defined as the ratio of impurity concentrations at the two sides of the interface, is

$$k_{seg} = N_{imp}(\text{A})/N_{imp}(\text{B}) = N_0(\text{A})/N_0(\text{B}) \exp(E_{seg}/k_\text{B}T), \tag{23}$$

where $E_{seg} = E_{form}$ (A) - E_{form} (B) is the segregation energy. Therefore, k_{seg} only depends on the difference between the formation energies and is independent of μ_{imp}, i.e., of the actual source of the impurity atoms. To describe quantitatively the interfacial segregation, the above analysis needs to be supplemented by first principles calculations of the total energies. They were done using the methods described in Sec. 4.

In general, dopants and defects in semiconductors are ionized. Ionization of dopants affects their distribution in GaN/AlN heterojunctions in two ways that are discussed in more detail below. First, electric fields present in a junction lead to field-driven diffusion and redistribution of ionized dopants. This effect was seen experimentally for B in a Si homojunction [34]. Second, in a homogeneous crystal, E_{form} depends on the charge state of the defect. The transfer of m electrons from a defect level E_{imp} to the Fermi level E_F changes the formation energy by $m(E_F - E_{imp})$, which should be added to the right-hand side of Eq. 21 [23, 24]. Consequently, since E_{seg} is the difference between E_{form} of the two materials, segregation depends on the charge state of the defect as well. It will be shown below that this dependence is in general weak for shallow defects, but may be substantial for deep ones.

9.3 Profile of H in AlN/GaN Superlattice

The first step towards analysis of the profile of H in a GaN/AlN SL is to find its equilibrium sites in bulk GaN and AlN. Because of electrostatic interactions between H and host atoms they depend on the charge state: H^+ is located close to negatively charged N anions, while H^- and H^0 prefer interstitial regions. The following four sites for H^+ are considered below: the bond center *c*-BC and the antibonding *c*-AB, with the H-N bond parallel to the *c*-axis, as well as the bond center *a*-BC and the antibonding *a*-AB, where the H-N bond is almost perpendicular to the *c*-axis. In AlN, the equilibrium site is *c*-BC, with the energy lower than those of *a*-AB and *c*-AB by 0.05 and 0.2 eV, respectively, see Table 2. The energy of *a*-BC is higher than that of *c*-BC by 0.7 eV, which demonstrates a strong non-equivalence between the two types of Al-N bonds in AlN. In agreement with Ref. [35], a similar sequence of energies is found for GaN, and is reported in Table 2. Accordingly, the superlattice calculations assumed that H^+ is at *c*-BC sites in both GaN and AlN layers. The most stable sites for both H^- and H^0 are interstitial at the center of the trigonal channel.

Figure 7 shows the total energy E_{tot} of hydrogen in the three charge states, H^+, H^0, and H^-, as a function of the position in $(GaN)_6(AlN)_6$ SL. As follows from this figure, there are three features characterizing the energetics and the spatial

Table 2 The total energy of H^+ in GaN and AlN relative to the *c*-BC bond-center site. See text for the explanation of the four sites.

	c-BC	*a*-BC	*c*-AB	*a*-AB
AlN	0.0	0.05	0.20	0.70
GaN	0.0	0.05	0.25	0.30

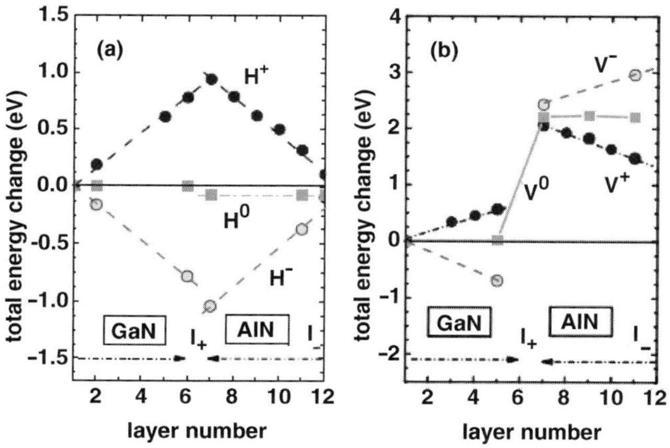

Fig. 7 Calculated total energy of (a) H^+, H^0, and H^-, and (b) V_N^+, V_N^0, and V_N^-, as a function of their location in a $(GaN)_6(AlN)_6$ superlattice. Bilayers 1–6 are GaN and 7–12 are AlN. The lines are only guides for the eye. Figure reprinted with permission from Ref. [33]. Copyright (2006) by the American Physical Society.

distribution of H in the SL, namely: (i) the total energy of the neutral H is constant within both GaN and AlN layers, while E_{tot} of charged H varies linearly with its location due to the electric field, (ii) at GaN/AlN interfaces, i.e., between the layers 6 and 7 or 1 and 12, there is a discontinuity in the total energy, which is the segregation energy E_{seg}, and (iii) E_{seg} depends on the charge state of H.

As follows from Fig. 7b, the same effects take place for the nitrogen vacancy V_N. In fact, these are universal features of interface segregation and we discuss them in detail below.

9.3.1 Electric Fields

Electric fields in an AlN/GaN SL act on charged defects by Coulomb forces and give rise to a linear dependence of the total energy on the defect's position within AlN or GaN layers, which is shown in Fig. 7. Neutral defects are of course not sensitive to the field. As was discussed in Sec. 6, periodic boundary conditions require that the average of the electric field over the superlattice period must vanish, and therefore the direction of the field changes sign between AlN and GaN. In agreement with Figs. 5 and 6, the sign of the excess charge density alternates at consecutive interfaces, which are labeled I_+ and I_-, respectively. The electric field leads to accumulation of H^+ in the vicinity of I_- interfaces, and of H^- in the vicinity of I_+. The most interesting case is that of p-doped wurtzite structures; in this case H^+ accumulates in GaN close to I_- interfaces with $E_{seg}(+) = 0.13\,\text{eV}$, see Fig. 7a. The value of the field deduced from Fig. 7, about $7 \times 10^6\,\text{V/cm}$, agrees with that obtained from material constants given in Ref. [36].

One should stress that the predicted accumulation of ionized H close to the appropriate interfaces leads to screening of the field, analogous to screening by free carriers. This effect may have an important impact on the electronic states and on optical and transport properties of GaN/AlN-based heterostructures, which are, to a large extent, determined by the fields [37].

One can also follow this process on a microscopic scale. Diffusion of H^- has an activated character and proceeds by jumps between local minima separated by energy barriers. The presence of a macroscopic electric field leads to left-right asymmetry of barrier heights, and thus to electromigration. In the present case, the migration is driven by *internal*, rather than external electric fields. This effect is revealed in Fig. 8, which shows the total energy of H^- as a function of its location along the c-axis in one unit cell of AlN in $(GaN)_6(AlN)_6$. The average barrier is $1.67\,\text{eV}$ [38] and the left barrier is lower than the right one by $0.16\,\text{eV}$.

9.3.2 Segregation Energies

As it is seen in Fig. 7b, the total energy of the nitrogen vacancy is lower by about 2 eV in GaN than in AlN. This energy difference is the segregation energy, defined earlier in this Section. The segregation is much less pronounced for H, see Fig. 7a.

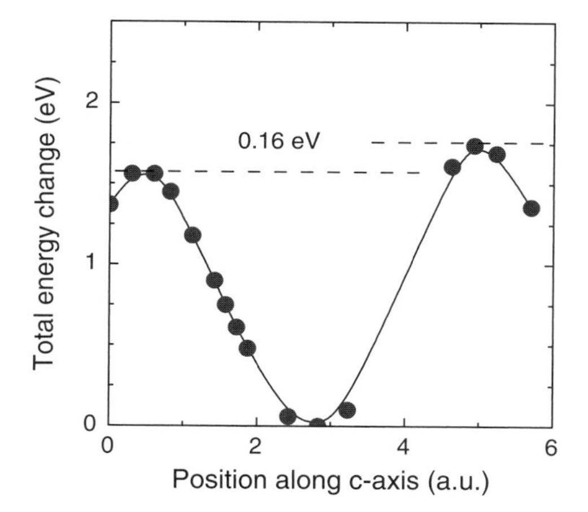

Fig. 8 Calculated total energy changes of H^- as a function of its location in an AlN unit cell in $(GaN)_6(AlN)_6$ superlattice. The lines are only guides for the eye. Figure reprinted with permission from Ref. [33]. Copyright (2006) by the American Physical Society.

Table 3 Segregation energies, $E_{seg} = E_{form}$ (AlN) - E_{form} (GaN), and their differences ΔE_{seg} between positive and negative charge states of the defect (in eV).

	E_{seg}	ΔE_{seg}
H in wurtzite	−0.07	0.40
H in cubic	−0.2	0.48
V_N	2.2	−1.55
Si_N	2.6	−0.35
Si_{cation}	0.15	−0.15
Mg	0.05	0.18

To illustrate the variation of the total energy with the H position, we show in Fig. 8 its values for negatively charged H along the AlN part of the GaN/AlN superlattice. The calculated values of E_{seg} are summarized in Table 3, which also includes two typical impurities, namely Mg, the most efficient (shallowest) acceptor, and Si, which has an amphoteric character in GaN [39].

Beginning the discussion with neutral defects, for which the effect is not influenced by electric fields, the calculated $E_{seg}(0)$ for H^0 is 0.07 eV in the wurtzite structure and ∼0.2 eV in the cubic one, respectively. The negative sign found in both structures indicates that H^0 has a tendency to accumulate in AlN layers. In contrast, all the remaining defects have a tendency to accumulate in GaN layers. Table 3 shows that this tendency is weak, since E_{seg} is small for cation-substituting defects (Mg, Si_{cation}, and V_{cation}), but it is very pronounced for N-substituting defects (Si_N, and V_N). This different behavior is caused by the fact that the nearest neighbors of a cation-substituting defect are N atoms at both sides of the interface, and thus the number of bonds of a given type before and after the segregation does not change. For example, in the case of Si_{cation}, there are 4 Si-N bonds independent of the actual location of Si either in AlN or in GaN layer. In contrast, after the diffusion of the

N-substituting Si_N from GaN to AlN, four Si-Ga bonds replace four Si-Al bonds, inducing corresponding changes in the total energy. This effect is further enhanced by the high bond energies of the nitrides compared with other III-V crystals. A similar effect occurs for vacancies in AlGaN and SiGe alloys, which are preferentially terminated by Ga and Ge atoms, respectively [40]. Finally, one should note that according to Eq. 23 even a modest value of E_{seg}, 0.1 eV, leads to defect equilibrium concentration that is three times higher in GaN than in AlN at 800° C.

9.3.3 Dependence of E_{seg} on the Defect Charge State

In Table 3 we list the values of $\Delta E_{seg} = E_{seg}(+) - E_{seg}(-)$, the difference between E_{seg} for positive and negative charge states. In certain cases, the charge dependence is pronounced. To simplify the discussion and to uncover the underlying physics, we also discuss an AlN/GaN (001)-oriented superlattice with GaN and AlN in the zinc-blende structure. In this case, there are no electric fields, not only because of the absence of spontaneous polarization in cubic crystals, but also the vanishing piezoelectric effect for strains oriented in the [1] direction. According to our results, E_{seg} of substitutional dopants and vacancies in the two structures are equal to within 0.05 eV. This stems from the same four fold tetrahedral defect coordination in both structures. In contrast, the number and coordination of neighbors of interstitial H differ between the two structures, resulting in a larger E_{seg} of 0.15 eV.

As follows from Table 3, ΔE_{seg} is about 0.2 eV for the two shallow dopants, Si_{cation} and Mg, and it is larger for the defects that induce deep states in the forbidden band gap (H, Si_N, and both vacancies). The stronger dependence of E_{seg} on the charge state of deep defects is caused by two factors: (i) in general, the energy levels of deep states are different at the two sides of the interface, and (ii) the energies of impurity levels depend on their occupation because of the electron-electron interaction. Both features are schematically illustrated in Fig. 9 for H. Hydrogen in the negative and the neutral charge states induces a singlet

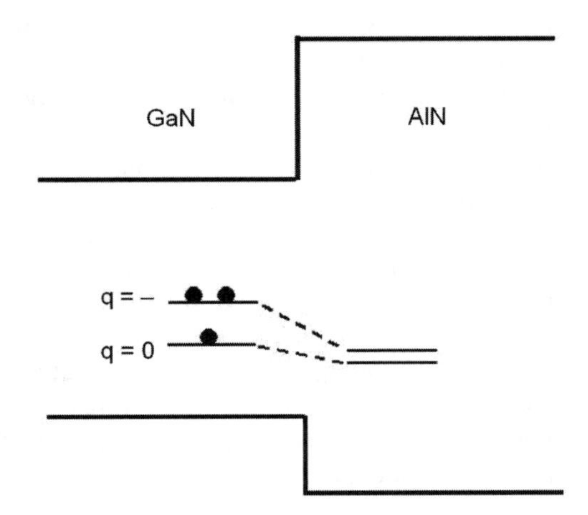

Fig. 9 Schematic of the energy levels of H at a GaN/AlN heterojunction. Figure reprinted with permission from Ref. [33]. Copyright (2006) by the American Physical Society.

about 1 eV above the top of the valence band. Consequently, when H migrates from GaN to AlN the energy of m electrons that occupy this level changes by $\delta = m\{E_{imp}(\text{H in AlN}) - E_{imp}(\text{H in GaN})\}$. We find that $\delta = E_{seg}$ to within 0.05 eV for both H^0 and H^-. Analogous results hold for the remaining deep defects. Thus, the electronic contribution to the total energy is the dominant factor that determines the charge state dependence of E_{seg}. On the other hand, in the case of a shallow dopant, the electron (or the hole) is located in the well independently of the defect's location in the junction. For this reason, the electronic contribution to the segregation energies of shallow dopants is smaller than in the case of deep ones.

10 Summary

The first part of this chapter aimed at presenting a comprehensive link between first principles calculations and basic classical electrostatics concepts used to find macroscopic electric fields in semiconductor heterostructures. The attention was focused on polarization, both spontaneous and of piezoelectric origin. The advanced and general theoretical methods developed in the last decade, which allow for accurate calculations of both the electronic structure and polarization, were introduced and briefly described. The formation of dipole layers at GaAs/AlAs interfaces was used an example emphasizing dipole-layer contributions to discontinuities in both the electrostatic potential and the band offsets in heterojunctions. The second example, wurtzite BeO, highlighted the direct link between its spontaneous polarization and charge densities accumulated at interfaces. In GaN/AlN heterostructures, discussed in Sec. 8, both effects are present, and these were studied in more detail.

The second part of this review has shown that the macroscopic electric fields present in polar structures based on, e.g., (0001)-oriented wurtzite or (111)-oriented zinc-blende crystals, lead to field-driven diffusion of H and its accumulation close to the appropriate interfaces. This electromigration screens electric fields similarly to free carrier screening, and thus affects the electronic structure of the junction. The theory of semiconductor doping was generalized in order to describe segregation and distribution of dopants and defects at heterojunctions. This theory shows that the profile of dopants depends strongly on the interface and it cannot be obtained from calculations for bulk components alone. The general features of interface segregation were illustrated by considering Si, Mg, and the nitrogen vacancy in GaN/AlN. The results reveal that the relative concentrations of dopants at both sides of an interface are determined by (i) the dopant charge state and the band offset at the interface, and (ii) the atomic site of a dopant or a defect. For example, a cation-substituting dopant has four N neighbors in either component of the nitride junction. Therefore, its segregation energy is an order of magnitude lower than that of an N-substituting dopant.

Acknowledgements This work was supported by U.S. DOE and DOD. We thank N. Gonzalez Szwacki, M. Buongiorno Nardelli and S. M. Nakhmanson for many discussions over the years on the fascinating topic of polarization. We also thank L. Yu for generating the results shown in Fig. 2. One of us (PB) thanks Alfonso Baldereschi for introducing him to first-principles methods and dedicates this paper to Prof. Baldereschi on the occasion of his 60th birthday.

References

1. For a review see, for example, S. J. Pearton, J. C. Zolper, R. J. Shul, and F. Ren, J. Appl. Phys. **86**, 1 (1999).
2. R. D. King-Smith and D. Vanderbilt, Phys. Rev. B **47**, 1651 (1993); D. Vanderbilt and R. D. King-Smith, ibid. **48**, 4442 (1993).
3. R. Resta, Rev. Mod. Phys. **66**, 899 (1994).
4. R. M. Martin, Phys. Rev. **B5**, 1607 (1972).
5. J. D. Jackson, Classical Electrodynamics, John Wiley & Sons, New York 1967.
6. For an excellent review see S. Baroni, S. de Gironcoli, A. Dal Corso, and P. Gianozzi, Rev. Mod. Phys. **73**, 515 (2001).
7. R. M. Dreizler and E. K. U. Gross, *Density Functional Theory* (Springer, Berlin, 1990).
8. See, for example, J. Bernholc, "Computational materials science: the era of applied quantum mechanics," Physics Today, **52**, September, p. 30 (1999).
9. E. L. Briggs, D. J. Sullivan, and J. Bernholc, Phys. Rev. B **54**, 14362 (1996).
10. R. M. Martin, Phys. Rev. B**9**, 1998 (1974).
11. D. Vanderbilt, J. Phys. Chem. Solids, **61**, 147 (2000).
12. S. M. Nakhmanson, V. Meunier, J. Bernholc, M. Buongiorno Nardelli, Phys. Rev. B **67**, 235406 (2003).
13. N. Marzari and D. Vanderbilt, Phys. Rev. B **56**, 12847 (1997).
14. F. Gygi, J.-L. Fattebert, and E. Schwegler, Computer Physics Comm., **155**, 1 (2003).
15. A. Baldereschi, S. Baroni, and R. Resta, Phys. Rev. Lett. **61**, 734 (1988).
16. M. Posternak, A. Baldereschi, A. Catellani, and R. Resta, Phys. Rev. Lett. **64**, 1777 (1990).
17. F. Bernardini and V. Fiorentini, Phys. Rev. B **57**, R9427 (1997).
18. A. Al-Yacoub and L. Bellaiche, Appl. Phys. Lett. **79**, 2166 (2001).
19. F. Bernardini, V. Fiorentini, and D. Vanderbilt, Phys. Rev. B **56**, R10024 (1997).
20. K. Shimada, T. Sota, K. Suzuki, and H.Okumura, Jpn. J. Appl. Phys. Part 2, **37**, L1421 (1998).
21. F. Bernardini and V. Fiorentini, Phys. Rev. B **64**, 85207 (2001); *ibid.* **65**, 129903 (2002).
22. M. Buongiorno Nardelli, K. Rapcewicz, and J. Bernholc, Phys. Rev. B **55**, R7323 (1997); Appl. Phys. Lett. **71**, 31315 (1997).
23. D. B. Laks *et al.*, Phys. Rev. Lett. **66**, 648 (1991); C. G. Van De Walle *et al.*, Phys. Rev. B **47**, 9425 (1993).
24. S. B. Zhang and John E. Northrup, Phys. Rev. Lett. **67**, 2339 (1991).
25. P. Bogusławski, E. Briggs, and J. Bernholc, Phys. Rev. B, Rapid Commun. **51**, 17255 (1995).
26. S. M. Hu *et al.*, Phys. Rev. Lett. **67**, 1450 (1991).
27. N. Moriya *et al.*, Phys. Rev. Lett. **75**, 1981 (1995).
28. T. T. Fang *et al.*, Appl. Phys. Lett. **68**, 791(1995).
29. S. Kobayashi *et al.*, J. Appl. Phys. **86**, 5480 (1999).
30. Be in InP/InGaAs: W. Haussler, J. W. Walter, and J. Muller, Mat. Res. Symp. Proc. vol. 147, 333 (1989); Be and Zn in AlGaAs/GaAs: T. Humer-Hager *et al.*, J. Appl. Phys. **66**, 181 (1989); Zn in InGaAsP/InP: R. Weber *et al.*, J. Electrochem. Soc. **138**, 2812 (1991).
31. A. Gaymann, M. Maier, and K. Kohler, J. Appl. Phys. **86**, 4312 (1999).
32. K. Kohler *et al.*, J. Appl. Phys. **97**, 104914 (2005).
33. P. Boguslawski, N. Gonzalez Szwacki, and J. Bernholc, Phys. Rev. Lett. **96**, 185501 (2006).
34. R. D. Chang, P. S. Choi, and D. L. Kwong, Appl. Phys. Lett. **72**, 1709 (1998).
35. A. F. Wright *et al.*, J. Appl. Phys. **94**, 2311 (2003); S. Limpijumnong and C. Van de Walle, Phys. Rev. B **68,** 235203 (2003).

36. O. Ambacher *et al.*, J. Appl. Phys. **87**, 334 (2000).
37. B. S. Kang *et al.*, Appl. Phys. Lett. **84**, 4635 (2004).
38. This value is very close to 1.7 eV obtained for H in GaN by S. M. Myers *et al.*, J. Appl. Phys. **88**, 4676 (2000).
39. P. Boguslawski and J. Bernholc, Phys. Rev. B **56**, 9496 (1997).
40. P. Boguslawski and J. Bernholc, Phys. Rev. B **59**, 1567 (1999).

Polarization Induced Effects in GaN-based Heterostructures and Novel Sensors

O. Ambacher and V. Cimalla

1 Introduction

The macroscopic non-linear pyroelectric polarization of wurtzite $Al_xGa_{1-x}N$, $In_xGa_{1-x}N$ and $Al_xIn_{1-x}N$ ternary compounds dramatically affects the optical and electrical properties of multilayered Al(In)GaN/GaN hetero-, nanostructures and devices, due to the huge built-in electrostatic fields and bound interface charges caused by gradients in polarization at surfaces and heterointerfaces. In modeling of polarization induced effects in GaN based devices it is often assumed that polarization in group-III-nitride alloys interpolates linearly between the limiting values determined by the binary compounds. In more advanced models it is taken into account that the macroscopic polarization in group-III-nitride alloys is a non-linear function of strain and composition. We have applied those results to reverse-model experimental data obtained in a number of InGaN/GaN quantum-wells (QW) as well as AlInN/GaN and AlGaN/GaN transistor structures (HEMTs). Thereby we find that the discrepancies of experiment and *ab initio* theory are almost completely eliminated for the AlGaN/GaN based heterostructures when polarization non-linearity is accounted for. The realization of undoped lattice matched AlInN/GaN heterostructures further allows the confirmation of the existence of a gradient in spontaneous polarization by the experimental observation of two dimensional electron gases (2DEGs). The confinement of 2DEGs in InGaN/GaN QWs in combination with the measured Stark shift of excitonic recombination is used to determine the polarization induced electric fields in nanostructures. To facilitate inclusion of the predicted non-linear polarization in future simulations, we give an explicit prescription to calculate polarization induced electric fields and bound interface charges for arbitrary x in each of the ternary III-N alloys. In addition, the theoretical and experimental results presented here allow a detailed comparison of the predicted electric fields and bound interface charges with the measured Stark shift and the sheet carrier concentration of polarization induced 2DEGs. This comparison provides an inside in the reliability of the calculated non-linear piezoelectric and spontaneous polarization of group-III-nitride ternary alloys.

It has been predicted by theory and confirmed by experiment, that InN, GaN and AlN with wurtzite crystal structure are pyroelectric materials [1–4]. Characteristic for these crystals, especially for AlN, are huge values of pyroelectric (spontaneous) polarizations up to $|P_{AlN}^{SP}| = 0.09\,\mathrm{C/m^2}$, which are present without any external electric field. In contrast to ferroelectric crystals, the orientation of this spontaneous polarization is always along the $\lfloor 000\overline{1} \rfloor$ direction and cannot be turned around by external electric fields. The orientation of the spontaneous polarization relative to the c-axis can only be controlled by adjusting the polarity of the material [1,5,6]. Although the spontaneous polarization is found to be very strong in group-III-nitrides, the pyroelectric constants, describing the change of the spontaneous polarization with temperature, are measured to be small (e.g. $dP_{AlN}^{SP}/dT = 7.5\,\mu\mathrm{C/Km^2}$, at room temperature [4,7]), which is supposed to be a big advantage in using these materials for high power and high temperature applications like surface acoustic wave devices (SAWs), high frequency, high power transistors (HEMTs), microwave amplifiers or very bright light emitting diodes (LEDs) and lasers.

Pyroelectric crystals are always piezoelectric. The piezoelectric constants of group-III-nitrides are a factor of 5 to 20 larger compared to those of InAs, GaAs and AlAs [1]. The high piezoelectricity of GaN and AlN in combination with the high stiffness and sound velocity has already been used to process surface acoustic wave devices for electric filters with transfer frequencies above 1 GHz [8,9]. In addition, SAW devices have been successfully applied to measure the elastic and piezoelectric constants of GaN and AlN [10,11]. Using the measured and predicted elastic and piezoelectric constants it has been previously shown that effects induced by spontaneous and piezoelectric polarization can exert a substantial influence on the concentration, distribution and recombination of free carriers in strained group-III-nitride heterostructures. Observations of Franz-Keldysh-oscillations and Stark-effects in InGaN/GaN and AlGaN/GaN quantum well structures provide direct evidence for the presence of strong electric fields caused by gradients in spontaneous and piezoelectric polarization [12–14]. These fields significantly influence the distribution and life time of free carriers in InGaN and GaN quantum wells and can have an important impact on the performance of AlGaN/GaN and InGaN/GaN based optoelectronic devices. Although these devices, e.g. continuous green light emitting LEDs and blue lasers have already reached outstanding performance like output powers of 10 and 30 mW, respectively, further improvement of design and efficiency seem to be possible only on the basis of a more detailed understanding of polarization in group-III-nitride nanostructures [15,16].

The importance of polarization induced effects is further emphasized in the case of pseudomorphic, wurtzite AlGaN/GaN based transistor structures [6], which have been subject of intense recent investigation and have emerged as attractive candidates for high-voltage, high-power operation at microwave frequencies [17–23]. Contributing to the outstanding performance of AlGaN/GaN based HEMTs is the ability to achieve polarization induced two dimensional electron gases (2DEGs) with sheet carrier concentrations up to $2{\cdot}10^{13}\,\mathrm{cm^{-2}}$, well in excess of those observed in other III-V material systems. These high sheet carrier concentrations enable the fabrication of AlGaN/GaN based HEMTs with channel currents above 1000 mA/mm, breakdown voltages of 140 V and cut off frequencies of $f_t = 68\,\text{GHz}$ and $f_{max} = 140\,\text{GHz}$. High output powers of about 11 W/mm together with high maximum power added efficiency of 40% have been measured at a frequency of 10 GHz [24]. To successfully model and further optimize the electronic devices, a precise prediction and control of the sheet carrier concentrations as well as electronic transport properties of the polarization induced 2DEGs is required. To meet these requirements, the spontaneous as well as piezoelectric polarizations of binary as well as of ternary group-III-nitrides have to be determined with a high accuracy.

The spontaneous polarization and piezoelectric constants of the ternary compounds have been often determined by linear interpolation of the physical properties of the relevant binary materials. A detailed comparison of theoretical predictions achieved by this approximation with experimental results has revealed a significant discrepancy. By example, the predicted sheet carrier concentration of polarization induced 2DEGs confined at interfaces of $Al_{0.3}Ga_{0.7}N/GaN$ HEMT-structures is about 20% higher than the experimental data [25]. The polarization induced electric fields calculated from the measured Stark-shift of excitonic recombination in AlGaN/GaN-based quantum wells are also significantly below the theoretical predictions [14]. The discrepancies between measured and calculated electric fields are even more pronounced if InGaN/GaN quantum well structures are investigated [12].

In this paper we review the theoretical and experimental results of the elastic and pyroelectric properties of binary group-III-nitrides with wurtzite crystal structure. We develop an improved method to calculate the piezoelectric and spontaneous polarization of the binary, and, more important, of the ternary group-III-nitrides taking non-linearities due to alloying and/or high internal strain into account. Polarization induced interface charges and sheet carrier concentrations of 2DEGs are predicted for pseudomorphic InGaN/GaN, AlGaN/GaN and AlInN/GaN quantum well and heterostructures on the basis of the improved theory and compared with experimental results achieved by a combination of elastic recoil detection, high resolution X-ray diffraction, X-ray standing wave, photoluminescence, C-V profiling, and Hall effect measurements. Based on the improved model of polarization induced surface and interface charges a review of novel sensors based on AlGaN/GaN-heterostructures is provided, enabling a detailed understanding of the detection mechanisms and new functionalities of these interesting devices.

2 First-Principles Prediction of Structural and Pyroelectric Properties

In order to predict the pyroelectricity of binary alloys as well as ternary random alloys, the structural, elastic and polarization properties of wurtzite group-III-nitrides are calculated from first principles within density-functional theory using the plane-wave ultra soft pseudopotential method, taking advantage of the generalized gradient approximation (GGA) to the exchange-correlation functional. The pseudopotentials for Ga and In include, respectively, the semicore 3d and 4d states in the valence. A plane wave basis is used to expand the wave functions and a cut-off of 350 eV is found to be sufficient to fully converge all properties of relevance here. For k-space summation at least a Monkhorst-Pack (888) grid is used. Lattice constants and internal parameters are calculated using standard total energy calculations. Polarization and related quantities are obtained using the Berry-phase approach as in previous works [1, 2, 26, 27]. Since we are mainly interested in the effects on polarization due to internal strain of ternary alloys related to size mismatch of the two alloyed group-III-nitrides, it is necessary that our calculations reproduce relative mismatches with the highest possible accuracy. Comparing density functional theory calculations, we found that GGA has advantages in comparison to LDA in reproducing the relative mismatch between the binary constituents. In addition, GGA is getting closer to the experimental lattice constants and cell-internal parameter. All forces and stress components have been converged to zero within 0.005 eV/Å and 0.01 kbar, respectively, for free standing alloys. The microscopic structure of a random alloy is represented by periodic boundary conditions using the special quasi-random structure method [28]. We enforce the periodic boundary conditions needed to predict the macroscopic polarization in the Berry phase approach. As a compromise between computational workload and the description of random structures, a 32-atom $2 \times 2 \times 2$ wurtzite super cell is adopted. It is possible to mimic the statistical properties of a random wurtzite alloy for a molar fraction $x = 0.5$ by suitably placing the cations on the 16 sites available in the cell. Since other molar fractions cannot be described as easily and also because non-linear effects are expected to be largest for this concentration, our theoretical approach on random alloys is restricted to $x = 0.5$. The predicted non-linearities of structural and polarization related properties of $A_xB_{1-x}N$ alloys are approximated by quadratic equations of the form:

$$Y_{ABN}(x) = Y_{AN}x + Y_{BN}(1-x) + bx(1-x), \tag{1}$$

where

$$b = 4Y_{ABN}(x = 0.5) - 2(Y_{AN} + Y_{BN}), \tag{2}$$

is the bowing parameter. In order to determine the quality of the theoretical predictions and the approximation of the non-linearities, the calculated physical properties are compared to experimental results discussed in chapters III and VIII.

3 Lattice Constants, Average Bond Length and Bond Angles in Ternary Compounds

The lattice constants $a(x)$ and $c(x)$ of wurtzite group-III-nitride random alloys $A_xB_{1-x}N(Al_xGa_{1-x}N, In_xGa_{1-x}N$ and $Al_xIn_{1-x}N)$ are predicted to follow the composition weighted average between the binary compounds AN and BN (Vegard's law) [2]:

$$a_{AlGaN}(x) = (3.1986 - 0.0891x)\text{Å}, \quad c_{AlGaN}(x) = (5.2262 - 0.2323x)\text{Å},$$
$$a_{InGaN}(x) = (3.1986 + 0.3862x)\text{Å}, \quad c_{InGaN}(x) = (5.2262 + 0.574x)\text{Å}, \qquad (3)$$
$$a_{AlInN}(x) = (3.5848 - 0.4753x)\text{Å}, \quad c_{AlInN}(x) = (5.8002 - 0.8063x)\text{Å}.$$

By a combination of high resolution X-ray diffraction (HRXRD), Rutherford and elastic recoil detection analysis, Vegard's law can be confirmed for the ternary group-III-nitrides [29–31].

The agreement between the predicted lattice constants and linear fits of the experimental data:

$$\begin{aligned}
a_{AlGaN}(x) &= ((3.189 \pm 0.002) - (0.086 \pm 0.004)x)\text{Å},\\
a_{InGaN}(x) &= ((3.192 \pm 0.001) + (0.351 \pm 0.004)x)\text{Å},\\
a_{AlInN}(x) &= ((3.560 \pm 0.019) - (0.449 \pm 0.019)x)\text{Å},\\
c_{AlGaN}(x) &= ((5.188 \pm 0.003) - (0.208 \pm 0.005)x)\text{Å},\\
c_{InGaN}(x) &= ((5.195 \pm 0.002) + (0.512 \pm 0.006)x)\text{Å},\\
c_{AlInN}(x) &= ((5.713 \pm 0.014) - (0.745 \pm 0.024)x)\text{Å},
\end{aligned} \qquad (4)$$

are better than 2% over the whole range of possible compositions (Fig. 1a,1b).

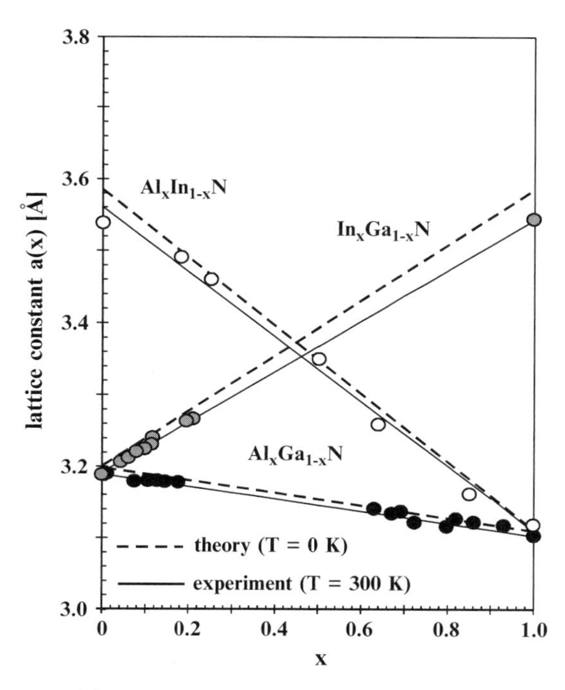

Fig. 1a Lattice constant $a(x)$ for random ternary alloys of group-III-nitrides measured by high resolution X-ray diffraction (HRXRD) at room temperature (solid line) and calculated by the method described in chapter II for $T = 0K$ (dashed line). The measured and calculated data confirm the validity of Vegard's law for wurtzite AlGaN, InGaN and AlInN alloys [29–31].

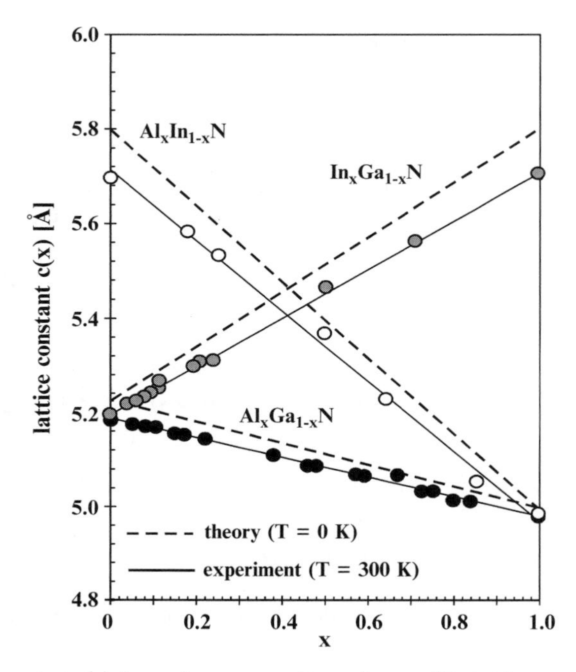

Fig. 1b Lattice constant $c(x)$ for random ternary alloys of group-III-nitrides measured by HRXRD at room temperature (solid line) and calculated by the method described in chapter II for $T = 0K$ (dashed line). The measured and calculated data confirm the validity of Vegard's law for wurtzite AlGaN. InGaN and AlInN alloys. The agreement between predicted and measured lattice constants is better than 2% over the whole range of possible compositions [29–31].

In contrast to the lattice constants, the nearest neighbor bond lengths, $A - N(x)$ and $B - N(x)$ change only slightly and in a non-linear way (especially for In-containing alloys) as function of alloy composition, resembling more their values in the binary constituents rather than an average value corresponding to the virtual crystal limit (see Fig. 2a, 2b [32–34]).
The wurtzite structure of interest here has two types of first neighbor metal-nitrogen bond distances: $M - N_c$ along the c-axis (one bond) and $M - N_b$ in the basal plane (three bonds):

$$M - N_{c1} = uc, \tag{5}$$

$$M - N_{b1} = \sqrt{\frac{1}{3}a^2 + \left(\frac{1}{2} - u\right)^2 c^2}, \tag{6}$$

and two bond angles $\alpha = \angle(M - N_{c1}; M - N_{b1})$, $\beta = \angle(M - N_{b1}; M - N_{b'1})$ (see Fig. 5b):

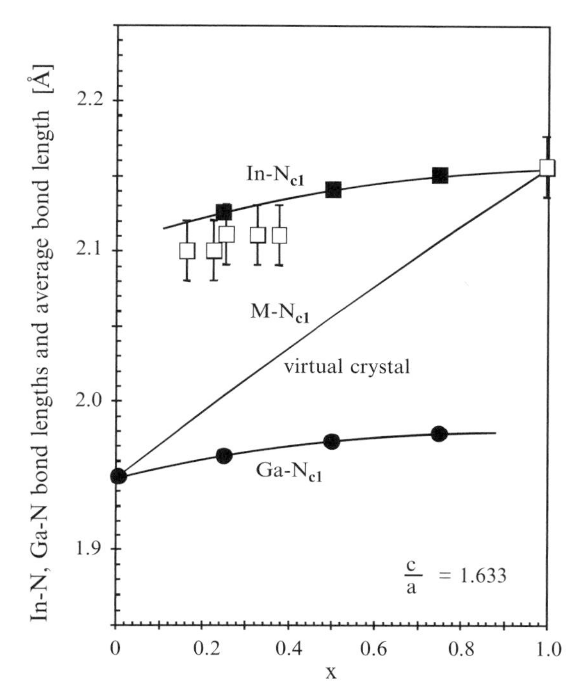

Fig. 2a Calculated (black symbols) and measured *In-N* and *Ga-N* bond lengths (open symbols) for relaxed, random InGaN alloys versus alloy composition [32–34]. The bond length along the c-axis is predicted, assuming a constant c/a-ratio of 1.633. In addition the average bond length *M-N* of wurtzite InGaN alloys is shown.

$$
\begin{aligned}
\alpha &= \frac{\pi}{2} + \arccos\left\{\left(\sqrt{1+3\left(\frac{c}{a}\right)^2\left(\frac{1}{2}-u\right)^2}\right)^{-1}\right\}, \\
\beta &= 2\arcsin\left\{\left(\sqrt{\frac{4}{3}+4\left(\frac{c}{a}\right)^2\left(\frac{1}{2}-u\right)^2}\right)^{-1}\right\},
\end{aligned}
\tag{7}
$$

where u denotes the cell-internal parameter. In addition three types of second neighbor cation-anion distances connecting the cation M to the anions N_{c2}, N_{b2}, and $N_{b'2}$ (see Fig. 3) are present:

$$M - N_{c2} = (1-u)c \quad \text{(one neighbor along the c-axis)}, \tag{8}$$

$$M - N_{b2} = \sqrt{a^2 + (uc)^2} \quad \text{(six neighbors)}, \tag{9}$$

$$M - N_{b'2} = \sqrt{\frac{4}{3}a^2 + \left(\frac{1}{2}-u\right)^2 c^2} \quad \text{(three neighbors)}. \tag{10}$$

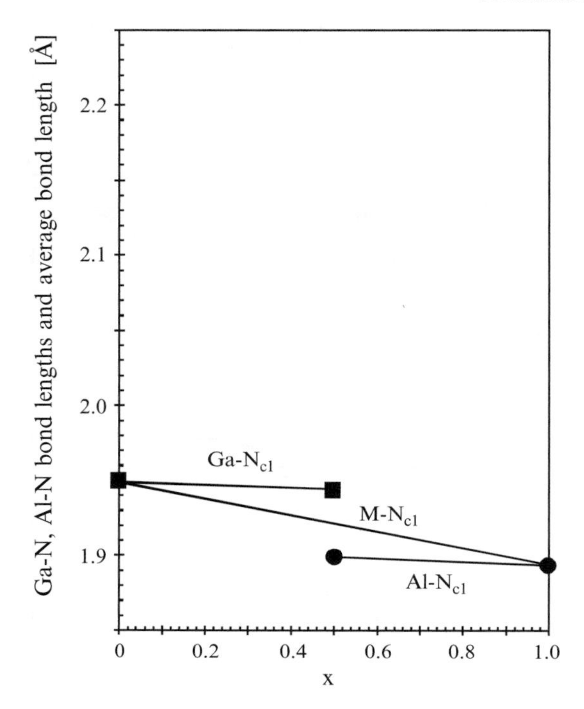

Fig. 2b Calculated *Ga-N* and *Al-N* bond lengths for relaxed, random AlGaN alloys versus alloy composition [32–34]. The bond length along the c-axis is predicted, assuming a constant c/a-ratio of 1.633. In addition the average bond length *M-N* of wurtzite AlGaN alloys is shown.

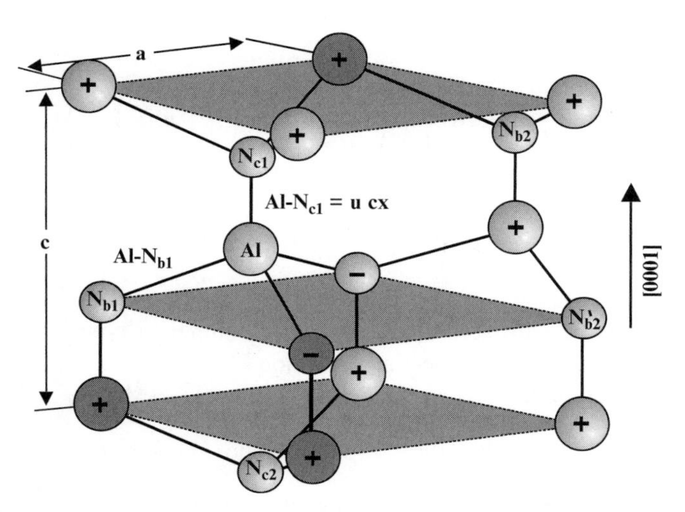

Fig. 3 Schematic drawing of a wurtzite AlN crystal with lattice constants a and c. The sign of the charge of metal and nitrogen atoms are indicated as well as the relevant positions of the nearest and second nearest neighbors to one Al-atom.

Table 1 Calculated cell-internal parameters, lattice constants, cation-anion distances between nearest and second nearest neighbors (given in Å) as well as bond angles (given in degree) of binary, wurtzite group-III-nitrides with ideal crystal structure or the predicted "real" structure.

	InN		GaN		AlN	
	ideal	real	ideal	real	ideal	real
u	0.375	0.379	0.375	0.377	0.375	0.382
a	3.585	3.585	3.199	3.199	3.110	3.110
c/a	1.633	1.618	1.633	1.634	1.633	1.606
$M-N_{c1}$	2.195	2.200	1.959	1.971	1.904	1.907
$M-N_{b1}$	2.195	2.185	1.959	1.955	1.904	1.890
$M-N_{c2}$	3.659	3.600	3.265	3.255	3.174	3.087
$M-N_{b2}$	4.204	4.206	3.751	3.757	3.646	3.648
$M-N_{b'2}$	4.204	4.198	3.751	3.749	3.646	3.639
α	109.47	108.69	109.47	109.17	109.47	108.19
β	109.47	110.24	109.47	109.78	109.47	110.73

It should be noticed that in the case of an ideal ratio of lattice constants $\frac{c}{a} = \sqrt{\frac{8}{3}} = 1.633$ and ideal cell-internal parameter $u = \frac{8}{3} = 0.375$ (ideal means most dense hexagonal sphere packing) it follows from equations (5–10) that the bond length and the bond angles between the nearest neighbors ($\alpha = \beta = 109.47°$, Table 1) are equal, but the distance to the second nearest neighbor along the c-axis is about 13% shorter than the distance to the other second nearest neighbors (this is not the case in the cubic structure [33]).

It is known from experiment as well as theoretical predictions that neither the cell-internal parameter u, nor the c/a-ratio is ideal in binary group-III-nitrides (see Fig. 4a, 4b) [2, 29–31]. In order to evaluate the consequences of the non-ideality of the wurtzite structure on polarization and polarization induced effects in group-III-nitrides we have calculated the average bond lengths and angles as well as the second neighbor distances (in the case of the virtual crystal limit) of ternary alloys taking advantage of the measured and calculated lattice constants and average u parameter. The average cell-internal parameter is defined as the average value of the projection of the connecting vector pointing from a nitrogen atom to its nearest neighbor metal atom in the [0001] direction (Fig. 3).

The u parameter has been calculated for randomly distributed $A_{0.5}B_{0.5}N$ alloys by the theoretical approach described above (more detailed information can be found in Ref. [35]) and can be approximated by the following quadratic equations:

$$u_{ABN}(x) = u_{AN}x + u_{BN}(1-x) + bx(1-x), \tag{11}$$

$$\begin{aligned} u_{AlGaN}(x) &= 0.3819x + 0.3772(1-x) - 0.0032x(1-x), \\ u_{InGaN}(x) &= 0.3793x + 0.3772(1-x) - 0.0057x(1-x), \\ u_{AlInN}(x) &= 0.3819x + 0.3793(1-x) - 0.0086x(1-x). \end{aligned} \tag{12}$$

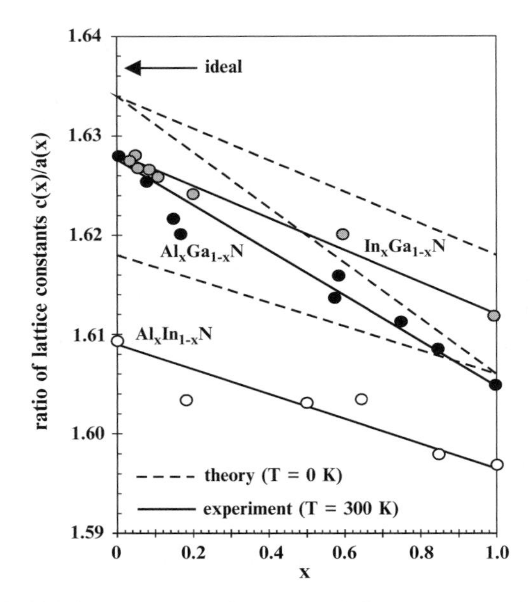

Fig. 4a c/a-ratio of random, ternary alloys of group-III-nitrides measured by high resolution X-ray diffraction (HRXRD) at room temperature (solid line) and calculated by the method described in chapter II for $T = 0K$ (dashed line). The measured and calculated data confirm that the c/a-ratio of wurtzite InGaN, AlGaN and AlInN crystals are always below the value of an ideal hexagonal crystal with $c/a = 1.633$ [29–31].

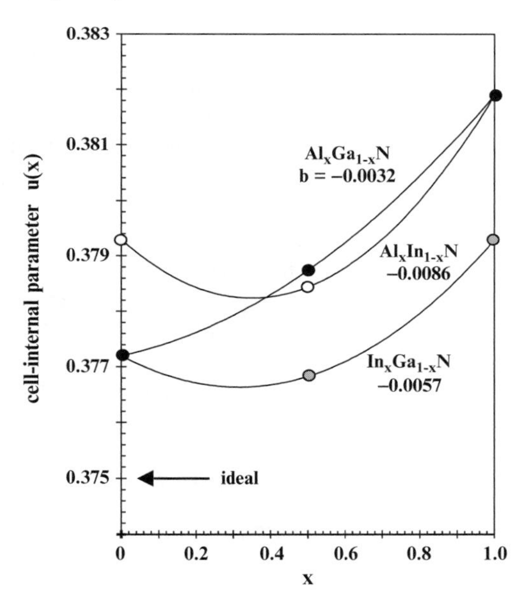

Fig. 4b The cell-internal parameter u, of random AlGaN, InGaN and AlInN alloys predicted by the theory described in chapter II and approximated by a quadratic equation (11). The non-linearity of the cell-internal parameter in dependence on alloy composition can be described by a negative bowing parameter b, given in the figure. The u parameter of the alloys is always above the value of an ideal hexagonal crystal ($u_{ideal} = 0.375$).

The cell-internal parameter is increasing from GaN to InN and, more significantly, to AlN. The non-linear dependence on alloy composition is described by a negative bowing parameter which value increases from AlGaN to InGaN and AlInN. It should be pointed out, that although the bowing parameters are negative, the average cell-internal parameter of random alloys is always above the ideal value ($u_{ABN} > u_{ideal} = 0.375$). If the lattice constants scale linearly with the alloy composition but the internal parameter does not, the bond angles and/or the bond lengths of the real and the virtual crystal must depend non-linearly on the alloy composition. The average bond lengths, angles and second nearest neighbor distances calculated by using equations (5–10) are shown in Fig. 5a-c and listed in Table 1 and 2. The average cation-anion distances to the nearest and second nearest neighbors scale nearly linearly with alloy composition for AlGaN, InGaN and AlInN. The average bond length along the c-axis is 0.7 to 0.9% longer than the nearest neighbor bonds in direction of the basal plane (Fig. 5a). This difference becomes smaller for In-containing alloys. The ratio r, of the distances to the second nearest neighbors along the c-axis ($M - N_{c2}$) and along the basal plane ($M - N_{b2}, M - N_{b'2}$ (Fig. 5c) is well above the ratio of the ideal structure ($r = \{1 - (M - N_{c2}/M - N_{b2})\}$ $100\% = 13\%$) and increases non-linearly from GaN to InN, reaching 15.4% for AlN (Fig. 5d). The difference between the ideal bond angle $\alpha_{\text{ideal}} = 109.47°$ and the

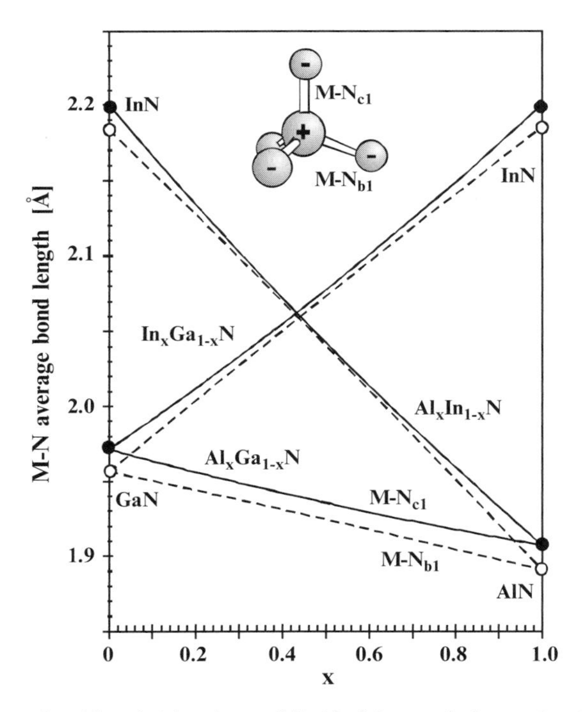

Fig. 5a The average bond length (virtual crystal limit) of the metal-nitrogen bond along the c-axis (solid line) and in the direction of the basal plane (dashed line) in dependence of alloy composition.

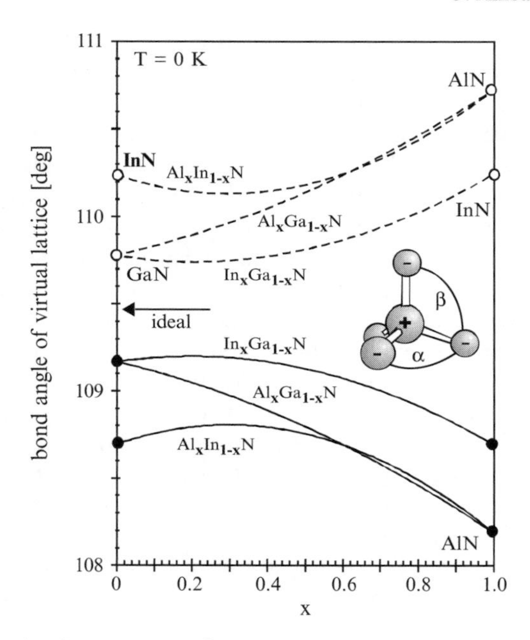

Fig. 5b The average bond angles α and β of random AlGaN, InGaN and AlInN alloys versus x. The average bond angles deviate significantly from the bond angle of an ideal hexagonal crystal, where $\alpha = \beta = 109.47°$. The deviation is increasing from GaN to InN and AlN in a lon-linear way.

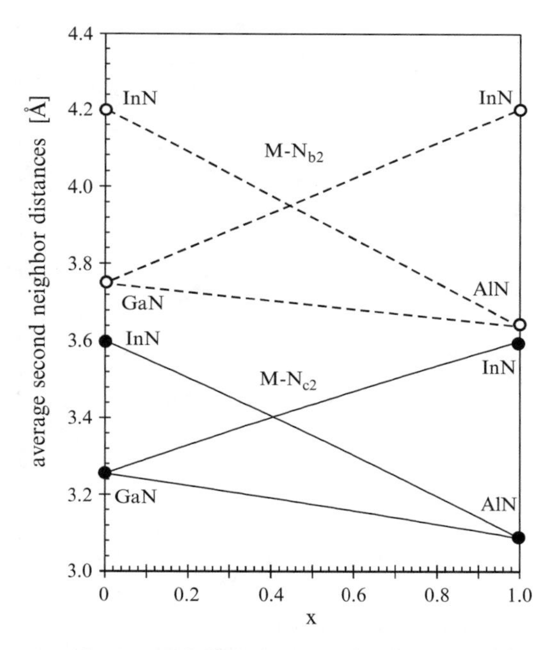

Fig. 5c The average second nearest neighbor distances of metal and nitrogen atoms in alloys with wurtzite crystal structure versus alloy composition. The second nearest neighbor distance along the c-axis is 13 to 15% shorter than the $M - N_{b2}$ distances along the basal planes (see also Table 1 and 2).

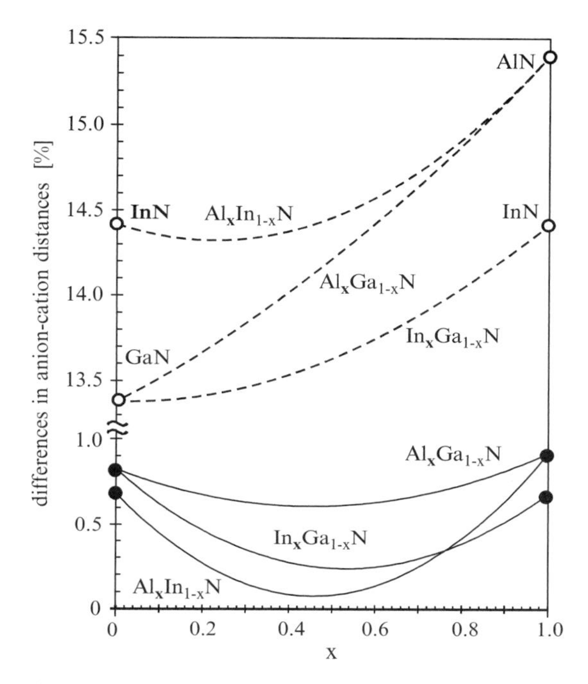

Fig. 5d The ratios $(r = \{1-(M-N_{c2}/M-N_{b2})\}\ 100\%)$ between the nearest and second nearest neighbor distances along the c-axis and in the direction of the basal plane are shown in dependence on alloy composition.

Table 2 Predicted average (virtual crystal limit) cell-internal parameters, lattice constants, cation-anion distances between nearest and second nearest neighbors (given in Å) as well as bond angles (given in degree) of ternary random alloys with $x = 0.5$.

	$Al_{0.5}Ga_{0.5}N$	$Al_{0.5}In_{0.5}N$	$In_{0.5}Ga_{0.5}N$
u	0.379	0.378	0.377
a	3.154	3.347	3.392
c/a	1.620	1.612	1.625
$M-N_{cl}$	1.935	2.042	2.078
$M-N_{bl}$	1.924	2.041	2.073
$M-N_{c2}$	3.175	3.354	3.436
$M-N_{b2}$	3.701	3.921	3.977
$M-N_{b'2}$	3.694	3.920	3.975
α	108.80	108.76	109.13
β	110.14	110.18	109.81

average bond angles α and β increases also non-linearly from GaN to InN and AlN reaching values of $\Delta\beta = \Delta\alpha = 1.25°$ (Fig. 5b).

Beside the significant deviation of the group-III-nitrides from the ideal wurtzite structure it should be noticed that the observed non-linearities in the cell-internal parameter, cation-anion distances and bond lengths of the random alloys always tend to decrease the difference between the "real" and ideal structure. An additional important structural property of group-III-nitrides with wurtzite structure, which strongly effects the orientation of the spontaneous and piezoelectric polarization, is the polarity of the crystals.

4 Polarity

Noncentrosymmetric compound crystals exhibit two different sequences of atomic layering in the two opposing directions parallel to certain crystallographic axes. Consequently, crystallographic polarity along these axes can be observed. For binary A-B compounds with wurtzite structure, the sequence of the atomic layers of the constituents A and B is reversed along the $[0001]$ and $[000\bar{1}]$ directions. The corresponding (0001) and $(000\bar{1})$ faces are the A-face and B-face, respectively. In the case of heteroepitaxial growth of thin films of a noncentrosymmetric compound, the polarity of the material cannot be predicted in a straightforward way, and must be determined by experiments. This is the case for group-III-nitrides with wurtzite structure and GaN-based heterostructures with the most common growth direction normal to the $\{0001\}$ basal plane, where the atoms are arranged in bilayers. These bilayers consist of two closely spaced hexagonal layers, one formed by cations and the other formed by anions, leading to polar faces. Thus, in the case of GaN, a basal surface should be either Ga- or N-faced. Ga-faced means Ga-atoms are placed on the top position of the $\{0001\}$ bilayer, corresponding to the $[0001]$ polarity (Fig. 4) (by convention, the $[0001]$ direction is given by a vector pointing from a Ga atom to a nearest-neighbor N atom). However, it is important to notice that (0001) and $(000\bar{1})$ surfaces of GaN are non-equivalent and differ in their chemical and physical properties [36].

Both types of polarity were reported to be found by ion channeling and convergent beam electron diffraction in GaN (0001) layers grown by metal-organic chemical vapor deposition (MOCVD) on c-plane sapphire if the layers exhibited rough morphology, while for smooth films Ga-face polarity was exclusively concluded from the experimental results [37]. This result was supported by a photoelectron diffraction study of MOCVD grown films [38]. Smith *et al.* reported on investigations of surface reconstructions of GaN grown by plasma-induced molecular-beam epitaxy (PIMBE) on c-plane sapphire and PIMBE homoepitaxy on a MOCVD grown GaN/sapphire substrate [39, 40]. They observed two structurally non-equivalent faces with completely different surface reconstructions attributed to the N-face for PIMBE growth on sapphire and to the Ga-face for PIMBE deposition on a MOCVD template.

Furthermore, the polarity of the investigated samples (see chapter V) was determined by the x-ray standing wave method (XSW) [41]. The advantage of this technique is the combination of the structural sensitivity of x-ray diffraction with the chemical elemental sensitivity inherent to x-ray spectroscopy. The method is based on generating an XSW field by x-ray Bragg diffraction and monitoring the x-ray fluorescence yield excited by this field as a function of glancing angle as the GaN layer is turned through the narrow region of Bragg reflection. In the case of the AlGaN/GaN, InGaN/GaN and AlInN/GaN hetero- and nanostructures grown on $c-Al_2O_3$ substrates covered with a thin AlN nucleation layer the XSW technique was successfully applied. The standing wave was generated by x-ray diffraction inside the GaN buffer and the GaK_α fluorescence yield was recorded as a function of the incident angle within the width of the (0002) reflection peak [42]. In these studies the samples grown by MOCVD as well as by PIMBE on AlN nucleation layer were determined to have Ga-face polarity exclusively.

5 Growth of Undoped AlGaN/GaN, InGaN/GaN and AlInN/GaN Hetero- and Nanostructures

The investigated pseudomorphic Ga-face AlGaN/GaN, InGaN/GaN and AlInN/GaN heterostructures were grown by low-pressure metal-organic chemical vapor deposition (MOCVD) or plasma-induced molecular beam epitaxy (PIMBE).

5.1 AlGaN/GaN Heterostructures

The MOCVD grown undoped AlGaN/GaN heterostructures were deposited at a pressure of 100 mbar, using triethylgallium (TEG), trimethylaluminium (TMA) and ammonia as precursors. A growth rate of about $0.5\,\mu m/h$ was achieved for a substrate temperature of 1040°C and V/III gas phase ratios of 1800 and about 900 for GaN and AlGaN, respectively [43]. The AlGaN/GaN heterostructures as well as the $Al_{0.1}Ga_{0.9}N$ nucleation layers were grown at a constant substrate temperature and without any growth interruptions to avoid the presence of excessively high free carrier background concentrations. $Al_xGa_{1-x}N$ barriers with thicknesses between 100 and 450 Å and alloy compositions of up to $x = 0.45$ were deposited on GaN buffer layers with thicknesses between 1 and $2.5\,\mu m$.

AlGaN/GaN heterostructures were also deposited by PIMBE using conventional effusion cells and a radio frequency plasma source for the generation of nitrogen radicals. The nitrogen flux through the plasma source was fixed at about 1 sccm causing a nitrogen partial pressure in the PIMBE chamber of $4\cdot 10^{-5}$ mbar during growth. The optimized growth temperature for GaN was determined to be between 780 and 800°C for a Ga flux of $1.5\cdot 10^{15}\,cm^{-2}s^{-1}$ leading to a growth rate of $0.5\,\mu m/h$, as discussed in more detail in Ref. [23, 44]. AlGaN barriers with thicknesses between 2 and 50 nm were grown on GaN buffer layers using a total flux of Al- and Ga-atoms equal to the Ga-flux optimized for the deposition of high quality GaN.

5.2 InGaN/GaN Heterostructures

MOCVD grown $In_{0.13}Ga_{0.87}N/GaN$ single quantum well structures (QWs) with well widths from 0.9 to 54 nm were embedded between a 2.7 μm thick GaN buffer layer, deposited on top of a low temperature nucleation layer, and a 130 nm thick GaN cap layer. The InGaN layers were grown at a substrate temperature of 770°C, which is 360°C lower than for the growth of GaN. The InGaN composition and layer thickness was determined by combination of secondary ion mass spectroscopy (SIMS), high resolution x-ray diffraction (HRXRD) and Raman spectroscopy [45,46]

5.3 AlInN/GaN Heterostructures

$Al_{1-x}In_xN$/GaN/AlN (50/540/80 nm) heterostructures with In-concentrations between $x = 0.12$ and 0.21 in the barrier were also grown by MOCVD. The AlN nucleation layer, GaN buffer layer and AlInN barrier were deposited at substrate temperatures of 900° C, 1000° C and 750° C, respectively. The In-concentration of the barrier was increased by increasing the flux of the Indium precursor (trimethylindium, TMIn) from 7.5 to 42 μmol/min keeping the flux of the Aluminum precursor (TMAl), the transport gas N_2 and the ammonia constant at 3.4 μmol/min, 2000 sccm and 4250 sccm, respectively [47].

Before we discuss the experimentally observed polarization induced effects in the grown hetero- and nanostructures we will provide the theoretical background for a detailed understanding of non-linear spontaneous and piezoelectric polarization in binary and ternary group-III-nitrides.

6 Non-Linear Spontaneous and Piezoelectric Polarization in Group-III-Nitrides

In modeling of polarization induced effects in GaN based electronic and optoelectronic devices it is often assumed that polarization in group-III-nitride alloys interpolates linearly between the limiting values determined by the binary compounds. It is known from earlier work on III-V alloys with zincblende structure that, although Vegard's law is found to be valid, physical properties like the band gap are non-linear functions of the alloy composition. These non-linearities can be caused by different cation electronegativity of the components, internal strain effects due to varying cation-anion bond length and disorder effects due to random distribution of the chemical elements on the cation side.

6.1 Spontaneous Polarization

Motivated by the observed non-linearities in the structural parameters, the pyroelectric properties of alloys have been checked for possible non-linearities. Zoroddu *et al.* pointed out that the spontaneous polarization of relaxed alloys for a given composition depends linearly on the average u parameter, which indicates that spontaneous polarization differences between alloys of the same composition are mainly due to varying cation-anion bond length, whereas disorder has a negligible influence [2]. This idea is supported by the fact that in binaries the polarization is strongly influenced by the relative displacement of the cation and anion sub-lattices in the [0001] direction [1,35]. In addition, theoretical calculations provided by N. Ashcroft *et al.* of the u parameter and spontaneous polarization of different binary compounds with wurtzite crystal structure, strongly indicate a nearly linear dependence between the cell-internal parameter and P_{SP} (Fig. 6). Because of the non-linear dependence of the cell-internal parameter on alloy composition, a non-linear behaviour of the spontaneous polarization versus x has to be expected for ternary alloys.

Beside the non-linearities which are caused by the structural properties, Bernardini *et al.* pointed out that the different cation electronegativities should contribute significantly to the non-linear behavior of spontaneous polarization in ternary random alloys. A moderate bowing is calculated for AlGaN which is increasing to InGaN and AlInN mainly due to the differences of the dynamical charges of Al-N, Ga-N

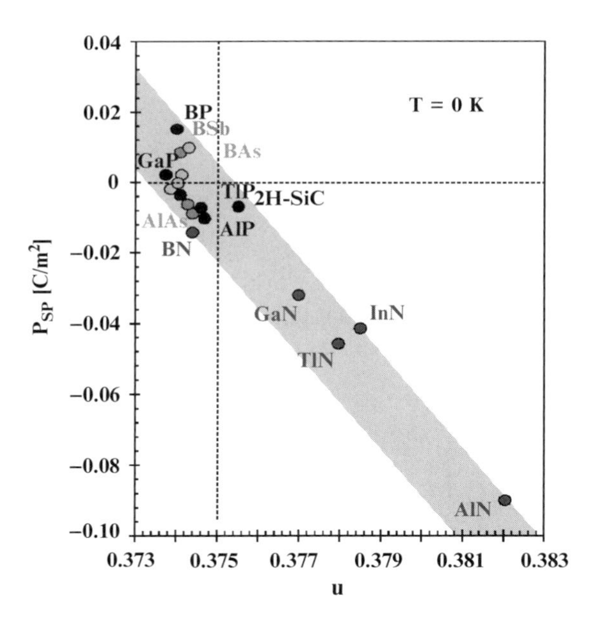

Fig. 6 Theoretical predictions of the spontaneous polarization in wurtzite compound semiconductors (private communication N. Ashcroft *et al.*).

and In-N bonds which are found to be 2.653, 2.670 and 3.105, respectively [1,2,35]. We have used the theoretical approach described in chapter II to calculate the spontaneous polarization of relaxed $A_{0.5}B_{0.5}N$ alloys (listed in Table 2) in order to determine the bowing parameter of P^{SP}_{ABN} as a function of x. The spontaneous polarization of the random ternary group-III-nitride alloys is given to second order in x by (in $\mathrm{C/m^2}$):

$$\begin{aligned} P^{SP}_{ABN}(x) &= P^{SP}_{AN}x + P^{SP}_{BN}(1-x) + bx(1-x), \\ P^{SP}_{AlGaN}(x) &= -0.090x - 0.034(1-x) + 0.021x(1-x), \\ P^{SP}_{InGaN}(x) &= -0.042x - 0.034(1-x) + 0.037x(1-x), \\ P^{SP}_{AlInN}(x) &= -0.090x - 0.042(1-x) + 0.070x(1-x). \end{aligned} \tag{13}$$

The first two terms in the equations are the usual linear interpolation between the binary compounds. The third term is embodying non-linearity to quadratic order (see also Fig. 7). Higher order terms are neglected as their effect was estimated to be less than 10%. Knowledge of the spontaneous polarization in alloys is not sufficient to describe the pyroelectric properties of alloys and GaN-based nanostructures or to predict the values of polarization induced interface and surface charges as well as electrostatic fields. GaN-based hetero- and nanostructures are usually grown

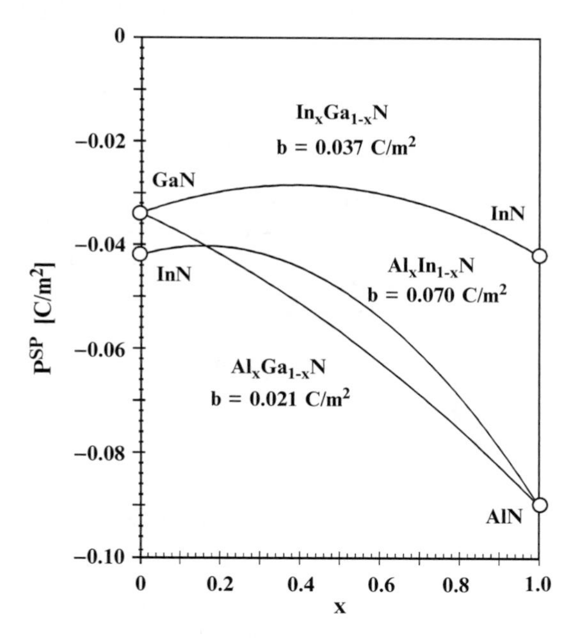

Fig. 7 Predicted spontaneous polarization for random, ternary alloys with wurtzite crystal structure. The dependence of P^{SP} on x can be approximated with high accuracy by quadratic equations (13). The non-linearity can be described by positive bowing parameters increasing from AlGaN to InGaN and AlInN.

pseudomorphically and strained on substrates and buffer layers with significant mismatch in lattice constants and thermal expansion coefficients.

The ensuing symmetry-conserving strain causes a change in the crystal structure and in polarization that amounts to a piezoelectric polarization. Therefore the polarization of every layer is a combination of spontaneous and piezoelectric polarization.

6.2 *Piezoelectric Polarization*

By Hook's law the deformation of a crystal ε_{kl}, due to external or internal forces or stresses σ_{ij} can be described by:

$$\sigma_{ij} = \sum_{k,l} C_{ijkl}\varepsilon_{kl}, \tag{14}$$

where C_{ijkl} is the elastic tensor. Due to spacial symmetry this fourth rank tensor can be reduced to a 6×6 matrix using the Voigt notation: xx $\rightarrow 1$, yy $\rightarrow 2$, zz $\rightarrow 3$, yz, zy $\rightarrow 4$, zx, xz $\rightarrow 5$, xy, yx $\rightarrow 6$. The elements of the elastic tensor can be rewritten as $C_{ijkl} = C_{mn}$ were i, j, k, l = x, y, z and m, n = 1,...,6. Using this notation Hook's law can be simplified to:

$$\sigma_i = \sum_j C_{ij}\varepsilon_j. \tag{15}$$

The 6×6 matrix of the elastic constants C_{ij}, (see Table 3 [48]) for crystals with wurtzite structure is given by:

$$C_{ij} = \begin{pmatrix} C_{11} & C_{12} & C_{13} & 0 & 0 & 0 \\ C_{12} & C_{11} & C_{13} & 0 & 0 & 0 \\ C_{13} & C_{13} & C_{33} & 0 & 0 & 0 \\ 0 & 0 & 0 & C_{44} & 0 & 0 \\ 0 & 0 & 0 & 0 & C_{44} & 0 \\ 0 & 0 & 0 & 0 & 0 & \frac{1}{2}(C_{11}-C_{12}) \end{pmatrix}. \tag{16}$$

Another useful way to describe the mechanical properties of the crystals is a complete set of the elastic compliance constants. These parameters are calculated taking advantage of the elastic constants listed in Table 3 and using the relations:

$$S_{11} = \frac{C_{11}C_{33} - C_{13}^2}{(C_{11}-C_{12})(C_{33}(C_{11}+C_{12}) - 2C_{13}^2)}, \tag{17}$$

$$S_{12} = -\frac{C_{12}C_{33} - C_{13}^2}{(C_{11}-C_{12})(C_{33}(C_{11}+C_{12}) - 2C_{13}^2)}, \tag{18}$$

$$S_{13} = -\frac{C_{13}}{C_{33}(C_{11}+C_{12}) - 2C_{13}^2}, \tag{19}$$

Table 3 Experimental and predicted elastic, elastic compliance and piezoelectric constants as well as the biaxial strain coefficient of wurtzite binary group-III-nitrides and 6H-SiC at room temperature (theory 1 [48], theory 2 [1,2], experimental results: a) [10], b) [7], c) [49] and d) [11]).

hexagonal crystal property	InN		GaN			AlN		
	theory 1	theory 2	theory 1	theory 2	exp.	theory 1	theory 2	exp.
C_{11} [GPa]	223		367		370[a)]	396		410[a)]
C_{12} [GPa]	115		135		145[a)]	137		140[a)]
C_{13} [GPa]	92	70	103	68	110[a)]	108	94	100[a)]
C_{33} [GPa]	224	205	405	354	390[a)]	373	377	390[a)]
C_{44} [GPa]	48		95		90[a)]	116		120[a)]
$\nu(0001)$	0.82	0.68	0.52	0.38	0.56	0.58	0.50	0.51
$S_{11}[10^{-12}\mathrm{N/m^2}]$	6.535		3.267		3.326	2.993		2.854
$S_{12}[10^{-12}\mathrm{N/m^2}]$	−2.724		−1.043		−1.118	−0.868		−0.849
$S_{13}[10^{-12}\mathrm{N/m^2}]$	−1.565		−0.566		−0.623	−0.615		−0.514
$S_{33}[10^{-12}\mathrm{N/m^2}]$	5.750		2.757		2.915	3.037		2.828
$S_{44}[10^{-12}\mathrm{N/m^2}]$	20.83		10.53		11.11	8.621		8.333
e_{31} $[\mathrm{C/m^2}]$		−0.41		−0.34			−0.53	−0.58[d)]
e_{33} $[\mathrm{C/m^2}]$		0.81		0.67			1.50	1.55[d)]
e_{15} $[\mathrm{C/m^2}]$			−0.22[b)]		−0.30[c)]			−0.48[d)]
$d_{31}[10^{-12}\mathrm{C/m^2Pa}]$		−3.147		−1.253			−2.298	−2.65
$d_{33}[10^{-12}\mathrm{C/m^2Pa}]$		6.201		2.291			5.352	5.53
$d_{15}[10^{-12}\mathrm{C/m^2Pa}]$		−2.292		−1.579			−2.069	−4.08

$$S_{33} = \frac{C_{11}+C_{12}}{C_{33}(C_{11}+C_{12})-2C_{13}^2} \tag{20}$$

and

$$S_{44} = \frac{1}{C_{44}}. \tag{21}$$

Using the compliance constants a helpful figure of merit to determine the directional hardness and the reciprocal Young's modulus as a function of orientation to the crystal axis, can be provided. For hexagonal materials the reciprocal Young's modulus S_{11}^* along an arbitrary direction at an angle Θ with respect to the $[0001]$-axis is given by [51]:

$$S_{11}^* = S_{11}\sin^4(\theta)+S_{33}\cos^4(\theta)+(S_{44}+2S_{13})\sin^2(\theta)\cos^2(\theta). \tag{22}$$

In Fig. 8a and 8b polar plots of S_{11}^* for InN, GaN and AlN as a function of the direction with respect to the $[0001]$-axis and in the basal plane ($\Theta = 0$) are shown.

It is seen that AlN and GaN are more than twice as hard as InN. The hardness of AlN is almost isotropic whereas GaN and InN show preferential "softness" along the $[0001]$ and $[2\bar{1}\bar{1}0]$-axis.

Of main interest here is the fact that the hardness of all binary compounds with wurtzite structure is isotropic in the basal plane. This is important because the strain in epitaxial layers of group-III-nitride heterostructures grown along the c-axis

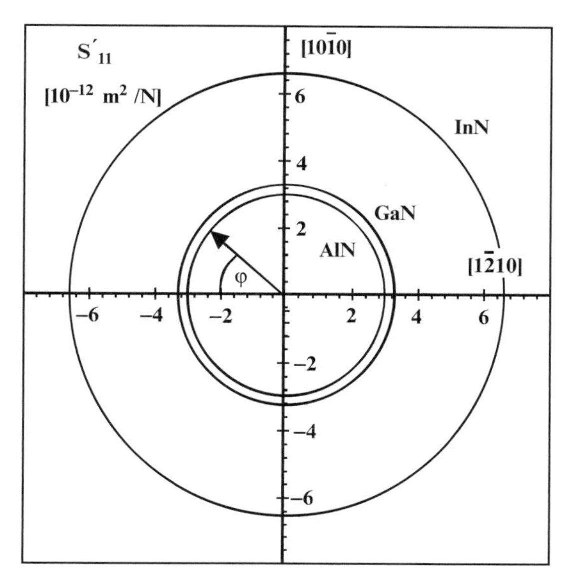

Fig. 8a The reciprocal Young's modulus in the basal plane of InN, GaN and AlN. The stiffness of AlN and GaN is more than twice as high as for InN. More important the hardness of the wurtzite crystals is isotropic in the basal plane.

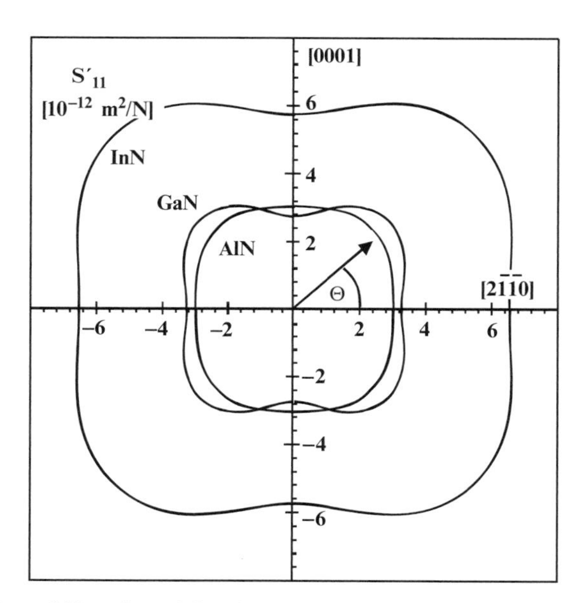

Fig. 8b The reciprocal Young's modulus along an arbitrary direction at an angle Θ with respect to the c-axis. In this plane the stiffness of AlN is isotropic whereas GaN and InN show preferential "softness" along the $[0001]$- and $[2\bar{1}\bar{1}0]$-directions.

([0001]-axis) caused by mismatch of the lattice constants a and/or a mismatch of the thermal expansion coefficients of layer and substrate, is directed along the basal plane (parallel to the substrate). No force is applied in growth direction and the crystal is free to move. The resulting biaxial strain ($\varepsilon_1 = \varepsilon_2$) causes stresses $\sigma_1 = \sigma_2$, whereas σ_3 has to be zero. Using equations (15) and (16) a relation between the strain along the c-axis and along the basal plane can be derived:

$$\varepsilon_3 = -2\frac{C_{13}}{C_{33}}\varepsilon_1, \tag{23}$$

where

$$\nu(0001) = 2\frac{C_{13}}{C_{33}} \text{ is the biaxial strain coefficient and} \tag{24}$$

$$\varepsilon_1 = \frac{a - a_0}{a_0}, \ \varepsilon_3 = \frac{c - c_0}{c_0} \tag{25}$$

are the relative changes of the lattice constants a and c with respect to the constants of the relaxed crystals a_0 and c_0. The stresses in the basal plane caused by the mismatch in the lattice constants can be calculated by:

$$\sigma_1 = \varepsilon_1 \left(C_{11} + C_{12} - 2\frac{C_{13}^2}{C_{33}} \right), \tag{26}$$

where

$$C_{11} + C_{12} - 2\frac{C_{13}^2}{C_{33}} > 0. \tag{27}$$

The piezoelectric polarization for hexagonal crystals belonging to the C_{6v} crystallographic point group [50] is given by:

$$P_i^{pz} = \sum_l d_{il}\sigma_l, i = 1,2,3, l = 1,\ldots,6, \tag{28}$$

where P_i^{pz} are the components of the piezoelectric polarization and d_{il} are the piezoelectric moduli (Tab. 3). Using the relations given by symmetry between the piezoelectric moduli, $d_{31} = d_{32}$, $d_{33} \neq 0$, $d_{15} = d_{24}$ and considering that all other components are $d_{il} = 0$, equation (28) can be expanded into:

$$P_1^{pz} = \frac{1}{2}d_{15}\sigma_5, \tag{29}$$

$$P_2^{pz} = \frac{1}{2}d_{15}\sigma_4, \tag{30}$$

$$P_3^{pz} = d_{31}(\sigma_1 + \sigma_2) + d_{33}\sigma_3. \tag{31}$$

Keeping in mind that for biaxial stress, as it is of interest here, $\sigma_1 = \sigma_2$, $\sigma_3 = 0$ and sheer stresses are assumed to be negligible ($\sigma_4 = \sigma_5 = 0$), the piezoelectric polarization has only one not vanishing component, which is directed along the growth direction and given by:

$$P_3^{pz} = 2d_{31}\sigma_1 = 2d_{31}\varepsilon_1\left(C_{11} + C_{12} - 2\frac{C_{13}^2}{C_{33}}\right). \tag{32}$$

More often than the piezoelectric moduli the piezoelectric constants e_{kl} are used to describe the piezoelectric properties of group-III-nitrides. They can be calculated by:

$$e_{kl} = \sum_j d_{kj}C_{jl}, \quad \text{where } k = 1,2,3, l = 1,\ldots,6, j = 1,\ldots,6. \tag{33}$$

For hexagonal crystals the relations between piezoelectric constants and modulus can be reduced to:

$$e_{31} = e_{32} = C_{11}d_{31} + C_{12}d_{32} + C_{13}d_{33} = (C_{11} + C_{12})d_{31} + C_{13}d_{33}, \tag{34}$$

$$e_{33} = 2C_{13}d_{31} + C_{33}d_{33}, \tag{35}$$

$$e_{15} = e_{24} = C_{44}d_{15}, \text{and for all other components:} \tag{36}$$

$$e_{kl} = 0. \tag{37}$$

The piezoelectric polarization as a function of strain can be written as:

$$P_k^{pz} = \sum_l e_{kl}\varepsilon_l, \quad \text{where } k = 1,2,3, l = 1,\ldots,6. \tag{38}$$

The non-vanishing component of the piezoelectric polarization caused by biaxial strain is:

$$P_3^{pz} = \varepsilon_1 e_{31} + \varepsilon_2 e_{32} + \varepsilon_3 e_{33}, \tag{39}$$

$$= 2\varepsilon_1 e_{31} + \varepsilon_3 e_{33}, \qquad \text{where} \quad \varepsilon_3 = -2\frac{C_{13}}{C_{33}}\varepsilon_1, \tag{40}$$

$$= 2\varepsilon_1\left(e_{31} - e_{33}\frac{C_{13}}{C_{33}}\right), \qquad \text{where} \quad \varepsilon_1 = \frac{a - a_0}{a_0}. \tag{41}$$

In analogy to the determination of piezoelectric polarization along the c-axis in dependence of biaxial strain, equations can be derived in case of uniaxial ($\sigma_1 = \sigma_2 = 0, \sigma_3 \neq 0$) as well as for hydrostatic strain ($\sigma_1 = \sigma_2 = \sigma_3$):

$$\text{uniaxial strain:} \qquad P_3^{pz} = \varepsilon_1\frac{C_{11} - C_{12}}{C_{11} - \frac{C_{13}^2}{C_{33}}}\left(e_{31} - e_{33}\frac{C_{13}}{C_{33}}\right), \tag{42}$$

$$\text{hydrostatic strain:} \qquad P_3^{pz} = \varepsilon_1\left(2e_{31} + e_{33}\frac{C_{11} + C_{12} - C_{13}}{C_{33} - C_{13}}\right). \tag{43}$$

It is interesting to notice that, beside the fact that the piezoelectric polarization along the c-axis is linearly dependent on the relative change of the lattice constant a, P_3^{pz} is

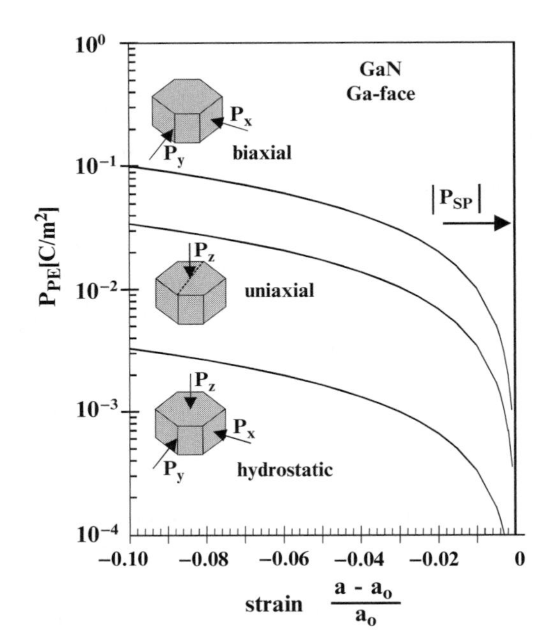

Fig. 9 Piezoelectric polarization along the c-axis versus strain of wurtzite GaN caused by biaxial, uniaxial or hydrostatic pressure.

always negative for GaN layers under tensile strain ε_1 and positive for crystals under compressive strain caused by uniaxial, biaxial or hydrostatic pressure (see Fig. 9). As a consequence, the spontaneous and the piezoelectric polarization are pointing along the $[000\bar{1}]$ direction for tensile strained GaN layers whereas in crystals under compressive strain the piezoelectric polarization is oriented along the [0001]-axis (anti-parallel to the spontaneous polarization).

Because of its relevance for group-III-nitride based devices, P_3^{pz} (we will skip the index 3 in the following discussion) is determined for alloys pseudomorphic grown on relaxed InN, GaN and AlN buffer layers. Under these assumptions the biaxial strain can be calculated by:

$$\varepsilon_1 = \frac{a_{buffer} - a(x)}{a(x)}. \tag{44}$$

As a first approach the lattice- (see equations (4)), piezoelectric- and elastic constants of the alloys are approximated by a linear interpolation between the constants of the relevant binary compounds and implemented into equation (41). For the following calculations we have used the physical properties of the binary compounds predicted by Zoroddu *et al.* (Tab. 3, theory 2) [2]. Fig. 10a–c show the piezoelectric polarization of the alloys pseudomorphically grown on binary buffer layers for different values of x. Very high values of piezoelectric polarization covering the range from -0.28 (AlN/InN) to $0.182\,\mathrm{C/m^2}$ (InN/AlN) are observed. In addition, it should be pointed out that the piezoelectric polarization calculated with the same piezoelectric constants as before but using the elastic constants predicted by Wright

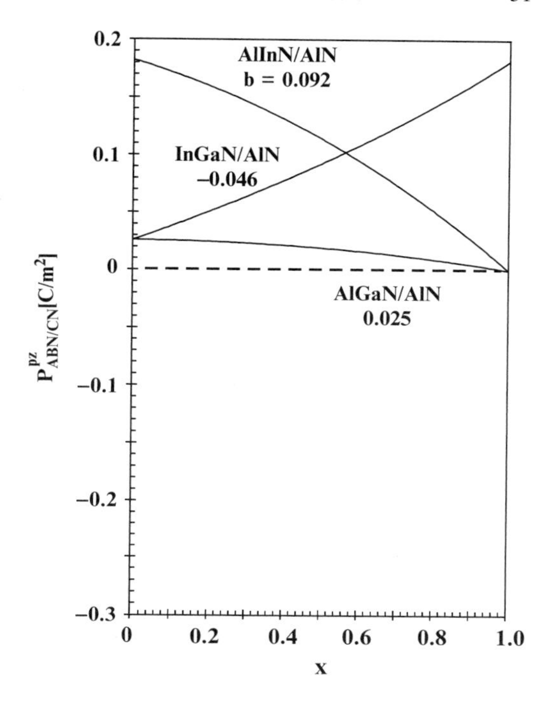

Fig. 10a The piezoelectric polarization of ternary alloys pseudomorphically grown on relaxed AlN buffer layers. For ternary alloys grown on AlN the positive piezoelectric polarization is oriented anti-parallel to the negative spontaneous polarization. The bowing parameters describing the non-linearity of P^{pz} in dependence of x are given in addition.

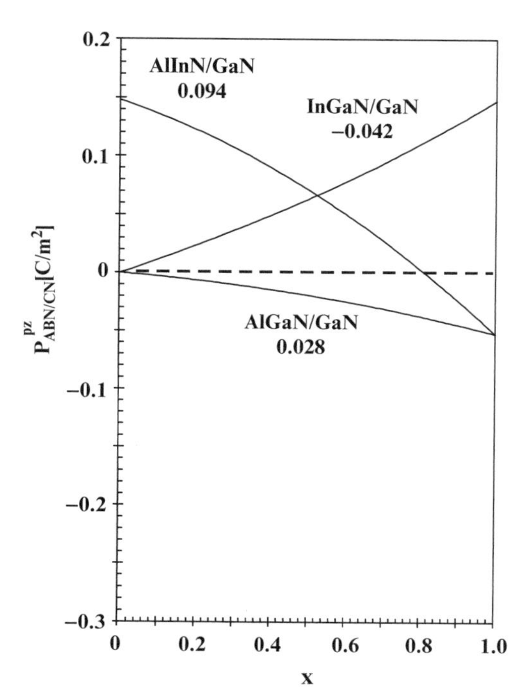

Fig. 10b The piezoelectric polarization of ternary alloys pseudomorphically grown on relaxed GaN buffer layers. For lattice matched $Al_{0.82}Ga_{0.18}N/GaN$ heterostructures the piezoelectric polarization equals zero. The bowing parameters describing the non-linearity of P^{pz} in dependence of x are given in addition.

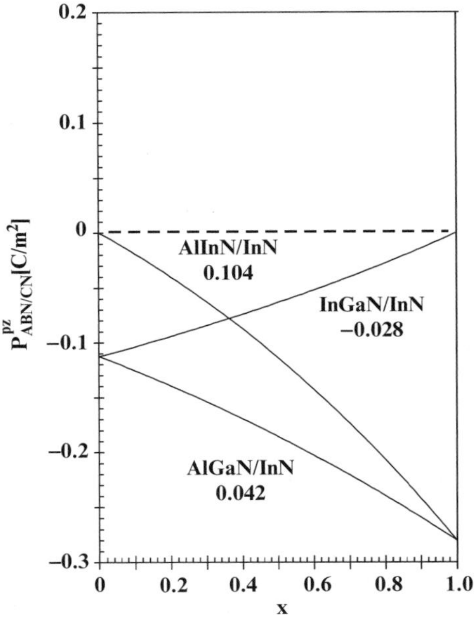

Fig. 10c The piezoelectric polarization of ternary alloys pseudomorphically grown on relaxed InN buffer layers. For ternary alloys grown on InN the negative piezoelectric polarization is oriented parallel to the spontaneous polarization along the c-axis. The bowing parameters describing the non-linearity of P^{pz} in dependence of x are given in addition.

(Table 3, theory 1) or measured by Deger *et al.* [10], P^{pz} deviates from the values shown in Fig. 8a–c by only 10% in the worst case (AlN/InN heterostructure).

As a consequence of equation (41) and the linear interpolation of the elastic and piezoelectric constants, the piezoelectric polarization is non-linear in terms of the alloy composition. The non-linear dependence of the piezoelectric polarization on the alloy composition can be approximated with an accuracy of better than 1% by the following quadratic equations:

$$\begin{aligned} P^{pz}_{AlGaN/InN}(x) &= [-0.28x - 0.113(1-x) + 0.042x(1-x)]C/m^2, \\ P^{pz}_{AlGaN/GaN}(x) &= [-0.0525x + 0.0282x(1-x)]C/m^2, \\ P^{pz}_{AlGaN/AlN}(x) &= [0.026(1-x) + 0.0248x(1-x)]C/m^2, \end{aligned} \tag{45}$$

$$\begin{aligned} P^{pz}_{InGaN/InN}(x) &= [-0.113(1-x) - 0.0276x(1-x)]C/m^2, \\ P^{pz}_{InGaN/GaN}(x) &= [0.148x - 0.0424x(1-x)]C/m^2, \\ P^{pz}_{InGaN/AlN}(x) &= [0.182x + 0.026(1-x) - 0.0456x(1-x)]C/m^2, \end{aligned} \tag{46}$$

$$\begin{aligned} P^{pz}_{AlInN/InN}(x) &= [-0.28x + 0.104x(1-x)]C/m^2, \\ P^{pz}_{AlInN/GaN}(x) &= [-0.0525x + 0.148(1-x) + 0.0938x(1-x)]C/m^2, \\ P^{pz}_{AlInN/AlN}(x) &= [0.182(1-x) + 0.092x(1-x)]C/m^2. \end{aligned} \tag{47}$$

For heterostructures with barriers under moderate strain, these equations can be used as an input, either directly as polarization or, as discussed in more detail later on, as interface bound sheet charge, depending on the implementation, in a self consistent Schrödinger-Poisson solver based e.g. on effective mass or tight binding theory.

Up to now, we have relied on Hooks law assuming that C_{ij} and e_{ij} are constant for a given binary or ternary crystal and that the piezoelectric polarization depends linearly on the strain. But it is known from other piezoelectric semiconductors that, if high forces or pressures are applied to the crystals, this relation can become non-linear [52]. In the following, we first demonstrate that the piezoelectricity of GaN, InN and AlN as well as of group-III-nitride alloys is non-linear in terms of strain, than we suggest how to use this understanding in practice for an improved prediction of the piezoelectric polarization of random alloys caused by high biaxial strain. We consider the technologically most relevant case of an alloy pseudomorphically grown on an unstrained GaN buffer layer.

Using the same theoretical approach as described in chapter II, the polarization based on the assumption $a_{buffer} = a_{GaN}$, is calculated and reoptimized for all structures. The piezoelectric polarization is computed as the difference of the total polarization obtained from this calculation and the spontaneous polarization.

Fig. 11a,b show the piezoelectric polarization versus strain calculated by equation (41) (linear in strain) and our improved theoretical approach for binary compounds.

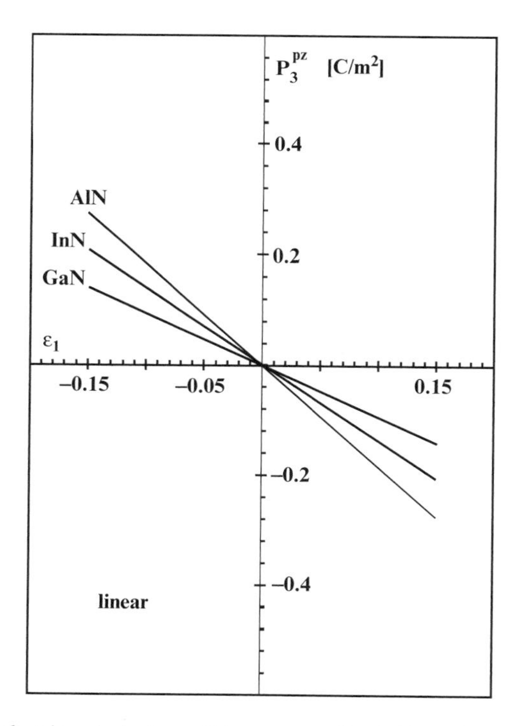

Fig. 11a The piezoelectric polarization of binary group-III-nitrides with wurtzite structure under biaxial tensile or compressive strain calculated by equation (41). The value of the piezoelectric polarization is increasing linearly with increasing strain and for a given biaxial strain from GaN to InN and AlN.

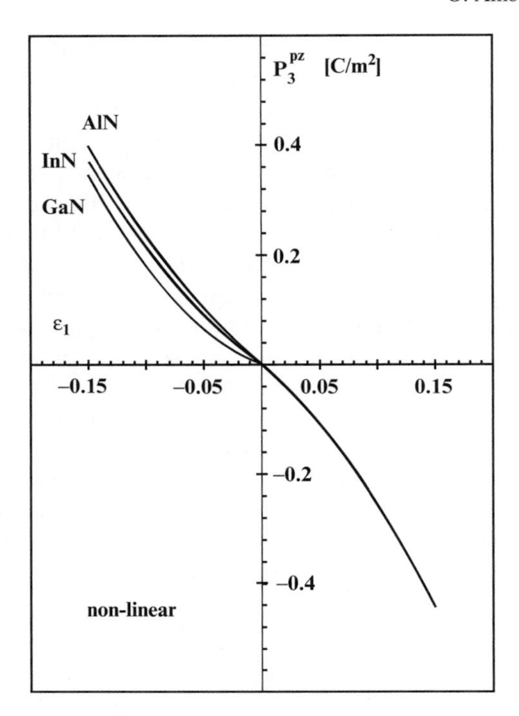

Fig. 11b The piezoelectric polarization of binary group-III-nitrides with wurtzite structure under biaxial tensile or compressive strain calculated by the theory described in chapter II. The value of piezoelectric polarization is increasing non-linearly with increasing strain, exceeding the values calculated by equation (41).

In both cases the piezoelectric polarization for a given strain is increasing from GaN to InN and AlN. It is important that calculations by our improved method lead to significantly higher piezoelectric polarizations, especially in cases of high strains (e.g. caused by high In-concentrations). The calculated non-linear piezoelectricity of the binary compounds can be described by the relations (in C/m^2):

$$\begin{aligned} P^{PZ}_{AlN} &= -1.761\varepsilon + 6.11\varepsilon^2, for\ \varepsilon < 0, \\ P^{PZ}_{AlN} &= -1.761\varepsilon - 8.00\varepsilon^2, for\ \varepsilon > 0, \\ P^{PZ}_{GaN} &= -0.775\varepsilon + 10.37\varepsilon^2, \\ P^{PZ}_{InN} &= -1.477\varepsilon + 6.837\varepsilon^2. \end{aligned} \tag{48}$$

The observed non-linearity in the bulk piezoelectricity exceeds any effects related to disorder or bond alternation effects which were taken into account [35]. Therefore, the calculation of the piezoelectric polarization of an $A_xB_{1-x}N$ alloy at any strain becomes straightforward. One can pick up a value for x, calculate the strain $\varepsilon_1 = \varepsilon(x)$ from Vegard's law, and the piezoelectric polarization by:

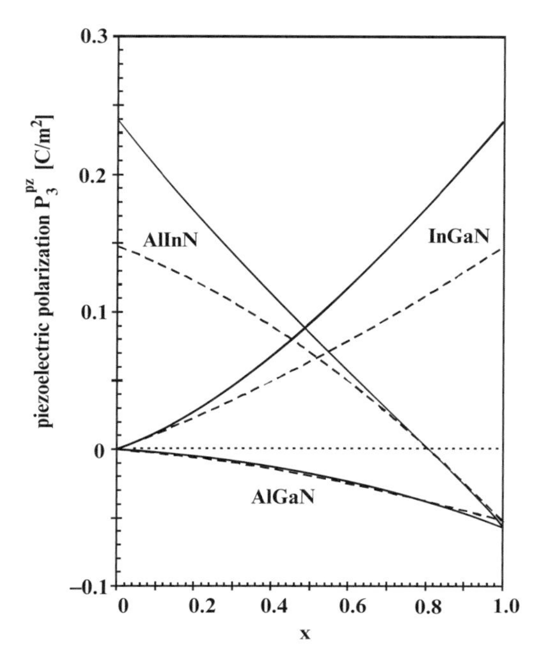

Fig. 12 The piezoelectric polarization of random ternary alloys pseudomorphically grown on relaxed GaN buffer layers calculated by: (i) equation (41) and using linear interpolations of the physical properties (e_{xy}, C_{xy}) for the relevant binary compounds (dashed lines) (ii) taking into account the non-linearity in the piezoelectric polarization in terms of strain (equations (46), solid lines). For alloys under high biaxial strain (layers with high Inconcentration) the piezoelectric polarization is underestimated by approach (i).

$$P_{ABN}^{PZ}(x) = xP_{AN}^{PZ}(\varepsilon(x)) + (1-x)P_{BN}^{PZ}(\varepsilon(x)), \tag{49}$$

where $P_{AN}^{PZ}(\varepsilon(x)), P_{BN}^{PZ}(\varepsilon(x))$ are the strain dependent bulk piezoelectric polarizations of the relevant binary compounds (given above).

The piezoelectric and spontaneous polarization of random ternary AlGaN, InGaN and AlInN alloys pseudomorphic grown on relaxed GaN buffer layers are shown in Fig. 12. The piezoelectric polarization is changing non-linearly with alloy composition reaching the lowest and highest value of -0.057 and $0.238\,\mathrm{C/m^2}$ for pseudomorphic AlN/GaN and InN/AlN heterostructures, respectively. Especially for alloys with high Indium concentrations, which are however difficult to realize in practice, the value of the piezoelectric polarization exceeds the values of the spontaneous polarization. It is interesting to notice that $\mathrm{Al_{0.82}In_{0.18}N/GaN}$ heterostructures are predicted to grow lattice matched and the piezoelectric polarization in the alloy should vanish in this case. By comparing the calculated values of P^{pz} using a linear interpolation of the elastic and piezoelectric constants or taking the non-linearity of the piezoelectric polarization into account, it becomes obvious that the linear interpolation leads to an underestimation of the piezoelectric polarization. Therefore the improved scheme represented by equations (48) and (49) is of interest in modeling hetero- and nanostructures with highly strained InGaN and AlInN layers.

Based on a theoretical understanding of the non-linear dependence of spontaneous and piezoelectric polarization on composition and strain we can provide an accurate prediction of polarization induced charges bound at surfaces and interfaces of $A_xB_{1-x}N/GaN$ heterostructures.

7 Polarization Induced Surface and Interface Charges

As mentioned above, the total polarization P, is the sum of the piezoelectric and spontaneous polarization,

$$P_{ABN} = P^{pz}_{ABN} + P^{SP}_{ABN} \tag{50}$$

Always associated with a gradient of polarization in space is a polarization induced charge density given by:

$$\rho_P = -\nabla P. \tag{51}$$

In analogy, at the surface of a relaxed or strained $A_xB_{1-x}N$ layer as well as at interfaces of a $A_xB_{1-x}N/GaN$ heterostructure, the total polarization can change abruptly, causing a fixed polarization charge σ, defined by:

$$\begin{aligned} \sigma_{ABN} &= P_{ABN} = P^{SP}_{ABN} + P^{pz}_{ABN} \qquad \text{for surfaces,} \\ \sigma_{ABN/GaN} &= P_{GaN} - P_{ABN}, \\ &= \left(P^{SP}_{GaN} + P^{pz}_{GaN}\right) - \left(P^{SP}_{ABN} + P^{pz}_{ABN}\right) \qquad \text{for interfaces,} \end{aligned} \tag{52}$$

respectively. If we assume just for simplification, that spontaneous polarization is negligible, we can use equation (52) and (41–43) to calculate the surface charges of ternary alloys caused by piezoelectric polarization only. Fig. 13a–c shows the surface sheet charge of ternary alloys in dependence on alloy composition in case of a constant tensile strain of 0.01. This strain is caused by uniaxial, biaxial and hydrostatic pressure. It becomes obvious that the surface sheet charge caused by biaxial strain is the highest and that the sign of the charge is the same for uniaxual and biaxial strain but can be opposite to the surface charge caused by hydrostatic strain.

In the following we concentrate on the more device relevant case of polarization induced sheet charges caused by biaxial strain and spontaneous polarization. Fig. 14a and 14b show the polarization induced surface and interface sheet charges $\sigma/e(e = -1.602 \cdot 10^{-19}\,\text{C})$ for relaxed and strained binary group-III-nitrides as well as for pseudomorphic $A_xB_{1-x}N/GaN$ heterostructures.

The (spontaneous) polarization induced bound surface charge of relaxed InN, GaN and AlN crystals is determined to be $2.62 \cdot 10^{13}$, $2.12 \cdot 10^{13}$ and $5.60 \cdot 10^{13}\,\text{cm}^{-2}$, respectively, which equals between 2 and 5 elementary electron charges per every

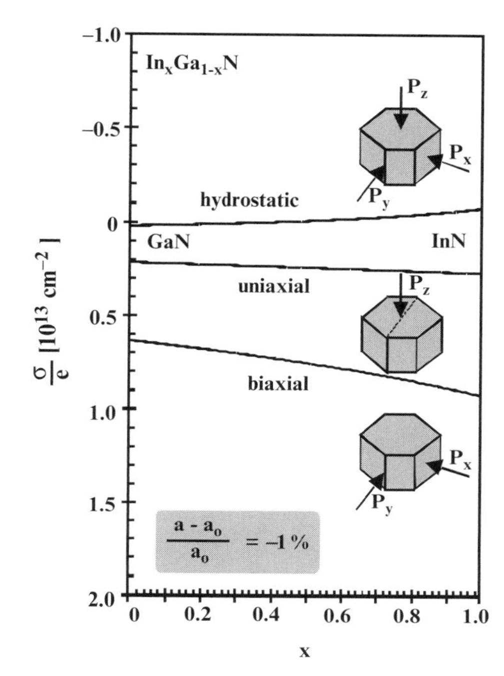

Fig. 13a Piezoelectric polarization induced surface charges of InGaN versus In concentration. The sheet charge is calculated for a constant strain of -1%, but caused by hydrostatic, uniaxial or biaxial pressure.

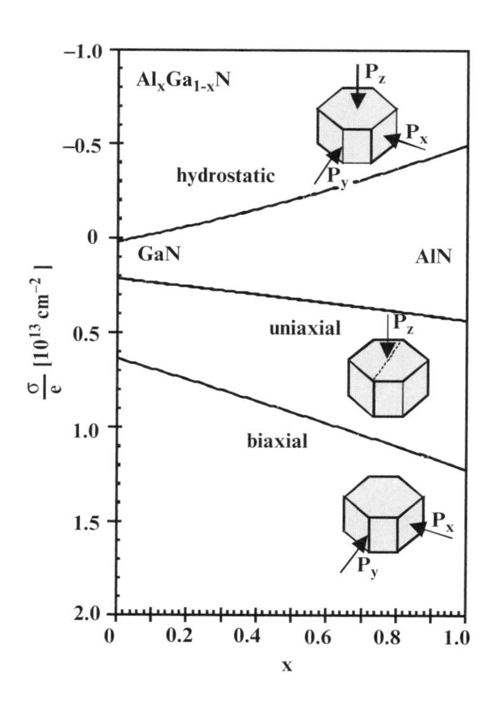

Fig. 13b Piezoelectric polarization induced surface charges of AlGaN versus Al concentration. The sheet charge is calculated for a constant strain of -1%, but caused by hydrostatic, uniaxial or biaxial pressure.

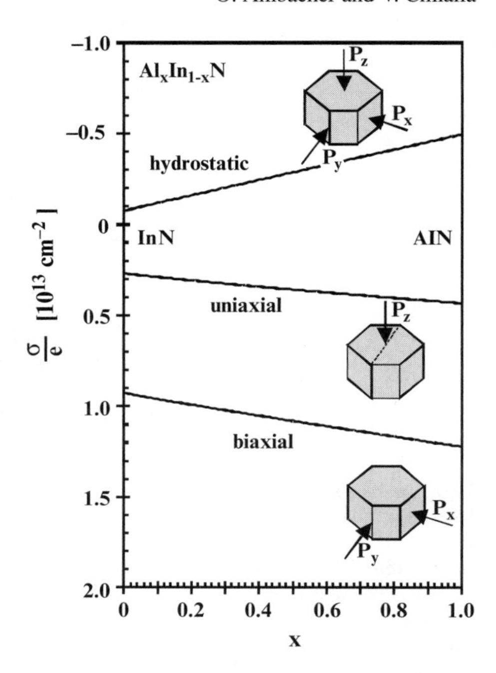

Fig. 13c Piezoelectric polarization induced surface charges of AlInN versus Al concentration. The sheet charge is calculated for a constant strain of −1%, but caused by hydrostatic, uniaxial or biaxial pressure.

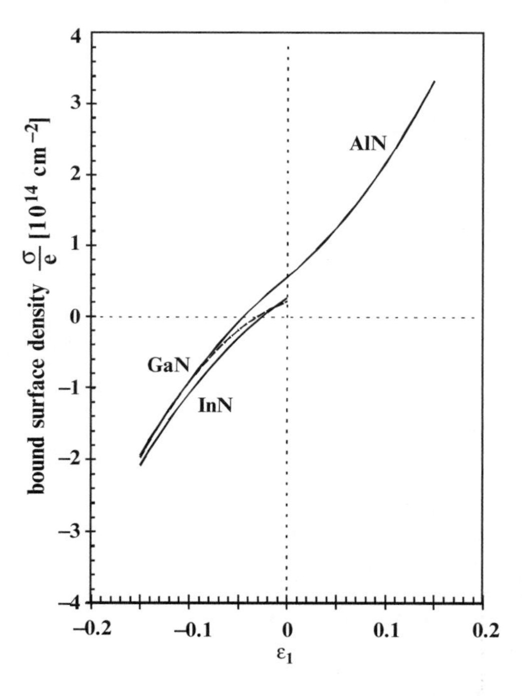

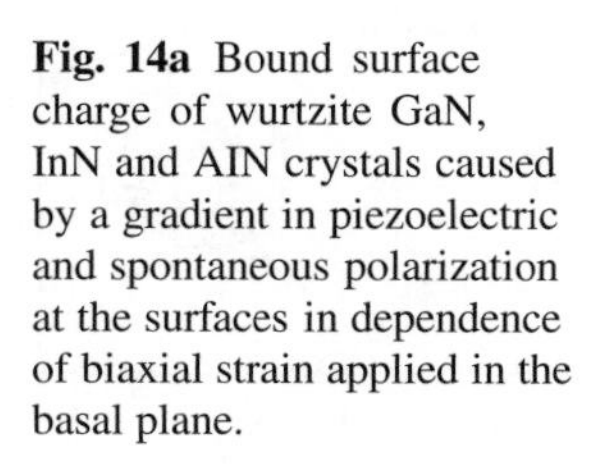
Fig. 14a Bound surface charge of wurtzite GaN, InN and AlN crystals caused by a gradient in piezoelectric and spontaneous polarization at the surfaces in dependence of biaxial strain applied in the basal plane.

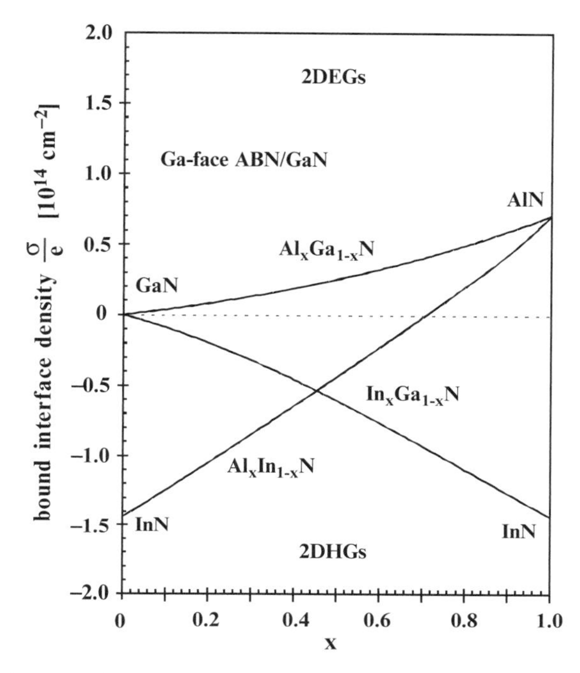

Fig. 14b Bound interface charges of pseudomorphic AlGaN/GaN, InGaN/GaN and AlInN/GaN heterostructures grown on relaxed GaN buffer layers with Ga-face polarity. Positive (negative) polarization induced interface charges are screened by electrons (holes) in n-type (p-type) samples, causing the formation of 2DEGs.

100 surface atoms. If a biaxial compressive strain of e.g. $\varepsilon_1 = 0.02$ is applied to the crystals the surface charges are reduced to $0.61 \cdot 10^{13}$, $0.90 \cdot 10^{13}$ and $3.27 \cdot 10^{13}\,\text{cm}^{-2}$, respectively (Fig. 14a). For compressive strains in InN, GaN and AlN of $\varepsilon_1 = 0.025$, 0.03 and 0.045, respectively, the piezoelectric compensates the spontaneous polarization and the strained crystals should be without surface charge and internal electric field. As can be seen from Fig. 14a, the compressive strain reduces whereas tensile strain enhances the bound surface charge. It should be mentioned, that for relaxed layers grown along the c-axis on a substrate of choice, the surface charge is positive (negative) for material with N-face (Ga-face) polarity.

Like in ferroelectrics, the bound surface charge can be screened by charged surface defects and adsorbed charges from the environment [53]. But it was shown that the effective barrier height of Schottky contacts on GaN can be influenced by the presence of polarization induced surface charges [54]. In addition, if the screening of the surface charge by defects and the ambient is not complete, the carrier concentration profiles inside the crystals will be affected. By example in slightly n-type doped Ga-face GaN grown on c-Al_2O_3 substrates free electrons are expected to accumulate at the GaN/Al_2O_3 interface compensating the positive bound (mainly spontaneous) polarization induced interface charge $+\sigma/e$, whereas the negative surface charge on

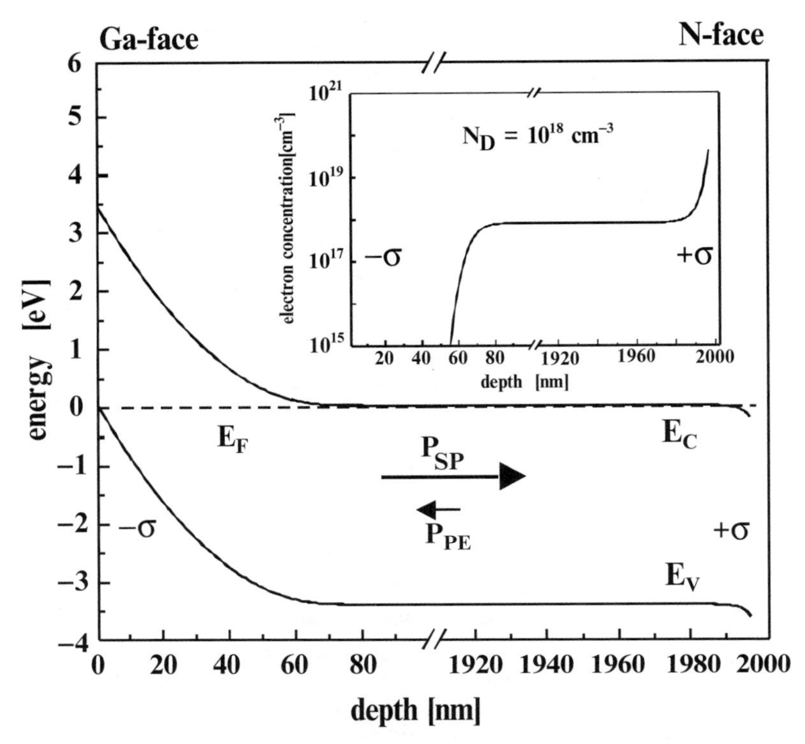

Fig. 15 Conduction and valence band edge profile of Si-doped GaN ($d = 2\,\mu$m) taken into account polarization induced surface charges induced by gradients in spontaneous (mainly) and piezoelectric polarization. The insert shows the electron concentration profile. At the Ga-face side a depletion zone is caused by negative polarization induced surface charges. It should be noticed, that screening of surface charges by the ambient is neglected.

top of the GaN layer $-\sigma/e$, is only partially compensated by the ionized donors. As a consequence, a significant band bending at the Ga-face GaN surface is expected if the surface charge is not compensated by defects or the environment (Fig. 15).

Since screening by charges from the ambient at $A_xB_{1-x}N/GaN$ interfaces can be excluded, polarization induced effects are much easier to study in hetero- and nanostructures in comparison to single epitaxial layers. For pseudomorphic Ga-face $Al_xGa_{1-x}N/GaN$ $(0 < x \leq 1)$ and $Al_xIn_{1-x}N/GaN$ $(0.71 < x \leq 1)$ heterostructures the polarization induced interface charges are predicted to be positive. In both cases the bound charge increases non-linearly with x up to $7.06 \cdot 10^{13}\,cm^{-2}$, determined for the AlN/GaN heterostructure. For Ga-face $In_xGa_{1-x}N/GaN$ $(0 < x \leq 1)$ and $Al_xIn_{1-x}N/GaN$ $(0 < x \leq 0.71)$ heterostructures σ/e is found to be negative. Again, the polarization induced interface charge changes in a non-linear manner if the alloy composition is varied. For pseudomorphic InN/GaN heterostructures a very high value of the bound sheet charge of $14.4 \cdot 10^{13}\,cm^{-2}$ is calculated.

In n-type heterostructures it has to be taken into account that free electrons will accumulate at interfaces with positive bound sheet charges, compensating $+\sigma/e$. As a consequence, a 2DEG with a sheet carrier concentration close to the concentration of the bound interface charge can be formed. It should be pointed out that these 2DEGs are realized without any need of a modulation doped barrier. This is of special interest for the fabrication of high frequency and high power HEMTs, as growth and processing of these devices is drastically simplified.

With a theoretical understanding of the polarization induced charge we now can predict the sheet carrier concentration of polarization induced 2DEGs and their dependence on alloy composition for pseudomorphic Ga-face AlGaN/GaN and AlInN/GaN $(0.71 < x \leq 1)$ heterostructures.

8 Sheet Carrier Concentration of Polarization Induced 2DEGs

For undoped pseudomorphic Ga-face AlGaN/GaN or AlInN/GaN HEMT structures, the sheet electron concentration, $n_s(x)$, can be approximated by taking advantage of the total bound sheet charge $\frac{\sigma_{ABN/GaN}(x)}{e}$ calculated above, and the following equation [55]:

$$n_s(x) = \frac{\sigma_{ABN/GaN}(x)}{e} - \frac{\varepsilon_0 E_F}{e^2}\left(\frac{\varepsilon_{ABN}(x)}{d_{ABN}} + \frac{\varepsilon_{GaN}}{d_{GaN}}\right) - \frac{\varepsilon_0 \varepsilon_{ABN}(x)}{e^2 d_{ABN}}\left(e\phi_{ABN}(x) + \Delta(x) - \Delta E^C_{ABN}(x)\right), \tag{53}$$

where ε_0 is the dielectric constant of the vacuum, ε_{GaN} and $\varepsilon_{ABN}(x)$ are the relative dielectric constants, d_{ABN} and d_{GaN} are the thicknesses of the barrier and the buffer layer, respectively. E_F is the position of the Fermi level with respect to the GaN conduction-band-edge close to the GaN/substrate interface, $e\phi_{ABN}(x)$ is the Schottky barrier height of the gate contact on top of the barrier, $\Delta E^C_{ABN}(x)$ is the conduction band offset and $\Delta(x)$ is the penetration of the conduction band edge below the Fermi level at the ABN/GaN interface. The penetration of the conduction band edge below the Fermi level is calculated by:

$$\Delta(x) = E_0(x) + \frac{\pi\hbar^2}{m^*_{GaN}} n_S(x), \tag{54}$$

where the ground subband level of the 2DEG is given as:

$$E_0(x) = \left\{\frac{9\pi\hbar e^2}{8\varepsilon_0\sqrt{8m^*_{GaN}}}\frac{n_S(x)}{\varepsilon_{GaN}}\right\}^{2/3}. \tag{55}$$

We used the following linear interpolations to describe the physical properties of the alloys in our calculations (values for InGaN are given for completeness and further calculations):
dielectric constants [56]:

$$\varepsilon_{AlGaN}(x) = 0.03x + 10.28, \quad (56)$$

$$\varepsilon_{InGaN}(x) = 4.33x + 10.28,$$

$$\varepsilon_{AlInN}(x) = -4.30x + 14.61,$$

Schottky barrier for Ni-contact [57]:

$$e\phi_{AlGaN}(x) = (1.3x + 0.84)\,eV, \quad (57)$$

$$e\phi_{InGaN}(x) = (-064x + 0.84)eV,$$

$$e\phi_{AlInN}(x) = (1.94x + 0.20)eV,$$

band gaps [35, 58–60]:

$$E^g_{AlGaN}(x) = [6.28x + 3.42(1-x) - 1.00x(1-x)]\,eV, \quad (58)$$

$$E^g_{InGaN}(x) = [0.70x + 3.42(1-x) - 1.67x(1-x)]\,eV,$$

$$E^g_{AlInN}(x) = [6.28x + 0.70(1-x) - 3.10x(1-x)]\,eV,$$

band offsets [61, 62]:

$$\Delta E^C_{ABN}(x) = 0.63(E^g_{ABN}(x) - E^g_{ABN}(0)) \quad (59)$$

and an effective electron mass of $m_e^* = 0.228\ m_e$ for GaN [63].

In order to determine the sheet carrier concentrations and carrier distribution profiles in undoped HEMT structures including spontaneous and piezoelectric polarization induced bound sheet charges and to verify the results obtained with the equations above, we have used a one-dimensional Schrödinger-Poisson solver [64].

To incorporate the effects of the non-linear spontaneous and piezoelectric polarization into the program, thin layers ($\approx 6\,\text{Å}$) of charge are added to the heterostructure interfaces equivalent to the bound sheet charge density σ/e.

For both kinds of calculations it is necessary to specify the boundary conditions at the surface and at the substrate interfaces. In our HEMT structures we assumed a Ni Schottky barrier contact at the surface, pinning the conduction band. At the interface to the substrate, the Fermi level was set to one half of the band gap of GaN. By changing the value of the conduction band level close to the substrate it turned out that this boundary condition has only very little impact on the sheet carrier concentration of the 2DEG for buffer thicknesses above 1 μm. For the same boundary conditions the calculations of the 2DEG sheet carrier concentrations using equation (53) or the Schrödinger-Poisson solver agree within 5% for all HEMT structures investigated.

8.1 2DEGs Confined at Interfaces of Undoped Ga-face AlGaN/GaN Heterostructures

In typical, undoped AlGaN/GaN HEMTs with insulating buffers ($N_d < 10^{16}\,\mathrm{cm}^{-3}$), GaN thicknesses of about 2 µm and barrier widths of more than 15 nm, the value of the sheet carrier concentration is dominated by the polarization induced sheet charge. This charge is increasing with increasing Al-concentration in the barrier, as the strain and therefore the piezoelectric as well as the spontaneous polarization are increasing. This offers the possibility to control the sheet carrier concentration of the 2DEG by the choice of the Al-concentration of the barrier instead of doping the barrier. In Fig. 16 the polarization induced bound interface charges and 2DEG sheet carrier concentrations are shown versus alloy composition of the barrier for AlGaN/GaN heterostructures with a barrier thickness of about 30 nm. The bound interface charge is calculated for barriers grown on relaxed GaN using again (i) a linear interpolation of the physical properties (C_{ij}, e_{ij}, and P^{SP}) of the binary compounds (upper curve) and (ii) considering the non-linear behaviour of piezoelectric and spontaneous polarization (lower curve). To enable a comparison between experimental (open symbols) and theoretical data, the sheet carrier concentration of the 2DEGs is calculated like described above, taking into account the depletion caused

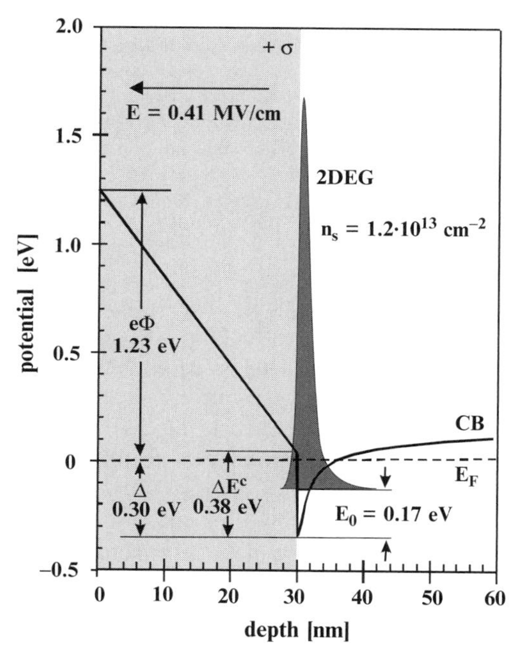

Fig. 16a Conduction band profile versus depth of an undoped $Al_{0.3}Ga_{0.7}N/GaN$ heterostructures (30 nm/2000 nm) calculated by a one-dimensional Schrödinger-Poisson solver. A polarization induced 2DEG confined at the interface with a sheet carrier concentration of $1.2 \cdot 10^{13}\,\mathrm{cm}^{-2}$ is predicted. The electrons confined are provided by surface donors.

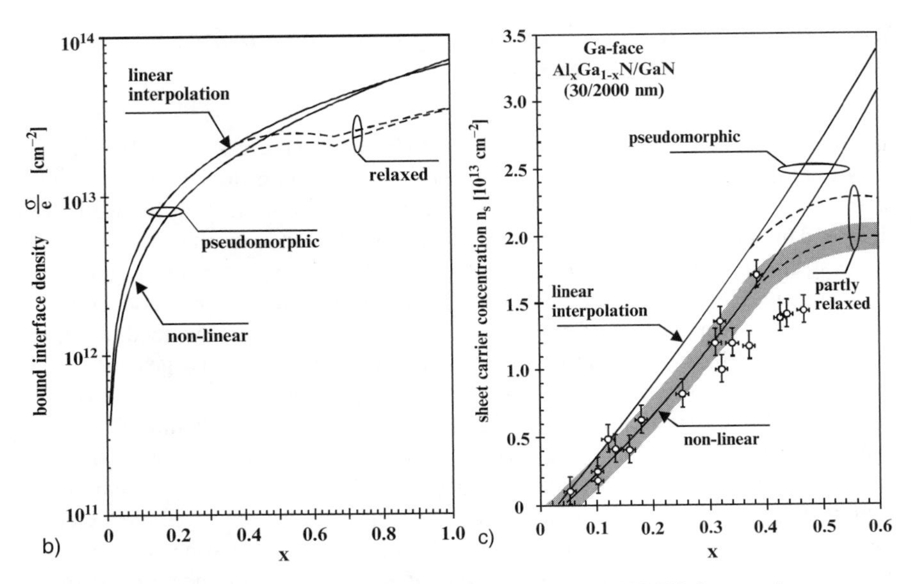

Fig. 16b,c b) The polarization induced bound interface charges and 2DEG sheet carrier concentrations are shown as a function of alloy composition of the barrier for AlGaN/GaN heterostructures ($x = 0.3, 30\,\text{nm}/2000\,\text{nm}$). The bound interface charge is calculated for barriers grown on relaxed GaN using again (i) a linear interpolation of the physical properties (C_{ij}, e_{ij}, and P^{SP}) of the binary compounds (upper curve) and (ii) considering the non-linear behavior of piezoelectric and spontaneous polarization (lower curve). c) The sheet carrier concentration of the 2DEGs is calculated, taking the depletion caused by the Ni Schottky contact into account (lower solid curve). The measured 2DEG carrier concentration measured by CV-profiling is shown as open symbols. The highest measured and calculated sheet carrier concentration of pseudomorphic AlGaN/GaN heterostructures is determined to be $2 \cdot 10^{13}\,\text{cm}^{-2}$ for $x = 0.37$. For higher Al-concentrations of the 30 nm thick barriers, the AlGaN layer starts to relax, diminishing the piezoelectric component (dashed lines) [25].

by the Ni Schottky contact. Experimentally the sheet carrier concentration of the 2DEGs confined in AlGaN/GaN heterostructures with alloy compositions of up to $x = 0.5$, is measured by CV-profiling using Ti/Al ohmic and Ni-Schottky contacts. The highest measured and calculated sheet carrier concentration of pseudomorphic AlGaN/GaN heterostructures is determined to be $2 \cdot 10^{13}\,\text{cm}^{-2}$ for $x = 0.37$. For higher Al-concentration, the AlGaN barrier starts to relax, diminishing the piezoelectric component [25]. The solid lines in Fig. 16 show the further increase of the sheet carrier concentration with increasing x for pseudomorphic structures with thinner barriers. For lower Al-concentrations of the 30 nm thick barriers the sheet carrier concentration decreases non-linearly down to $x \approx 0.06$, where the 2DEG disappears.

For pseudomorphic HEMT structures with barrier thicknesses of more than 15 nm the non-linear dependence of the 2DEG carrier concentration on the alloy composition can be approximated very well by:

$$n_s(x) = (-0.169 + 2.61x + 4.50x^2)10^{13}cm^{-2}, \quad x > 0.06. \tag{60}$$

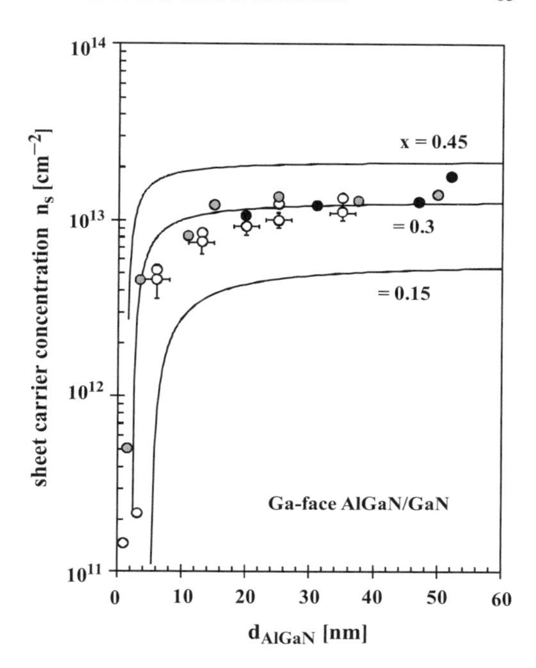

Fig. 17 The dependences of $n_s(x)$ on barrier thickness calculated using equation (53) for $x = 0.15$, 0.30 and 0.45 are shown. A drastic decrease in sheet carrier concentration is predicted for thicknesses below 12, 8 and 6 nm, respectively. For $x = 0.3$ the sheet carrier concentration of AlGaN/GaN HEMTs was measured by CV-profiling for barrier thicknesses between 1 and 50 nm. These data (white open symbols) and experimental results of other groups (black and gray symbols [65, 66]) are in good agreement with the theoretical predictions.

For thinner barriers the depletion by the Schottky contact or the influence of the surface charge of the heterostructure becomes important (for more information see Ref. [55]). In Fig. 17 the dependences of $n_s(x)$ on the barrier thickness calculated using equation (53) for $x = 0.15$, 0.30 and 0.45 are shown. A drastic decrease in sheet carrier concentration is predicted for thicknesses below 12, 8 and 6 nm, respectively. For $x = 0.3$ the sheet carrier concentration of AlGaN/GaN HEMTs was measured by CV-profiling for barrier thicknesses between 1 and 50 nm. These data and experimental results of other groups are presented in Fig. 17 in addition [65, 66]. A good agreement of our theoretical predictions and the experimental data is observed.

The comparison of the calculated bound sheet charge and the 2DEG sheet carrier concentration to the measured interface-accumulated free charge is clearly more direct than if it is compared to the Stark shift of excitonic radiative recombination due to the depolarizing field as discussed later. More important, the non-linear prediction reproduces the measured data accurately, recovering almost all of the residual deviations from experiment observed in former approaches [6, 67, 68]. The predicted piezoelectric and spontaneous polarization emerges therefore as rather reliable for AlGaN/GaN heterostructures.

8.2 2DEGs Confined at Interfaces of Undoped, Ga-face AlInN/GaN Heterostructures

In order to determine the precise composition x and the strain of AlInN barriers grown on Ga-face buffer layers, reciprocal space maps of the symmetric (200) and asymmetric (205) reflexes are measured by HRXRD.

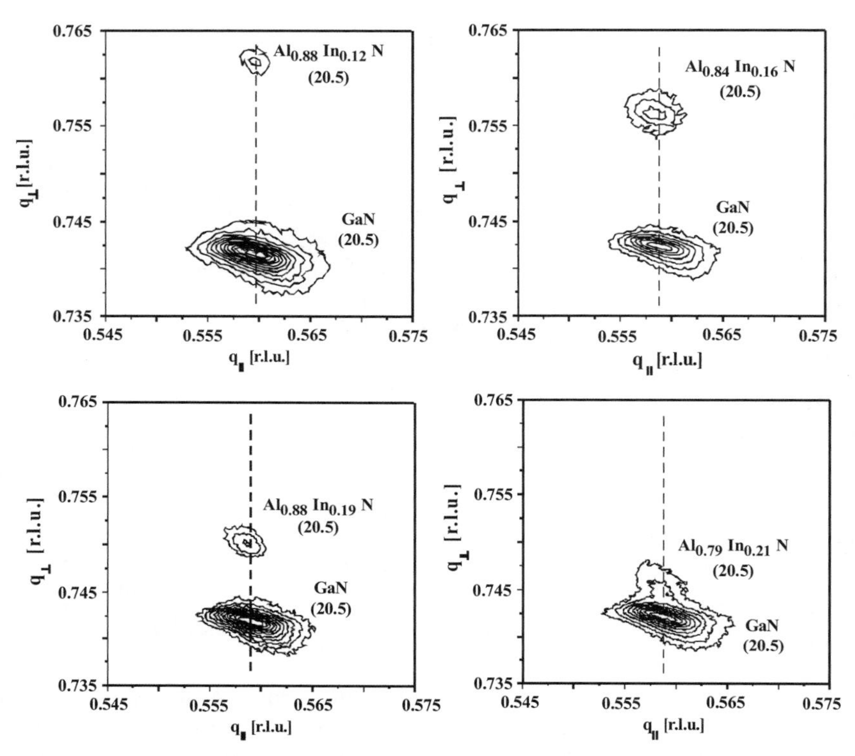

Fig. 18 Reciprocal space maps of the (20.5) reflections of undoped AlInN/GaN-heterostructures measured by HRXRD. AlInN barriers with Al-concentrations between $x = 0.79$ and 0.88 are grown pseudomorphic.

From the reciprocal space maps (Fig. 18) it can be immediately seen that the $Al_xIn_{1-x}N/GaN$ (50/540 nm) heterostructures with Al-concentrations between 0.78 and 0.88 are grown pseudomorphically. The GaN buffer layer is under weak compressive strain ($\varepsilon_1 = -1.9 \cdot 10^{-3}$) causing a piezoelectric polarization of 0.0015 C/m^2. More important, by calculating the strain of the AlInN along the c-axis and in the basal plane using the measured lattice constants, the Al-concentration at which the AlInN can be grown lattice matched to the GaN buffer layer is determined to be $x = (0.83 \pm 0.01)$. Close to this composition the piezoelectric polarization in the AlInN layer vanishes. Although the AlInN barrier is intentionally undoped and the residual piezoelectric polarization of the insulating GaN buffer layer can cause a maximum negative bound sheet of only $-\frac{\sigma}{e} \approx -10^{12}\,\text{cm}^{-2}$ (Fig. 19), sheet carrier concentrations of accumulated electrons of more than $10^{13}\,\text{cm}^{-2}$ are measured by Hall-effect and CV-profiling (Fig. 20). In Fig. 21 the bound sheet charge caused by the gradient in piezoelectric (lower solid line) and spontaneous polarization (upper dashed line) as well as the calculated 2DEG sheet carrier concentration (upper solid line) are shown. It becomes obvious that the calculated positive bound sheet

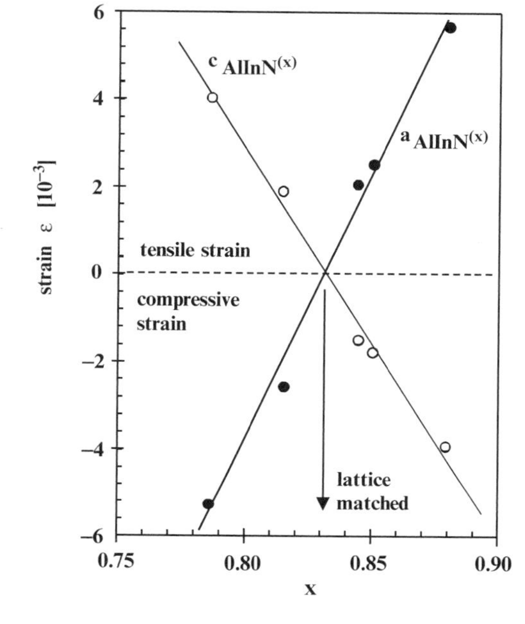

Fig. 19 Strain of the AlInN barrier along the c-axis and in the basal plane, determined from the measured lattice constants $c_{AlInN}(x)$ and $a_{AlInN}(x)$. For $x = 0.83$ the pseudomorphic AlInN layer is grown lattice matched on the GaN buffer layer.

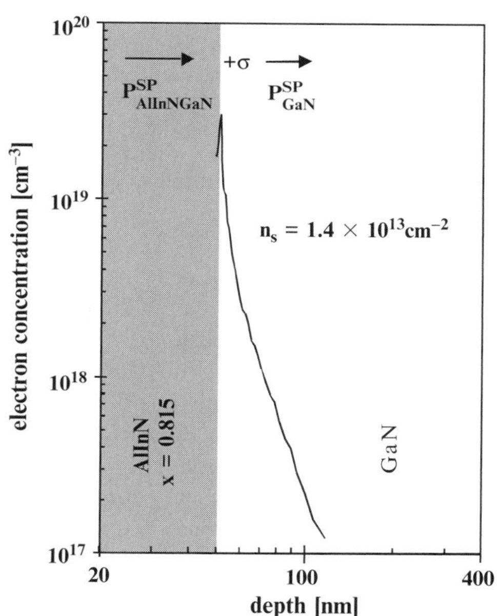

Fig. 20 Electron concentration profile measured by CV-profiling of a nearly lattice matched AlInN/GaN heterostructure (50/540 nm) with Ga-face polarity. A very high concentration of electrons accumulated at the AlInN/GaN interface can be measured, proving the presence of a strong gradient in spontaneous polarization at the heterointerface.

charge and the measured 2DEG sheet carrier concentrations of the lattice matched heterostructures can only be caused by a gradient in spontaneous polarization at the AlInN/GaN interface.

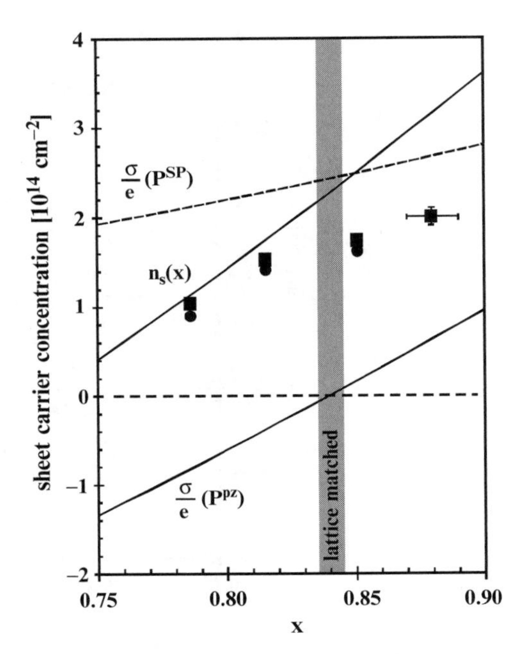

Fig. 21 Calculated bound sheet charges located at Al_xIn_{1-x} N/GaN interfaces, induced by a gradient in piezoelectric polarization (lower solid line) and/or a gradient in spontaneous polarization (dashed line) in dependence on Al-concentration of the barrier. The predicted 2DEG carrier density (upper solid line) is compared to sheet carrier concentrations determined by Hall effect (solid circles) and CV-profiling (solid squares) measurements. For lattice matched heterostructures (gray area) sheet carrier concentrations of more than $10^{13}\,\mathrm{cm}^{-2}$ are predicted and measured, proving the presence of a large gradient in spontaneous polarization at the heterostructure interface.

The observation of high concentrations of accumulated electrons in theses samples provide direct evidence for the presence of a strong gradient in spontaneous polarization at the heterostructure interface. The predicted sheet carrier concentrations for lattice matched AlInN/GaN heterostructures achieved by a linear interpolation of the physical properties (C_{ij}, e_{ij}, and P^{SP}) of the binary compounds and taking the non-linearity of P^{pz} and P^{SP} into account are $n_s(0.82) = 2.95 \cdot 10^{13}$ and $2.28 \cdot 10^{13}\,\mathrm{cm}^{-2}$, respectively. Again, our improved non-linear theory matches better with the measured sheet carrier concentration of about $(1.5 \pm 0.1) \cdot 10^{13}\,\mathrm{cm}^{-2}$. The discrepancy between the measured and calculated values for n_s could be caused by an underestimation of the bowing parameters used to describe the non-linearity of P^{pz} and P^{SP}, but more likely charged defects located close to the AlInN/GaN interface screen part of the bound polarization induced interface charge lowering the measured concentration of free electrons.

Also In-fluctuations causing a smear out and a reduction of the polarization induced interface charge can be present in the AlInN barriers. A further improvement in the agreement between experimental and theoretical results has to be expected if the structural and interface quality of the heterostructures is optimized.

8.3 2DEGs Confined in InGaN/GaN Single Quantum Wells

InGaN/GaN single quantum wells offer two possibilities to obtain information about polarization induced interface charges. First, in analogy to the AlGaN/GaN and AlInN/GaN HEMT structures, the formation of 2DEGs confined in the InGaN SQW can be measured. Second, the Stark shift of the excitonic recombination can be used to determine the polarization induced electric field inside the quantum well and the polarization induced bound charge located at the InGaN/GaN interfaces. We have investigated undoped, n-type $GaN/In_{0.13}Ga_{0.87}N/GaN$ nanostructures with Ga-face polarity, where the width of the pseudomorphic SQW was varied between $d_{InGaN} = 0.9$ and 54 nm [69, 70]. The spontaneous polarization of the InGaN layer ($P^{SP}_{InGaN}(x = 0.13) = -0.031\,\mathrm{C/m^2}$), pointing in the $[000\overline{1}]$ direction, is predicted to be slightly smaller than the polarization of the relaxed GaN-buffer and cap layer. The piezoelectric polarization is calculated to be $P^{pz}_{InGaN}(x = 0.13) = 0.016\,\mathrm{C/m^2}$. Because the InGaN well is under compressive strain, the piezoelectric polarization is anti-parallel to P^{SP}_{InGaN}.

The gradient in total polarization at the InGaN/GaN interface should cause a polarization induced interface charge of:

$$\begin{aligned}\frac{\sigma}{e} &= \frac{1}{e}\left\{\left(P^{SP}_{GaN} + P^{pz}_{GaN}\right) - \left(P^{SP}_{InGaN} + P^{pz}_{InGaN}\right)\right\} \\ &= \frac{1}{e}\{(-0.034+0) - (-0.031+0.016)\}\,\frac{C}{m^2} = 1.18 \cdot 10^{13} cm^{-2}.\end{aligned} \quad (61)$$

It should be mentioned that the bound polarization induced charge in these heterostructures is positive at the interface close to the surface, whereas the interface below the InGaN QW has to be negatively charged. Therefore, in analogy to the AlGaN/GaN and AlInN/GaN heterostructures, the formation of a 2DEG is expected at the upper interface. As a consequence, 2DEGs with sheet carrier concentrations equal to the bound interface charge should be observed 130 nm below the surface for wells with widths of more than 20 nm. By CV-profiling the carrier concentration profiles of the nanostructures are measured and shown in Fig. 22. For well widths above 4 nm an accumulation of electrons can be confirmed. By increasing the thickness of the InGaN QWs from 4.3 to 54 nm the sheet carrier concentration is increasing from $n_s = 3.6 \cdot 10^{10}$ to $5 \cdot 10^{12}\,\mathrm{cm^{-2}}$ (Fig. 23). The increase in carrier concentration is proportional to the volume of the InGaN, indicating that the source for the electrons could be a density of shallow donors of about $10^{18}\,\mathrm{cm^{-3}}$. But even for thick InGaN SQWs the measured electron density of the 2DEGs is below 50% of the predicted bound interface charge.

Are the polarization induced interface charges in InGaN/GaN SQWs not completely screened by electrons and ionized donors? In this case an electric field causing a red shift of the radiative recombination of generated electron hole pairs should be present in the InGaN QW, scaling with the thickness of the well. Assuming that the polarization induced charge at the surface and the GaN/substrate interface is completely screened by charge from the ambient or charged surface defects (which

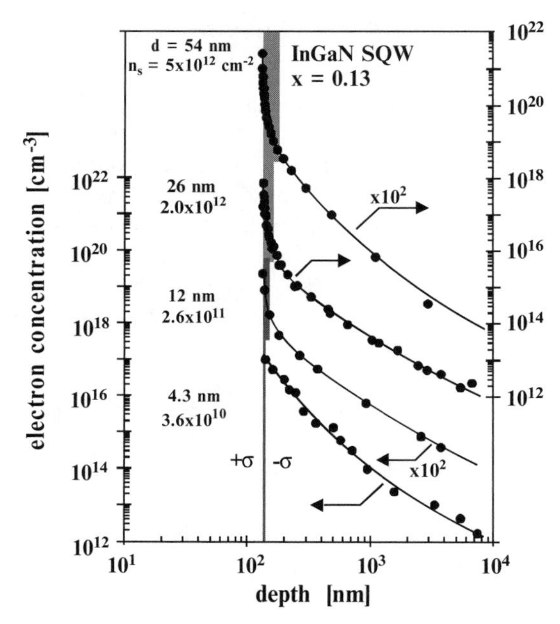

Fig. 22 Electron concentration profiles measured by CV-profiling of undoped, n-type $GaN/In_{0.13}Ga_{0.87}N$-/GaN QWs with well widths of 4.3, 12, 26 and 54 nm. For comparison, the carrier profiles are shifted against each other. For QW's (gray area) with widths of more than 4.3 nm, the formation of 2DEGs at the upper interface, compensating partly the positive polarization induced charge, can be confirmed.

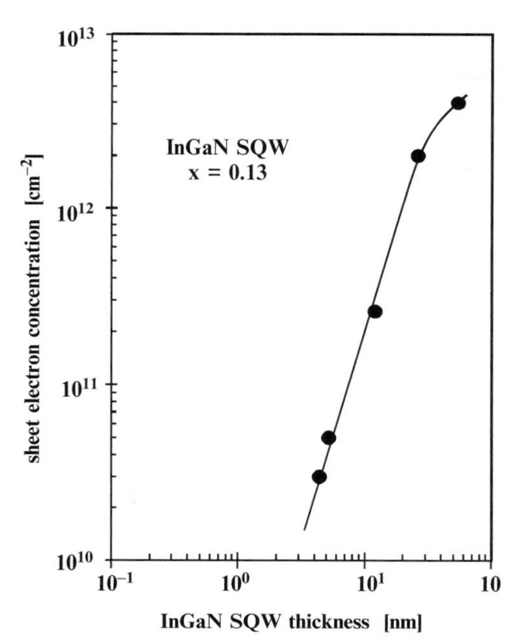

Fig. 23 The increase of sheet carrier concentration inside the InGaN layers in dependence on the widths of the QWs.

is in agreement with our experiments discussed in Ref. [53] the electric field inside an insulating well can be calculated by:

$$E_{InGaN} = -\frac{1}{\varepsilon_0 \varepsilon_{InGaN}} \left\{ \left(P^{SP}_{GaN} + P^{pz}_{GaN} \right) - \left(P^{SP}_{InGaN} + P^{pz}_{InGaN} \right) \right\}. \qquad (62)$$

If free carriers confined in the well are not homogeneously distributed, like it is the case in the presence of a 2DEG, the slope of the conduction band edge and the electric field inside the InGaN are not constant (for more detailed information please see Ref. [71]).

For this case the average electric field can be approximated by:

$$\overline{E_{InGaN}} = -\frac{e}{\varepsilon_0 \varepsilon_{InGaN}} \left\{ \frac{\sigma}{e} - n_s \right\}. \qquad (63)$$

For InGaN QWs with well widths below 6 nm the screening by free carriers can be neglected and a very high electric field of 1.97 MV/cm is predicted by equation (63). For the thickest quantum well of 54 nm, the observed 2DEG should only partially screen the predicted interface charge causing an average field of about 1.14 MV/cm. In the presence of an electric field and in the absence of free carriers, the rectangular cross section of the quantum well has to be replaced by a triangular shape (Fig. 24) in which the energy of the radiative recombination between the hole and electron ground level is described by:

$$E_{1e1h} = E^g_{InGaN} - eE_{InGaN} d_{InGaN} + \left(\frac{9\pi\hbar e E_{InGaN}}{8\sqrt{2}} \right)^{2/3} \left(\frac{1}{m_e^*} + \frac{1}{m_h^*} \right)^{1/3}, \qquad (64)$$

where $m_e^* = 0.228\ m_e$, $m_h^* = 0.9\ m_e$, are the effective electron and hole mass of GaN, respectively, which we use as an approximation for effective masses of the $In_{0.13}Ga_{0.87}N$ QW [72]. The energy of the radiative recombination E_{1e1h}, as well as the band gap of the InGaN QW, as a function of the well width, has been measured using the same samples by Ramakrishan *et al.* earlier [70]. The band gap is determined by spectroscopic ellipsometry and room-temperature photoluminescence (PL) using an excitation energy of 3.41 eV, only absorbed by the InGaN well. The QW is pumped with high intensity in order to generate a high density of electron-hole pairs, completely screening the polarization induced electric field. The Stark shift is measured also by PL, but using the lowest possible pumping power and a photon energy of 3.81 eV, well above the band gap of GaN. By this experimental procedure the screening of the polarization induced charge by photogenerated carriers is negligible.

For $In_{0.13}Ga_{0.87}N$ QWs with widths of more than 26 nm we measured a band bap of (2.902 ± 0.012) eV in good agreement with the value of bulk InGaN of the same composition ($E^g_{InGaN}(x = 0.13) = 2.946\,eV$).

The spectroscopic ellipsometry as well as PL data recorded with direct excitation at 3.41 eV show a monotonous shift to higher energies of up to 3.21 eV if the well

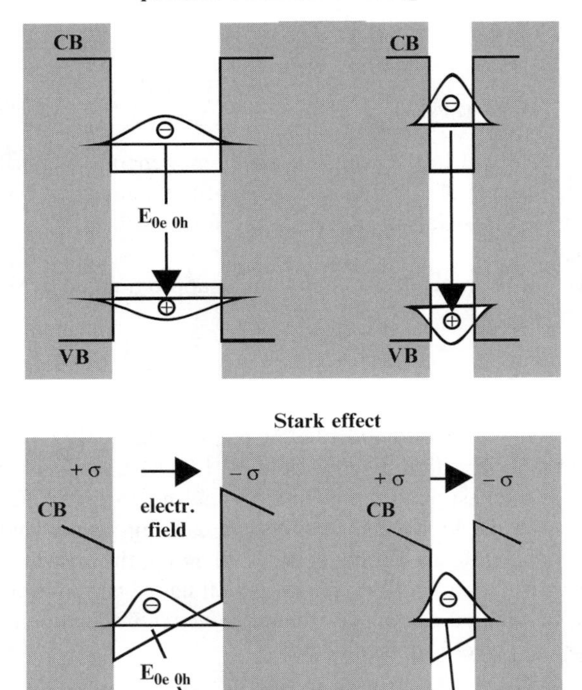

Fig. 24 Schematic drawing of the band edge profiles of two quantum wells with different thicknesses. The radiative recombination between the hole and electron ground level inside the rectangular QW is increasing in energy with decreasing well width due to quantum confinement. The optical transition is blue shifted. In the presence of a polarization induced electric field, the shape of the rectangular well has to be replaced by a triangular quantum well in which the recombination of the photogenerated electron-hole pairs is red shifted due to a Stark shift.

width is decreased to 0.9 nm, as expected for pure quantum confinement (Fig. 25). PL spectra measured by excitation at 3.81 eV with a pumping power density of $1\,\mathrm{kW/cm^2}$, yield a PL peak position which is shifted to lower energies if the QW widths is decreased from 0.9 to 5.3 nm (Stark effect). From the measured red shift of the PL peak position and equation (64) electric field strength of 0.83 MV/cm can be estimated for thin QW's. This field strength corresponds to an unscreened polarization induced bound interface charge of about $\frac{\sigma}{e} = 5 \cdot 10^{12}\,cm^{-2}$. The measured electric fields and calculated interface charges are in good agreement with the

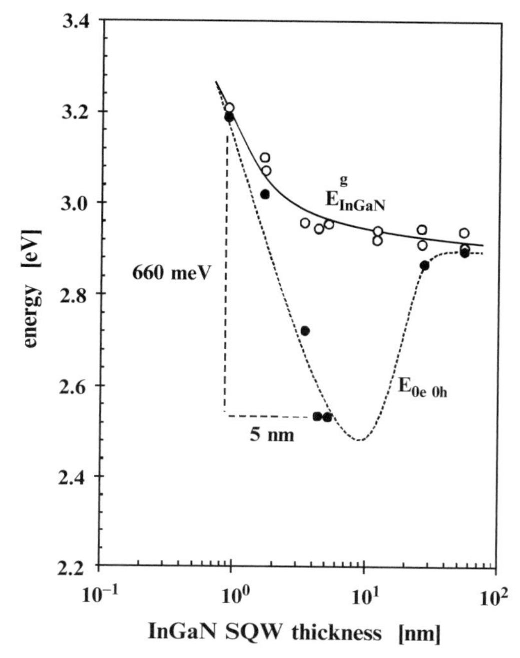

Fig. 25 The band gap of the $In_{0.13}Ga_{0.87}N$ QWs and the energy of the radiative recombination in dependence on the well width. The band gap is determined by spectroscopic ellipsometry and room-temperature PL using excitation energy of 3.4 eV, only absorbed by the InGaN well. The QW is pumped with high intensity in order to generate a high density of electron-hole pairs screening completely the polarization induced electric field. The Stark shift is measured also by PL, but using the lowest possible pumping power and a photon energy of 3.81 eV, well above the band gap of GaN [70]. By this experimental procedure the screening of the polarization induced charge by photogenerated carriers is negligible.

data obtained by Wetzel *et al.* for $In_{0.12}Ga_{0.88}N/GaN$ MQW's of 0.62 MV/cm [73], corresponding to $\frac{\sigma}{e} = 3.7 \cdot 10^{12} cm^{-2}$.

The polarization induced charge deduced from the measured red shift fits very well to our experimental observation, that the Stark shift is not detectable for quantum wells with a width of more than 26 nm (Fig. 25). For these samples, the measured band gap of InGaN and the PL peak position agree within 40 meV. Especially for the thickest QW it is obvious that the polarization induced interface charge must be completely screened by the presence of a 2DEG (Fig. 22) with a sheet carrier concentration of $5 \cdot 10^{12}\,cm^{-2}$. Both the measured Stark shift and the 2DEG sheet carrier density indicate an unscreened bound interface charge of $5 \cdot 10^{12}\,cm^{-2}$ located at $In_{0.13}Ga_{0.87}N/GaN$ interfaces. A significant discrepancy of about 50% is observed between the predicted and measured polarization induced sheet charge in the case of InGaN/GaN heterostructures. In addition our data prove that this discrepancy is not caused by screening due to free carriers.

What are the possible explanations for the deviation of our theory from the experimental data in the case of InGaN/GaN heterostructures? One reason for the measured low polarization induced interface charge could be the presence of Indium fluctuations which were observed on a scale of about 10 nm inside thick wells. Indium fluctuations will smear out the polarization induced charge throughout the well causing a complex profile of the electric field. These effects are not included in our one dimensional model of polarization induced effects. Another explanation could be the presence of charged, fixed point defects caused by impurities, vacancies or charged structural defects, located at grain boundaries and screening the polarization induced charge but not showing up in our electronic characterization.

The influence of In fluctuations can be understood by simulating the corresponding strain fields and piezoelectric polarization profiles in a two dimensional theory, combining finite element and tight binding calculations. Information of charged defects in InGaN can be obtained by a combination of temperature dependent Hall measurements and deep level transient spectroscopy.

9 Sensors Based on Polarization Induced 2DEGs

9.1 Overview

The 2DEG in AlGaN/GaN heterostructures have demonstrated extremely promising results for the fabrication of HEMTs with applications in high frequency power amplifiers [3, 74]. Together with optoelectronics, these implemented topics are the most considered and investigated application fields for group-III-nitrides. However, it has been shown above that the 2DEG is influenced by several intrinsic and extrinsic properties. This suggests that GaN based HEMTs are promising structures to sense those properties through modulation of the carrier density of the 2DEG. Sensing applications received an increased attention and wide band gap materials like the group-III-nitrides exhibit several excellent materials properties to be implemented in devices operating at high temperatures, in corrosive media or high energetic impact [75]. Up to date, several sensing principles were investigated; some of them are summarized in the first reviewing papers of pioneering activities at the Technical University Munich [76, 77] and the University of Florida [78]. Generally, the origin of modulation of the 2DEG, i.e., the sensing principles can be classified into three groups (Fig. 26).

1. In the first case, the polarization induced, surface-near 2DEG responds on any change of the charges on the free, unpassivated gate surface. It has been shown that the polarization induced 2DEG at the AlGaN/GaN heterointerface is balanced by positive charges on the surface. Any additional charges on the surface will compensate or amplify the field inside the AlGaN barrier, i.e. positive or negative charges on the surface will increase or decrease the carrier concentration in the 2DEG, respectively. Such changes can be induced by ions or polar molecules

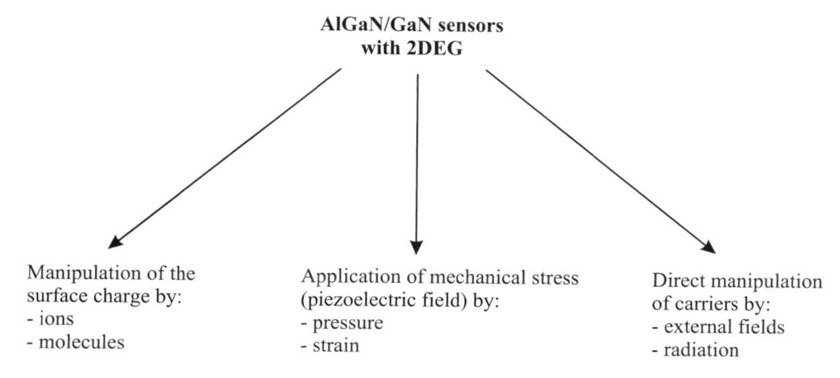

Fig. 26 Possible sensing principles through manipulation of the carrier density of a 2DEG in AlGaN/GaN heterostructures.

and their interaction with the surface of the AlGaN/GaN heterostructure can obey very different mechanism. This basic principle can be used for sensing of gases, ions, (bio-) molecules and the determination of the pH-value in solutions.

2. In the second application field it will be used that, unlike conventional III–V-based HEMTs (i.e. AlGaAs/GaAs), all the layers of the typical group-III-nitride-based HEMT structure are undoped and the carriers in the 2DEG channel are induced by piezoelectric and spontaneous polarization. As has shown above, the influence of the piezoelectric polarization on the carrier concentration in the 2DEG can be up to 50%. Consequently, such heterostructures will be sensitive to mechanical impact, which makes group-III-nitride HEMTs promising candidates for pressure or strain sensors.
3. Finally, the electron concentration or the transport in the 2DEG can be influenced directly by external fields, i.e. magnetic fields or electromagnetic radiation, which can be used for Hall sensors or ultraviolet light (UV) detectors.

These mechanisms and their practical realization in sensors will be described in more detail in the following part.

9.2 Surface Sensitive Sensors

9.2.1 Basic Sensor Structure and Technology

Although many different designs for group-III-nitride based sensors are possible, a very attractive feature of the transistors discussed in this section is the principal operation with gain and thus, the potentially higher sensitivity than the "simple" bulk material, and that they all are derived from the same undoped AlGaN/GaN heterostructure shown in Fig. 27. Moreover, the same heterostructure also can be used for the fabrication of high frequency, high power HEMTs as well as passive

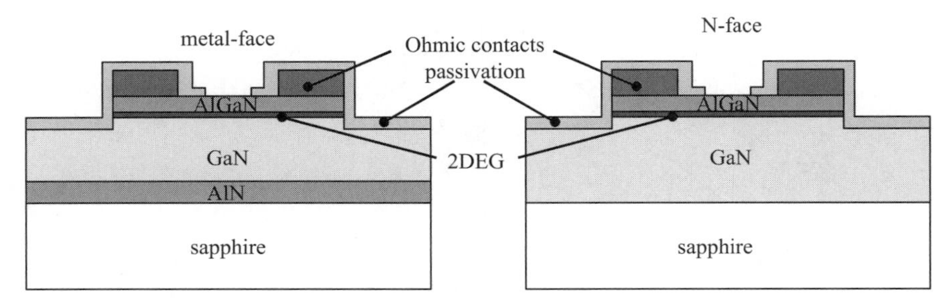

Fig. 27 Schematic cross section of the basic sensor structure using an open-gate HEMT on a metal face and N-face heterostructure.

surface acoustic wave devices, which opens up a possible route for monolithic integration of both, sensor functions and analog as well as digital data processing and transmission. Thus, integrated devices with multifunctional sensor capability, on-chip amplification and computation capability, and even remote wireless readout capability would be possible by suitable lithographic processing and metallization.

The simplest configuration to use a 2DEG confining AlGaN/GaN heterostructure is a planar open-gate HEMT (Fig. 27). The size of the active sensor area can be a few millimeter [53] for single devices or scaled down to a few micrometer for sensor arrays [79] or integrated structures. The majority of the investigations is concentrated on the use of metal-face HEMT structures (Fig. 27, left) since for a long time the growth of metal-face group-III-nitrides has been better controlled resulting in a materials quality superior to N-face structures (Fig. 27, right). However, first studies [80] have shown no qualitative difference in the sensing behaviour of heterostructures based on the two polarities except a longer response time on the N-face sensor, which might be attributed to polarity dependent differences in the surface chemistry. However, general conclusions are not possible from a single experiment, and in the following we concentrate on the sensors based on metal-face heterostructures.

AlGaN/GaN heterostructures are grown by PIMBE [81–83] or MOCVD [79, 84, 85]. An additional 2 nm GaN cap layer improves the chemical stability of the sensor device. The majority of sensing application was investigated on AlGaN/GaN heterostructures on sapphire substrates. This substrate offers different advantages. First, the thermal expansion coefficient is close to common packaging materials like alumina or aluminum nitride ceramics, which simplifies the packaging technologies for high temperature sensors and chemical sensors in aggressive media. Second, sapphire is transparent to visible light, which enables the implementation of additional optical analysis techniques (transmission or fluorescence spectroscopy) for more complex biological investigations (Fig. 28 [86]).

Usually, a mesa etching is required to isolate the device [89]. Alternatively, ion implantation [74] or UV laser cuts [86] can be used to isolate the 2DEG channels of the active sensor area [86]. A conventional Ti/Al/Au based metallization used for the HEMT technology [90, 91] is forming the ohmic contacts for source and drain. The

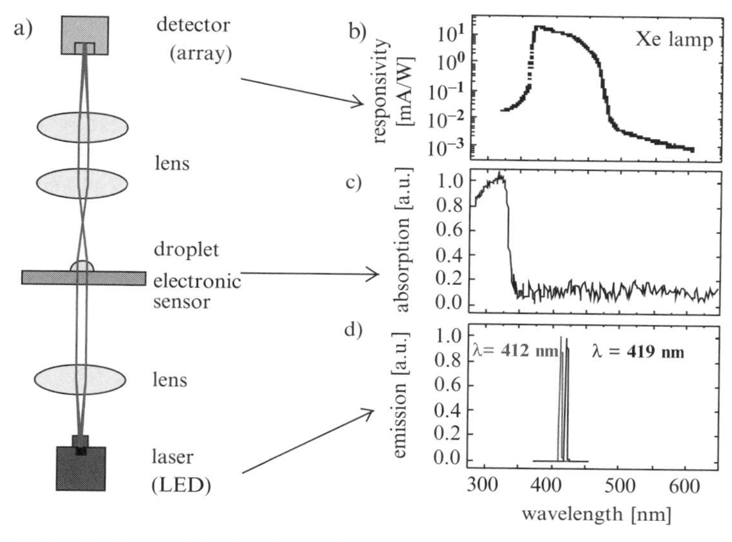

Fig. 28 (a) Setup for the combined optical and electrical analysis of bioreactions in small amounts of liquids [87, 88]. A laser diode or LED emits light with specific wavelength (b). At this wavelength, the sensor is completely transparent (c). A detector with spectral selectivity (d) records the transmitted light for colorimetric measurements of the droplet simultaneously to the electrical measurements with the sensor.

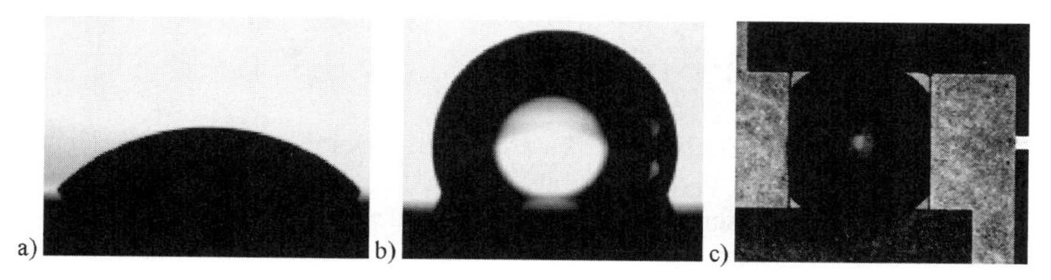

Fig. 29 (a) Hydrophilic and (b) hydrophobic AlGaN surface obtained by oxidation and CF_x coating, respectively [83]. (c) Nanodroplet confined on the hydrophilic active area of an AlGaN/GaN sensor with hydrophobic surrounding [86, 87].

active area is confined and the contacts are passivated by additional isolating layers as SiN_x or Sc_2O_3 [84], which are often used for high performance rf-HEMT passivation [90], CF_x [83], SiO_2 [85], silicon rubber [81], polyimide [82], epoxy [79], polymethyl methacrylate [92] or other materials depending on the actual sensing application. For fluidic applications the gate area and the surrounding parts can be modified to achieve hydrophilic or hydrophobic behavior (Fig. 29). This treatment will confine the liquid on the active area, which becomes important for the measurements in droplets [83, 86, 87]. Finally, the active gate might be modified or coated to activate it or to achieve certain selectivity, which will be described in the context of the application. For the mechanical sensing (pressure, strain) additional technological steps are necessary, which will be described in section C.

9.2.2 Ion Sensors

A direct manipulation of the conductivity of the 2DEG channel was achieved by exposure to ions generated by an ion spray technique [53, 80]. The ion detectors were PIMBE grown GaN/$Al_{0.3}Ga_{0.7}N$/GaN heterostructures with open gate, while all other parts are covered by a SiN_x or silicone passivation. Positive and negative ions were generated at ambient conditions by a high voltage cascade plasma spray ionizer, which allowed ion fluxes up to $10^{13}\,cm^{-2}\,s^{-1}$. By switching the polarity of the bias voltage, either positive or negative ions were directed towards the sensor surface and the channel current was recorded. Fig. 30 shows the observed modulation of the 2DEG channel current I_D induced by alternating flux of positive and negative ions. An incident negative ion flux onto the gate area results in a drastic decrease of the channel current corresponding to a reduction in sheet carrier concentration by nearly 4 orders of magnitude, i.e., the channel is effectively depleted. Exposure to positive ions recovers the 2DEG and increases the channel current with respect to the initial state. Cyclic changes have been found to be completely reversible.

With a 2DEG sheet carrier concentration of $1.2^{*}10^{13}\,cm^{-2}$, and an ion flux of about $10^{13}\,cm^{-2}\,s^{-1}$, the response time to negative ions is about 1 s, indicating that for every negative ion reaching the surface one electron is lost in the 2DEG. With these observations, AlGaN/GaN based ion sensors can be fabricated with a dynamic range of 4 orders of magnitude $(10^9 - 10^{13}\,cm^{-2}\,s^{-1})$ and an immediate response.

9.2.3 Electrolyte Gate HEMT Sensors

The high sensitivity of AlGaN/GaN based sensors towards ions in air anticipate a response to ions in electrolytic solutions, too. In the case of Electrolyte Gate field FETs (EGFETs) or ion sensitive FETs (ISFETs), the open gate area is exposed directly to an electrolyte whose concentration of specific ions shall be determined. The

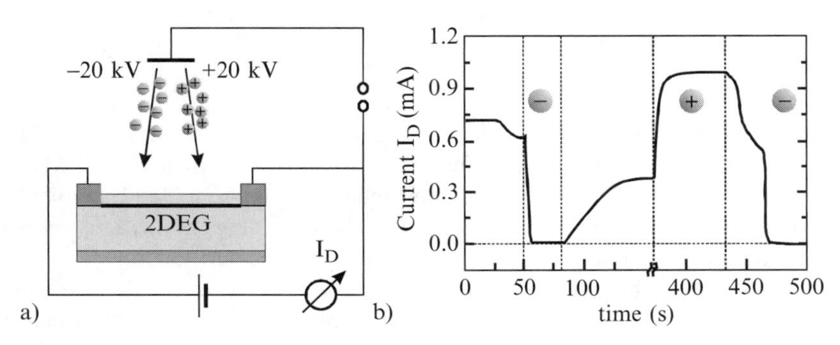

Fig. 30 (a) Measurement setup exposing the gate area to ions of both signs generated by the plasma spray technique. (b) Time response of the channel current I_D upon exposure to an ion spray: rapid increase (positive ions) or decrease (negative ions) upon switching the sign of the ion flux; slow return of I_D to its initial value after turning off the ion spray [80].

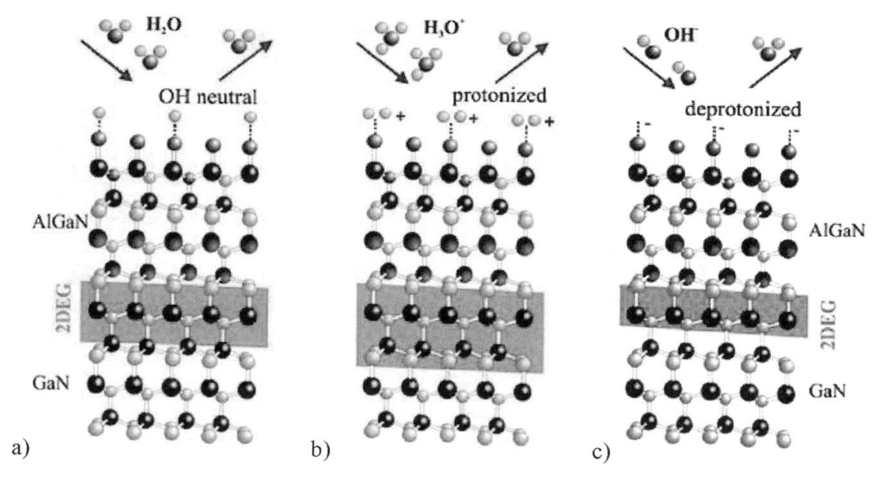

Fig. 31 (a) Model for the oxidized GaN top layer capping a HEMT structure in water and the reaction with (b) positive H_3O^+- and (c) negative OH^- ions, as well as the resulting impact on the 2DEG sheet carrier density.

influence of the ions on the surface potential can be explained with the site-binding model [93]. It proposes that atoms in the surface are acting as amphoteres, when they are in contact with an electrolyte. They can release protons into the electrolyte ("Donors") and thus get negatively charged, form neutral OH sites or bind protons from the electrolyte ("Acceptors"), resulting in a negative surface charge (Fig. 31). In the case of a high concentration of H_3O^+ (low pH, Fig. 31b), the M-OH groups tend to accept a proton and act as acceptors, and the oxide surface becomes positively charged. In contrast, if the concentration of H_3O^+ is low, i.e. the concentration of OH^- is high (high pH, Fig. 31c), most of the M-OH groups release a proton, and the surface charge is getting negative. These changes in surface charge, due to the change of pH in the electrolyte, directly affect the surface potential and consequently the sheet carrier density of the 2DEG [94, 95].

An alternative model proposes the direct interaction of the electron deficient gallium in Ga-face GaN with anions from the electrolyte [96–98]. Thus, the pH value response is originated from the interaction with OH^- ions rather than H_3O^+-ions and does not require an oxidized surface. The observed (weak) response of the AlGaN/GaN HEMT on anions at constant pH supports this assumption [99]. However, the opposite changes of I_D upon exposure by KCl and KF [99] as well as the observed drift of the sensors after the first use let expect that both mechanisms may contribute to the sensing effect.

The chemical response of an open gate HEMT to changes in the electrolyte composition can be monitored by recording of the source-drain current I_D at fixed V_{GS} (or $V_{GS} = 0$), or by adjusting the gate voltage via an Ag/AgCl reference electrode (Fig. 32) in order to keep I_D constant. This reference electrode can be integrated onto the sensor chip and connected with the active gate area via an agarose gel [100].

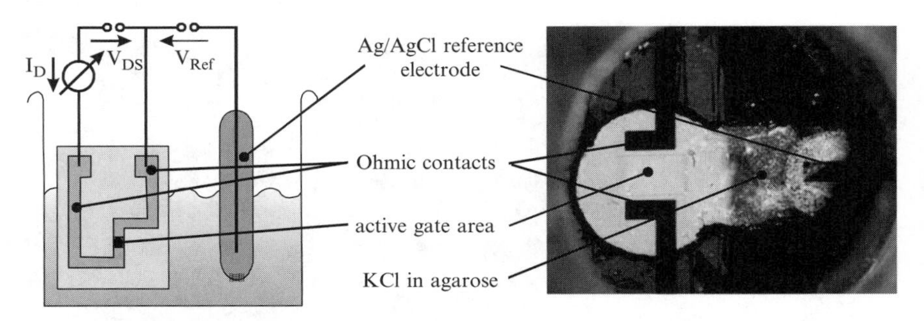

Fig. 32 Schematic setup for sensing of ions in electrolytes and experimental realizations of an integrated Ag/AgCl reference electrode on an AlGaN/GaN sensor chip using an agarose gel [100].

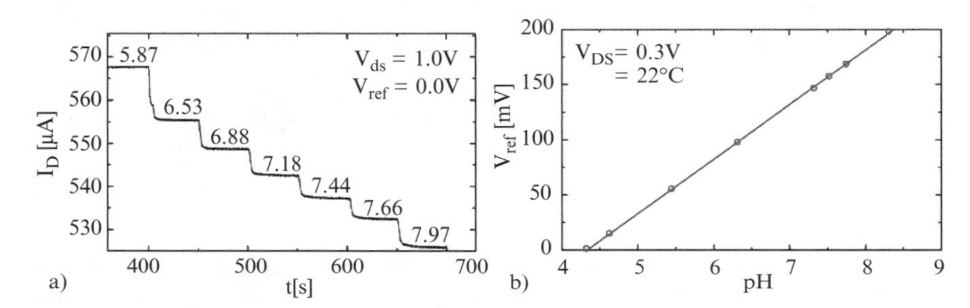

Fig. 33 (a) Transient behavior of the source–drain current I_D during changes of the electrolyte pH by titration with NaOH. Numbers correspond to pH values measured by a calibrated pH meter. (b) Reference potential as a function of pH [100].

The transient behavior of the GaN-based ISFET devices was investigated by time-resolved measurements of the I_D for V_{GS} = constant. The pH of the electrolyte was changed in steps between 0.1 and 0.3 pH by titration with diluted NaOH every 50 s (Fig. 33a, [100, 101]). All investigated GaN-based devices showed immediate response to changes in the pH. Non-optimized ISFETs already provide stable operation in the range from pH 2 to 12 with a resolution better than 0.05 pH [95]. Surprisingly, Kang *et al.* [102, 103] observed an opposite response on HCl, i.e. a decreasing of I_D with increasing HCl concentration. However, this controversy was not discussed and the underlying mechanisms are not understood up to date. The reference potential for constant I_D as a function of pH is shown in Fig. 33b. A linear behaviour over the entire investigated range from pH 4 to 10 is observed and a sensitivity of 55.0 mV/pH was measured [100]. Similar values were obtained also by other groups (56 mV/pH [95], 57.5 mV/pH [85]). These sensitivities are close to the Nernstian response to H^+ ions, which is 58.7 mV/pH at 23° C. As surfaces with native oxide and thermally oxidized samples showed almost no difference in sensitivity, it was proposed that a Ga_xO_y surface layer is responsible for the observed behaviour [95]. All experimental work was accomplished on metal face AlGaN/GaN heterostructures. It should be noted that theoretical considerations of the sensor-electrolyte interface proposed a higher sensitivity for N-face heterostructures [94].

Table 4 Achieved sensitivity to specific ions with polymer membrane coated HEMTs [104].

ion	detection limit	linear range	sensitivity
K^+	$3.1 \cdot 10^{-6}$ M	$10^{-5} - 10^{-2}$ M	-52.4 mV/pK^+
NH_4^+	$5.4 \cdot 10^{-6}$ M	$10^{-5} - 10^{-2}$ M	-55.5 mV/pNH_4^+
Na^+	$\sim 10^{-6}$ M	$10^{-6} - 10^{-2}$ M	-45.6 mV/pNa^+
NO_3^-	$9.4 \cdot 10^{-6}$ M	$10^{-5} - 10^{-2}$ M	61.1 mV/pNO_3^+

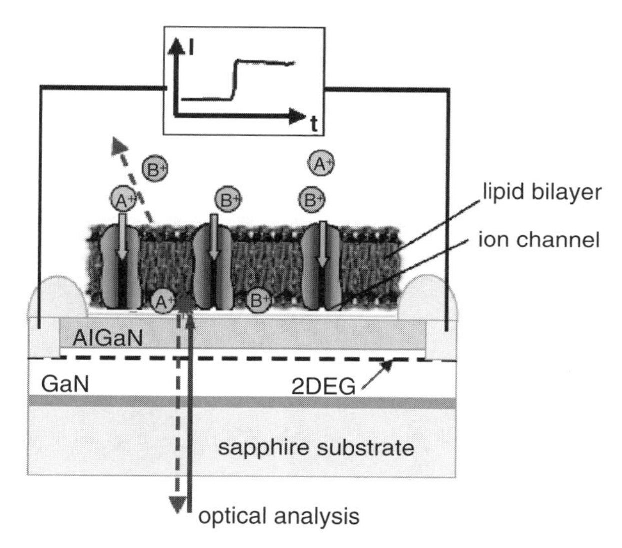

Fig. 34 Schematic design of a biosensor on a transparent AlGaN/GaN heterostructure as read-out device for selective ion transport across a lipid membrane via trans-membrane ion channels [105].

The selectivity to specific ions can be improved by functionalization of the gate area. In the simplest case it was achieved by a thin Au film [99], which resulted in a slightly increased response on halide ions. Higher selectivity can be achieved by covering the active gate area with polymeric membranes. This technique was successfully employed for the selective sensing of potassium, ammonium, sodium, and nitrate (Tab. 4, [104]). Other techniques could be the use of selectice ion channels in lipid membranes as proposed by Steinhoff *et al.* (Fig. 34, [106] and Kang *et al.* [78, 102], or ion selective chalcogenide glasses [106].

9.2.4 Sensors for Polar Liquids

The adsorption of polar molecules on the surface of semiconductors leads to a corresponding variation of the surface potential. These dipoles superimposing the electrical field inside the AlGaN barrier and modulate the 2DEG channel carrier density

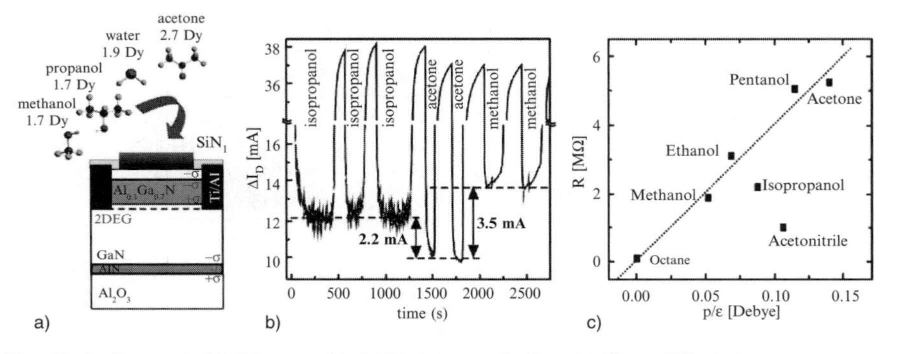

Fig. 35 (a) Ungated HEMT exposed to different polar liquids. (b) The source-drain current versus time during an alternating exposure by isopropanol, acetone, and methanol. By monitoring the channel current it is possible to distinguish between these different polar liquids [81, 107]. (c) Saturation values of the resistance of an AlGaAs/GaAs HEMT as a function of p/ε [108].

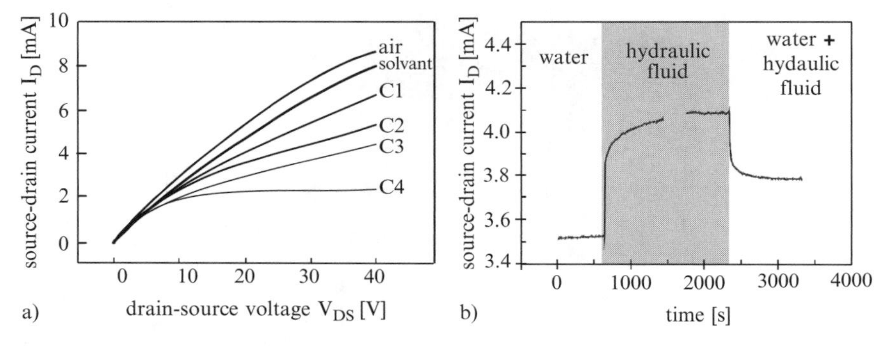

Fig. 36 (a) Drain I–V characteristics for different concentrations of copolymer solutions ($C1 = 1.0917\,mg/ml, C2 = 0.7612\,mg/ml, C3 = 0.5504\,mg/ml, C4 = 0.08734\,mg/ml$) [78]. (b) Soure-drain current of an ungated HEMT exposed to water and (polar) hydraulic liquid (unpolar) and a mixture of both [81].

and current. For the sensing of three different polar liquids, an unpassivated HEMT structure as described above was used [77, 81, 102, 103], (Fig. 35).

A decrease in the 2DEG channel current I_D by a factor of about three due to the interaction of polar liquids with the AlGaN surface is observed (Fig. 35b). Acetone with the largest dipole moment (2.7 Dy) leads to the strongest decrease in the channel current. Isopropanol and methanol adsorption results in different sensor response despite a similar dipole moment (1.7 Dy) which might be caused by size effects. A more systematic study was accomplished on AlGaAs/GaAs HEMT based sensors [108] which were protected by self-assembled monolayers of 4'-substituted 4-mercaptobiphenyls due to the chemical instability of GaAs surfaces in electrolytes. An empiric relationship between the sensors response and the normalized dipole moment p/ε was observed (Fig. 35c).

This concept was used for the sensing of different concentrations of block copolymers (Fig. 36a [78, 109]) and for the sensing of polar contaminations in

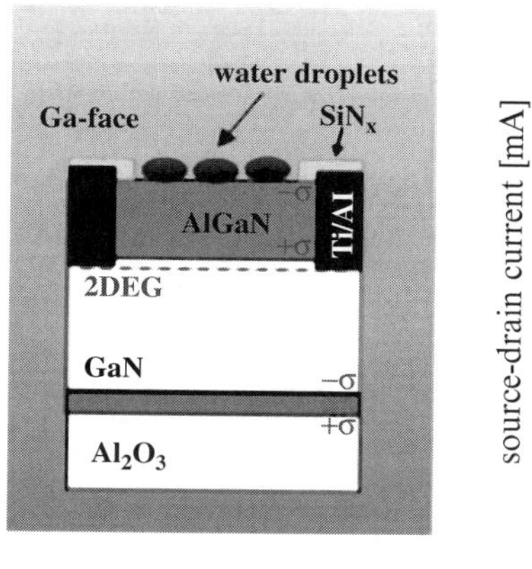

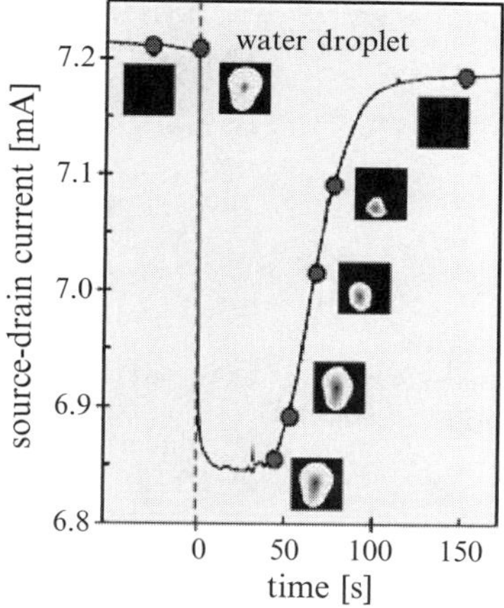

Fig. 37 Source-drain current of an ungated HEMT upon water droplet deposition and re-evaporation as a function of time.

nonpolar liquids [81, 107]. The sensor was exposed to water (polar liquid) hydraulic liquid (unpolar liquid) and a mixture with 10% water (Fig. 36b). The obvious sensor response demonstrates the possibility to use AlGaN/GaN heterostructures for applications like monitoring the water content of hydraulic oils.

Finally, also pure water droplets can be monitored with an open-gate HEMT sensor (Fig. 37). Polar water molecules deplete the 2DEG after the exposure onto the active surface area. For small volumes of water, i.e. small droplet sizes, the re-evaporation of the droplet can be monitored by measuring the recovery of the 2DEG channel current. Thus, with appropriate dimensions of the active area, the AlGaN/GaN heterostructure effectively acts as a size sensor for the water droplets.

9.2.5 Biosensors

With the mechanisms described above and the excellent stability, AlGaN/GaN heterostructures bear a high potential for biosensing applications. Such biosensors can be used for direct monitoring of the cell activity, i.e. measuring the cell action potential [79] or by indirect monitoring of bioreaction by recording of the pH value [110]. Moreover, by appropriate functionalization, selective sensing of conjugated biomolecules can be achieved [111] (see Fig. 38).

(1) Direct monitoring of cell activity

A first critical issue for the realization of cell-on-chip concepts is the survival of living cells on the substrate or the sensor surface. Silicon is known to be attacked by many biological important agents; however, an appropriate cell growth can be

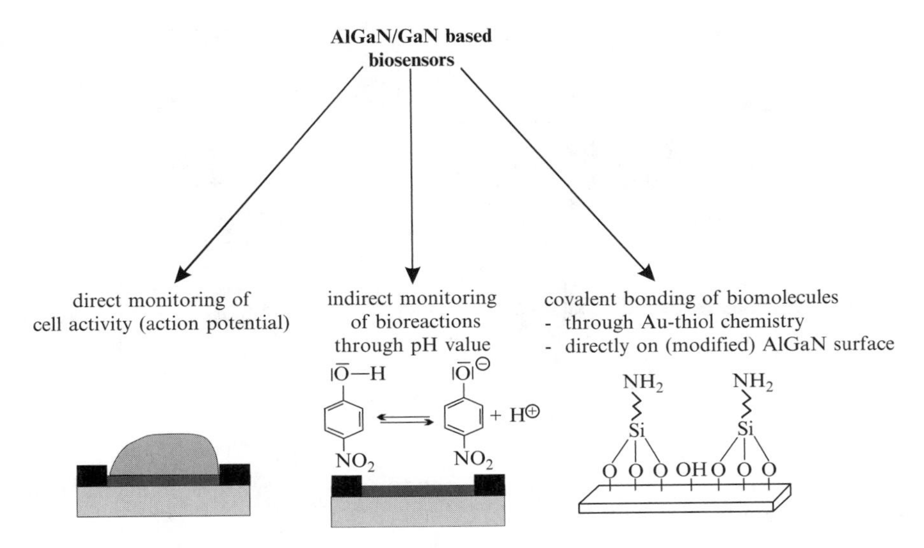

Fig. 38 Mechanisms for biosensing with AlGaN/GaN heterostructures.

achieved on an oxidized surface. Other semiconductors require a passivation (e.g. GaAs [112]). In contrast, AlGaN based alloys are chemically inert. First studies with rat-fibroblasts (3T3 cells) have demonstrated good adhesion properties of the group-III-nitrides independent of the aluminum concentration in the $Al_xGa_{1-x}N$ alloy ($x = 0, 0.22, 1$) or a pre-treatment by oxidation [105]. Recent investigations on cerebellar granule neurons prepared from 7-day-old Wistar rats [113] and mammalian cell lines HEK 293FT and fibroblastoid cells CHO-K1 [114] clearly demonstrated that adhesion and growth of living cells on GaN is superior to silicon.

The recording of an electrical cell activity with planar devices is a promising approach for the study of biological networks. Such cell-sensor hybrids are suitable for a variety of applications, such as drug screening in pharmacology, detection of toxins, and environmental monitoring. These applications require a noninvasive analyzing system suitable for long-term measurements under physiological conditions, which can be achieved by microelectrode arrays [115] or FET arrays [116]. The main drawbacks of these devices are their long-term drift in electrolytes due to the electrochemical instability.

Group-III-nitrides based HEMTs are chemically stable under physiological conditions, nontoxic to living cells and exhibit promising properties for sensor applications in liquid and electrolyte environments as shown above. The concept was demonstrated with cardiac myocyte cells of embryonic Wistar rats on HEMT arrays (Fig. 39a) [79]. After 5 to 6 days in culture a confluent monolayer of cells developed, which spontaneously contracted with stable frequency and the electrical signals of the cells were recorded. In Fig. 39b, voltage V_J in the junction area between cell and transistor gate is shown. Transistor signals were 100–150 ms in duration, firing at a stable frequency for several minutes. The signal shapes recorded by the AlGaN/GaN

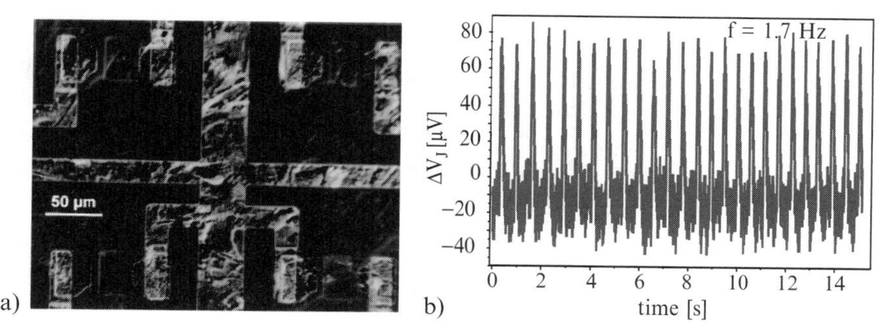

Fig. 39 a) Cardiac mycoyte syncytium cultivated on the device surface of an AlGaN/GaN EGFET array. b) Extracellular potential of a spontaneously beating cardiac myocyte syncytium recorded with an AlGaN/GaN EGFET. The cells were cultivated on the device surface [79, 117].

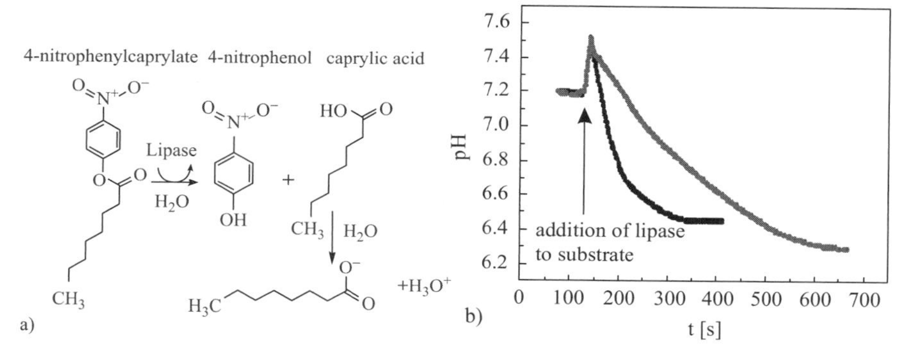

Fig. 40 (a) Model reaction, which changes the pH value by secession of a proton, and (b) change of the pH-value versus time during the reaction measured by an open gate HEMT sensor [88].

transistors seem to consist mainly out of the potassium signal part; however, the exact reason for this signal shape remains to be clarified.

(2) Indirect monitoring of bioreactions through pH value

A second approach uses the change of the pH-value by several bioreactions, for example enzymatic reactions. Lipase enzymes and their reactions with different substrates were monitored [88, 110] by continuous pH-measurement and transmission spectroscopy (see Fig. 28). This enzyme is responsible for breaking down lipids. In the experiments the enzymes catalyzed the reaction of 4-nitrophenylcaprilate to capryl acid and 4-nitrophenol. The resulting acid causes a change in pH (Fig. 40a). The AlGaN/GaN sensors were used to continuously monitor the pH value of the volume (Fig. 40b). After adding the lipase the pH increases from 7.2 to 7.5 followed by an expected decreasing of around 0.85 pH within the next 10 min. From the measured pH value and a comparative titration of caprylic acid the metabolic

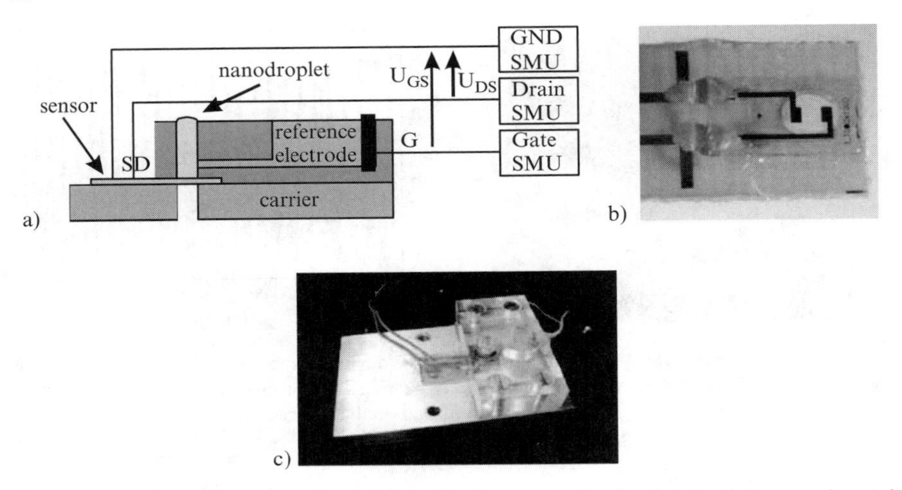

Fig. 41 (a) Schematic cross section of the pH-value sensor for droplets, and the experimental realization of (b) a single sensor chip on carrier, and (c) a chip integrated into the microfluidic system for nanodroplets [88].

rate can be determined. Dilution of the enzyme causes a different reaction kinetic as expected [88, 110]. The setup for the measurements is shown in Fig. 41.

(3) Site-specific, selective detection of biomolecules on functionalized surfaces

For the electronic detection of specific biomolecular processes based on molecular recognition, a covalent attachment of specific molecules on the sensor surface is a basic requirement. A first approach is the well known Au-thiol group chemistry. For this purpose, a thin Au film is deposited on top of the sensor surface. This approach is realized to bound thiol-modified oligonucleotides with 5 nm Au for HEMT based DNA sensing [92]. When the HEMT is exposed into the solution containing matched target DNA, after around 100 s, the source-drain current abruptly decreases, as illustrated in the inset of Fig. 42b. The source-drain current continuously decreases until the hybridization is completed after 20 min in reasonable agreement with previous reports of time scales of tens of minutes. It was suggested that the thiol modified probe DNA in the gate region hybridizing with target DNAs led to a double layer that alters the surface charge on the HEMT and changes the source-drain current. The device shows a good repeatability after a regenerating process.

The direct covalent coupling of biomolecules to the surface is investigated in detail for silicon based sensors using oxidized or hydrogen-terminated surfaces. In contrast, for the reaction with group-III-nitrides only a few investigations have been published. The first studies were accomplished by Bermudez [118, 119] who has analyzed the adsorption of aniline and pyrroline from the gas phase and proposed a

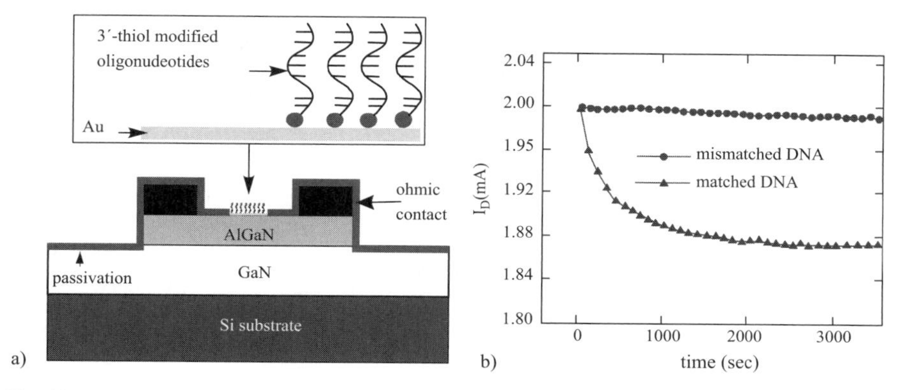

Fig. 42 (a) Schematic of the Au-gated HEMT. The Au coated gate area was functionalized with 15-mer 3′-thiol-modified oligonucleotides. (b) Change in HEMT source-drain current at $V_{DS} = 0.5\,\text{V}$ as a result of hybridization between immobilized thiol-modified DNA and matched or mismatched target DNA.

bonding model. The functionalization of a HEMT surface with receptors to sense the binding of target molecules in a liquid phase was demonstrated using aminopropyltriethyoxysilane (APTES) [111, 120] and octadecyltrimethoxysilane [120] and the covalent bounding was proven by electron spectroscopy [120]. This approach was used to demonstrate the electronic detection of streptavidin proteins [111]. First, the oxidized AlGaN surface was treated with APTES to functionalize the surface with amine groups. By covalent attaching of the antigen biotin to the surface of the HEMT gate region, the device can detect the conjugation of the streptavidin-a protein. The immobilization of the protein on these sites was detected as a change in HEMT source-drain current (Fig. 43, [111]).

These few examples demonstrate the high potential of AlGaN/GaN HEMTs for the rapidly developing application field of biosensing.

9.2.6 Gas sensors

Gas sensing based on the adsorption on a catalytic metal surface [121] on top of chemically sensitive semiconductor is a widely used principle. The first demonstration of this mechanism has been reported for hydrogen gas exposed to a silicon FET with catalytic Pd gate [122]. In the meantime, the mechanism was applied for the detection of many oxidizing and reducing gases. The operation requires the dissociative chemisorption induced by the catalyst, and the transport of reactive or charged species to the semiconductor surface, both processes are dependent on temperature. Thus, high temperature operation not only widens the application field, it also alters the selectivity to specific gases and reduces the response time to milliseconds [75]. However, the operation of widely investigated silicon based gas sensors is limited to temperatures below 150° C. Wide band gap electronics potentially keep

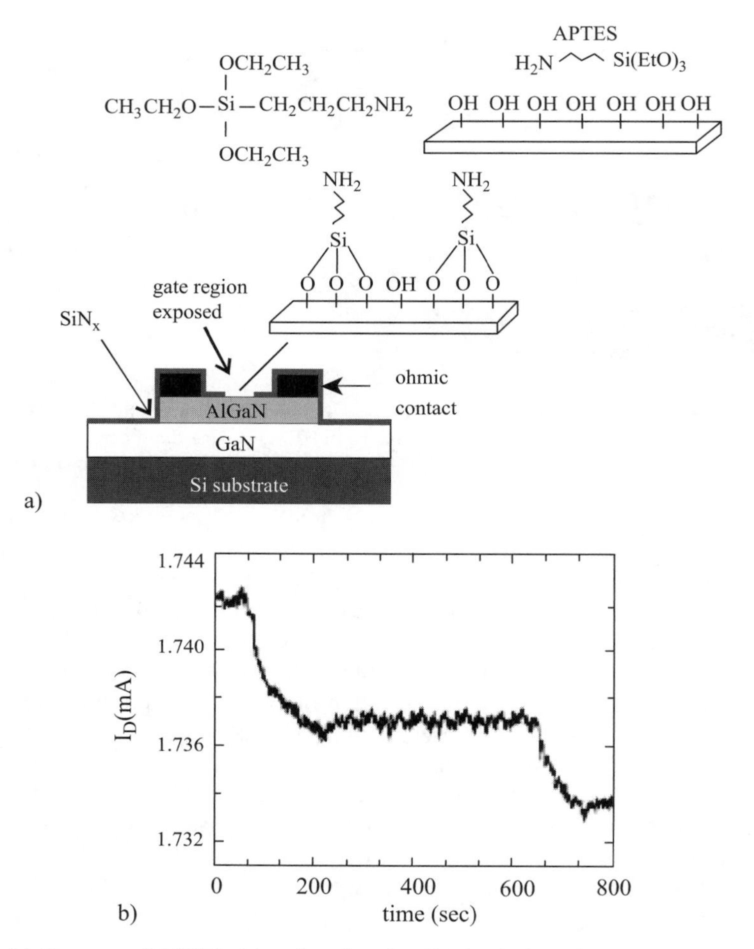

Fig. 43 (a) Structure of APTES, (b) surface functionalization before chemical modification, and (c) after chemical modification. (d) Scheme of the gateless HEMT whose surface is functionalized in the gate region. (e) Change in the source-drain current as a result of interaction between biotin and streptavidine introduced to the gateless HEMT surface [111].

their functionality at temperatures above 400° C. For example, SiC based sensors are able to operate at temperatures up to 800° C [75]. Thus, the wide band gap of the AlN-GaN system as well as the high chemical stability enables the application for high temperature electronic and sensing devices.

The most used catalytic metal is Pt, other materials are Pd [123], Rh [124], and Pd/Au [124]. These metals form effective Schottky barriers to AlGaN and, in particular Pt is one of the metals implemented into the HEMT technology to create the gate contact [54]. These GaN Schottky diodes exhibit strong changes in current upon exposure to hydrogen and hydrogen containing molecules [125]. The effect is explained by the lowering of the effective Schottky barrier height [126] when molecular hydrogen is catalytically cracked by the catalyst and hydrogen atoms diffuse to the GaN-metal interface. This mechanism was first investigated on Pt/GaN-Schottky diodes [125]. Recently, Pt-AlGaN/GaN heterostructure Schottky diodes were used to fabricate hydrogen sensors with improved performance [127], which operate at temperatures up to 800° C [128]. The heterostructure effectively alters the Schottky barrier; the influence of the present 2DEG was not investigated.

At the metal-nitride interface, atomic hydrogen forms a dipole layer. This dipole layer is effectively created by O-H-bonds (Fig. 44a) and consequently strongly dependent on the presence of oxygen on the nitride surface [126, 129]. The exposure of the GaN surface to air prior to the catalyst deposition is sufficient to accumulate oxygen for effective sensor operation. In contrast, complete *in vacuo* processes resulted in decreased sensing performance [129]. The concept of an oxidized interface between the AlGaN/GaN heterostructure and the catalytic metal was further developed by the deposition of additional oxide. For this purpose Sc_2O_3 was grown by PIMBE on ozone exposed AlGaN/GaN HEMTs [130]. The device effectively sensed hydrogen [130] and ethylene [131], and in the case of hydrogen an improved sensitivity compared to conventional Pt/GaN Schottky diodes was demonstrated.

However, Pt/GaN sensors exhibit sensitivity to both oxidizing and reducing gases [132]. Since hydrogen is the only specie that can diffuse through the catalyst, an alternating transport mechanism was proposed [132] taken the porosity of Pt into account (Fig. 44b). Here, the gas molecules diffuse through pores directly to the (oxidized) GaN surface leading to polarized adsorbates. This model was confirmed by the investigation of the CO sensitivity of Pt/GaN Schottky diodes [133]. By decreasing the catalyst thickness and thermal annealing, the response on CO was increased due to the surface restructuring of the Pt film, which opens channels (pores) for gas transport towards the free GaN surface.

The manipulation of the effective barrier at the surface of GaN predestines the effect to be used in HEMT structures in order to alter the 2DEG channel. The decrease of the electric field in the AlGaN barrier and the resulting drop in the carrier

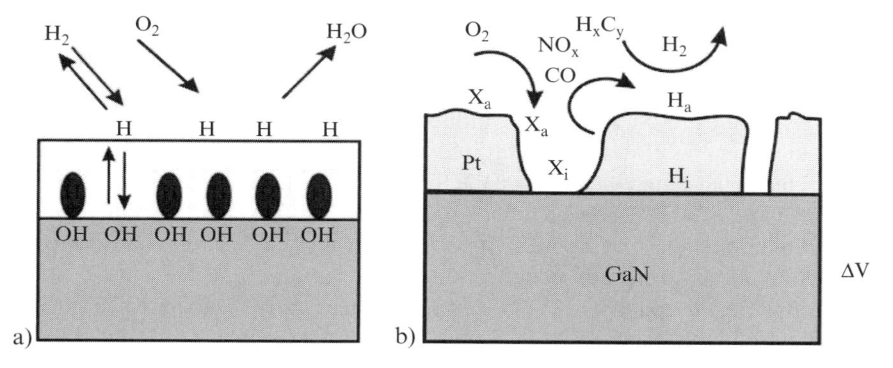

Fig. 44 (a) Schematic representation of the sensing mechanism of catalyst-GaN-Schottky heterojunctions. (b) Gas-sensing mechanisms underlying the function of Pt/GaN Schottky barrier devices (H_a, H_i: adsorbed and interfacial hydrogen; X_a, X_i: adsorbed and interfacial molecules).

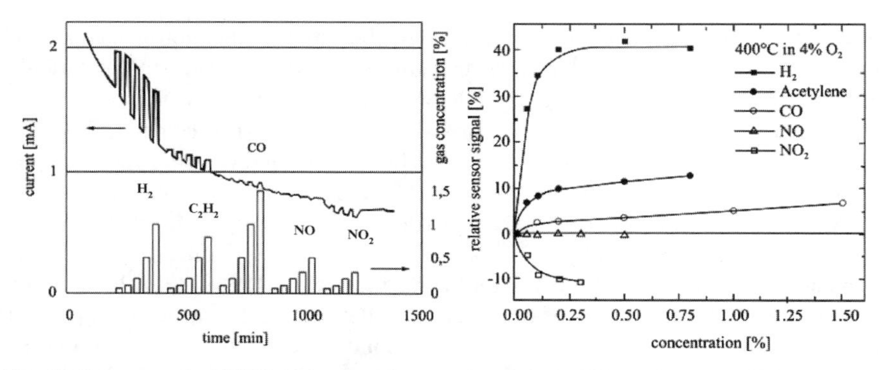

Fig. 45 Response of a HEMT device to various concentrations (500 ppm to 1.5%) of reducing and oxidizing gases diluted in a background of 4% O_2 in N_2 (operation temperature: 400° C).

density of the 2DEG after exposure to hydrogen was recently determined by modulation spectroscopy [134]. In saturation, a decrease of the Schottky barrier by 0.85 and 0.65 eV was determined directly for $Al_xGa_{1-x}N/GaN$ heterostructures with $x = 0.15$ and 0.20, respectively. The mechanism was successfully applied on AlGaN/GaN sensors [135], and hydrogen was reproducibly detected at concentration of 500 ppm in synthetic air [132]. In addition, the response of the 2DEG channel conductivity on further gaseous molecules was investigated (Fig. 45 [132, 136]).

The results in Fig. 45 show that similar to the H_2 case, reducing agents such as acetylene and CO led to an increased source-drain current, whereas, no response was detected upon exposure to NO. Upon exposure to oxidizing NO_2, a decrease in the channel current was observed. In comparison, the Pt-HEMT sensor exhibited a relatively good selectivity towards hydrogen with a sensor response around four times higher in comparison to other test gases. This relatively high selectivity arises from the lower porosity of the Pt based catalytic gate in this study as a result of a lower annealing temperature compared to Schottky diodes.

A direct comparison of a Schottky diode and an identically processed HEMT sensor has shown the higher output signal (source-drain current) for the transistor, which is expected since this device works with gain [137–139]. However, for small forward bias the relative sensitivity of the Schottky diode was higher. This phenomenon might be caused by the strong influence of hydrogen on the Schottky barrier, and at bias voltages close to the threshold, the effect of small changes in the barrier is expected to have high influence on the forward current. Reliability issues were not discussed in [137–139], however, one might argue that this effect is highly instable since contamination or cross selectivity effects can alter the Schottky barrier slightly with the observed strong consequences on the current.

Meanwhile, the concept of Pt-Oxide-FET was transferred onto an AlGaAs/GaAs HEMT with high sensitivity to hydrogen at room temperature [140]. Also on AlGaAs/GaAs based sensors a three-terminal device was demonstrated where the Pd gate was biased to improve the linearity of the response on hydrogen [141]. Similar to the above described electrolyte gate HEMTs, improved sensing behavior can be expected also for AlGaN/GaN gas sensor by optimization of the operation point.

9.2.7 Stability of the Sensors

Up to date mainly the sensing principles are investigated. Most of the devices and demonstrators exhibited an excellent stability at conditions present in laboratories; however, reliability and stability issues at practical conditions were only rarely addressed. However, a few conclusions can be drawn.

The temperature dependence of a HEMT device is very small compared to other semiconductor sensors (100 ppm/K [142]). For high temperature sensing the stability of the contacts and / or passivation layers are major problems [143]. The operation up to 400° C has no influence on the 2DEG, even in oxidizing environment [143]. Above 600° C a degradation of the electrical properties and thus, of the sensor performance occurs.

Despite the high chemical stability of the group-III-nitrides, AlGaN/GaN based heterostructures can be attacked by reagent in (bio-) chemical sensing applications. Most critical issue is the stability at extreme pH-values, which requires a careful passivation of all parts except the active gate area (see section A1). The active open gate has shown to be very stable at low pH-values, except in HF. However, for high pH-values ($pH > 10$) a non-stability and a deviation from the linear (Nernstian) dependence of the sensor response on the pH-value was observed [144], which might be caused by the instability of the oxidized surface in alkaline solutions [145]. Moreover, for biological applications the sensors has to be carefully sterilized. The first common method, a treatment in steam at temperatures of about 120° C was shown to not affect the sensor properties [114]. In contrast, the treatment in NaOH at 80° C tend to attack both the passivation layers and crystal defects such as dislocations, which degrades the sensor performance or even destroys the structure [101]. Appropriate passivation layers to be integrated into the sensor, which withstand the above mentioned treatments, are not available and the stability of the sensor and the protection of the active open gate area are still technological challenges. MOCVD grown sensors have generally shown higher stability than heterostructures grown by PIMBE [101]. It is futher improved by a thin GaN cap layer of 2-3 nm. Moreover, it has been shown that a thin SiN_x film (10 nm) on the gate shifts the operational point, but has very little effect on the sensitivity to the pH-value [85].

Finally, the electrical behavior is influenced by several sources of outer noise (light, temperature variation, electrical noise). The influence can be minimized by appropriate selection of the operation point [100] or design of the heterostructure, for example, by doping [101]. For the reduction of such influences, the well-known bridge techniques can be used, which in the case of the HEMT sensors exhibit very simple integrable geometry with four transistors (Fig. 46 [101, 146]).

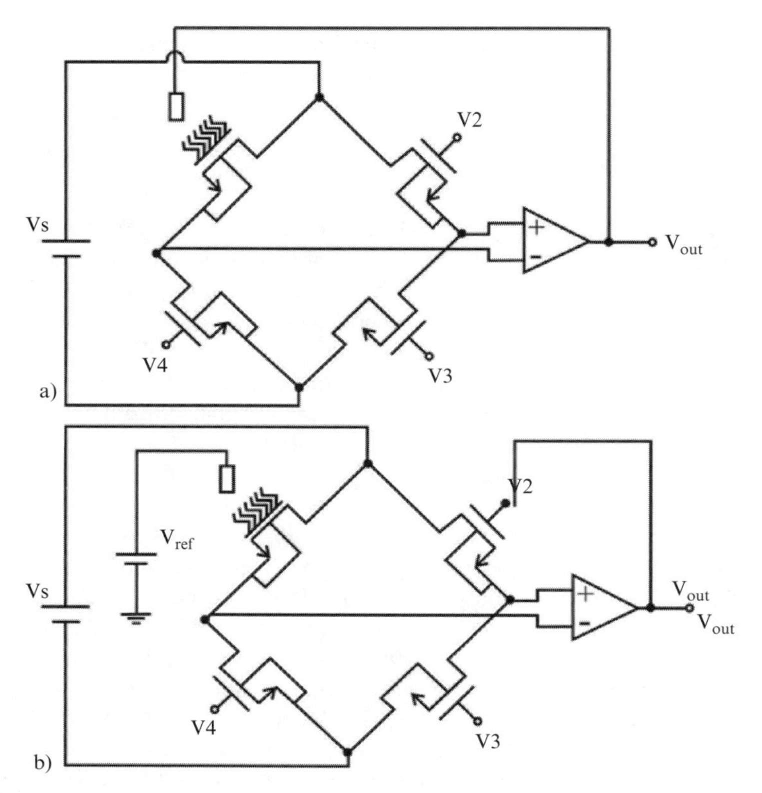

Fig. 46 Possible configuration of (a) direct and (b) indirect Wheatstone bridges for improved stability in pH-sensing [146].

9.3 Mechanical Sensors

Mechanical sensors like pressure sensors are based on the measurement of the externally induced strain in the heterostructure. The pyrolelectric properties of group-III-nitrides allow two mechanisms for strain determination: the piezoelectric and the piezoresistive transduction. The direct piezoelectric effect is used for dynamical pressure sensing. For measuring static pressure such sensors are not suited due to leakage of electric charges under constant conditions. For static operation, the piezoresistive operation is preferable. Piezoresistive sensor using wide band gap materials have been mainly employed using hexagonal SiC bulk materials for high temperature operation or heteroepitaxial cubic SiC on silicon for integration into the silicon micromachining technology [75]. The piezoresistivity of GaN [147] and AlGaN/GaN heterostructures [148] is comparable to SiC, however, it can be amplified by a HEMT structure [149]. For piezoresistive strain sensing two basic configurations are possible (Fig. 47).

The first configuration is a planar device and requires basically the same processing steps as the sensors described above. The function is based on the action of

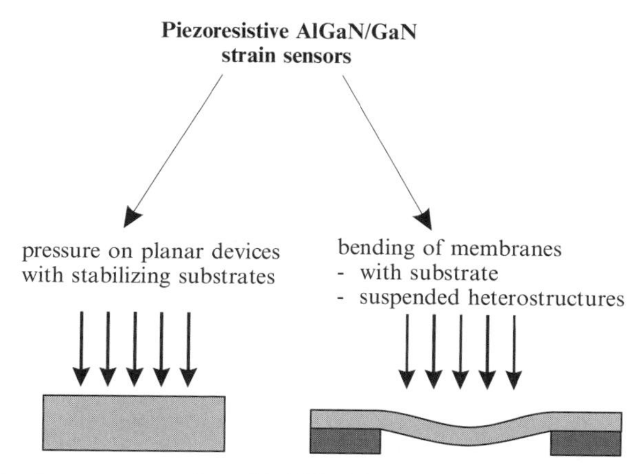

Fig. 47 Principles for pressure sensing with AlGaN/GaN heterostructures.

hydrostatic pressure, which alters the Ni/AlGaN Schottky barrier height [150] and the polarization in the AlGaN/GaN heterostructure [151]. These effects are relatively small compared to the strain sensors exposed to bending. Thus, the realized sensor is sensitive to high pressure (kbar) and the high mechanical stability of this structure favors such applications. The performance can be controlled by the gate voltage and highest sensitivity is achieved close to the pinch-off [151].

For lower pressure (differences) the application of diaphragm or membranes is preferable where the external pressure is transferred into a changed internal strain caused by the bending (Fig. 48). The resulting change in the polarization alters the 2DEG channel current, which will be measured. The first experiments were accomplished on N-face AlGaN/GaN/AlGaNGaN HEMT structures, where the 2DEG is confined between the upper GaN and the AlGaN barrier [149]. The heterostructure was supported by the sapphire substrate and the full structure was exposed to bending.

A well controlled uniaxial strain is applied by a blade driven by a servo-controlled piezomanipulator (Fig. 48a). The blade dynamically deflects the center of the sample with a frequency of 30 Hz. The resulting voltage modulation is measured at a current of 1 mA between two ohmic contacts. Fig. 54c demonstrates the linear dependence of the channel resistivity on the applied strain. Moreover, a direct comparison to cubic (3C) SiC and a single AlGaN layer clearly demonstrates the superior piezoresistive properties of the AlGaN/GaN heterostructure. From these results it can be concluded that the interaction of piezoelectric and piezoresistive properties improves the sensitivity of pressure sensors by using AlGaN/GaN heterostructures with confined 2DEG. Moreover, the weak temperature dependence of the piezoelectric constants enables the fabrication of pressure sensors, which are relatively independent of the temperature without further compensation mechanism.

However, the manufacture of applicable sensors requires a micromachining to realize freestanding membranes containing the AlGaN/GaN heterostructure. It was achieved by epitaxial deposition onto silicon substrates and etching of circular holes from the backside [152–154]. Fig. 49a shows the concept of such devices with

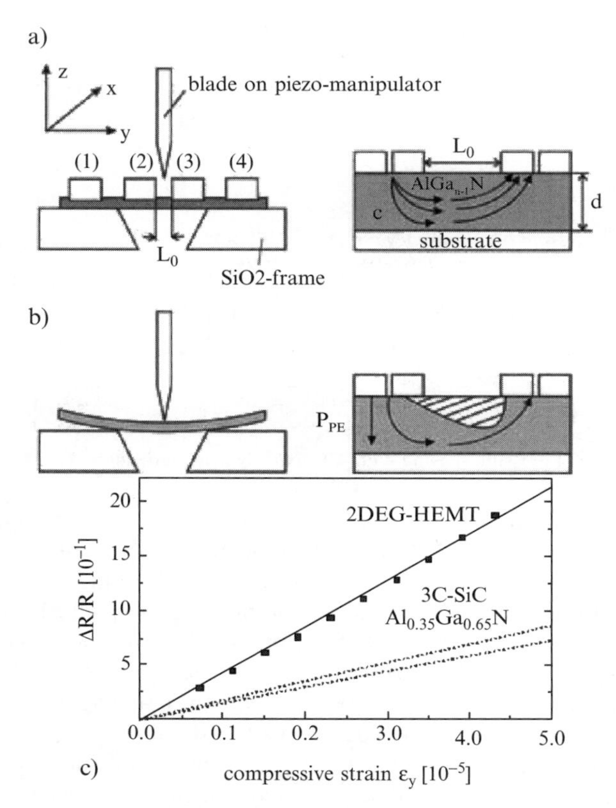

Fig. 48 Schematic drawing of the experimental setup. The current is measured between contact (1) and (4), the voltage drop between (2) and (3). The influence of the strain induced piezoelectric field on the current is shown on the right: (a) sample without strain, (b) compressive strain leads to narrowing of the channel and an increase in resistance due to the piezoelectric field. (c) Relative change in conductivity as a function of the applied compressive strain for two HEMTs. For comparison the corresponding results for $Al_{0.35}Ga_{0.65}N$ and 3C–SiC are given [149].

inter-digitated finger pattern. The 2DEG channel conductance was determined from I-V characteristics and reveals a linear dependence on the pressure (Fig. 49b).

Despite the low thermal drift of the AlGaN/GaN heterostructures, a higher stability is expected by capacitive read-out of the sensor (Fig. 50, [152]). Here, the 2DEG at the AlGaN/GaN interface acts as counter electrode to the silicon substrate bonded to the backside of the wafer with suspended membrane. The realized device was sensitive to changes of pressure ranging from -0.5 to $+1$ bar with a hysteresis less than 0.4 %.

Finally, a novel concept for strain sensing using a 2DEG was theoretically investigated by Wu and Singh [155]. The device is based on the use of a thin oxide with high piezoelectric coefficients under the gate region of a HEMT. Thus, the strain induced piezoelectricity is superimposed to the gate voltage and modulates the 2DEG sheet carrier concentration and as a consequence the channel current in

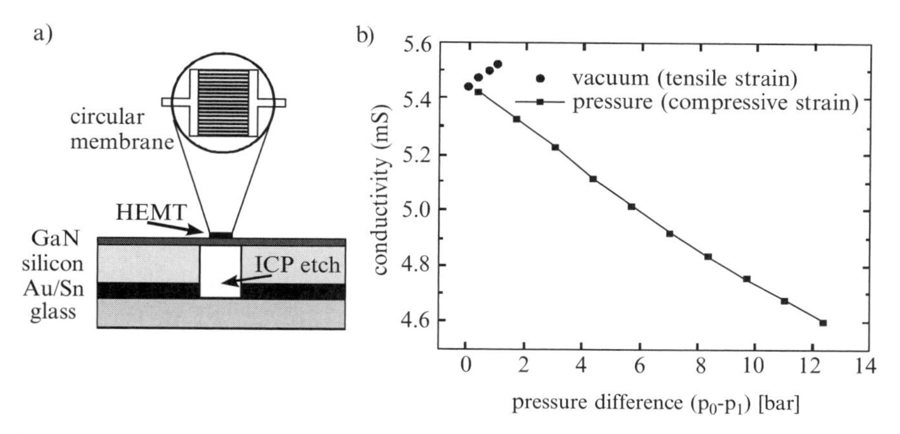

Fig. 49 (a) Schematic diagram of differential pressure sensor device with 2DEG channel and (b) channel conductivity of the AlGaN/GaN HEMT membrane as a function of differential pressure [153].

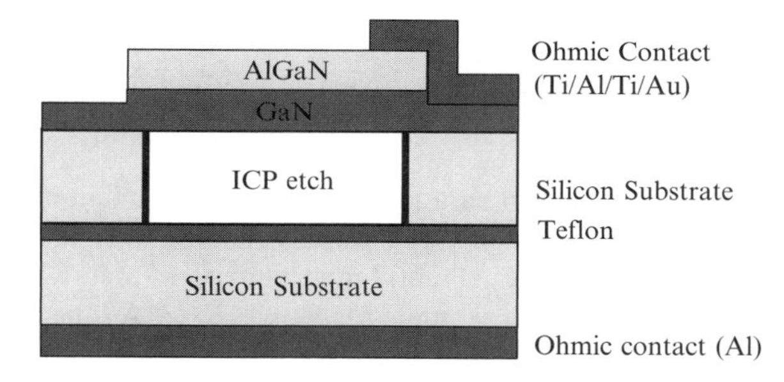

Fig. 50 Schematic diagram of the capacitive pressure sensor [152].

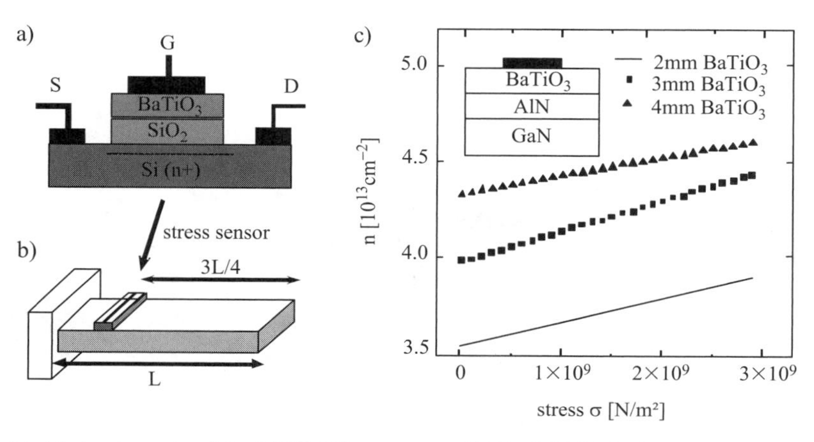

Fig. 51 (a) A schematic of the MPISFETs structure. (b) The MPISFETs is grown on top of the cantilever. (c) Calculated sheet charge densities, (nS), for the GaN/AlN heterostructure interface. The thickness of AlN layer is 30 Å [155].

dependence of the external stress (Fig. 51, [155]). A direct comparison with a silicon FET structure shows the superior performance of the AlGaN/GaN HEMT. $BaTiO_3$ was used as piezoelectric material due to the high piezoelectric constants, which are about one order of magnitude higher than AlN, which enhances the sensitivity compared to pure group-III-nitride based sensors. The optimal thickness of the piezoelectric oxide is about a few nm and therefore will not alter the mechanical properties and stability of the structure.

9.3.1 Micro- and Nanoelectromechanical Devices (MEMS / NEMS)

The potential of group-III-nitrides for micro- and nano-electromechanical systems (MEMS / NEMS) is not completely evaluated up to date. In particular, the activities on the integration of a 2DEG to improve the performance or implement new sensing principles are just at the beginning. MEMS and NEMS devices require the freedom of movement in different directions and consequently a three-dimensional geometry. The basic structures are single-clamped cantilevers and double-clamped bridges (Fig. 52 [156]). To use such structures as sensors, the effect of external influence on either the deflection of a static device or a shift of the resonant frequency in dynamic devices is measured. In the latter case, the cantilever / bridge oscillates as mechanical resonator.

First studies implemented AlN as a pure mechanically functional layer due to the excellent properties (low density, high elasticity module) for the manufacture of NEMS. There, a single crystalline AlN bridge was prepared and oscillations about 100 MHz were achieved [157]. An example for an AlN based resonator array, which

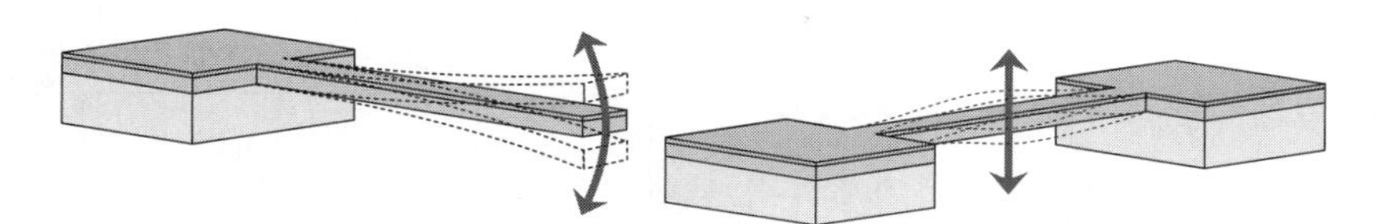

Fig. 52 Basic configuration of resonant MEMS devices [156].

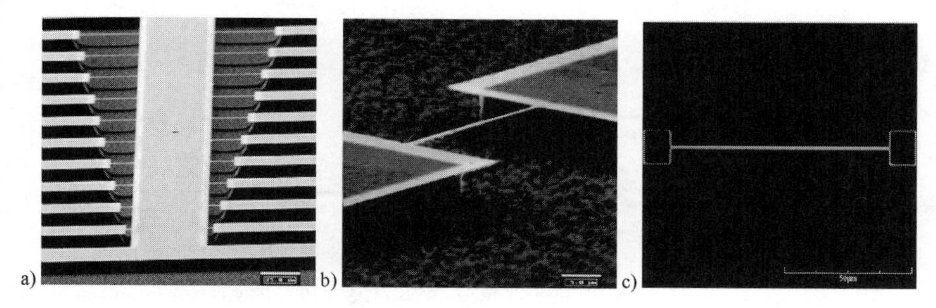

Fig. 53 AlN resonator beams on Si-subtrates: (a) arrays with a fixed width w and different lengths L, (b) magnification of a single resonator bridge (c) e-beam patterned structure with $w \sim 150\,\text{nm}$ [156]

was used to evaluate the sensing behavior in air [156] is shown in Fig. 53 [158]. These devices can be scaled down to submicron dimensions (Fig. 53c) in order to increase the sensitivity. A step beyond was the use of the piezoelectric properties of AlN for actuation of the resonators. This principle was employed for the stimulation of waves in AlN membranes [159] and was transferred onto pure AlN bridges [160] as well as MEMS based on AlN/SiC heterostructures (Fig. 54 [161]). With the latter device sensing properties have been demonstrated by showing the shift of the resonant frequency in dependence on temperature, pressure and mass loading. The mass loading was investigated by covering the resonator with an analyte-specific polymer and the concentration of a nerve agent stimulant (dimethyl methylphosphonate) as well as an explosive stimulant (3-nitrotoluene) was measured.

The integration of an AlGaN/GaN based 2DEG in MEMS or NEMS devices was only rarely investigated [162,163]. The first attempt used stripes of AlGaN/GaN heterostructures on sapphire substrates clamped between two blocks of Lucite (Fig. 55a [162]). This rather macroscopic device was tested at compressive as well as tensile strain and exhibited a linear response (Fig. 55b).

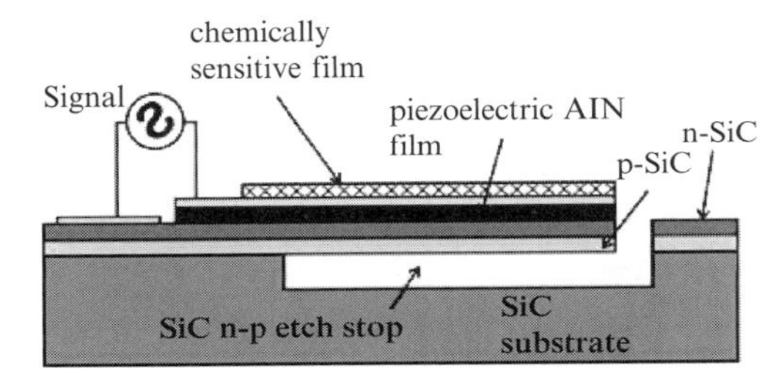

Fig. 54 Cross section of an active MEMS sensor based on an AlN/SiC heterostructure [161].

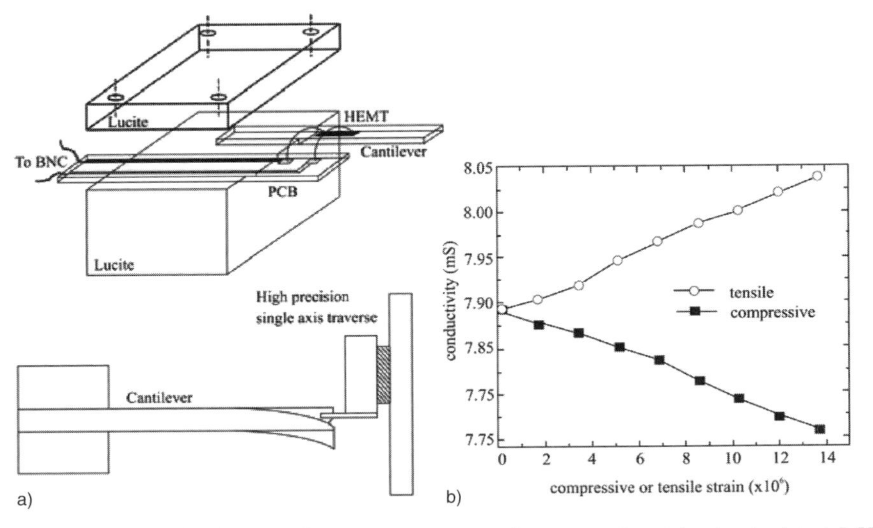

Fig. 55 (a) Schematic diagram of a pressure sensor package: experimental setup to detect I–V characteristics connected to the BNC cable according to various mechanical stresses (top) and mechanical stressor with cantilever (bottom). (b) The effect of tensile (top) or compressive (bottom) stress on the conductivity of the AlGaN/GaN HEMT with mesa etching.

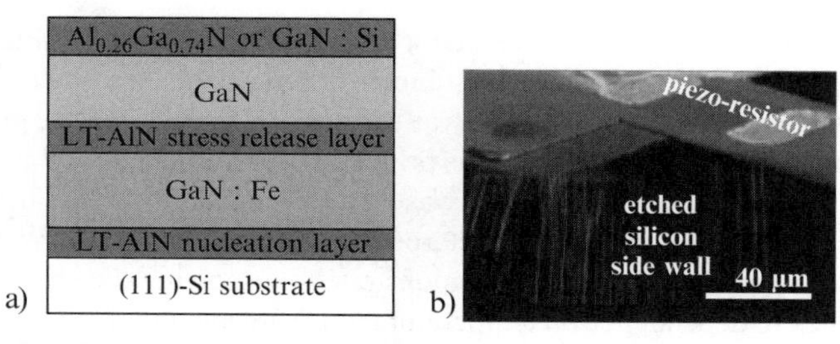

Fig. 56 (a) Schematic of the cross section of used GaN structures, and (b).realized cantilever with piezoresistor located at the pivot point [163].

The difficulty to fabricate suspended AlGaN/GaN heterostructure arises from the high chemical stability of group-III-nitrides. This complicates the necessary undercutting techniques, which require selective etching of either the substrate or a sacrificial layer. A promising approach is the deposition of AlGaN/GaN heterostructures on silicon which can be easily etched [164] without damaging the AlGaN/GaN heterostructure [114]. A critical issue is the control of the polarity of the group-III-nitride, and early investigations on suspended AlGaN/GaN heterostructures exhibited no 2DEG due to the existence of inversion domain boundaries [165]. However, the growth of single crystalline AlGaN/GaN heterostructures on silicon by MOCVD is now well controlled [166] and cantilever with integrated HEMT structure could be processed (Fig. 56 [163]).

The undercut of the heterostructure was achieved by selective CF_4 based dry etching of the silicon. The strain sensing element was located at the pivot point at the cantilever. The strain was induced using a needle displacing the edge of the cantilever. Fig. shows that the device is reacting on both tensile and compressive strain. An additional gate was implemented into the HEMT structure in order to control the operational point of the device. Highest sensitivity is observed when the gate is biased with a voltage close to pinch-off (Fig. 57).

More investigations on suspended 2DEG were accomplished on the AlGaAs/GaAs system, where novel concepts were demonstrated for strain sensing in a resonant device (i.e. the measurement of the resonant frequency and the quality factor) [167], for sensing of magnetic fields [168], as well as for actuation and the read-out of the device (Fig. 58 [169]). These few realized examples demonstrate the huge potential of AlGaN/GaN based heterostructures implemented in MEMS devices for sensing application with high sensitivity.

9.4 Sensor for Electromagnetic Fields

The method of Hall measurements is an established technique for the electrical characterization of the 2DEG confined in AlGaN/GaN heterostructures. Thus, the Hall effect can be used contrariwise for the sensing of magnetic fields. The concept was

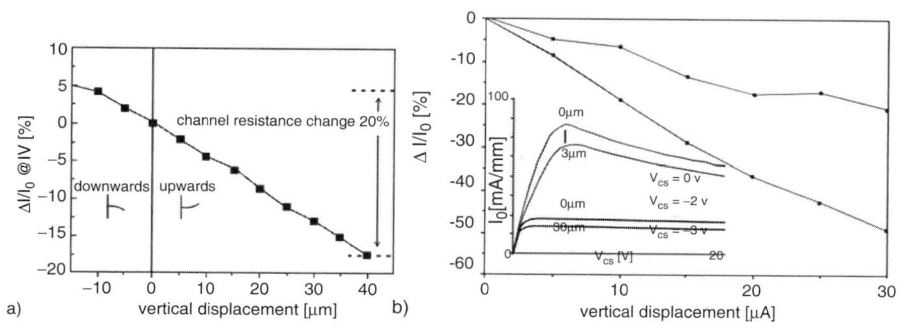

Fig. 57 (a) Response of a n-type doped channel piezoresistor ($N_s = 3 \times 10^{11}\,cm^{-2}$) for vertical displacement of the cantilever ($L = 200\,\mu m, w = 100\,\mu m$) at 1 V applied voltage. The initial current level $I_{D,0}$ was $24\,\mu A$, and the HEMT resistor dimensions were $L = 80\,\mu m$ and $w = 30\,\mu m$, gate length $L_g = 2.0\,\mu m$. The current decreases with upward bending as indicated. (b) Relative change in HEMT output current for two gate bias points at a displacement of up to $30\,\mu m$. The inset shows the change of output characteristic at $30\,\mu m$ cantilever bending, $I_{Dsat} = 80\,mA/mm$ for $V_{GS} = 0\,V, V_{DS} = 3.0\,V$, and $V_{th} = -4.0\,V$.

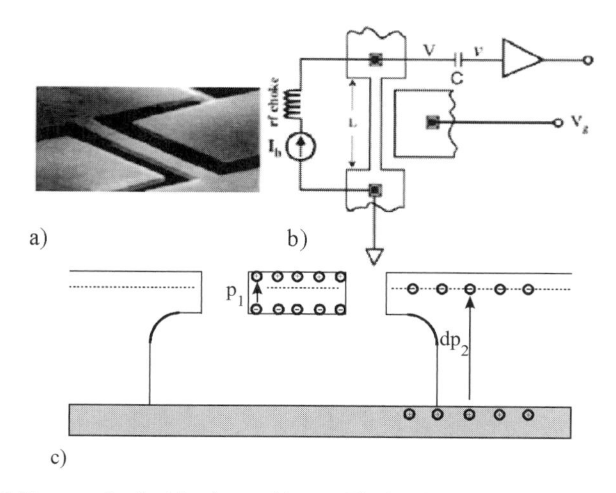

Fig. 58 (a) SEM image of a doubly clamped beam. The in-plane gates are formed by the 2DEG. (b) Sketch of measurement setup. A constant dc bias current I_b is sent through a large rf choke (10 mH). Gate drive voltage consists of both dc and rf components. (c) A cross-sectional schematic of the dipolar actuation mechanism, showing dipole formation on the beam (p_1) and on the driving gate (dp_2) [169].

demonstrated by Lu *et al.* [142]. An ultralow thermal drift, a reasonable magnetic sensitivity, and a good linearity versus magnetic field and bias current have been obtained (Fig. 59).

These Hall sensors offer the possibility to be combined with devices optimized to monitor biochemical reactions in liquids and cells in order to study the influence

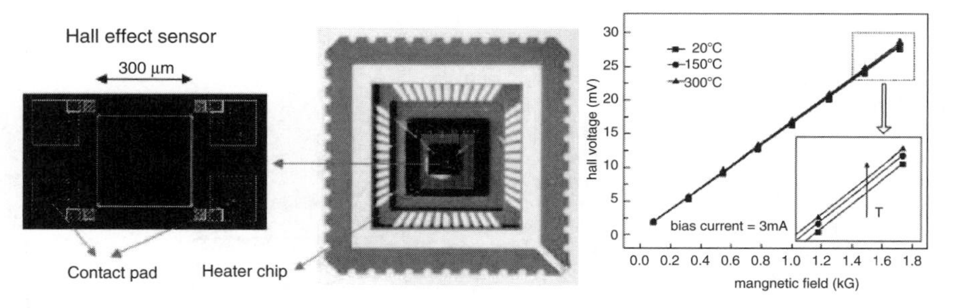

Fig. 59 (a) Schematic of an $Al_{0.3}Ga_{0.7}N/GaN$ heterojunction based Hall effect sensor mounted on a heater chip inside a carrier. (b) Temperature dependence of output Hall voltage as a function of magnetic induction. The inset of this figure shows the enlarged portion of the original output curves [142].

of electromagnetic fields on biological systems which becomes more and more an safety issue because of the increasing power of transmitters needed e.g. for telephone networks.

10 Summary

The application of GaN-based hetrostructures for (bio-) chemical, mechanical, gas and further sensors has been reviewed and the basic mechanisms and functionalities are described. The realized sensors published up to date are summarized in Tab. 5 to the convenience of the reader. These examples show the high feasibility to fabricate robust sensors in various application fields. An attractive feature especially of AlGaN/GaN heterostructure based sensors is the often realized transistor structure, i.e. the operation with gain, and the possibility for integration with electronic, optoelectronic or surface acoustic wave devices for data processing and contact less read out. An optimized device design and the more basic understanding of the sensing mechanisms require a comprehensive knowledge of the piezoelectric and spontaneous polarization of binary and ternary group-III-nitride with wurtzite crystal structure.

Therefore, we have presented simulated and experimental evidence for the non-linearity of spontaneous and piezoelectric polarization in group-III-nitride alloys as a function of strain and composition. We have applied our improved theory to reverse-model experimental data obtained from a number of InGaN/GaN quantum-wells as well as AlInN/GaN and AlGaN/GaN transistor structures. We have found that the discrepancies of experiment and ab-initio theory so far present are almost completely eliminated for the AlGaN/GaN based heterostructures when polarization non-linearity is taken into account. To facilitate inclusion of the predicted non-linear polarization in future simulations, we gave an explicit prescription to calculate polarization induced surface and interface charges as well as electric fields for arbitrary x in each of the random ternary III-N alloys.

Table 5 Summary of realized 2DEG based sensor structures and devices reported up to date.

Sensor	Referenz	Device	Epitaxy	Area	Contacts *(Gate)*	Passivation	Range
Ions	53, 80	GaN/AlGaN/GaN	PIMBE	$1 \times 3\,mm^2$	Ti/Au	silicone	10^9–$10^{13}\,cm^{-2}s^{-1}$
pH	85	AlGaN/GaN	MOCVD	$500 \times 10\,\mu m^2$	Ti/Al/Ti/Au *(SiN_x)*	SiO_2	pH 4–10
pH	95	GaN/AlGaN/GaN	PIMBE				pH 2–12
pH	101, 110	GaN/AlGaN/GaN	MOCVD	$600 \times 2400\,\mu m^2$	Ti/Al/Ti/Au	polyimide	pH 3–10
pH	101, 170	AlGaN/GaN	PIMBE	$600 \times (500$–$1000)\,\mu m^2$	Ti/Al/Ti/Au	solder stop resist	pH 3–10
HCl	102, 103	AlGaN/GaN	MOCVD	$3\,mm^2$	Ti/Al/Pt/Au	SiN_x	5–10% HCl
pH, anions	82	GaN/AlGaN/GaN	PIMBE	$100 \times (10$–$100)\,\mu m^2$	Ti/Al/Ni/Au	polyimide	pH 3.3–12.4
anions	99	Si:AlGaN/GaN	MOCVD	$100 \times (5$–$40)\,\mu m^2$	Ti/Al/Pt/Au *(Au)*	SiN_x	
K^+, Na^+, NH_4^+, NO_3^-	104	GaN/AlGaN/GaN	PIMBE		Ti/Al/Ni/Au *(polymer membrane)*	polyimide	10^{-6}–$10^{-2}\,M$
polar molecules	81, 107	GaN/AlGaN/GaN	PIMBE	$3 \times 1\,mm^2$	Ti/Au	silicone	
polar molecules	85	AlGaN/GaN	MOCVD	$500 \times 10\,\mu m^2$	Ti/Al/Ti/Au	SiO_2	
block copolymer	102, 109	AlGaN/GaN	MOCVD	$20 \times 150\,\mu m^2$	Ti/Al/Pt/Au	SiN_x	
polar molecules	103	AlGaN/GaN	MOCVD	$3\,mm^2$	Ti/Al/Pt/Au	SiN_x	50–75% acetone
cell action potential	79	GaN/AlGaN/GaN	MOCVD	$35 \times 35\,\mu m^2$		epoxy	
DNA	92	Si:AlGaN/GaN	MOCVD	$10 \times 4\,\mu m^2$	Ti/Al/Pt/Au *(Au)*	polymethyl methacrylate	

(Continued)

Table 5 (Continued)

Sensor	Referenz	Device	Epitaxy	Area	Contacts *(Gate)*	Passivation	Range
protein	111	Si:AlGaN/GaN	MOCVD	$100 \times 5\,\mu m^2$	Ti/Al/Pt/Au *(functionalized)*	SiN_x	
H_2	84, 137–139	AlGaN/GaN	MOCVD	$1 \times 50\,\mu m^2$	Ti/Al/Pt/Au (Pt/Sc_2O_3)	Sc_2O_3/SiN_x	
H_2, NO	124, 143	Si:AlGaN/GaN	PIMBE	$10 \times 200\,\mu m^2$	Ti/Al/Cr/Au $(Pt, Pd/Ag, Rh)$		
H_2, C_2H_4, CO, NO, NO_2	132, 135, 136	GaN/AlGaN/GaN	PIMBE	$500 \times 70\,\mu m^2$	Ti/Al/Ti/Au *(Pt)*		
pressure / strain	149	N-face AlGaN/GaN/ AlGaN/GaN	PIMBE	$5000 \times (50–500)\,\mu m^2$			
pressure	151	GaN/AlGaN/GaN	MOCVD/ MBE	$2 \times 40/150\,\mu m^2$	Ti/Al *(Ni/Au)*		0.4–2 kbar
pressure (capacitive)	152	AlGaN/GaN	MOCVD		Ti/Al/Pt/Au		−0.5–1 bar
pressure	153, 154	AlGaN/GaN	MOCVD	4 µm IT	Ti/Al/Pt/Au		−1–12 bar
strain	162	Si:AlGaN/GaN	MOCVD	$100 \times (5–40)\,\mu m^2$	Ti/Al/Pt/Au		
strain	163	AlGaN/GaN	MOCVD	$80 \times 30\,\mu m$	Ti/Al/Ni/Au *(Ni/Au)*		
magnetic field	142	AlGaN/GaN	MOCVD	$300 \times 300\,\mu m^2$	Ti/Al/Mo/Au		0.1–1.7 kG

With the realization of undoped lattice matched AlInN/GaN heterostructures we were able to prove the existence of a gradient in spontaneous polarization predicted by our theory. The confinement of 2DEGs in InGaN/GaN QWs in combination with the measured Stark shift of excitonic recombinations was used to determine the polarization induced electric fields and interface charges. A significant discrepancy between the predicted and measured bound interface charges is found, stimulating the development of a two dimensional model for polarization induced effects in QWs with In-fluctuations and a more detailed characterization of charged defects. This will provide further stimulation for the development of novel sensors based on GaN hetero- and nanostructures.

Acknowledgements The authors would like to thank J. Wagner (Fraunhofer-Institute, Freiburg, Germany) as well as F. Scholz (University Ulm, Germany) for providing high quality InGaN/GaN- and AlInN/GaN-heterostructures. F. Bernardini and V. Fiorentini provided important results and theoretical predictions related to the piezoelectric and spontaneous polarization of group-III nitrides. We thank all partners of the European project "GaNano" for helpful and stimulating discussions. In addition we would like to thank IOP Publishing Limited to provide us with the permission to reproduce parts of the paper: O. Ambacher "Pyroelectric properties of Al(In)GaN/GaN hetero- and quantum well structures", J. Phys.: Condensed Matter 14 (2002) 3399.

The work was funded by: the Office of Naval Research (MURI N00014-99-10714, Project Officer C. Wood), the European Commission (GaNano, NMP4-CT-2003-505641), the German Science Foundation (π-NEMS, AM105/2-1; Mechnano CI148/2-1) and the Thurigian Ministry of Culture (B678-03001).

References

1. F. Bernardini, V. Fiorentini, and D. Vanderbilt, Phys. Rev. B **56** (1997) R10024.
2. A. Zoroddu, F. Bernardini, P. Ruggerone, and V. Fiorentini, Phys. Rev. B. **64** (2001) 45208.
3. O. Ambacher, J. Phys. D: Appl. Phys. **31** (1998) 2653.
4. M.-A. Dubois and P. Muralt, Appl. Phys. Lett. 74 (1999) 3032.
5. R. Dimitrov, A. Mitchell, L. Wittmer, O. Ambacher, M. Stutzmann, J. Hilsenbeck, and W. Rieger, Jpn. J. Appl. Phys. **38** (1999) 4962.
6. O. Ambacher, J. Smart, J.R. Shealy, N.G. Weimann, K. Chu, M. Murphy, W.J. Schaff, L.F. Eastman, R. Dimitrov, L. Wittmer, M. Stutzmann, W. Rieger, and J. Hilsenbeck, J. Appl. Phys. **85** (1999) 3222.
7. M.S. Shur, A.D. Bykhovski, and R. Gaska, MRS Internet J. Nitride Semicond. Res. **4S1** (1999) G1.6.
8. K. Tsubouchi, and N. Mikoshiba, IEEE Ultrason. Symp. **90** (1983) 299.
9. T. Palacios, F. Calle, E. Monroy, J. Grajal, M. Eickhoff, O. Ambacher, and C. Prieto, Mater. Sci. Eng. B **93** (2002) 154.
10. C. Deger, E. Born, H. Angerer, O. Ambacher, M. Stutzmann, J. Hornsteiner, E. Riha and G. Fischerauer, Appl. Phys. Lett. **72** (1998) 2400.
11. K. Tsubouchi, K. Sugai, and N. Mikoshiba, IEEE Ultrason. Symp. **90** (1981) 375.
12. C. Wetzel, T. Takeuchi, S. Yamaguchi, H. Katoh, H. Amano, and I. Akasaki, Appl. Phys. Lett. **73** (1998) 1994.
13. J.S. Im, V. Härle, F. Scholz, A. Hangleiter, MRS Internet J. Nitride Semicond. Res. **1** (1996) 37.
14. N. Grandjean, J. Massies, and M. Leroux, Appl. Phys. Lett. **74** (1999) 2361.

15. S. Nakamura, G. Frasol, The Blue Laser Diode – GaN based Light Emitters and Lasers, Springer Verlag, Berlin, 1998.
16. S. Nakamura, Semicond. Sci. Technol. **14** (1999) R27.
17. L.F. Eastman, V. Tilak, J. Smart, B.M. Green, E.M. Chumbes, R. Dimitrov, H. Kim, O. Ambacher, N. Weimann, T. Prunty, M. Murphy, W.J. Schaff, and J.R. Shealy, IEEE Trans. Electr. Dev. **48** (2001) 479.
18. M. Higashiwaki, T. Matsui, and T. Mimura, IEEE Electron Device Letters **27** (2006) 16.
19. M.A. Khan, Q. Chen, M.S. Shur, B.T. MsDermott, J.A. Higgins, J. Burm, W.J. Schaff, and L.F. Eastman, IEEE Electron Device Lett. **17** (1996) 584.
20. S.C. Binari, J.M. Redwing, G. Kelner, and W. Kruppa, Electron. Lett. **33** (1997) 242 .
21. R. Gaska, Q. Chen, J. Yang, A. Osinsky, M.A. Khan, and M.S. Shur, IEEE Electron Device Lett. **18** (1997) 492.
22. T. Palacios, A. Chakraborty, S. Heikmann, S. Keller, S.P. DenBaars, and U.K. Mishra, IEEE Electron Device Letters **27** (2006) 13.
23. R. Dimitrov, L. Wittmer, H.P. Felsl, A. Mitchell, O. Ambacher, and M. Stutzmann, phys. stat. sol. (a) **168** (1998) R7.
24. L.F. Eastman, V. Tilak, J. Smart, B. Green, A. Vertiatchikh, N. Weimann, O. Ambacher, E. Chumbes, H. Kim, T. Prunty, J.H. Hwang, W.J. Schaff, B.K. Ridley, V. Kaper and J.R. Shealy, Proc. of 28th International Symposium on Compound Semiconductors, ISCS 2001, Tokyo, October, 2001.
25. O. Ambacher, B. Foutz, J. Smart, J.R. Shealy, N.G. Weimann, K. Chu, M. Murphy, A.J. Sierakowski, W.J. Schaff, and L. F. Eastman, R. Dimitrov, A. Mitchell, and M. Stutzmann, J. Appl. Phys. **87** (2000) 334.
26. R.D. King-Smith and D. Vanderbilt, Phys. Rev. B **47** (1993) 1651.
27. R. Resta, Rev. Mod. Phys. **66** (1994) 899.
28. S.-H. Wei, L.G. Ferreira, J.E. Bernard, and A. Zunger, Phys. Rev. B **42** (1990) 9622.
29. L. Görgens, O. Ambacher, M. Stutzmann, C. Miskys, F. Scholz, and J. Off, Appl. Phys. Lett. **76** (2000) 577.
30. H. Angerer, D. Brunner, F. Freudenberg, O. Ambacher, M. Stutzmann, R. Höpler, T. Metzger, E. Born, G. Dollinger, A. Bergmaier, S. Karsch, and H.-J. Körner, Appl. Phys. Lett. **71** (1997) 1504.
31. T. Peng, J. Piprek, G. Qiu, J.O. Olowolafe, K.M. Unruh, C.P. Swann, E.F. Schubert, Appl. Phys. Lett. **71** (1997) 2439.
32. L. Bellaiche, S.-H. Wei, and A. Zunger, Phys. Rev. **56** (1997) 13872.
33. T. Mattila and A. Zunger, J. Appl. Phys. **85** (1999) 160.
34. K.P. O'Donell, J.F.W. Mosselmans, R.W. Martin, S. Pereira, and M.E. White, J. Phys.: Condensed Matter **13** (2001) 6977.
35. F. Bernardini and V. Fiorentini, Phys. Rev. B **64** (2001) 85207.
36. E.S. Hellman, MRS Internet J. Nitride Semicond. Res. **3** (1998) 11.
37. B. Daudin, J.L. Rouviére, and M. Arley, Appl. Phys. Lett. **69** (1996) 2480.
38. M. Seelmann-Eggebert, J.L. Weyher, H. Obloh, H. Zimmermann, A. Rar, and S. Porowski, Appl. Phys. Lett. **71** (1997) 2635.
39. A.R. Smith, R.M. Feenstra, D.W. Greve, J. Neugebauer and J.E. Northrup, Phys. Rev. Lett. **79** (1997) 3934.
40. A.R. Smith, R.M. Feenstra, D.W. Greve, M.S. Shin, M. Skowronski, J. Neugebauer, J.E. Northrup, Appl. Phys. Lett. **72** (1998) 2114.
41. B.W. Batterman, Phys. Rev. Lett. **22** (1969) 703.
42. A. Kazimirov, G. Scherb, J. Zegenhagen, T.-L. Lee, M.J. Bedzyk, M.K. Kelly, H. Angerer, and O. Ambacher, J. Appl. Phys. **84** (1998) 1703.
43. J.A. Smart, A.T. Schremer, N.G. Weimann, O. Ambacher, L.F. Eastman, and J.R. Shealy, Appl. Phys. Lett. **75** (1999) 388.
44. M.J. Murphy, K. Chu, H. Wu, W. Yeo, W.J. Schaff, O. Ambacher, J. Smart, J.R. Shealy, L.F. Eastman, and T.J. Eustis, J. Vac. Sci. Technol. B **17** (1999) 1252.
45. J. Wagner, A. Ramakrishnan, H. Obloh, and M. Maier, Appl. Phys. Lett. **74** (1999) 3863.

46. A. Ramakrishnan, J. Wagner, M. Kunzer, H. Obloh, K. Köhler, and B. Johs, Appl. Phys. Lett. **76** (2000) 79.
47. A. Kasic, M. Schubert, J. Off, and F. Scholz, Appl. Phys. Lett. **78** (2001) 1526.
48. A.F. Wright, J. Appl. Phys. **82** (1997) 2833.
49. G.D. O'Clock and M.T. Duffy, Appl. Phys. Lett. **23** (1973) 55.
50. J.F. Nye, Physical Properties of Crystals; Their Representation by Tensors and Matrices, Clarendon, Oxford, (1985).
51. B. Jogai, J. Appl. Phys. **90** (2001) 699.
52. M. Rotter, A. Wixforth, A.O. Govorov, W. Ruille, D. Bernklau, and H. Riechert, Appl. Phys. Lett. **75** (1999) 965.
53. R. Neuberger, G. Müller, O. Ambacher, and M. Stutzmann, phys. stat. sol. (a) **183** (2001) R10.
54. U. Karrer, O. Ambacher, and M. Stutzmann, Appl. Phys. Lett. **77** (2000) 2012.
55. B.K. Ridley, O. Ambacher, and L.F. Eastman, Semicond. Sci. Technol. **15** (2000) 270.
56. F. Bernardini, V. Fiorentini, Phys. Rev. Lett. **79** (1997) 3958.
57. L.S. Yu, D.J. Qiao, Q.J. Xing, S.S. Lau, K.S. Boutros, and J.M. Redwing, Appl. Phys. Lett. **73** (1998) 238.
58. T. Onuma *et al.*, J. Appl. Phys. **94** (2003) 2449.
59. B. Lee, L.W. Wang, J. Appl. Phys. **100** (2006) 93717.
60. D. Brunner, H. Angerer, E. Bustarret, R. Höpler, R. Dimitrov, O. Ambacher, and M. Stuzmann, J. Appl. Phys. **82** (1997) 5090.
61. G. Martin, S. Strite, A. Botchkaev, A. Agarwal, A. Rockett, H. Morkoç, W.R.L. Lambrecht, B. Segall, Appl. Phys. Lett. **65** (1994) 610.
62. G. Martin, A. Botchkarev, A. Rockett, H. Morkoç, Appl. Phys. Lett. **68** (1996) 2541.
63. L.W. Wong, S.J. Cai, R. Li, K. Wang, H.W. Jiang, and M. Chen, Appl. Phys. Lett. **73** (1998) 1391.
64. B.E. Foutz, M.J. Murphy, O. Ambacher, V. Tilak, J. Smart, J.R. Shealy, W.J. Schaff, and L.F. Eastman, Mat. Res. Soc. Proc. **572** (1999) 501.
65. I.P. Smorchkova, C.R. Elsass, J.P. Ibbetson, R. Vetury, B. Heying, P. Fini, E. Haus, S.P. DenBaars, J.S. Speck, and U.K. Mishra, J. Appl. Phys. **86** (1999) 4520.
66. Z. Bougrioua, J.-L. Farvacque, I. Moerman, and F. Carosella, phys. stat. sol. (b) **228** (2001) 625.
67. P.M. Asbeck, E.T. Yu, S.S. Lau, G.J. Sullivan, J. Van Hove, and J.M. Redwing, Electron Lett. **33** (1997) 1230.
68. E.T. Yu, G.J. Sullivan, P.M. Asbeck, C.D. Wang, D. Qiao, and S.S. Lau, Appl. Phys. Lett. **71** (1997) 2794.
69. J. Wagner, A. Ramakrishnan, H. Obloh, and M. Maier, Appl. Phys. Lett. **74** (1999) 3863.
70. A. Ramakrishnan, J. Wagner, M. Kunzer, H. Obloh, and K. Köhler Appl. Phys. Lett. **76** (2000) 79.
71. F.D. Sala, A. di Carlo, P. Lugli, F. Bernardini, V. Fiorentini, R. Scholz, J.-M. Januc, Appl. Phys. Lett. **74** (1999) 2002.
72. L. Eckey, A. Hoffmann, P. Thurian, I. Broser, B.K. Meyer, K. Hiramatsu, Mat. Res. Soc. Symp. Proc. **482** (1998) 555.
73. C. Wetzel, T. Takeuchi, H. Amano and I. Akasaki, Jpn. J. Appl. Phys. **38** (1999) L163.
74. S.C. Binari, H.B. Dietrich, G. Kelner, L.B. Rowland, K. Doverspike, and D.K. Wickenden, J. Appl. Phys. **78** (1995) 3008.
75. G. Müller, G. Krötz, and J. Schalk, phys. stat. sol. (a) **185** (2001) 1.
76. M. Eickhoff, J. Schalwig, G. Steinhoff, O. Weidemann, L. Görgens, R. Neuberger, M. Hermann, B. Baur, G. Müller, O. Ambacher, and M. Stutzmann, phys. stat. sol. (c) **0** (2003) 1908.
77. M. Stutzmann, G. Steinhoff, M. Eickfoff, O. Ambacher, C.E. Nebel, J. Schalwig, R. Neuberger, and G. Müller, Diam. Rel. Mater. **11** (2002) 886.
78. S.J. Pearton, B.S. Kang, S. Kim, F. Ren, B.P. Gila, C.R. Abernathy, J. Lin, and S.N.G. Chu, J. Phys.: Condens Matter **16** (2004) R961.

79. G. Steinhoff, B. Baur, G. Wrobel, S. Ingebrandt, A. Offenhäuser, A. Dadgar, A. Krost, M. Stutzmann, and M. Eickhoff, Appl. Phys. Lett. **86** (2005) 033901.
80. R. Neuberger, G. Müller, M. Eickhoff, O. Ambacher, and M. Stutzmann, Mater. Sci. Eng. B **93** (2002) 143.
81. A. Neuberger, G. Müller, O. Ambacher, and M. Stutzmann, phys. stat. sol. (a) **185** (2001) 85.
82. Y. Alifragis, A. Georgakilas, G. Konstantinidis, E. Iliopoulos, A. Kostopoulos, and N.A. Chaniotakis, Appl. Phys. Lett. **87** (2005) 253507.
83. C. Buchheim, G. Kittler, V. Cimalla, V. Lebedev, M. Fischer, S. Krischok, V. Yanev, M. Himmerlich, G. Ecke, J.A. Schaefer, and Oliver Ambacher, IEEE Sensors J. **6** (2006) 881.
84. B.S. Kang, R. Mehandru, S. Kim, F. Ren, R.C. Fitch, J.K. Gillespie, N. Moser, G. Jessen, T. Jenkins, R. Dettmer, D. Via, A. Crespo, B.P. Gila, C.R. Abernathy, and S.J. Pearton, Appl. Phys. Lett. **84** (2004) 4635.
85. T. Kokawa, T. Sato, H. Hasegawa, and T. Hashizume, J. Vac. Sci. Technol. B **24** (2006) 1972.
86. A. Schober, G. Kittler, C. Buchheim, A. Majdeddin, V. Cimalla, M. Fischer, V. Yanev, M. Himmerlich, S. Krischok, J.A. Schaefer, H. Romanus, T. Sändig, J. Burgold, F. Weise, H. Wurmus, K.H. Drüe, M. Hintz, H. Thust, M. Gebinoga, M. Kittler, A. Spitznas, E. Gottwald, K-F. Weibezahn, D. Wegener, A. Schwienhorst, and O. Ambacher, Techn. Proc. 2005 NSTI Nanotechn. Conf. Trade Show, Vol. 1, (2005) 489.
87. A. Schober, G. Kittler, B. Lübbers, C. Buchheim, A. Majdeddin, V. Cimalla, M. Fiscger, A. Spitznas, M. Gebinoga, V. Yanev, M. Himmerlich, T. Kerekes, M. Kittler, K.H. Drüe, M. Hintz, S. Krischok, J. Burgold, F. Weise, O. Ambacher, E. Gottwald, K.F. Weibezahn, D. Wegener, G. Schlingloff, A. Schwienhorst, 7. Dresdner Sensor-Symposium, Dresden, 12.–14. 12. 2005, Dresden, TUD Press, Dresden, 2005, p. 143.
88. B. Lübbers, G. Kittler, V. Cimalla, M. Gebinoga, C. Buchheim, D. Wegener, Schober and O. Ambacher, 51st Internationales Wissenschaftliches Kolloquium Technische Universität Ilmenau, September 11–15, (2006) 249.
89. S.J. Pearton, J.C. Zolper, R.J. Shul, and F. Ren, J. Appl. Phys. **86** (1999) 1.
90. U.K. Mishra, P. Parikh, and Y.F. Wu, Proc. IEEE **90** (2009) 1022.
91. Z. Fan, S.N. Mohammand, W. Kim, O. Aktas, A.E. Botchkarev, and H. Morkoç, Appl. Phys. Lett. **68** (1996) 1672.
92. B.S. Kang, S.J. Pearton, J.J. Chen, F. Ren, J.W. Johnson, R.J. Therrien, P. Rajagopal, J.C. Roberts, E.L. Piner, and K.J. Linthicum, Appl. Phys. Lett. **89** (2006) 122102.
93. D. E. Yates, S. Levine, and T.W. Healy. J. Chem. Soc. Fraday Trans. **1** (1974) 1807.
94. M. Bayer, C. Uhl, and P. Vogl, J. Appl. Phys. **97** (2005) 033703.
95. G. Steinhoff, M. Hermann, W.J. Schaff, L.F. Eastman, M. Stutzmann, and M. Eickhoff, Appl. Phys. Lett. **83** (2003) 177.
96. N.A. Chaniotakis, Y. Alifragis, G. Konstantinidis, and A. Georgakilas, Anal Chem. **76** (2004) 5552.
97. N.A. Chaniotakis, Y. Alifragis, A. Georgakilas, and , G. Konstantinidis, Appl Phys. Lett. **86** (2005) 164103.
98. Y. Alifragis, G. Konstantinidis, A. Georgakilas, and N.A. Chaniotakis, Electroanalysis **17** (2005) 527.
99. B.S. Kang, F. Ren, M.C. Kang, C. Lofton, W. Ran, S.J. Pearton, A. Dabiran, A. Osinsky, and P.P. Chow, Appl. Phys. Lett. **86** (2005) 173502.
100. A. Spitznas, Diploma thesis, Technical University Ilmenau, 2005.
101. G. Kittler, A. Spitznas, B. Lübbers, V. Lebedev, D. Wegener, A. Schober, M. Gebinoga, F. Schwierz, V. Polyakov, F. Weise, O. Ambacher, European Workshop on III-Nitride Semiconductor Materials and Devices EW3NS, September 18-20, 2006, Heraklion, Crete, Greece.
102. B.S, Kang, S. Kim, F. Ren, B.P. Gila, C.R. Abernathy, and S.J. Pearton, IEEE Sensors J. **5** (2005) 677.
103. R. Mehandru, B. Luo, B.S. Kang, S. Kim, F. Ren, S.J. Pearton, C.C. Pan, G.T. Chen, and J.I. Chyi, Solid State Electron. **48** (2004) 351.
104. A. Alifragis, N.A. Chaniotakis, G. Konstantinidis, A. Volosirakis, A. Adikimenakis, and A. Georgakilas, Biosens. Bioelectron., submitted.

105. G. Steinhoff, O. Purrucker, M. Tanaka, M. Stutzmann, and M. Eickhoff, Adv. Funct. Mater. **13** (2003) 841.
106. Y.G. Mourzina, J. Schubert, W. Zander, A. Legin, Y.G. Vlasov, H. Lüth, and M.J. Schöning, Electrochimica Acta **47** (2001) 251.
107. M. Eickhoff, O. Ambacher, G. Steinhoff, J. Schalwig, R. Neuberger, T. Palacios, E. Monroy, F. Calle, G. Müller, and M. Stutzmann, MRS Symp. Proc. **693** (2002) I12.1.
108. S.M. Luber, K. Adlkofer, U. Rant, A. Ulman, A. Gölzhauser, M. Grunze, D. Schuh, M. Tanaka, M. Tornow, and G. Abstreiter, Physica E **21** (2004) 1111.
109. B.S. Kang, G. Louche, R.S. Duran, Y. Gnanou, S.J. Pearton, and F. Ren, Solid State Comm. 48 (2004) 851.
110. G. Kittler, A. Spitznas, B. Lübbers, V. Lebedev, D. Wegener, A. Schober, M. Gebinoga, F. Schwierz, V. Polyakov, F. Weise, and O. Ambacher, 51st Internationales Wissenschaftliches Kolloquium Technische Universität Ilmenau, September 11–15, (2006) 251.
111. B.S. Kang, F. Ren, L. Wang, C. Lofton, W.W. Tan, S.J. Pearton, A. Dabiran, A. Osinsky, and P.P. Chow, Appl. Phys. Lett. **87** (2005) 023508.
112. K. Ozasa, S. Nemoto, M. Hara, and M. Maeda, phys. stat. sol. (a) **203** (2006) 2287.
113. T.H. Young and C.R. Chen, Biomaterials **27** (2006) 3361.
114. I. Cimalla, F. Will, K. Tonisch, M. Niebelschütz, V. Cimalla, V. Lebedev, G. Kittler, M. Himmerlich, S. Krischok, J. A. Schaefer, M. Gebinoga, A. Schober, Th. Friedrich, and O. Ambacher, Sens. Actuators B: Chem. (2006), doi:10.1016/j.snb.2006.10.030.
115. F. Heer, W. Franks, A. Blau, S. Taschini, C. Ziegler, A. Hierlemann, and H. Baltes, Biosens Bioelectr. **20** (2004) 358.
116. A. Offenhäuser, C. Sprössler, M. Matsutawa, and W. Knoll, Biosens. Bioelectron. **12** (1997) 819.
117. G. Steinhoff, B. Baur, G. Wrobel, S. Ingebrandt, A. Offenhäuser, A. Dadgar, A. Krost, M. Stutzmann, and M. Eickhoff, Appl. Phys. Lett. **89** (2006) 011901.
118. V.M. Bermudez, Surf. Sci. **499** (2002) 109.
119. V.M. Bermudez, Surf. Sci. **499** (2002) 124.
120. B. Baur, G. Steinhoff, J. Hernando, O. Purrucker, M. Tanaka, B. Nickel, M. Stutzmann, and M. Eickhoff, Appl. Phys. Lett. **87** (2005) 263901.
121. I. Lundstrom, M. Armgarth, and L.G. Petersson, CRC Critical Reviews in Solid State and Material Sciences **15** (1989) 201.
122. I. Lundstrom, M.S. Shivaraman,, C. Svensson, and L. Lundquist, J. Appl Phys. **26** (1975) 55.
123. J. Kim, B.P. Gila, C.R. Abernathy, and S.J. Pearton, Appl. Phys. Lett. **82** (2003) 739.
124. S.C. Pyke, J.H. Chern, R.J. Hwu, and L.P. Sadwick, Annual Meeting of the DOE Hydrogen and Fuel Cell Program, San Ramon, CA, May 2000.
125. B.P. Luther, S.D. Wolter, and S.E. Mohney, Sens. Act. B **56** (1999) 164.
126. J. Schalwig, G. Müller, U. Karrer, M. Eickhoff, O. Ambacher, M. Stutzman, L. Görgens, and G. Dollinger, Appl. Phys. Lett. **80** (2002) 1222.
127. G. Zhao, W. Sutton, D. Pavlidis, E.L. Piner, J. Schwank, and S. Hubbard, IEICE Trans. Electron. E86-C (2003) 2027.
128. J. Song, W. Lu, J.S. Flynn, and G.R. Brandes, Appl. Phys. Lett. **87** (2005) 133501.
129. O. Weidemann, M. Hermann, G. Steihoff, H. Wingbrant, A. Lloyd-Spetz, M. Stutzmann, and M. Eickhoff, Appl. Phys. Lett. **83** (2003) 773.
130. B.S. Kang, F. Ren, B.P. Gila, C.R. Abernathy, and S.J. Pearton, Appl. Phys. Lett. **84** (2004) 1123.
131. B.S. Kang, S. Kim, F. Ren, K. Ip, Y.W. Heo, B. Gila, C.R. Abernathy, D.P. Norton, and S.J. Pearton, J. Electrochem. Soc. **151** (2004) G468.
132. J. Schalwig, G. Müller, M. Eickhoff, O. Ambacher, and M. Stutzmann, Mater. Sci. Eng. B **93** (2002) 207.
133. E.J. Cho, D. Pavlidis, G.Y. Zhao, S.M. Hubbard, and J. Schwank, IEICE Trans. Electron. E89-C (2006) 1047.
134. A.T. Winzer, R. Goldhahn, G. Gobsch, A. Dadgar, A. Krost, O. Weidemann, M. Stutzmann, and M. Eickhoff, Appl. Phys. Lett. **88** (2006) 024101.

135. J. Schalwig, G. Müller, O. Ambacher, and M. Stutzmann, phys. stat. sol. (a) **185** (2001) 39.
136. J. Schalwig, G. Müller, M. Eickhoff, O. Ambacher, and M. Stutzmann, Sens. Actuat. B **87** (2002) 425.
137. H.T. Wang, B.S. Kang, F. Ren, R.C. Fitch, J.K. Gillespie, N. Moser, G. Jessen, T. Jenkins, R. Dettmer, D. Via, B.P. Gila, C.R. Abernathy, and S.J. Pearton, Appl. Phys. Lett. **87** (2005) 172105.
138. B.S. Kang, R. Mehandru, S. Kim, F. Ren, R.C. Fitch, J.K. Gillespie, N. Moser, G. Jessen, T. Jenkins, R. Dettmer, D. Via, A. Crespo, K.H. Baik, B.P. Gila, C.R. Abernathy, and S.J. Pearton, phys. stat. sol. (c) **2** (2005) 2672.
139. B.S. Kang, H.T. Wang, L.C. Tien, F. Ren, B.P. Gila, D.P. Norton, C.R. Abernathy, J. Lin, and S.J. Pearton, Sensors **6** (2006) 643.
140. C.C. Cheng, Y.Y. Tsai, K.W. Lin, H.I. Chen, and W.C. liu, Appl. Phys. Lett. **86** (2005) 112103.
141. C.W. Hung, H.L. Lin, Y.Y. Tsai, P.H. Lai, S.I. Fu, H.I. Chen, and W.C. Liu, Electronics Letters, **42** (2006) 578.
142. H. Lu, P. Sandvik, A. Vertiatchikh, J. Tucker, and A. Elasser, J. Appl. Phys. **99** (2006) 114510.
143. S.C. Pyke and L.P. Sadwick, Proc. 2002 U.S. DOE Hydrogen Program Review, NREL/CP-610-32405.
144. Y. Alifragis, A. Geogakilas, G. Konstantinidis, E. Iliopoulos, M. Zervos, and N.A. Chaniotakis, Biosens. Bioelectron., submitted.
145. K. Prabhakaran, T.G. Andersson, and K. Nozawa, Appl. Phys. Lett. **69** (1996) 3212.
146. A. Morgenshtein, L. Sudakov-Boreysha, U. Dinnar, C.G. Jakobson, and Y. Nemirovsky, Sens. Actuat. B **98** (2004) 18.
147. A. D. Bykhovski, V. V. Kaminski, M. S. Shur, Q. C. Chen, and M. A. Khan, Appl, Phys. Lett. **68** (1996) 818.
148. R. Gaska, A. D. Bykhovski, M. S. Shur, V. V. Kaminskii, and S. M. Soloviov, J. Appl. Phys. **85** (1999) 6932.
149. M. Eickhoff, O. Ambacher, G. Krötz, and M. Stutzmann, J. Appl. Phys. **90** (2001) 3383.
150. Y. Liu, M.Z. Kauser, P.P. Ruden, Z. Hassan, Y.C. Lee, S.S. Ng, and F.K. Yam, Appl. Phys. Lett. **88** (2006) 022109.
151. Y. Liu, P.P. Ruden, J. Xie, H. Morkoç, and K.A. Son, Appl Phys. Lett. **88** (2006) 013505.
152. B. S. Kang, S. Kim, S. Jang, F. Ren. J. W. Johnson, R. J. Therrien, P. Rajagopal, J. C. Roberts, E. L. Piner, K. J. Linthicum, S. N.G. Chu, K. Baik, B. P. Gila, C. R. Abernathy, and S. J. Pearton, Appl. Phys. Lett. **86** (2005) 253502.
153. B. S. Kang, S. Kim, F. Ren. J. W. Johnson, R. J. Therrien, P. Rajagopal, J. C. Roberts, E. L. Piner, K. J. Linthicum, S. N.G. Chu, K. Baik, B. P. Gila, C. R. Abernathy, and S. J. Pearton, Appl. Phys. Lett. **85** (2004) 2962.
154. S.N.G. Chu, F. Ren, S.J. Pearton, B.S. Kang, S. Kim, B.P. Gila, C.R. Abernathy, J.I. Chyi, W.J. Johnson , and J. Lin, Mater. Sci. Eng. A **409** (2005) 340.
155. Y.R. Wu and J. Singh, Appl. Phys. Lett. **85** (2004) 1223.
156. V. Cimalla, F. Will, K. Tonisch, Ch. Foerster, K. Brueckner, I. Cimalla, T. Friedrich, J. Pezoldt, R. Stephan, M. Hein, and O. Ambacher, Sens. Actuators B: Chem. (2006), doi:10.1016/j.snb.2006.10.049.
157. A.N. Cleland, M. Pophristic, and I. Ferguson, Appl. Phys. Lett. **79** (2001) 2070.
158. V. Cimalla, Ch. Foerster F. Will, K. Tonisch, K. Brueckner, R. Stephan, M.E. Hein, O. Ambacher, and E. Aperathitis, Appl. Phys. Lett. **88** (2006) 253501.
159. M. Akiyama, N. Uono, K. Nonaka, and H. Tateyama, Appl. Phys. Lett. **82** (2003) 1977.
160. J. Olivares, E. Iborra, M. Clement, L. Vergara, J. Sangrador, and A. Sanz-Hervas, Sens. Actuat. A **123–124** (2005) 590.
161. D. Doppalapudi, R. Mlcak, J. Chan, H.L. Tuller, J. Abell, W. Li, and T.D. Moustakas, Electrochem. Soc. Proc. **2004–06** (2004) 287.
162. B.S. Kang, S. Kim, J. Kim, F. Ren, K. Baik, S. J. Pearton, B. P. Gila, C. R. Abernathy, C.-C. Pan, G.-T. Chen, J.-I. Chyi, V. Chandrasekaran , M. Sheplak, T. Nishida, and S. N. G. Chu, Appl. Phys. Lett. **83** (2003) 4845.

163. T. Zimmermann, M. Neuburger, P. Benkart, F. J. Hernández-Guillén, C. Pietzka, M. Kunze, I. Daumiller, A. Dadgar, A. Krost, and E. Kohn, IEEE Electron Lett. **27** (2006) 309.
164. Ch. Förster, V. Cimalla, K. Brueckner, V. Lebedev, R. Stephan, M. Hein, and O. Ambacher, phys. stat. sol. (a) **202** (2005) 671.
165. S. Davies, T.S. Huang, M.H. Gass, A.J. Papworth, T.B. Joyce, P.R. Chalker, Appl. Phys. Lett. **84** (2004) 2566.
166. A. Dadgar, *et al.* phys. stat. sol. (c) **0** (2003) 1940.
167. R.G. Beck, M.A. Eriksson, M.A. Topinka, R.M. Westervelt, K.D. Maranowski, and A.C. Gossard, Appl. Phys. Lett. **73** (1998) 1149
168. M.P. Schwarz, D. Grundler, I. Meinel, Ch. Heyn, and D. Heitman, Appl. Phys. Lett. **76** (2000) 3564.
169. H.X. Tang, X.M.H. Huang, M.L. Roukes, M. Bichler, and W. Wegschneider, Appl. Phys. Lett. **81** (2002) 3879.
170. G. Kittler, A. Spitznas, C. Buchheim, V. Lebedev, D. Wegener, A. Schober, and O. Ambacher., 7. Dresdner Sensor-Symposium, Dresden, 12.–14. 12. 2005, Dresden, TUD Press, Dresden, 2005, p. 207.

Lateral and Vertical Charge Transport in Polar Nitride Heterostructures:

Applications for HEMTs, Novel Vertical Junction and Sensors

Yuh-Renn Wu, Madhusudan Singh, and Jasprit Singh

1 Polar Heterostructures: What Do They Offer?

Information processing devices driving the modern technology revolution are based on materials such as semiconductors, ferroelectrics, piezoelectrics, ferromagnetics, etc. In these materials there is a strong change in one or more physical property in response to an external perturbation. Semiconductors where there is a strong change in electrical conductivity or optical properties are the most important materials in today's technology. However, most semiconductors don't have strong response to external stress or magnetic fields or temperature changes. Also most traditional semiconductors breakdown in the presence of strong electrical fields and cannot be used for very high power generation. The realization of new devices based on GaN, InN and AlN has increased the capabilities of the semiconductor family allowing not only traditional devices to operate better but also allowing new devices to be conceived. In this chapter we will focus on transport and charge control devices based on nitrides.

Most traditional semiconductors are based on the diamond or zinc blende structure. This limits the polar properties of these materials. The nitride family has an underlying wurtzite structure which provides many of the unusual properties exploited in nitride devices. In addition to being able to provide materials with a large bandgap range (bandgap can range from 0.7 to 6.2 eV) the nitride system has strong piezoelectric response and has significant spontaneous polarization. In heterostructures, piezoelectric related polarization and spontaneous polarization can be exploited to create strong built-in electric fields (reaching more than a megavolt per cm) which can result in induction of free charge that can exceed 10^{13} cm^{-2} . These issues have been explored in other chapters in this volume and we will summarize the relevant concepts in this chapter as well. As discussed below these unique polar properties can be exploited for undoped electronics and tailorable vertical junction transport. Additionally the very strong polar response along with very good transport properties allow for smart-FETs which can respond to external inputs such as strain and temperature change.

1.1 Polar Heterostructures: Undoped Electronics

As noted above, the nitrides have a very large spontaneous polarization arising from the underlying wurtzite structure. In Fig.1, we show the GaN wurtzite structure. The shift between cation and anion sublattices results in a large spontaneous polarization in the system. In hetero-interface, the net spontaneous polarization of two materials appears as polar charge. When three of the bonds on a Ga atom with tetrahedral coordination face towards the substrate, the polarity is traditionally called Ga-face. In contrast, when three bonds face in the growth direction, the material is termed N-face. The polar structure with the three bonds of III-atom facing toward the substrate is defined as +c polarity and the opposite as −c polarity. In pseudomorphic combinations using GaN/AlGaN heterojunctions, etc., strain induced piezoelectric effect contributes high sheet charge densities at the interfaces in $[000\bar{1}]$ direction [3,4,11]. As shown in Fig. 2, a large net polarization at the heterojunction (sum of piezoelectric and spontaneous polarization) can introduce large electric fields ($\sim 10^6$ V$\cdot$ cm^{-1}) and band bending, and induce a two dimensional electron gas (2DEG) with sheet charge of $\sim 10^{13}$ cm^{-2} in the heterostructure interface [80]. For certain designs, a two dimensional hole gas (2DHG) can also be induced.

Strong polarization effects are also present in other materials such as ZnO, PZT, $BaTiO_3$, $LiNbO_3$, etc. Materials like PZT, $BaTiO_3$, and $LiNbO_3$ are not only piezoelectric but also ferroelectric and polarization can be altered by external electric field and temperature. Therefore, ferroelectric materials are widely used as sensors, memory, and optoelectronic materials. At present ferroelectric materials have poor material quality resulting in poor transport properties. While the polar effects of

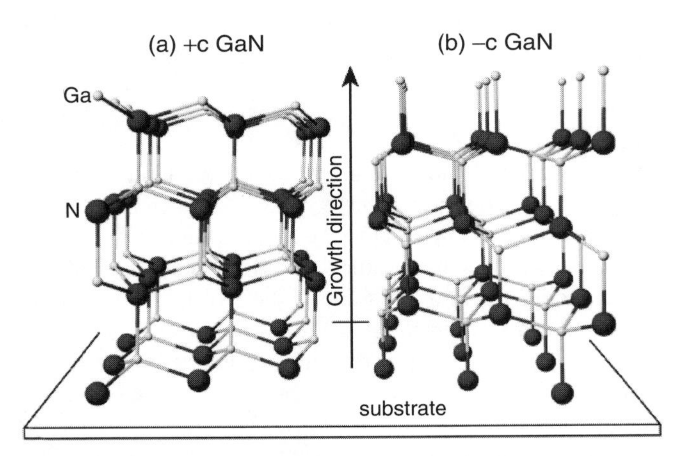

Fig. 1 Schematic illustration of GaN wurtzite crystal structure exhibiting the polarity along the c-axis. The small and large spheres indicate Ga and N, respectively. GaN with Ga-face (+c) polarity on left side and GaN with N-face (−c) polarity on right side. When the direction of the three bonds of the III-element is towards the substrate, the polar structure defined has having a +c polarity. On the other hand, when that of the bonds is upward against substrate, it is defined as having −c polarity. [83]

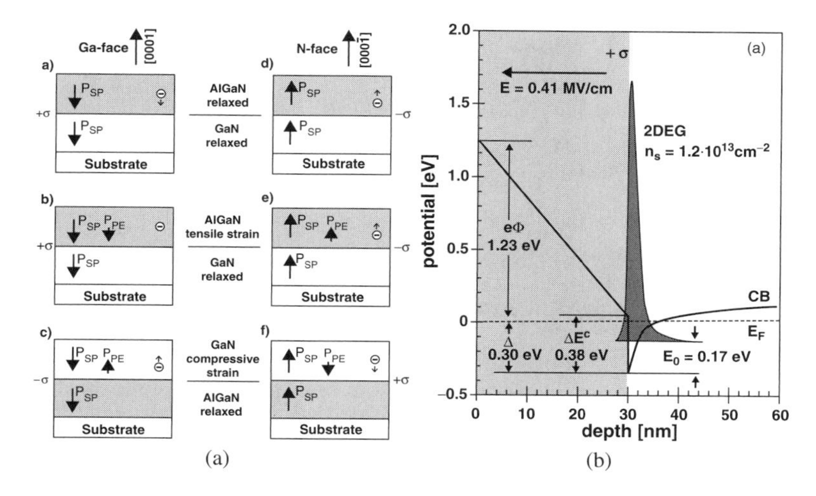

Fig. 2 (a) Polarization induced sheet charge density and directions of the spontaneous and piezoelectric polarization in Ga- and N-face strained and relaxed AlGaN/GaN heterostructures. (b) Self-consistent calculation of the conduction band and electron concentration profile of an undoped Ga-face $Al_{0.3}Ga_{0.7}N$/GaN (30/2000 nm) heterostructure including the Schottky barrier of the Ni contact on top and the polarization-induced surface and interface charges. [3, 4]

nitrides are not as strong as in some ferroelectrics,there is no other system that has such good transport properties and good polar effects.

1.2 *The Applications of Nitrides*

Nitride related compound semiconductors have been the most attractive materials in optoelectronic devices [1, 16, 54, 75, 86, 91] for short wavelength emission. The bandgap of nitride compounds including AlN, GaN, InN, and their alloys cover a wide ranges from 0.7eV to 6.2eV, which cover infrared to ultraviolet(UV) emission spectrum. Therefore, they have been widely applied in blue and ultraviolet (UV) emission optoelectronic devices such as laser diodes. Nitride alloys and heterostructures are also important as white light sources, especially for room lighting and LCD back light modules.

In addition to applications in optoelectronics, nitrides are also very important materials for power electronics. A variety of power amplifier technologies, such as GaAs MESFETs, HBT, Si bipolar transistors, SiC MESFET and GaN HEMTs have been used in commercial products. Table 1 shows a material properties comparison for the current available technologies. Si and GaAs are the basis of the most matured technologies in the market. However, as shown in Table 1, Si and GaAs have relatively small band gaps and have low saturation velocities in high field region. The breakdown fields for Si and GaAs are also smaller than other materials listed in

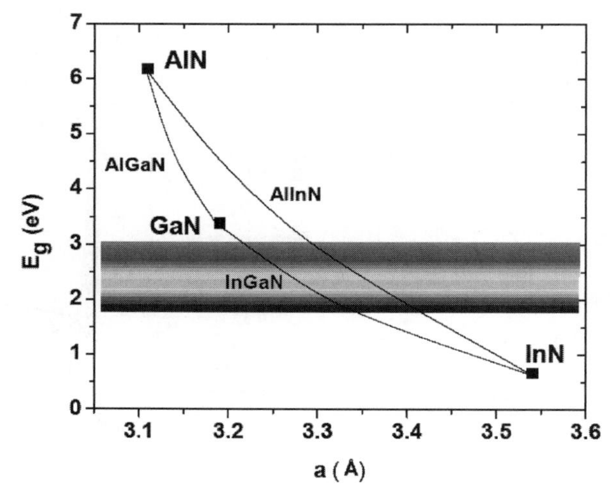

Fig. 3 Bandgaps of group III-nitride alloys as a function of in-plane lattice constant. [90]

Table 1 Material Comparison for high power high frequency operation

	Si	GaAs	Diamond	SiC	GaN
μ (cm^2/Vs)	600	8500	2200	1000	1600
Eg (eV)	1.12	1.43	5.45	3.26	3.40
v_{sat} (10^7 cm/s)	1.0	1.0	2.7	2.0	2.0
E_{break} (kV/cm)	300	400	10,000	2,200	2,000
Thermal conductivity (W/cmk)	1.5	0.46	22.0	4.40	2.30

Table 1 and therefore, these devices are only suitable for low power and high speed device applications. Diamond compared to other materials has the best features. However, due to the difficulties in material growth and fabrication limitations [88], it is not close to device applications. SiC and GaN have very close features and can be operated at very high voltages (up to 42V at gate length equals of 1 μm) due to their wide bandgap and high breakdown electric field. The higher efficiency that results from the high operating voltage reduces power requirements and simplifies cooling systems. GaN and SiC have competed against each other for a while owing to their similar properties. However, GaN has some important advantages due to its strong spontaneous and piezoelectric polarization which can induce a large amount of two dimensional electron gas (2DEG) in the channel without intentional doping. The maximum sheet charge density for GaN HEMTs can be as high as $1-2\times10^{13}$ cm^{-2}, which can provide large current for the signal amplification. Because of these advantages, FETs based on III-V nitride technology have become increasingly important in applications of microwave transistors for high power, high frequency applications.

1.3 Transport Issues in Nitride Device

In polar heterojunction devices (as in non-polar devices), both vertical transport and lateral transport are of concern. Vertical junction transport controls two terminal (diode like) devices as well as gate leakage issues. In the case of polar heterojunctions, the strong electric field due to interface polar charge creates effects that can be both beneficial and deleterious depending upon applications. It is possible to tailor the I–V characteristics of junctions. However, gate leakage effects can be important and may limit FET performance. In this chapter we will examine vertical transport in polar nitrides in section 3.

In nitride HEMTs, lateral (source to drain) transport is critical. It is important to identify not only the velocity in device channels but how scaling limits behave and what the role of self-heating [6, 50, 51, 61, 72] (under high power operation) is in transport. These issues will be examined in section 4.

1.4 Polar Materials: Use in Sensor Technology–Potential of Merging Polar Materials with Semiconductors

We have noted above that ferroelectric materials are widely used for sensor applications but their transport properties are poor. Is it possible to combine ferroelectric materials with semiconductors (including nitrides) to create novel devices? Ferroelectric materials have many properties which make them interesting. They usually have high values for the dielectric constants, strong piezoelectric, and pyroelectric effects and strong electron-optical interactions useful in double frequency generation in optical regime. The dielectric responses also have strong temperature dependence. They exhibit polarization in the absence of an external electric field, with hysteresis in polarization. The plot of polarization versus electric field for ferroelectric state shows a hysteresis loop and can be expressed as:

$$P_{sat}^{+}(E) = P_s \tanh\left[\frac{E - E_c}{2\delta}\right] \tag{1}$$

$$P_{sat}^{-}(E) = P_s \tanh\left[\frac{-E - E_c}{2\delta}\right] \tag{2}$$

$$\delta = \frac{E_c}{\ln\left(\frac{1+P_r/P_s}{1-P_r/p_s}\right)}, \tag{3}$$

where the $\pm$ signs refer to the direction of the sweep. P_s and E_c are material parameters, as is the remnant polarization ratio, P_r/P_s. In some crystals, we can observe changes in the spontaneous polarization when temperature is changed. Such crystals are known as pyroelectric materials. When the temperature is higher than the Curie temperature, ferroelectric materials lose their spontaneous polarization and

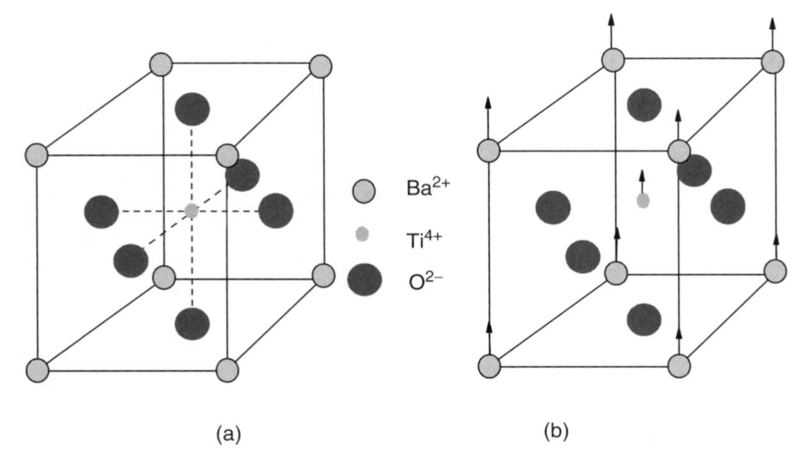

Fig. 4 The crystal structure of $BaTiO_3$ (a) above the Curie point the cell is cubic; (b) below the Curie point the structure is tetragonal with Ba^{2+} and Ti^{4+} ions displaced relative to O^{2-} ions.

reach the most stable state. This is the paraelectric phase. Figure 4 shows the crystal structure of $BaTiO_3$ above the Curie temperature and below Curie temperature.

Ferroelectric materials have been used in many areas [55, 71]. For example, the hysteresis properties have been used in making memory cells. The strong Kerr effects in ferroelectrics can be used to double the frequency of input wave. It can also be used to detect the polarization of crystal itself. Piezoelectric effects have been widely used for stress and thermal sensors. Modulation of refractive index with electric field is widely used to design optical switches. Their overall superior dielectric properties are widely being considered for use as gate dielectrics in FET technology. In section 5, we propose novel devices which incorporate ferroelectric materials in FETs to create stress sensors, thermal sensors.

1.4.1 Smart Sensor FET

If we examine high performance electronic and optoelectronic devices, we find that they are made from semiconductors (such as Si, GaAs, InGaAs, etc.). On the other hand, devices that fall in the general category of sensors are made from insulators (a variety of oxides). The dominance of semiconductors in electronic and optoelectronic devices can be traced to two properties (see Fig. 5): (i) A small change in the Fermi level position can alter density of "free" carriers and conductivities by orders of magnitude; (ii) Charge injection can alter the conductivities and optical properties by orders of magnitude. Semiconductors have also been used in many sensor applications, such as silicon strain sensors [85], GaAs FET strain sensors [10, 20], and P-N diode temperature sensors [39]. However, the sensitivity to such perturbations is relatively poor.

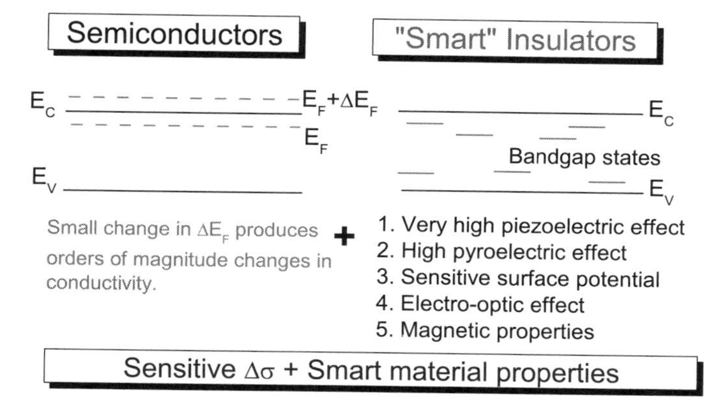

Fig. 5 Semiconductors have the remarkable properties of conductivity and *e-h* injection control. Smart materials based on various oxides have remarkable polarization, magnetic, and electro-optic properties. The heterostructures between the two classes would lead to novel devices. [93]

Materials such as $BaTiO_3$, PLZT, KDP, $LiNbO_3$, etc. have excellent piezoelectric, pyroelectric, ferroelectric, and electro-optic properties [22,25,44,56,60], which make them attractive for many applications. Sensor devices based on these materials rely upon the fact that a small perturbation (stress, temperature, charge, etc.) produces a large change in the polarization of the material. This polarization change can then be detected as current or a voltage signal. Ideally one would like to create heterostructures which have the properties of semiconductors and also the superior polar properties of certain oxides.

2 Theoretical Approach

To understand the issues mentioned in the introductory section and to examine potential of polar heterostructure devices, we need to develop formalisms for the following: i) How does polar charge depend upon strain created through epitaxy? ii) How does vertical junction transport occur and how is the current influenced by the presence of polar charge and defects? iii) How does lateral transport occur in HEMTs and how does scaling influence device response? How do self-heating effects influence device response? iv) How do polar insulator-semiconductor structures behave in term of smart FET applications? In this section, we will outline these formalisms.

2.1 Polarization by Strain

For nitride and ferroelectric materials, piezoelectric effects affect the channel charge density, and are expected to significantly influence the operation of HEMTs. To study how polarization is induced by strain, we need to use the basic formalism linking strain and polarization. We will assume that the strain values are small ($< 3\%$) so that the linear equations can be used. Strain can be expressed by a strain tensor $\epsilon_{\alpha\beta}$:

$$\epsilon_{\alpha\beta} = \begin{bmatrix} \epsilon_{xx} & \epsilon_{yx} & \epsilon_{zx} \\ \epsilon_{xy} & \epsilon_{yy} & \epsilon_{zy} \\ \epsilon_{xz} & \epsilon_{yz} & \epsilon_{zz} \end{bmatrix} . \tag{4}$$

The tensor has nine components. The elastic constants can be expressed by stress train relation

$$\sigma_i = \sum_{j=1}^{6} C_{ij}\epsilon_j, \tag{5}$$

where $1 = xx$, $2 = yy$, $3 = zz$, $4 = yz$, $5 = zx$, $6 = xy$, and C_{ij} is the elastic constant. The polarization induced by strain can be expressed as

$$P_i = \sum_j e_{ij}\epsilon_j \tag{6}$$

For c-axis GaN/AlGaN heterostructures grown on GaN buffered substrate (actual substrate may be sapphire or SiC), the AlGaN is under strain. We can use the equation given above to calculate the polarization in the vertical direction, z. Nitrides are wurtzite structures, and with the symmetry reduction, the matrix elements, C_{ij}, can be simplified to,

$$C = \begin{bmatrix} C_{11} & C_{12} & C_{13} & 0 & 0 & 0 \\ C_{12} & C_{11} & C_{13} & 0 & 0 & 0 \\ C_{13} & C_{13} & C_{33} & 0 & 0 & 0 \\ 0 & 0 & 0 & C_{44} & 0 & 0 \\ 0 & 0 & 0 & 0 & C_{44} & 0 \\ 0 & 0 & 0 & 0 & 0 & C_{66} \end{bmatrix} . \tag{7}$$

Since there is no stress in the z direction, $\sigma_z = C_{13}\epsilon_x + C_{13}\epsilon_y + C_{33}\epsilon_z = 0$, the strain in z direction is

$$\epsilon_z = -\frac{C_{13}}{C_{33}}(\epsilon_x + \epsilon_y). \tag{8}$$

The polarization in the z direction P_z is given by

$$P_z = P_{sp} + \left(e_{13} - e_{33}\frac{C_{13}}{C_{33}} \right) (\epsilon_x + \epsilon_y), \tag{9}$$

where e_{ij} is the piezoelectric constant and P_{sp} is the spontaneous polarization of the nitrides. The ϵ_x and ϵ_y can be obtained by lattice mismatch of AlGaN and GaN. The material parameters can be directly interpolated from the value listed in Table 2 and in reference [3].

Table 2 Some parameters used for nitride related materials [3]

Parameter Name	GaN	AlN	InN
Effective mass (m_0)	0.20	0.48	0.05
lattice constant, a_0 Å	3.189	3.112	3.54
Bandgap (eV) Γ	3.4	6.2	0.9
Gap (eV) L	5.29	6.9	4.09
Gap (eV) Γ_3	5.49	7.2	4.49
e_{33} ($C{\cdot}m^{-2}$)	0.73	1.46	0.97
e_{31} ($C{\cdot}m^{-2}$)	−0.49	−0.60	−0.57
P_{sp} ($C{\cdot}m^{-2}$)	−0.029	−0.081	−0.032
ϵ_s	10.4	9.2	15.3
ϵ_∞	5.35	4.77	8.4
mass density ($g{\cdot}cm^{-3}$)	6.095	3.230	6.810
c_{11} (10^{12} $dyn{\cdot}cm^{-2}$)	3.742	4.105	2.231
L-O Phonon energy	0.091	0.110	0.073
Acoustic Deformation Potential	10.1	9.5	7.1
Non-parabolicity (eV^{-1})	0.189	0.044	0.491

2.2 Vertical Junction Transport

2.2.1 One-dimensional Drift-Diffusion Charge Control Model

We will now examine the mathematical and numerical underpinnings of the formalisms that are used to model polar heterostructure devices that exploit built-in polarization charges. To understand how such devices work, we need to develop basic formalisms to simulate the induction of piezoelectric polarization, potential at any point, and carrier wavefunctions, which yield the local charge density. The task of solving the coupled system of equations for potential and wavefunctions is referred to as the charge control calculation [34, 49, 68, 99, 102]. Figure 6 shows the self-consistent loop for the charge control model. To develop the charge control model, we need to address the following

- Solution of Poisson equation to get the band profiles [80, 100],
- solution of Schrödinger equation to obtain the wave function and charge density, and
- use the drift-diffusion current model to get quasi-Fermi levels.

The quasi-Fermi levels are then fed back into Poisson equation until the solution converges. We will now describe the charge control model to examine how changes in fixed polar charge influence 2DEG. For this model, the semiconductor interface is treated as an ideal interface with abrupt transition from the channel material to the barrier material. Piezoelectric and spontaneous polarization are accounted for through the application of the boundary condition for the heterojunction interface,

$$\varepsilon_1 \mathbf{E_1} + \mathbf{P_1} = \varepsilon_2 \mathbf{E_2}, \tag{10}$$

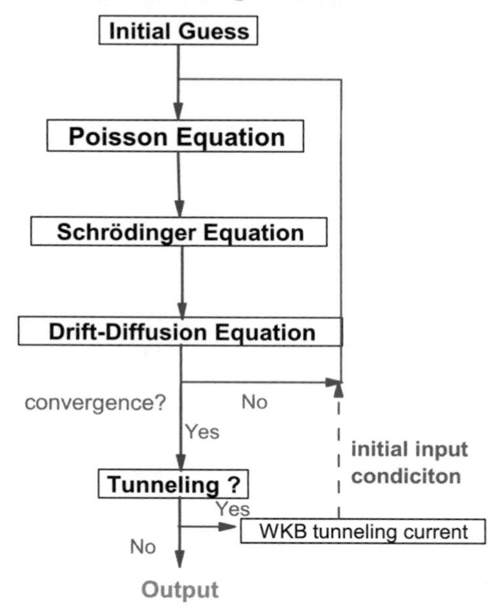

Fig. 6 The self-consistent loop for solving the formalism needed in charge control model.

where ε_1 and ε_2 are the dielectric constants and $\mathbf{E_1}$ and $\mathbf{E_2}$ are electric fields at the interface. $\mathbf{P_1}$ are the polarization. It also needs to satisfy the band discontinuity boundary condition at the interface expressed by

$$E_{c1} + \triangle E_c = E_{c2}, \tag{11}$$

where E_c is conduction band energy. $\triangle E_c$ is the conduction band edge discontinuity at the abrupt heterojunction interface. Once this boundary condition is applied, we write down the Poisson equation,

$$\nabla^2 E_c = -\frac{\rho}{\varepsilon}, \tag{12}$$

where ρ is the total charge density and ε is the dielectric constant. Since the variation of fixed polar charges, dielectric constant, free electron charge distribution, and doping is along the z direction. We can use the one-dimensional Poisson equation,

$$\frac{d^2}{dz^2} E_c(z) = -\frac{\rho(z)}{\varepsilon(z)} \tag{13}$$

The total charge $\rho(z)$ is

$$\begin{aligned}\rho(z) = q(N_d^*(z) - N_a^*(z) - n_{free}(z) + p_{free}(z) \\ - \sum_i n_i \psi_i^*(z) \psi_i(z)),\end{aligned} \tag{14}$$

where N_d^* and N_a^* are the effective doping concentrations and $n_{free}(z)$ and $p_{free}(z)$ are the free carrier concentrations in the bulk region. The sum over i in Eq. (14) is over 2D confined sub-bands with normalized envelope functions, ψ_i, and occupation, n_i (p_i for the hole case). In the electron case, we can write the occupation

$$n_i = \frac{m_i k_B T}{\pi \hbar^2} \ln \left[1 + \exp\left(\frac{E_i - E_f}{k_B T} \right) \right], \tag{15}$$

where m_i is the in-plane effective mass and E_i is the sub-band level. To determine the 2D confined sub-band envelope functions, ψ_i, and sub-band levels, E_i, the Schrödinger equation must be solved. The one dimensional Schrödinger equation can be written using the perpendicular part of the effective mass tensor as follows:

$$\frac{d^2}{dz^2}\psi_i(z) + \frac{2m_e}{\hbar^2}\left[E_i - E_c(z)\right]\psi_i(z) = 0, \tag{16}$$

where m_e is the effective mass along the quantum confinement direction. Once the Poisson and Schrödinger equations have been solved, the drift-diffusion equation must be considered if the charges are injected into the devices, or the gate is under bias. When current flows in the device, the Fermi level splits into two different quasi-Fermi levels. The quasi-Fermi levels can be obtained by solving the drift-diffusion equations, which can be expressed as:

$$J_n = -\mu_n n(z)\frac{\partial E_c}{\partial z} + qD_n\frac{\partial n(z)}{\partial z} \tag{17}$$

$$J_p = \mu_p p(z)\frac{\partial E_v}{\partial z} - qD_p\frac{\partial p(z)}{\partial z}, \tag{18}$$

where μ_n and μ_p are the mobilities of the electrons and holes, and D_n and D_p are the diffusion coefficients of electrons and holes. Using the continuity equation,

$$\frac{\partial J_n}{\partial z} = qR \tag{19}$$

$$\frac{\partial J_p}{\partial z} = -qR, \tag{20}$$

where R represents the Shockley-Read-Hall [76, 84] rate and can be expressed as

$$R = \frac{pn - n_i^2}{\tau_{n0}\left(p + n_i exp\left(\frac{E_i - E_t}{K_B T}\right)\right) + \tau_{p0}\left(n + n_i exp\left(\frac{E_t - E_i}{K_B T}\right)\right)} \tag{21}$$

where $\tau_{n,p}$ are electron and hole lifetimes, n_i is the intrinsic carrier concentration, and E_t is the trap energy. Once quasi-Fermi levels are obtained, they are fed into the Poisson equations and the equations are solved self-consistently.

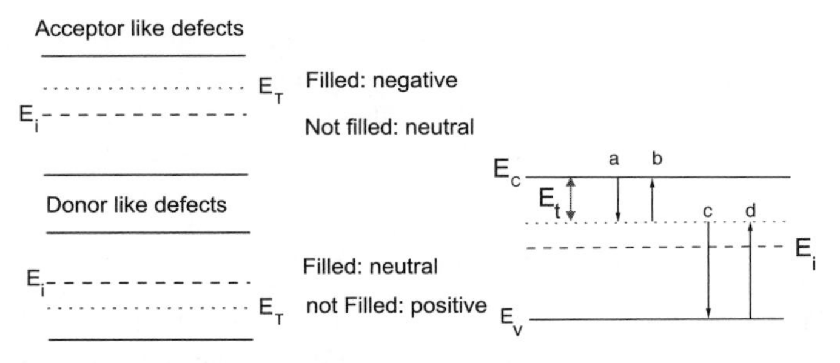

Fig. 7 (a) The two basic types of traps which are call donor-like and acceptor-like traps. (b) The mechanism of generation and recombination at trap states for acceptor like traps.

2.2.2 Incorporation of Defects

To consider the effect of defects in the system, we need to consider two basic types of traps: acceptor-like and donor-like traps [26, 74], as shown in Fig. 7 (a). An acceptor-like trap is negatively charge so it might only release an electron. Therefore, acceptor-like traps are negative when filled but are neutral when empty. On the other hand, a donor-like trap is positively charged and therefore can only capture an electron. This means that donor-like traps are positive when vacant but neutral when filled.

The fraction of trapped electron charges can be expressed as:

$$F_n = \frac{\upsilon_n \sigma_n n + e_p}{\upsilon_n \sigma_n n + \upsilon_p \sigma_p p + e_n + e_p}. \tag{22}$$

Similarly, we can obtain the probability of occupation, F_p, for donor-like traps,

$$F_p = \frac{\upsilon_p \sigma_p p + e_n}{\upsilon_n \sigma_n n + \upsilon_p \sigma_p p + e_n + e_p}. \tag{23}$$

The electron and hole lifetimes τ_n and τ_p are related to carrier capture cross sections σ_n and σ_p through equations

$$\tau_n = \frac{1}{\sigma_n \upsilon_n N_t} \tag{24}$$

$$\tau_p = \frac{1}{\sigma_p \upsilon_p N_t}. \tag{25}$$

The electron and hole charges are calculated by the equations

$$n_t = F_n N_t \tag{26}$$

$$p_t = F_p N_t. \tag{27}$$

The effect of trapped charge is then added to the charge term of Poisson's equation. Thus Eq. (14) becomes,

$$\rho(z) = q(N_d^*(z) - N_a^*(z) - n_{free}(z) - n_t + p_{free}(z) + p_t - \sum_i n_i \psi_i^*(z)\psi_i(z)), \tag{28}$$

The effect of defects is then considered in the Poisson equation so that we can examine how the defects influence electrical characteristics of the device.

2.2.3 WKB Approximation and Tunnelling Current

The basic equations for the calculation of tunneling probability are based on an Wentzel-Kramers-Brillouin approximation [76]. The transmission probability is given by

$$T(E_1) = e^{-\int_a^b \frac{\sqrt{2m^*(V(z)-E_1)}}{\hbar}dz} \tag{29}$$

To obtain the tunnelling current, we consider the density of states in the GaN substrate. The density of states can be divided into two regions: With the region of quantum confinement, which is the two dimensional density of states, the corresponding current density J_{2D} is defined below:

$$J_{2D} = q\sum_{n=1}^{n=n_{2D}} \frac{v(E_n, E_f)D_{2D}(E_n)}{L_n} \int_0^{\infty} (f(E_{tot}, E_{fs}, T) - f(E_{tot}, E_{fm}, T)) \times T(E_n, E_t)dE_t, \tag{30}$$

where n_{2D} is the total number of bound levels in the 2DEG. E_f is given by

$$E_f = \min(E_{fs}, E_{fm}) \tag{31}$$

At energies greater than confinement energy, we have three dimensional density of states, the corresponding current density J_{3D} is defined as:

$$J_{3D} = q\int_{E_{n_{2D}}}^{\infty} \int_0^{\infty} v(E_l, E_f)D_{1D}(E_l)D_{2D}(E_t)(f(E_{tot}, E_{fs}, T) - f(E_{tot}, E_{fm}, T)) \times T(E_l, E_t)dE_t dE_l. \tag{32}$$

Thermionic emission is accounted for by extending the range of energies above the height classically forbidden barriers.

2.3 *Lateral Transport in Undoped HEMTs*

To study the carrier transport issues, Monte Carlo method has been the most popular way to realistically model the transport characteristics of devices [29, 33, 41, 97]. In the Monte Carlo process, the electron is considered as a point particle whose scattering rates are given by the Fermi Golden Rule expressions. The simulation involves the following stages:

1. Initial calculation of the scattering rates for intra- and inter-sub-band scattering.
2. Particle injection into the region of study: The particles are injected with a pre-chosen distribution of carrier momenta.
3. Free flight of the carriers: Scattering events are considered to be instantaneous. Between scattering processes the carriers simply move in the electric field according to the free-particle equation of motion.
4. Scattering Events: A specific prescription is used in the Monte Carlo methods to determine the time between scattering events. At the end of a free flight, scattering occurs which alters the carrier flight direction.
5. Selection of Scattering events: Random choice of scattering events. This choice is again based on the Monte Carlo method.
6. State of electron after scattering: To determine the final state of carriers, we need detailed information on the scattering process.

Steps 3-6 are repeated until the stopping condition is met. This condition usually corresponds to the end of the channel. Figure 8 shows the flowchart of the Monte Carlo program. The process takes the following scattering mechanisms into account:

- Polar optical phonon absorption and emission;
- acoustic phonon scattering;

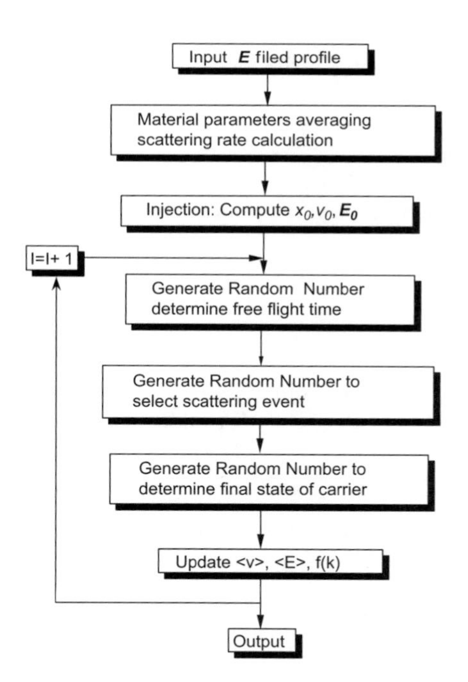

Fig. 8 A flowchart of the Monte Carlo program.

- interface roughness scattering;
- equivalent and non-equivalent inter-valley scattering;
- alloy scattering, and;
- charged dislocation scattering.

According to Fermi Golden Rule, a scattering process is described to first order by

$$M_{k,k'} = \langle k|H_{pert}|k'\rangle \tag{33}$$

$$W_i(k,k') = \frac{2\pi}{\hbar}\sum_{k'}|M_{k,k'}|^2\delta\left(E_f - E_i\right), \tag{34}$$

where H_{pert} is the perturbation Hamiltonian for the interaction, and k' is the index over all final states. The resultant wave functions from solution of 16 are then used to calculate overlap integrals, $I_{k,k'}$, which are used in the calculation of the inter-sub-band matrix elements.

$$I_{k,k'} = \int_{\text{Entiredevice}} \phi_k^*(z)\phi_{k'}(z)dz. \tag{35}$$

Monte Carlo simulation is carried out in two modes - a) Steady state Monte Carlo and b) Ensemble Monte Carlo. For the steady state simulation, one electron is injected into the device for a very long period of simulation for certain electric field until it reaches the steady state velocity. The resultant v–E curve is used to deduce the low field drift mobility, μ_l and be used to solved the drift-diffusion equation. For the ensemble Monte Carlo method, a large number of electrons (typically $>$ 100,000) are injected into real device potential profiles to calculate the average transient time of the electrons.

In confined three-dimensional or quasi - two-dimensional structures, the matrix elements used in eqn.(33) would be slightly different. The matrix element is calculated by integrating an overlap integral term with the three dimensional matrix element. For example, for phonon scattering, we have

$$I_{kk'}(q) = \int_{\text{Entiredevice}} \phi_k^*(z)\phi_{k'}(z)dz. \tag{36}$$

$$|M_{kk'}|^2 = \int |M(Q,q)|^2|I_{kk'}(q)|^2dq, \tag{37}$$

where Q and q are components of phonon wave vectors parallel and perpendicular to the hetero-interface, respectively.

In the quasi-2D Monte Carlo approach, the electron is treated as a particle in the parallel direction. In the vertical direction, the electron is treated as an envelope wave function. For the heterostructure like AlGaN/GaN or InN/GaN, the wave function is localized at the heterojunction interface. For higher sub-band, the envelope will spread into GaN or AlGaN region. To calculate the scattering rate for a state which overlaps several different regions, we need to use a reasonable averaging procedure. For different sub-bands, the wave function has its own fraction over different material region. Therefore, for each state i, we can obtain the fraction in the different region to be

$$f_{ij} = \int_{Z_j}^{Z_{j+1}} \phi_i^*(z)\phi_i(z)dz, \tag{38}$$

where j denotes the number of layers N located at z_j, $j = 1, ..., N$, and the average material parameters, G_i, for each sub-band can be calculated by

$$D_i = \sum_{j=1}^{N} f_{ij} D_j \tag{39}$$

Eqn.(39) is used for all GaN parameters shown in Table 2 except effective masses, which are averaged by,

$$\frac{1}{m_i^*} = \sum_{j=1}^{N} \frac{f_{ij}}{m_j^*}. \tag{40}$$

These space-averaged subband-dependent parameters are then used in all calculation of the simulation.

2.3.1 Heat Dissipation Issues

As mentioned earlier, GaN based devices are starting to dominate power electronics. When the device is operated at very high voltage and large current, the self-heating problem would become very important. To model heating effects, we need to consider the thermal conduction equation given by

$$C\frac{\partial T}{\partial t} = H + \nabla(\kappa \nabla T), \tag{41}$$

where κ is the thermal conductivity ($\mathrm{W \cdot cm^{-1} \cdot K^{-1}}$), H is the heat generation term ($\mathrm{W \cdot cm^{-3}}$), C is the heat capacity ($\mathrm{Joul \cdot K \cdot cm^{-3}}$), and T is the temperature. If we only consider the steady state thermal equilibrium condition, the time variation term is zero. Therefore, the equation can be simplified as:

$$H = -\nabla(\kappa \nabla T). \tag{42}$$

The equation can be easily solved by numerical methods such as finite element method (FEM) in 2D or 3D cases.

2.4 $k \cdot p$ Method for Strained Nitride Quantum Wells and Quantum Dots

Polarization effects at nitride heterojunctions are known to exert strong influence on junction properties, especially for nitride quantum wells and dots. For instance, the large Stokes shift and non-linear electro-optic effects are directly related to the piezoelectric effect. The built in field causes a separation of the electrons and holes in a quantum well. As carriers are injected, field is screened. Electron hole separation and screening effects play a significant role in the optical gain and light emission. In section 2.2.1, we have introduced the self-consistent Poisson and Schrodinger equation method to consider the polarization effect in the conduction band. In this section, we will introduce a six-band $k \cdot p$ method [14] to calculate valence band structure of nitride quantum well and quantum dot.

In zinc blende type crystals, the Kohn-Luttinger Hamiltonian is used to described the electronic structure around the valence band edge. However, because of the different symmetric for wurtzite GaN materials, the valence-band edge cannot be described by Kohn-Luttinger Hamiltonian. For the GaN wurtzite strained structures, the new Hamiltonian [14] can be written as:

$$H = \begin{bmatrix} F & -K^* & -H^* & 0 & 0 & 0 \\ -K & G & H & 0 & 0 & \Delta \\ -H & H^* & \lambda & 0 & \Delta & 0 \\ 0 & 0 & 0 & F & -K & H \\ 0 & 0 & \Delta & -K^* & G & -H^* \\ 0 & \Delta & 0 & H^* & -H & \lambda \end{bmatrix} \begin{matrix} |u_1\rangle \\ |u_2\rangle \\ |u_3\rangle \\ |u_4\rangle \\ |u_5\rangle \\ |u_6\rangle \end{matrix}, \tag{43}$$

where

$$\begin{aligned} F &= \Delta_1 + \Delta_2 + \lambda + \theta, \\ G &= \Delta_1 - \Delta_2 + \lambda + \theta, \\ \lambda &= \frac{\hbar^2}{2m_0}[A_1 k_z^2 + A_2(k_x^2 + k_y^2)] + \lambda_\epsilon, \\ \lambda_\epsilon &= D_1\epsilon_{zz} + D_2(\epsilon_{xx} + \epsilon_{yy}), \\ \theta &= \frac{\hbar^2}{2m_0}[A_3 k_z^2 + A_4(k_x^2 + k_y^2)] + \theta_\epsilon, \\ \theta_\epsilon &= D_3\epsilon_{zz} + D_4(\epsilon_{xx} + \epsilon_{yy}), \\ K &= \frac{\hbar^2}{2m_0}A_5(k_x + ik_y)^2 + D_5\epsilon_+, \\ H &= \frac{\hbar^2}{2m_0}A_6(k_x + ik_y) + D_6\epsilon_{z+}, \\ \Delta &= \sqrt{2}\Delta_3, \end{aligned} \tag{44}$$

and

$$\epsilon_{\pm} = \epsilon_{xx} - \epsilon_{yy} \pm 2i\epsilon_{xy}, \tag{45}$$
$$\epsilon_{z\pm} = \epsilon_{zx} \pm i\epsilon_{yz}. \tag{46}$$

The $\epsilon_{i,j}$ are the strain tensors and the detailed derivation of the strain tensor can be found in section 2.1. The paramters A and δ can be found in [14, 21, 70, 73]. The basis functions of $|u_i\rangle$ are

$$\begin{aligned}
|u_1\rangle &= \frac{-1}{\sqrt{2}}|(X+iY)\uparrow\rangle \\
|u_2\rangle &= \frac{1}{\sqrt{2}}|(X-iY)\uparrow\rangle \\
|u_3\rangle &= |Z\uparrow\rangle \\
|u_4\rangle &= \frac{1}{\sqrt{2}}|(X-iY)\downarrow\rangle \\
|u_5\rangle &= \frac{-1}{\sqrt{2}}|(X+iY)\downarrow\rangle \\
|u_6\rangle &= |Z\downarrow\rangle
\end{aligned} \tag{47}$$

For the quantum well problem, the wavefunctions are quantized in the z direction. Therefore, the operator k_z should be replaced by $-i\partial/\partial z$. For the quantum dot problem, The wavefunctions are all quantized in three directions. The k_x, k_y, and k_z are replaced by $-i\partial/\partial x$, $-i\partial/\partial y$, and $-i\partial/\partial z$, respectively.

3 Tailoring of Vertical Junctions

Advances in nitride heterostructure technology have allowed careful studies into junction properties of metal-polar heterostructure combinations where dopants are replaced by built-in polar fixed charges. Very large band bending caused by interface fixed charges permits the tailoring of the current-voltage relations [77] by simply controlling the AlGaN layer thickness.

To apply a vertical transport model [77] to the general case, we need to determine the current flow inside the entire device, which can be expressed by the drift-diffusion current model, as well as effects arising out of hysteresis exhibited by ferroelectric materials [78]. The transport model described in section 2.2 is used here.

In this section, we will discuss three classes of junctions that are important for device technologies:

1. The alloy $Al_{0.3}Ga_{0.7}N$ constitutes the barrier layer with a GaN channel and substrate. The exact Al fraction can of course vary in general but the 30% value is found to be an optimum value. This structure can offer tailorable I–V characteristics for junction diodes as well as for fabricating ohmic contacts.

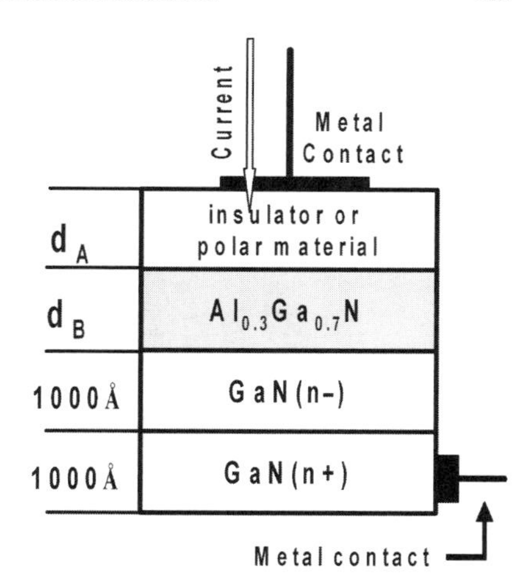

Fig. 9 A schematic of the structures considered is shown. GaN(n-) is doped 2×10^{16} cm^{-3} and GaN(n+) substrate is doped 5×10^{16} cm^{-3}. The contact regions shown here are merely schematics, as is the rest of the figure, and do not represent any actual contact structures. The arrow represents current flow direction for positive gate bias.

2. A high dielectric constant insulator layer is added on top of the $Al_{0.3}Ga_{0.7}N$ layer. Such a structure would be useful for gate tunneling suppression.
3. A high dielectric constant polar layer (with a charge density at 1×10^{14} cm^{-2}) is added on top of the $Al_{0.3}Ga_{0.7}N$ layer instead. Such structures can be useful for tailorable I–V characteristics.

The general structure simulated by us is shown in Fig. 9. It consists of a layer of an insulator or a generic polar material grown on a GaN/$Al_{0.3}Ga_{0.7}N$ structure. A fixed polar charge density, 1.68×10^{13} cm^{-2}, exists at the heterointerface between AlGaN and GaN. The GaN (n-) layer is doped at 2×10^{16} cm^{-3}, and the GaN (n+) layer is doped at 5×10^{16} cm^{-3}. The thicknesses of both layers are 1000Å .

3.1 Gate Leakage Suppression

Figure 10(a) shows the sheet charge density and corresponding tunnelling I-V characteristic. We compared two barrier thickness, 50Å and 70Å. As expected, the tunnelling current decreases, and the sheet charge density increases with increase in barrier thickness. In Si technology, a "rule of thumb" relating gate oxide thickness to gate length [87] is given by:

$$L_G \approx 45 d_{\mathrm{ox}}. \tag{48}$$

In the nitride system, no such clear rule has been established. However, we can see from Fig. 10(a) that for AlGaN thickness of ~50Å, gate tunnelling becomes very significant. Its been seen that there is a trade-off between induced 2DEG charge and tunnelling probability [77]. For reverse biases $\sim -1V$, the tunnelling current of 70Å barrier thickness is close to that for 50Å barrier thickness. A possible reason for this could be that the increasing free carrier density cancels part of the contribution to 2DEG from a thicker barrier.

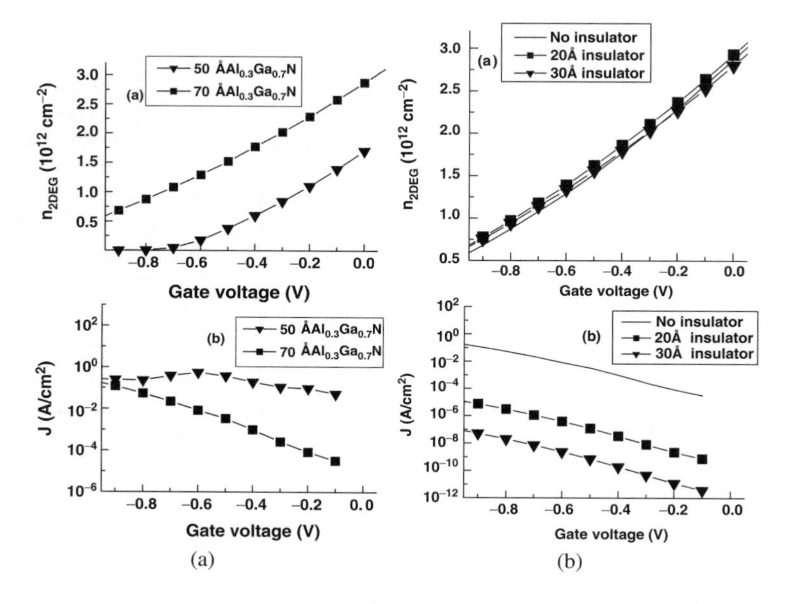

(a) (b)

Fig. 10 (a) The sheet charge density and I-V characteristics for the device structure GaN(n+)/GaN (n-)/$Al_{0.3}Ga_{0.7}N$. Schottky barrier height, $\phi_{SB} = 1.45$ eV; (b) I-V characteristics and sheet charge densities for the device structure incorporating the insulator. The thickness of the $Al_{0.3}Ga_{0.7}N$ is taken to be 70Å. Insulator parameters chosen: dielectric constant $\kappa = 29$, band gap, $E_g = 3.74$ eV and Schottky barrier height, $\phi_{SB} = 1.15$ eV. [92]

It is well known that one way to reduce gate tunnelling in FETs is to add a high dielectric constant region between the gate and the channel [17, 18]. Such approaches have been successfully demonstrated in Si MOSFET where $BaTiO_3$ has been added. High-κ dielectrics such as TiO_2 have been grown on nitride heterostructures [28]. In Fig. 10(b) we show results of calculation for structures where a material with $E_g = 3.74$ eV and $\kappa = 29\epsilon_0$ has been added between the metal and AlGaN. The high-κ material is assumed to be non-polar. Our results indicate that 2DEG sheet charge density is relatively insensitive to increase in insulator thickness. However, tunnelling current is reduced at least by a factor of 10^4 with introduction of an insulator layer of thickness, $d_A = 30$Å. Thus, a high-κ insulator layer can be used to decouple the relationship between barrier layer thickness and 2DEG sheet charge density.

It is also worthwhile to note that the gate control of channel properties of a field effect transistor described above (such as transconductance, g_m) is relatively unaffected owing to the high capacitance of the polar layer. The structure shown in Fig. (9) is composed of two capacitors in series with the total capacitance C_g given by

$$\frac{1}{C_g} = \frac{1}{C_A} + \frac{1}{C_B} \approx \frac{1}{C_B}, \tag{49}$$

where C_A is the capacitance of either the polar or the insulator layer and C_B is the capacitance of a similar FET structure without additional polar or insulator layer. This makes it clear that the introduction of the polar or insulator layer does not affect the transconductance, which is still adequately described by the relation,

$$g_m = C_g v_{sat} \approx C_B v_{sat}, \tag{50}$$

which may be seen to be the same transconductance as that of a structure without either a polar or an insulator layer(v_{sat} is the saturation velocity of the carriers in FET channel region). Since g_m is the main figure of merit in description of FET's, its apparent that characteristics of circuits employing the proposed structure which has the aforementioned advantages, would not be degraded significantly by the use of high-κ gate insulators.

3.1.1 Experimental Work on Gate Leakage Reduction

In the previous section, we have demonstrated that using high-κ dielectric layer can significantly reduce the gate leakage. There are many experimental studies on different gate insulators for GaN HEMTs to reduce the gate leakage. Gate insulators, such as SiO_2 [12, 13, 37], ZrO_2 [82], Si_3N_4 [59], Al_2O_3/Si_3N_4 [46–48], BST [27] and HfO_2-Al_2O_3 [67] have been demonstrated to reduce the gate leakage by 3 to 5 orders of magnitudes. Table 3 shows the experimental results on gate insulators for AlGaN/GaN HEMTs. As shown in the table, some materials such as SiO_2 and SiON are low κ materials where the threshold voltage would significantly shift if the gate insulator is too thick. Si_3N_4 and Al_2O_3 have very good properties in adhesion and surface passivation, and have the lowest defect density values at the interface. However, their dielectric constant is very similar to GaN and therefore there is a significant voltage shift in the device. As our simulation suggests, using high-κ dielectric gate insulators, such as BST in table 3, can have the smallest voltage shift. As shown in Table 3, the layer thickness of BST is the largest. However, due to the high-dielectric constant, the V_t shifts slightly around 14% to 25%. Therefore,

Table 3 The experimental results on gate insulator used in AlGaN/GaN HEMTs to reduce the leakages

Material	ϵ (ϵ_0)	thickness (Å)	leakage reduction	V_t shift (V)
SiO_2 [13]	3.9	320	10^{-5}	−4V to −8V
Si_3N_4 [59]	7.9	100	10^{-5}	−2V to −5V
SiON [7]	4∼20	100	10^{-3}–10^{-4}	−2.1V to −9.1V
ZrO_2 [8]	20	300	10^{-3}–10^{-4}	−2.5V to −7V
Al_2O_3-Si_3N_4 [48]	9-7.9	40-10	10^{-2}–10^{-3}	−10V to −12.5V
BST [27]	22	1250	10^{-5}	−8V to −10V
	66	400	10^{-4}	−8V to −9V
HfO_2-Al_2O_3 [67]	25-9	100	10^{-4}	−2V to −4V

to reduce the gate leakage and maintain the same transconductance behavior, it is necessary to explore new high-κ materials as the gate insulator.

3.2 Forming Ohmic Contacts by Using Polarization Effects

An important issue in semiconductor, especially in large gap semiconductors, is Ohmic contact formation. It is well known that in semiconductor technology [84], one way to convert a rectifying contact into an Ohmic contact is to heavily dope the semiconductor in the region of the contact. Another way to achieve the same result is to use polar materials [80] in the region of the contact. In the nitride system, the interfacial polar charge results from spontaneous polarization and piezoelectric effect, which cannot be altered once the crystal is grown. In ferroelectrics such $LiNbO_3$, $SrTiO_3$ etc., polarization values can be much larger than what is possible in nitrides and can also be controlled by external fields. Polar charge densities in the range of 10^{14} cm^{-2} are present in a number of ferroelectrics. For example, $LiNbO_3$ has a polar charge of $\sim 2 \times 10^{14}$ cm^{-2} and has a very high coercive field [38].

Ferroelectrics have been grown on nitrides and other semiconductors although their junction properties have not been reported. To examine the potential of polar materials for ohmic junction formation, we examine the I-V characteristics of a structure where a thin layer of a polar material with $\sigma = +10^{14}$ cm^{-2} is present between a metal and the AlGaN/GaN structure. Results are shown in Fig. (11). As we can see, a very strong band bending can be induced which leads to a very high tunnelling current. Fig. (11) shows the band profiles and sheet charge distribution in the contact heterojunction in absence of "gate" bias. It shows that the barrier becomes very thin and the tunnelling current density is increased by a factor of nearly 10^5 (in either bias condition) when compared to the case without any polar layer. It essentially forms the Ohmic junction without heavy doping, traditionally difficult to obtain in large band gap materials [40, 42, 89, 98].

We also examine the influence of thickness, d_A, of the polar material. As can be seen in Fig. (12), an increase in the polar material's thickness initially increases the tunnelling current and the 2DEG sheet charge density. It can be seen that the positive polar charges at the AlGaN/polar material heterointerface lead to higher band bending when d_A increases from 0Å to 20Å. However, when the thickness d_A is over 20Å, the induced 2DEG charge at the AlGaN/polar material heterointerface starts to annul the band bending. Then, the barrier thickness, $d_A + d_B$, dominates the tunnelling probability. Hence, there exists an optimal thickness ($\simeq$20Å) that corresponds to the highest tunnelling current.

A study of variation of AlGaN thickness with fixed polar material thickness indicates that while the charge density induced in the channel remains practically unchanged, the tunnelling current exhibits an expected decrease with increase in that thickness.

For p-type nitride based materials, it is more difficult to form an ohmic contact due to the large Schottky barrier height and difficulty in achieving for heavy p type

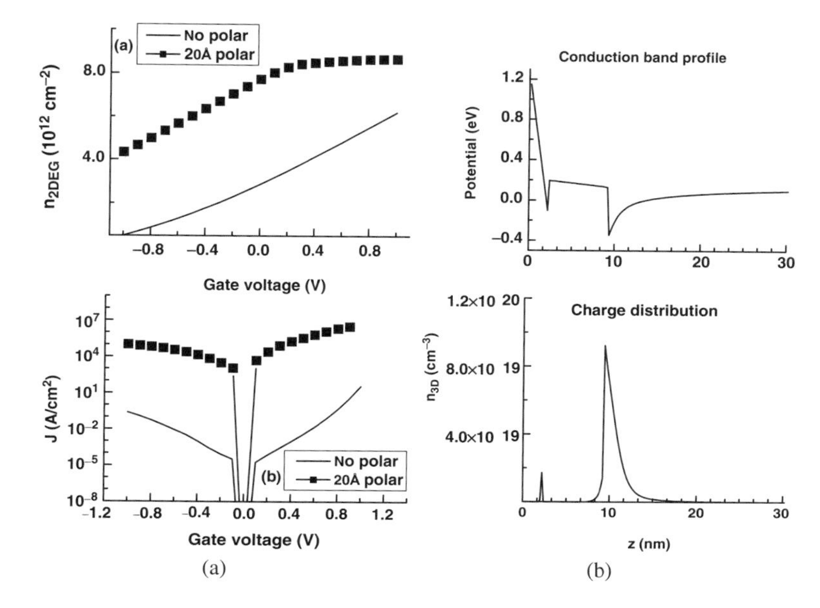

(a) (b)

Fig. 11 (a) I-V characteristics and sheet charge densities for the device structure incorporating the polar material layer. The thickness of the $Al_{0.3}Ga_{0.7}N$ is taken to be 70Å. Polar charge density at the interface between AlGaN and the polar material = 1×10^{14} cm^{-2}. Other parameters of the polar layer are taken to be the same as that of the insulator layer in Fig. 10(b); (b) Conduction band profile and charge density distribution for the structure incorporating a polar material layer of thickness 20Å. Gate voltage, $V_g = 0V$. [92]

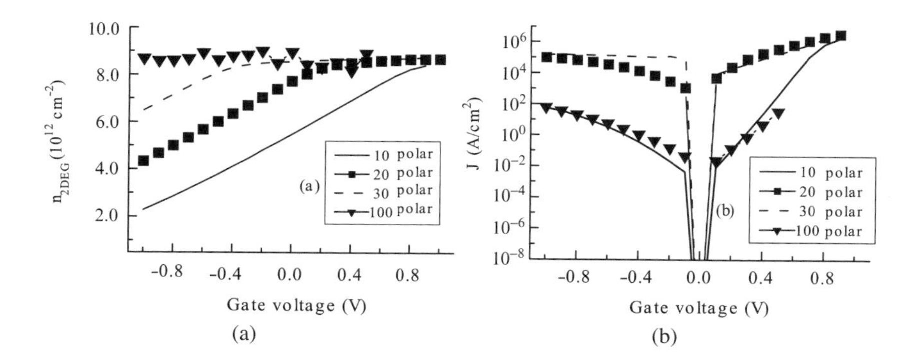

(a) (b)

Fig. 12 Dependence of I-V characteristics and sheet charge densities on thickness, d_A, of the polar layer. [92]

doping. One way to overcome this problem is to use a metal with high work function. However, such high work function metals are difficult to find. Therefore, by adding a polar material to enhance the valence band bending would be a choice to form an ohmic contact. It is well known that for ferroelectric materials, we can easily

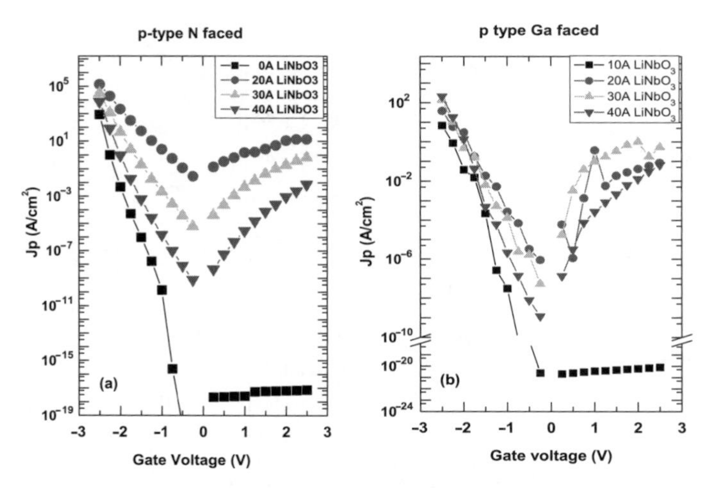

Fig. 13 (a) Dependence of $I-V$ characteristics on thickness, d_A, of the polar layer for the p-type N-faced nitride based devices; (b) Dependence of $I-V$ characteristics on thickness, d_A, of the polar layer for the p-type Ga-faced nitride based devices. The barrier thickness of $Al_{0.3}Ga_{0.7}N$ is 50 Å. [94]

reverse the polarization by applying electrical field. Figure 13 shows the $I-V$ characteristic of p-type Ga-faced and N-faced with different $LiNbO_3$ layer thickness. It is known that Ga-faced nitride based materials would have higher mobility in the channel region. However, a fixed positive polar charge is introduced at the Ga-faced GaN/AlGaN heterostructure interface, thereby forming a large barrier height for the hole current. Figure 13(a) shows that by adding a $LiNbO_3$ layer, we can significantly increase the tunneling current with a predicted optimal thickness of 30 Å. However, in comparison with the n-type Ga-faced GaN/AlGaN devices, the current density is still lower. Unlike Ga-faced nitride based materials, N-faced nitride based materials have a large fixed negative polar charge at the AlGaN/GaN heterostructure interface, which enhances the band bending of the valence band.

Figure 13(b) shows the results of N-faced GaN devices. It is shown that the current density increases a factor of 100 compared to the Ga-faced GaN devices. In this cases, the optimal predicted $LiNbO_3$ layer thickness is 20 Å. It is shown that by adding a polar material for p-type nitride based materials, one can be more easily to form an ohmic contact, which would be very useful to design optoelectronic devices.

4 Nitride HFETS: Transport Issues

In recent years, significant progress has been made in GaN HEMT devices, which represent the most promising class of devices for microwave and millimeter-wave power applications. Transport in GaN is strongly related to the unique properties of GaN such as high LO optical phonon energy, large inter-valley separations, polarization charges, etc.

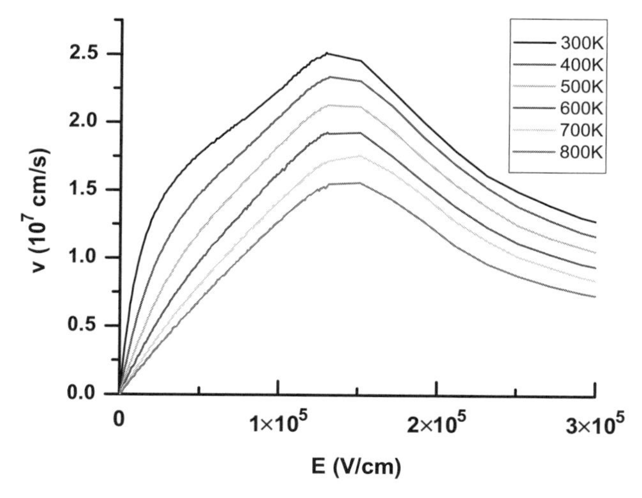

Fig. 14 Velocity field curves for GaN under different temperatures.

Figure 14 shows the velocity-field characteristics of AlGaN/GaN Heterostructures calculated by Monte Carlo method. As shown in the figure, the mobility of nitrides is strongly effected by the temperature. Transport in GaN has some unusual factors which can be traced to a very high optical phonon energy and very large phonon scattering rates ($\sim 10^{14}$ s^{-1}) for electron energies larger than phonon energy. The transport can be described by (i) Ohmic in low field region; (ii) Nonlinear transport characterized by a hump in the velocity-field curve at fields of $\sim$10 to 50 KV/cm; (iii) peak velocity region occurring at $\sim 1.5 \times 10^5$ V/cm; (iv) negative resistance region followed by and (V) saturation and until breakdown. In addition, for very short channel devices ($L_G <$ than 0.1 μm), we can have non-local transport.

4.1 Nonlinear Access Resistance and GaN Device Operation

The presence of a *hump* region in the velocity field characteristic has interesting origins and has important repercussions on device performance. To understand the source of this *hump* we show the phonon scattering rates and velocity as a function of electric field in Fig. 15. As shown in Fig. 15, in the very low field region, steady state Monte Carlo simulation shows that the average electron energy is lower than LO phonon energy, which implies a low probability for the 2D variant of the polar optical phonon emission process (POPEM2D) [97, 101]. As the electric field increases, the probability of an electron acquiring enough energy to surmount the LO phonon energy is increased. Therefore, the gradual increase in the strength of the POPEM2D process slows down the rate of increase of electron velocity and results in the nonlinear *hump*.

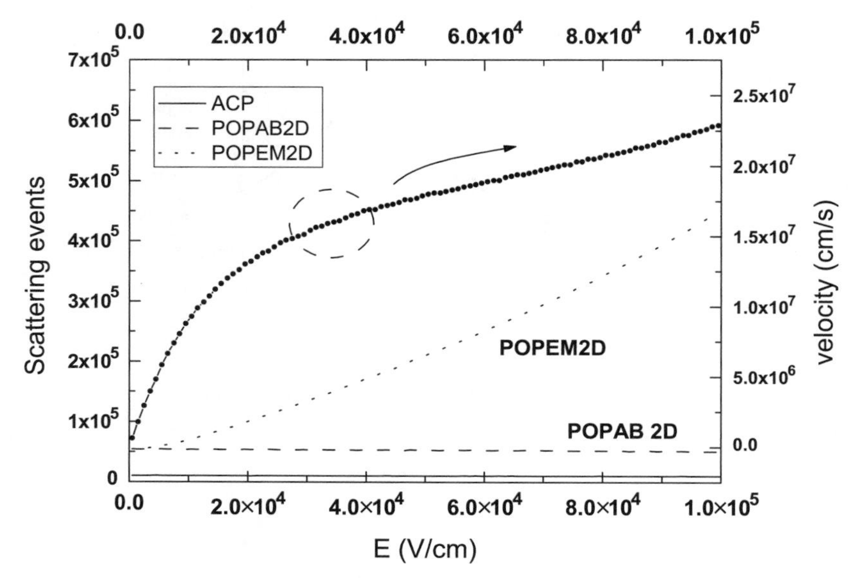

Fig. 15 The number of scattering events corresponding to the 2D polar optical emission (POPEM2D) and 2D polar optical absorption (POPAB2D) processes as a function of the electric field. The marked curve is a portion of the velocity-field characteristic. [95]

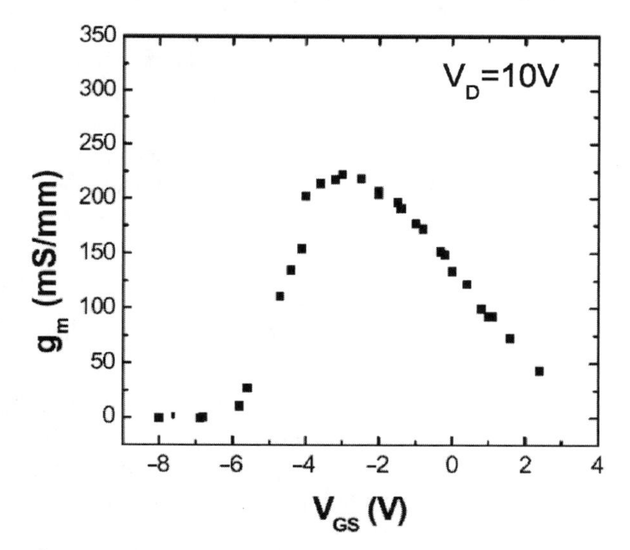

Fig. 16 The transconductance collapse behavior of GaN HEMTs. [66]

The nonlinear behavior in the low field region directly impacts the performance of GaN HEMTs. It has been shown that the GaN based HEMTs exhibit a fairly large gate bias dependent transconductance as shown in Fig.16. This poses a problem for circuit design as many FET circuit models are predicated upon constant g_m (see for instance [24]). This gate bias dependent g_m also limits power linearity and large signal operation at high frequencies.

There are several proposed reasons for the variation in transconductance:

- interface roughness scattering [35]; and
- *hot* phonon scattering and self-heating [6, 50, 51];
- non-linear increase of source-gate resistance with I_{DS} [66, 95]

According to Monte Carlo simulations, interface roughness scattering is not the main source of g_m collapse due to better interface quality and much smaller barrier potential difference at AlGaN/GaN interface compared to SiO_2/Si interface. The self-heating issues are undoubtedly important factors for these transconductance decrease at higher current density since the electron velocity will decrease significantly as shown in Fig. 15. However, short pulse measurements to remove the self-heating effects still show the presence of transconductance decrease [66]. Further, iterative Monte Carlo calculations [9, 79] seem to suggest that hot phonon effects are not the major cause for self-heating induced variations in g_m. It has been suggested [66, 93] that the non-linear increase in the source-gate resistance with increase in I_{DS} arising from the peculiar shape of the GaN velocity-field characteristic is the cause for this collapse in transconductance. Detailed Monte Carlo studies show that this to be the case and the reason are discussed below.

For GaN HEMTs, because of the lack of compatible dielectrics/insulators, and their use in high power applications, there are large spacings between source-gate and drain-gate to drop potential in high bias operation and to prevent current leakage. With the high sheet charge density caused by polarization effects, the overall sheet resistance of GaN HEMTs is comparable to GaAs based HEMTs although they have lower mobility. When the bias current increases, the electric field in the access channel increases. However, because of nonlinear v–E characteristics, the resistance increases significantly as current increases and as a result the transconductance decreases. In other materials such as GaAs and InAs, this non-linearity does occur but at much lower energies as the LO phonon energy is much lower for those systems.

To verify if the nonlinear access resistance in the low-field region is the reason for the collapse in the transconductance, modeling results based on 2D Poisson and Drift-diffusion solver are applied for many different v–E characteristics depicted in Fig. (17). The solid line in Fig. (17) corresponds to the actual v–E characteristics in the HEMT (Fig. 14). Three different test v–E characteristics are considered. In test model 1, a monotonic relation for v–E is used and then we impose a saturation value in the velocity. In test model 2, the velocity increases linearly and the behavior at high fields is taken to be the same as it is for GaN. In test model 3, a much faster increase in the velocity is used, corresponding to a much lower effective mass (for materials like GaAs, InN, InAs, etc.). In the modelling of the three cases, the transconductance is much more insensitive to change in gate bias in the region of interest. Thus, the source of the collapse in the transconductance observed for the actual v–E characteristics is a result of the non-linearity in the low-field v–E characteristic.

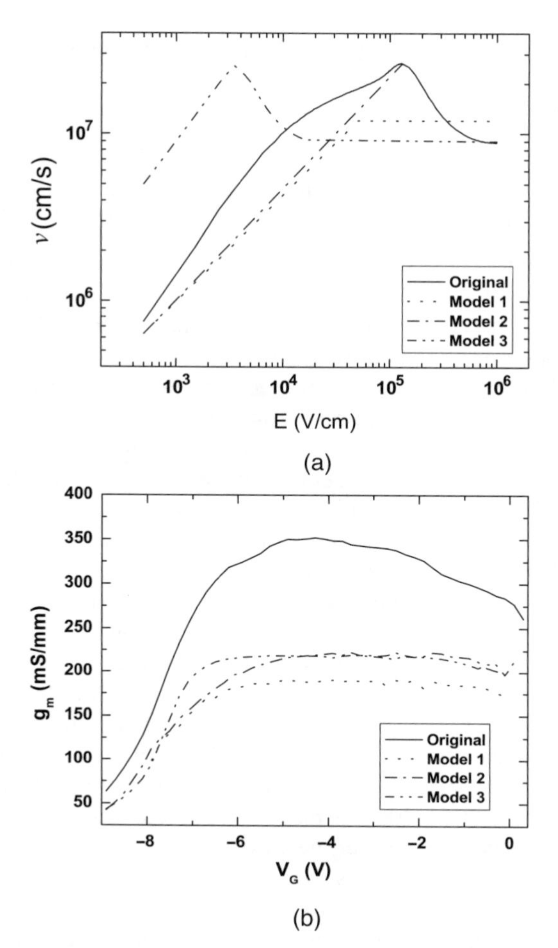

Fig. 17 (a) The actual and 3 test v–E characteristics in the low-field region. (b) Corresponding bias dependence of transconductance. [93]

Since the non-linear access resistance is an important factor affecting transconductance, it is logical to investigate the effect of reducing the magnitude of the resistance as a possible solution for improving the bias dependence of g_m. There are two possible approaches [93] to achieve that goal:

1. Making the structure more self-aligned. The access resistance arises from the source-gate and drain-gate spacings. If we reduce these spacings, the overall effect of access resistance can be reduced. However, as mentioned earlier, the reduction of channel spacing will lead to higher leakage because of the increase in channel peak electric field. Also the breakdown voltage will decrease.
2. Increasing the sheet charge density in the channel access regions. This will significantly reduce the potential drop in the access region and reduce the non-linear effects. There are several ways to increase the channel sheet charge density. As in the GaAs HEMTs, one can apply modulation doping to increase sheet charge

density. Buried doping by ion implantation to the channel is also an effective way to increase the channel sheet charge density. Using high Al concentration AlGaN layer could also increase the net polarization charge in the channel to increase the sheet charge density [31]. However, higher Al concentration might cause larger lattice mismatch and limit the layer thickness. The other way is using AlInN as the top barrier, which is shown to have higher sheet charge density [3] and reduce the lattice mismatch [52].

4.2 Scaling Issues in Nitride HEMTs

Over the last several years, AlGaN/GaN HEMTs have been demonstrated with unity current gain frequency f_T of as high as 180 GHz for 30 nm gate length [30] and f_{max} as high as 230 GHz [64] at $V_{DS} = 18$ V. An experimental picture of the scaling of f_T with gate length has emerged over the last few years [19, 30–32, 43, 63, 65]. Cut-off frequencies range from ~86 GHz for a 0.25 μm gate length [65] device to 180 GHz for a 30 nm gate length [30] device and 163 GHz for a 90 nm gate length [64] with an $In_{0.05}Ga_{0.95}N$ back-barrier. It has been suggested that due to the high optical phonon scattering rates, build-up of optical phonons causes high scattering rates and suppresses carrier velocities. However, Monte Carlo studies [79] and latest experimental reports on very high frequency devices [64, 65] suggest this may not occur. Recently, average velocities in the channel have been extracted through measurement and have reached $1.6 \sim 2.0 \times 10^7$ cm/s [62]. Furthermore, even if the low velocity is due to the build-up of optical phonons, the f_T of the device should still scale inversely with gate length, which is not observed experimentally.

Several questions need to be addressed to fully exploit AlGaN-GaN HEMT technology. These include: i) What is the source of sublinear f_T dependence on inverse gate length? ii) Is the velocity in GaN channel suppressed below values expected from Monte Carlo studies? iii) What is the highest cut-off frequency possible in AlGaN-GaN HEMTs; and iv) How do scaling issues in AlGaN-GaN devices compare with other high frequency devices?

In FETs, carrier transport plays an important role in device performance. We have noted the various regions of transport in GaN. The ohmic region determines the source resistance, the *hump* impacts the gate access resistance, and the saturation velocity impacts channel transit time. How do all these regions control scaling of GaN devices and is the scaling different from what is seen in Si MOSFETs? Si MOSFET technology benefits from a very large gap insulator SiO_2 allowing for self-aligned technology down to gate length of 10 nm. No other technology allows this possibility so far. In Fig. 19 we show scaling of gate delay as a function of gate length in Si MOSFET. We see the linear relation between gate delay and gate length. In GaN device, it has been seen that linear scaling occurs up to $L_G \sim 0.35$ μm and then the cutoff frequency starts to saturate. Theoretically was shown that this behavior is related to the difference between metallurgical gate length and actual effective gate length. For this purpose, we examine the following issues: i) the

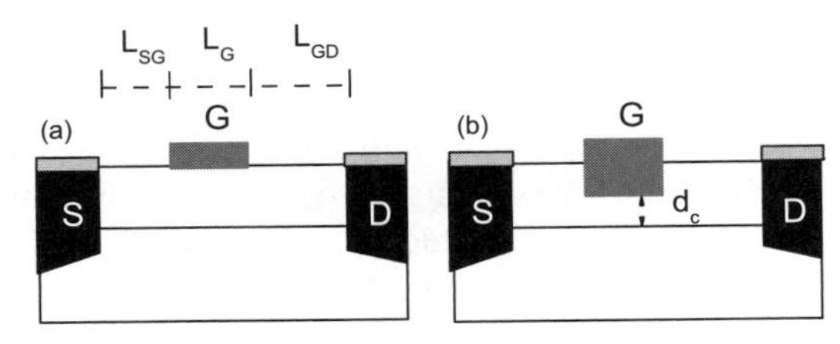

Fig. 18 The schematic cross section of AlGaN/GaN HEMT structures. (a) The device without recessed gate, and (b) the device with recessed gate.

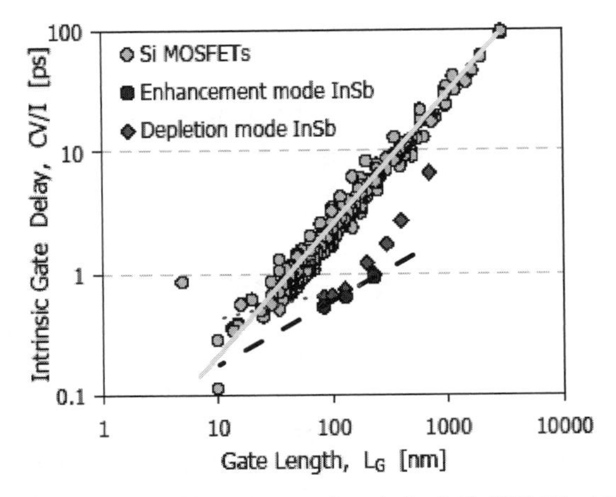

Fig. 19 Transistor Gate delay (CV/I) versus gate length for InSb HEMTs at V_{DS} = 0.5V benchmarked against state-of-the-art silicon MOSFET transistors. [15]

charge distribution in the channel and its variation with bias conditions; ii) the velocity distribution in the channel; iii) a comparison of the effective velocity in the channel using the effective gate length and with that using the lithographic gate length and resulting transit time; and iv) cutoff frequency determined by effective transit time that includes effects of depletion length on the source and drain side.

Figure 18 shows the schematic cross section of a typical nitride HEMT. The barrier layer thickness of $Al_{0.35}Ga_{0.65}N$ is 290 Å. The temperature is 300K and the L_{GS} is 0.35 μm in all cases, unless otherwise specified. The total channel length ($L_{GS}+L_G+L_{GD}$) is 1.6 μm. For very short gate lengths (<0.15μm), the gate loses control over the channel unless it is recessed. Thus, we examine the recessed gate device as well. Fig. 18(b) shows a schematic of the recessed gate case. The gate to channel distance d_c is 100 Å for our recessed gate studies.

Figure 20(a) shows the calculated charge distribution along the channel for different gate biases for a 0.1μm gate length device. The actual size and position of the real gate is labelled in Fig. 20(a). The drain bias V_{DS} is 10V. The estimated effective gate length is also marked in the figure for the gate bias $V_{GS} = -5V$. The effective gate length is determined by the length of the depletion region for different gate biases individually. The increased effective gate length σ on the source side is

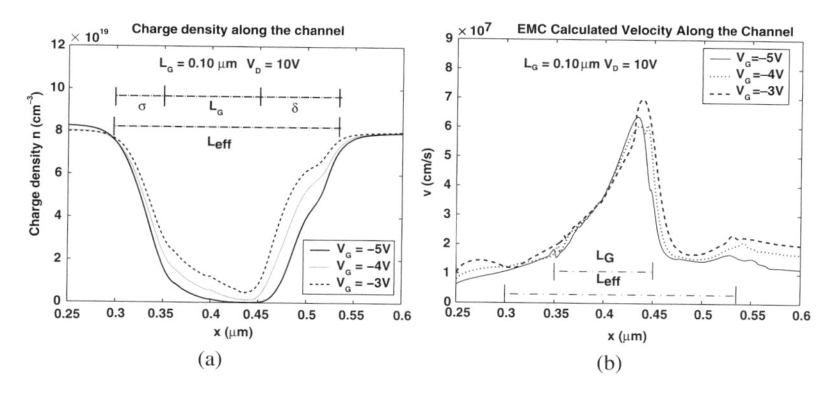

Fig. 20 (a) Charge distribution in the channel with different gate voltages. L_G is 0.1μm and V_D is 10V. The label of effective gate length in the figure is marked for V_{GS}=-5V. (b) The velocity along the channel calculated by ensemble Monte Carlo method (EMC). [96]

likely due to the fringing effect at the edges of the gate capacitor formed between the gate and the channel region. This effect can be reduced by decreasing the gate to channel distance and increasing the aspect ratio of gate length to gate to channel distance. However, for the smaller d_c (<100 Å), there might be tunneling-related gate leakage problems [92] in the system. The threshold voltage of the recessed gate will shift to ~ -1.5V for $d_c = 10$ nm and the device will need to operate at positive gate voltage to prevent the channel from becoming fully depleted. However, our studies [92] show that when the gate bias is larger than 1.0 V, the leakage current density is higher than 10^8 A/cm^2. Therefore, the range of recessed thickness is limited and modulation doping in the AlGaN layer may be needed to increase the charge density in the channel. The increased effective gate length at the drain end, δ, is much larger than the value at the source end. This is due to the additional strong horizontal electric field E_x induced by the drain bias. As the drain bias increases, the larger E_x to E_y ratio leads to larger asymmetric channel depletion length δ. For zero drain bias, the L_{eff} reduces to $L_G + 2\sigma$, which is the fundamental limiting value of L_{eff}.

Figure 21 shows a comparison of calculated maximum f_T with experimental data from [30,31,63,65]. The maximum f_T of GaN is calculated for $V_{DS} = 7$V and 10V. The simulation shows a less than 10% difference with the maximum f_T reported by experiments [30,31,63,65]. In comparing the recessed gate and non-recessed gate cases, we find that the predicted increase of f_T is less than 15%. However, a recessed gate allows us to turn the device off since for short channels ($\leq 0.10\,\mu$m), the gate loses control on the channel if a recessed structure is not employed. These results suggest that the maximum achievable f_T for nitrides would be close to 220 GHz $\pm$ 5%, with errors in estimating L_{eff} at V_{DS}=7V. It is noted that most GaN HEMTs are passivated with Si_3N_4 layer at the AlGaN surface to remove the surface traps effects. In our simulation, the effect of passivation layer such as Si_3N_4 is not considered. The passivated layer might increase the effective gate length significantly

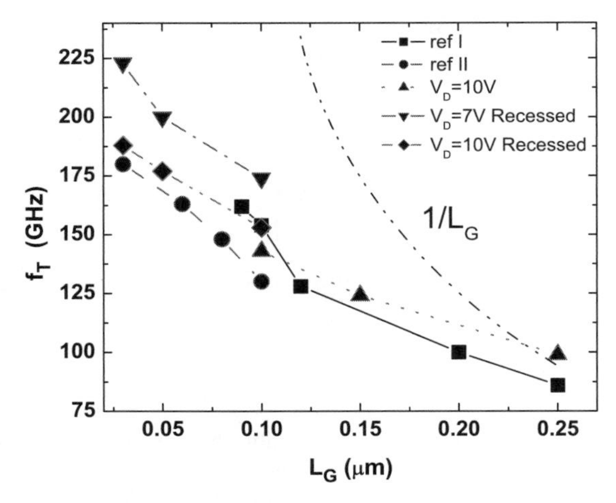

Fig. 21 Comparison of maximum f_T between simulations and experimental results. The drain bias V_{DS} is 10V for nitride HEMT. The experimental results of ref I and ref II are from different references [30, 31, 63, 65].

by increasing the drain-gate and source-gate capacitance. The device performance might be improved by the use of thin surface passivation layer [63] or with a smaller dielectric constant materials.

5 Smart HFETs: Multi-Functional Devices

As mentioned in section 1.4, there are many potential ways of incorporating polar materials into semiconductor structures to create smart HFETs. One can imagine the gate oxide of a field effect transistor (FET) to be replaced by a smart oxide leading to a "FET-sensor". It is important to note that in the near term, the oxides will have a high level of defects and therefore in the device design, the free charge in the device should be spatially separated from the polar region and should reside in the semiconductor part. This has been the approach in the Si/SiO_2 technology, where the two–dimensional electron charge (2DEG) is in the semiconductor with minimal penetration of the carrier wave function into the SiO_2 region.

We will explore the charge control of a FET where the gate oxide (or large band gap material) incorporates a material with very high polar response. The structure will be examined as a FET and as a stress or a thermal sensor. In Fig. 22, we show a generic description of these devices. The first structure represents a traditional FET where a gate voltage V_G controls the channel current I_D. A common measure of the device performance is the transconductance,

$$g_m = \frac{dI_D}{dV_G}. \tag{51}$$

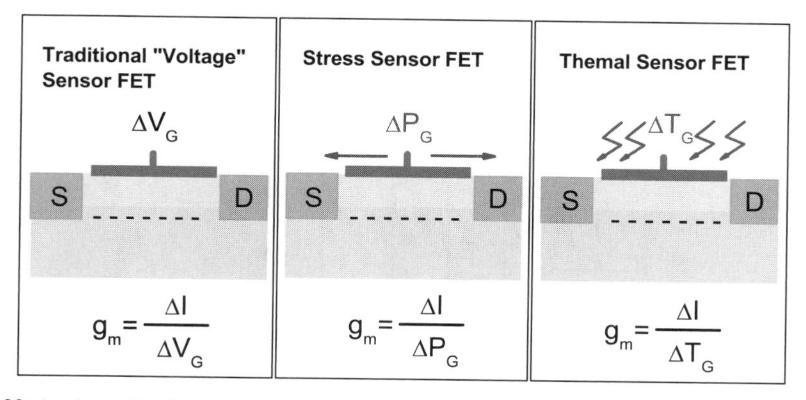

Fig. 22 A schematic of smart sensors based on oxide/semiconductor heterostructures.

In Fig. 22, we also show example devices with thin layers of polar gate oxides deposited on top of the semiconductor structure. As noted above, it is important that the region where the free carriers reside is separated from the oxide.

In the smart-FET devices shown in Fig. 22, we consider devices in which the two dimensional (2D) channel charge is modulated by stress and temperature change. An appropriate measure of the device performance is a new "transconductance" describing how the perturbation of interest controls the channel current,

$$\text{“}g_m\text{”} = \frac{dI_D}{dP}, \tag{52}$$

where dP is the appropriate perturbation such as pressure, temperature, and electric field. The synthesis of polar oxides with traditional semiconductor is in its infancy. A number of structures have been grown experimentally and have been studied structurally and electrically. The results presented in this section are to provide guidance on the potential benefit of such structures in the area of smart-FETs.

We will examine some typical materials with high piezoelectric coefficients placed above a 2D electron channel in several semiconductor systems (Si, GaAs, and GaN). Our results will focus on answering the following questions: (i) What kind of layer thickness is needed to allow a polar gate oxide to influence and control channel charge? and (ii) What kind of sensitivity is possible in such a functional device? Our calculations reveal that for a stress-sensor-FET to operate, the thickness of the polar oxide needs to be ~ 30 Å. We will also show results of a calculation for temperature-sensor-FETs exploiting pyroelectric properties.

The general structure is shown in Fig. 23. For Si based device, an oxide, SiO_2, with large bandgap would in general be needed on top of Si semiconductor as an insulator to stop the leakage current from the gate. In the cases of nitrides or arsenides, AlN and AlGaAs are grown on top of GaN and GaAs, respectively. Then, a polar oxide, such as $BaTiO_3$, is placed on top of the insulator. In Fig. 23(a), we show the key energy levels associated with the metal, the polar oxide, the insulator oxide, and the semiconductor. When the capacitor is formed, as shown in Fig 23(b), the Fermi

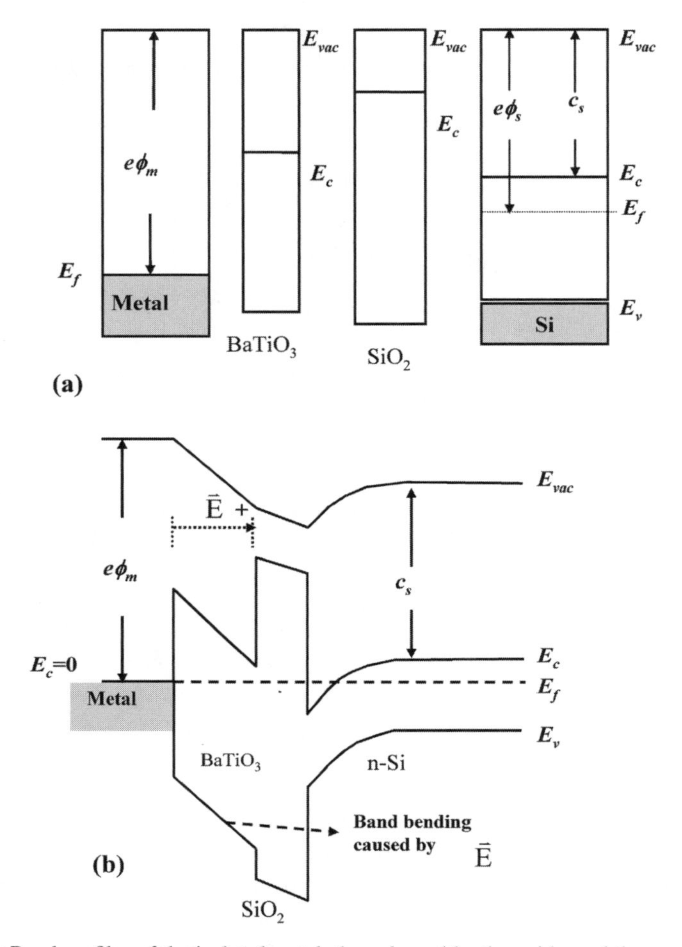

Fig. 23 (a) Band profiles of the isolated metal, the polar oxide, the oxide, and the semiconductor. ϕ_m is metal work function, χ_s is the electron affinity, E_{vac} is the vacuum energy level, E_c is the conduction energy band, E_v is valence band, and E_f is the Fermi-level. (b) Band profile of the heterostructure junctions. [93]

level is aligned so that there is no gradient. The gate bias is set to zero in the whole simulation such that we assume no externally applied electric bias. At the metal / polar material interface, the fixed negative polarization charge is balanced by the metal contact. At the polar material / insulator interface, there exists a fixed positive polarization, which will cause band bending as shown in Fig. 23(b). The degree of band bending will affect the 2DEG formed at the insulator / semiconductor or polar oxide/insulator interface. To examine the device, we need to understand how stress and temperature influence polar charge.

5.1 Stress and Strain Calculation

To use the piezoelectric equation to calculate the change in polar charge as a function of strain, we need to obtain the strain tensor in a structure shown in Fig. 24. Such a generic cantilever structure can be used as a force sensor. As shown, a "smart oxide" (in our simulation, this is $BaTiO_3$) is placed between the gate and 2D channel of a FET. We note that since oxides such as $BaTiO_3$ are likely to have high defect density and very poor mobility, we design the smart FET so that the free carrier density is essentially at the high quality Si / SiO_2 interface. The metal–piezoelectric–insulator–semiconductor FET (MPISFET) sensor is assumed to be grown on a Silicon cantilever as shown in Fig. 24(b).

When a force **F** is applied on the side of the cantilever, the deflection dz in the z direction is given by

$$dz = \frac{L^3}{3EI}\mathbf{F}, \tag{53}$$

where E is Young's modulus of the cantilever, I is the momentum of inertia, and L is the length of the cantilever. The momentum of inertia, I, of a rectangular cantilever is $hd^3/12$, where h is the cantilever width and d is the cantilever height. For the results presented here, the values of L, h, and d are set to be 200 μm, 200 μm, and 10 μm, respectively. The gate length of sensor-FET is 1 μm in this case. The cantilever under a stress given by

$$\sigma_x(x,z) = -\frac{xz}{I}\mathbf{F}, \tag{54}$$

$$\sigma_y = \sigma_z = 0, \tag{55}$$

$$\sigma_{xz}(z) = -\frac{1}{2I}\left(\frac{d^2}{4} - z^2\right)\mathbf{F}, \tag{56}$$

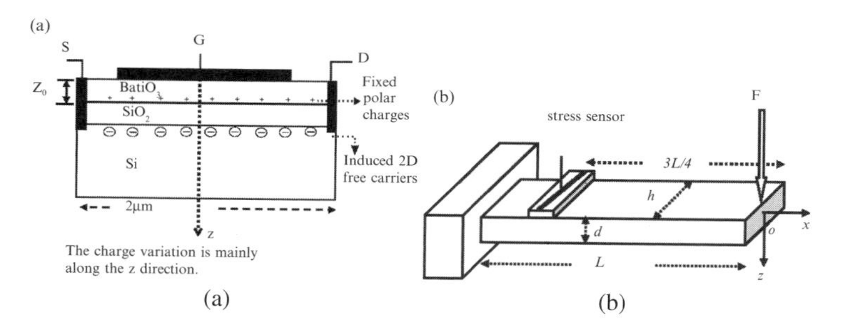

Fig. 24 (a) A schematic of the MPISFETs structure. The gate length is 1 μm. (b) The MPISFETs is assumed to be grown on top of the cantilever. L is the total length of the cantilever which is set to be 200 μm. d is the height of the cantilever, which is set to be 10 μm. O is the origin of the cartesian coordinate. The sensor FET is at $x = 3/4L$. [93]

Table 4 Material Parameters for piezoelectric materials – elastic constant and piezoelectric constant

Parameter	$BaTiO_3$ [45]	AlN	GaN [3]
C_{11} (Gpa)	211	410	370
C_{12} (Gpa)	107	140	145
C_{13} (Gpa)	114	100	110
C_{33} (Gpa)	160	390	390
e_{13} (C/m^2)	−3.88	−0.58	−0.34
e_{33} (C/m^2)	5.48	1.55	0.67
e_{15} (C/m^2)	32.6	−0.48	−0.30

where σ_x, σ_y and σ_z are the stress in the x, y, and z direction, respectively. The σ_{xz} is the shear stress in $x-y$ direction. When the sensor is grown on top of the cantilever, $z = -d/2$, and σ_{xz} would be zero for $z = -d/2$. Therefore, the shear force is not considered in this case. From the Eq. (53) and Eq. (54), the strain ϵ_x is shown below,

$$\epsilon_x(x, -d/2) = -\frac{xz}{IE}\mathbf{F} \tag{57}$$

$$= \frac{3x(d/2)}{L^3}dz, \tag{58}$$

where $z = -d/2$ since the MPISFET is grown on top of the cantilever. In addition the strain ϵ_y is equal to $-\tau\epsilon_x$, where τ is the Poisson ratio. The detailed formalism can be found in [56]. Here, we have assumed that MPISFETs are much thinner than the height of the cantilever, and therefore the strain, ϵ_x and ϵ_y, on top of cantilever is equal to the strain on the piezoelectric material. The piezoelectric constants and elastic constants of $BaTiO_3$ are listed on Table 4.

5.2 Pyroelectricity

We now consider how the fixed polar charge (and consequently the 2DEG) is influenced by the temperature changes. Polar oxides are not only good piezoelectric materials, but also have good pyroelectric properties, especially near their Curie temperatures. Pyroelectricity results from the temperature dependence of spontaneous polarization, P_s, of polar materials, whether they are single domain single crystals or poled ceramics. Therefore, the pyroelectric coefficient is given by

$$p_g(T) = p + E\frac{\partial \varepsilon}{\partial T}, \tag{59}$$

where p is the real pyroelectric coefficient and p_g is called generalized pyroelectric coefficient. As the temperature changes from T_0 to T, the spontaneous polarization of the oxide can be expressed by

$$P_{sp}(T) = P_{sp}(T_0) + p\Delta T. \quad (60)$$

Therefore, for thermal sensors, one needs to consider the changes in the spontaneous polarization when the temperature changes. Eq.(9) will be modified to $P_z = P_{sp}(T)$. Once the polarization P_z is obtained, the 2DEG induced by the polarization can be calculated.

5.3 Strain Sensor FETs: Results

To study the performance of the sensor-FETs, we first begin with Si / SiO_2 / $BaTiO_3$ heterostructure junctions. In the simulation, the SiO_2 layer thickness is fixed at 8 Å. Table 5 lists the parameters relevant for this material combination and for several other systems. Traditionally, silicon is used for mechanical sensors, because it combines well-established electronic properties with excellent mechanical properties [69]. Additionally, it would be very easy to integrate with other electronic circuits and microprocessors. Therefore, the material of the cantilever is assumed to be silicon in this case. Young's modulus and Poisson ratio for silicon are 130 Gpa and 0.28, respectively [5]. The Schottky barrier height for metal/ $BaTiO_3$ is assumed to be 1.5 eV.

The purpose of the simulation is to shed light on two issues: (i) What is the effect of strain on 2DEG with strain? (ii) What is the thickness of the $BaTiO_3$ layer for an optimal sensor-transistor? In Fig. 25, we show how the sheet charge density in the channel changes as stress increases. In Fig. 25(a), we show the 2DEG at the $BaTiO_3$ / SiO_2 junction while in Fig. 25(b), it is shown at the Si / SiO_2 junction. For both 50 Å and 30 Å $BaTiO_3$ films, most of electron free charge is at the $BaTiO_3$ / SiO_2 interface. This is not desirable as discussed earlier. Figure 26 shows the conduction band profile and charge density distribution for 30 Å and 50 Å $BaTiO_3$ layer thickness under a stress of 1×10^9 N/m^2. In Fig. 26(a), the charge density at the SiO_2/ $BaTiO_3$ interface starts to increase, but for the 50 Å $BaTiO_3$ shown in Fig. 26(b), most of the charge is accumulated at the SiO_2 / $BaTiO_3$ interface. Since

Table 5 Material Parameters used in the calculation – Bandgap, electron effective mass, heavy hole effective mass, light hole effective mass, dielectric constant, and spontaneous polarization

Parameter	$BaTiO_3$	Si	SiO_2	AlN	GaN	GaAs	AlGaAs
E_g (eV)	3.1	1.12	9.0	6.2	3.4	1.424	1.79
m_e (m_0)	1.0	0.19	0.58	0.48	0.20	0.67	0.77
m_{hh} (m_0)	1.0	0.16	0.6	0.60	0.60	0.45	0.50
m_{lh} (m_0)	1.0	0.49	0.6	0.60	0.60	0.082	0.92
ε_{33} (ε_0)	48	11.9	3.9	9.14	10.4	13.18	12.244
P_{sp} (C/m^2)	−0.25	0	0	−0.081	−0.029	0	0

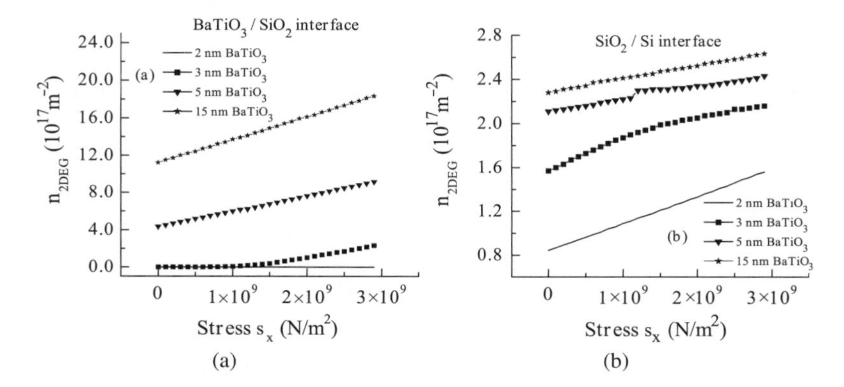

(a) (b)

Fig. 25 Calculated sheet charge density, (n_{2DEG}), at the Si / SiO_2 / $BaTiO_3$ heterostructure junctions. (a) The n_{2DEG} at the SiO_2 / $BaTiO_3$ interface; and (b) The n_{2DEG} at the Si/SiO_2 interfaces. The SiO_2 layer thickness is 8 Å. The gate voltage is zero and Schottky barrier height is taken to be 1.5 eV. [93]

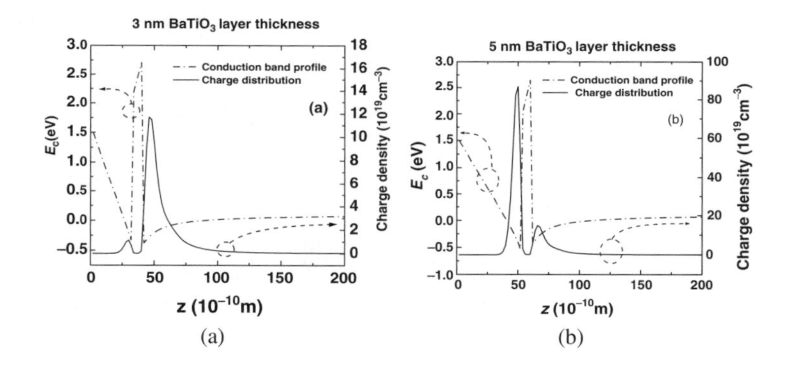

(a) (b)

Fig. 26 Calculated charge density and conduction band profiles of Si / SiO_2 / $BaTiO_3$ heterostructure junctions. (a) 30 Å $BaTiO_3$ layer thickness under stress $\sigma_x = 10^9$ N/m^2, and (b) 50 Å $BaTiO_3$ layer thickness under the same stress $\sigma_x = 10^9$ N/m^2. The SiO_2 layer thickness of is 8 Å. The gate voltage is zero and Schottky barrier height is taken to be 1.5 eV. [93]

mobile charges are expected to have very poor transport properties with deleterious trap-related problems, it is important that the electron charge should reside at the Si / SiO_2 interface. Once the $BaTiO_3$ thickness reaches 20 Å, most of the channel charge is still at the Si / SiO_2 interface and the device acts as a MOSFET but with a very large stress response.

To understand and compare the sensitivity of the stress sensor-FET and other stress sensors, we define the sensitivity of the cantilever to be

$$S = \frac{dn_{2DEG}}{n_{2DEG}} \frac{1}{dz} \tag{61}$$

$$= \frac{dn_{2DEG}}{d\sigma_x} \frac{3Exd}{2L^3 n_{2DEG}}, \tag{62}$$

For the optimal configuration, the slope of $dn/d\sigma_x$ is found to be 2.57×10^7 N^{-1}. Therefore, we can obtain the sensitivity, S, as 1.6×10^{-5} Å^{-1}. Compared to silicon piezoresistive cantilever [85], where the sensitivity, $\triangle R/Rz^{-1}$ is equal to 2.4×10^{-7} Å^{-1}, the sensor-FET has a superior potential performance. The change in channel charge shown in Fig. 25(b) will result in a change in the source-drain current if a standard FET is fabricated.

We next consider GaN / AlN / $BaTiO_3$ heterojunctions. Since the AlN is grown pseudomorphically on the GaN layer, a static strain in the AlN layer would affect the polarization [4, 102]. In this calculation, The cantilever is assumed to be GaN. The Schottky barrier height is assumed to be 1.2 eV [2]. Since the piezoelectric constants of GaN and AlN are similar, part of the strain induced fixed polar charge variation at the GaN / AlN interface is cancelled. Therefore, a structure with a thin $BaTiO_3$ layer on top would improve the sensor properties. Figure 27 shows a schematic of the structure and calculations for GaN / AlN / $BaTiO_3$ heterojunctions. Once again, a large thickness of $BaTiO_3$ layer leads to higher band bending in $BaTiO_3$ layer, and accumulation of sheet charge density at the AlN / $BaTiO_3$ interface. This would lower the mobility in FETs channel, and its consequent sensor performance. The

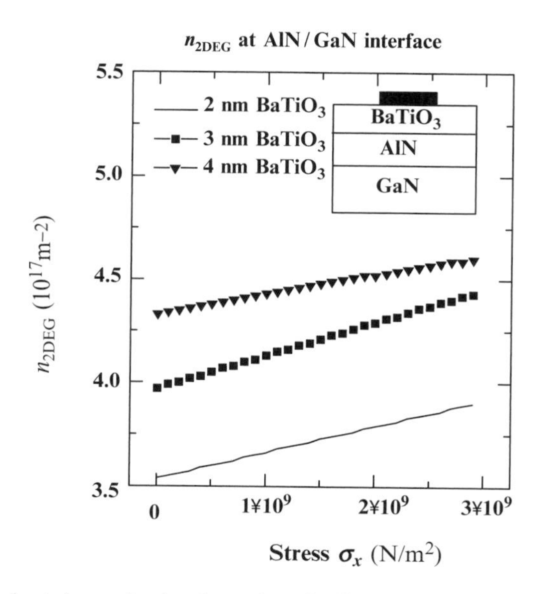

Fig. 27 Calculated sheet charge density, (n_{2DEG}), at the GaN / AlN heterostructure interface. The thickness of AlN layer is 30 Å. The gate voltage is zero and Schottky barrier height is taken to be 1.2 eV, respectively. [93]

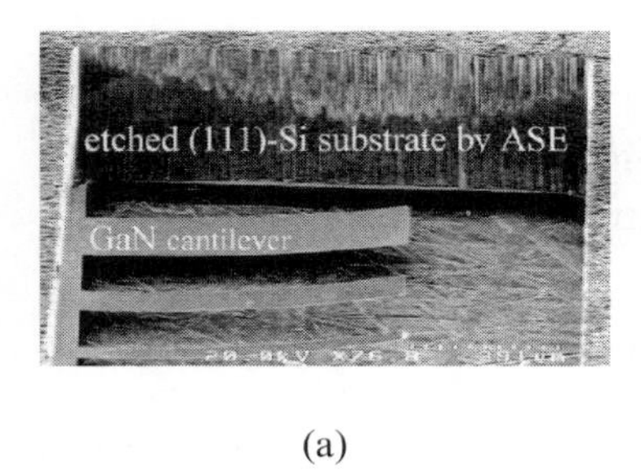

(a)

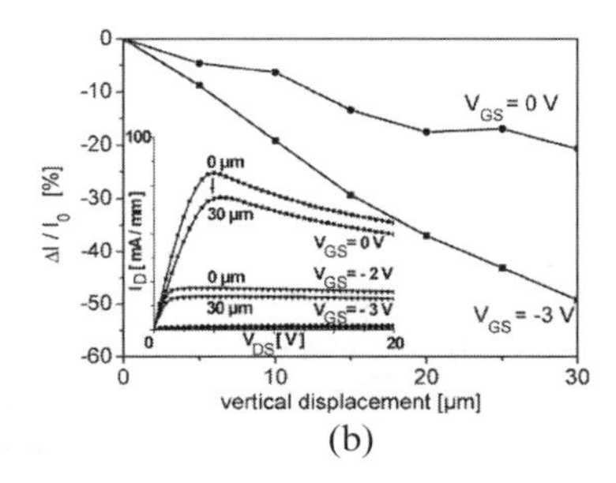

(b)

Fig. 28 (a) GaN cantilever structures fabricated by membrane etching of the Si substrate by ASE from the rear side; and (b) Relative change in HEMT output current for two gate bias points for a displacement of up to 30 μm. The inset shows the change of output characteristic by 30 μm cantilever bending. Cantilever length L = 200 μm. HEMT structure data: W = 80 μm, L_{SD} = 30 μm, L_g = 2.0 μm, I_{Dsat} = 80 mA/mm for V_{GS} = 0 V, V_{DS} = 3.0 V, and $V_{th} = -4.0$ V. [103]

optimal thickness of $BaTiO_3$ is found to be 30 Å, for which the slope $dn/d\sigma_x$ is equal to 1.608×10^7 N^{-1}. The sensitivity of the cantilever is found to be 4.95×10^{-6} $Å^{-1}$.

Due to the difficulties in fabricating cantilever based FET piezoelectric sensors, there have been few published results on these devices. However, recently Zimmermann et. al. [103] have published results on free-standing GaN and GaN / AlGaN contilevers fabricated on (111) silicon substrates. As shown in Fig. 28, 2 μm gate length GaN based HFET is placed on the pivot point of a cantilever. Changes in I_{DS} for two different gate bias points $V_{GS} = 0$ V and $V_{GS} = -3$ V are plotted in Fig.28(b). As seen, a 50% relative change in current is obtained near pinch-off voltage ($V_{GS} = -3$V). The result shows very promising sensitivity even though an additional ferroelectric material has not been added in.

5.4 Thermal Sensor FETs

We will now discuss the use of pyroelectric effect for the thermal sensor-FET. We choose the Si / SiO_2 / $BaTiO_3$ heterostructure. The pyroelectric coefficient of $BaTiO_3$ at room temperature, T (300K), has been reported to be -243μ C/m^{2}K [23]. This value is relatively small in comparison to BST ($\sim$7000μ C/m^{2}K) or PST($\sim$3500μ C/m^{2}K) [57]. The pyroelectric coefficient and dielectric constant change as the temperature increases. For temperatures near the Curie temperature, ($T_c\sim 130°C$), the pyroelectric coefficient has a peak and the spontaneous polarization becomes zero at Curie temperature. To consider contributions of variation in both the pyroelectric coefficient and dielectric constant variation, we directly fit the spontaneous polarization published in [53] as a function of temperature. The fitted pyroelectric coefficient for large scale temperature variation is ~ -700 μ C/m^{2}K.

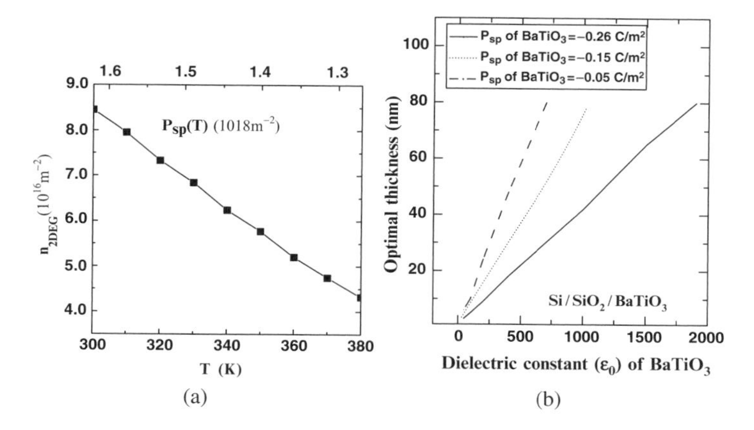

Fig. 29 (a) Calculated sheet charge density, (n_{2DEG}), versus temperature for a Si / SiO_2 / $BaTiO_3$ heterostructure junction. The thickness of $BaTiO_3$ layer is 20Å and the thickness of SiO_2 layer is 8Å; (b)Calculated optimal $BaTiO_3$ layer thickness of Si / SiO_2 / $BaTiO_3$ heterostructure sensor-FETs for different dielectric constants and spontaneous polarization. The thickness of SiO_2 layer is 8 Å. The gate voltage is zero and Schottky barrier height is taken to be 1.5 eV. As discussed in the text, at the optimal thickness the 2DEG is located at the SiO_2 / Si interface. [93]

As shown in Fig. 29(a), temperature changes are predicted to reduce the spontaneous polarization, thereby decreasing the sheet charge density at the Si / SiO_2 interface. It is found that the sheet charge density decreases from 8.45×10^{12} to 4.32×10^{12} (cm^{-2}), for temperature change of 80K, revealing a high temperature sensitivity of the device. To compare the performance to other thermal sensors, the sensitivity of the thermal sensor-FET is defined as,

$$S = \frac{dn_{2DEG}}{n_{2DEG}} \frac{1}{dT} = \frac{dn_{2DEG}}{dT} \frac{1}{n_{2DEG}}. \tag{63}$$

Using the results shown in Fig.29(a), we can obtain $dn_{2DEG}/dT = 5.24 \times 10^{14}$ (m^{-2}K^{-1}), and sensitivity, S is found to be 6.2×10^{-3} (K^{-1}). P-N diode thermal sensors typically have a variation in voltage of 2 mV/K. Thus the pyroelectric-FET should have better performance. To enhance the performance, larger pyroelectric coefficient materials such as BST and PST would be better choices. Otherwise, operating the FET thermal sensor near Curie temperature would be another choice. An issue that needs to be kept in mind is that pyroelectric materials with very high dielectric constants would reduce the band bending. Therefore, a larger layer thickness would be needed to obtain the same performance.

The parameters for oxides used in the calculations are assumed to be those for perfect bulk materials. However, in reality, the properties of thin film piezoelectric or pyroelectric material may have differences from the bulk crystal. Several theoretical and experimental reports [36, 81] indicate that the spontaneous polarization of $BaTiO_3$ would decrease when the film thickness is smaller than 100 Å. There are reports of a critical thickness, 24 Å for $BaTiO_3$ thin layer to keep its spontaneous

Table 6 Material Parameters for related piezoelectric materials

Parameter	$BaTiO_3$ [45]	$LiNbO_3$ [44]	$LiTaO_3$ [44]	ZnO [58]
P_{sp} (C/m^2)	−0.26	−0.71	−0.50	−0.057
ε_{33} (ε_0)	48	28.6	43.7	9.9
ε_{11} (ε_0)	1600	82.9	52.6	
E_g (eV)	3.1	4.0	4.6	3.3
p (μC/m^{2}K)	238 [23]		230 [57]	
e_{13} (C/m^2)	−3.88	0.332	−0.143	−0.534
e_{33} (C/m^2)	5.48	1.896	1.804	1.200
e_{15} (C/m^2)	32.6	3.631	2.609	−0.48 [25]
e_{22} (C/m^2)		2.394	1.818	

polarization. The spontaneous polarization decreases to -0.05 C/m^2 at the critical thickness. Also, some studies of MFISFET structures suggest that much lower polarization polarization is observed due to the leakage issues and difficulties to apply large electric on ferroelectric materials because of their high dielectric constant. Most electric field is dropped in the semiconductor layer so that the ferroelectric domain is hardly switched. Therefore, many experimentally obtained parameters of polar materials need to be determined. To examine the influences of parameter changes, we also examine the results for cases where the dielectric constant, ε_{33}, of $BaTiO_3$ varies from 48 [45] to 1900 [56] and spontaneous polarization P_{sp} varies from -0.26 C/m^2 to -0.05 C/m^2. Figure 29(b) shows the calculated optimal $BaTiO_3$ layer thickness versus the dielectric constant for different spontaneous polarization. When the spontaneous polarization decreases to -0.05 C/m^2, the optimum $BaTiO_3$ changes from 30 Å to 60 Å. It is also clear that optimal polar material layer thickness is roughly proportional to $1/\varepsilon$. Further more when ε is assumed to be 1900 ε_0, we obtain an optimum layer thickness of $\sim$800 Å. Thus, if ε increases or spontaneous polarization decreases, the optimum thickness of polar materials increases. The parameter study of Fig. 29 is useful for device design.

Table 6 lists the parameters of other polar materials that have potential to be used for sensor-FET applications. One must note that the use of polar oxides can also influence tunneling related to the gate current [92]. Another point to be kept in mind is that as a stress sensor, the sensitivity to temperature variation is undesirable. Therefore, choosing a high piezoelectric material with low pyroelectric coefficient would be suggested. Also, as for most thermal sensor technologies [71], one more reference calibration sensor can be designed to cancel the variation caused by thermal variation.

5.5 *Effects of Defects*

In this section, we have discussed theoretical studies which show the potential of integrating polar oxides with semiconductors. However, so far we have not addressed

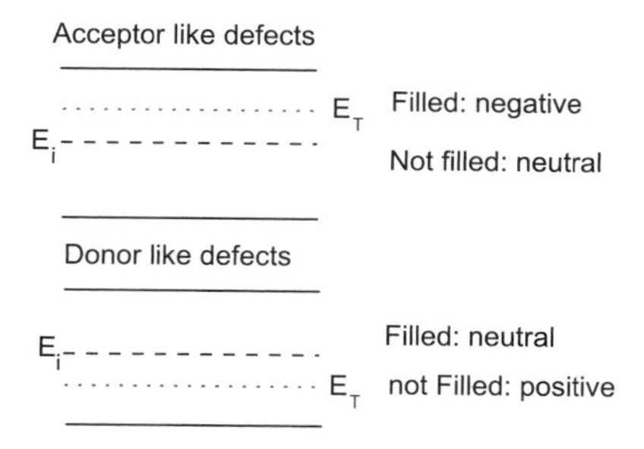

Fig. 30 The mechanism of generation and recombination of trap states for acceptor like traps.

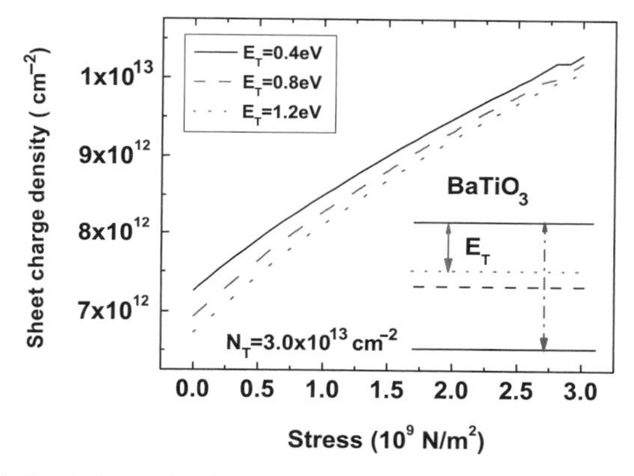

Fig. 31 Calculated sheet charge density versus stress at different defect energy levels for the $BaTiO_3$-SiO_2-Si stress sensors. N_t is equal to 3.0×10^{13} cm^{-2}.

the issues of defects. It is expected that polar oxides will have defects at the oxide-semiconductor interface. In this section, we will examine how trap states due to defects would influence devices. Trap centers, whose energy levels lie in the forbidden gap, exchange charge with the conduction and valence bands through the emission and recombination of electrons. The trap centers change the density of space charge in semiconductor bulk and influence the recombination statics. In the chapter 2.2.2, we discussed how to calculate the defect effects in the semiconductor devices. In this section, we will use this model to calculate the effects of defects.

Figure 31 shows the sheet charge density versus stress for a variety of defect densities and energy levels for the $BaTiO_3$-SiO_2-Si stress sensors. The device is operated at inversion mode where the substrate is p-type Si. As shown in the figure, the deep level defects are already filled before applying stress because Fermi level is already higher than the defect energy level. Therefore, it has little impact on the sensitivity. For the shallow level defects are empty in absence of stress. While the Fermi level starts to pin at the defect levels, most induced electrons are trapped at the defect states and affect sheet charge density in the channel. Therefore we observe the changes of sensitivity with different trap level. Defect effects become more clear in simulation results when their density rises above 10^{12} cm^{-2}. Although we do not observe a strong influence of defects at lower densities, we need to consider that this result is valid only for the steady state case.

6 Conclusions

In this chapter we have examined how polar charge can be incorporated into heterostructures through the use of spontaneous polarization and piezoelectric effect. The charge can be used to incorporate built-in fields that in turn can allow induction of free carriers in a controlled manner. The polar charge can act as dopants for electronic devices without the usual ionized impurity scattering. The field profiles created by polar charge can be used to tailor vertical junction transport leading to novel junction characteristics. We have also examined transport in the nitride structures focusing on GaN based HFET devices. In these undoped transistors it is possible to reach very high frequency performance. We have also examined how the incorporation of polar structures with traditional semiconductors can lead to sensitive smart transistors capable of sensing voltages, pressure or temperature. For these devices, experimental work needs to be done to grow high quality structures and to measure their material properties at lower thicknesses relevant for devices.

There are many issues that need to be addressed before sensor FETs become feasible. Most importantly, Most importantly, the thickness dependence of parameters like the dielectric constant and polarization need to be explored experimentally. We also need to examine the ferroelectric, piezoelectric, and pyroelectric properties for the oxide thin film. We need to make C-V measurements to determine interface properties. We also need to study to the defect states and the leakage issues by the PL and I-V measurements. Although there are some promising results for growing the oxide thin film, most films are grown on another oxides substrates. Therefore, finding a suitable material which has very good interface quality with semiconductor will be critical.

References

1. O. Aktas, Z. F. Fan, A. Botchkarev, S. N. Mohammad, M. Roth, T. Jenkins, L. Kehias, and H. Morkoç. Microwave performance of AlGaN/GaN inverted MODFET's. *IEEE Electron Device Lett.*, 18(6):293–295, June 1997.
2. O. Ambacher, B. Foutz, J. Smart, J. R. Shealy, N. G. Weimann, K. Chu, M. Murphy, A. J. Sierakowski, W. J. Schaff, L. F. Eastman, R. Dimitrov, A. Mitchell, and M. Stutzmann. Two dimensional electron gases induced by spontaneous and piezoelectric polarization in undoped and doped AlGaN/GaN heterostructures. *J. Appl. Phys.*, 87(1):334–344, January 2000.
3. O. Ambacher, J. Majewski, C. Miskys, A. Link, M. Hermann, M. Eickhoff, M. Stuzmann, F. Bernardini, V. Fiorentini, V. Tilak, B. Schaff, and L. F. Eastman. Pyroelectric properties of Al(In)GaN/GaN hetero- and quantum well structures. *J. Phys.: Condens. Matter*, 14:3399–3434, 2002.

4. O. Ambacher, J. Smart, J. R. Shealy, N. G. Weimann, K. Chu, M. Murohy, W. J. Schaff, L. F. Eastman, R. Dimitrov, L. Wittmer, M. Stutzmann, W. Rieger, and J. Hilsenbeck. Two-dimensional electron gases induced by spontaneous and piezoelectric polarization charges in N- and Ga-face AlGaN/GaN heterostructures. *J. Appl. Phys.*, 85(6):3222–3233, March 1999.
5. E. Anastassakis and M. Siakavellas. Elastic properties of textured diamond and silicon. *J. Appl. Phys.*, 90(1):144–152, July 2001.
6. L. Ardaravicius, A. Matulionis, J. Liberis, O. Kiprijanovic, M. Ramonas, L. F. Eastman, J. R. Shealy, and A. Vertiatchikh. Electron drift velocity in AlGaN/GaN channel at high electric fields. *Appl. Phys. Lett.*, 83(19):4038–4040, 2003.
7. K. Balachander, S. Arulkumaran, T. Egawa, Y. Sano, and K. Baskar. Demonstration of AlGaN/GaN metal-oxide-semiconductor high-electron-mobility transistors with silicon-oxy-nitride as the gate insulator. *Materials Sci. Engg. B*, 119(1):36–40, May 2005.
8. K. Balachander, S. Arulkumaran, H. Ishikawa, K. Baskar, and T. Egawal. Studies on electron beam evaporated ZrO_2/AlGaN/GaN metal-oxide-semiconductor high-electron-mobility transistors. *Physica Status Solidi A - Applied Research*, 202(2):R16–R18, January 2005.
9. J. M. Barker, D. K. Ferry, S. M. Goodnick, D. D. Koleske, A. Allerman, and R. J. Shul. High field transport in GaN/AlGaN heterostructures. *J. Vac. Sci. Technol. B*, 22(4):2045–2050, 2004.
10. R. G. Beck, M. A. Eriksson, M. A. Topinka, and R. M. Westervelt. GaAs/AlGaAs self-sensing cantilevers for low temperature scanning probe microscopy. *Appl. Phys. Lett.*, 73(8):1149–1151, 1998.
11. Fabio Bernardini, Vicenzo Fiorentini, and David Vanderbilt. Spontaneous polarization and piezoelectric constant of III-V nitrides. *Phys. Rev. B*, 56(16):10024–10027, October 1997.
12. J. Bernat, D. Gregusova, G. Heidelberger, A. Fox, M. Marso, H. Luth, and P. Kordos. SiO2/AlGaN/GaN MOSHFET with 0.7 μn gate-length and f_{max}/f_T of 40/24 GHz. *Electronics Lett.*, 41(11):667–668, May 2005.
13. Y.Z. Chiou, S.J. Chang, Y.K. Su, C.K. Wang, T.K. Lin, and Bohr-Ran Huang. Photo-CVD SiO_2 layers on AlGaN and AlGaN-GaN MOSHFET. *IEEE Transactions on Electron Devices*, 50:1748–1752, 2003.
14. S. L. Chuang and C. S. Chang. k·p method for strained wurtzite semiconductors. *Physics Review B*, 54:2491–2504, 1996.
15. S. Datta, T. Ashley, J. Brask, L. Buckle, M. Doczy, M. Emeny, D. Hayes, K. Hilton, R. Jefferies, T. Martin, T. J. Phillips, D. Wallis, P. Wilding, and R. Chau. 85nm Gate Length Enhancement and Depletion mode InSb Quantum Well Transistors for Ultra High Speed and Very Low Power Digital Logic Applications. *IEDM*, page Sec 32.1, 2005.
16. R. Dimitrov, L. Wittmer, H. Felsl, A. Mitchell, O. Ambacher, and M. Stutzmann. Carrier Confinement in AlGaN/GaN Heterostructures Grown by Plasma Induced Molecular Beam Epitaxy. *Phys. Status Solidi A*, 168:R7–R8, 1998.
17. R. Droopad, Z. Yu, J. Ramdani, L. Hilt, J. Copturless, C. Overgaard, J. L. Edwards, J. Finder, K. Eisenbeiser, and W. Ooms. Development of high dielectric constant epitaxial oxides on silicon by molecular beam epitaxy. *Mater. Sci. Engineering B*, 87(3):292–296, December 2001.
18. Ravi Droopad, Zhiyi Yu, Jamal Ramdani, Lyndee Hilt, Jay Copturless, Corey Overgaard, John L. Edwards, Jeff Finder, Kurt Eisenbeiser, Jun Wang, V. Kaushik, B. Y. Ngyuen, and Bill Ooms. Epitaxial oxides on silicon grown by molecular beam epitaxy. *J. Cryst. Growth*, 227-228:936–947, July 2001.
19. L. F. Eastman, V. Tilak, J. Smart, B. M. Green, E. M. Chumbes, R. Dimitrov, Hyungtak Kim, O. S. Ambacher, N. Weimann, T. Prunty, M. Murphy, W. J. Schaff, and J. R. Shealy. Undoped AlGaN/GaN HEMTs for microwave power amplification. *IEEE Trans. Electron Devices*, 48(3):479–485, Mar. 2001.
20. A. K. Fung, L. Cong, J. D. Albrecht, M. I. Nathan, and P. P. Ruden. Linear in-plane uniaxial stress effects on the device characteristics of AlGaAs/GaAs modulation doped field effect transistors. *J. Appl. Phys.*, 81:502–505, 1997.

21. Sandip Ghosh, P. Waltereit, O. Brandt, H. T. Grahn, and K. J. Ploog. Electronic band structure of wurtzite gan under biaxial strain in the m plane investigated with photoreflectance spectroscopy. *Physical Review B*, 65:075202, 2002.
22. Venkatraman Gopalan and Mool C. Gupta. Observation of internal field in $LiTaO_3$ single crystals: Its origin and time-temperature dependence. *Appl. Phys. Lett.*, 68(7):888–890, February 1996.
23. A. Govindan, A. K. Tripathi, and T. C. and Goel. Pyroelectric and piezoelectric studies on $BaTiO_3$: silica glass composites. In *Proceedings., 7th International Symposium on 25-27 Sept. 1991*, pages 524–529, September 1991.
24. Paul R. Gray and Robert G. Meyer. *Analysis and Design of Analog Integrated Circuits*. John Wiley & Sons, 1993.
25. J. G. Gualtieri, J. A. Kosinski, and A. Ballato. Piezoelectric materials for acoustic wave applications. *IEEE Trans. Ultrason., Ferroelect. Freq. Contr.*, 41(1):53–59, January 1994.
26. R. N. Hall. Electron-hole recombination in Germanium. *Phys. Rev.*, 87:387, 1952.
27. P. J. Hansen, L. Shen, Y. Wu, A. Stonas, Y. Terao, S. Heikman, D. Buttari, T. R. Taylor, S. P. DenBaars, U. K. Mishra, R. A. York, and J. S. Speck. AlGaN/GaN metal-oxlde-semiconductor heterostructure field-effect transistors using barium strontium titanate. *Journal of Vacuum Science & Technology B*, 22(5):2479–2485, 2004.
28. Peter Hansen, Y. Wu, L. Shen, S. Heikman, S. P. Denbaars, R. A. York, U. K. Mishra, and J. S. Speck. Oxide / *GaN* Integration. In *Center for Advanced Nitride Electronics CANE / MURI REVIEW*. University of California, Santa Barbara, May 2003.
29. D. C. Herbert, P. A. Childs, R. A. Abram, G. C. Crow, and M. Walmsley. Monte carlo simulations of high-speed insb-inalsb fets. *Electron Devices, IEEE Transactions on*, 52(6):1072–1078, 2005.
30. M. Higashiwaki, T. Matsui, and T. Mimura. 30-nm-gate AlGaN/GaN MIS-HFETs with 180 GHz f_T. In *2006 Device Research conference*, 2006.
31. M. Higashiwaki, T. Matsui, and T. Mimura. AlGaN/GaN MIS-HFETs With f_T of 163 GHz Using Cat-CVD SiN Gate-Insulating and Passivation Layers. *IEEE Electron Device Lett.*, 27(1):16–18, January 2006.
32. Masataka Higashiwaki and Toshiaki Matsui. AlGaN/GaN Heterostructure Field-Effect Transistors with Current Gain Cut-off Frequency of 152 GHz on Sapphire Substrates. *Jpn. J. Appl. Phys.*, 44(16):475–478, April 2005.
33. Carlo Jacobini and Lino Reggiani. The Monte Carlo method for the solution of charge transport in semiconductors with applications to covalent materials. *Rev. Mod. Phys.*, 55(3):645–705, 1983.
34. B. Jogai. Free electron distribution in AlGaN/GaN heterojunction field-effect transistors. *J. Appl. Phys.*, 91(6):3721–3729, March 2002.
35. J. R. Juang, Tsai Y. Huang, Tse M. Chen, Ming G. Lin, Gil H. Kim, Y. Lee, C. T. Liang, D. R. Hang, Y. F. Chen, and Jen I. Chyi. Transport in a gated $Al_{0.18}Ga_{0.82}N$/GaN electron system. *J. Appl. Phys.*, 94(5):3181–3184, 2003.
36. Javier Junquera and Phulippe Ghosez. Critical thickness for ferroelectricity in perovskite ultrathin films. *Nature*, 422:506–509, April 2003.
37. D. Kikuta, R. Takaki, J. Matsuda, M. Okada, X. Wei, J. P. Ao, and Y. Ohno. Gate leakage reduction mechanism of AlGaN/GaN MIS-HFETs. *Jpn. J. Appl. Phys. Part 1*, 44(4B):2479–2482, April 2005.
38. Sungwon Kim, Venkatraman Gopalan, and Alexei Gruverman. Coercive fields in ferroelectrics: A case study in lithium niobate and lithium tantalate. *Appl. Phys. Lett.*, 80(15):2740–2742, April 2002.
39. Mitsuteru Kimura and Katsuhisa Toshima. Thermistor-like pn junction temperature-sensor with variable sensitivity and its combination with a micro-air-bridge heater. *Sensors and Actuators A: Physical*, 108:239–243, 2003.
40. R. Y. Korotkov, J. M. Gregie, and B. W. Wessels. Codoping of wide gap epitaxial III-Nitride semiconductors. *Opto-Elec. Rev.*, 10(4):243–249, 2002.
41. H. Kosina and S. Selberherr. A hybrid device simulator that combines Monte Carlo and drift-diffusion analysis. *IEEE Trans. Computer-Aided Design*, 13(2):201–210, Feb. 1994.

42. Peter Kozodoy, Huili Xing, Steven P. Denbaars, Umesh K. Mishra, A. Saxler, R. Perrin, S. Elhamri, and W. C. Mitchel. Heavy doping effects in Mg-doped GaN. *J. Appl. Phys.*, 87(4):1832–1835, February 2000.
43. V. Kumar, W. Lu, R. Schwindt, A. Kuliev, G. Simin, J. Yang, Asif M. Khan, and I. Adesida. AlGaN/GaN HEMTs on SiC with f_T of over 120 GHz. *IEEE Electron Device Lett.*, 23(8):455–457, August 2002.
44. J. Kushibiki, I. Takanaga, S. Komatsuzaki, and T. Ujiie. Chemical composition dependences of the acoustical physical contants of $LiNbO_3$ single crystal. *J. Appl. Phys.*, 91(10):6341–6349, May 2002.
45. Z. Li, S. K. Chan, M. H. Grimsditch, and E. S. Zouboulis. The elastic and electromechanical properties of tetragonal $BaTiO_3$ single crystals. *J. Appl. Phys.*, 70(12):7327–7332, December 1991.
46. N. Maeda, T. Makimura, T. Maruyama, C. X. Wang, M. Hiroki, H. Yokoyama, T. Makimoto, T. Kobayashi, and T. Enoki. DC and RF characteristics in Al_2O_3/Si_3NA_4 insulated-gate AlGaN/GaN heterostructure field-effect transistors. *Jpn. J. Appl. Phys. Part 2*, 44(20-23):L646–L648, 2005.
47. N. Maeda, T. Makimura, C. X. Wang, M. Hiroki, T. Makimoto, T. Kobayashi, and T. Enoki. Al_2O_3/Si_3N_4 insulated gate channel-doped AlGaN/GaN heterostructure field-effect transistors with regrown ohmic structure: Low gate leakage current with high transconductance. *Jpn. J. Appl. Phys. Part 1*, 44(4B):2747–2750, April 2005.
48. N. Maeda, C. X. Wang, T. Enoki, T. Makimoto, and T. Tawara. High drain current density and reduced gate leakage current in channel-doped AlGaN/GaN heterostructure field-effect transistors with Al_2O_3/Si_3N_4 gate insulator. *Appl. Phys. Lett.*, 87(7):073504, August 2005.
49. Narihiko Maeda, Tadashi Saitoh, Kotaro Tsubaki, Toshio Nishida, and Naoki Kobayashi. Enhanced effect of polarization on electron transport properties in AlGaN/GaN double-heterostructure field-effect transistors. *Appl. Phys. Lett.*, 76(21):3118–3120, May 2000.
50. A. Matulionis, J. Liberis, L. Ardaravicius, L. F. Eastman, J. R. Shealy, and A. Vertiatchikh. Hot-phonon lifetime in AlGaN/GaN at a high lattice temperature. *Semicond. Sci. Technol.*, 19(4):S421–S423, 2004.
51. A. Matulionis, J. Liberis, L. Ardaravicius, J. Smart, D. Pavlidis, S. Hubbard, and L. F. Eastman. Hot-phonon limited electron energy relaxation in aln/gan. *Int. J. High Speed Electron. Systems*, 12(2):459–468, December 2002.
52. F. Medjdoub, J.-F. Carlin, M. Gonschorek, E. Feltin, M.A. Py, D. Ducatteau, C. Gaquire, N. Grandjean, and E. Kohn. Can InAlN/GaN be an alternative to high power / high temperature AlGaN/GaN devices? *Electron Devices Meeting, 2006. IEDM Technical Digest. IEEE International Dec. 11-13*, page 35.7, 2006.
53. W. J. Merz. Double hysteresis of $BaTiO_3$ at the Curie point. *Phys. Rev.*, 91(3):513–517, August 1953.
54. U. K. Mishra, Yi F. Wu, B. P. Keller, S. Keller, and S. P. Denbaars. GaN microwave electronics. *IEEE Trans. Microwave Theory Tech.*, 46(6):756–761, June 1998.
55. A. J. Moulson and J. M. Herbert. *Eletroceramics*. Wiley, England, 2 edition, 2003.
56. A. J. Moulson and J. M. Herbert. *Eletroceramics*, chapter 6, pages 381–402. Wiley, England, 2 edition, 2003.
57. A. J. Moulson and J. M. Herbert. *Eletroceramics*, chapter 7. Wiley, England, 2 edition, 2003.
58. Y. Noel, C. M. Zicovich-Wilson, B. Civalleri, Ph. DArco, and R. Dovesi. Polarization properties of ZnO and BeO: An ab *initio study* through the Berry phase and Wannier functions approaches. *Phys. Rev. B*, 65:014111, December 2002.
59. M. Ochiai, M. Akita, Y. Ohno, S. Kishimoto, K. Maezawa, and T. Mizutani. AlGaN/GaN heterostructure metal-insulator-semiconductor high-electron-mobility transistors with Si_3N_4 gate insulator. *Jpn. J. Appl. Phys. Part 1*, 42(4B):2278–2280, April 2003.
60. Hirotsugu Ogi, Yasunori Kawasaki, and Masahiko Hirao. Acoustic spectroscopy of lithium niobate: Elastic and piezoelectric coefficients. *J. Appl. Phys.*, 92(5):2451–2456, September 2002.

61. C. H. Oxley and M. J. Uren. Measurements of unity gain cutoff frequency and saturation velocity of a GaN HEMT transistor. *IEEE Trans. Electron Devices*, 52(2):165–169, February 2005.
62. T. Palacios, A. Chakraborty, S. Heikman, S. Keller, S. P. Denbaars, and U. K. Mishra. AlGaN/GaN High Electron Mobility Transistors With InGaN Back-Barriers. *IEEE Electron Device Lett.*, 27(1):13–15, January 2006.
63. T. Palacios, A. Chakraborty, S. Keller, S. P. Denbaars, and U. K. Mishra. AlGaN/GaN HEMTs with an InGaN-based back-barrier. *63rd Device Research Conference*, IEEE DRC Technical Digest, pages 181–182, 2005.
64. T. Palacios, E. Snow, Y. Pei, A. Chakraborty, S. Keller, S. P. Denbaars, and U. K. Mishra. Ge-Spacer Technology in AlGaN/GaN HEMTs for mm-Wave Applications. *IEEE IEDM Digest*, Dec. May-July 2005.
65. Tomas Palacios and U. K. Mishra. Improved technology for high frequency AlGaN/GaN HEMTs. *ONR CANE/MURI Review*, April 2005.
66. Tomas Palacios, S. Rajan, S. Heikman, S. Keller, S. P. Denbaars, and U. K. Mishra. Influence of the access resistance in the RF performance of mm-wave AlGaN/GaN HEMTs. In *62nd Device Research Conference*, pages 75–76, June 2004.
67. K. Y. Park, H. I. Cho, H. C. Choi, Y. H. Bae, C. S. Lee, J. L. Lee, and J. H. Lee. Device characteristics of AlGaN/GaN MIS-HFET using Al_2O_3-HfO_2 laminated high-k dielectric. *Jpn. J. Appl. Phys. Part 2*, 43(11A):L1433–L1435, November 2004.
68. Seoung H. Park and Shun L. Chuang. Spontaneous polarization effects in wurtzite GaN/AlGaN quantum wells and comparison with experiment. *Appl. Phys. Lett.*, 76(15):1981–1983, April 2000.
69. K. E. Peterson. Silicon as a mechnical material. *Proc. IEEE*, 70:420–457, 1992.
70. S. K. Pugh, D. J. Dugdale, S. Brand, and R. A. Abram. Electronic structure calculations on nitride semiconductors. *Semicond. Sci. Technol.*, 14:23–31, 1999.
71. R. Ramesh. *Thin Film Ferroelectric Materials and Devices*. Kluwer International Series in Electronic Materials : Science and Technology. Kluwer Academic Publishers, 1997.
72. P. Regoliosi, A. Reale, A. Dicarlo, P. Romanini, M. Peroni, C. Lanzieri, A. Angelini, M. Pirola, and G. Ghione. Experimental Validation of GaN HEMTs Thermal Management by Using Photocurrent Measurements. *Electron Devices, IEEE Transactions on*, 53(2):182–188, 2006.
73. A. E. Romanov, P. Waltereit, and J. S. Speck. Buried stressors in nitride semiconductors: Influence on electronic properties. *J. Appl. Phys.*, 97:043708, 2005.
74. W. Shockley and W. T. Read. Statistics of the recombinations of holes and electrons. *Phys. Rev.*, 87:835–842, 1952.
75. M. S. Shur and M. A. Khan. GaN/AlGaN heterostructure devices: Photodetectors and field-effect transistors. *MRS Bulletin*, 22(2), February 1997.
76. Jasprit Singh. *Physics of Semiconductors and Their Heterostructures*. McGraw-Hill, Inc., 1993.
77. Madhusudan Singh, Jasprit Singh, and Umesh Mishra. Current-voltage characteristics of polar heterostructure junctions. *J. Appl. Phys.*, 91(5):2989–2993, 2002.
78. Madhusudan Singh, Yuh R. Wu, and Jasprit Singh. Examination of $LiNbO_3$ / nitride heterostructures. *Solid-State Electron.*, 47(12):2155–2159, 2003.
79. Madhusudan Singh, Yuh R. Wu, and Jasprit Singh. Velocity overshoot effects and scaling issues in III-V nitrides. *IEEE Trans. Electron Devices*, 52(3):311–316, March 2005.
80. Madhusudan Singh, Yifei Zhang, Jasprit Singh, and Umesh Mishra. Examination of tunnel junctions in the AlGaN/GaN system : Consequences of polarization charge. *Appl. Phys. Lett.*, 77(12):1867–1869, 2000.
81. M. G. Stachiotti. Ferroelectricity in $BaTiO_3$ nanoscopic structures. *Appl. Phys. Lett.*, 84(2):251–253, January 2004.
82. T. Sugimoto, Y. Ohno, S. Kishimoto, K. Maezawa, J. Osaka, and T. Mizutani. AlGaN/GaN MIS-HEMTs with ZrO_2 gate insulator. *Compound Semiconductors 2004, Proceedings*, 184:279–282, 2005.

83. M. Sumiya and S. Fuke. Review of polarity determination and control of GaN. *MRS Internet Journal of Nitride Semiconductor Research*, 9(1), 2004.
84. S. M. Sze. *Semiconductor Devices Physics and Technology*. John Wiley & Sons, 1985.
85. J. Thaysen, A. Boisen, O. Hansen, and S. Bouwstra. Atomic force microscopy probe with piezoresistive read-out and a highly symmetrical wheatstone bridge arrangement. *Sensors and Actuators A: Physical*, 83:47–53, 2000.
86. R. Therrien, S. Singhal, J. W. Johnson, W. Nagy, R. Borges, A. Chaudhari, A. W. Hanson, A. Edwards, J. Marquart, P. Rajagopal, C. Park, I. C. Kizilyalli, and K. J. Linthicum. A 36mm GaN-on-Si HFET Producing 368W at 60V with 70Efficiency. *IEEE IEDM Digest*, Dec. May-July 2005.
87. S. Thompson, P. Packan, and M. Bohr. Mos scaling: transistor challenges for the twentyfirst century. *Intel. Technol. J.*, 2, 1998.
88. K. Ueda, M. Kasu, Y. Yamauchi, T. Makimoto, M. Schwitters, D.J. Twitchen, G.A. Scarsbrook, and S.E. Coe. Diamond FET Using High-Quality Polycrystalline Diamond With f_T of 45 GHz and f_{max} of 120 GHz. *Electron Device Letters, IEEE*, 27(7):570–572, 2006.
89. W. Walukiewicz. Intrinsic limitations to the doping of wide-gap semiconductors. *Physica B: Condensed Matter*, 302-303:123–134, June 2001.
90. J. Wu, W. Walukiewicz, K. M. Yu, J. W. Ager III, S. X. Lib, E. E. Hallerb, Hai Lud, and William J. Schaff. Universal bandgap bowing in group-III nitride alloys. *Solid State Comm.*, 127:411–414, 2003.
91. Y. F. Wu, S. Keller, P. Kozodoy, B. P. Keller, P. Parikh, D. Kapolnek, S. P. Denbaars, and U. K. Mishra. Bias dependent microwave performance of AlGaN/GaN MODFET's up to 100 V. *IEEE Electron Device Lett.*, 18(6):290–292, June 1997.
92. Yuh R. Wu, Madhusudan Singh, and Jasprit Singh. Gate leakage suppression and contact engineering in nitride heterostructures. *J. Appl. Phys.*, 94(9):5826–5831, November 2003.
93. Yuh-Renn Wu and Jasprit Singh. Polar heterostructure for multi-function devices: Theoretical studies. *IEEE Trans. Electron Devices*, 52(2):284–293, February 2005.
94. Yuh-Renn Wu, Madhusudan Singh, and Jasprit Singh. Gate leakage suppression and contact engineering in nitride heterostructures. *Mat. Res. Soc.*, 798:Y11.1, 2004.
95. Yuh-Renn Wu, Madhusudan Singh, and Jasprit Singh. Sources of transconductance collapse in III-V nitrides - Consequences of velocity-field relations and source-gate design. *IEEE Trans. Electron Devices*, 52(6):1048–1054, June 2005.
96. Yuh Renn Wu, Madhusudan Singh, and Jasprit Singh. Device Scaling Physics and Channel Velocities in AlGaN-GaN HFETs: Velocities and Effective Gate Length. *IEEE Trans. Electron Devices*, 53(4):588–593, April 2006.
97. Kiyoyuki Yokoyama and Karl Hess. Monte Carlo study of electronic transport in $Al_{1-x}Ga_xAs/GaAs$ single-well heterostructures. *Phys. Rev. B*, 33(8):5595–5606, 1986.
98. S. B. Zhang, S. H. Wei, and Alex Zunger. Microscopic origin of the phenomenological equilibrium "doping limit rule" in n-type III-V semiconductors. *Phys. Rev. Lett.*, 84(6):1232–1235, February 2000.
99. Y. Zhang and J. Singh. Charge control and mobility studies for an AlGaN/GaN high electron mobility transistor. *J. Appl. Phys.*, 85(1):587–594, January 1999.
100. Y. Zhang and J. Singh. Charge control and mobility studies for an AlGaN/GaN high electron mobility transistor. *J. Appl. Phys.*, 85(1):587–594, 1999.
101. Yifei Zhang and Jasprit Singh. Monte Carlo studies of two dimensional transport in GaN/AlGaN transistors : Comparison with transport in AlGaAs/GaAs channels. *J. Appl. Phys.*, 89(1):386–389, 2001.
102. Yifei Zhang, I. P. Smorchkova, C. R. Elsass, Stacia Keller, James P. Ibbetson, Steven Denbaars, Umesh K. Mishra, and Jasprit Singh. Charge control and mobility in AlGaN/GaN transistors: Experimental and theoretical studies. *J. Appl. Phys.*, 87(11):7981–7987, June 2000.
103. T. Zimmermann, M. Neuburger, P. Benkart, F.J. Hernandez-Guillen, C. Pietzka, M. Kunze, I. Daumiller, A. Dadgar, A. Krost, and E. Kohn. Piezoelectric GaN sensor structures. *IEEE Electron Device Letters*, 27:309–312, 2006.

Polarization Effects on Low-Field Transport & Mobility in III-V Nitride HEMTs

Debdeep Jena

1 Introduction

III-V Nitride heterostructures are very attractive for high-power RF power amplifiers, among a host of other applications. The large bandgap of GaN, in addition to a number of intrinsic material properties that include high electron saturation velocity, high breakdown fields, and high thermal conductivity make the material system especially suitable for mm-wave high-electron mobility transistors (HEMTs).

The speed of operation of HEMTs is limited by two major factors:

a) intrinsic speed limits determined by how fast electrons can traverse through the active region of the device, and
b) extrinsic factors, loosely called parasitics, which can comprise series resistances, capacitances, etc.

Figure 1 shows the structure of a typical AlGaN/GaN HEMT. A high-density 2-dimensional electron gas (2DEG) forms at the AlGaN/GaN heterojunction due to the discontinuity in electronic polarization across the heterointerface. The electrostatics, band diagrams, and other details of the HEMT have been described in detail in the chapters by Ambacher & Cimalla, and Morkoç & Leach, in this book. The 2DEG constitutes the conductive channel of the HEMT. The sheet density of the 2DEG is modulated capacitively by the gate contact, while the current flows through the channel between the source and the drain ohmic contacts. The electric field component directed along the channel $E(x)$ during typical device operation is shown in the same figure. The integrated area under the electric field curve is equal to the drain-source bias applied ($-\int_S^D E(x)dx = V_{DS}$). Due to the gate bias, the field in the channel is non-uniform, and it peaks at the drain-end of the gate. Under typical device operation, the field in this region exceeds E_{sat}, the field beyond which the electron velocity saturates. Even in this situation which is attained at high bias conditions, a major fraction of the channel has an electric field lower than the saturation field, where the electron velocity $v(x)$ is linearly related to the

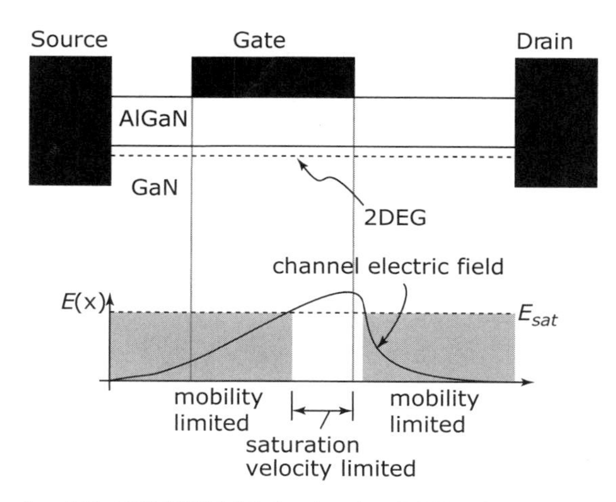

Fig. 1 Structure of an AlGaN/GaN HEMT, showing the 2DEG at the heterointerface. The electric field along the channel under typical bias conditions is sketched, demarcating the regions in the channel where electron mobility is important, and those where the saturation velocity is important.

field through 'Ohm's law' : $v(x) = \mu E(x)$, μ being the electron mobility. The section of the channel between the source and the gate (the so-called source-access region) is an unwanted series resistance which slows the device operation. The resistivity of this access region is limited by the product of the 2DEG sheet density and the mobility. Thus, the transport of electrons through a part of the HEMT channel is always mobility-limited, as indicated in the figure.

A high electron mobility, therefore, is of paramount importance for achieving high speeds. What limits the mobility of electrons in such polarization-induced 2DEGs? What can one do to achieve high mobilities? These are the questions that are answered in the rest of this chapter. It is organized in the following fashion - in section 2, AlGaN/GaN 2DEGs are studied from the point of view of charge densities and spatial distributions, and the crucial role of polarization is highlighted. Experimental electron mobilities reported in the literature are discussed.

In section 3, the various scattering mechanisms responsible for limiting electron mobility in III-V nitride HEMTs are discussed in detail. Some scattering mechanisms are not unique to the III-V nitrides, but assume increased importance in this material system - dislocation scattering is a good example. Some scattering mechanisms arise entirely due to the large polarization - dipole scattering is an example. They have no analogues in non-polar and weakly polar semiconductors. Such scattering mechanisms are also described in section 3.

In section 4, the theoretical tools developed are used to explain experimentally measured AlGaN/GaN 2DEG mobility data. Mobility-limiting scattering mechanisms are identified, and strategies to make further progress are outlined. A summary of the major concepts covered in the chapter, and conclusions are listed in section 5.

For those readers who do not have a firm grounding on the basics of electron transport theory, a detailed treatment of this topic is provided in section 6. For such readers, it is recommended that they cover section 6 right after section 2 to develop their understanding before returning to section 3, where many concepts of transport theory are used.

2 Polarization-Induced 2DEGs in AlGaN/GaN HEMTs

When a thin layer of AlGaN is epitaxially grown on Ga-face GaN, a two-dimensional electron gas (2DEG) results at the heterojunction. The need for modulation doping as in AlGaAs/GaAs heterostructures does not exist. The 2DEG carrier density can be modulated by changing the thickness of the AlGaN barrier layer as well as the aluminum composition in it. In addition, extremely high 2DEG densities can be achieved ($n_{2d} \approx 2 \times 10^{13}\text{cm}^{-2}$) with reasonably high room temperature mobility $\mu \approx 1500$ cm^2/V·s. Such properties make the AlGaN/GaN heterojunction 2DEG very attractive for field-effect transistors [12]. High Electron Mobility Transistors (HEMTs) utilizing the AlGaN/GaN structure have demonstrated record high breakdown and power performance [2, 3]. Of especial interest are the transport properties of the 2DEG - identification of the scattering processes that limit the mobility.

The transport properties of the AlGaN/GaN 2DEGs form the major part of this chapter. A simple charge control model is presented that accurately captures experimentally observed behavior. The major difference of charge control in the AlGaN/GaN 2DEG system is the absence of modulation dopants and the appearance of polarization sheet charges at the heterojunction. A suitable analytical model for the 2DEG is then chosen for use in the analysis of transport properties.

Various defects are identified in a realistic AlGaN/GaN structure. The effects of the large number of dislocations on transport properties is evaluated. The effect of polarization disorder in the AlGaN barrier on carrier transport in the 2DEG is studied by modelling it as scattering from dipoles. The traditional scattering mechanisms in 2DEGs - interface roughness scattering, alloy scattering, impurity scattering and phonon (optical and acoustic) scattering are analyzed. The theoretical results are compared with available experimental data. The chapter ends with a summary of the relative effects of various scattering processes, the intrinsic limits on low-field mobility and a brief discussion of high field effects.

2.1 Polarization Effects on Charge Transport and Scattering

At first sight, the 2DEG at the AlGaN/GaN heterojunction is similar in many aspects to 2DEGs in modulation-doped AlGaAs/GaAs heterostructures, or even in the Si MOSFET. However, the fact that the AlGaN/GaN 2DEG is induced due to polarization makes it markedly different from the others in many aspects.

Densities: The 2DEG densities for polarization-induced 2DEGs are strongly dependent on the spontaneous and piezoelectric polarization coefficients. Since the polarization coefficients are very high in the nitride semiconductors, the 2DEG densities achieved in nitride heterojunctions ($\sim 10^{13}$ cm^{-2}) are typically an order of magnitude higher than in the zinc-blende heterojunctions.

Control Parameters: The 2DEG density in polarization-induced 2DEGs is directly determined by two parameters: *composition* and *thickness* of layers. The strength of the polarization field is controlled by the composition, and the magnitude of the band-bending that results from the polarization charges is directly dependent on the thickness of layers. The importance of these two parameters have been emphasized in other chapters, and will be reinforced later in this chapter as well.

Polarization-induced modification of 2DEG wavefunctions: Due to the very high electric fields that result from the large polarization ($\sim$1 MV/cm), electrons in the 2DEG are electrostatically pushed close to the AlGaN/GaN interface, and the centroid of the wavefunction is brought closer to the heterointerface. This directly leads to an increased sensitivity to alloy disorder and interface roughness scattering, which turn out to be the dominant scattering processes at low temperatures, and even at room temperature for very high density 2DEGs.

Dipole scattering due to Polarization: A new scattering mechanism arises in highly polar semiconductors that have alloy layers (such as AlGaN, AlInN, or AlInGaN). Due to the microscopic disorder in an alloy layer, the dipole moment in each unit cell is no more periodic with the crystal lattice, and leads to what is now known as 'dipole-scattering'. This interesting novel scattering mechanism has no analogue in tradiational non-polar and weakly-polar semiconductors, and is described in detail later in this chapter.

2.2 Charge Control

The free-charge distribution of a [0001]-oriented AlGaN/GaN heterostructure can be calculated exactly from a self-consistent numerical solution of Schrödinger and Poisson equations in the effective mass approximation. However, a simpler model based on the band diagram is useful for obtaining physical insight. In figure 2 is shown a schematic band diagram and charges at the AlGaN/GaN heterojunction.

The AlGaN surface potential is pinned at a level $\Phi_S(x) = (1+x)$ eV below the AlGaN conduction band edge; here x is the aluminum composition of the alloy [4]. The fixed polarization sheet charge $\sigma_\pi(x)$ at the heterojunction is calculated using the polarization constants. The polarization coefficients $e_{ij}(x)$, the elastic coefficients $c_{ij}(x)$ and the band discontinuity $\Delta E_c(x)$ for the alloy are linearly interpolated (Vegard's law). The total (spontaneous and piezoelectric) polarization-induced sheet charge at the heterojunction at the AlGaN/GaN heterojunction is given by

$$\sigma_\pi(x) = \Delta P_{sp}(x) + 2(e_{31}(x) - e_{33}(x)\frac{c_{13}(x)}{c_{33}(x)}) \times \left(\frac{a(x) - a_{GaN}}{a_{GaN}}\right), \quad (1)$$

where ΔP_{sp} is the difference in spontaneous polarization of the barrier and GaN. The thickness of the AlGaN barrier is t_b and $a(x), a_{GaN}$ are the relaxed lattice constants of AlGaN and GaN respectively. The ground-state energy of the triangular quantum well formed at the heterojunction is given by a variational solution to be [5]

$$E_0(n_s) \approx \left(\frac{9\pi\hbar e^2 n_{2d}}{8\epsilon_0\epsilon_b(x)\sqrt{8m^\star}} \right)^{2/3}, \tag{2}$$

where $\hbar$ is the reduced Planck's constant, e is the electron charge, n_{2d} is the 2DEG sheet density, ϵ_0 is the permittivity of free space, $\epsilon_b(x)$ is the alloy-composition dependent relative dielectric constant, and $m^\star$ is the electron effective mass.

AlGaN is assumed to be coherently strained on GaN. The *mobile* 2DEG charge at the AlGaN-GaN interface can be related to the aluminum composition and the barrier thickness. From figure 2 it follows that

$$e\Phi_s(x) - E \times t_b - \Delta E_c(x) + E_0 + (E_F - E_0) = 0, \tag{3}$$

where $E = e(\sigma_\pi(x) - n_{2d})/(\epsilon_0\epsilon(x))$ is the electric field in the barrier pointing along the growth direction, and

$$E_F - E_0 = \frac{\pi\hbar^2}{m^\star} n_{2d}, \tag{4}$$

where it is assumed that only one subband of the quantum well is filled. The roots of Equation 3 yield the mobile 2DEG densities n_{2d} as a function of the alloy

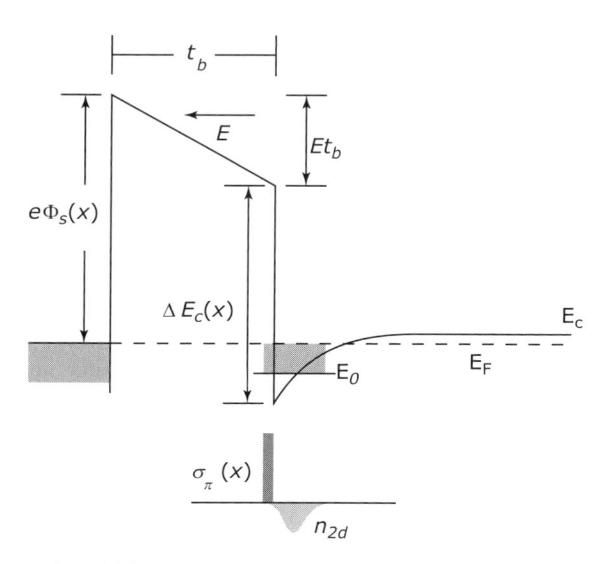

Fig. 2 Charge control model for evaluating 2DEG density at the AlGaN/GaN interface.

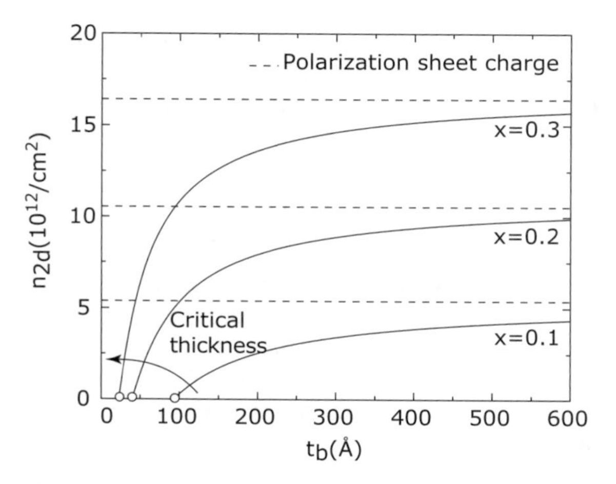

Fig. 3 Changing Aluminum composition and changing AlGaN thickness. Note that the critical thickness reduces as the Aluminum composition increases.

composition x and the barrier thickness t_b. Solutions are plotted as a function of t_b for $x = 0.1, 0.2, 0.3$ in figure 3.

The simple charge control model sets a cutoff critical thickness t_{cr} of the barrier below which a 2DEG is not formed. Such a cutoff has indeed been observed [8]. The polarization-induced mobile 2DEG density can be tuned by changing the AlGaN barrier thickness. The mobile 2DEG charge comes from donor-like surface states. As the barrier thickness is increased, the 2DEG density approaches the polarization sheet charge density $\sigma_\pi(x)$ as seen in figure 3. Normally, the strained AlGaN will relax before the 2DEG density becomes equal to $\sigma_\pi(x)$. The 2DEG carrier densities for high aluminum composition ($x \approx 0.3$) and easily achievable thicknesses ($t_b \approx 30$ nm) are extremely high ($n_{2d} > 10^{13}/\text{cm}^2$) compared to similar 2DEGs in modulation-doped AlGaAs/GaAs structure or piezoelectric-doped $[111]$-oriented zinc-blende III-V quantum wells [15]. The reason is the large polarization, and the large band-offsets in the material system.

Figure 4 shows the numerically evaluated conduction band-diagram and the free-carrier distribution for a $Al_{0.3}Ga_{0.7}N$/GaN heterojunction with a changing barrier thickness. The calculation was done using a computer program [24]. The program uses a self-consistent iterative procedure to solve Schrödinger and Poisson equations in the effective-mass approximation. The polarization sheet charge is modelled by an extremely thin ($t = 0.1$ nm) junction layer appropriately doped ($N_D = \sigma_\pi(x)/t$) to mimic polarization charge. The region around the heterojunction is more finely meshed than the bulk material for ensuring good convergence of the self-consistent routine. Also calculated in this manner are the two lowest subband energy eigenvalues (E_0, E_1) for different barrier thicknesses. The subband edges are shown as small ticks in the band diagram. The first subband of the quantum well at the heterojunction appears at the critical thickness for 2DEG formation. The state grows deeper ($(E_F - E_0)$ increases) allowing more 2DEG carriers as the barrier thickness

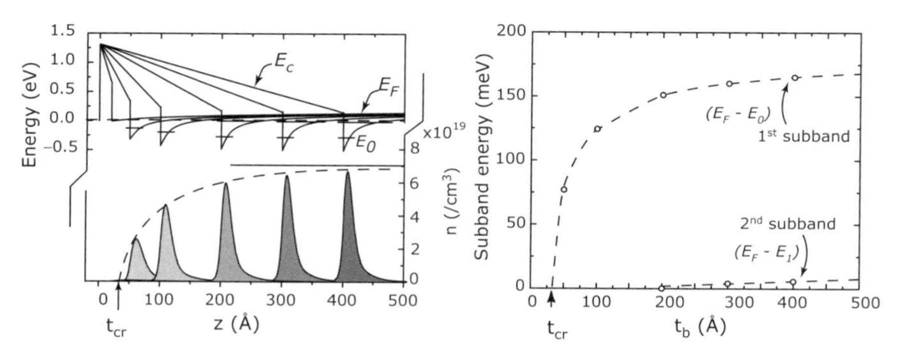

Fig. 4 A self-consistently solved band-diagram with only the conduction band shown in the picture.

increases and a second subband appears at a barrier thickness of $t_b = 20$ nm. The second subband is extremely shallow and is neglected for the charge control analysis.

For accurate evaluation of transport properties and scattering rates, the finite extent of the 2DEG along the z direction must be accounted for. The exact form of the wavefunction from the self-consistent Schrödinger - Poisson solution is very useful in determining the 2DEG sheet density and the shape of the wavefunction.

However, for analytic evaluation of scattering rates, the Fang-Howard variational wavefunction is a better candidate, and has been used successfully for transport calculations in the past [5]. The form of the wavefunction is

$$\chi(z) = 0, z < 0$$

$$\chi(z) = \sqrt{\frac{b^3}{2}} z e^{-\frac{bz}{2}}, z \geq 0. \tag{5}$$

where b is a variational parameter (see section 6 in this chapter). The variational carrier density $\rho(z) = en_s|\chi(z)|^2$ and the numerical Hartree solution are plotted for a (30 nm) 30% AlGaN/GaN heterostructure in figure 5 for comparison. The variational wavefunction does not take into account the wavefunction penetration into the alloy barrier - this is the price paid in the process of obtaining analytical results for scattering rates. However, the *shape* of the variational wavefunction is accurate, though there is a rigid shift away from the heterojunction as compared to the Hartree wavefunction. This property is not important in many scattering processes and as and when it is, it will be pointed out.

The total carrier sheet-density for the 2DEG with single subband occupation can be written as (see section 6)

$$n_s = n_{2d} = \frac{m^\star k_B T}{\pi \hbar^2 e} \ln(1 + e^\zeta), \tag{6}$$

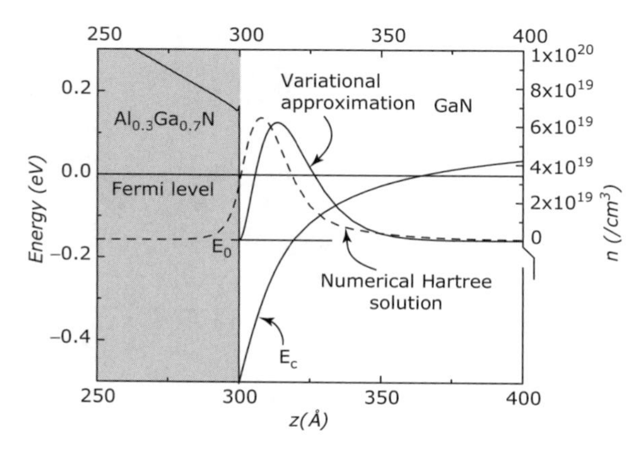

Fig. 5 Figure illustrating the difference between the exact Hartree-Fock wavefunction of the 2DEG and the variational (Fang-Howard) approximation. Note that the variational approximation loses information of the wavefunction penetration into the barrier.

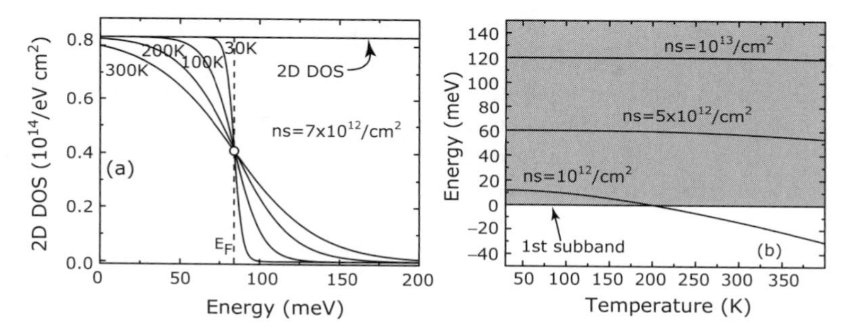

Fig. 6 Charge statistics and Fermi level movement with temperature for a 2DEG.

where $\zeta = (\varepsilon_F - \varepsilon_0)/k_B T$, and ε_0 is the first subband energy. With the knowledge of the carrier density and the effective mass of the carriers the Fermi level variation with temperature is given by

$$\varepsilon_F = k_B T \ln(e^{\frac{\pi\hbar^2 n_{2d}}{m^\star k_B T}} - 1). \tag{7}$$

Figure 6(a) shows the 2DEG density of states and the energy-occupation of carriers for different temperatures for $n_s = 7 \times 10^{12}$/cm^2. Figure 6(b) shows the movement of the Fermi level (in meV) with temperature for three values of the AlGaN/GaN 2DEG sheet-densities $n_s = 1, 5, \& 10 \times 10^{12}$ cm^{-2}. The 2DEG carriers are heavily degenerate at temperatures $T \leq 100$K since $\zeta \gg 1$ is satisfied. At higher temperatures, 2DEGs with higher carrier densities maintain their degenerate nature; non-degeneracy sets in only for the low-density 2DEGs. The degenerate nature of the high density carriers simplifies the evaluation of transport scattering rates and mobility by avoiding cumbersome Fermi-Dirac integrals (section 6). The expression for momentum scattering rates would depend only on carriers at the Fermi-surface, which simplifies the averaging for calculating mobility.

For low carrier densities ($n_s \sim 10^{12}\text{cm}^{-2}$), though the low-temperature behavior is strongly degenerate, the 2DEG becomes non-degenerate for higher temperatures. This requires a proper averaging of the mobility using a generalized expression for arbitrary degeneracy. However, since we will not be interested in the high temperature behavior of low-density 2DEGs, that topic is not treated in this chapter.

2.3 Survey of Experimental 2DEG Mobility Data

Figure 7 shows the highest reported mobilities for Al(Ga)N/GaN heterojunction 2DEGs as a function of the 2DEG density n_s. The dashed lines are guides to the eye, showing the prevailing trend. It is obvious that the low-temperature mobility reduces with increasing carrier density. The highest mobilities reported are for the lowest 2DEG densities. Thus the dominating scattering mechanisms at low temperatures should have a strong dependence on the 2DEG carrier density. Though the turnaround of the trend for mobility at the lowest carrier concentrations has not

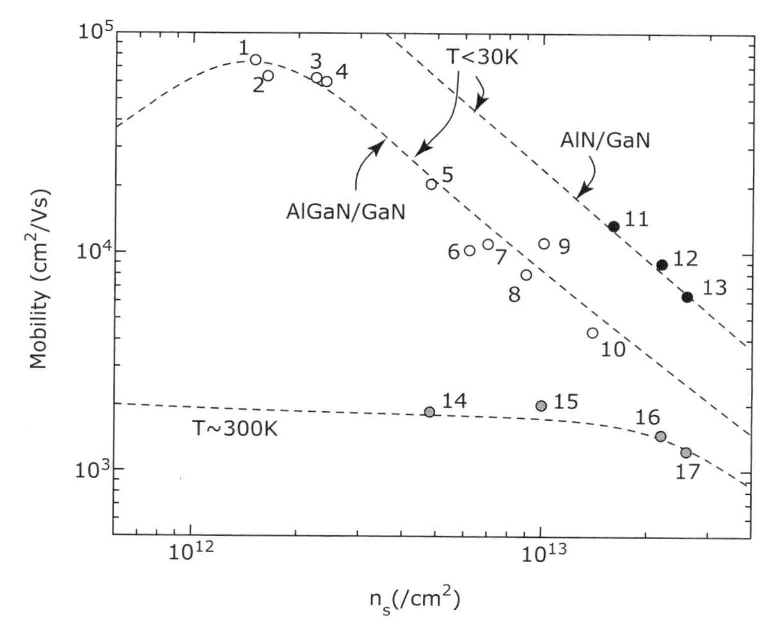

Fig. 7 A collection of highest mobility data reported till date in AlGaN/GaN 2DEGs. For references, refer to Table 1. The dashed lines are guides to the eye.

Table 1 AlGaN/GaN 2DEG Hall data

Points	Year	Group	Growth	μ $cm^2/V{\cdot}s$	n_s 10^{12} cm^{-2}
1,2	2002	Manfra et. al. [12]	MBE	75000, 62000	1.5,1.7
3	1999	Ioulia et. al. [13]	MBE	62000	2.23
4	2000	Frayssinet et. al. [14]	MBE	60000	2.4
5,14	2000	Elsass et. al. [15]	MBE	20500	4.8
6	1999	Wang et. al. [16]	MOCVD	10300	6.2
7,9,15	1999	Gaska et. al. [17]	MOCVD	11000,10300,2019	7,10,13
8	2000	Jena et. al. [18]	MBE	8000	9
10-13,16,17	2001	Smorchkova et. al. [19]	MBE	Range	Range

been reported, there are indications of such an effect [9]. Carrier densities lower than $n_s = 10^{12}$ cm^{-2} are difficult to achieve in polar AlGaN/GaN heterojunctions due to the large polarization discontinuity at the heterointerface, and are accessible only by gated Hall measurements. Using such gated Hall samples, very high mobilities exceeding 100,000 $cm^2/V.s$ has been achieved for very low-density 2DEGs ($n_{2d} \sim 10^{12}$ cm^{-2} and less) [10, 11]. Integral Quantum Hall Effect (IQHE) has also been observed in such 2DEGs [10, 11]. Since these samples have low 2DEG densities, they are very useful for fundamental transport studies, but have not found wide applications in HEMT technology due to the low 2DEG densities. Therefore, the discussion here is limited to relatively high density 2DEGs.

The highest low-temperature mobilities are in the range of $\mu \approx 7 \times 10^4$ $cm^2/V{\cdot}s$ at carrier densities in the range of $n_s \approx 10^{12}$ cm^{-2}. This is orders of magnitude lower than the highest mobilities reported for AlGaAs/GaAs modulation-doped 2DEGs, where the highest mobilities reported [20] are in the range $\mu \approx 10^7$ $cm^2/V{\cdot}s$ for carrier densities $n_s \approx 2 \times 10^{11}$ cm^{-2}. The highest mobilities in AlGaAs/GaAs 2DEGs are remote ionized-impurity scattering limited; thus, the mobility *increases* with 2DEG density [20] - this is opposite in trend to AlGaN/GaN 2DEGs. Of special interest in AlGaN/GaN 2DEGs is the fact that there is a *large improvement* in the low-temperature mobility seen if a thin layer of AlN is sandwiched between the AlGaN/GaN layers (or the barrier is entirely AlN) [19]. The insertion of AlN causes the removal of alloy scattering, and thus shows that alloy scattering is a *major* scattering mechanism at low temperatures.

The highest room-temperature mobility reported for the AlGaN/GaN 2DEG is $\mu \approx 2000$ $cm^2/V{\cdot}s$ as compared to 2DEGs in the arsenide system that reaches ≈ 5000 $cm^2/V{\cdot}s$. However, the carrier densities in the AlGaN/GaN 2DEGs are typically an order of magnitude higher than that in AlGaAs/GaAs 2DEGs, making nitride structures more suited for field-effect device applications [12].

The rest of this chapter is devoted to a study of various defects and scattering mechanisms that limit the mobility of AlGaN/GaN 2DEGs, i.e., to piece together the story behind figure 7.

2.4 Theoretical Tools to Address AlGaN/GaN 2DEG Mobilities

Transport of electrons in response to an applied electric field may occur either in the conduction band by drift-diffusion processes, or by hopping between localized states in a heavily disordered material. Thus, it is essential to first determine the transport regime to choose the correct theoretical approach. Low-temperature transport in the III-V nitride heterostructure 2DEGs is characterized by short Fermi-wavelengths ($\lambda_F = 2\pi/k_F \approx 8$nm) and long mean free paths ($L \approx 0.5\mu$m) for a 2DEG of density $n_{2d} = 10^{13}$ cm^{-2} and mobility $\mu = 10,000$cm^2/V·s. Electron wavefunctions will be *localized* around the defects if $L < \lambda_F$ (the Ioffe-Regel criterion [21]), and are *extended* if $L \gg \lambda_F$.

Clearly, electrons in AlGaN/GaN 2DEGs experience band-transport. This simplifies the problem at hand enormously, since the theoretical approach to transport in disordered materials is much more complicated and requires results from many-body theory and percolation theory. The problem can then be attacked in the single-particle approximation. The only many-body effect needed is in the phenomena of screening. The theoretical formalism for 2DEG transport in the drift-diffusion regime can be found in textbooks [5]; the important results are treated comprehensively in section 6.

The momentum scattering rate for degenerate 2DEG electrons is given by

$$\frac{1}{\tau_m(k_F)} = n_{2D}^{imp} \frac{m^\star}{2\pi\hbar^3 k_F^3} \int_0^{2k_F} |V(q)|^2 \frac{q^2}{\sqrt{1-(\frac{q}{2k_F})^2}}, \tag{8}$$

where n_{2D}^{imp} is the areal density of scatterers and $k_F = \sqrt{2\pi n_s}$ is the Fermi wavevector, n_s being the 2DEG density. $|V(q)|$ is the Fourier-transform of the screened scattering potential. The experimental probe for the momentum-scattering rate $\tau_m(k_F)$ is the electron mobility. The mobility is related to the momentum scattering time through the Drude relation $\mu_{2d} = e\tau_m(k_F)/m^\star$. This relation is the result of the solution of the Botlzmann transport equation, solved in the relaxation-time approximation, as discussed in section 6.

3 Scattering Mechanisms

3.1 Typical AlGaN/GaN 2DEG Structures

A high-resolution Transmission-Electron-Microscope (HRTEM) picture of a typical AlGaN/GaN heterostructure is shown in figure 8. Electrons moving in the 2DEG experience interface-roughness scattering due to the non-abrupt interface between AlGaN and GaN. The 2DEG wavefunction is mostly confined in GaN, but there is a finite part that penetrates the AlGaN barrier, leading to alloy-disorder

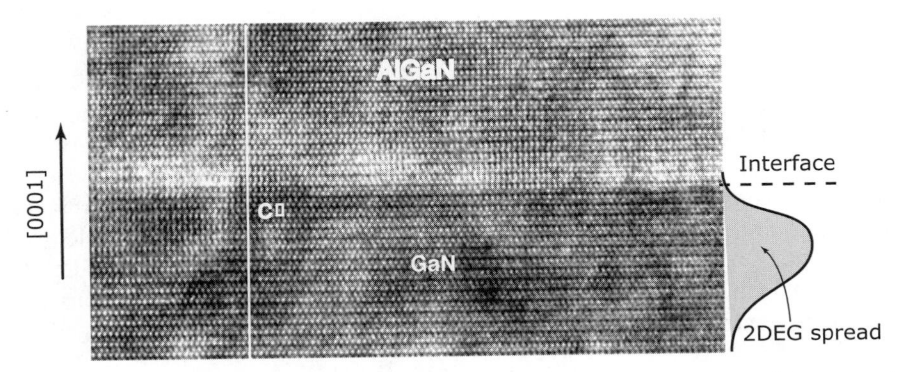

Fig. 8 HRTEM picture of the AlGaN/GaN interface structure - provided by Wuyuan and J. S. Speck (UCSB).

scattering. Interface-roughness scattering and alloy scattering are short-range scattering sources [22]. Charged impurities (remote and residual) are always present in the samples, and constitute a form of long-range Coulombic scattering source. The lattice vibrates at finite temperatures and phonons form a potent scattering mechanism at high temperatures.

The scattering mechanisms listed above are 'traditional', since their effects have been studied for AlGaAs/GaAs and Si-MOSFET systems in a fair amount of detail [21, 22]. They are important in AlGaN/GaN 2DEG transport as well; so the results for scattering rates from the existing literature is quoted and used for calculations. The scattering mechanisms with specific relevance to AlGaN/GaN 2DEG transport mentioned below are treated in far more detail, and some new results are derived.

An important form of Coulombic scattering in AlGaN/GaN 2DEGs is dislocation scattering, owing to the large density of dislocations in the material. The cores of threading edge dislocations have dangling bonds that introduce states in the gap of the semiconductor, causing a dislocation to become a line of charge. Such charged dislocations scatter conduction electrons. Dislocations also scatter from strain-fields that develop around them.

Finally, *dipole scattering* originates in the AlGaN/GaN system due to the coupling of alloy disorder in the barrier and the strong polarization of the material system. This is also long-range (Coulombic) in nature, though it is considerably weaker than the single-impurity scattering.

3.2 Traditional Scattering Mechanisms

Phonon scattering limits electron mobility at temperatures $T \geq 80K$ for 2DEGs. Scattering by three types of phonons are important for our study - acoustic phonons by the deformation potential coupling and the piezoelectric coupling, and polar optical phonons.

3.2.1 Acoustic Phonons

The solution of Boltzmann equation in the relaxation time approximation requires the scattering processes to be elastic (see section 6). The acoustic phonon linear dispersion $\omega = v_s k$ makes the acoustic phonon energy very low, and scattering is thus essentially elastic. Thus, a relaxation time may be defined. The coupling of electron transport to acoustic phonons can be through deformation potential or piezoelectric components. Since the acoustic branch of dispersion has both longitudinal and transverse components, ideally one should consider both branches for finding the scattering rates. However, the transverse modes are weaker than the longitudinal modes for deformation potential scattering. For piezoelectric coupling, both have to be considered; however, it has been shown [46] that the piezoelectric component of acoustic phonon scattering in AlGaN/GaN 2DEGs is weaker than deformation potential scattering and may be safely neglected in comparison.

So, scattering by only the longitudinal-mode acoustic phonon is considered. The Γ-valley conduction band deformation potential is $a_C = 9.1\text{eV}$. Since acoustic phonon energy $\hbar v_s k$ is very small, one can assume that the Bose-Einstein distribution reduces to $N_B \approx k_B T / \hbar v_s k$ which is the number of acoustic scatterers. With the Fang-Howard wavefunction, the momentum scattering rate of the 2DEG is [5]

$$\frac{1}{\langle \tau_m^{ac} \rangle} = \frac{3m^\star b a_C^2 k_B T}{16 \rho v_s^2 \hbar^3}, \tag{9}$$

where the mass density ρ and the sound velocity v_s are given in Table 2.

3.2.2 Optical Phonons

Polar optical phonon (POP) energy for the wurtzite GaN crystal is higher than other III-Vs ($\hbar\omega_{op} = 92\text{meV}$)[1]. Scattering by polar optical phonons is highly inelastic; such a case demands the solution of Boltzmann equation by coupled equations for both emission and absorption thus making the relaxation-time approximation invalid [25].

Table 2 Material properties for transport calculations [1]

Property	Symbol	GaN	AlN	InN	Units
Effective mass (Γ valley)	$m^\star$	0.2	0.5	0.1	m_0
Mass density	ρ	6.15	3.23	6.81	g/cm^{-3}
Static dielectric constant	$\epsilon(0)$	8.9	8.5	15.3	-
High frequency dielectric constant	$\epsilon(\infty)$	5.35	4.77	8.4	-
Optical phonon energy	$\hbar\omega_{op}$	92	100	89	meV
Deformation potential	Ξ	8.3	9.5	7.1	meV
Sound velocity (Longitudinal)	v_s	8	11	5.2	10^5 cm/s

[1] Polar optical phonon energy is large since the Ga-N bond is much stronger than bonds in other (zinc blende) III-Vs. In fact, it turns out to be the general trend in all wide-bandgap materials (including SiC).

An analytic expression for the momentum relaxation rate in 2DEGs was nevertheless derived by Gelmont, Shur, and Stroscio [26], which is able to match experimental data over a wide temperature range rather accurately. Their theory takes advantage of the fact that the optical phonon energy is large; $\hbar\omega_{op} \gg k_B T$ (the *thermal* energy of carriers for a wide range of temperatures), and $\hbar\omega_{op} > E_F$ (*kinetic* energy of carriers for 2DEG densities $n_{2d} \leq 10^{13}$ cm^{-2} in AlGaN/GaN 2DEGs). This means most carriers have energies lower than the optical phonon energy, thus *blocking* the emission of optical phonons. The absorption process dominates, and is used to find the momentum relaxation time; it is given by

$$\frac{1}{\tau_{pop}} = \frac{e^2 \omega_0 m^* N_B(T) G(k_0)}{2\varepsilon^\star q_0 \hbar^2 F(y)}. \tag{10}$$

Here, $q_0 = \sqrt{2m^*(\hbar\omega_{pop})/\hbar^2}$ is the polar optical phonon wavevector, N_B is the Bose-Einstein distribution function $N_B(T) = 1/(\exp(\hbar\omega_{pop}/k_B T) - 1)$, and $F(y)$ is given by

$$F(y) = 1 + \frac{1 - e^{-y}}{y}, \tag{11}$$

y being the dimensionless variable $y = \pi\hbar^2 n_{2d}/m^\star k_B T$. $G(k_0)$ is the form factor for the 2DEG wavefunction (see section 6).

Figure 9(a) shows the dependence of mobility on carrier density at 300K due to optical and acoustic phonon scattering. 9(b) shows the contributions of acoustic and optical phonon scattering to the mobility as a function of temperature for two

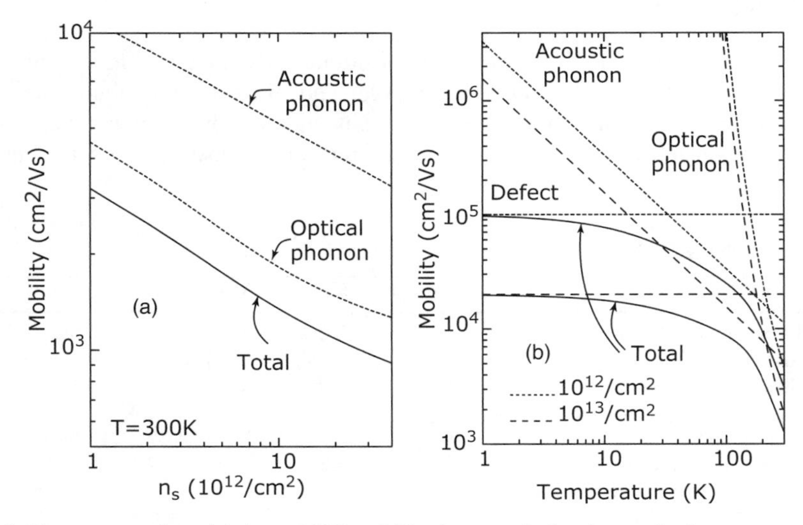

Fig. 9 Phonon scattering - (a) shows 300K mobility due to optical and acoustic phonon scattering. (b) shows the mobility dependence on temperature for scattering by Phonons and (unspecified) defects.

carrier densities. At the lowest temperatures, it is assumed some form of scattering (alloy, interface) limits the electron mobility - the defect related mobility is temperature independent, but decreases with increasing carrier density (from figure 7). One might identify three distinct regions for low density samples. At high temperatures (around and above 300K), the mobility is limited by polar optical phonon scattering. In the intermediate regime ($10K \leq T \leq 80K$), the mobility is limited by acoustic deformation potential *only* for low-density 2DEGs. For high-density 2DEGs, this part is washed out by defect scattering (Figure 9(b)). Finally, at the lowest temperatures, mobility is determined entirely by defect and disorder scattering, and is sensitive to the growth methods. The mobility at room-temperature limited by optical phonon scattering is $\mu_{300K} \sim 2000$ cm^2/V.s, close to the highest room temperature mobility data reported [17] for reasonably large 2DEG densities.

3.2.3 Alloy Disorder Scattering

Alloy disorder scattering originates from the randomly varying alloy potential in the barrier. This form of scattering is known (Bastard, [27]) to be the mobility-limiting mechanism for 2DEGs confined in an alloy channel such as in InGaAs/GaAs heterostructures. In 2DEGs confined in binary wells, alloy scattering occurs as a result of the finite penetration of the 2DEG wavefunction into the barrier. Since the Fang-Howard type of wavefunction does not take the penetration of the wavefunction into the barrier into account, one has to resort to other methods to find the 'volume' of the 2DEG wavefunction residing in the barrier.

One way to do this is to obtain the exact Hartree-form and find the penetration numerically. However, a modified Fang-Howard wavefunction can be used with sufficient accuracy for the same problem [27]. The modified wavefunction is

$$\chi(z) = Me^{\frac{\kappa_b z}{2}}, z < 0$$
$$\chi(z) = N(z+z_0)e^{-\frac{bz}{2}}, z \geq 0. \tag{12}$$

Here $\kappa_b = 2\sqrt{2m^\star \Delta E_c(x)/\hbar^2}$, the wavevector characterizing the wavefunction penetration into the barrier, $\Delta E_c(x)$ is the conduction band-offset between AlGaN and GaN, and N, M are normalization constants. Normalization and continuity conditions yield the parameters

$$z_0 = \frac{2}{b + \kappa_b \frac{m_A}{m_B}}, \tag{13}$$

where m_A and m_B are the effective masses of the electron in the barrier and the well respectively,

$$N = \sqrt{\frac{b^3}{2}} \frac{1}{\left(1 + bz_0 + \frac{1}{2}b^2 z_0^2 (1 + \frac{b}{\kappa_b})\right)^{1/2}}, \tag{14}$$

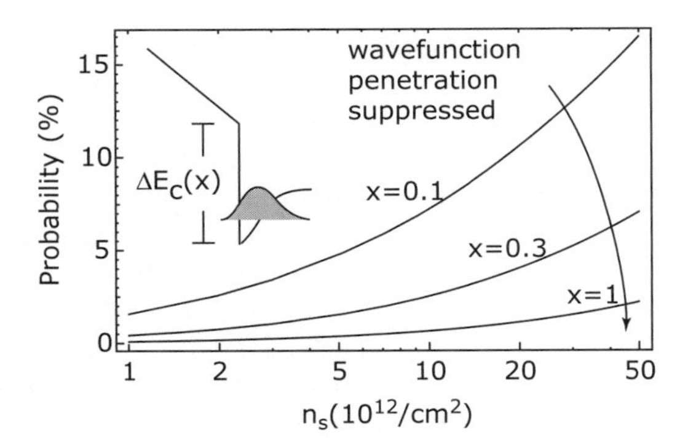

Fig. 10 Integrated probability of the wavefunction penetration in to the AlGaN barrier as a percentage. The probability can be as high as 10% for typical AlGaN/GaN structures. Note that the penetration is suppressed strongly with increasing Al composition, owing to the increase in the barrier height.

and

$$M = Nz_0. \tag{15}$$

With this modified Fang-Howard algebra, one obtains the integrated probability of finding the particle in the barrier region to be

$$P_b = \frac{N^2 z_0^2}{\kappa_b}. \tag{16}$$

Figure 10 shows the probability P_b as a percentage of the total 2DEG density.

The momentum scattering rate due to alloy disorder is given by (Bastard, [27])

$$\frac{1}{\tau_m^{alloy}} = \frac{m^* \Omega_0 (V_A - V_B)^2 x(1-x)}{e^2 \hbar^3} \times \frac{\kappa_b P_b^2}{2}, \tag{17}$$

where Ω_0 is the volume associated with each Al(Ga) atom, $(V_A - V_B)$ is the alloy scattering potential that results on replacing a Ga atom by Al. The exact value of this potential is a reason for some controversy, and is best determined by experimental techniques. The general rule of thumb is $(V_A - V_B) = \Delta E_c = (E_c^{AlN} - E_c^{GaN})$, i.e., the conduction band offset between AlN and GaN.

There has been opposition to the usage of such a form of the alloy scattering potential [28]. However, transport measurements for electron gases housed entirely in the AlGaN alloy has made it possible to *measure* the alloy scattering potential for the AlGaN system, and it is found to be $(V_A - V_B) = 1.8$ eV. This is not very different from the conduction band offset of $\Delta E_c = 2.1 eV$ between AlN and GaN. This measured value is used in all calculations in this chapter. We also note that the difference of the two values can cause an error of $\approx 25\%$ in the calculated mobility, and the earlier results on mobility in AlGaN/GaN 2DEGs [29] should be interpreted in that light.

In AlGaAs/GaAs heterostructures, alloy disorder scattering is weak, and often negligible when the 2DEG is located in GaAs. However, in AlGaN/GaN heterostructures, the large electron effective mass, the high 2DEG density and the large alloy scattering potential all combine to make this form of scattering quite strong in spite of the confinement in the binary semiconductor. Figure 11 shows the alloy- scattering-limited electron mobility for a range of 2DEG densities and alloy compositions. Part (a) of the figure shows a strong dependence of mobility on the carrier density. More importantly, mobility decreases with increasing 2DEG density, which is identical to the experimentally observed trend, suggesting that this form of scattering is an important one. Alloy-scattering-limited mobilities are also of the same magnitudes as the measured low-temperature mobilities (figure 7). When the final mobilities are calculated by considering all scattering mechanisms, this form of scattering will be found to be very strong at low temperatures. Part (b) of figure 11 shows the dependence of 2DEG mobility on the alloy-composition of Al in the AlGaN barrier for a number of 2DEG densities. The low mobilities at low alloy concentrations are due to large penetration of the 2DEG wavefunction into the barrier due to reduced conduction-band discontinuity. At large band discontinuities, the wavefunction penetration is strongly suppressed and the mobility rises. The dependence on the carrier density is important; as carrier density increases, the wavefunction gets pushed closer to the junction, leading to more penetration into the alloy barrier and hence stronger scattering for the same alloy composition. This situation in figure 11(a) is met during the gate-modulation in a HEMT: when the channel is in the 'on' state ($V_{gs} \geq V_p$, i.e., gate voltage larger than the pinch-off voltage), the 2DEG density is high, and alloy scattering is severe. As the gate bias approaches the pinch-off voltage, the 2DEG density reduces, and thus alloy scattering limited mobility will increase, and one moves along the constant composition lines in figure 11(a).

3.2.4 Interface Roughness Scattering

Scattering at rough interfaces can be severe if the 2DEG density is high, since the 2DEG tends to shift closer to the interface as the density increases. The roughness at heterojunction interfaces has been traditionally modelled by a Gaussian autocovariance function. The scattering rate by a rough interface with a root mean square roughness height Δ and a correlation length L is given by (Ferry and Goodnick, [30])

$$\frac{1}{\tau_{IR}} = \frac{\Delta^2 L^2 e^4 m^\star}{2\epsilon^2 \hbar^3} \left(\frac{1}{2} n_{2d}\right)^2 \int_0^1 du \frac{u^4 e^{-k_F^2 L^2 u^2}}{(u + G(u)\frac{q_{TF}}{2k_F})^2 \sqrt{1-u^2}}, \tag{18}$$

where the substitution $u = q/2k_F$ is used to make the integral dimensionless. Here q_{TF} is the Thomas-Fermi screening wavevector for the 2DEG, and $G(u)$ is a form-factor that arises due to finite extend of the 2DEG wavefunction along the growth direction, as described in section 6.

Figure 12(a) shows how the distance of the centroid of the 2DEG distribution from the heterojunction interface varies with the 2DEG sheet density for different

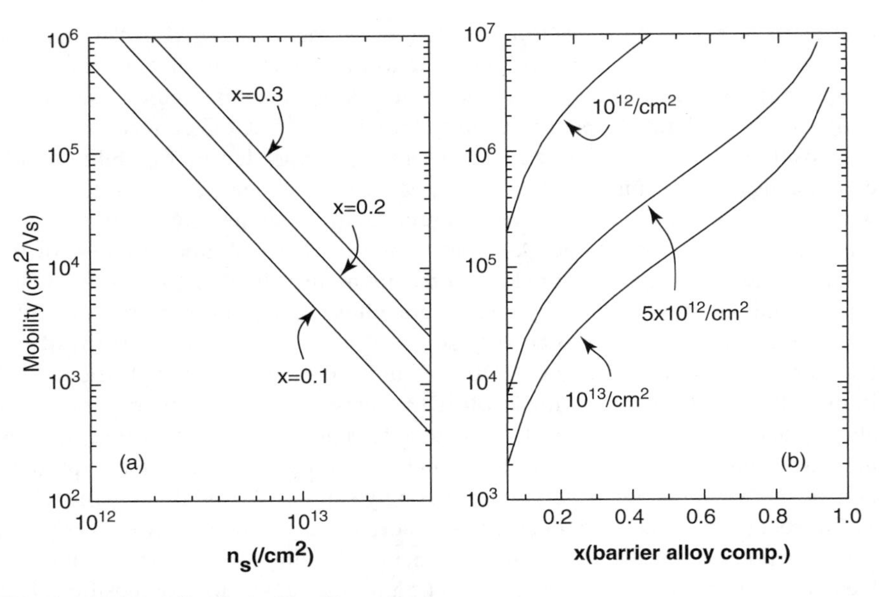

Fig. 11 2DEG mobility limited by alloy scattering.

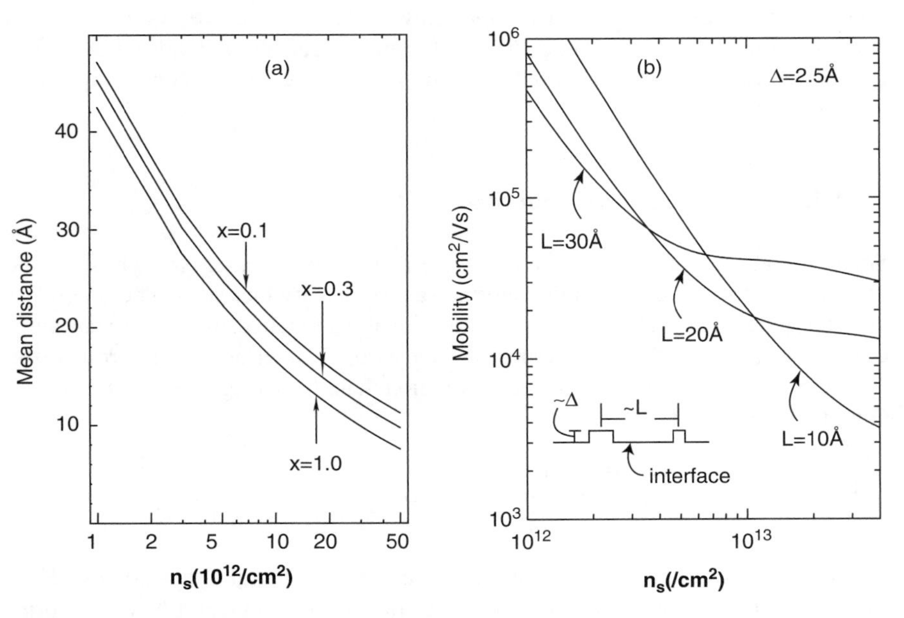

Fig. 12 Mobility limited by interface roughness scattering - (a) shows the mean distance of the 2DEG wavefunction from the interface, and (b) is the mobility for different carrier densities.

alloy concentrations in the AlGaN barrier layer. The dependence was calculated from the self-consistent Fang-Howard variational wavefunction. The dependence on the 2DEG density is characteristically much stronger than on the alloy composition. Interface roughness scattering affects transport even in the presence of a binary barrier - for example, in an AlN/GaN heterojunction, where alloy scattering is absent by definition.

Figure 12(b) shows the calculated mobility limited by interface roughness scattering for the AlGaN/GaN 2DEG. The correlation length between islands is varied and the dependence changes when the correlation length approaches the Fermi-wavelength. This is the reason for the switching of dependence on carrier densities between $10^{12}-10^{13}\text{cm}^{-2}$. The effect of interface roughness scattering on mobility is also quite strong, as the mobilities are of the order of the highest reported, only slightly higher than alloy scattering limited mobility.

A very rough interface can localize electrons at the 2DEG; this limit was analyzed by Zhang and Singh, who proposed that in such a case, transport will require phonon assisted hopping [31]. However, since the highest reported mobilities are for low-density samples ($n_{2D} \approx 10^{12}\text{cm}^{-2}$) with no temperature dependence of conductivity for $T \leq 30K$, transport in the best samples is by band conduction.

Interface roughness-scattering-limited mobility has a characteristic L^{-6} dependence [32] for 2DEGs in quantum wells (of thickness L), which can be observed by transport measurements on quantum wells of different thicknesses. An interesting feature of the III-V nitrides is that due to the unscreened polarization fields in thin epitaxial layers, there is a large band bending inside the well even under no external bias. Hence the 2DEG samples one interface much more than the other; this acts as an built-in mechanism to restrict interface roughness effects on 2DEGs confined in thin unscreened quantum wells.

3.2.5 Remote Ionized Impurities

The typical AlGaN/GaN heterostructure 2DEG is polarization-doped, and the surface donor-like states [8] are positively charged. The donor charge-density is equal to the 2DEG sheet density to maintain charge neutrality. Polarization sheet charges exist at the heterojunction as well as the surface. However, charges on these sheets assume the lattice periodicity, causing no scattering[2]. Due to the atomic origin of polarization-doping, the III-V nitride semiconductors can be expected to scale to very small dimensions easily, without any 'discrete-dopant' effects as is experienced in MOSFETs.

An ionized charge at a distance z_0 from the heterointerface has a Coulomb-potential $V_{uns}(r, z_0) = 1/4\pi\epsilon_0\epsilon(0)\sqrt{r^2+z_0^2}$ where r is the in-plane distance in the 2DEG. The *screened* matrix element for this potential for a perfect 2DEG is given by [5]

[2] Note that for a random alloy, this is not strictly valid since the alloy is disordered, and the polarization sheet charge should replicate the disorder and deviate from periodicity.

$$V(q) = \frac{V(q,z_0)}{\epsilon_{2d}(q)} = \int_0^\infty r dr \int_0^{2\pi} d\theta \frac{e^2}{4\pi\epsilon(q)\sqrt{r^2+z_0^2}} e^{iqr\cos\theta}$$
$$= \frac{e^2}{2\epsilon_0\epsilon(0)} \frac{e^{-qz_0}}{q+q_{TF}}, \quad (19)$$

where q_{TF} is the Thomas-Fermi screening wavevector (see section 6). The e^{-qz_0} term damps the remote scattering potential, enhancing mobility. This result for the matrix element for remote ionized impurity is used as the backbone for much of the further calculations of scattering rates from Coulombic charge centers in different configurations such as dipoles and charged dislocations. If the sheet density of the remote donors is N_D, and they are at a distance t_b form the heterostructure interface (t_b is the thickness of the AlGaN barrier), the scattering rate is (see Davies, [5])

$$\frac{1}{\tau_{rem}(k_F)} = N_s \frac{m^\star}{2\pi\hbar^3 k_F^3} \left(\frac{e^2}{2\epsilon_0\epsilon(0)}\right)^2 \int_0^{2k_F} dq \frac{F(q)e^{-2qt_b}}{(q+q_{TF}G(q))^2} \frac{q^2}{\sqrt{1-(\frac{q}{2k_F})^2}}, \quad (20)$$

where the Fang-Howard algebra results in the form factors $F(q), G(q)$. This may be evaluated by changing the variable using $q = 2k_F \sin(\theta/2)$ whereupon the integral depends only upon the 2DEG density n_{2d}. If the carrier mobility in the 2DEG was limited by remote ionized impurity scattering alone, one can evaluate it as $\mu_{rem} = e\tau_m(k_F)/m^\star$. This has been evaluated for a 2DEG densities $n_{2d} = 1, 5, \& 10 \times 10^{12}$ cm^{-2} and barrier thicknesses $0.1 < t_b < 50$ nm. It is shown in figure 13(a).

For typical barrier thicknesses of $t_b = 30$ nm, scattering by remote ionized impurities is seen to be relatively weak, causing a drift mobility of $\mu_{rem} \approx 10^6$ cm^2/V·s. From the survey of experimental 2DEG mobilities, we see that this form of scattering is relatively weak, unless the barrier is too thin. Low-temperature mobility will be limited by this form of scattering only if the barrier thickness is less than $\approx$10 nm.

3.2.6 Background Residual Impurities

The advantage of modulation-doping is a spatial separation of the 2DEG from the ionized donors, thus reducing scattering and improving electron mobility. State of the art AlGaN/GaN structures have $N_{back} \approx 10^{16}$cm^{-3} unintentional residual background donors. These donors are believed to be unwanted oxygen and silicon atoms (or vacancies) that incorporate during the growth process.

The scattering rate for background residual impurities may be calculated from the result for the matrix element of remote ionized impurity scattering. The matrix element for background impurity scattering is calculated by passing from the 2-dimensional remote impurity distribution to a three-dimensional impurity distribution (Davies, [5]) by the transformation

$$n_{imp}^{2d} e^{-2qz} \to n_{imp}^{3d} \int_{-\infty}^{\infty} dz e^{-2qz} = \frac{n_{imp}^{3d}}{q}. \tag{21}$$

The momentum scattering rate due to a homogeneous background donor density of N_{imp} is thus given by

$$\frac{1}{\tau_m^{imp}(k_F)} = N_{imp} \frac{m^\star}{2\pi\hbar^3 k_F^3} \left(\frac{e^2}{2\epsilon_0\epsilon(0)}\right)^2 \int_0^{2k_F} dq \frac{P_0^2}{(q+q_{TF}G(q))^2} \frac{q}{\sqrt{1-(\frac{q}{2k_F})^2}}. \tag{22}$$

which can be approximated to a form useful for numerical estimates

$$\frac{1}{\tau_m^{imp}} \approx N_{imp} \frac{m^\star}{2\pi\hbar^3 k_F^3} \left(\frac{e^2}{2\epsilon_0\epsilon(0)}\right)^2. \tag{23}$$

Hence background impurity limited 2DEG mobility is given by

$$\mu_{imp} \approx \frac{4(2\pi)^{5/2}\hbar^3(\epsilon_0\epsilon(0))^2}{(m^\star)^2 e^3} \times \frac{n_s^{3/2}}{N_{imp}}, \tag{24}$$

which has a $n_s^{3/2}/N_{imp}$ dependence. The calculated mobility is shown in figure 13(b). The effect of scattering by background residual impurities is rather weak for AlGaN/GaN 2DEGs. From figure 13(b) it is seen that background impurity scattering is strong for low sheet densities, but still more than an order of magnitude higher than the observed highest mobilities for typical background concentrations of $N_{imp} \approx 10^{16}\text{cm}^{-3}$. For very high background doping density $N_{imp} > 10^{18}\text{cm}^{-3}$, the

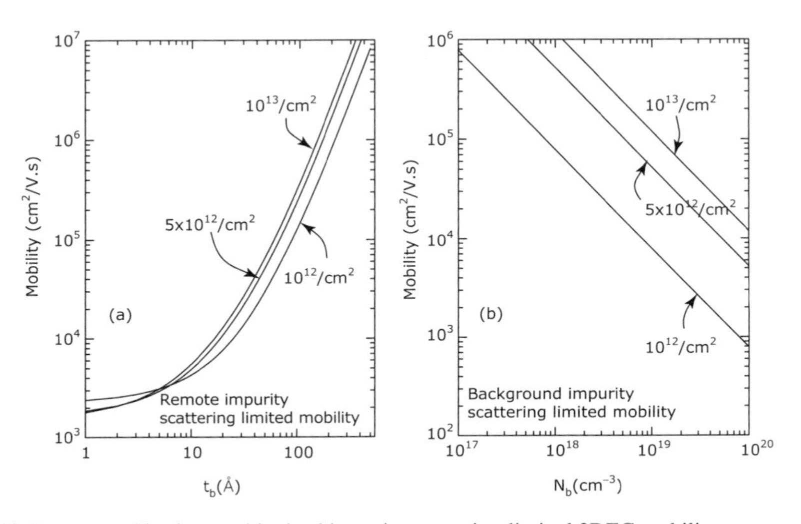

Fig. 13 Remote and background ionized impurity-scattering-limited 2DEG mobility.

mobility of low-density 2DEGs is affected. However, the mobility of high-density 2DEGs is very resistant to background impurity scattering due to strong screening. The analysis shows that background impurity scattering is relatively unimportant in current state-of-the-art AlGaN/GaN 2DEGs.

3.3 Novel Scattering Mechanisms in AlGaN/GaN 2DEGs

3.3.1 Dipole Scattering

For the perfectly periodic III-V nitride crystal, the classical microscopic picture of polarization is a dipole in each primitive cell aligned along the [0001] axis. The dipole moment $\mathbf{p_0} = e \cdot d_0$ (d_0 is the effective charge separation) is related to the macroscopic polarization $\mathbf{P}$ by the relation $\mathbf{P} = \mathbf{p_0}/\Omega$, where Ω is the volume of the primitive cell [33]. $\mathbf{P}$ is the total polarization, which includes the spontaneous and piezoelectric components,

$$\mathbf{P} = \mathbf{P_{sp}} + \mathbf{P_{pz}}. \tag{25}$$

A perfect binary polar lattice thus has a periodically arranged array of dipoles in every unit cell with equal dipole moments. Such a periodic arrangement of similar dipoles has a characteristic wavevector, and hence does not contribute to the scattering of carriers.

However, the 2DEG in AlGaN/GaN heterostructures is confined by a barrier due to the undoped $Al_xGa_{1-x}N$ ternary alloy barrier. The alloy is a disordered system with Al and Ga atoms arranged in a random array such that the overall composition over any plane is constant over Al(Ga) planes. The difference in spontaneous and piezoelectric polarizations between AlN and GaN implies that we have a dipole moment of randomly fluctuating magnitude in the barrier. A method similar to the treatment of disordered alloys by virtual crystal approximation is used here to treat dipoles in disordered polar semiconductor alloys.

The dipole moments in a unit cell of coherently strained AlN and GaN binary wurtzite crystals is first calculated. The piezoelectric field in a binary wurtzite primitive cell coherently strained to a $x-y$ lattice constant $a(x)$ from its unstrained lattice constant a_0 and[3] $c(x)$ from c_0 in the z direction is [3]

$$P_{pz}(x) = 2 \cdot \left(\frac{a(x) - a_0}{a_0}\right) \cdot \left(e_{31}(x) - e_{33}(x)\frac{c_{13}(x)}{c_{33}(x)}\right), \tag{26}$$

where $e_{31}(x)$ and $e_{33}(x)$ are the piezoelectric coefficients and $c_{13}(x)$ and $c_{33}(x)$ are the elastic constants of the crystal structure. The volume of the unit cell of the wurtzite structure is

$$\Omega(x) = \frac{\sqrt{3}}{2} c_0(x) \cdot a_0^2(x). \tag{27}$$

[3] The c/a ratio for ideal hexagonal close packed structure is $\sqrt{8/3} \approx 1.63$. For calculations, the known c and a values are used.

Thus the dipole moment in a strained binary crystal is given by

$$p_{dip}(x) = (P_{sp} + P_{pz}(x)) \cdot \Omega(x). \tag{28}$$

This dipole moment is calculated for both binary semiconductors as p_{dip}^{AlN} and p_{dip}^{GaN}.

The disordered $Al_xGa_{1-x}N$ barrier is modelled as a perfect crystal superposed with a randomly fluctuating dipole moment at each primitive cell. Such a virtual crystal has a dipole moment of magnitude

$$p_{dip}(av) = x \cdot p_{dip}^{AlN} + (1-x) \cdot p_{dip}^{GaN}. \tag{29}$$

The deviation from the perfect virtual crystal at all Al sites is $(1-x) \cdot \Delta p_{dip}$ where

$$\Delta p_{dip} = p_{dip}^{AlN} - p_{dip}^{GaN}. \tag{30}$$

The deviation at Ga sites is $(-x) \cdot \Delta p_{dip}$. Since there are x Al sites and $(1-x)$ Ga sites on average on a Al(Ga) plane, the average randomly fluctuating dipole moment at each site is

$$\delta p_{dip} = e \cdot d_0 = 2 \cdot x \cdot (1-x) \cdot |\Delta p_{dip}|. \tag{31}$$

The absolute value is used in adding the dipole contributions since the direction of the dipole is immaterial in the scattering matrix element, which involves the square of the dipole potential.

The number of such dipoles present on each Al(Ga) plane is given by

$$n_{dip}^{2D} = \frac{1}{\frac{\sqrt{3}}{4} a_0^2(x)}, \tag{32}$$

where the in plane lattice constant $a_0(x)$ is interpolated for the alloy. Since all constants are known, the sheet-density n_{dip}^{2D} and the effective dipole-length d_0 of such dipoles is easily calculated depending on the alloy composition of the barrier. It is instructive to look at the dependence of the dipole strength term d_0 on the barrier alloy composition x, since it determines the strength of the scattering. This will be done when the scattering rate due to dipoles is calculated.

Scattering by dipoles and their effects on electron transport in bulk semiconductor samples has been studied, albeit not extensively owing to it's insignificance in the non-polar Si and relatively weakly polar GaAs material systems [35, 36]. However, the effect of dipole scattering on 2DEG electron transport has not been studied. We derive the scattering rate due to dipoles for a semiconductor two dimensional electron gas.

We consider the 2DEG to be perfect (i.e., the extent along the z direction to be zero) for our derivation. Extension to the more physical case of a 2DEG with finite extent along the growth direction involves incorporation of the relevant form factors.

Figure 14 shows the model for the system under consideration. The dipole charges are separated from each other by distance d_0, and the center is a distance

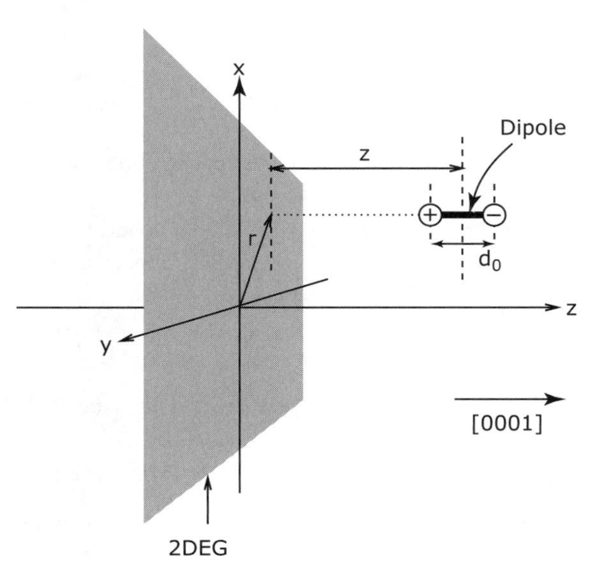

Fig. 14 The location of the dipole with respect to the 2DEG is shown. The dipole axis is taken to be perpendicular to the plane of the 2DEG, keeping with the direction of the polarization field in the AlGaN barrier of AlGaN/GaN HEMTs. The distances used in the text in the derivation of the scattering rate are defined.

z from the plane containing the 2DEG. Spontaneous and piezoelectric polarization fields ($\mathbf{P_{sp}}$ and $\mathbf{P_{pz}}$ respectively) in wurtzite AlGaN/GaN is directed perpendicular to the 2DEG plane [25]. The dipole axis is thus chosen to be aligned in the $[0001]$ direction.

The unscreened Coulomb potential seen by a 2DEG electron at $\mathbf{r}$ due the dipole is written as

$$V_{uns}(r,z) = \frac{e^2}{4\pi\epsilon} \cdot \left[\frac{1}{\sqrt{r^2 + (z - \frac{d_0}{2})^2}} - \frac{1}{\sqrt{r^2 + (z + \frac{d_0}{2})^2}} \right]. \tag{33}$$

The screened matrix element is easily written down in analogy to the remote ionized impurity matrix element

$$V(q,z) = V_+(q, z + \frac{d_0}{2}) + V_-(q, z - \frac{d_0}{2}) = \frac{e^2}{2\epsilon_0\epsilon(0)} \cdot \frac{2e^{-qz}\sinh(\frac{qd_0}{2})}{q + q_{TF}}, \tag{34}$$

where q is the $x - y$ in-plane wavevector. This is the scattering potential experienced by an electron in the 2DEG due to a *single* dipole at a distance z from the 2DEG plane. We have to add the effect of all dipoles present for evaluating the scattering rate.

Figure 15 illustrates the physical location of the dipoles in AlGaN/GaN HEMT structures. Due to the interface roughness, there are dipoles located at the interface too; however, their effect on the 2DEG mobility is not considered in light of the far denser distribution of dipoles in the barrier. We consider the 2DEG to be physically located at the centroid z_0 of the spatially extending quasi-2DEG for illustrating the role of dipoles.

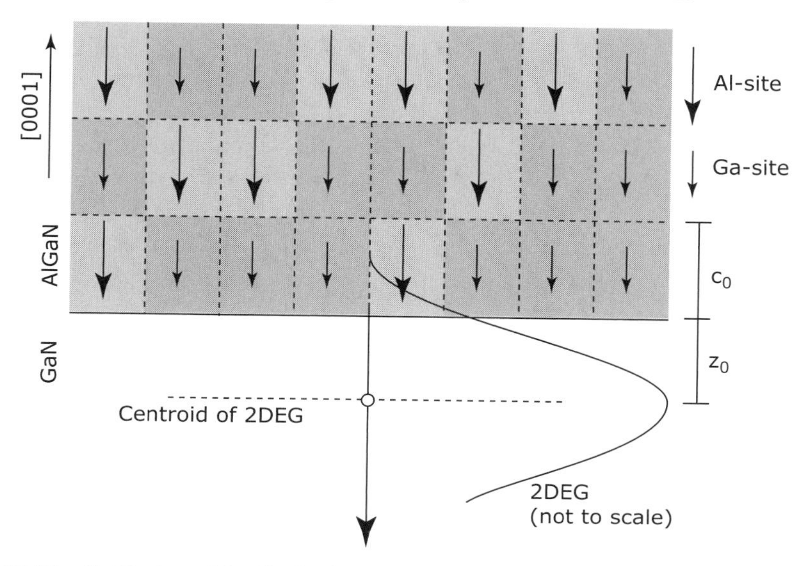

Fig. 15 The distribution of the dipoles in the AlGaN barrier is shown. The rectangular boxes depict unit cells. The dipole moment at Al sites is higher than that at the Ga sites owing to the higher spontaneous polarization and piezoelectric constants in AlN than in GaN. This fluctuation leads to a random distribution of dipole moments which leads to scattering of the electrons in the 2DEG. The 2DEG is assumed to be located entirely at the centroid of the quasi-2DEG distribution for simplicity.

The total screened scattering potential due to the distribution of dipoles in the barrier is hence given by a Fourier-weighted sum over all spatial locations of dipoles

$$V_{dip}(q) = \sum_i e^{i\mathbf{q}\cdot\mathbf{r_i}} \frac{V(q,z_i)}{\epsilon_{2d}(q)}. \tag{35}$$

If we assume that the dipole distribution on each Al(Ga) plane is completely uncorrelated, the cross-terms arising in the sum cancel, and we are left with a sum over different planes. This calls for the alloy to be disordered with no clustering of any form. The complex exponential can then be factored out and therefore does not contribute to the matrix element. For a thick AlGaN barrier, this evaluates to

$$V_{dip}(q) = \frac{e^2}{2\epsilon_0\epsilon(0)} \cdot \frac{2e^{-q(z_0+c_0)}}{1-e^{-qc_0}} \cdot \frac{\sinh(\frac{qd_0}{2})}{q+q_{TF}}, \tag{36}$$

where z_0 is the distance of the centroid of the 2DEG from the interface (figure 15), and c_0 is the separation of the planes containing the dipoles in the barrier, which is the lattice constant in the [0001] direction.

The momentum scattering rate is now evaluated by using the dipole-scattering matrix element.

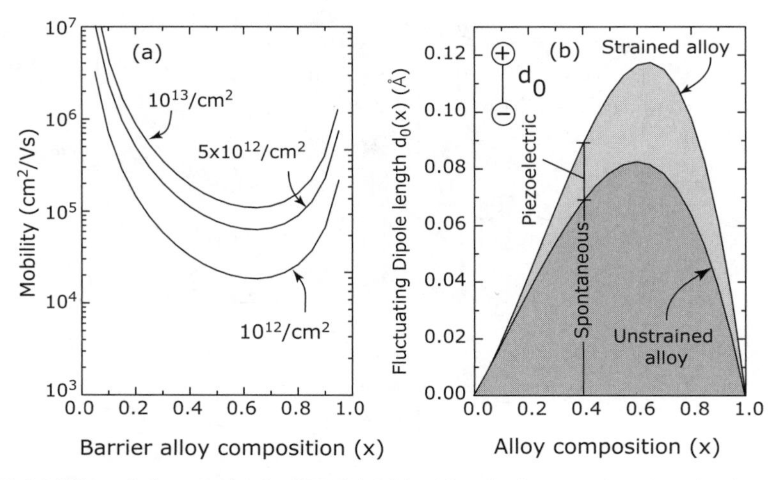

Fig. 16 Mobility of electrons in the 2DEG inhibited by dipole scattering alone is plotted as a function of alloy composition. Also shown is the fluctuating dipole length d_0.

$$\frac{1}{\langle \tau_m^{dip} \rangle} = n_{dip}^{2D} \frac{m^\star}{2\pi\hbar^3 k_F^3} \int_0^{2k_F} |V_{dip}(q)|^2 \frac{q^2 dq}{\sqrt{1-(\frac{q}{2k_F})^2}}. \tag{37}$$

where n_{dip}^{2D} is the sheet density of dipoles in any AlGaN plane. This is in the form that we can evaluate the momentum scattering time due to dipoles numerically.

The mobility inhibited by dipole scattering alone is evaluated for different alloy compositions and different 2DEG carrier densities. The results are plotted in figure 16(a). Figure 16(b) depicts the dipole length d_0 vs x for the fluctuating dipole moment in the alloy - this determines the strength of scattering. The piezoelectric and spontaneous parts are depicted separately[4].

An expected increase in mobility with the increase in the binary nature of the alloy barrier is seen. It is well worth noticing that the mobility limited by this form of scattering is much lower than the record low-temperature mobilities ($\approx 10^7$ cm^2/V·s) of AlGaAs/GaAs modulation doped heterostructures, and an order of magnitude higher than the record high mobilities in AlGaN/GaN HEMTs for the respective carrier densities.

Dipole scattering can assume increased importance if the electrons were physically located inside the alloys. Such situations can be encountered in graded base bipolar transistors, or in UV detectors that employ AlGaN active regions. The effect of dipole scattering in such cases has been treated in a recent work, where it is shown that dipole scattering can become the dominant scattering mechanism under certain situations [30]. Thus, future devices employing InGaN, AlGaN, or AlInGaN

---

[4] There is a shift from the $x(1-x)$ variation (typical in alloy scattering) due to the *larger* polarization for larger Al compositions of the barrier.

disordered alloys in active regions of devices should pay special attention to the effects of dipole scattering on the electron mobility.

3.3.2 Dislocation Scattering

A good lattice-matched substrate for the growth of III-V nitride semiconductors still remains elusive. Due to the large lattice mismatch with the present substrates of choice (sapphire or SiC), state of the art AlGaN/GaN HEMTs have a two-dimensional-electron-gas (2DEG) which has $1 - 100 \times 10^8$ cm^{-2} line dislocations passing through it. Look and Sizelove [39] analyzed the effect of dislocation scattering on the mobility of bulk GaN structures. However, dislocation scattering effects on the transport of AlGaN/GaN 2DEGs has not received much attention. An effort was made to treat dislocation scattering in a AlGaAs/InGaAs/AlGaAs quantum well before the advent of AlGaN/GaN heterostructures [40]. The scope of the work was limited, and the reasons will be explained. A theory is developed that shows that 2DEG mobility is affected strongly by a high density of dislocations. The effect is weaker, however, than that in 3D bulk; the reasons for this are pointed out.

The dislocations in AlGaN/GaN heterostructures are observed to be oriented in the direction of crystal growth, i.e., along the $[0001]$ direction [41]. Thus, electrons in the 2DEG would see the dislocation as a line perpendicular to the plane, as in figure 17.

Charged dislocation scattering in 2DEGs

A perfect 2DEG with no spatial spread in the $z-$axis is considered. The line charge density on the dislocation is assumed to be $\lambda_L = ef/c_0$, where c_0 is the lattice

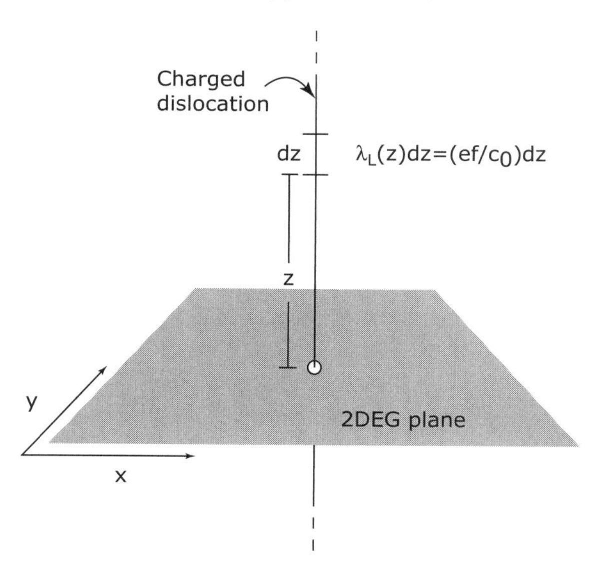

Fig. 17 Schematic of the line-charge model of a dislocation.

constant in the [0001] direction. An edge dislocation will have a dangling bond every lattice constant along its axis. These dangling bonds introduce states in the gap. f is the fraction of the states introduced by the dislocation in the energy gap that are filled. The filling factor, as well as whether edge dislocations introduce states in the bandgap has been a topic of considerable debate [41–43]. From a comparison of experimental data with the theoretical results of this work, it is possible to estimate the bounds on f, as will be seen later.

The differential contribution to the matrix element of the scattering potential of a slice of the charged line of length dz at distance z from the 2DEG plane (figure 17) is same as the matrix element due to a point charge given by (Equation 19)

$$\delta V(q,z) = \frac{e}{2\epsilon_0\epsilon(0)} \cdot \frac{e^{-qz}\lambda_L dz}{q+q_{TF}}, \tag{38}$$

whence for a dislocation that has a large length along the z-direction the total matrix element is

$$V(q) = \int dz \delta V(q,z) = \frac{e\lambda_L}{\epsilon_0\epsilon(0)q(q+q_{TF})}. \tag{39}$$

Zhao and Kuhn [40] arrived at a scattering potential which models an in-plane charged impurity rather than the spatially extending dislocation line, and did not consider the strong screening contribution in the highly degenerate 2DEG. The model used here overcomes these difficulties of the previous model.

If there are N_{dis} line dislocations piercing the 2DEG per unit area, the momentum scattering rate for the 2DEG is given by

$$\frac{1}{\langle\tau_m^{dis}\rangle} = N_{dis} \cdot \left(\frac{m^\star}{2\pi\hbar^3 k_F^3}\right) \cdot \int_0^{2k_F} |V(q)|^2 \frac{q^2 dq}{\sqrt{1-(\frac{q}{2k_F})^2}}. \tag{40}$$

Using the screened potential and the substitution $u = q/2k_F$, the scattering rate for a perfect 2DEG is

$$\frac{1}{\tau_{2d}^{dis}} = \frac{N_{dis} m^\star e^2 \lambda_L^2}{\hbar^3 \epsilon_0^2 \epsilon(0)^2} \cdot \left(\frac{I(n_s)}{4\pi k_F^4}\right) \cdot \int_0^1 \frac{du}{(u+\frac{q_{TF}}{2k_F})^2\sqrt{1-u^2}}. \tag{41}$$

The dimensionless integral

$$I(n_s) = \int_0^1 \frac{du}{(u+\frac{q_{TF}}{2k_F})^2\sqrt{1-u^2}} \tag{42}$$

can be evaluated exactly for the perfect 2DEG. It depends on the 2DEG density. For the perfect 2DEG, with $a = q_{TF}/2k_F$, the integral factor reduces to

$$I(a(n_s)) = \frac{\sqrt{1-a^2} + a^2 ln(\frac{1-\sqrt{1-a^2}}{a})}{a(1-a^2)^{\frac{3}{2}}}. \tag{43}$$

For a more realistic 2DEG, the wavefunction spread introduces the form factors of the Fang-Howard wavefunction. Using the Fang-Howard function, and re-evaluating the momentum scattering rate for the 2DEG, the charged dislocation-scattering limited 2DEG mobility (in cm^2/V·s) for AlGaN/GaN 2DEGs may be cast in the form

$$\mu_{disl} = 43365 \left(\frac{10^8 cm^{-2}}{N_{disl}}\right) \left(\frac{n_s}{10^{12} cm^{-2}}\right)^{1.34} \left(\frac{1}{f^2}\right), \tag{44}$$

where the dislocation density and the 2DEG density are in cm^{-2}, and f is the dislocation filling factor. The result in this form is convenient for numerical estimates of the strength of charged dislocation scattering for 2DEGs.

Figure 18 depicts the dislocation-scattering limited 2DEG electron mobility for (a)-changing dislocation density for three 2DEG sheet charges, (b)-changing occupation fraction of the states introduced by dislocations for $N_{dis} = 10^9$cm^{-2}, and (c)-changing sheet density for three dislocation densities. Scattering by charged dislocations is seen to be strongly affected by the fraction of filled stated at the dislocation core - it goes as $1/f^2$. Dislocation scattering limited mobility for $f = 1$ would lead to a *lower* mobility than the highest reported mobilities. Besides, the trend of mobility in 18(c) is *opposite* with increasing carrier density than what is observed (figure 7). This points towards lower charge on the dislocations ($f < 1$).

f, the fraction of filled states was first calculated by Read [44] using a simple thermodynamic model. There has been some controversy regarding the electrical activity of dislocations, as well as for the value of the occupation function f, given that dislocations are charged. Through scanning capacitance microscopy measurements, Hansen et. al. [45] showed that dislocations in GaN are electrically charged. Brazel et. al. [46] and recently, Hsu et. al. [47] have shown that dislocations offer highly preferential localized current paths. Additionally, Kozodoy et. al. [48]

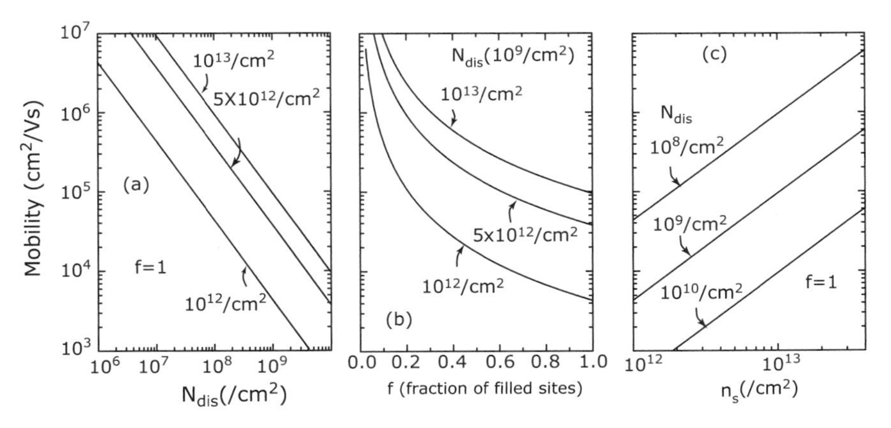

Fig. 18 Charged-dislocation scattering limited 2DEG mobility, dependence on various properties of the dislocations.

showed a direct relationship of reverse leakage currents in GaN junction diodes to the number of dislocations. Schaadt et. al. [49] confirm the notion of charged dislocations from their scanning capacitance voltage measurements, and are also able to predict the amount of charge on the dislocations and the screening lengths around the charged lines.

Hence, experimental evidence strongly suggests that dislocations introduce states in band gap. However, the controversy has been in theoretical studies. Elsner et. al. calculated the electronic properties of dislocations in III-V nitrides from both ab-initio local-density-functional methods and density-functional tight binding methods [43] for both pure edge and pure screw-type dislocations. For screw-type dislocations, the authors found deep states in the bandgap. They found no deep states in the gap for pure edge type dislocations. Calculations of Wright and Furthmüller [50] and Wright and Grossner [51], however, show that for both AlN and GaN, edge dislocations introduce electronic states in the gap. Leung et. al. did an energetics study [42] of the occupation probabilities for the states introduced by threading edge dislocations, drawing upon the theoretical results of Wright et. al. [50,51]. They find that electronic states introduced in the bandgap by threading edge dislocations can be multiply occupied, and the probability of occupation of sites is a function of the background doping density.

The energy calculations by Leung et. al [42] show that typically only $10-50\%$ of the states will be occupied $(f = 0.1-0.5)$ for a background donor density of $N_d \approx 10^{16}\text{cm}^{-3}$ and dislocation densities in the $10^8 - 10^{10}\text{cm}^{-2}$ range (which is typical of high purity molecular beam epitaxy samples). Schaadt et. al. [49] reported $f = 0.5$ from scanning capacitance-voltage measurements. This makes dislocations much more benign as scatterers than if $f = 1$, i.e., all dangling bonds were charged.

The scattering time arrived at highlights the metallic nature of the 2DEG electrons. The screening length for a 2DEG depends on q_{TF} and k_F. The Thomas Fermi wavevector q_{TF} is constant. As the free carrier density is increased, k_F increases, and λ_F, the Fermi wavelength gets shorter, leading to better screening. The 2DEG carrier density does not freeze out at low temperatures as in 3D. These factors contribute to the observed high mobilities in a 2DEG. In contrast, 3D screening and scattering is controlled by the Debye screening factor $q_D = \sqrt{e^2 n'/\epsilon k_B T}$, where n' is the effective screening concentration, involving both free and bound carriers. At low temperatures, free carriers freeze out exponentially in a semiconductor. An elongation of the Debye screening length $\lambda_D = 1/q_D$ leads to weaker screening. In addition, the carriers are less energetic, leading to strong scattering, and hence to lower mobilities. Thus, scattering from charged dislocations in AlGaN/GaN 2DEGs is not strong enough to limit electron mobilities at present, and there are stronger scattering mechanisms in operation. This insensitivity is aided by incomplete occupation of the states introduced by the dislocations.

Strain scattering from dislocations

Localized strain fields exist around point and extended defects in semiconductors. Traditionally in electronic transport theory one considers charge scattering by

Coulombic interaction of mobile carriers with charged defects; strain fields associated with defects is generally neglected. This approximation is justified for substitutional donors/acceptors for example, since the lattice distortion around them is minimal. However, for dislocations, which may or may not be charged, the strain fields can contribute substantially to scattering of mobile carriers in semiconductors, just as in metals [52–55]. Electron-strain field interaction will affect transport properties for vacancies/interstitials as well. It is important to note that this form of scattering arises due to the presence of dislocations regardless of the presence or absence of charges at the core.

Dislocations set up a strain field around them with atoms displaced from their equilibrium positions in a perfect crystal. The band extrema (conduction band minimum, valence band maximum) shift under influence of the strain fields. The magnitude of spatial variation of the band extrema to linear order in strain is given by the deformation potential theorem of Bardeen and Shockley [56].

It is necessary to start with a suitable model for behavior of quantum well band-edges in the presence of a localized strain field around a dislocation. A flat quantum well is assumed, with no built in fields, which houses a 2DEG (the case of finite extension will be taken into account with the Fang-Howard factor in the numerical evaluation). The problem of hole transport can be formulated in a similar fashion, though the strain splitting of various bands of holes makes it a more complicated affair. The effect of a strain in the quantum well is to shift the conduction and valence band edges. The shift in the conduction band edge was shown by Chuang [57] to be given by

$$\Delta E_C = a_C \; Tr(\epsilon), \tag{45}$$

where a_C is the conduction band deformation potential, and $Tr(\epsilon) = \epsilon_{xx} + \epsilon_{yy} + \epsilon_{zz} = \delta\Omega/\Omega$ is the trace of the strain matrix. The trace is also equal to the fractional change in the volume of unit cells ($\delta\Omega/\Omega$).

Dislocations perpendicular to the quantum well (2DEG) plane are considered. As an electron in the 2DEG approaches a dislocation, it experiences a potential due the strain around the dislocation, which causes scattering (see figure 19 for a schematic). The strain distribution radially outward from an edge dislocation is well known [58]. Combined with the preceding equation, we get the necessary scattering potential responsible for electron scattering

$$\delta V = \Delta E_C = a_C Tr(\epsilon) = -\frac{a_C b_e}{2\pi} \frac{1-2\gamma}{1-\gamma} \frac{\sin\theta}{r}. \tag{46}$$

Here b_e is the magnitude of the Burgers vector of the edge dislocation, and γ is the Poisson's ratio for the crystal. $\epsilon_{zz} = 0$ for an edge dislocation, and nonzero for a screw dislocation. For a screw dislocation in a cubic crystal, the strain field has purely shear strain, causing no dilation/compression of the unit cells. This means there can be no deformation potential scattering for screw dislocations in cubic crystals. However, for uniaxial crystals such as GaN, the argument does not hold, and there is a deformation potential coupling even for screw dislocations for bulk transport. Screw dislocation strain scattering is not considered.

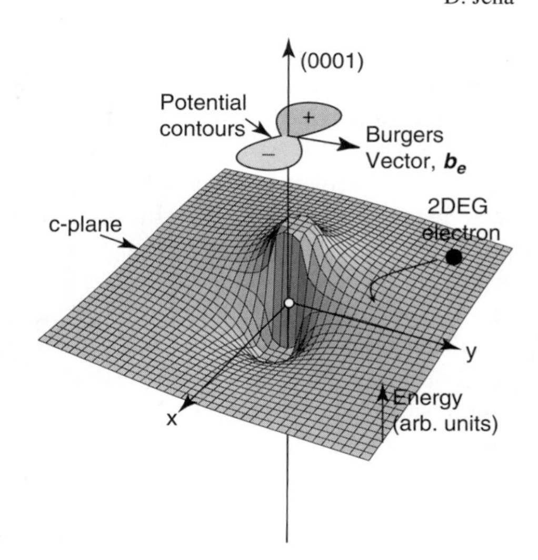

Fig. 19 Strain fields surrounding an edge dislocation and the band-edge shift caused by it as a scattering potential.

The screened matrix element for the scattering potential is

$$V(q,\phi) = \frac{b_e a_C}{2\pi S}\frac{1-2\gamma}{1-\gamma}\frac{sin(\phi)}{q+q_{TF}}, \tag{47}$$

where ϕ is the angle between **q** and $\mathbf{b_e}$, the Burgers vector. Summing the square of the matrix element over all scatterers in the dilute scatterers limit requires an average of the angular dependence over random orientations of the burger's vectors for different dislocations; averaging yields $< sin^2(\phi) >= \frac{1}{2}$. Momentum scattering rate is thus given by

$$\frac{1}{\tau_m^{str}} = \frac{N_{disl} m^\star b_e^2 a_C^2}{2\pi k_F^2 \hbar^3}(\frac{1-2\gamma}{1-\gamma})^2 I(n_s). \tag{48}$$

The dimensionless integral $I(n_s)$ given by

$$I(n_s) = \int_0^1 \frac{u^2}{(u+\frac{q_{TF}}{2k_F})^2\sqrt{1-u^2}du} \tag{49}$$

is again dependent only on the sheet density n_s, and can be evaluated explicitly.

Finally, we arrive at the dislocation strain field scattering limited electron mobility given by the Drude result $\mu = e\tau_{disl}^{strain}/m^*$

$$\mu_{disl}^{strain} = \frac{2e\hbar^3\pi k_F^2}{N_{disl} m^{\star 2} b_e^2 a_C^2}(\frac{1-\gamma}{1-2\gamma})^2\frac{1}{I(n_s)}. \tag{50}$$

Quantities needed for a numerical evaluation are the magnitude of the Burger's vector $b_e = a_0 = 3.189$Å, the conduction electron effective mass $m^* = 0.2m_0$ (m_0 is free electron mass), Poisson's ratio for the crystal, $\gamma = 0.3$ [58], and the conduction band deformation potential a_C.

For uniaxial crystals such as the wurtzite crystal, the second-rank deformation potential tensor Ξ_{ij} has two independent components, Ξ_1 and Ξ_2 at the Γ point in the E-k diagram. The volume change (compression or dilatation) leads to a shift in the band gap

$$\Delta E_G = \Xi_1 \epsilon_{zz} + \Xi_2 \underbrace{(\epsilon_{xx} + \epsilon_{yy})}_{\epsilon_\perp}, \tag{51}$$

where $\Xi_1 = a_1 = -6.5\text{eV}$ and $\Xi_2 = a_2 = -11.8\text{eV}$ for GaN ([44]). For an edge dislocation, there is no strain along the z [0001] axis ($\epsilon_{zz} = 0$); thus only Ξ_2 will be required in our analysis. The deformation potential has contributions from both the CB and the VB, $\Xi_2 = \Xi_2^{CB} + \Xi_2^{VB}$. We require only the conduction band deformation potential for our calculation. Knap et.al. measured the conduction band deformation potential from transport analysis to be $a_C = (9.1 \pm 0.7)\text{eV}$ [46]. Their measured value is used for calculations here.

Figure 20 shows the dislocation strain-field scattering-limited electron mobility for three sheet densities. Strain scattering for the AlGaN/GaN 2DEG is insensitive to strain-field scattering from dislocations at present. The strain-field scattering limited

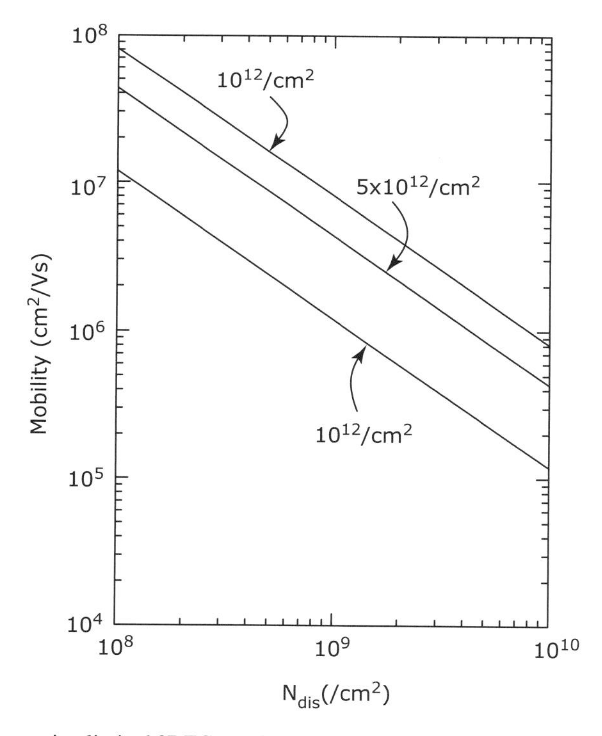

Fig. 20 Strain scattering limited 2DEG mobility.

mobility is higher than the highest reported data. At very high dislocation densities of $N_{dis} = 10^{10}\text{cm}^{-2}$, the mobility starts affecting low temperature mobility for low-density 2DEGs. However, it is still not the dominant scattering mechanism.

In addition to the deformation potential scattering from the strain fields, in non-centrosymmetric crystals such as GaN there is also a possibility of piezoelectric fields associated with dislocations. Scattering from such a field is expected to be negligible [58]. The effect of screw dislocations on transport in uniaxial crystals is a more subtle question, and is not considered.

4 Using Theory to Explain Experimental Data

Low Temperature

At low temperatures, the different scattering processes act independently; Mathiessen's rule offers a simple way of combining the effect of all scatterers. Figure 21

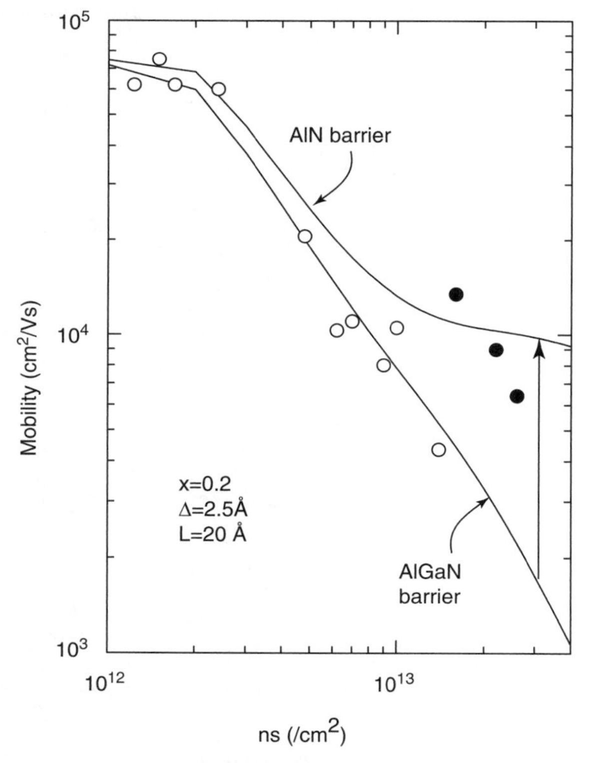

Fig. 21 Experimental versus theoretical plots of low temperature mobility of highest reported mobilities.

shows the total low temperature mobility, calculated by considering all scattering mechanisms, as a function of the 2DEG sheet density n_s. Also shown in the same figure are the experimental highest values tabulated earlier in this chapter. Mobility at typical AlGaN/GaN sheet densities is limited by short-range scatterers due to alloy disorder and interface roughness. In the range of 2DEG densities $n_{2D} \geq 10^{12}/cm^2$, alloy scattering or interface roughness scattering dominate, depending on the nature of the barrier. Alloy scattering dominates mobility for AlGaN barriers for *all* Aluminum compositions. The effects at high carrier densities is significant - even at room temperature, since at very high carrier densities, alloy-scattering limited mobility approaches the limits set by optical phonon scattering.

Introduction of a thin AlN interlayer at the AlGaN/GaN interface suppresses the penetration of the wavefunction into the barrier and effectively removes alloy scattering. This is illustrated in figure 22, where the band diagram and 2DEG distribution is shown with and without the AlN interlayer. Introducing a 1nm AlN interlayer pushes out most of the 2DEG from the barrier, and the small finite part that still penetrates the barrier does not experience alloy scattering due to the presence of AlN. This was predicted [60] by Hsu and Walukiewicz and later verified by Shen et. al. and Smorchkova et. al. [19, 61]. The introduction of a thin AlN barrier layer in a HEMT structure results in several advantages, including better confinement, and reduced alloy scattering and enhanced conductivity. These are attractive qualities for the design of HEMTs.

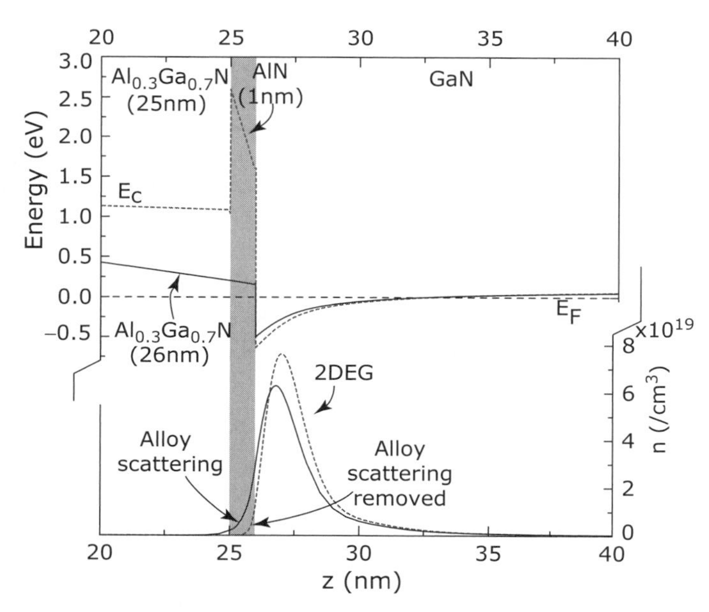

Fig. 22 The effect of introduction of a thin AlN interlayer to remove alloy scattering, enhancing the 2DEG conductivity.

The mobility limit for $n_s \geq n_{cr} = 10^{12}$ cm^{-2} is 'intrinsic' in the sense that removal of charged defects (dislocations, background impurities, etc) will not be useful in improving the mobility than those reported. The critical density n_{cr} can be used as a guideline for designing high mobility 2DEG structures. It can be predicted that highest low temperature mobilities will be achieved for lowest density ($n_s \approx n_{cr}$) 2DEGs. For the same carrier density, a barrier of AlN will have a larger mobility than an AlGaN barrier.

If carrier densities are lowered below the critical density n_{cr}, the effects of charged impurities can be probed. Reduction of carrier density in the AlGaN/GaN 2DEG is not straightforward, since the density is not controlled by intentional modulation doping (indeed, almost all high-mobility AlGaN/GaN 2DEGs are in undoped structures). Gating is one way of achieving low carrier densities; another way is the growth of GaN (or low composition AlGaN) cap layers on top of the AlGaN barrier layer. Introduction of acceptors in the barrier can also be used to reduce the 2DEG density by compensation, though p-doping in the nitrides is currently not under good control. Growth along non-polar faces, as pioneered by Waltereit et. al. [10] may be a good alternative for exercising precise control over 2DEG densities by intentional modulation doping.

Variable temperature

At high temperatures ($T > 100$K), the approximation that various scattering processes are independent breaks down due to strong optical phonon scattering. However, since the total scattering rate is dominated by optical phonon scattering, using Mathiessen's rule will not cause significant deviations from a more accurate calculation [22].

For comparing theory to experimental mobility data, three samples with carrier densities of (a) $n_s = 2.23 \times 10^{12}$ cm^{-2} [13], (b) $n_s = 5 \times 10^{12}$ cm^{-2} [63], and (c) $n_s = 10^{13}$ cm^{-2} [17] were chosen. The temperature dependent mobility data and the calculated curves are shown in figure 23(a), (b), and (c) respectively. The

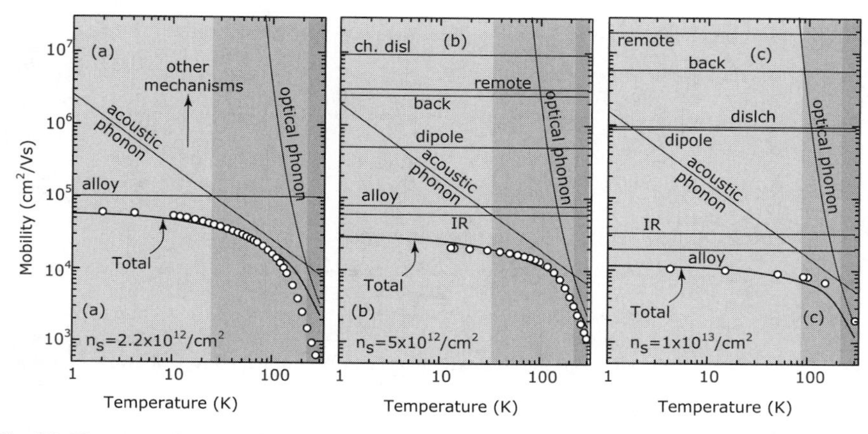

Fig. 23 Three samples with different 2DEG densities studied for the behavior of alloy scattering.

characteristic three regions predicted for low-density 2DEGs is shown in different shades of grey.

For sample (a), the mobility reaches $\mu = 62,000$ cm^2/V$\cdot$s at the lowest temperatures, limited by alloy scattering. The region of mobility for temperature 10K $<$ $T <$ 100K is acoustic phonon scattering limited. The value of acoustic phonon-scattering can be *extracted* by fitting the calculated mobility to the measured value; this yield $a_C = 9.1$eV, close to the value found by Knap [46]. At still higher temperatures $T >$ 150K, sample (a) showed parallel conduction through the underlying GaN layer, and the measured mobility dropped from what is theoretically predicted due to phonon scattering.

The second sample (b) shows a rather good fit of theory to the experimental values. The sample exhibited no parallel conduction, and thus the high-temperature values are accurate. Alloy scattering gets more severe than acoustic phonon scattering - so does interface roughness scattering; they cut into the region dominated by acoustic phonon scattering for lower carrier density samples. This effect becomes even more pronounced in figure 23(c), where the 2DEG density is higher (10^{13} cm^{-2}). The acoustic phonon-scattering limited mobility region is completely suppressed by severe alloy scattering.

Since most HEMT structures require high 2DEG sheet densities for high conductivity, alloy scattering has to be reduced for improving the conductivity. Beneficial properties of AlN layers are thus a subject of much current interest [19].

High Field effects

At high electric fields, the electrons gain more energy from the field than the optical phonon energy before they can scatter off defects, and end up emitting optical phonons copiously [64]. This causes a saturation of the drift velocity, thus entering into a non-linear regime of transport. Saturation velocity of 2DEG carriers is thus an important design parameter for HEMTs, since it limits the maximum drain-source current achievable. A high saturation velocity is highly desirable for high-power devices.

A crude estimate of the saturation velocity can be made by equating the kinetic energy the electrons gained from the electric field to the optical phonon energy. Assuming that the electron loses all its kinetic energy gained from the field by emitting optical phonons, the time-averaged saturated drift-velocity is

$$v_{sat} \approx \frac{1}{2}\sqrt{\frac{2\hbar\omega_{op}}{m^{\star}}}, \tag{52}$$

which yields $v_{sat} \approx 2 \times 10^7$ cm/s. This is close to experimental values [65], and that calculated by a Monte-Carlo simulation [66]. Inter-valley transfer of electrons in GaN is predicted to cause negative differential conductivity [66, 67], which has not been experimentally observed.

In spite of the large effective mass of electrons in GaN ($0.2m_0$) compared to GaAs ($0.067m_0$), the optical phonon energy of GaN (92 meV) is much larger than of GaAs (36 meV), resulting in similar saturation velocities. However, the 2DEG densities achievable by polarization doping in AlGaN/GaN structures is much larger than that can be achieved by modulation doping in AlGaAs/GaAs structures. The larger *conductivity* makes AlGaN/GaN structures more suited for high-power HEMTs.

5 Summary and Conclusions

Results that are relevant for the design of HEMTs structures (or any other structures requiring AlGaN/GaN 2DEGs) with high conductivity are -

(a) Mobility of low-density AlGaN/GaN 2DEGs ($n_{2d} \leq 10^{12}$ cm^{-2}) is limited by scattering from charged defects (dislocations, dipoles, residual impurities).
(b) Mobility of high-density AlGaN/GaN 2DEGs is *insensitive* to scattering by various *charged impurities* (dislocations, dipole, residual impurities).
(c) Alloy disorder scattering limits the mobility for AlGaN/GaN 2DEGs at low temperatures. At extremely high carrier densities, alloy scattering is as severe as scattering from phonons, even at room temperature.
(d) Alloy scattering can be removed by the introduction of a thin AlN interlayer at the AlGaN/GaN heterojuntion, or with a AlN barrier. In such a case, interface roughness scattering is the mobility limiting scattering mechanism.

6 Appendix on the Theory of Low-Field Transport & Mobility

This section is intended to be pedagogical. Much of the material in this section is collected from textbooks and research articles. The main references for this section are Seeger [25], Wolfe et. al. [64], and Davies [5]. No claim to originality is made for much of the material. The subsection on *generalization* of mobility expressions for arbitrary dimensions and arbitrary degeneracy is original, though much of it is inspired from the references.

We begin our study of charge transport with a brief outline of the microscopic picture. Even in the absence of any electric or magnetic fields, the charge carriers (electrons or holes) in a semiconductor are in a constant state of motion. The carriers (say electrons) possess kinetic energy of the order of $k_B T$, where k_B is the Boltzmann constant, and T the absolute temperature. This energy is exchanged continually between the electrons and the crystal lattice, and the two systems are in thermal equilibrium. Since the thermal velocities of an ensemble of electrons point in random directions, the net current flowing across any surface is zero, and therefore the thermal kinetic energy of electrons does not result in a net current if the spatial density is uniform. However, if there exist gradients in spatial density of electrons, the random thermal kinetic energy drives a diffusion current density, given by $J_n^{diff} = eD_n dn(x)/dx$, where D_n is the diffusion constant, and e is the electron charge. If there are no external fields, the diffusion current is exactly balanced by an opposing drift current that results from the electric field due to charge imbalance caused by the diffusive flow of electrons. This exact cancellation of the drift and diffusion current components leads to the Einstein relation, $D_n/\mu_n = k_B T/e$, where μ_n is the electron mobility.

Under the application of an external electric field, electrons acquire an additional 'drift' component to the velocity in addition to the random thermal component. The drift component points in the direction opposite to the electric field. To understand the motion of a large number of electrons, one needs to resort to results from statistical thermodynamics - in particular, one needs to understand how the *distribution function* behaves in the presence of external perturbations. A formal theory that achieves this goal was first proposed by Ludwig Boltzmann to understand the kinetic theory of gases. The whole theory is encapsulated in one formal equation, the much-used Boltzmann Transport Equation (BTE).

6.1 The Boltzmann Transport Equation

The distribution-function $f(\mathbf{k}, \mathbf{r}, t)$ is defined as the probability that an electron at time t resides in space in the location $(\mathbf{r}, \mathbf{r} + \mathbf{dr})$ with wavevectors lying between $(\mathbf{k}, \mathbf{k} + \mathbf{dk})$. Under thermal equilibrium ($\mathbf{E} = \mathbf{B} = \nabla_r f = \nabla_T f = 0$, i.e., no external electric ($\mathbf{E}$) or magnetic ($\mathbf{B}$) field and no spatial and thermal gradients), the distribution function for fermionic particles (both electrons and holes) is found from quantum-statistical analysis to be the Fermi-Dirac function -

$$f_0(\varepsilon) = \frac{1}{1 + e^{\frac{\varepsilon_{\mathbf{k}} - \varepsilon_F}{k_B T}}}, \tag{53}$$

where $\varepsilon_{\mathbf{k}}$ is the momentum-dependent energy of the electron, and ε_F is the Fermi energy. The energy of the electron is related to its momentum ($\hbar\mathbf{k}$) through the dispersion relation, which for the semiconductor is the bandstructure $\varepsilon(\mathbf{k})$. The bandstructures for most semiconductors are well-known.

Any external perturbation drives the distribution function away from the equilibrium. The Boltzmann-transport equation (BTE) is a book-keeping construct that relates the shift of the distribution function from equilibrium to the driving forces. It is written formally as [64]

$$\frac{df}{dt} = -\Big[\frac{\mathbf{F_t}}{\hbar} \cdot \nabla_{\mathbf{k}} f(\mathbf{k})\Big] - \big[\mathbf{v} \cdot \nabla_{\mathbf{r}} f(\mathbf{k})\big] + \Big[\frac{\partial f}{\partial t}\Big|_{coll.}\Big], \tag{54}$$

where on the right hand side, the first term tracks the change in the distribution function due to the total Lorentz force $\mathbf{F_t} = \mathbf{E} + \mathbf{v} \times \mathbf{B}$, the second term is the change due to concentration gradients, and the last term is the change due to collisions (scattering by defects & lattice vibrations, i.e., phonons). One way to visualize the BTE is that while the force and the diffusion terms (the first two) try to drive the system away from the equilibrium state (reduce f), the electron distribution tries to relax back to equilibrium through collisions (the collision term increases the weight of f).

Since the total number of carriers in the semiconductor is constant in the steady state, the total rate of change of the distribution function is identically zero by Liouville's theorem. Hence, one gets

$$\frac{\partial f}{\partial t}|_{coll} = \frac{\mathbf{F_t}}{\hbar} \cdot \nabla_{\mathbf{k}} f(\mathbf{k}) + \mathbf{v} \cdot \nabla_{\mathbf{r}} f(\mathbf{k}). \tag{55}$$

Scattering of electrons by defects and lattice vibrations changes their momentum. Denoting the scattering rate from state $\mathbf{k} \rightarrow \mathbf{k}'$ as $S(\mathbf{k},\mathbf{k}')$, the collision term is given by

$$\frac{\partial f(\mathbf{k})}{\partial t}|_{coll} = \sum_{\mathbf{k}'}[S(\mathbf{k}',\mathbf{k})f(\mathbf{k}')[1-f(\mathbf{k})] - S(\mathbf{k},\mathbf{k}')f(\mathbf{k})[1-f(\mathbf{k}')]]. \tag{56}$$

Figure 24 provides a visual representation of the collision term. The increase of the distribution function in the small volume $\Delta\mathbf{k}$ by particles flowing in by the field term is balanced by the net flow out by the two collision terms.

At equilibrium ($f = f_0$), the principle of detailed balance enforces the condition

$$S(\mathbf{k}',\mathbf{k})f_0(\mathbf{k}')[1-f_0(\mathbf{k})] = S(\mathbf{k},\mathbf{k}')f_0(\mathbf{k})[1-f_0(\mathbf{k}')], \tag{57}$$

which translates to

$$S(\mathbf{k}',\mathbf{k})e^{\frac{\varepsilon_{\mathbf{k}}}{k_B T}} = S(\mathbf{k},\mathbf{k}')e^{\frac{\varepsilon_{\mathbf{k}'}}{k_B T}}. \tag{58}$$

In the special case of *elastic* scattering, $\varepsilon_{\mathbf{k}} = \varepsilon_{\mathbf{k}'}$, and as a result, $S(\mathbf{k}',\mathbf{k}) = S(\mathbf{k},\mathbf{k}')$ irrespective of the nature of the distribution function. Using this, the collision term is written as

$$\frac{\partial f(\mathbf{k})}{\partial t}|_{coll} = \sum_{\mathbf{k}'} S(\mathbf{k},\mathbf{k}')(f(\mathbf{k}') - \mathbf{f}(\mathbf{k})). \tag{59}$$

A characteristic time scale of collisions can be defined by writing this collision equation in another form:

$$\frac{\partial f(\mathbf{k})}{\partial t}|_{coll} + \frac{f(\mathbf{k})}{\tau_q(\mathbf{k})} = \sum_{\mathbf{k}'} S(\mathbf{k},\mathbf{k}')f(\mathbf{k}'), \tag{60}$$

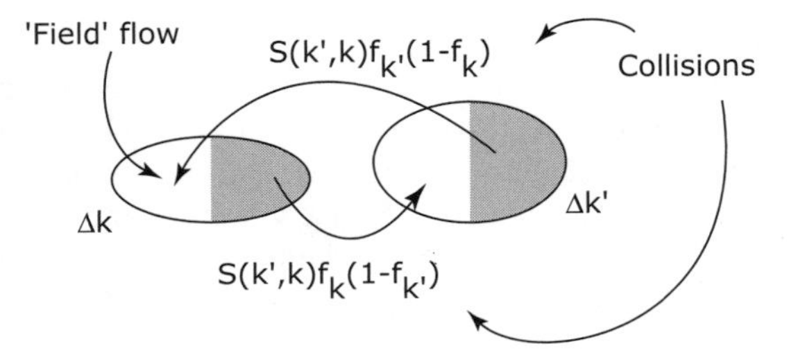

Fig. 24 Scattering term of Boltzmann transport equation depicting the inflow and outflow of the distribution function. The shaded regions represent occupied states, and the unshaded parts represent unoccupied states.

where the *quantum scattering time* is defined as

$$\frac{1}{\tau_q(\mathbf{k})} = \sum_{\mathbf{k}'} S(\mathbf{k},\mathbf{k}'). \tag{61}$$

A particle in state $|\mathbf{k}\rangle$ at time $t = 0$ will be scattered into other states $|\mathbf{k}'\rangle$ due to collisions, and the distribution function in that state will approach the equilibrium distribution exponentially fast with the time constant $\tau_q(\mathbf{k})$ upon the removal of the applied field. The quantum scattering time $\tau_q(\mathbf{k})$ may be viewed as a 'lifetime' of the particle in the state $|\mathbf{k}\rangle$.

Let us now assume that the external fields and gradients have been turned on for a long time. They have driven the distribution function to a *steady state* value f from f_0. The perturbation is assumed to be small, i.e., distribution function is assumed not to deviate far from its equilibrium value of f_0. Under this condition, the collision term in the BTE may be approximated by

$$\frac{\partial f}{\partial t}|_{coll} \approx -\frac{f - f_0}{\tau}, \tag{62}$$

where τ is a time scale characterizing the relaxation of the distribution due to collisions, and is intimately connected to the electron mobility. This is the relaxation time approximation (RTA), which is crucial for obtaining a solution of the Boltzmann transport equation.

When the distribution function reaches a steady state, the complete BTE (equation 54) may be written as

$$\frac{df}{dt} = -\left[\frac{\mathbf{F_t}}{\hbar} \cdot \nabla_{\mathbf{k}} f(\mathbf{k})\right] - \left[\mathbf{v} \cdot \nabla_{\mathbf{r}} f(\mathbf{k})\right] + \left(-\frac{f - f_0}{\tau}\right) = 0, \tag{63}$$

where the relaxation time approximation (RTA) to the collision term has been used. In the absence of concentration gradients, the distribution function simplifies to

$$f(\mathbf{k}) = f_0(\mathbf{k}) - \tau \frac{\mathbf{F_t}}{\hbar} \cdot \nabla_{\mathbf{k}} f(\mathbf{k}). \tag{64}$$

Using the definition of the velocity $\mathbf{v} = 1/\hbar(\partial \varepsilon_{\mathbf{k}}/\partial k)$, the distribution function becomes

$$f(\mathbf{k}) = f_0(\mathbf{k}) - \tau \mathbf{F_t} \cdot \mathbf{v} \frac{\partial f(\mathbf{k})}{\partial \varepsilon}. \tag{65}$$

With the RTA, the complete BTE has been reduced to a much simpler form. However, we have not yet obtained a 'solution' since the unknown distribution function f occurs on both sides of equation 65. At this point, it can be argued that since f and f_0 are not too different, $\partial f/\partial \varepsilon$ and $\partial f_0/\partial \varepsilon$ cannot be too different either. Then, the replacement $f(\mathbf{k}) \to f_0(\mathbf{k})$ yields

$$f(\mathbf{k}) = f_0(\mathbf{k}) - \tau \mathbf{F_t} \cdot \mathbf{v} \frac{\partial f_0(\mathbf{k})}{\partial \varepsilon}, \tag{66}$$

which is the *solution* of the BTE for a perturbing Lorentz force $\mathbf{F_t}$. It is a solution since everything on the right hand side is known, the only unknown is $f(\mathbf{k})$.

BTE in the presence of an Electric field

The external force $\mathbf{F_t}$ may be due to electric or magnetic fields. We first look for the solution in the presence of only the electric field; thus, $\mathbf{F_t} = -e\mathbf{E}$.

Using Equation 66, for elastic scattering processes one immediately obtains

$$f(\mathbf{k}') - f(\mathbf{k}) = \underbrace{e\tau\frac{\partial f_0}{\partial \varepsilon}\mathbf{E}\cdot\mathbf{v}}_{f(\mathbf{k})-f_0(\mathbf{k})}(1 - \frac{\mathbf{E}\cdot\mathbf{v}'}{\mathbf{E}\cdot\mathbf{v}}) \tag{67}$$

for a parabolic bandstructure ($\mathbf{v} = \hbar\mathbf{k}/m^\star$). Using this relation, the collision term in the form of the relaxation time approximation becomes

$$\frac{\partial f(\mathbf{k})}{\partial t} = \sum_{\mathbf{k}'} S(\mathbf{k},\mathbf{k}')(f(\mathbf{k}') - f(\mathbf{k})) = -\frac{(f(\mathbf{k}) - f_0(\mathbf{k}))}{\tau_m(\mathbf{k})}, \tag{68}$$

where a new relaxation time is defined by

$$\frac{1}{\tau_m(\mathbf{k})} = \sum_{\mathbf{k}'} S(\mathbf{k},\mathbf{k}')(1 - \frac{\mathbf{E}\cdot\mathbf{k}'}{\mathbf{E}\cdot\mathbf{k}}). \tag{69}$$

This is the *momentum relaxation time*.

Let the vectors $\mathbf{k}, \mathbf{k}', \mathbf{E}$ be directed along random directions in the 3−dimensional space. We fix the z−axis along $\mathbf{k}$ and the y−axis so that $\mathbf{E}$ lies in the $y-z$ plane. From figure 25, we get the relation

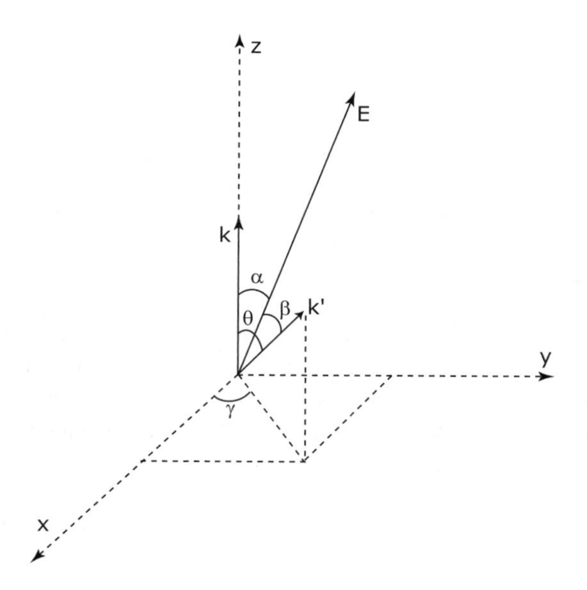

Fig. 25 Angular relations between the vectors in Boltzmann transport equation.

$$\frac{\mathbf{k}' \cdot \mathbf{E}}{\mathbf{k} \cdot \mathbf{E}} = \cos\theta + \sin\theta \sin\gamma \tan\alpha, \tag{70}$$

where the angles are shown in the figure.

When the sum over *all* $\mathbf{k}'$ is performed for the collision term, the $\sin(\gamma)$ sums to zero and the momentum relaxation time $\tau_m(\mathbf{k})$ becomes

$$\frac{1}{\tau_m(\mathbf{k})} = \sum_{\mathbf{k}'} S(\mathbf{k}, \mathbf{k}')(1 - \cos\theta). \tag{71}$$

We note here that this relation can be generalized to an arbitrary number of dimensions, the three-dimensional case was used as a tool. This is the general form for momentum scattering time, which is used heavily in the text for finding scattering rates determining mobility. It is related to mobility by the Drude relation $\mu = e\langle\tau(\mathbf{k})\rangle/m^\star$, where the momentum scattering time has been averaged over all energies of carriers.

The quantum scattering rate $1/\tau_q(\mathbf{k}) = \sum_{\mathbf{k}'} S(\mathbf{k}, \mathbf{k}')$ and the momentum scattering rate $1/\tau_m(\mathbf{k}) = \sum_{\mathbf{k}'} S(\mathbf{k}, \mathbf{k}')(1 - \cos\theta)$ are both experimentally accessible quantities, and provide a valuable method to identify the nature of scattering mechanisms. The momentum scattering time $\tau_m(\mathbf{k})$ measures the average time spent by the particle moving along the external field. It differs from the quantum lifetime due to the $\cos\theta$ term. The angle θ is identified from figure 25 as the angle between the initial and final wavevectors upon a scattering event. Thus for scattering processes that are isotropic $S(\mathbf{k}, \mathbf{k}')$ has no angle dependence, the $\cos\theta$ term sums to zero, and $\tau_q = \tau_m$. However, for scattering processes that favor small angle ($\theta \to 0$) scattering, it is easily seen that $\tau_m > \tau_q$.

6.2 Mobility-Basic Theory

We will now arrive at a general expression for the drift mobility of carriers of arbitrary degeneracy confined in d spatial dimensions. d may be 1,2 or 3; for $d = 0$, the carrier in principle does not move in response to a field. Let the electric field be applied along the ith spatial dimension, ($\mathbf{E} = E_i\mathbf{i}$) and the magnetic field $\mathbf{B} = 0$. We assume an isotropic effective mass $m^\star$. Starting from the Boltzmann equation for the distribution function of carriers $f(k, r, t)$, and using the relaxation-time approximation solution, we write the distribution function as

$$f(\mathbf{k}) = f_0(\mathbf{k}) + eF_i\tau(k)v_i\frac{\partial f_0}{\partial \varepsilon}, \tag{72}$$

where $\tau(k)$ is the momentum relaxation time and v_i is the velocity of carriers in the ith direction in response to the field.

The total number of carriers per unit 'volume' in the $d-$dimensional space is

$$n = \int \frac{d^d k}{(2\pi)^d} f(k) = \int d\varepsilon f(\varepsilon) g_d(\varepsilon), \tag{73}$$

where the generalized $d-$dimensional DOS expressed in terms of the energy of carriers is given by

$$g_d(\varepsilon) = \frac{1}{2^{d-1}\pi^{\frac{d}{2}}\Gamma(\frac{d}{2})}(\frac{2m^{\star}}{\hbar^2})^{\frac{d}{2}}\varepsilon^{\frac{d}{2}-1}. \tag{74}$$

Here $\hbar$ is the reduced Planck's constant and $\Gamma(...)$ is the gamma function. Using this, and the parabolic dispersion we can switch between the k-space and energy-space.

The current in response to the electric field along the ith direction is given by

$$\mathbf{J} = 2e\int \frac{d^d k}{(2\pi)^d}\mathbf{v}f(\mathbf{k}). \tag{75}$$

Using the distribution function from the solution of the BTE, we see that the f_0 term integrates out to zero, and only the second term contributes to a current.

For a particle moving in $d-$dimensions the total kinetic energy ε is related to the average squared velocity $\langle v_i^2\rangle$ along *one* direction by the expression $\langle v_i^2\rangle = 2\varepsilon/dm^{\star}$. Using this result, we re-write the current as

$$J_i = en\underbrace{(-\frac{2e}{dm^{\star}}\frac{\int d\varepsilon\tau_m\varepsilon^{\frac{d}{2}}\frac{\partial f_0}{\partial\varepsilon}}{\int d\varepsilon f_0(\varepsilon)\varepsilon^{\frac{d}{2}-1}})}_{\mu_d}F_i, \tag{76}$$

where the mobility in the $d-$dimensional case is denoted by the underbrace. τ_m, the momentum relaxation time due to scattering events calculated in the Born approximation by Fermi's golden rule using the scattering potential, turns out to depend on the energy of the mobile carrier and the temperature. Let us assume that it is possible to split off the energy dependence of the relaxation time in the form

$$\tau_m = \tau_0\left(\frac{\varepsilon}{k_B T}\right)^n, \tag{77}$$

where τ_0 does not depend upon the energy of the carriers. Using this, and the fact that $f_0(\varepsilon)\to 0$ as $\varepsilon\to\infty$ and $\varepsilon^m\to 0$ as $\varepsilon\to 0$, the expression for mobility can be converted by an integration by parts to

$$\mu_d = \frac{e\tau_0}{m^{\star}}\cdot\left(\frac{\Gamma(\frac{d}{2}+n+1)}{\Gamma(\frac{d}{2}+1)}\right)\cdot\left(\frac{F_{\frac{d}{2}+n-1}(\zeta)}{F_{\frac{d}{2}-1}(\zeta)}\right), \tag{78}$$

where $F_j(\zeta)$ are the traditional Fermi-Dirac integrals of the jth order defined as

$$F_j(\zeta) = \frac{1}{\Gamma(j+1)}\int_0^{\infty} dx\frac{x^j}{1+e^{x-\zeta}}. \tag{79}$$

Equation 78 may be viewed as a *generalized* formula for mobility of carriers in $d-$dimensional space.

Note that this is a *general* expression that holds true for an *arbitrary* degeneracy of carriers that are confined in *arbitrary* (d) dimensions. We now proceed to use this form of the expression for determining the mobility for two extreme cases. The strongly non-degenerate ('ND') case, where $\zeta \ll -1$ and the strongly degenerate ('D') case, where $\zeta \gg +1$.

For the non-degenerate case, the Fermi integrals can be shown to reduce to $F_j(\zeta) \approx e^{\zeta}$. This reduces the expression for mobility to the simple form

$$\mu_d^{ND} \approx \frac{e\tau_0}{m^\star}\left(\frac{\Gamma(\frac{d}{2}+n+1)}{\Gamma(\frac{d}{2}+1)}\right). \tag{80}$$

For the strongly degenerate case, we make another approximation of the Fermi-Dirac integral. We re-write it as

$$F_j(\zeta) = \frac{e^{\zeta}}{\Gamma(j+1)}\left[\int_0^{\zeta} dx \frac{x^j}{e^{\zeta}+e^x} + \int_{\zeta}^{\infty} dx \frac{x^j}{e^{\zeta}+e^x}\right]. \tag{81}$$

In the first integral, $e^x \ll e^{\zeta}$ and in the second integral, $e^x \gg e^{\zeta}$ since $\zeta \gg 1$. Using this and retaining only the *leading* power of ζ, the Fermi-Dirac integral can be approximated as $F_j(\zeta) \approx \zeta^{j+1}/\Gamma(j+2)$. Further, $\zeta = \varepsilon_F/k_BT$ where ε_F is the Fermi-energy that is known for the degenerate case if one knows *only* the carrier density. So the expression for degenerate carrier mobility finally reduces to the simple form

$$\mu_d^D \approx \frac{e\tau_0}{m^\star}\left(\frac{\varepsilon_F}{k_BT}\right)^n. \tag{82}$$

The validity of the degenerate and non-degenerate limits rests on the accuracy of the approximations made to the Fermi-Dirac integrals. For strong degeneracy and non-degeneracy, the approximations for the three-dimensional case are shown with the exact Fermi-Dirac integrals in figure 26.

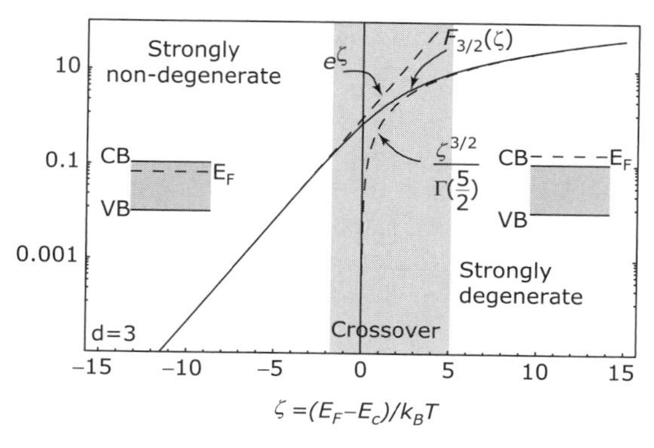

Fig. 26 The accuracy of the approximations to the Fermi-Dirac integral in extreme degeneracy and extreme non-degeneracy.

6.3 Statistics for Two- and Three-Dimensional Carriers

The concentration of free carriers in the conduction band determines the location of the Fermi level. The carrier density for the $d-$dimensional case is given by

$$n = \int d\varepsilon g_d(\varepsilon) f(\varepsilon), \tag{83}$$

where $g_d(\varepsilon)$ is the $d-$dimensional density of states and $f(\varepsilon)$ is the distribution function. The distribution function is the solution of the Boltzmann transport equation. From the Boltzmann transport equation, the perturbation term in the distribution function has a $\partial f_0/\partial k$ term that is odd in k and integrates to zero. So the only term contributing to the carrier density is $f_0(\varepsilon)$, the equilibrium value of the distribution function given by the Fermi-Dirac function. This is saying nothing more than the fact that the carrier density does not change from the equilibrium value upon application of a field. Thus, the carrier density for the $d-$dimensional case is evaluated using the generalized $d-$dimensional density of states to be

$$n = \frac{1}{2^{d-1}} \left(\frac{2m^\star k_B T}{\pi \hbar^2} \right)^{d/2} F_{\frac{d}{2}-1}(\zeta), \tag{84}$$

where $F_{d/2-1}(\zeta)$ is the Fermi-Dirac integral.

For the three-dimensional case, it reduces to the form

$$n_{3d} = \underbrace{2 \left(\frac{m^\star k_B T}{2\pi \hbar^2} \right)^{3/2}}_{N_c^{3d}} F_{1/2}(\zeta), \tag{85}$$

where N_c^{3d} is the 3-d band-edge density of states. The result holds true for arbitrary degeneracy. Sometimes, ζ is needed as a function of the carrier density and temperature; this is achieved by inverting the above expression by a numerical technique (the Joyce-Dixon approximation [68])

$$\zeta \simeq \ln(\frac{n}{N_c}) + \sum_{m=1}^{4} A_m (\frac{n}{N_c})^m, \tag{86}$$

where the constants $A_m = 3.536 \times 10^{-1}, -4.950 \times 10^{-3}, 1.484 \times 10^{-4}, -4.426 \times 10^{-6}$ for $m = 1, 2, 3, 4$ respectively. The Joyce-Dixon approximation holds good for the *entire* range of degeneracies that are achievable in semiconductors.

Similarly, for the two-dimensional case, we get immediately

$$n_{2d} = \underbrace{\frac{m^\star k_B T}{\pi \hbar^2}}_{N_c^{2d}} \ln(1 + e^\zeta), \tag{87}$$

which is a well known result for 2-d carrier density. For the 2-d case, $\zeta = (\varepsilon_F - \varepsilon_i)/k_BT$ where ε_i is the lowest subband energy.

6.4 Screening by Two- and Three-Dimensional Carriers

An important effect of the presence of mobile carriers in a semiconductor is screening. Since we are interested in scattering of mobile carriers from various defect potentials in the III-V nitrides, we summarize the theoretical tool used to attack the problem of screening in the presence of free carriers in the semiconductor.

The permittivity of vacuum is denoted as ϵ_0. If a material has no free carriers, an external d.c. electric field E will be scaled due to screening by movement of electron charge clouds of the atoms and the nuclei themselves - this yields the dielectric constant of the material, $\epsilon(0)$. The electric field inside the material is accordingly scaled down to $E/\epsilon(0)\epsilon_0$. If the electric field is oscillating in time, the screening by atomic polarization becomes weaker since the nuclei movements are sluggish, and in the limit of a very fast changing field, only the electron charge clouds contribute to screening, resulting in a reduced dielectric constant $\epsilon(\infty) < \epsilon(0)$. These two material constants are listed for the III-V nitrides in Table 2, and are related to the transverse and longitudinal modes of optical phonons by the Lyddane-Sachs-Teller equation $\epsilon(0)/\epsilon(\infty) = \omega_{LO}^2/\omega_{TO}^2$ [33].

The situation is more lively in the presence of mobile carriers in the conduction band [70]. In the situation where the perfect periodic potential of the crystal lattice is disturbed by a most general perturbing potential $V(r)e^{i\omega t}e^{-\Gamma t}$ (the potential may be due to a defect, impurity, or band variations due to phonons), *additional* screening of the potential is achieved by the flow of the mobile carriers. Lindhard first attacked this problem and with a random-phase approximation (RPA), arrived at a most-general form of the relative dielectric constant $\epsilon(q,\omega)$ given by [71]

$$\epsilon(q,\omega) = \epsilon(\infty) + (\epsilon(0) - \epsilon(\infty))\frac{\omega_{TO}^2}{\omega_{TO}^2 - \omega^2} + \epsilon(0)V_{uns}(q)\sum_{\mathbf{k}}\frac{f_{\mathbf{k}-\mathbf{q}} - f_{\mathbf{k}}}{\hbar\omega + i\Gamma + \varepsilon_{\mathbf{k}-\mathbf{q}} - \varepsilon_{\mathbf{k}}}. \tag{88}$$

Here, the first two terms take into account the contributions from the nuclei, the core electron clouds, and the valence electron clouds. The last term has a sum running over the free carriers only, and is zero for an intrinsic semiconductor. With this form of the dielectric function, the unscreened spatial part of the perturbation $V_{uns}(q)$ gets screened to $V_{scr} = V_{uns}(q)/\epsilon(q,\omega)$. Here $V_{uns}(q)$ is the Fourier-coefficient of the perturbing potential $V(q) = \int d^d r e^{iqr} V(r)$.

We are interested exclusively in *static* perturbations (defects in the material), and thus the time dependent part $\omega, \hbar\omega + i\Gamma \to 0$. With the approximations $f_{\mathbf{k}-\mathbf{q}} - f_{\mathbf{k}} \approx -\mathbf{q}\cdot\nabla_{\mathbf{k}} f_{\mathbf{k}}$ and $\varepsilon_{\mathbf{k}-\mathbf{q}} - \varepsilon_{\mathbf{k}} \approx -\hbar^2\mathbf{q}\cdot\mathbf{k}/m^\star$, the dielectric function may be converted to [33]

$$\epsilon(q) = \epsilon_0\left(1 + V(q)\sum_k \frac{\partial f}{\partial \varepsilon}\right), \tag{89}$$

which is a very useful form that applies regardless of the dimensionality of the problem.

For 2-dimensional carriers, a Coulombic potential $V(r)$ which has the well-known Fourier transform $V_{2d}(q) = e^2/L^2\epsilon(0)\epsilon_0 q$, where L^2 is the 2DEG area and q is the 2DEG wavevector [5], the dielectric function may be written as

$$\epsilon_{2d}(q) = \epsilon(0)(1 + \frac{e^2}{q\epsilon(0)\epsilon_0} \frac{\partial(\sum_k f_k/L^2)}{\partial \varepsilon}) = \epsilon(0)(1 + \frac{q_{TF}}{q}). \tag{90}$$

Since the factor in brackets is the sheet density $\sum_k f_k/L^2 = n_s$, we get the 'Thomas-Fermi' screening wavevector q_{TF} given by

$$q_{TF} = \frac{m^\star e^2}{2\pi\epsilon(0)\epsilon_0\hbar^2} = \frac{2}{a_B^\star}, \tag{91}$$

$a_B^\star$ being the effective Bohr-radius in the semiconductor. Thus, the screening in a perfect 2DEG is surprisingly *independent* of the 2DEG density, and depends only on the basic material properties, within limits of the approximations made in reaching this result [21]. For quasi-2DEGs, where there is a finite extent of the wavefunction in the third dimension, the dielectric function acquires form-factors that depend on the nature of the wavefunction. Finally, the screened 2-d Coulomb potential is given by

$$V_{scr}(q) = \frac{e^2}{\epsilon_0\epsilon(0)(q + q_{TF})}. \tag{92}$$

Similarly, for the 3-d case, the Coulomb potential $V(q) = e^2/L^3\epsilon_0\epsilon(0)q^2$ leads to a dielectric function

$$\epsilon_{3d}(q) = \epsilon(0)(1 + \frac{q_D^2}{q^2}), \tag{93}$$

where Debye screening-wavevector q_D is given by

$$q_D = \sqrt{\frac{e^2 N_c F_{-1/2}(\zeta)}{\epsilon_0\epsilon(0)k_B T}}, \tag{94}$$

for an arbitrary degeneracy of carriers.

6.5 Mobility of 2DEGs

Two-dimensional carriers

The wavefunction of electrons for band-transport[5] in 2DEG is

[5] In the presence of heavy disorder, the wavefunctions are localized and transport occurs by hopping and activation. For such cases, we cannot assume plane-wave eigenfunctions for electrons. All samples studied here are sufficiently pure, localization effects are neglected.

$$\langle \mathrm{r}|\mathrm{k}\rangle = \frac{1}{\sqrt{A}} e^{i\mathbf{k}\cdot\mathbf{r}} \chi(z) u_{n\mathbf{k}}(\mathbf{r}), \tag{95}$$

where the wavefunction is decomposed into a plane-wave part in the 2-dimensional $x-y$ plane of area A and a finite extent in the $z-$direction governed by the wave-function $\chi(z)$. $\mathbf{k}, \mathbf{r}$ are both two-dimensional vectors in the $x-y$ plane. $u_{n\mathbf{k}}(\mathbf{r})$ are the unit-cell-periodic Bloch-wavefunctions, which are generally not known exactly. The Kane model of bandstructure presents an analytical approximation for the Bloch-functions [72], which is not presented in anticipation of the cancellation of the Bloch-function for transport in parabolic bands.

Assuming that the defect potential is given by $V(\mathbf{r}, z)$, which depends on both the in-plane two-dimensional vector $\mathbf{r}$ and z perpendicular to the plane, time-dependent perturbation theory provides the solution for the scattering rate of electrons in the 2DEG. Scattering rate from a state $|\mathbf{k}\rangle$ to a state $|\mathbf{k}'\rangle$ is evaluated using Fermi's Golden Rule [5]. The use of Fermi's Golden rule in the $\delta-$function form is justified since the typical duration of a collision in a semiconductor is much less than the time spent between collisions [71, 73]. The scattering rate is written as

$$S(\mathbf{k}, \mathbf{k}') = \frac{2\pi}{\hbar} |H_{\mathbf{k},\mathbf{k}'}|^2 \delta(\varepsilon_{\mathbf{k}} - \varepsilon_{\mathbf{k}'}), \tag{96}$$

where $H_{\mathbf{k},\mathbf{k}'} = \langle \mathbf{k}'|V(\mathbf{r}, z)|\mathbf{k}\rangle \cdot I_{\mathbf{k},\mathbf{k}'}$ is the product of the matrix element $\langle \mathbf{k}'|V(\mathbf{r}, z)|\mathbf{k}\rangle$ of the scattering potential $V(\mathbf{r}, z)$ between states $|\mathbf{k}\rangle, |\mathbf{k}'\rangle$ and the matrix element $I_{\mathbf{k}\mathbf{k}'}$ between lattice-periodic Bloch functions. Owing to the wide bandgap of the III-V nitrides, the matrix element $I_{kk'} \approx 1$, the approximation holding good even if there is appreciable non-parabolicity in the dispersion [72].

By writing the scattering term in the form of Equation 96, we reach a point of connection to the Boltzmann-transport equation. Once the matrix element is determined, the momentum relaxation time $\tau_m(\mathbf{k})$ of the single particle state $|\mathbf{k}\rangle$ is evaluated from the solution of the Boltzmann-transport equation as

$$\frac{1}{\tau_m(\mathbf{k})} = N_{2D} \sum_{k'} S(k', k)(1 - \cos\theta), \tag{97}$$

where N_{2D} is the total number of scatterers in the 2D area A and θ is the angle of scattering. Implicit in this formulation is the assumption that all scatterers act *independently* of each other, which is true if they are in a dilute concentration. If this does not hold (as in heavily disordered systems), then one has to take recourse to interference effects from multiple scattering centers by the route of Green's functions [74]. The impurity concentration in AlGaN/GaN 2DEGs is dilute due to good growth control - this is confirmed by the band transport characteristics.

We write $\mathbf{q} = \mathbf{k} - \mathbf{k}'$ as depicted in figure 27. Since states for the subband with $\varepsilon < \varepsilon_F$ are filled, they do not contribute to transport. Transport then occurs by scattering in the Fermi circle shown in the figure, and $|\mathbf{k}| = |\mathbf{k}'| \approx k_F$. From the figure, the magnitude of $\mathbf{q}$ is $q = 2k_F \sin(\theta/2)$ where θ is the angle of scattering. This makes

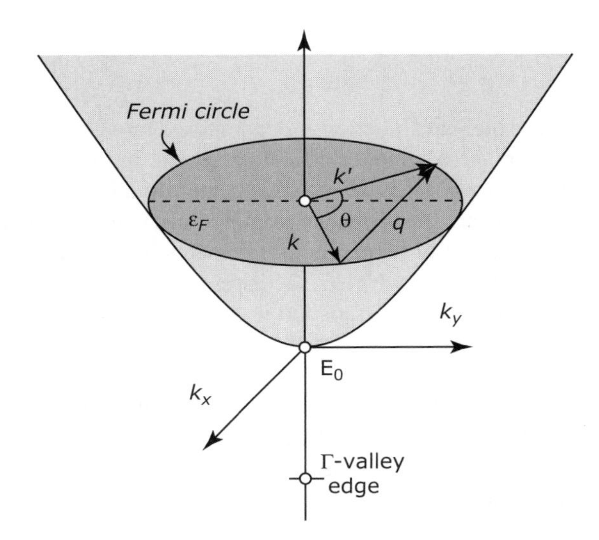

Fig. 27 Visualization of the scattering process on the 2DEG Fermi-circle.

$1-\cos\theta = q^2/2k_F^2$. As a result, all integrals in the vector $\mathbf{q}$ reduce to integrals over angle θ.

Any *measurement* of transport properties samples over all state $|\mathbf{k}\rangle$ values. Converting the summation to an integral over the quasi-continuous wavevector states and exploiting the degenerate nature of the carriers for averaging $\tau_m(\mathbf{k})$, the measurable momentum scattering rate $\langle 1/\tau_m\rangle$ reduces to the simple form [5]

$$\frac{1}{\langle\tau_m\rangle} = n_{2D}^{imp}\frac{m^*}{2\pi\hbar^3 k_F^3}\int_0^{2k_F}|V(q)|^2\frac{q^2}{\sqrt{1-(\frac{q}{2k_F})^2}}, \tag{98}$$

where $n_{2D}^{imp} = N_{2D}/A$ is the areal density of scatterers and $k_F = \sqrt{2\pi n_s}$ is the Fermi wavevector, n_s being the 2DEG density.

The perturbation potential matrix element is given by

$$V_{nm}(q) = \frac{1}{A}\int dz\left(\chi_n^\star(z)\chi_m(z)\int d^2\mathbf{r}V(\mathbf{r},z)e^{i\mathbf{q}\cdot\mathbf{r}}\right), \tag{99}$$

where n,m are the subband indices. This reduces to

$$V(q) = V_{00}(q) = \frac{1}{A}F(q)V(q,z_0) \tag{100}$$

when only the lowest subband ($n=m=0$) is occupied. Here, $F(q)$ is a form factor that is unity when the 2DEG spread in the $z-$direction is a delta function. The scattering potential

$$V(q,z_0) = \frac{V_{uns}(q,z_0)}{\epsilon_{2d}(q)} \tag{101}$$

is the screened two-dimensional Fourier transform of the scattering potential of a scatterer located at z_0 for a perfect 2DEG (no $z-$spread), where the screened-dielectric function (Equation 90) was used.

For accurate evaluation of transport properties and scattering rates, the finite extent of the 2DEG along the z direction must be accounted for. The exact form of the wavefunction from the self-consistent Schrödinger - Poisson solution is very useful in determining the 2DEG sheet density and the shape of the wavefunction.

However, for analytic evaluation of scattering rates, the Fang-Howard variational wavefunction is a better candidate, and has been used successfully for transport calculations in the past [5]. The form of the wavefunction is

$$\chi(z) = 0, z < 0$$
$$\chi(z) = \sqrt{\frac{b^3}{2}} z e^{-\frac{bz}{2}}, z \geq 0, \tag{102}$$

where b is a variational parameter. The parameter is chosen such that it minimizes the energy; this is achieved when $b = \left(33m^* e^2 n_s / 8\hbar^2 \epsilon_0 \epsilon_b\right)^{1/3}$, where n_s is the 2DEG density. The centroid of the distribution (which is also a measure of the spread) is $\langle z \rangle = \int_0^\infty z|\chi(z)|^2 = 3/b$. Thus, if the 2DEG $z-$dependence in not important in a scattering mechanism, the Fang-Howard function reduces to an ideal sheet charge $\rho(z) \rightarrow e n_s \delta(z)$ when $b \rightarrow \infty$.

The Fang-Howard variational wavefunction leads to a form factor

$$F(q) = \eta^3 = (\frac{b}{b+q})^3. \tag{103}$$

Screening by free carriers in the 2DEG is also affected due to the finite extent. This is reflected in another form factor $G(q)$ entering the 2D dielectric function

$$\epsilon_{2d}(q) = \epsilon(0)\left(1 + \frac{q_{TF}}{q} G(q)\right). \tag{104}$$

The screening form factor $G(q)$ is given by

$$G(q) = \frac{1}{8}(2\eta^3 + 3\eta^2 + 3\eta), \tag{105}$$

and q_{TF} is the Thomas-Fermi screening wavevector [21]. For a perfect 2DEG with no $z-$spread, $\eta \rightarrow 1$, and both form factors $F(q), G(q)$ reduce to unity. Thus, the Fang-Howard approximation along with particular scattering potentials can be used for evaluating the scattering rates of electrons in AlGaN/GaN 2DEGs.

6.6 Material Properties of III-V Nitrides Relevant to Transport

For the calculation of transport properties of the III-V nitride semiconductors, it is essential to know various properties of the material. The bandstructure is of utmost importance, determining the effective mass of carriers and the allowed energy and momentum eigenvalues for the carriers. The band-alignment is essential for analyzing transport of quantum-confined carriers at heterojunctions. The electromechanical properties such as deformation potentials of the bands and acoustic wave (sound) velocities are essential in the electron-acoustic phonon coupling study. Optical deformation potential and optical phonon energies are essential in calculating optical phonon scattering rates. In what follows, these properties are surveyed for the III-V nitride semiconductors with an eye for application for transport studies.

Bandstructure and Alignment

Since we are interested in analyzing transport in GaN, a close look at the bandstructure is in order. In figure 28, the Brillouin zone, the theoretically calculated Bandstructure [75] and the simplified bandstructure [67] with only the lowest three conduction band valleys are shown.

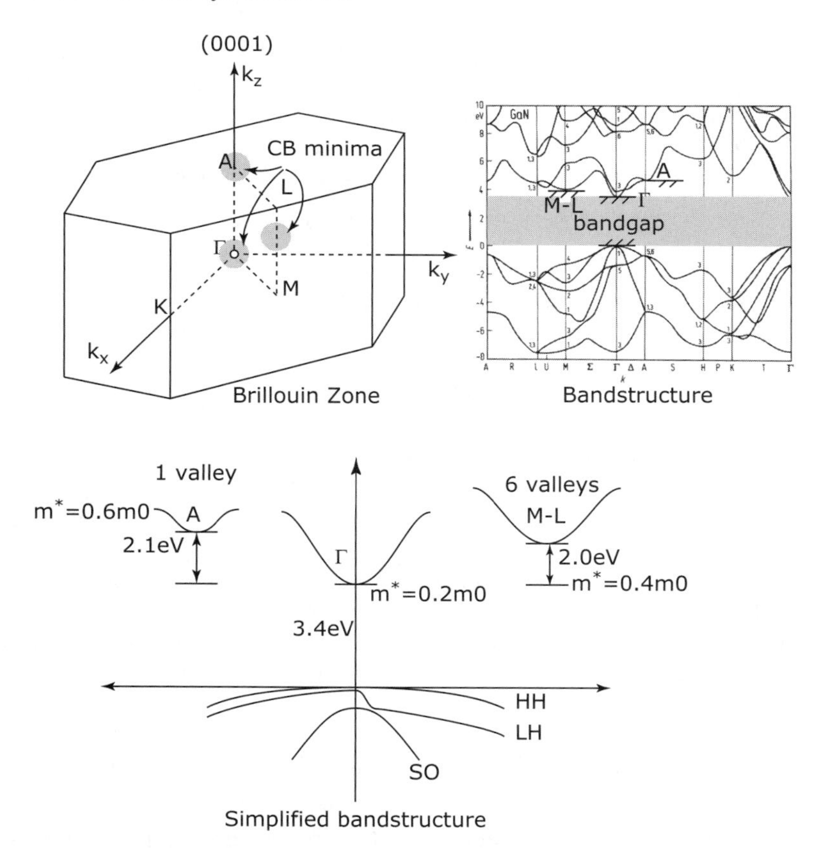

Fig. 28 Bandstructure, Brillouin zone and simplified bandstructure for wurtzite GaN.

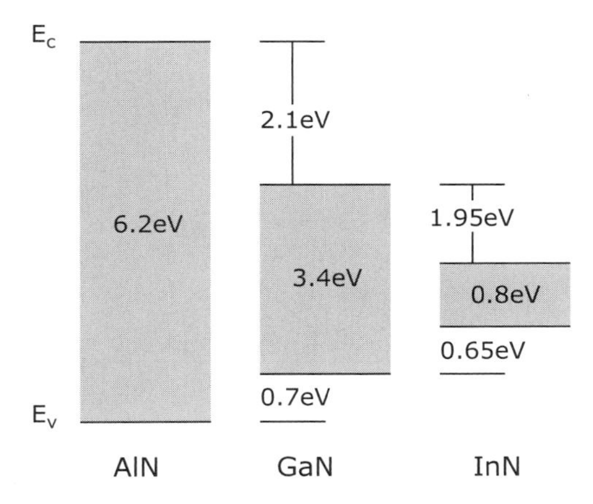

Fig. 29 The band alignment in the III-V nitrides. Note that the fundamental bandgap of InN was previously believed to be 1.9eV - measured on heavily doped polycrystalline films. However, developments in growth techniques only recently made it possible to grow pure single crystals which lead to the new bandgap. The band alignment of InN is assumed to follow the same 3:1 ratio as in the case of GaN and AlN; it has not been experimentally measured yet.

Carriers residing in the lowest conduction-band valley (Γ-valley, direct gap) have an effective mass of $m^{\star} = 0.2m_0$. The next band minimum is in the $M-L$ direction of the Brillouin zone at 2.1 eV from the Γ point. The effective mass of carriers in this valley is $m^{\star}_{ML} = 0.4m_0$. From the Brillouin-zone picture, there are six such equivalent minima. The next minimum is close to the $M-L$ minimum in energy - it is at the A-point in the Brillouin-zone. The effective mass in this valley is $m^{\star}_{A} = 0.6m_0$. We note here that the effective masses and energy separations of the $M-L$ and A valleys are from theoretical bandstructure calculations [67] whereas that of the Γ valley has been verified experimentally ([42,46]).

The large separation of the direct-gap Γ valley minima from the other minima at $A, L-M$ ($\Delta\varepsilon \geq 1$ eV) bodes well for low-field transport analysis; a parabolic dispersion $\varepsilon = \hbar^2 k^2/2m^{\star}$ is a good approximation. The effective mass is $m^{\star} = 0.2m_0$, and has been reported to be isotropic [44], which simplifies the transport analysis.

Phonon dispersion

The phonon dispersion curves of GaN have been calculated theoretically and measured experimentally as well. The experimental determination of phonon dispersion curves is traditionally done by the technique of neutron scattering. However, this technique requires unstrained samples of large sizes, which has not been possible till date. Bulk unstrained GaN samples of GaN can be grown by high pressure techniques, however, these samples are small in size (5×5mm). Ruf et. al. [77] overcame the size limitation by using the similar technique of inelastic X-Ray scattering for determining the phonon-dispersion of the material. They found very close agreement between the measured phonon-dispersion and the calculated dispersions by ab-initio techniques.

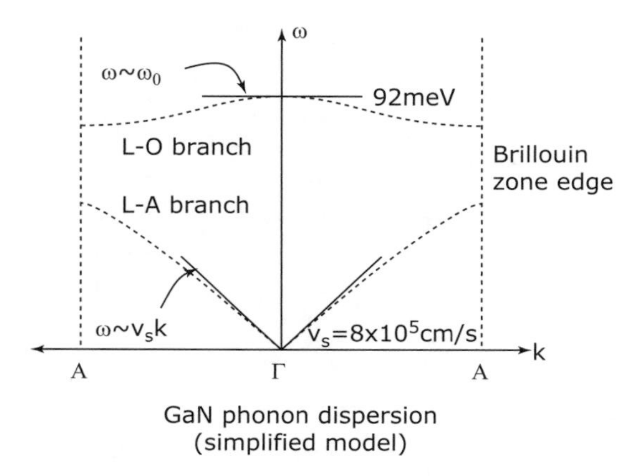

Fig. 30 Phonon dispersion of GaN with the two parameters needed for transport studies.

Since our interest in this work is aimed towards determining the electron-phonon coupling for transport calculations, we use the simplified phonon-dispersion shown in figure 30. The presence of both longitudinal and transverse modes of phonon propagation complicates the usage - we will assume only the longitudinal components since the effect of the transverse modes is much weaker.

The longitudinal sound velocities are given by $v_s = \sqrt{c_{11}/\rho}$ and $v_s = \sqrt{c_3/\rho}$ for the [1000] and [0001] directions of the wurtzite lattice. Since $c_{11} \sim c_{33}$ for GaN, AlN, and InN, it is a good approximation to assume that sound velocity is isotropic along the two directions; the sound velocities are tabulated in Table 2. Since acoustic phonons couple to electrons in the semiconductor through the deformation potential, the deformation potentials of the lowest conduction band valley are also listed in Table 2.

Acknowledgements The author of this chapter acknowledges U. Mishra, A. C. Gossard, J. Speck, and H. Kroemer at UCSB for discussions and guidance during the course of the research presented here.

References

1. U. K. Mishra, P. Parikh, and Y. F. Wu *Proceedings of the IEEE.*, vol. 90, p. 1022, 2002.
2. Y. F. Wu, B. P. Keller, P. Fini, S. Keller, T. J. Jenkins, L. T. Kehias, S. P. DenBaars, and U. K. Mishra *IEEE Electron Device Lett.*, vol. 19, p. 50, 1998.
3. J. R. Shealy, V. Kaper, V. Tilak, T. Prunty, J. A. Smart, B. M. Green, and L. F. Eastman *J. Phys.: Condens. Matter*, vol. 14, p. 3499, 2002.
4. A. Rizzi, R. Lantier, F. Monti, H. Lüth, F. D. Sala, A. Di Carlo, and P. Lugli *J. Vac. Sci. Tech. B*, vol. 17, p. 1674, 1999.
5. J. H. Davies, *The Physics of Low-Dimensional Semiconductors*. Cambridge, United Kingdom: Cambridge University Press, 1st ed., 1998.

6. J. P. Ibbetson, P. T. Fini, K. D. Ness, S. P. DenBaars, J. S. Speck, and U. K. Mishra *Appl. Phys. Lett.*, vol. 77, p. 250, 2000.
7. E. S. Snow, B. V. Shanabrook, and D. Gammon *Appl. Phys. Lett.*, vol. 56, p. 758, 1990.
8. G. L. Snider *1DPoisson*, http://www.nd.edu/g̃snider/.
9. M. J. Manfra and C. R. Elsass *personal communication*, 2002.
10. M. J. Manfra, K. W. Baldwin, A. M. Sergent, K. W. West, R. J. Molnar, and J. Caissie *Appl. Phys. Lett.*, vol. 85, p. 5394, 2004.
11. C. Skierbiszewski, K. Dybko, W. Knap, J. Lusakowski, Z. R. Wasilewski, Z. R. Maude, T. Suski, and S. Porowski *Appl. Phys. Lett.*, vol. 86, p. 102106, 2005.
12. M. J. Manfra, N. G. Weimann, J. W. P. Hsu, L. N. Pfeiffer, K. W. West, S. Syed, H. L. Stormer, W. Pan, D. V. Lang, S. N. G. Chu, G. Kowach, A. M. Sergent, J. Caissie, K. M. Molvar, L. J. Mahoney, and R. J. Molnar *J. Appl. Phys.*, vol. 92, p. 338, 2002.
13. I. P. Smorchkova, C. R. Elsass, J. P. Ibbetson, R. Vetury, B. Heying, P. Fini, E. Haus, S. P. DenBaars, J. S. Speck, and U. K. Mishra *J. Appl. Phys.*, vol. 86, p. 4520, 1999.
14. E. Frayssinet, W. Knap, P. Lorenzini, N. Grandjean, J. Massies, C. Skierbiszewski, T. Suski, I. Grzegory, S. Porowski, G. Simin, X. Hu, M. A. Khan, M. S. Shur, R. Gaska, and D. Maude *Appl. Phys. Lett.*, vol. 77, p. 2551, 2000.
15. C. R. Elsass, I. P. Smorchkova, B. Heying, E. Haus, P. Fini, K. Maranowski, J. P. Ibbetson, S. Keller, P. Petroff, S. P. DenBaars, U. K. Mishra, and J. S. Speck *Appl. Phys. Lett.*, vol. 74, p. 3528, 1999.
16. T. Wang, Y. Ohno, M. Lachab, D. Nakagawa, T. Shirahama, S. Sakai, and H. Ohno *Appl. Phys. Lett.*, vol. 74, p. 3531, 1999.
17. R. Gaska, J. W. Yang, A. Osinsky, Q. Chen, M. A. Khan, A. O. Orlov, G. L. Snider, and M. S. Shur *Appl. Phys. Lett.*, vol. 72, p. 707, 1998.
18. D. Jena Unpublished.
19. I. P. Smorchkova, L. Chen, T. Mates, L. Shen, S. Heikman, B. Moran, S. Keller, S. P. DenBaars, J. S. Speck, and U. K. Mishra *J. Appl. Phys.*, vol. 90, p. 5196, 2001.
20. L. Pfeiffer, K. W. West, H. L. Stormer, and K. W. Baldwin *Appl. Phys. Lett.*, vol. 55, p. 1888, 1989.
21. T. Ando, A. B. Fowler, and F. Stern *Rev. Mod. Phy.*, vol. 54, p. 437, 1982.
22. W. Walukiewicz, H. E. Ruda, J. Lagowski, and H. C. Gatos *Phys. Rev. B*, vol. 30, p. 4571, 1984.
23. W. Knap, S. Contreras, H. Alause, C. Skierbiszewski, J. Camassel, M. Dyakonov, J. L. Robert, J. Yang, Q. Chen, M. A. Khan, M. L. Sadowski, S. Huant, F. H. Yang, M. Goian, J. Leotin, and M. S. Shur *Appl. Phys. Lett.*, vol. 70, p. 2123, 1997.
24. E. archive New Semiconductor Materials Characteristics and Properties *http://www.ioffe.rssi.ru/SVA/NSM/.*
25. K. Seeger, *Semiconductor Physics, An Introduction*. Berlin: Springer Verlag, 6th ed., 1999.
26. B. L. Gelmont, M. Shur, and M. Stroscio *J. Appl. Phys.*, vol. 77, p. 657, 1995.
27. G. D. Bastard, *Wave-Mechanics applied to Semiconductor Heterostructures*. Les Ulis Cedex, France: Les Editions de Physique, 1st ed.
28. B. K. Ridley *physica status solidi*, vol. 176, p. 359, 1999.
29. L. Hsu and W. Walukiewicz *J. Appl. Phys.*, vol. 89, p. 1783, 2001.
30. D. K. Ferry and S. M. Goodnick, *Transport in Nanostructures*. Cambridge, UK: Cambridge University Press, 1st ed., 1999.
31. Y. Zhang and J. Singh *J. Appl. Phys.*, vol. 85, p. 587, 1999.
32. A. Gold *Phys. Rev. B*, vol. 35, p. 723, 1987.
33. N. W. Ashcroft and D. N. Mermin, *Solid State Physics*. Philadelphia: Saunders College, 1st ed., 1976.
34. O. Ambacher, B. Foutz, J. Smart, J. R. Shealy, N. G. Weimann, K. Chu, M. Murphy, A. J. Sierakowski, W. J. Schaff, L. F. Eastman, R. Dimitrov, A. Mitchell, and M. Stutzmann *J. Appl. Phys.*, vol. 87, p. 334, 2000.
35. R. Stratton *J. Phys. Chem. Solids*, vol. 23, p. 1011, 1962.
36. B. K. Ridley, *Quantum Processes in Semiconductors*. Great Clanderon St. Oxford: Clanderon Press, 4th ed., 1999.

37. F. Bernardini, V. Fiorentini, and D. Vanderbilt *Phys. Rev. B*, vol. 56, p. R10 024, 1997.
38. W. Zhano and D. Jena *J. Appl. Phys.*, vol. 96, p. 2095, 2004.
39. D. C. Look and J. R. Sizelove *Phys. Rev. Lett.*, vol. 82, p. 1237, 1999.
40. D. Zhao and K. J. Kuhn *IEEE Trans. Electron Devices*, vol. 38, p. 1520, 1991.
41. J. S. Speck and S. J. Rosner *Physica B*, vol. 273-274, p. 24, 1999.
42. K. Leung, A. F. Wright, and E. B. Stechel *Appl. Phys. Lett.*, vol. 74, p. 2495, 1999.
43. J. Elsner, R. Jones, P. K. Sitch, V. D. Porezag, M. Elstner, T. Frauenheim, M. I. Heggie, S. Öberg, and P. R. Briddon *Phys. Rev. Lett.*, vol. 79, p. 3672, 1997.
44. W. T. Read *Philos. Mag.*, vol. 45, p. 775, 1954.
45. P. J. Hansen, Y. E. Strausser, A. N. Erickson, E. J. Tarsa, P. Kozodoy, E. Brazel, J. P. Ibbetson, V. Narayanamurti, S. P. DenBaars, and J. S. Speck *Appl. Phys. Lett.*, vol. 72, p. 2247, 1998.
46. E. G. Brazel, M. A. Chin, and V. Narayanamurti *Appl. Phys. Lett.*, vol. 74, p. 2367, 1999.
47. J. W. P. Hsu, M. J. Manfra, D. V. Lang, S. Richter, S. N. G. Chu, A. M. Sergent, R. N. Kleinman, L. N. Pfeiffer, and R. J. Molnar *Appl. Phys. Lett.*, vol. 78, p. 1685, 2001.
48. P. Kozodoy, J. P. Ibbetson, H. Marchand, P. T. Fini, S. Keller, J. S. Speck, S. P. DenBaars, and U. K. Mishra *Appl. Phys. Lett.*, vol. 73, p. 975, 1998.
49. D. M. Schaadt, E. J. Miller, E. T. Yu, and J. M. Redwing *Appl. Phys. Lett.*, vol. 78, p. 88, 2001.
50. A. F. Wright and J. Furthmüller *Appl. Phys. Lett.*, vol. 72, p. 3467, 1998.
51. A. F. Wright and U. Grossner *Appl. Phys. Lett.*, vol. 73, p. 2751, 1998.
52. J. S. Koehler *Phys. Rev.*, vol. 75, p. 106, 1949.
53. J. K. Mackenzie and E. H. Sondheimer *Phys. Rev.*, vol. 82, p. 264, 1950.
54. R. Landauer *Phys. Rev.*, vol. 82, p. 520, 1951.
55. D. L. Dexter *Phys. Rev.*, vol. 86, p. 770, 1952.
56. J. Bardeen and W. Shockley *Phys. Rev.*, vol. 80, p. 72, 1950.
57. S. L. Chuang *Phys. Rev. B*, vol. 43, p. 9649, 1991.
58. C. Shi, P. M. Asbeck, and E. T. Yu *Appl. Phys. Lett.*, vol. 74, p. 573, 1999.
59. I. Vurgaftman, J. R. Meyer, and L. R. Ram-Mohan *J. Appl. Phys.*, vol. 89, p. 8815, 2001.
60. L. Hsu and W. Walukiewicz *J. Appl. Phys.*, vol. 89, p. 1783, 2001.
61. L. Shen, S. Heikman, B. Moran, R. Coffie, N. Q. Zhang, D. Buttari, I. P. Smorchkova, S. Keller, S. P. DenBaars, and U. K. Mishra *IEEE Electron. Dev. Lett.*, vol. 22, p. 457, 2001.
62. P. Waltereit, O. Brandt, A. Trampert, H. T. Grahn, J. Menniger, M. Ramsteiner, M. Reiche, and K. Ploog *Nature*, vol. 406, p. 865, 2000.
63. S. data supplied by Chris Elsass (UCSB)
64. C. M. Wolfe, N. Holonyak Jr., and G. E. Stillman, *Physical Properties of Semiconductors*. Englewood Cliffs, New Jersey: Prentice Hall, 1st ed., 1989.
65. M. Wraback, H. Shen, J. C. Carrano, T. Li, J. C. Campbell, M. J. Schurman, and I. T. Ferguson *Appl. Phys. Lett.*, vol. 76, p. 1155, 2000.
66. T.-H. Yu and K. F. Brennan *J. Appl. Phys.*, vol. 91, p. 3730, 2002.
67. U. Bhapkar and M. S. Shur *J. Appl. Phys.*, vol. 82, p. 1649, 1997.
68. W. B. Joyce and R. W. Dixon *Appl. Phys. Lett.*, vol. 31, p. 354, 1977.
69. C. Hamaguchi *Basic Semiconductor Physics*, p. 280, 2001.
70. J. Ziman *Theory of Solids*, Cambridge University Press, 1972.
71. D. K. Ferry, *Semiconductor Transport*. London: Taylor & Francis, 1st ed., 2000.
72. B. M. Askerov, *Electron Transport Phenomena in Semiconductors*. Singapore: World Scientific, 1st ed., 1994.
73. H. Kroemer, *Quantum Mechanics for Engineering, Materials Science, and Applied Physics*. Englewoods Cliff, New Jersey: Prentice Hall, 1st ed., 1994.
74. G. D. Mahan, *Many Particle Physics*. New York: Kluwer Academic/Plenum Publishers, 3rd ed., 2000.
75. S. Bloom, G. Harbeke, E. Meier, and I. B. Ortenburger *Phys. Stat. Solidi*, vol. 66, p. 161, 1974.
76. A. F. Brana, C. Diaz-Paniagua, F. Batallan, J. A. Garrido, E. Munoz, and F. Omnes *J. Appl. Phys.*, vol. 88, p. 932, 2000.
77. T. Ruf, J. Serrano, M. Cardona, P. Pavone, M. Pabst, M. Krisch, M. D'Astuto, T. Suski, I. Grzegory, and M. Leszczynski *Phys. Rev. Lett.*, vol. 86, p. 906, 2001.

Local Polarization Effects in Nitride Heterostructures and Devices

E. T. Yu and P. M. Asbeck

1 Introduction

1.1 Basic Physics of Polarization Effects

As described in the accompanying chapters of this volume, spontaneous and piezoelectric polarization effects play an extremely prominent role in Group III-nitride semiconductors, influencing potential and charge distributions and carrier transport in a broad range of nitride-based semiconductor heterostructure devices. In this chapter we focus on the existence, nature, and consequences of inhomogeneities in polarization fields and polarization charge distributions that arise from factors such as defects, non-uniform strain fields, or nanoscale compositional and layer thickness variations in basic heterostructure materials, as well as from process-induced defects, non-uniform stress due to metallization or etching, and related issues in electronic device structures. Possibilities for the intentional introduction of polarization fields to enhance device performance are also highlighted.

A detailed understanding of these effects requires first a clear understanding of the physical origin and nature of polarization fields generally in nitride semiconductor materials. In brief, the wurtzite crystal structure characteristic of nitride semiconductors leads to the existence of a large spontaneous polarization field $\mathbf{P}_{sp}$ aligned along the $[000\overline{1}]$ direction. The nonzero spontaneous polarization field is a consequence of the lower symmetry of the wurtzite crystal structure compared to that of the cubic zincblende structure, for which the spontaneous polarization vanishes. Nitride semiconductors are also characterized by large piezoelectric coefficients that, in strained materials, lead to the generation of a piezoelectric polarization field $\mathbf{P}_{pz}$ that can be comparable in magnitude to the spontaneous polarization field. These polarization fields can then give rise to a polarization-induced electrostatic charge density, ρ_{pol}, given by

$$\nabla \cdot \mathbf{P} = \nabla \cdot (\mathbf{P}_{sp} + \mathbf{P}_{pz}) = -\rho_{pol} \tag{1}$$

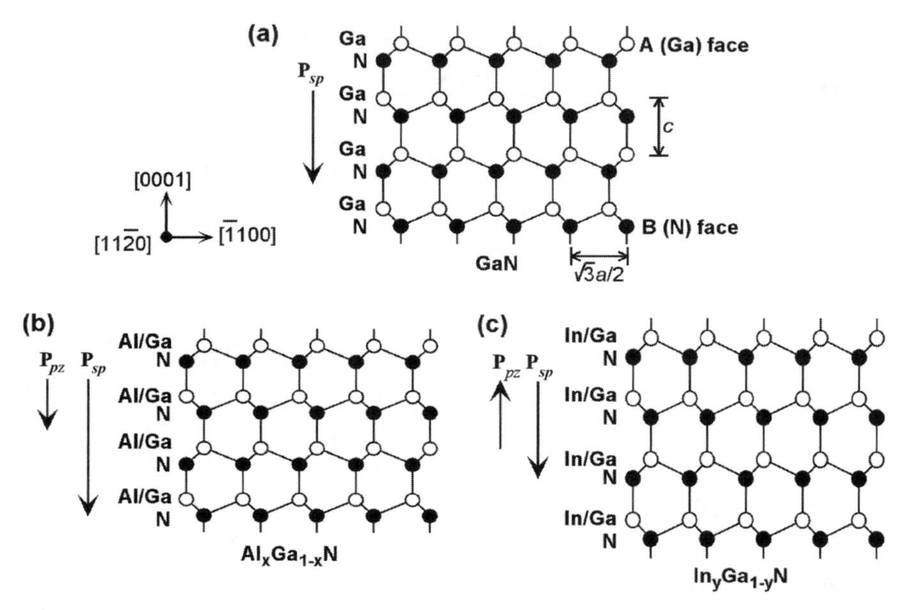

Fig. 1 Schematic diagram of crystal structure and spontaneous and piezoelectric polarization fields for (a) GaN, (b) $Al_xGa_{1-x}N$, and (c) $In_yGa_{1-y}N$, assumed in all cases to be coherently strained to a GaN (0001) substrate.

A schematic diagram of the nitride crystal structure and of the spontaneous and piezoelectric polarization fields for GaN, AlGaN, and InGaN coherently strained to GaN (0001) is shown in Figure 1.

In spatially uniform, structurally perfect bulk material, it is generally assumed that rearrangement of surface charges will nullify the spontaneous polarization and any piezoelectric polarization field that may be present. In heterostructures or inhomogeneous alloy layers, or in the presence of defects, variations in composition or local strain may create nonvanishing and spatially varying spontaneous and piezoelectric polarization fields and associated charge densities that can dramatically influence material properties and device behavior. The piezoelectric aspects of these phenomena were, in fact, first predicted and analyzed for cubic zincblende semiconductors to explain orientation-dependent device characteristics in GaAs MESFETs, and to design quantum well structures with built-in fields using (111) grown epitaxial layers, [1–5], but the magnitude of the polarization fields present in nitride semiconductors dramatically increases their prominence and significance in these materials.

Numerical values for the spontaneous polarization field along the [0001] direction in binary Group III-nitride semiconductors have been calculated theoretically and are given in Table 1. The method for computation of the piezoelectric polarization, given values of the appropriate piezoelectric moduli, the elasticity tensor, and the stress and/or strain tensor, is well established [6] and summarized here. In brief, the piezoelectric moduli d_{ijk} and e_{ijk} form third-rank tensors that relate $\mathbf{P}_{pz}$ to the stress tensor σ_{jk} and the strain tensor ε_{jk}, respectively, via the relations

$$P_{pz,i} = d_{ijk}\sigma_{jk} = e_{ijk}\varepsilon_{jk}, \sigma_{ij} = c_{ijkl}\varepsilon_{kl}, \tag{2}$$

where c_{ijkl} is the elastic tensor of the crystal and summation over repeated indices is assumed. Because the tensors d_{ijk}, e_{ijk}, σ_{jk}, and ε_{jk} are symmetric with respect to the indices j and k, we may define new matrix (rather than tensor) indices $j = xx$, yy, zz, yz, zx, $xy \equiv 1, \ldots, 6$ and write the tensor quantities in Eq. (2) in the more concise matrix notation with $d_{ijk} \rightarrow d_{ij}$, $e_{ijk} \rightarrow e_{ij}$, $c_{ijkl} \rightarrow c_{ij}$, $\sigma_{jk} \rightarrow \sigma_j$, and $\varepsilon_{jk} \rightarrow \varepsilon_j$ with, for example, e_{ij} and ε_j being given in the matrix notation by

$$e_{ij} = \begin{pmatrix} e_{11} & e_{12} & e_{13} & e_{14} & e_{15} & e_{16} \\ e_{21} & e_{22} & e_{23} & e_{24} & e_{25} & e_{26} \\ e_{31} & e_{32} & e_{33} & e_{34} & e_{35} & e_{36} \end{pmatrix}, \tag{3}$$

$$\varepsilon_j = \begin{pmatrix} \varepsilon_1 \\ \varepsilon_2 \\ \varepsilon_3 \\ \varepsilon_4 \\ \varepsilon_5 \\ \varepsilon_6 \end{pmatrix} \equiv \begin{pmatrix} \varepsilon_1 & \varepsilon_6 & \varepsilon_5 \\ \varepsilon_6 & \varepsilon_2 & \varepsilon_4 \\ \varepsilon_5 & \varepsilon_4 & \varepsilon_3 \end{pmatrix}. \tag{4}$$

For the wurtzite crystal structure, the piezoelectric tensors each contain three independent nonzero coefficients, with e_{ij} consequently being given by

$$e_{ij} = \begin{pmatrix} 0 & 0 & 0 & 0 & e_{15} & 0 \\ 0 & 0 & 0 & e_{15} & 0 & 0 \\ e_{31} & e_{31} & e_{33} & 0 & 0 & 0 \end{pmatrix}, \tag{5}$$

and d_{ij} by an analogous expression with independent coefficients d_{31}, d_{33}, and d_{15}.

Computation of the spontaneous and piezoelectric polarization fields, and of the associated polarization-induced charge densities, in nitride semiconductors requires the determination of values for the relevant physical constants, e.g., spontaneous polarization, lattice constants, piezoelectric moduli, and elastic constants. A summary of values, compiled from a variety of sources, [7–23], for selected physical parameters for GaN, AlN, and InN is shown in Table 1.

1.2 Experimental Determination of Polarization Charge Densities

The experimental determination of values for the polarization fields and charge densities in nitride heterostructures presents a particular challenge in several respects. First, because the spontaneous and piezoelectric polarization fields, while having different physical origins, are equivalent electrostatically, what is measured in

Table 1 Selected physical constants for GaN, AlN, and InN

	GaN	AlN	InN	Reference
a (Å)	3.189	3.112	3.548	[7]
c (Å)	5.185	4.982	5.760	[7]
P_{sp} (C/m^2)	−0.034	−0.090	−0.042	[8]
e_{31} (C/m^2)	−0.32			[9]
e_{31} (C/m^2)	−0.36			[10, 11]
e_{31} (C/m^2)	−0.22			[12]
e_{31} (C/m^2)		−0.58		[13]
e_{31} (C/m^2)	−0.49	−0.60	−0.57	[14]
e_{33} (C/m^2)	0.65			[9]
e_{33} (C/m^2)	1			[10, 11]
e_{33} (C/m^2)	0.44			[12]
e_{33} (C/m^2)		1.55		[13]
e_{33} (C/m^2)	0.73	1.46	0.97	[14]
c_{13} (GPa)	158			[15]
c_{13} (GPa)	70			[16]
c_{13} (GPa)	106			[17]
c_{13} (GPa)	114			[18]
c_{13} (GPa)	110	100		[19]
c_{13} (GPa)		120		[20]
c_{13} (GPa)		99		[21]
c_{13} (GPa)	100	127	94	[22]
c_{13} (GPa)	103	108	92	[23]
c_{33} (GPa)	267			[15]
c_{33} (GPa)	379			[16]
c_{33} (GPa)	398			[17]
c_{33} (GPa)	209			[18]
c_{33} (GPa)	390	390		[19]
c_{33} (GPa)		395		[20]
c_{33} (GPa)		389		[21]
c_{33} (GPa)	392	382	200	[22]
c_{33} (GPa)	405	373	224	[23]

strained structures is typically a combination of both spontaneous and piezoelectric polarization effects. Second, the determination of polarization fields or associated charge densities in nitride heterostructures is complicated by mobile-carrier screening effects, and any experimental determination of polarization-related fields or charge densities must account for the inevitable presence of mobile carriers in the materials and device structures being studied. And third, accurate extraction of the dependence of spontaneous and piezoelectric fields or charge densities on composition and strain requires an accurate experimental determination of composition and strain in the heterostructures for which the measurements are performed – a significant challenge in many typical nitride heterostructures, e.g., structures incorporating thin quantum-well layers.

In the face of these complications, a variety of optical [24–33] and electrical [34,34–36] methods emerge as viable approaches to the experimental determination

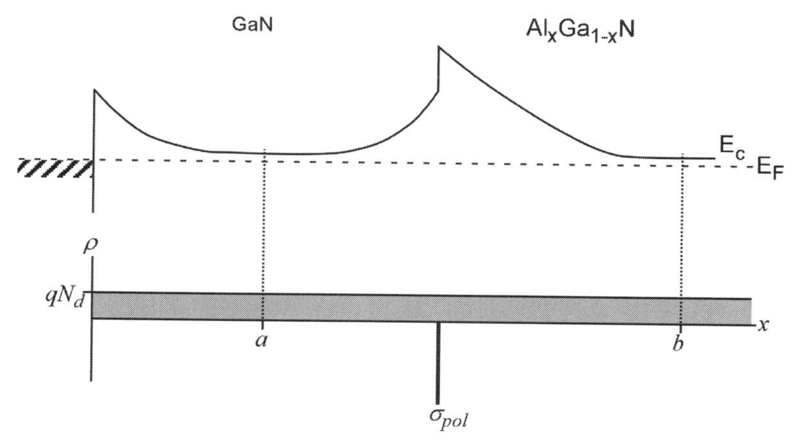

Fig. 2 Conduction band-edge energy diagram and charge density profile for $Al_xGa_{1-x}N/GaN$ heterostructure used in capacitance-voltage measurement of polarization charge density at $Al_xGa_{1-x}N/GaN$ heterojunction interface.

of polarization fields and charge densities in nitride heterostructures. Among these, capacitance-voltage profiling has yielded particularly detailed information concerning polarization charge density as a function of composition at $Al_xGa_{1-x}N$/GaN and $In_yGa_{1-y}N$/GaN heterojunction interfaces. In this approach, capacitance-voltage profiling of a Schottky diode is performed through a heterojunction interface, [35, 36] shown schematically in Figure 2, with integration of the apparent carrier concentration profile and net charge density yielding a measure of the polarization charge density at the heterojunction interface. Although the carrier profile $\hat{n}(x)$ obtained from capacitance-voltage profiling deviates from the actual carrier concentration profile $n(x)$ due to the intrinsic Debye length limitation, the number of charges depleted in profiling through a heterojunction interface is accurately reflected in the integrated apparent carrier concentration, and in the absence of trap-related interface charges can be related to the polarization charge at the interface, σ_{pol}, and the ionized dopant concentration, $N_d(x)$, according to [32, 36]

$$\sigma_{pol} = \int_a^b [\hat{n}(x) - N_d(x)]\, dx. \tag{6}$$

The apparent carrier concentration profile measured by capacitance-voltage profiling for a $GaN/Al_xGa_{1-x}N$ heterostructure, clearly revealing the depletion layer created by the negative polarization charge at the $GaN/Al_xGa_{1-x}N$ interface, is shown in Figure 3. Combined with Eq. (6), the apparent carrier concentration profile determined in this manner allows the polarization charge density σ_{pol} to be calculated. Figure 4 shows the polarization charge density determined in this manner for a series of $Al_xGa_{1-x}N/GaN$ (0001) heterojunctions as a function of composition of the $Al_xGa_{1-x}N$ layer. Also plotted in the figure are polarization charge densities estimated either by linear interpolation using a polarization charge density

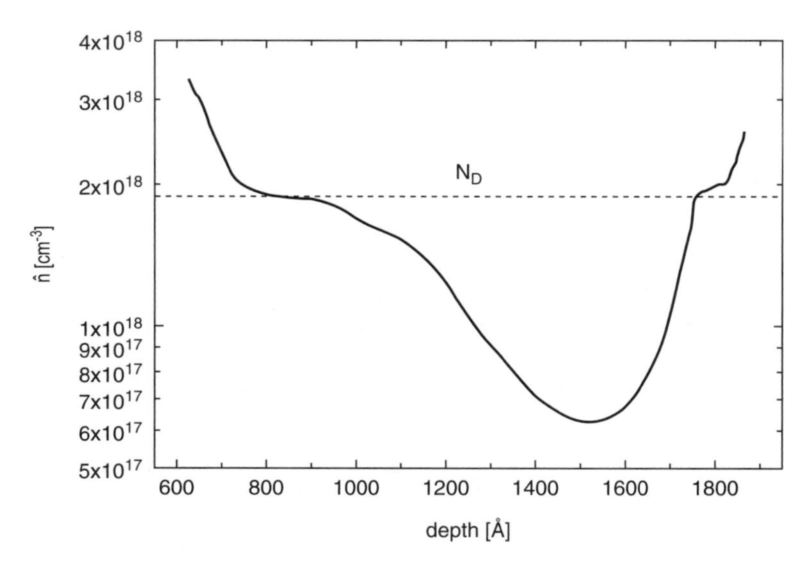

Fig. 3 Apparent carrier concentration profile extracted from capacitance voltage profiling of an $Al_{0.12}Ga_{0.88}N/GaN$ heterojunction as shown schematically in [32]. Also shown is the dopant concentration N_d.

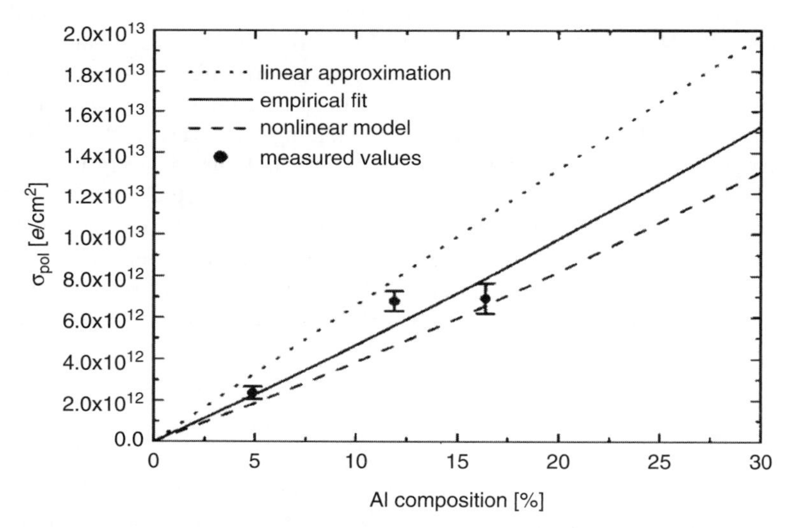

Fig. 4 Experimentally measured $Al_xGa_{1-x}N/GaN$ interface polarization charge density as a function of Al composition. Also shown are values obtained via linear interpolation to the theoretically calculated value for a AlN/GaN heterojunction coherently strained to the GaN (0001) lattice constant (dotted line), and via direct calculation using a theoretical model incorporating nonlinear dependence on composition (dashed line). The empirical fit corresponding to Eq. (7) is shown by the solid line.

calculated for a pure AlN/GaN (0001) interface with the AlN coherently strained to match the unstrained GaN lattice constant, or by direct calculation including nonlinear effects in the dependence on composition. [8] An empirical fit to the experimentally measured values plus the calculated value for the AlN/GaN heterojunction yields values intermediate between the linear and nonlinear models, with the polarization charge density magnitude for an $Al_xGa_{1-x}N/GaN$ (0001) heterojunction coherently strained to the GaN lattice constant being given by

$$\sigma_{pol}\,(Al_xGa_{1-x}N/GaN) = 4.45\times10^{13}x + 2.13\times10^{13}x^2\,[e/cm^2]. \qquad (7)$$

Analogous measurements have been performed for $In_yGa_{1-y}N/GaN$ (0001) heterojunctions coherently strained to the GaN lattice constant. [33] Figure 5 shows the resulting polarization charge densities as a function of composition for $In_yGa_{1-y}N/GaN$ (0001) heterojunctions obtained from these measurements and derived from a variety of optical measurements of internal electric fields performed on $In_yGa_{1-y}N/GaN$ quantum-well structures. Also shown is the theoretically calculated polarization charge density as a function of composition. [8] As is evident from the figure, the majority of these measurements yield polarization charge values significantly lower than those calculated theoretically. This observation has substantial implications for the design and analysis of nitride heterostructures incorporating In. On the basis of the experimentally determined polarization charge values from Refs. 25, 26, 28, 31, and 33 and the theoretically calculated value of $1.38\times10^{14}e/cm^2$ for a binary InN/GaN (0001) heterojunction coherently strained to the GaN lattice

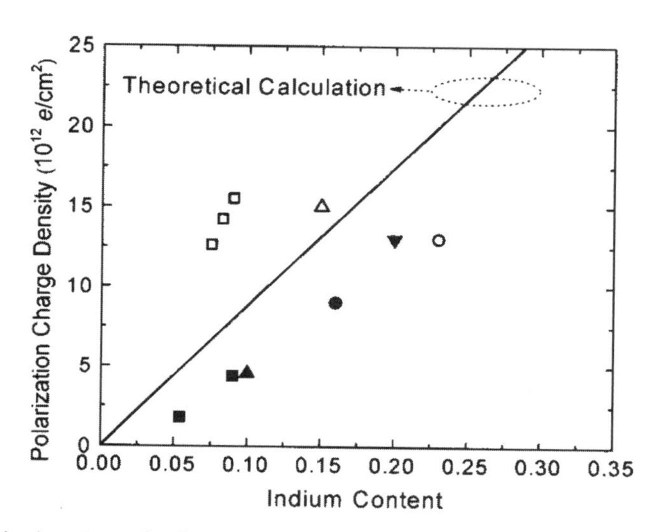

Fig. 5 Polarization charge density at the $In_yGa_{1-y}N/GaN$ heterojunction interface, measured by capacitance-voltage profiling in Ref. 33 (■), and inferred from Ref. 25 (▲), Ref. 26 (●), Ref. 28 (▼), Ref. 29 (△), Ref. 30 (□), and Ref. 31 (O). Theoretically calculated polarization charge densities (Ref. 8) are also shown.

constant, we obtain a dependence of the polarization charge density magnitude for an $In_yGa_{1-y}N$/GaN (0001) heterojunction coherently strained to the GaN lattice constant given by

$$\sigma_{pol}\,(\mathrm{In}_y\mathrm{Ga}_{1-y}\mathrm{N}/\mathrm{GaN}) = 3.81\times 10^{13}x + 9.99\times 10^{13}x^2[e/\mathrm{cm}^2]. \tag{8}$$

1.3 Consequences for Heterostructures, Defects, and Devices

Given the very sensitive dependence of polarization fields and charge densities on local alloy composition variations and strain, and the resulting intimate and complex interaction of structural morphology, electrostatic charge and potential distributions, and mobile carrier distributions, it is not unexpected that local, nanoscale variations in alloy composition or strain – introduced deliberately as part of a device engineering strategy or incidentally via the incorporation of defects or other structural inhomogeneities – can have a profound impact on electronic and optical material properties and on device performance.

In the remainder of this chapter we explore various aspects of these effects. We begin in the next section with a review of various approaches for engineering barrier structures based on incorporation of polarization-based charges in a heterostructures. It is shown how inclusion of negative polarization-induced sheet charges can be used to generate large electrostatic barriers in Schottky contacts, thereby achieving dramatic reductions in reverse-bias leakage current. We discuss localized polarization effects that are prominent in various electronic device structures, most notably high-electron mobility transistor (HEMT) structures and heterostructure bipolar transistors (HBT's) realized in nitride semiconductor material systems. The influence of polarization fields and charge densities on charge and mobile carrier behavior in the channel and buffer layer regions of nitride HEMT's, the consequences of localized stress fields created by device processing or by metal contact and passivation layers, and dynamic polarization effects are described. In nitride-based HBT structures, polarization effects in emitter-up and in collector-up device geometries are discussed.

We then discuss theoretical and experimental studies of localized, nanoscale variations in electronic structure that can arise in the presence of defects such as dislocations, leading, for example, to substantial variations at the nanoscale in local electron density in two-dimensional electron gas structures. Similarly, nanoscale and even monolayer fluctuations in heterostructure layer thickness are seen to lead to pronounced variations in local mobile carrier density via the close connection between structure and electronic behavior arising from large polarization fields and their influence on mobile carrier distributions. The energetics of piezoelectric fields in ternary alloy layers, and their possible influence on the occurrence of nanoscale clustering in nitride alloys, is also discussed.

2 Polarization-Based Engineering of Nitride Heterostructures

The structure of a representative AlGaN/GaN HFET is shown in Fig. 6, together with the corresponding charge density as a function of depth, for a FET produced on (0001) GaN (Ga-face). The polarization charge at the AlGaN/GaN interface is largely compensated by the presence of electrons in the 2DEG, while the surface polarization charge is compensated by charge in the Schottky metal gate; polarization charge associated with the buffer layer is compensated by shallow or deep acceptors within the nucleation or buffer layers. More generally, the bound charge associated with the divergence of the polarization may be viewed as a source of charge analogous to ionized impurities. The incorporation of the polarization charge represents another design variable that may be used for the optimization of structures. Such "polarization doping" obeys several rules, including the following:

1) the overall system is constrained to be neutral. Thus polarization-produced donors and acceptors occur in equal numbers. The doping effects of polarization engineering thus can be better described as the controlled incorporation of dipoles rather than of monopoles.
2) the sign of the charge obtained is dependent on the orientation of the crystal. Epitaxial layers grown with Ga-terminated faces ([0001] orientation) contain polarization dipoles opposite in orientation to those grown with N-terminated faces ($[000\bar{1}]$ orientation).
3) the polarization "dopants" are always fully ionized. There is no possibility of binding of electrons or holes to the polarization-induced charge, since it is dispersed over surfaces or volumes. This feature is of significant benefit in the nitrides, for which impurity acceptors have large binding energies, and, consequently, relatively low ionization ratios at room temperature.
4) the polarization "dopants" do not produce ionized impurity scattering. Since there are no localized centers for electrons to scatter from, with ideal planar

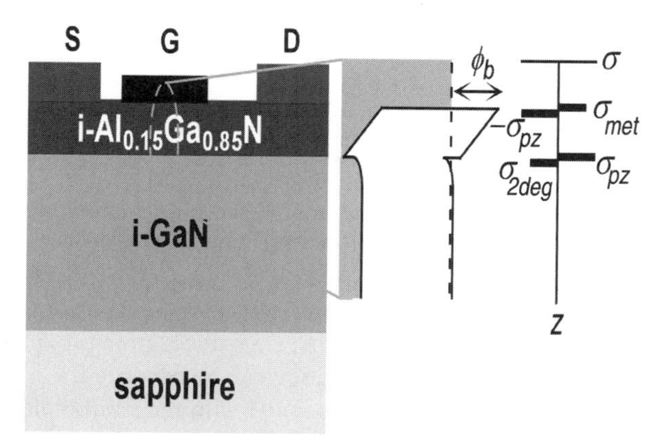

Fig. 6 Schematic structure of AlGaN/GaN HFET, together with corresponding band diagram and charge densities.

interfaces and uniform materials the presence of polarization doping should not decrease mobility. This feature is shared with "modulation doping" in AlGaAs/GaAs FETs. In real crystals in which interfaces are rough and alloy compositions fluctuate, however, there will be distributions of polarization-induced charge that can cause scattering and mobility reduction.

There are a variety of limits to the application of polarization-based doping in devices. The most significant ones are associated with stress relaxation and compensation.

Limitations due to Stress Relaxation

The maximum value of polarization doping that can be incorporated into a device is limited by the maximum difference in alloy composition that can be maintained, and in most cases this is fixed by the maximum strain that can be induced in lattice-mismatched layers. In the growth of heteromorphic layers with many III-V compounds such as GaAs and InP, it is well established that if a critical layer thickness is exceeded in lattice-mismatched growth, misfit dislocations will form and result in partial to complete strain relaxation. Experimental results in the nitride-based semiconductors typically show different behavior, however. The hexagonal wurtzite crystals do not permit glide of dislocations as readily as the zincblende structures. As a result, mechanical equilibrium is not reached in most samples. Bykhovski et al. [37] have reported a computation of critical thickness for AlGaN layers on GaN buffers, indicating approximately 200A critical thickness for 15% aluminum. Experimental data inferred from C-V measurements of polarization charge indicate that much thicker layers can be produced at that aluminum concentration while still incorporating substantial strain. Strain relaxation mechanisms representative of nitride heterostructures are illustrated schematically in figure 7. For AlGaN layers on GaN buffers, for which strain is typically tensile, the strain levels can typically be built up until the onset of cracking. It is noteworthy that in highly strained samples, cracking may not be immediate, but can occur over a period of days to months (after which a dramatic decrease in the 2DEG density of HFET structures is observed, as a result of strain relaxation and the loss of piezoelectric polarization). For InGaN layers on GaN buffers, for which compressive strain is observed, the strain levels appear to be limited by the onset of crystallographic defects, such as V-defects, during the growth process; these may or may not be accompanied by misfit dislocation formation, or the onset of phase segregation [38]. Large densities of threading dislocations, whose line direction is along the c-axis, are present in most nitride samples. These dislocations do not contribute to stress relaxation between the layers (for any orientation of the burgers vector). If threading edge dislocations are inclined slightly from the c axis, however, then stress relaxation can be produced.

There are polarization components associated with the strain fields of threading dislocations, as discussed later in this chapter. However, the divergence of the polarization vanishes for c-axis oriented dislocations almost everywhere. Charges arise from the dislocation strain fields only at surfaces and interfaces, in the case of c-axis

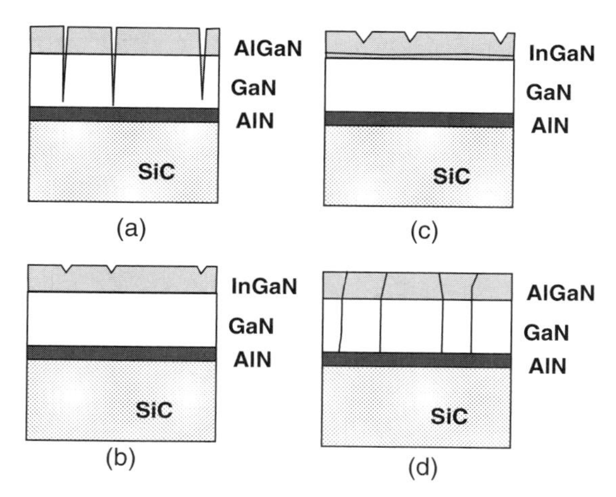

Fig. 7 Schematic representation of the response of nitride-based epitaxial structure to high levels of stress: a) formation of cracks; b) introduction of V defects; c) introduction of misfit dislocations in addition to the above; d) inclination or annihilation of threading dislocations.

directed dislocations with an edge component. Inclined threading dislocations produce polarization charge, and can be source of effective doping in nitride layers. For example, in figure 8 are shown experimental curves for the average strain in a GaN layer grown on a SiC substrate as a function of layer thickness [39]. The results differ according to whether the nucleation layer used at the SiC/GaN interface was composed of AlN or AlGaN. For each case, it is possible to infer the variation of strain vs depth for the layer. It is likely that, on a microscopic basis, the relaxation of strain occurs via formation of inclined threading dislocations, or via annihilation of edge dislocations. In either case, the variation of strain leads to the existence of a bound charge ρ_{bound}, which can be determined from equation (1). Figure 9 shows values computed for the bound charge in the GaN layers, as a function of layer thickness. The relatively large values indicate that GaN buffer layers used for HFET applications may have substantial intrinsic n-type "doping" that can contribute to source-drain leakage currents, to an extent that is dependent on the nucleation and growth procedures.

Limitations due to Compensation

It can be expected that polarization "doping" will be partially compensated by native defects and impurities during crystal growth and subsequent processing. The polarization-induced charge is not localized, and will not create bonding interactions and undergo complex formation (such as Mg incorporation in Mg-H complexes); however, the effect of the polarization charge on the fermi level will contribute to decreasing the energy of formation of defects, thereby increasing their density. Walukiewicz [40] has simulated the increase of native donors in AlGaN/GaN

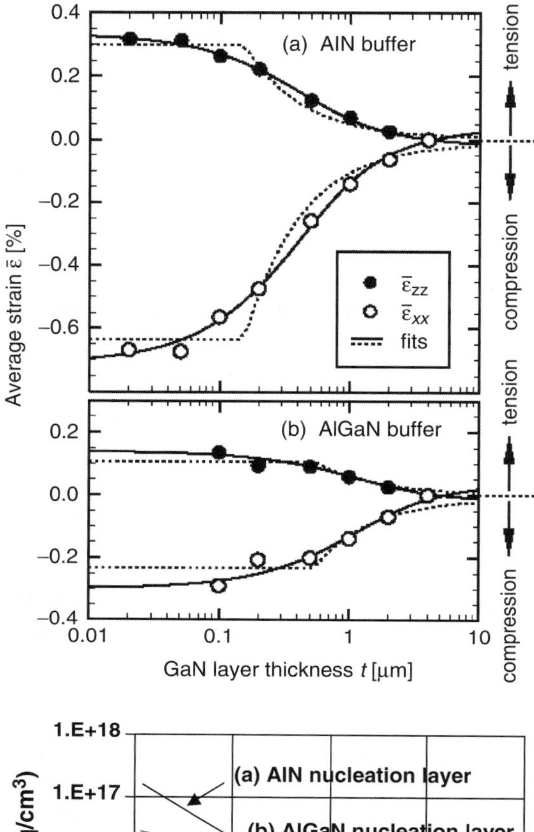

Fig. 8 Experimental measurements of average strain of layers of GaN grown on SiC substrates, as a function of GaN layer thickness: a) for AlN buffer layer; b) for AlGaN buffer layer [39].

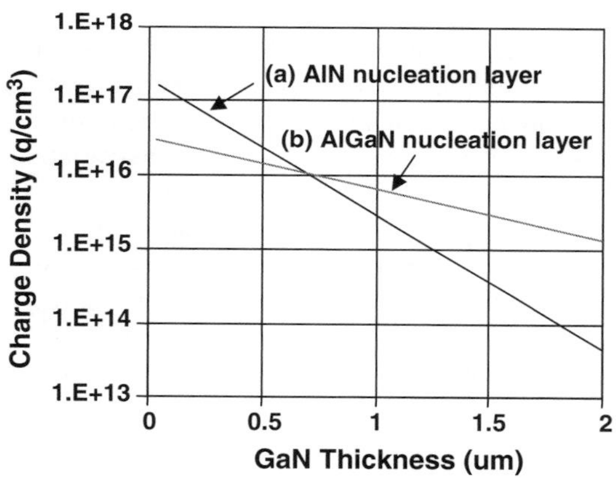

Fig. 9 Computed polarization-inducted charge density for the GaN layers described in figure 3: a) for AlN buffer layer; b) for AlGaN buffer layer.

HFETs as a result of the presence of polarization-induced negative charge at the surface of growing heterostructures. It is also possible that hydrogen incorporation as H+ entities will be increased at the surface as polarization acceptors are incorporated into the material, since they are energetically strongly favorable as the fermi level in the semiconductor approaches the valence band. The degree of compensation obtained under different growth conditions is not yet known. Under worst case scenarios, the compensation will negate the formation of p-type conductivity. An attractive scenario, already proposed by Van Vechten [41] to explain the p-type conductivity resulting from Mg doping, is that H+ may compensate the acceptors during growth to an extent that discourages the incorporation of more stable donor-like centers. Subsequently, the hydrogen may be expelled during relatively low temperature anneals to activate the acceptors.

In the following, we consider additional effects of polarization doping in FET structures, both intentional and unintentional.

2.1 Enhancement of Schottky Barrier Height in HFET Structures

An important characteristic for HFET structures is the effective height of the Schottky barrier between metal gate and underlying semiconductor. By maximizing this barrier height, the limiting electron concentration of the channel can be increased; the breakdown voltage of the Schottky junction can be increased; and the turn-on voltage for forward current of the gate-channel junction can be increased. With conventional structures, the barrier height is a characteristic of the semiconductor chosen for the top layer of the FET epitaxial stack, and the gate metal. An additional increase in barrier height can in principle be obtained by the incorporation of acceptor-like charge near the Schottky gate. In the conventional camel diode, this charge is produced by incorporation of acceptor impurities (culminating in the production of a p-n junction gate as in junction FETs). The use of polarization design principles provides, for the nitrides, an alternative powerful technique for maximizing the barrier height. The overall layer structure consists of a layer of GaN on top of an AlGaN layer, which is on top of a GaN buffer layer. Figure 10 depicts the structure, charges, and expected band diagram for an HFET structure with polarization-enhanced barriers. The barrier height is increased by the amount of potential drop

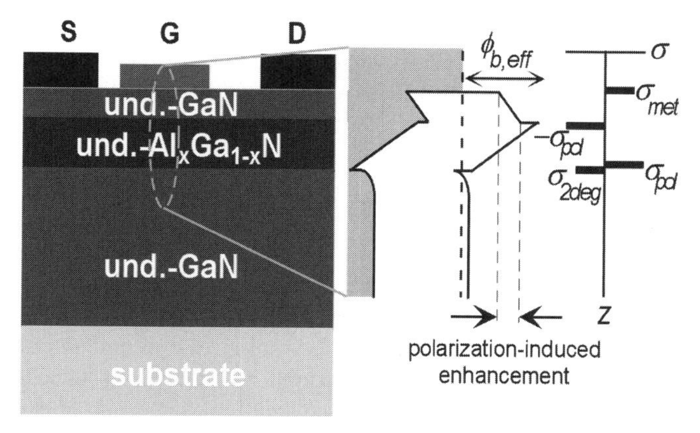

Fig. 10 (a) Structure with enhanced Schottky barrier structure via polarization engineering. (b) Corresponding band diagram. (c) Schematic charge densities [42].

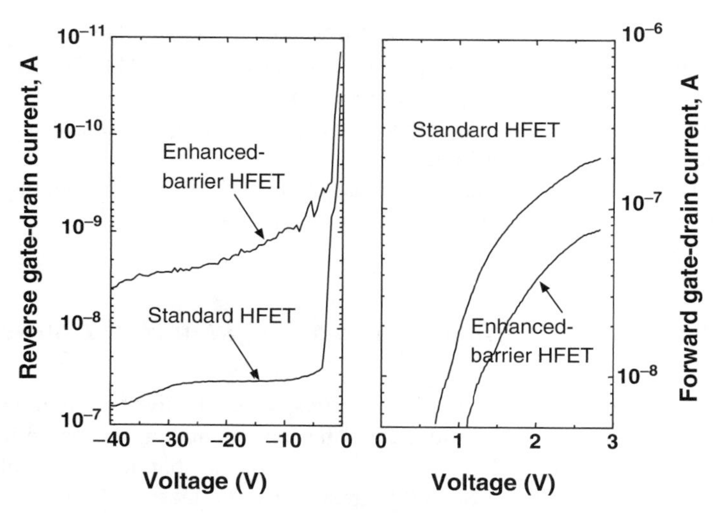

Fig. 11 Experimental I-V characteristics for Schottky barrier diode, with and without polarization enhanced barrier: a) reverse current; b) forward current [42].

in the surface GaN layer, which in turn is dependent on the polarization-induced acceptor charges at the top GaN/AlGaN interface. Photoresponse measurements of barrier height have been made for samples prepared in this manner. The observed results indicated that the barrier height was enhanced by up to 0.37 eV, for samples employing AlGaN layers of 25% aluminum mol fraction, at zero bias [42]. As is expected for camel diodes, the barrier height is somewhat dependent on the voltage applied to the Schottky diode, and increases for forward bias. The measured barriers are in approximate accord with the values predicted with a simple analysis (which assumes that the material has residual unintentional donors at a level of $1 \times 10^{18}\,\text{cm}^{-3}$). HFETs have been demonstrated with barrier enhancement produced in this manner. The HFET characteristics show, as expected, a decrease in the leakage current associated with the Schottky gate, and an increase in the forward voltage needed to turn on the device has been measured, as illustrated in figure 11.

2.2 *Polarization-Based Energy Barrier Engineering*

For GaAs and InP-based heterostructures, the energy barriers at heterojunctions are given largely by conduction and valence band offset energies between different materials. With the strong polarization effects in nitride semiconductors, additional approaches are available to form energy barriers. A thin layer of AlGaN embedded in a GaN matrix produces bound charges of opposite sign at its upper and lower interfaces, for example. The resultant charge dipole produces an electric field which is confined to the AlGaN layer (provided it is sufficiently thin and compensating

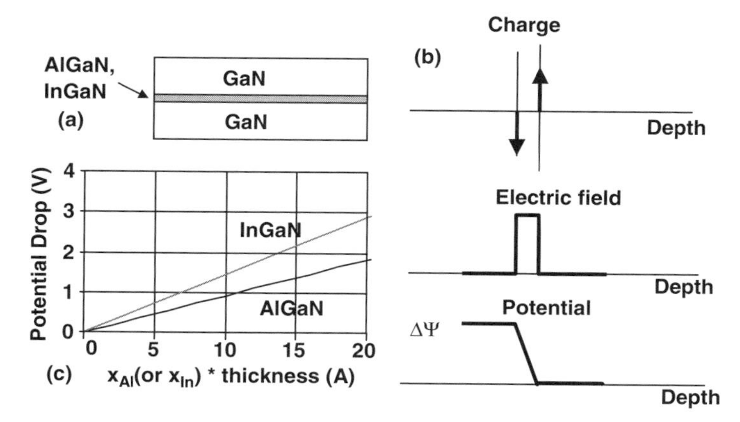

Fig. 12 Polarization-induced energy barrier: a) schematic layer structure; b) charge, electric field and potential distributions; c) approximate barrier voltage as a function of product of alloy mol fraction and barrier thickness (in A).

charges are not induced), and an electrostatic potential variation is formed which is nearly step-like across the AlGaN layer, as pictured in figure 12a. The effective variation in conduction or valence band energy is nearly identical to what can be provided by band offsets, and as a result, the polarization-based barriers can be used to confine electrons or holes [43]. The magnitude of the energy offset produced via polarization charges can be tailored by the aluminum concentration or thickness of the barrier. A potential step of opposite sign can be produced by a thin InGaN layer embedded in GaN. Figure 12b illustrates the approximate electrostatic potential variation associated with the barrier layer, as a function of the product of AlGaN or InGaN mol fraction and layer thickness (a linear variation is assumed, which is valid for "small" mol fractions only). The figure indicates that very large potential variations can be established, for example, a 1 eV barrier results from only a 25A thick AlGaN layer with $x_{Al} = 0.40$.

It is noteworthy that to form an energy barrier with an accurately specified shape, the alloy composition profile of the AlGaN or InGaN layer can be appropriately tailored. The conduction band energy $E_c(x)$ varies in accordance with the conduction band offset $\Delta E_c(x)$ in combination with the electrostatic potential contribution $-q\Psi(x)$. If the mol fraction of the embedded alloy varies as c(x) as a function of depth x, then

$$
\begin{aligned}
-q\Psi(x) &= \beta \int_0^x c(x')dx' \\
E_c(x) &= \Delta E_c(x) - q\Psi(x) = \alpha c(x) + \beta \int_0^x c(x')dx'
\end{aligned}
\tag{9}
$$

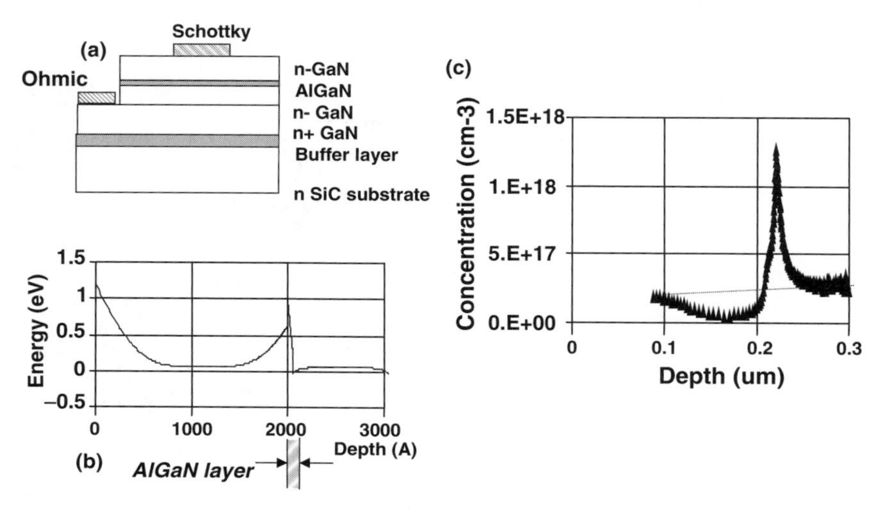

Fig. 13 Experimental verification of polarization barrier for AlGaN layers: a) sample structure; b) simulated conduction band diagram; c) electron density vs depth computed from C-V measurements.

Here α represents the value of dEc(x)/c(x), and β corresponds to $q/\varepsilon P_z(x)/c(x)$. To achieve a step-like variation in Ec(x) for example, the desired form of c(x) follows step-like behavior on one side, and follows a decaying exponential on the other side. The exponential decay constant for the case of AlGaN is of the order of 340A.

The formation of energy barriers analogous to band offsets, via polarization effects, is straightforward to confirm for AlGaN and InGaN layers embedded in GaN. C-V profiling allows estimates to be made of the effective barrier height, analogous to measurement of band offsets for heterojunctions in GaAs-based materials [36]. Figure 13a depicts a structure containing an AlGaN-induced polarization barrier in GaN, suitable for C-V profiling. Figure 13b shows a corresponding calculated band diagram, showing the large potential barrier expected as a result of the AlGaN layer, which has the appearance of a conventional conduction band offset. Experimental C-V curves yielded the calculated values of electron density vs position shown in figure 13c, which are similar in nature to those produced at AlGaAs/GaAs heterobarriers. The barrier heights inferred from C-V measurements were 1.43 eV (for a 100 A AlGaN film with $x_{Al} = 0.13$) and 0.60 eV for a 50 A AlGaN film with $x_{Al} = 0.13$) in reasonable accord with expectations from simple theory. Corresponding structures could be produced with InGaN thin layers embedded in GaN, as shown in figure 14. The calculated band diagram and measured carrier density vs position via C-V are also shown. For the InGaN structure, the inferred barrier height was 0.37 eV, for a 50 A InGaN film with $x_{In} = 0.08$, in good agreement with polarization coefficients measured with separate layers reported below.

Polarization-induced barriers may be applied to various devices. An important application is in HFET structures, for which barriers between the channel and the

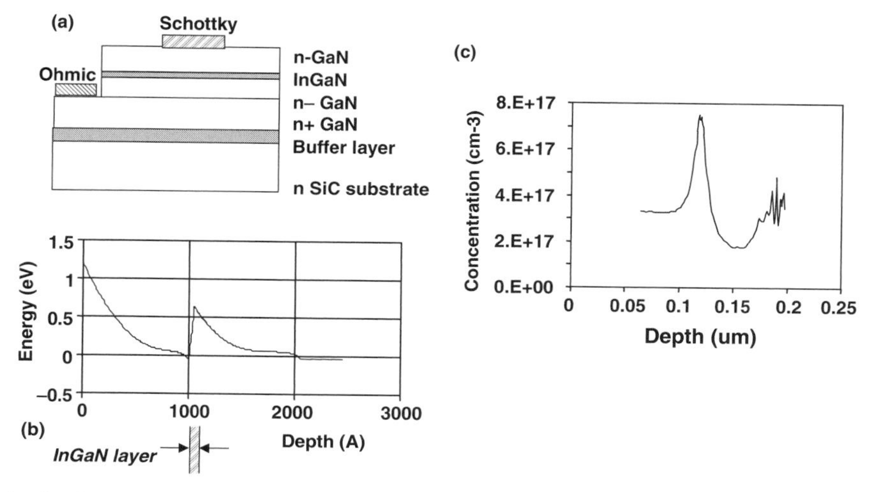

Fig. 14 Experimental verification of polarization barrier for InGaN layers: a) sample structure; b) simulated conduction band diagram; c) electron density vs depth computed from C-V measurements.

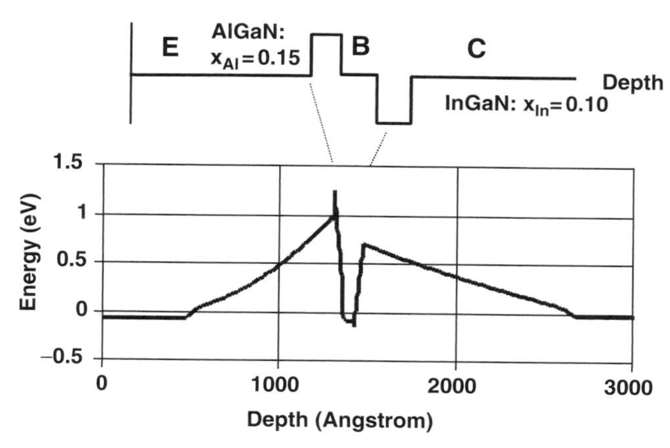

Fig. 15 Alloy composition vs depth and simulated conduction band profile for Hot Electron Transistor engineered with polarization-induced barriers.

underlying buffer layer are desirable in order to confine electrons to the near-surface region, in order to increase the degree of control over the current by the gate (and thereby reduce the "short channel" effects), as well as to reduce leakage of carriers through the buffer and to avoid trapping of electrons in the buffer layer. Potential layer structures to achieve this are shown in figure 15. In part (a) of the figure, a thin InGaN layer is used to provide the barrier at the backside of the channel, using a conventional buffer layer of GaN [43–45]. In the second structure, a buffer layer based on AlGaN of low Al mol fraction is employed, and the channel is GaN or an AlGaN/GaN double layer. Experimental results for HFETs have been obtained with both of these structures, which demonstrate the benefits of the confinement barrier [44, 45]. An additional interesting structure can be produced, in which the backside barrier is used in order to form the 2DEG channel layer, in the absence of a top layer of AlGaN. Figure 15c illustrates such a structure, in which an InGaN thin layer provides the necessary barrier. This structure, an "inverted HFET", benefits from a lower barrier for ohmic contact formation than the conventional HFET, as well as lower buffer layer leakage, and, perhaps most importantly, an absence of anomalous transient effects determined by surface charging [46].

An additional possible application of the polarization barriers for high speed transistors has been suggested, in the formation of a Hot Electron Transistor (HET). The HET epitaxial layer structure and calculated band structure is shown in figure 16. Polarization barriers of opposite sign are formed at the top and bottom on an n-type base layer. When the topmost "emitter" layer is biased appropriately, electrons are

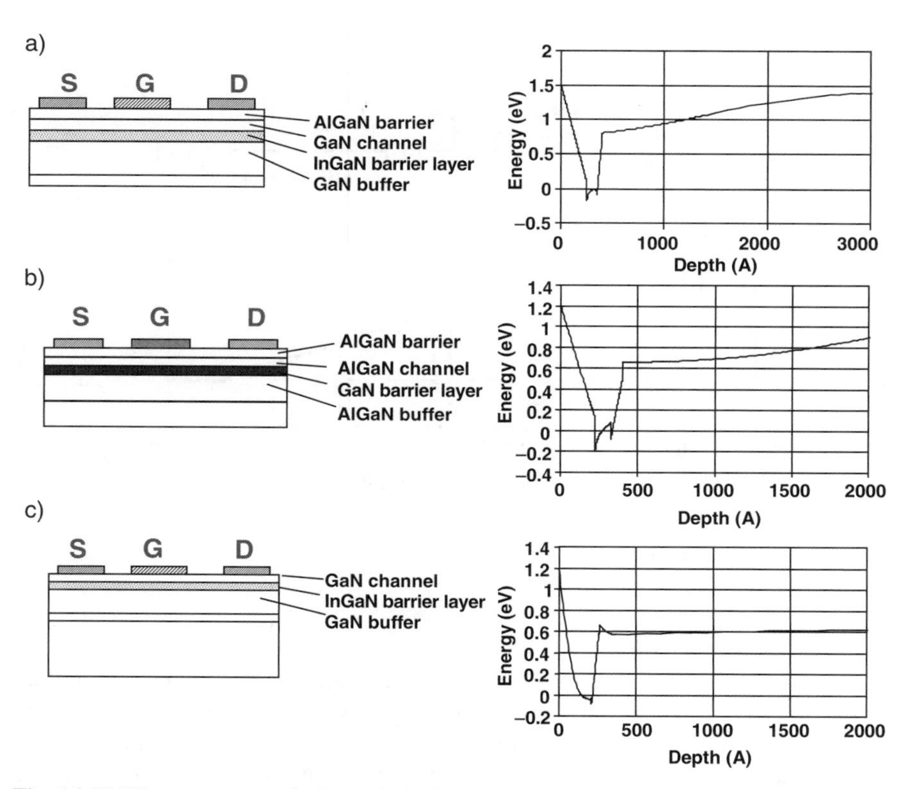

Fig. 16 HFET structures employing polarization barriers at channel / buffer layer interface: a) InGaN barrier layer on GaN buffer; b) GaN barrier layer on AlGaN buffer; c) inverted HFET structure with InGaN barrier layer.

injected at high energy into the base, and if their energy loss as they traverse the base layer is sufficiently small, they may surmount the base-collector barrier, and give rise to collector current. Such a transistor could potentially operate at very high speed, since the base transit time is expected to be minimal. The base conductivity is potentially far higher than achievable with HBT structures, because of the relatively high mobility of the base electrons. In work to date, the HET structures have not provided adequate current gain, principally because of leakage currents found experimentally between the base and collector. These leakage currents are potentially due to nonuniformity of the polarization-induced barrier over the device area.

2.3 Residual Stress and Piezoelectric Effects in GaN HFETs

The etching or deposition of films during the fabrication of HFETs often leads to the introduction of stresses into the device structure. These stresses can significantly affect the electrical characteristics of the device. In Si MOSFETs, for example, recent work has illustrated that process-induced stresses can produce a major enhancement in mobility and channel current, and thus stress has been intentionally designed into many modern CMOS transistors [47]. In GaAs-based FETs, stresses produced unintentionally by gate metals or dielectrics produce a shift in threshold voltage that leads to a significant difference between FETs oriented along different crystal directions [1]. A corresponding stress effect is expected in the case of nitride-based HFETs, as a result of piezoelectric polarization. Stress can be introduced into device structures by the deposition of dielectric layers such as silicon nitride, typically employed for surface passivation. The silicon nitride layers are under compressive or tensile stress, depending on the technique and deposition parameters used for their preparation. Etching of AlGaN during gate recess also introduces stress in the vicinity of the gate, due to the absence of the forces associated with the stressed AlGaN removed from the gate region. The addition or removal of stressed layers produces a marked enhancement of the stress in the vicinity of the edges of the film. If the stressed layers can be approximated as thin sheets, and the underlying nitride semiconductor is approximated as elastically isotropic, then well-known solutions for the stress distribution can be applied [48]:

$$\sigma_{xx} = -\frac{2}{\pi}\sigma_f d_f\left(\frac{x_1^3}{r_1^4} - \frac{x_2^3}{r_2^4}\right)$$

$$\sigma_{yy} = -\frac{2}{\pi}\sigma_f d_f\left(\frac{x_1}{r_1^2} - \frac{x_2}{r_2^2}\right)$$

$$\sigma_{zz} = -\frac{2}{\pi}\sigma_f d_f\left(\frac{x_1 z^2}{r_1^4} - \frac{x_2 z^2}{r_2^4}\right)$$

$$\sigma_{xz} = -\frac{2}{\pi}\sigma_f d_f\left(\frac{x_1^2 z}{r_1^4} - \frac{x_2^2 z}{r_2^4}\right)$$

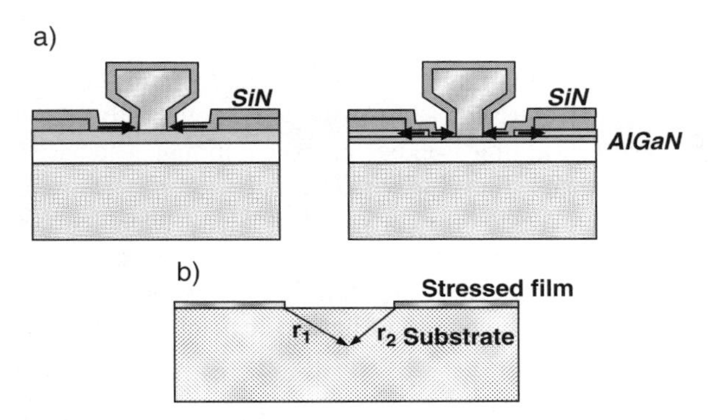

Fig. 17 Representative HFET structures, showing origin of surface stress near gate region: a) stress due to Si_3N_4 passivation layer; b) stress due to recess etch through stressed AlGaN layer. The geometry for computing stress distribution in the underlying GaN is also shown.

Here σ_{ij} are components of the stress tensor in the semiconductor, σ_f is the film stress and d_f is its thickness; the coordinate system is shown in figure 17, along with representative experimental structures. There is a corresponding piezoelectrically-induced polarization P_i, and associated bound charge density ρ_b, given by

$$[P_i] = -\frac{2}{\pi}\sigma_f d_f \begin{bmatrix} d_{15}\left(\frac{x_1^2 z}{r_1^4} - \frac{x_2^2 z}{r_2^4}\right) \\ 0 \\ d_{31}\left(\frac{x_1^3}{r_1^4} - \frac{x_2^3}{r_2^4}\right) + d_{31}\nu\left(\frac{x_1}{r_1^2} - \frac{x_2}{r_2^2}\right) + d_{33}\left(\frac{x_1 z^2}{r_1^4} - \frac{x_2 z^2}{r_2^4}\right) \end{bmatrix}$$

$$\rho_b = -\nabla \bullet P \tag{11}$$

$$Q_b = P_z\,(z=0)$$

The charge density computed from (5) is pictured in figure 18 for a representative HFET [49]. The charge density can reach appreciable magnitude ($>10^{18}\,\mathrm{cm}^{-3}$) for film stress of the type magnitude produced in many silicon nitride deposition processes (-5×10^9 to $5 \times 10^9\,\mathrm{dynes/cm}^2$). Computations have been made to estimate the effect of the silicon nitride stress-induced charges on device characteristics. The simulations were carried out using 2-dimensional device solver (Silvaco Atlas) in which the additional charge densities were manually introduced into an otherwise ideal AlGaN/GaN HFET structure. It was found that according to the sign of the stress, substantial changes in threshold voltage could be obtained. Figure 19a shows representative computed changes in the Id-Vgs curves of the FETs with varying silicon nitride stress, a gate length of 0.25 um, and a Si_3N_4 film thickness of 1000A. The corresponding changes in threshold voltage V_t are strongly dependent on gate

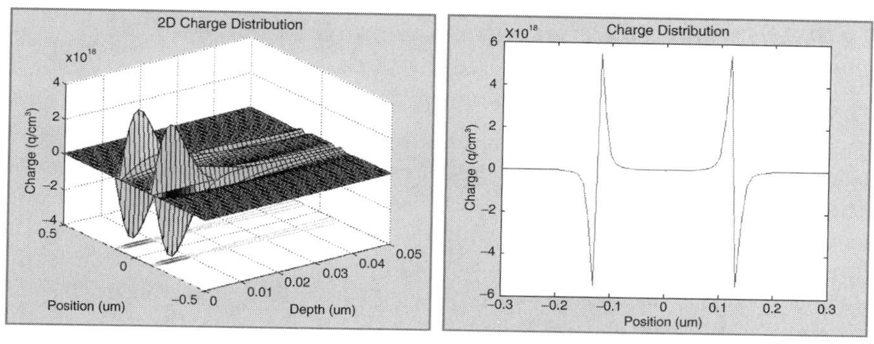

Fig. 18 Computed distribution of polarization-induced charge distribution near the channel region of an AlGaN/GaN HFET, due to Si_3N_4 layer (assumed to have thickness 100 nm and stress 0.5 GPa).

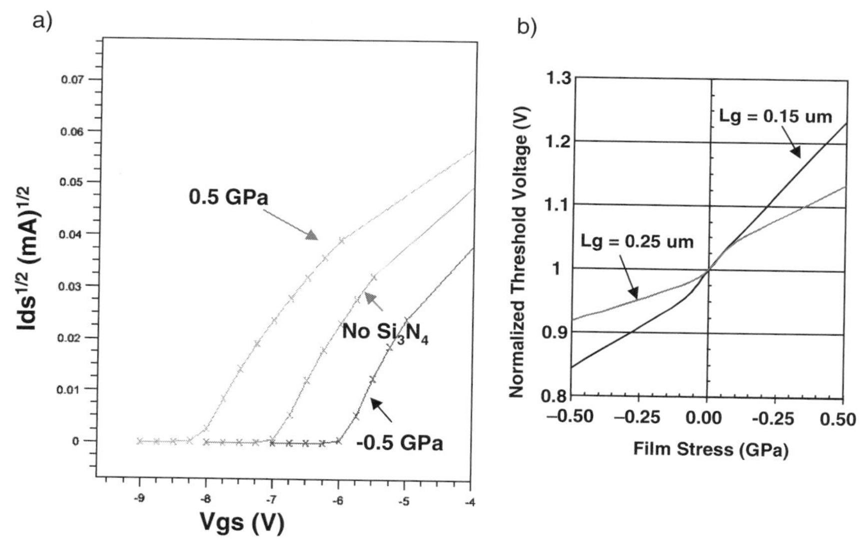

Fig. 19 a) Variation of Id-Vgs characteristics for AlGaN/GaN HFETs, with different values of stress in the Si_3N_4 passivation layers. A pronounced shift in threshold voltage Vt is apparent. b) Simulated variation of Vt for HFETs with varying gate length.

length (since the charge densities have only a small spatial extent, and V_t is sensitive primarily to the charge density at the mid-point of the channel). Figure 19b shows computed changes in threshold voltage as a function of silicon nitride stress. The simulations show that the breakdown voltage of the FETs is also dependent on the stress-induced charge density, and can be increased by employing a silicon nitride layer under compressive stress.

2.4 Polarization Effects in Nitride-Based HBTs and p-Type Structures

Another device for which polarization effects have a pronounced role is the Heterojunction Bipolar Transistor (HBT). HBTs employing nitride semiconductors can achieve much higher breakdown voltage and higher temperature operation than their counterparts based on GaAs, InP or SiGe semiconductors. In comparison with nitride-based HFETs, HBTs offer the possibility of higher transconductance, due to the exponential relationship between output current and input voltage; they are inherently normally off-devices, a highly desirable characteristic for power circuits; and they tend to exhibit fewer trap-related effects, since their intrinsic region is shielded from surfaces, so they can be free from anomalous transients exhibited by HFETs. As a result of these potential advantages, nitride-based HBTs are promising for application in power conditioning and amplification, at frequencies ranging from power-line frequency to the microwave regime, and numerous research efforts to develop nitride-based HBTs have been reported around the world [50, 51]. Representative material systems employed for HBTs include structures with AlGaN emitters, GaN bases and GaN collectors; or GaN emitters, InGaN bases, and GaN collectors, as shown schematically in figure 20. In both cases, a wide bandgap emitter is employed, in order to avoid minority carrier injection into the emitter and maximize dc current gain. The structures also feature graded composition in the base, in order to provide a built-in quasi-electric field to propel minority carriers across the base by drift as well as diffusion. Transistors with acceptable values of dc current gain have been demonstrated. Figure 21 illustrates, for example, the dc I-V characteristics of an experimental HBT with GaN / InGaN / GaN structure, exhibiting dc current gain of 20. There are significant challenges in the development of high performance HBTs in the nitrides, however, associated in large part with the difficulty of achieving adequate conductivity in the p-type based of the npn transistors. Available acceptor dopants in GaN all have large activation energy. The preferred dopant, Mg, has an acceptor depth of 170 meV, such that only a small fraction of the Mg acceptors incorporated in the material are ionized at room temperature. Even with Mg concentrations of order $10^{20}\,\mathrm{cm}^{-3}$, the hole concentration achieved is in the neighborhood of $10^{18}\,\mathrm{cm}^{-3}$. Additionally, the mobility of holes in

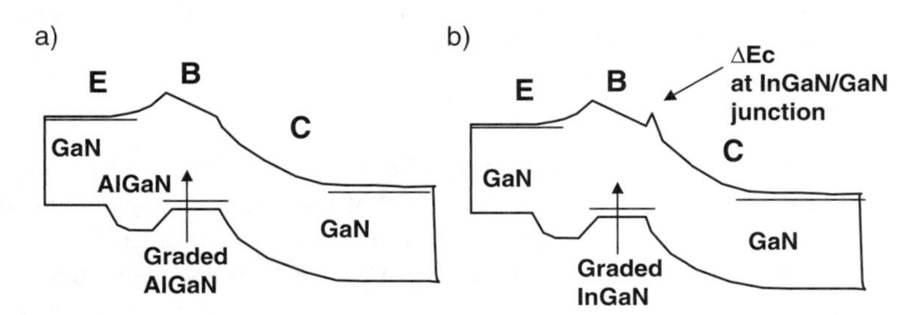

Fig. 20 Band diagrams for representative nitride-based HBTs. a) AlGaN emitter, AlGaN graded base, GaN collector; b) GaN or AlGaN emitter, InGaN graded base, GaN collector.

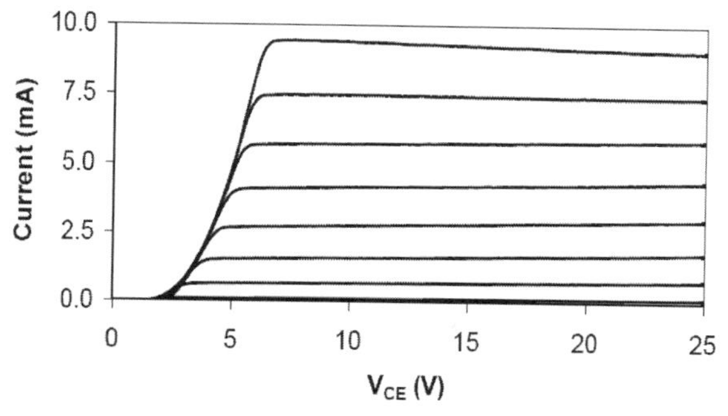

Fig. 21 Experimental Ic-Vce curves for representative GaN/InGaN/GaN HBT. The base current is varied in 62.5 uA steps.

nitride materials is low, with typical values of order 5–10 cm^2/Vsec, and the contact resistance between metals and p-type nitrides is very high. It is of considerable interest to employ polarization-based strategies to enhance the performance of HBTs. Potential approaches are discussed in the following.

2.4.1 Polarization Enhancement of Base Contacts

The high specific contact resistance of ohmic contacts to p-type GaN and nitride heterostructure layers has been a persistent problem for the development of HBTs as well as LEDs and lasers. Values of the order of 10^{-4} ohm cm^2 are typical of the best reported in HBT structures, and frequently much higher values are obtained. A proposed mechanism to improve the contact resistance is to create polarization-induced negative (acceptor-like) charge in the vicinity of the metal-semiconductor interface, where it can contribute to thinning of the depletion region, and enhance the tunneling probability of holes [52]. The polarization charge can be established with the use of a layer of InGaN deposited on a GaN substrate. A graded composition may be used in order to distribute the charge over a finite width, as illustrated in figure 22. The surface layer is also doped with Mg. A conventional ohmic contact may be represented in approximate fashion as an ideal Schottky barrier, in which the depletion region is thin enough to permit tunneling of electrons whose initial energy is near the fermi level (or somewhat close to it, corresponding to thermionic field emission current). Figure 22c shows the predicted valence band diagrams for a structure with and without polarization effects. The In composition of the InGaN contact layer is graded from 0 to 15% at the surface of a layer 60A thick. A Schottky barrier height of 2.2eV has been assumed. The figure shows that significant thinning of the depletion layer occurs, which should lead to enhanced current flow at a given bias. In the simulation, Mg has been assumed not to be complexed with hydrogen. Ionization of the Mg is enhanced in the vicinity of the contact as a result of band-bending. Experiments carried out by Gessman et al. have shown that significantly improved linearity of ohmic contacts to p-GaN results from the use of such cap layers [53].

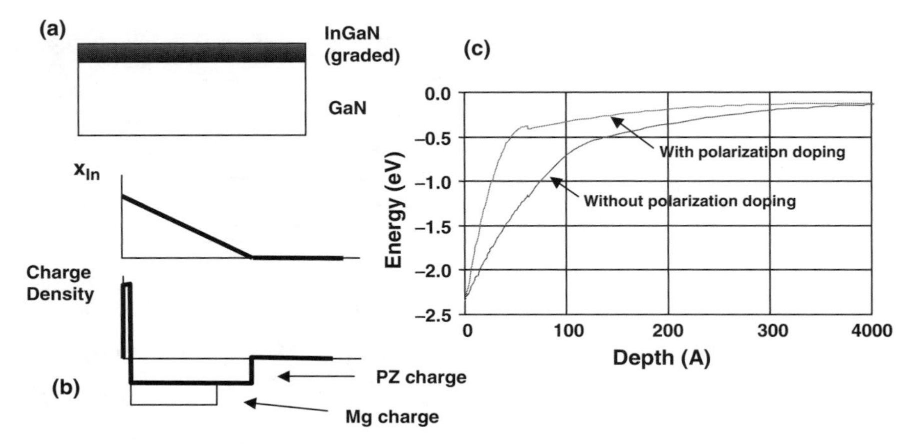

Fig. 22 (a) Layer structure for p-type ohmic contact enhanced via polarization doping. (b) Corresponding Indium alloy composition vs depth, and charge density. (c) Calculated valence band profile with and without polarization doping [52].

2.4.2 Polarization Enhancement of HBT Base Layer Conductivity

The variation of alloy composition across the HBT structure used to implement the wide bandgap emitter, graded base, and potentially wide bandgap collector have associated values of spatially varying polarization and bound charge ("polarization doping"). Figure 23 illustrates schematically the distribution of this bound charge within the HBT structure, along with the remaining doping contributions. Values of bound charge within the base region can be appreciable, for example, with a 20% Al alloy composition grading over a 500A base, the charge density reaches approximately $2 \times 10^{18}\,\text{cm}^{-3}$. Even higher densities can be achieved at the emitter-base and collector-base interfaces. As noted previously, however, the sign of the polarization doping is dependent on the crystallographic direction of epitaxial growth and the wafer surface. It is an unfortunate that for the conventional "emitter-up" device structure, and the [0001] (Ga face) growth direction, the sign of the polarization-induced charge is opposite to that which is desirable for a high performance HBT, that is, the charge in the base is donor-like for the graded base, rather than acceptorlike. One possible approach to remedy this situation is to carry out growth in the opposite direction, on the N face of the GaN. Figure 23 also shows the polarization doping for this case. Although it has been shown to be possible to reliably induce N face growth via appropriate choice of nucleation layers, the growth techniques have not yet been perfected for this direction, and no HBT structures have yet been grown with this geometry.

To assess the impact of the polarization-induced charge on the hole concentration in the base it is necessary to take into account the degree of ionization of the deep Mg acceptor concentration that may also be incorporated in the layer, as well as the possible presence of unintentionally incorporated residual donors. The hole density may be determined from the following:

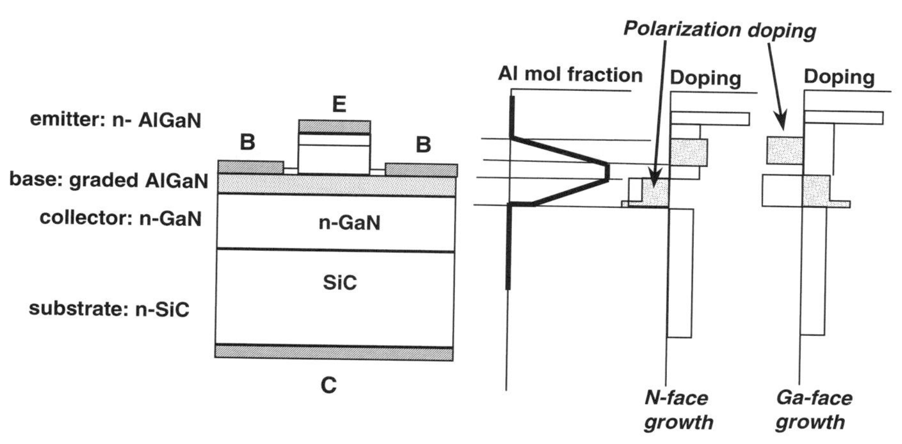

Fig. 23 Schematic structure of emitter-up AlGaN/GaN HBT, along with aluminum mol fraction as a function of depth, and schematic charge densities. The charge densities are shown for N-face growth and Ga-face growth, illustrating the changing role of the polarization charge [52].

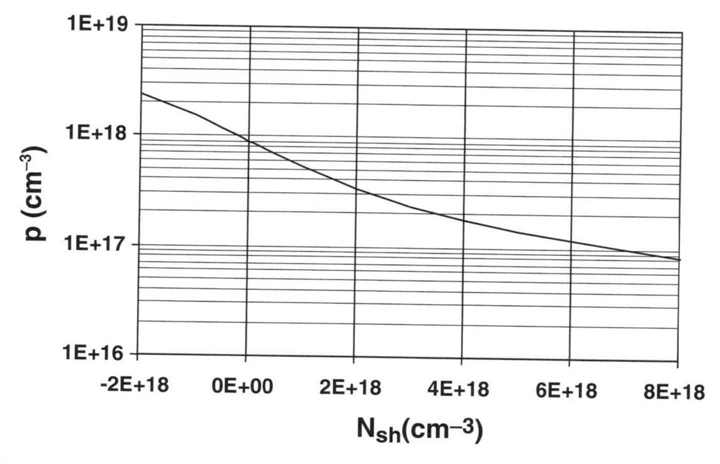

Fig. 24 Computed hole concentration vs shallow dopant density Nsh, for a Mg doping level of $5 \times 10^{19}\,\mathrm{cm}^{-3}$. Here Nsh is the combined result of residual shallow donor concentration and polarization-induced charge.

$$\frac{p(p+N_{sh})}{N_A - N_{sh} - p} = \frac{N_V}{g} \exp\left(-\frac{E_A}{KT}\right) \tag{12}$$

Here N_A is the concentration of deep acceptors (with ionization energy E_A), N_{sh} is the sum of polarization charge and residual donor concentration, and N_V is the effective valence band density of states. Figure 24 illustrates that variation of hole concentration p with N_{sh} for Mg doping $N_A = 5 \times 10^{19}\,\mathrm{cm}^{-3}$. The variation of p with polarization charge is significantly less than "one-for-one": as N_{sh} varies between

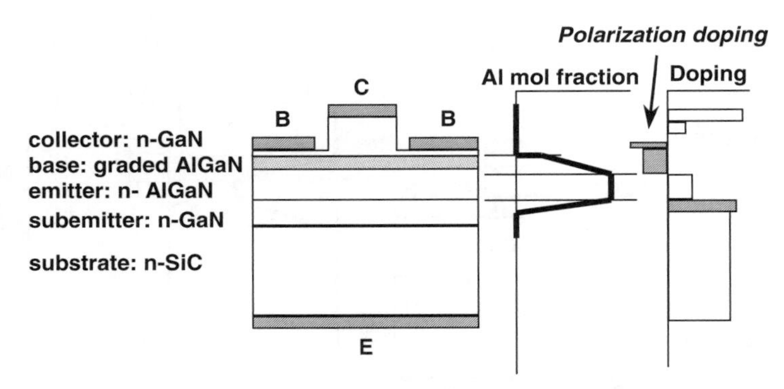

Fig. 25 Schematic structure of collector-up AlGaN/GaN HBT, along with aluminum mol fraction vs depth, and schematic charge densities. Ga-face growth is assumed [52].

$2\times10^{18}\,\mathrm{cm}^{-3}$ and $-2\times10^{18}\,\mathrm{cm}^{-3}$, p is found to vary between $3\times10^{17}\,\mathrm{cm}^{-3}$ and $2.2\times10^{18}\,\mathrm{cm}^{-3}$.

Alternate strategies to insure that the polarization charge within an HBT structure adds to, rather than detracts from, the hole density in the base, rely on altering the position of the emitter and collector within the device structure. High performance "collect-up" HBTs can be fabricated with the collector at the wafer surface. Kroemer [54] pointed out that this geometry has the desirable feature of dramatically reduced base-collector capacitance (which is the most critical parasitic limiting high frequency performance). A schematic structure for a collector-up AlGaN/GaN HBT with polarization-enhanced base conductivity is shown in figure 25; the structure is a mirror-image of the emitter-up design. A well-known problem with the collector-up design, however, is the fact that regions of the emitter not covered by the device collector ("extrinsic emitter" regions), may inject minority carriers into the base, which will not be collected, but rather will contribute to base current. It is critical to suppress this extrinsic base current contribution in order to have adequate current gain. The problem is particularly severe for devices with narrow collectors, since the collector current scales as the collector area, while the extrinsic emitter base current scales as the collector periphery. There are a number of strategies to suppress the extrinsic current, including the use of compensating implants in the extrinsic emitter regions; the use of SiO_2 buried layers, such as used in lateral epitaxial overgrowth; and the use of transferred substrate technology [55], in which processing is done on both emitter and collector sides of the device (made possible by transferring the device-containing film from one substrate to another). These structure have yet to be demonstrated with nitride-based HBTs. However, experiments have been conducted in order to demonstrate the role of polarization enhancement of doping in a structure analogous to a collector-up HBT [56]. MOCVD grown structures with Mg doping incorporated into GaN layers or into graded AlGaN/GaN layers were compared. Figure 26 illustrates the sample structure and the resultant sheet conductivity. It was found that the samples which contained graded AlGaN (providing negative polarization charge) had sheet conductance more than $2\times$ higher than that of the

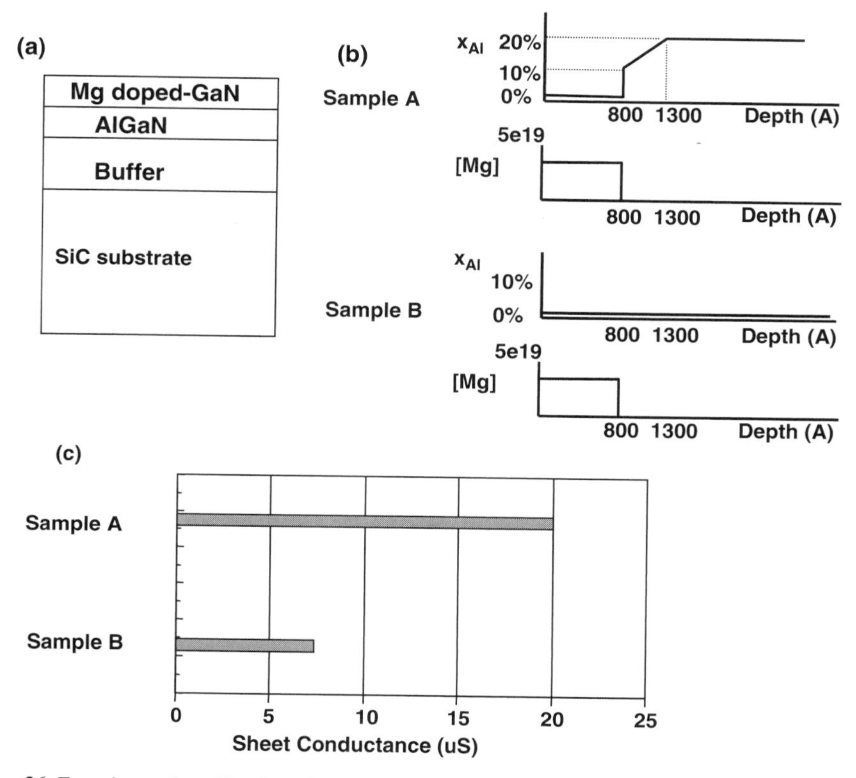

Fig. 26 Experimental verification of polarization enhancement of hole conductivity in GaN layers: a) sample structure; b) sample alloy composition and Mg doping profiles; c) measured sheet conductance.

GaN alone, although the Mg concentration and doping profile grown into the material were identical.

3 Localized Effects of Polarization

The strong coupling between structural morphology and composition and electronic properties in nitride semiconductors and heterostructures that arises from polarization effects increases, in many circumstances, the influence of defects and material inhomogeneities on both basic electronic and optical properties of nitride heterostructures and on various aspects of electronic and optoelectronic device performance. In this section we discuss the nature and characterization of a variety of such effects, and in subsequent sections the influence of these and related phenomena in electronic devices is addressed.

3.1 Dislocation-induced Polarization Fields

The heteroepitaxial nature of most nitride semiconductor epitaxial crystal growth leads to the presence of very high dislocation densities (typically in the range of $10^8 - 10^{10}\,\mathrm{cm}^{-2}$) in nitride semiconductor thin films and heterostructures. An understanding of the electronic properties of dislocations in the nitrides, and their consequences for devices, is therefore essential. There have been extensive studies showing that threading dislocations in GaN exhibit very prominent electrical behavior, often containing high densities of negative charge in the dislocation core [57–61] and, particularly for material grown by molecular-beam epitaxy, very high conductivity. [62–64] These properties, however, are not directly consequences of or related to polarization fields and therefore are not discussed further here.

In contrast, it has been shown theoretically [65] that the local strain fields associated with dislocations in nitride semiconductors can lead to the presence of large piezoelectric polarization fields in the vicinity of both screw and edge threading dislocations, and in the case of edge dislocations these polarization fields can generate substantial electrostatic charge densities in the vicinity of dislocation line intersections with surfaces and heterojunction interfaces. Specifically, computation of piezoelectric polarization fields in GaN for threading dislocation lines running along the [0001] direction, using Eqs. (2) and (5) and the established forms for strain fields associated with screw and edge dislocations, [66] yields polarization fields of the forms represented in Figures 27(a) and 27(b), respectively.

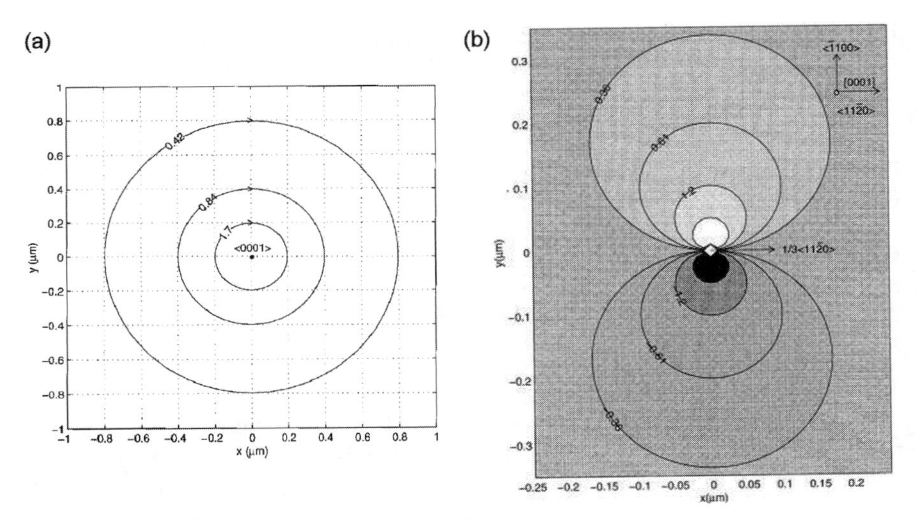

Fig. 27 (a) Polarization vector (in $10^{-4}\,\mathrm{C/m^2}$) associated with a pure screw dislocation in GaN running along the [0001] direction. (b) Contours of constant z component of polarization (in $10^{-4}\,\mathrm{C/m^2}$) associated with a pure edge dislocation running along the [0001] direction in GaN [65].

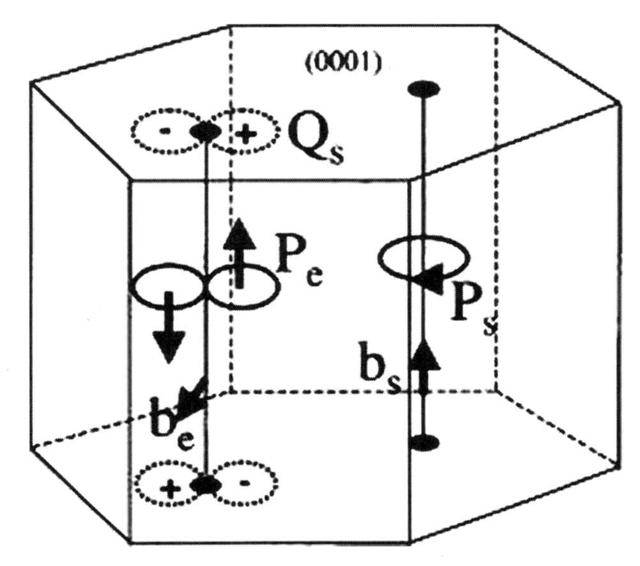

Fig. 28 Schematic diagram illustrating edge and screw dislocations along the [0001] direction in GaN, along with their Burgers vectors $\mathbf{b}_e$ and $\mathbf{b}_s$, polarization fields $\mathbf{P}_e$ and $\mathbf{P}_s$, and, in the case of the edge dislocation, polarization-induced charge densities Q at the upper and lower surfaces [65].

The polarization fields associated with pure screw dislocations running in the [0001] direction are easily seen to be divergence-free in bulk nitride material and planar heterostructures; polarization-induced charge densities are therefore not present in the vicinity of such dislocations. For edge dislocations running along the [0001] direction, the polarization fields also have zero divergence in bulk material. However, at intersections of the dislocation line with a free surface or a heterojunction interface, the discontinuity in the z-component of the dislocation-induced piezoelectric polarization field leads to the formation of dipole-like charge densities in the plane of the surface or interface, with magnitudes in the range of $\sim 10^{11}\, e/\text{cm}^2$ within $\sim 0.1\,\mu\text{m}$ of the dislocation core at a free surface, or within ~ 0.03–$0.05\,\mu\text{m}$ of the dislocation core at a heterojunction interface. A schematic summary of the polarization fields and charge densities associated with these dislocations is shown in Figure 28.

3.2 Scanning Capacitance Microscopy

Scanning probe techniques have emerged as a powerful approach for characterization of these and related phenomena at the nanoscale. In particular, scanning capacitance microscopy (SCM) enables charge and mobile carrier distributions to be imaged with spatial resolution of tens of nm laterally and, under favorable circumstances, approaching monoatomic layer resolution in depth. As shown in

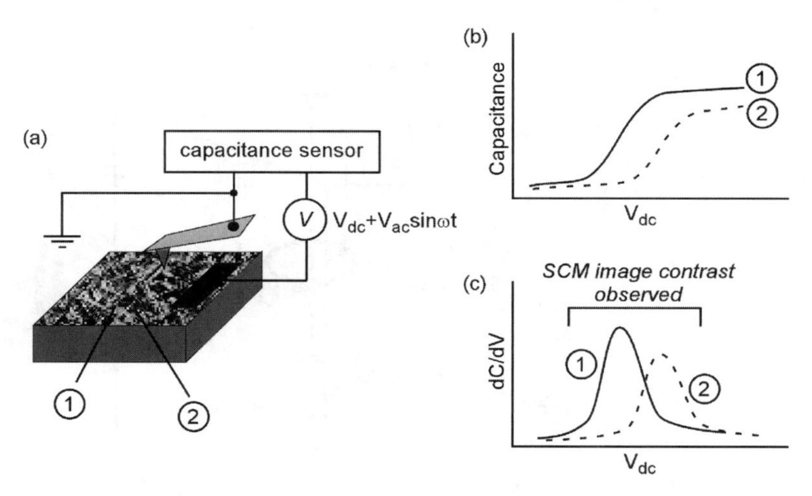

Fig. 29 (a) Schematic diagram of experimental apparatus and sample geometry for characterization of nitride heterostructures by scanning capacitance microscopy. (b) Schematic representation of tip-sample capacitance C for locations labeled "1" and "2" in (a). (c) Schematic representation of SCM signal, proportional to dC/dV, indicating range of voltage for which SCM image contrast between locations "1" and "2" is expected to be observed.

Figure 29, the SCM measurement entails the attachment of a very sensitive, resonant capacitance sensor to a contact-mode atomic force microscope to detect variations in the tip-sample capacitance arising from localized depletion of charge in the sample. The nature of the capacitance sensor is such that the signal detected is proportional not to the capacitance C itself, but instead to dC/dV; thus, local, nanoscale variations in capacitance are detected but an absolute measure of the capacitance magnitude itself is not obtained. The lateral spatial variation attainable in this measurement is determined by the probe tip radius, a "fringing" capacitance that depends on the detailed shape of the probe tip, and the extent of the depletion layer in the sample, as shown in Figure 30, and is generally in the range of a few tens of nm upward, depending on the sample structure under investigation. The resolution in depth depends on the depletion layer width and its sensitivity to voltage; for structures such as 2DEG layers or quantum wells, depth resolution can approach a single monoatomic layer, as demonstrated below in studies of monolayer thickness fluctuations in InGaN/GaN quantum-well structures.

3.3 Threshold Voltage Variations in AlGaN/GaN HEMT Structures

While the strain fields associated with threading dislocations in nitride semiconductors can generate moderate charge densities at interfaces in nitride heterostructures, the negative charge present in the core of most threading dislocations in n-type nitride semiconductors, due to its very large magnitude – variously estimated [58, 67–76] to be in the range of $0.1 - 2e/\mathrm{cm}$ – can have much more pronounced effects on mobile carrier distributions and transport. The nature and extent of these

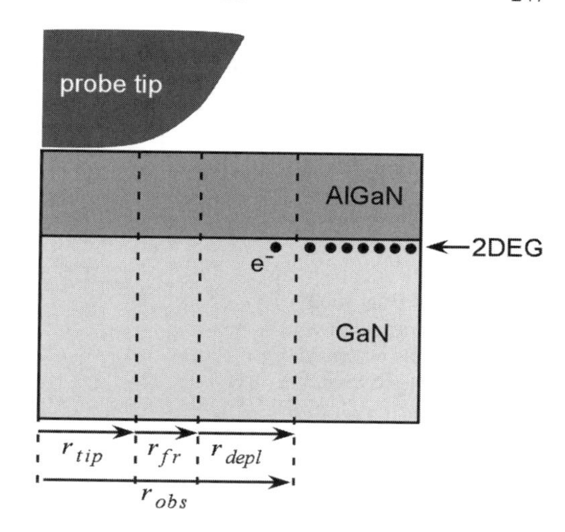

Fig. 30 Schematic diagram illustrating factors contributing to lateral spatial resolution in SCM measurement: probe tip radius r_{tip}, increased area due to "fringing" capacitance r_{fr}, and depletion layer width r_{depl}. The lateral spatial resolution or, similarly, the observed radius r_{obs} of a pointlike feature in the sample electronic structure, is given very approximately by the sum of these contributions.

effects has been studied in particular detail, with emphasis on elucidation of charge distributions at the nanoscale, in $Al_xGa_{1-x}N/GaN$ two-dimensional electron gas (2DEG) structures, in which nanoscale variations in carrier density can be observed very effectively by techniques such as scanning capacitance microscopy. [76–81] Studies of these effects in 2DEG structures have also revealed pronounced micron-scale variations in electron concentration within the 2DEG arising from small variations in thickness of the $Al_xGa_{1-x}N$ barrier layer. Because of the prominent role of polarization fields and polarization-induced charges in determining the electron concentration in an $Al_xGa_{1-x}N/GaN$ 2DEG structure, the local electron concentration is highly sensitive to the thickness of the $Al_xGa_{1-x}N$ layer in the immediate vicinity; variations of even a few nm in thickness can lead to changes in electron concentration that shift the threshold voltage for carrier depletion – a key characteristic in a field-effect transistor structure – by several tenths of a volt. Detailed characterization and understanding of these and related effects are therefore essential in device engineering and optimization.

Figure 31 shows a schematic diagram of an $Al_xGa_{1-x}N/GaN$ 2DEG heterostructure, with fluctuations in the $Al_xGa_{1-x}N$ layer thickness and traversed by a single, negatively charged threading dislocation line, along with the corresponding electron concentration profile in the 2DEG. Previous studies [82] have shown that, for a fixed Schottky barrier height or pinning position of the Fermi level at the $Al_xGa_{1-x}N$ surface, the electron concentration in the 2DEG should increase with increasing $Al_xGa_{1-x}N$ barrier layer thickness. This occurs because, as the $Al_xGa_{1-x}N$ layer increases in thickness, the potential drop across the $Al_xGa_{1-x}N$ layer remains approximately constant and the electric field within the $Al_xGa_{1-x}N$ layer therefore decreases. By Gauss's Law, the net charge at the $Al_xGa_{1-x}N/GaN$ interface – due to the combination of positive polarization-induced charge and negative charge due to electrons in the 2DEG – must decrease (but typically remain positive). The electron concentration therefore increases compared to that at lower $Al_xGa_{1-x}N$ barrier

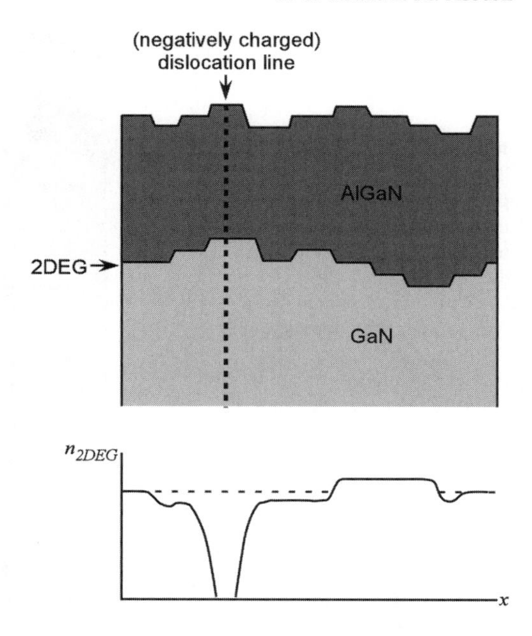

Fig. 31 Schematic diagrams of an AlGaN/GaN (0001) heterostructure within which a 2DEG forms at the AlGaN/GaN interface, and of the corresponding electron concentration profile. Fluctuations in AlGaN layer thickness giving rise to corresponding variations in electron concentration, and a negatively charged dislocation line that depletes carriers from the 2DEG are also shown.

layer thicknesses, reducing the net charge at the interface. Superimposed on these effects is local carrier depletion in the 2DEG in the vicinity of the dislocation line, as shown in Figure 31, due to the negative charge in the dislocation line core.

Figure 33(a) shows large-area atomic force microscope (AFM) and SCM images, obtained simultaneously, of an $Al_xGa_{1-x}N/GaN$ 2DEG epitaxial layer structure. The sample structure consisted of a 23 nm undoped $Al_{0.26}Ga_{0.74}N/1.2\,\mu m$ undoped $GaN/0.1\,\mu m$ AlN heterostructure grown by metalorganic chemical vapor deposition (MOCVD) on 4H-SiC (0001) The data shown in Figure 32(a) were obtained at a dc bias voltage of -7 V applied to the probe tip relative to the sample. Micron-scale variations in SCM signal, indicative of corresponding variations in local electron concentration in the 2DEG, are evident. To quantify the degree of this variation, or equivalently the amount by which the threshold voltage – the applied voltage required to deplete the 2DEG locally – varies, a spectroscopic approach is required. Figure 32(b) shows a map of local threshold voltage V_T required to deplete the 2DEG locally; clear patterns in threshold voltage at the micron scale are evident. The threshold voltages were determined in the manner shown in Figure 32(c). An SCM signal spectrum, i.e., a signal proportional to dC/dV, was measured as a function of dc bias voltage at each point in Figure 33(b) as shown in the top plot in Figure 33(c). Integration of the signal spectra with respect to voltage yielded capacitance-voltage curves associated with carrier modulation in the 2DEG, as shown in the middle plot of Figure 33(c). Finally, integration of this capacitance signal yielded a measure of the 2DEG charge at each point, from which the local threshold voltage V_T was extracted as shown in the bottom plot of Figure 33(c). As is evident from Figure 33(b), there are variations of a volt or more in local V_T over micron length

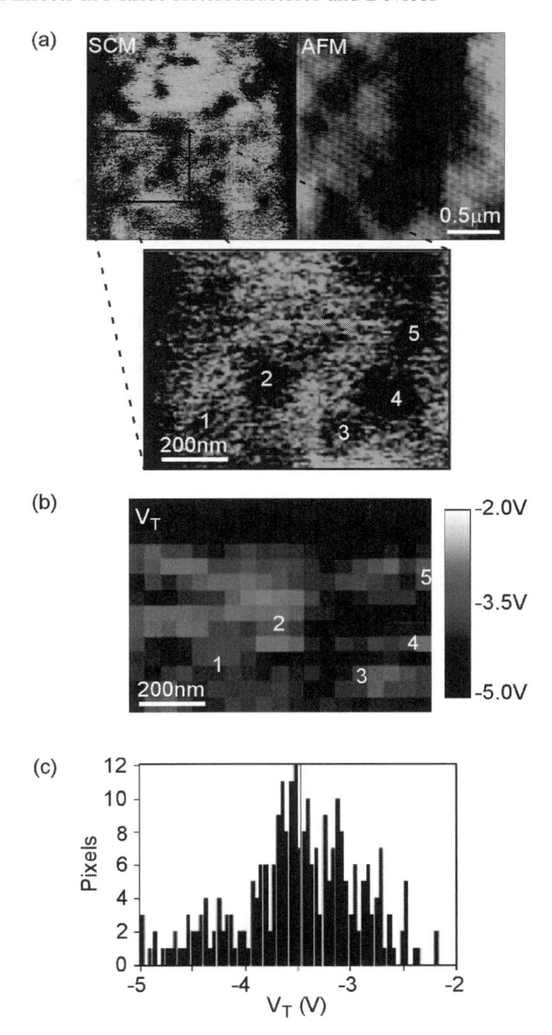

Fig. 32 (a) $2\,\mu\text{m} \times 2\,\mu\text{m}$ SCM and AFM images of AlGaN/GaN 2DEG epitaxial layer structure, along with a magnified view of part of the SCM image. (b) V_T map of area corresponding approximately to the magnified view in part (a), with corresponding locations in each marked '1'–'5'. (c) Histogram of V_T values mapped in part (b).

scales; Figure 33(d) shows a histogram of values of V_T for this region, confirming that variations in V_T of $\pm$ 1 V or more exist at these length scales. A detailed analysis of the SCM image contrast as a function of voltage has revealed that these variations arise predominantly from local inhomogeneities in AlGaN barrier layer thickness; [77] simple calculations have shown that variations in $Al_xGa_{1-x}N$ layer thickness of a few nm can lead to threshold voltage shifts of several tenths of a volt or more. [80]

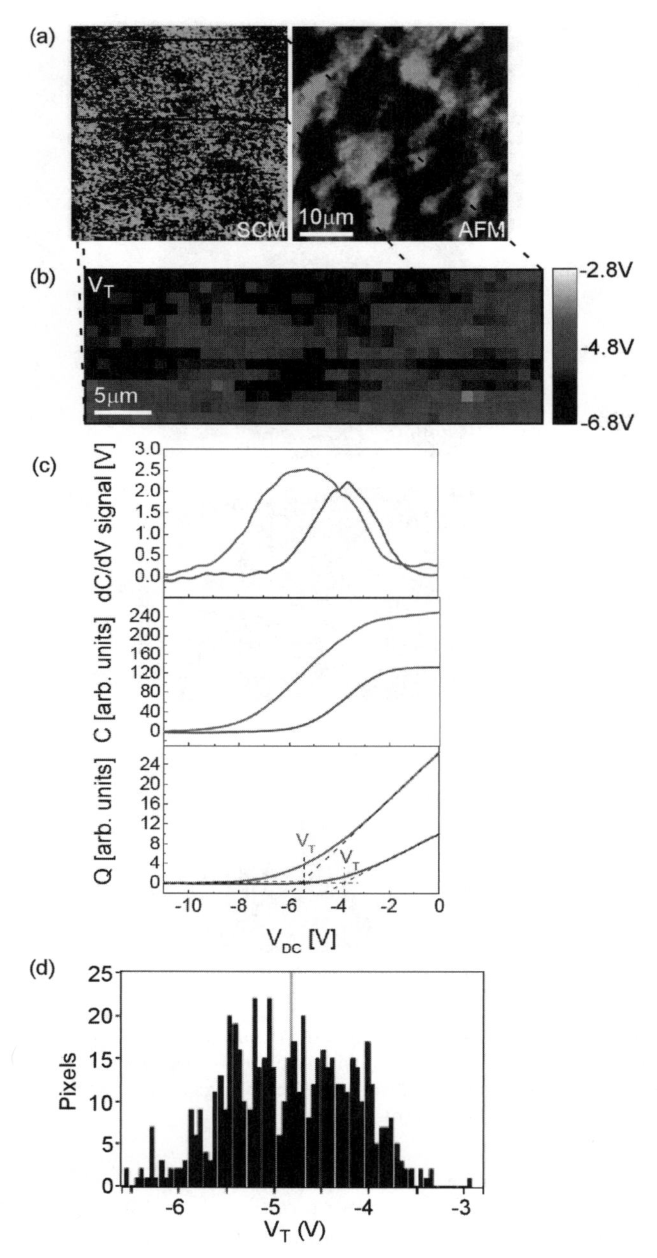

Fig. 33 (a) 40 μm × 40 μm SCM and AFM images, obtained simultaneously, of an AlGaN/GaN 2DEG epitaxial layer structure. (b) Map of local threshold voltage V_T for 2DEG depletion for the boxed region of part (a). (c) Local SCM signal spectra (dC/dV, top), and plots of 2DEG capacitance (middle) and 2DEG charge (bottom) obtained by successive integration of the SCM

Figure 32(a) shows SCM and AFM images of the same sample structure at higher lateral spatial resolution, with a further magnified view of one portion of the SCM image; these images were obtained simultaneously with a dc bias voltage of $-5\,\mathrm{V}$ applied to the probe tip relative to the sample. In the SCM image, a number of approximately circular features 100–150 nm in diameter are visible. Figure 32(b) shows a map of the local value of V_T for the area corresponding approximately to the magnified view of Figure 32(a); corresponding locations in the two images are labeled '1' –'5' in each. The circular features visible in Figure 32(a) can be seen to correspond to regions of less negative V_T, i.e., locally reduced electron concentration in the 2DEG. Figure 32(c) shows a histogram of values of V_T for the threshold voltage map shown in Figure 32(b), revealing that variations of up to approximately $\pm 1\,\mathrm{V}$ are present.

A detailed analysis [76, 80, 81] of these data indicates that the features observed in Figure 32 correspond to local regions of carrier depletion from the 2DEG associated with single threading dislocations. Specifically, the negative charge present in the core of the threading dislocation depletes electrons from the 2DEG in the vicinity of the intersection of the dislocation line with the $Al_xGa_{1-x}N/GaN$ interface; electrostatic simulations indicate that the shift in V_T observed is consistent with linear charge densities in the dislocation core of $\sim 0.1 - 0.5\,e/c$, where c is the c-axis lattice constant of GaN. The observation of such a phenomenon is notable both from the perspective of understanding the effects of extended defects such as dislocation on carrier behavior in the channel region of nitride-based FET structures, and also as a demonstration of imaging and characterization capability at the nanoscale for subsurface regions of realistic device epitaxial layer structures.

3.4 Nanoscale Electronic Structure in InGaN/GaN Quantum Wells

Polarization effects also play a prominent role in influencing electronic structure and carrier distributions in $In_yGa_{1-y}N/GaN$ quantum-well structures of interest for visible light emitter applications. The coupling between, on the one hand, local structural morphology and composition and, on the other, electronic structure and carrier behavior via polarization effects then leads to a very prominent influence of local, nanoscale inhomogeneities on carrier distributions in such structures. Once again, scanning probe techniques and in particular SCM offers a powerful method of elucidating the nature and extent of these effects.

Figure 34 shows a schematic diagram of an $In_yGa_{1-y}N/GaN$ quantum-well structure and of the corresponding charge density profile and conduction band-edge energy diagram. Scanning probe studies, as shown schematically in the figure, have been employed to elucidate the nature and influence of monolayer fluctuations in quantum-well thickness and nanoscale fluctuations in In composition of the $In_yGa_{1-y}N$ quantum-well region on carrier behavior, and in the latter case to correlate the presence of these fluctuations with increased luminescence efficiency. As discussed below, monolayer fluctuations and compositional variations can be distinguished from each other by SCM imaging at different bias voltages, with bias

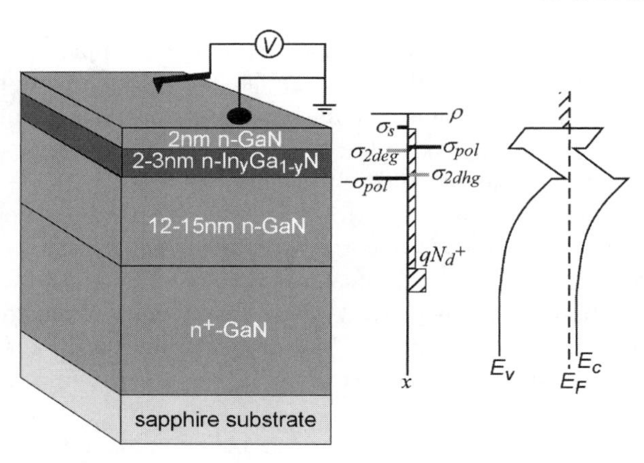

Fig. 34 Schematic diagram of InGaN/GaN quantum-well sample structure, scanning probe experiment geometry, and corresponding charge density and band-edge energy profiles.

conditions leading to electron accumulation in the quantum well allowing imaging of compositional fluctuations, [83] and bias conditions leading to hole accumulation providing sensitivity to quantum-well thickness fluctuations. [84] The very close proximity of the quantum-well region to the sample surface enables very high spatial resolution to be achieved since modulation of the carrier density in the quantum well occurs at only a very small depletion depth into the sample, and in addition improves sensitivity to the quantum-well electronic structure since the shallow quantum-well depth leads to a high tip-sample capacitance.

3.4.1 Monolayer Quantum-well Thickness Fluctuations

Figure 35 shows an AFM topographic image and a series of SCM images, obtained in the geometry shown in Figure 34, of a sample consisting of a 3 nm $In_{0.3}Ga_{0.7}N$ quantum-well layer grown on GaN and capped with 2nm n-type GaN. The sample was grown by MOCVD on a *c*-plane sapphire substrate with the n^+-GaN layer being 300nm in thickness and doped to a donor concentration of $\sim 1 \times 10^{18}\,cm^{-3}$, and the remaining layers doped to a donor concentration in the mid-$10^{16}\,cm^{-3}$ range. The images were obtained with negative bias voltage applied to the sample, leading to the reverse-bias conditions, and hole accumulation in the quantum-well region, shown in Figure 34(a). As seen in Figure 34(b), the SCM image contrast varies dramatically with bias voltage in this range. Specifically, at bias voltages in the range $-2.25\,V$ to $-2.50\,V$, near the threshold voltage for hole accumulation in the quantum well, contrast in the SCM images similar in nature and length scale to the topography observed in the AFM image, becomes evident. The dependence of the SCM image contrast on bias voltage is a clear indication that the features observed are associated with electronic structure, rather than being topographically induced.

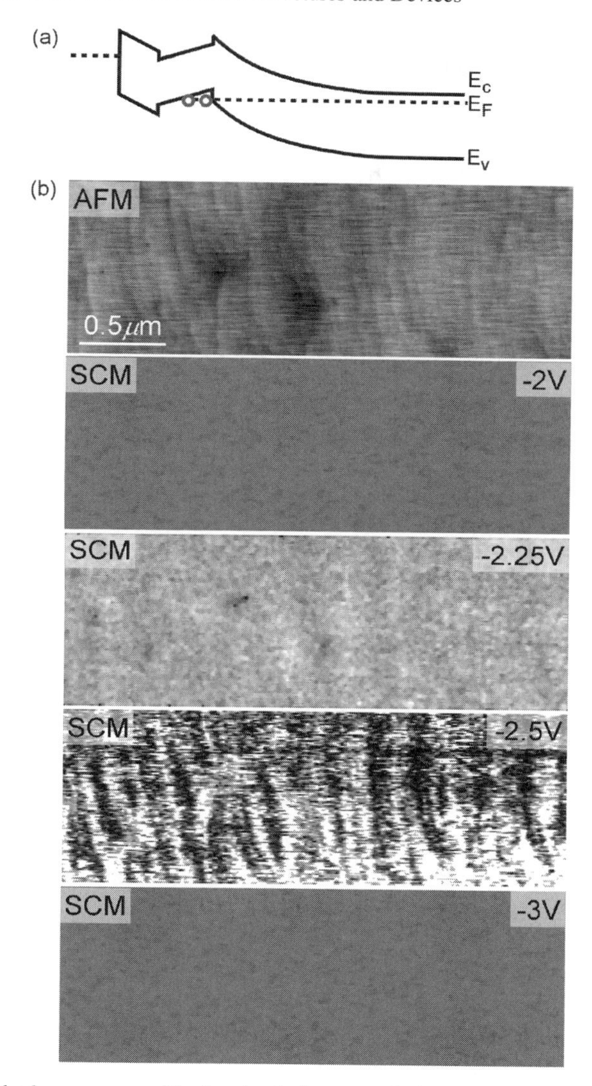

Fig. 35 (a) Band-edge energy profile showing hole accumulation under SCM imaging conditions employed here. (b) AFM topograph and series of SCM images at bias voltages of −2 V to −3 V.

Figure 36(a) shows topographic and corresponding SCM signal line scans extracted from the images shown. From the line scans it is evident that while the surface topography and SCM signal profile exhibit variations over comparable length scales, there is not any consistent correlation between specific topographic features and the SCM signal in the corresponding locations. This observation suggests that the SCM signal contrast is related to, but not a direct consequence of, the surface step structure during and after completion of growth. The most obvious hypothesis

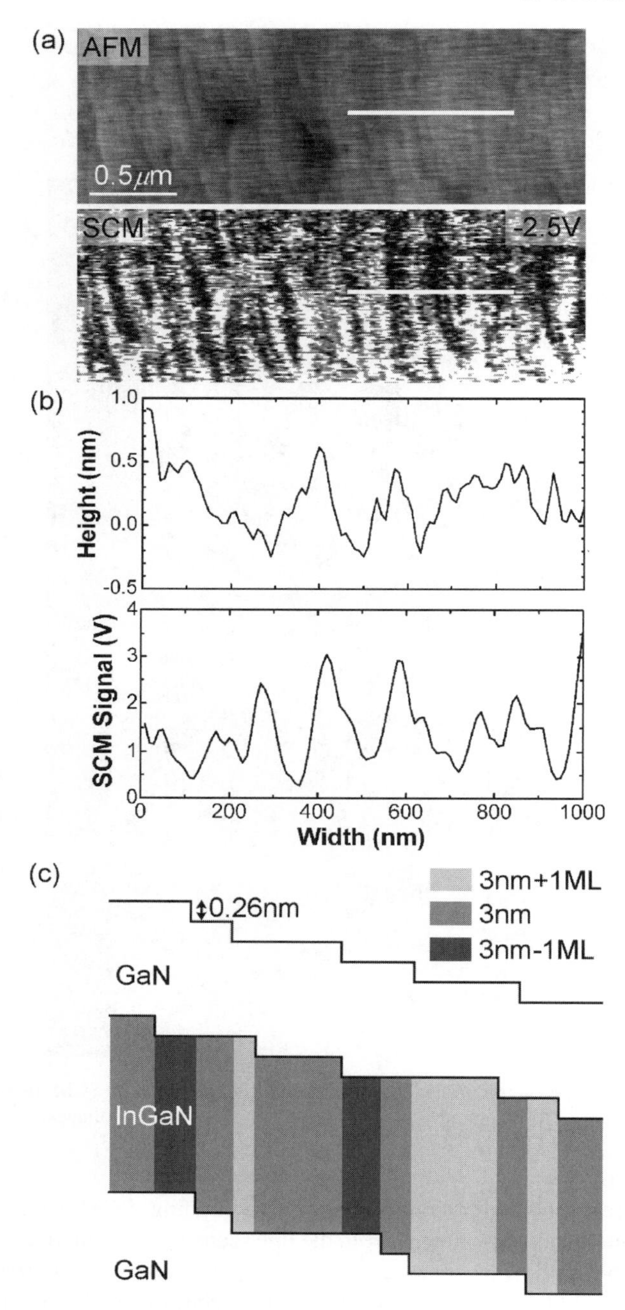

Fig. 36 (a) AFM topograph and SCM image of InGaN/GaN quantum-well sample at $-2.5\,\mathrm{V}$ bias. (b) Topographic and SCM signal line profiles extracted from images in (a) along lines indicated. (c) Schematic diagram of relationship between surface/interface roughness and monolayer fluctuations in InGaN quantum-well thickness.

is that the observed SCM signal contrast arises from monolayer fluctuations in thickness of the $In_{0.3}Ga_{0.7}N$ quantum well. As shown in Figure 36(c), monoatomic steps on the surface during growth will lead to the presence of monolayer fluctuations in quantum-well thickness, with a lateral spatial distribution related but not identical to the surface step structure. However, the issue of whether the observed SCM signal contrast arises from variations in the thickness of the $In_{0.3}Ga_{0.7}N$ quantum-well layer or the GaN capping layer remains.

This issue can be resolved by examination of spectroscopic data and results of numerical simulations. Figure 38 shows an SCM image along with experimental SCM signal spectra obtained at two locations marked on the image, and results of a numerical simulation of capacitance as a function of voltage for structures with $In_{0.3}Ga_{0.7}N$ layers 3 nm $\pm$ 1 ML in thickness. The experimental SCM spectra show that the voltage at which hole accumulation in the $In_{0.3}Ga_{0.7}N$ quantum

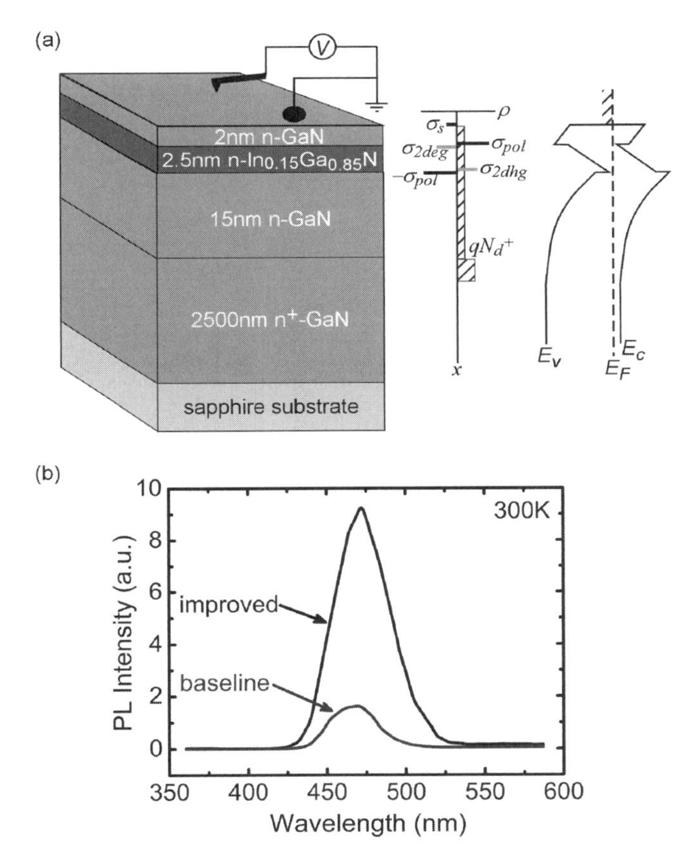

Fig. 37 (a) Schematic diagram of InGaN/GaN quantum-well sample structure, scanning probe experiment geometry, and corresponding charge density and band-edge energy profiles for studies of nanoscale In clustering. (b) Photoluminescence spectra for "baseline" and "improved" samples.

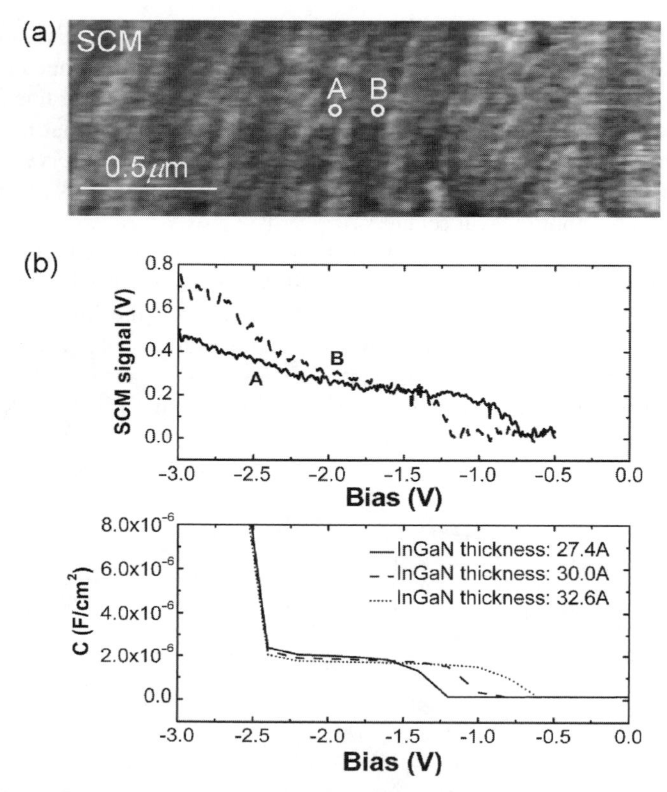

Fig. 38 (a) SCM image obtained at bias voltage of $-2.5\,\mathrm{V}$. (b) Experimental SCM signal spectra from locations marked "A" and "B" in (a), and numerical simulation of capacitance-voltage spectra for different InGaN layer thicknesses.

well commences, corresponding to the increase in SCM signal level observed at approximately $-0.7\,\mathrm{V}$ to $-1.2\,\mathrm{V}$, shifts significantly with SCM signal level in a constant-voltage image. The numerical simulations clearly show that even a change in $In_{0.3}Ga_{0.7}N$ layer thickness by a single monolayer causes a substantial shift – approximately 0.2-0.3 V – in this voltage. This very sensitive dependence of voltage on layer thickness is largely a consequence of the strong coupling of layer structure and internal electric field that arises from the large magnitude of the polarization fields and charges in these structures. Similar simulations of structures in which only the thickness of the top GaN layer is varied demonstrate that variations in GaN layer thickness have only a negligible effect on the voltages at which hole accumulation begins to occur. Thus, the combination of voltage-dependent SCM imaging, spatially resolved SCM spectroscopic measurements, and numerical simulations confirms the observation of nanoscale variations in carrier accumulation arising from monolayer fluctuations in quantum-well thickness in an $In_{0.3}Ga_{0.7}N/GaN$ quantum-well structure.

3.4.2 Nanoscale In Clustering

SCM imaging also enables local nanoscale variations in In composition in $In_yGa_{1-y}N$/GaN quantum-well structures to be imaged and characterized via the connection between carrier density and In concentration in the quantum well. In contrast to the imaging of quantum-well layer thickness fluctuations, however, compositional variations are most evident in imaging at bias voltages for which electron accumulation, rather than hole accumulation, occurs in the quantum-well region. This capability has proven to be particularly fruitful in elucidating the connection between In clustering at the nanoscale and improved luminescence efficiency in $In_yGa_{1-y}N/GaN$ quantum-well structures for light emitter applications.

To illustrate these points, we consider a sample structure and experimental geometry as shown in Figure 37(a). The sample structure consists of a 2.5 nm $In_{0.15}Ga_{0.85}N$ quantum well grown atop 15 nm GaN with a 2 nm GaN cap layer, all doped to a donor concentration in the mid-$10^{16}\,cm^{-3}$ range, deposited by MOCVD on a 2500 nm n^+-GaN buffer layer on a *c*-axis sapphire substrate. Two samples of nominally identical structure as shown in Figure 37(a), but differing in the detailed reactor conditions employed during quantum-well growth, yield very different luminescence efficiencies, as shown in Figure 37(b) comparing photoluminescence from a "baseline" structure with that from an "improved" structure exhibiting markedly increased luminescence efficiency.

SCM imaging of these samples reveals very different compositional structure at the nanoscale. Figure 38 shows an AFM topograph and a series of SCM images, of the same area as the AFM image, of the "baseline" sample. As expected, a general shift in signal level associated with the onset of electron accumulation in the quantum well at approximately 1 V to 1.5 V bias is observed, but the carrier accumulation characteristics appear to be essentially homogeneous within the resolution of the imaging system. In contrast, the SCM images shown in Figure 39 reveal marked contrast at voltages in the range of $\sim$1.25 V to $\sim$2.25 V. From the SCM image obtained at a bias of 1.25 V, we see that the onset of electron accumulation in the quantum well occurs not uniformly throughout the quantum-well layer, but in highly localized regions corresponding to the circular bright spots in the image. As the bias voltage increases, the density of such features increases, as does their average size, until at approximately 2.00 V bias electron accumulation is present throughout the quantum-well layer.

Numerical simulations and spectroscopic data provide information essential to the interpretation of these images. Figure 41(a) shows an SCM image of the "improved" sample structure obtained at a bias voltage of 1.75 V, and Figure 41(b) shows SCM signal spectra constructed at locations marked "A", "B", and "C" in Figure 41(a) constructed from images obtained at a series of bias voltages. Figure 41(c) shows a complete SCM signal spectrum obtained directly at a single point, confirming the validity of the image-constructed spectra shown in Figure 40(b). Figure 40(d) and (e) show numerically simulated SCM signal spectra computed for structures with, respectively, varying In concentration and varying $In_{0.15}Ga_{0.85}N$ layer thickness. One can see from the simulations that the variations

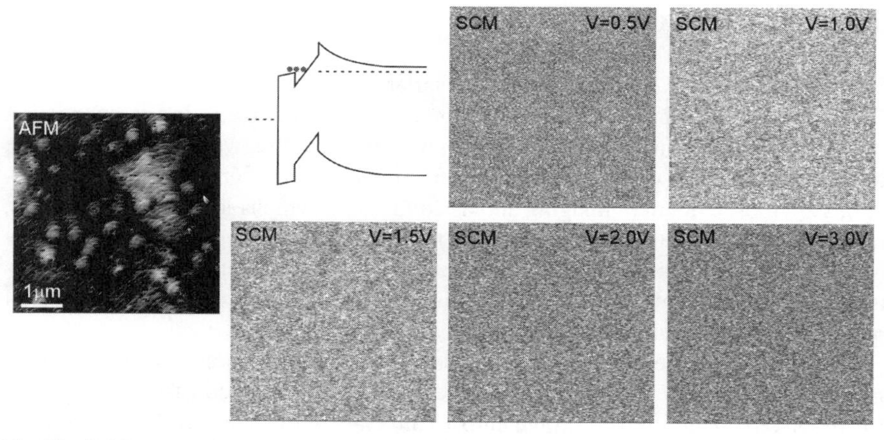

Fig. 39 AFM topograph, schematic band-edge energy diagram showing electron accumulation in quantum well for bias conditions employed here, and series of SCM images at 0.5 V to 3.0 V bias for "baseline" InGaN/GaN quantum-well sample.

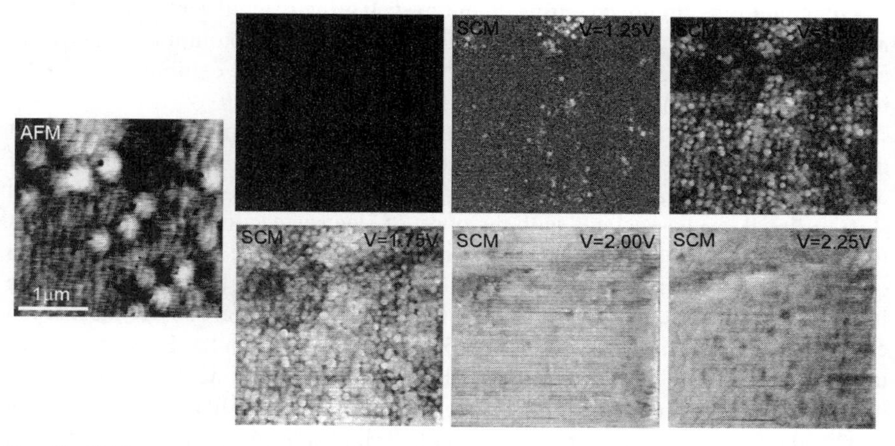

Fig. 40 AFM topograph and series of SCM images at 0.50 V to 2.25 V bias for "improved" InGaN/GaN quantum-well sample. Inhomogeneity in quantum-well electronic structure is clearly visible.

in measured local SCM signal spectra shown in Figure 41(b) are much more consistent with variations in In concentration within the quantum well than with local variations in $In_{0.15}Ga_{0.85}N$ layer thickness. Physically, this is reasonable as well as a locally higher In concentration, as would be present at location "A", should lead to electron accumulation, and hence a higher SCM signal level, at lower bias voltage – as we observe both experimentally and via simulation. Thus, we interpret the SCM contrast visible in Figure 40 as a consequence of local, nanoscale variations in In concentration with the circular features visible at intermediate bias voltages corresponding to local regions of increased In concentration within the quantum

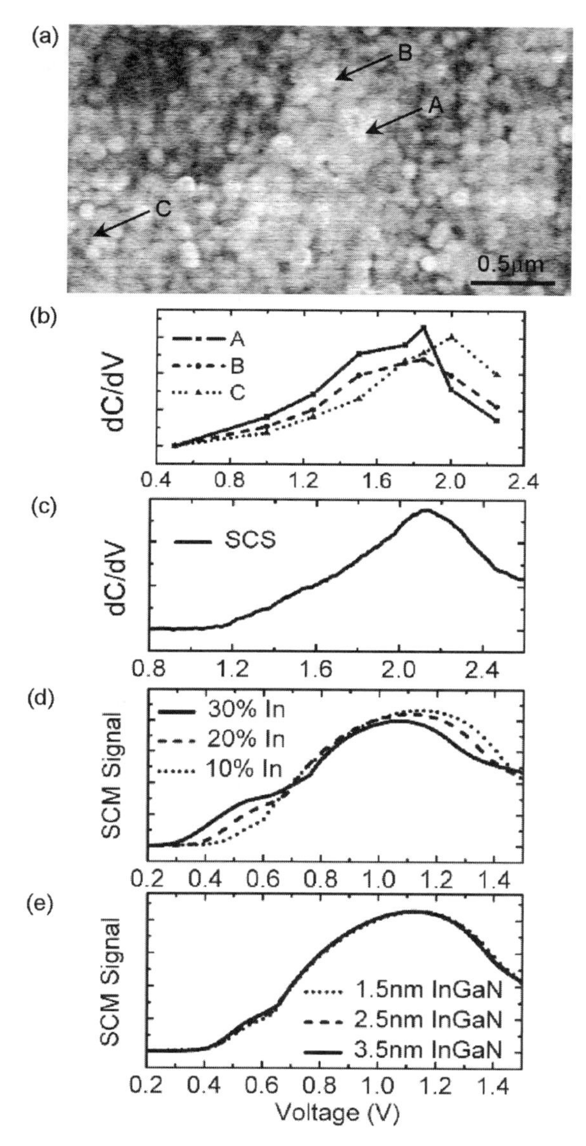

Fig. 41 (a) SCM image of "improved" InGaN/GaN quantum-well structure obtained at 1.75 V bias. (b) SCM signal spectra constructed at locations "A", "B", and "C" marked in (a) from bias-dependent series of SCM images. (c) SCM signal spectrum obtained directly at a single point. Numerical simulations of SCM signal spectrum for structures with (a) varying In concentration in InGaN quantum well and (b) varying InGaN layer thickness.

well. The evolution of the SCM image contrast with bias voltage suggests that these In-rich nanoscale clusters are present at concentrations in the vicinity of $\sim 10^{10}\,\mathrm{cm}^{-2}$ and with a range of In concentrations.

These observations demonstrate an approach for direct imaging of nanoscale compositional variations in subsurface semiconductor quantum-well structures via their influence on local electron structure and carrier density – a significant advance in scanning probe-based nanoscale imaging methodology. In addition, these experiments also provide direct evidence and confirmation of the hypothesis that formation of these In-rich nanoscale clusters is correlated with increased luminescence efficiency – a key issue for optimization of nitride-based light emitters.

Acknowledgements The authors are grateful to the Office of Naval Research for funding a portion of the work reported here, and to Dr. Colin Wood for guidance and support throughout the research. The authors are also indebted to numerous colleagues for stimulating discussions, particularly to S. Lau, L. Eastman, W. Schaff, U. Mishra, J. Speck, R. Davis and O. Ambacher, as well as to numerous graduate students whose research contributed to the material presented in this chapter.

References

1. P. Asbeck, C. P. Lee and M. F. Chang, "Piezoelectric effects in GaAs FETs and their role in orientation dependent device characteristics", IEEE Trans. on Electron Devices 31, 1377 (1984).
2. D. L. Smith. Strain-generated electric-fields in [111] growth axis strained-layer superlattices. *Solid-State Commun.* **57**, 919–21 (1986).
3. D. L. Smith and C. Mailhiot. Piezoelectric effects in strained-layer superlattices. *J. Appl. Phys.* **63**, 2717–9 (1988).
4. T. F. Kuech, R. T. Collins, D. L. Smith, and C. Mailhiot. Field-effect transistor structure based on strain-induced polarization charges. *J. Appl. Phys.* **67**, 2650–2 (1990).
5. E. S. Snow, B. V. Shanabrook, and D. Gammon. Strain-induced 2-dimensional electron-gas in [111] growth-axis strained-layer structures. *Appl. Phys. Lett.* **56**, 758–60 (1990).
6. J. F. Nye. *Physical Properties of Crystals: Their Representation by Tensors and Matrices* (Oxford University Press, Oxford, 1998).
7. S. Strite, M. E. Lin, and H. Morkoç. Progress and prospects for GaN and the III-V-nitride semiconductors. *Thin Solid Films* **231**, 197–210 (1993).
8. O. Ambacher, J. Majewski, C. Miskys, A. Link, M. Hermann, M. Eickhoff, M. Stutzmann, F. Bernardini, V. Fiorentini, V. Tilak, B. Schaff, and L. F. Eastman. Pyroelectric properties of Al(In)GaN/GaN hetero- and quantum well structures. *J. Phys.: Condens. Matter* **14**, 3399–3434 (2002).
9. M. A. Littlejohn, J. R. Hauser, and T. H. Glisson. Monte-Carlo calculation of velocity-field relationship for gallium nitride. *Appl. Phys. Lett.* **26**, 625–7 (1975).
10. A. D. Bykhovski, B. L. Gelmont, and M. S. Shur. Elastic strain relaxation and piezoeffect in GaN-AlN, GaN-AlGaN, and GaN-InGaN superlattices. *J. Appl. Phys.* **81**, 6332–8 (1997).
11. G. D. O'Clock, Jr. and M. T. Duffy. Acoustic surface-wave properties of epitaxially grown aluminum nitride and gallium nitride on sapphire. *Appl. Phys. Lett.* **23**, 55–6 (1973).
12. A. D. Bykhovski, V. V. Kaminski, M. S. Shur, Q. C. Chen, and M. A. Khan. Piezoresistive effect in wurtzite n-type GaN. *Appl. Phys. Lett.* **68**, 818–9 (1996).
13. J. G. Gualtieri, J. A. Kosinski, and A. Ballato. Piezoelectric materials for acoustic-wave applications. *IEEE Trans. Ultrason. Ferroelectr. Freq. Control* **41**, 53–9 (1994).
14. F. Bernardini, V. Fiorentini, and D. Vanderbilt. Spontaneous polarization and piezoelectric constants of III-V nitrides. *Phys. Rev. B* **56**, R10024–7 (1997).
15. V. A. Savastenko and A. U. Sheleg. Study of elastic properties of gallium nitride. *Phys. Status Solidi A* **48**, K135–9 (1978).

16. Y. Takagi, M. Ahart, T. Azuhata, T. Sota, K. Suzuki, and S. Nakamura. Brillouin scattering study in the GaN epitaxial layer. *Physica B* **219&220**, 547–9 (1996).
17. A. Polian, M. Grimsditch, and I. Grzegory. Elastic constants of gallium nitride. *J. Appl. Phys.* **79**, 3343–4 (1996).
18. R. B. Schwarz, K. Khachaturyan, and E. R. Weber. Elastic moduli of gallium nitride. *Appl. Phys. Lett.* **70**, 1122–4 (1997).
19. C. Deger, E. Born, H. Angerer, O. Ambacher, M. Stutzmann, J. Hornsteiner, E. Riha, and G. Fischerauer. Sound velocity of $Al_xGa_{1-x}N$ thin films obtained by surface acoustic-wave measurements. *Appl. Phys. Lett.* **72**, 2400–2 (1998).
20. K. Tsubouchi and N. Mikoshiba. Zero-temperature-coefficient SAW devices on AlN epitaxial-films. *IEEE Trans. Sonics Ultrason.* **SU-32**, 634–44 (1985).
21. L. E. McNeil, M. Grimsditch, and R. H. French. Vibrational spectroscopy of aluminum nitride. *J. Am. Ceram. Soc.* **76**, 1132–6 (1993).
22. K. Kim, W. R. L. Lambrecht, and B. Segall. Elastic constants and related properties of tetrahedrally bonded BN, AlN, GaN, and InN. *Phys. Rev. B* **53**, 16310–26 (1996).
23. A. F. Wright. Elastic properties of zinc-blende and wurtzite AlN, GaN, and InN. *J. Appl. Phys.* **82**, 2833–9 (1997).
24. S. Chichibu, T. Azuhata, T. Sota, and S. Nakamura. Spontaneous emission of localized excitons in InGaN single and multiquantum well structures. *Appl. Phys. Lett.* **69**, 4188–90 (1996).
25. S. F. Chichibu, A. C. Abare, M. S. Minsky, S. Keller, S. B. Fleischer, J. E. Bowers, E. Hu, U. K. Mishra, L. A. Coldren, and S. P. DenBaars, and T. Sota. Effective band gap inhomogeneity and piezoelectric field in InGaN/GaN multiquantum well structures. *Appl. Phys. Lett.* **73**, 2006–8 (1998).
26. T. Takeuchi, C. Wetzel, S. Yamaguchi, H. Sakai, H. Amano, I. Akasaki, Y. Kaneko, S. Nakagawa, Y. Yamaoka, and N. Yamada. Determination of piezoelectric fields in strained GaInN quantum wells using the quantum-confined Stark effect. *Appl. Phys. Lett.* **73**, 1691–3 (1998).
27. J. A. Garrido, J. L. Sanchez-Rojas, A. Jimenez, E. Munoz, F. Omnes, and P. Gibart. Polarization fields determination in AlGaN/GaN heterostructure field-effect transistors from charge control analysis. *Appl. Phys. Lett.* **75**, 2407–9 (1999).
28. P. Lefebvre, A. Morel, M. Gallart, T. Taliercio, J. Allegre, B. Gil, H. Mathieu, B. Damilano, N. Grandjean, and J. Massies. High internal electric field in a graded-width InGaN/GaN quantum well: accurate determination by time-resolved photoluminescence spectroscopy. *Appl. Phys. Lett.* **78**, 1252–4 (2001).
29. Y. D. Jho, J. S. Yahng, E. Oh, and D. S. Kim. Measurement of piezoelectric field and tunneling times in strongly biased InGaN/GaN quantum wells. *Appl. Phys. Lett.* **79**, 1130–2 (2001).
30. F. Renner, P. Kiesel, and G. H. Dohler, M. Kneissl, C. G. Van de Walle, and N. M. Johnson. Quantitative analysis of the polarization fields and absorption changes in InGaN/GaN quantum wells with electroabsorption spectroscopy. *Appl. Phys. Lett.* **81**, 490–2 (2002).
31. C. Y. Lai, T. M. Hsu, W.-H. Chang, and K.-U. Tseng. Direct measurement of piezoelectric field in $In_{0.23}Ga_{0.77}N/GaN$ multiple quantum wells by electrotransmission spectroscopy. *J. Appl. Phys.* **91**, 531–3 (2002).
32. E. J. Miller, E. T. Yu, C. Poblenz, C. Elsass, and J. S. Speck. Direct measurement of the polarization charge in AlGaN/GaN heterostructures using capacitance-voltage carrier profiling. *Appl. Phys. Lett.* **80**, 3551–3 (2002).
33. H. Zhang, E. J. Miller, E. T. Yu, C. Poblenz, and J. S. Speck. Measurement of polarization charge and conduction band offset at InxGa1-xN/GaN heterojunction interfaces. *Appl. Phys. Lett.* **84**, 4644–6 (2004).
34. E. T. Yu, G. J. Sullivan, P. M. Asbeck, C. D.Wang, D. Qiao, and S. S. Lau. Measurement of piezoelectrically induced charge in GaN/AlGaN heterostructure field-effect transistors. *Appl. Phys. Lett.* **71**, 2794–6 (1997).
35. R. People, K. W. Wecht, K. Alavi, and A. Y. Cho. Measurement of the conduction-band discontinuity of molecular-beam epitaxial grown $In_{0.52}Al_{0.48}As/In_{0.53}Ga_{0.47}As$ n-n heterojunction by C-V profiling. *Appl. Phys. Lett.* **43**, 118–20 (1983).

36. H. Kroemer, Wu-Yi Chien, J. S. Harris, Jr., and D. D. Edwall. Measurement of isotype heterojunction barriers by C-V profiling. *Appl. Phys. Lett.* **36**, 295–7 (1980).
37. A.Bykhovski, R.Gaska, M.S. Shur, "Piezoelectric doping and elastic strain relaxation in AlGaN-GaN heterostructure field effect transistors", Appl. Phys. Letts. 73, .3577–9 (1998).
38. F. A. Ponce, S. Srinivasan, A. Bell, L. Geng, R. Liu, M. Stevens, J. Cai, H. Omiya, H. Marui, and S. Tanaka, "Microstructure and electronic properties of InGaN alloys", Phys. Stat. Sol. (b) 240, 273–284 (2003)
39. S. Einfeldt, Z.J.Reitmeier and R.F.Davis, "Strain of GaN Layers Grown Using 6H-SiC(0001) Substrates with Different Buffer Layers", International J. of High Speed Electronics and Systems 14, 39 (2004).
40. L. Hsu, W.Walukiewicz, "Effects of piezoelectric field on defect formation, charge transfer, and electron transport at GaN/AlxGa1-xN interfaces", Appl. Phys. Letts., .73, 339–41 (1998).
41. J.A.Van Vechten, J.D. Zook, R.D.Horning, B. Goldenberg, "Defeating compensation in wide gap semiconductors by growing in H that is removed by low temperature de-ionizing radiation", Japanese Journal of Applied Physics, Part 1, 31, 3662 (1992).
42. E. T. Yu, X. Z. Dang, L. S. Yu, D. Qiao, P. M. Asbeck, and S.S. Lau, "Schottky Barrier Engineering in III-V Nitrides via the Piezoelectric Effect", *Appl. Phys. Lett.*, **73**(13) 1880 (1998).
43. P. Asbeck, "Polarization Barriers for GaN-Based Devices", MRS 2001, Symposium E.
44. Liu, Y. Zhou, J. Zhu, K. M. Lau, and K. J. Chen, "AlGaN/GaN/InGaN/GaN DH-HEMTs With an InGaN Notch for Enhanced Carrier Confinement", IEEE Electron Device Letters 27, 10, 2006.
45. T.Palacios, A. Chakraborty, S. Heikman, S. Keller, S. P. DenBaars, and U. K. Mishra, "AlGaN/GaN High Electron Mobility Transistors With InGaN Back-Barriers", IEEE Electron Device Letters 27, 13, 2006.
46. E. Kohn, I. Daumiller, M. Kunze, M. Neuburger, M. Seyboth, T. Jenkins, J. Sewell, J. Van Norstand, Y. Smorchkova, and U.K. Mishra, "Transient Characteristics of GaN-Based Heterostructure Field-Effect Transistors", IEEE Trans. Microwave Theory and Techniques, 51, 634, 2003.
47. K.Rim, J. L. Hoyt, and J. F. Gibbons, "Fabrication and Analysis of Deep Submicron Strained-Si N-MOSFET's" IEEE Trans. on Electron Devices 47, 1406 (2000).
48. P. Kirkby, P. Selway and L. Westbrook, "Photoelastic waveguides and their effect on stripe-geometry GaAs/GaAlAs lasers", J. Appl. Phys 50, 4567 (1979).
49. A. Conway, P. Asbeck and J. Moon,"The Effects of Processing Induced Stress on AlGaN/GaN HFET Characteristics," Electronics Materials Conference, 2003.
50. L.McCarthy, P.Kozodoy, S. DenBaars, M.Rodwell and U. Mishra, "First Demonstration of an AlGaN/GaN Heterojunction Bipolar Transistor", Int. Symp. Comp. Semiconductors, 1998.
51. F. Ren, C. Abernathy, J. Van Hove, P. Chow, R.Hickman, J. Klaasen, R. Kopf, H.Cho, K.Jung, J. La Roche, R. Wilson, J.Han, R.Shul, A. Baca, S.Pearton, "300C GaN/AlGaN Heterojunction Bipolar Transistor", Internet Jour. of Nitride Sem. Res. 3, 41 (1998).
52. P. M. Asbeck, E. T. Yu, S. S. Lau, W. Sun, X. Dang and C. Shi, "Enhancement of base conductivity via the piezoelectric effect in AlGaN/GaN HBTs", Solid-State Electronics 44, 211 (2000).
53. Th. Gessmann, J. W. Graff, Y.-L. Li, E. L. Waldron, and E. F. Schubert, "Ohmic contact technology in III nitrides using polarization effects of cap layers", J. Applied Physics 92, 3740, 2002.
54. H. Kroemer, "Heterostructure Bipolar Transistors and Integrated Circuits", Proc. IEEE 70, 13 (1982).
55. Q. Lee, B. Agarwal, D.Mensa, R.Pullela, J. Guthrie, L. Samoska, M.J.W. Rodwell, "A >400 GHz fmax transferred-substrate heterojunction bipolar transistor IC technology. IEEE Electr. Dev.Letts., 19, 77 (1998).
56. A.Michel, D. Hanser, R.F. Davis, D. Qiao, S.S. Lau, L.S. Yu, W. Sun, P. Asbeck, "Growth and Characterization of Piezoelectrically Enhanced Acceptor-Type AlGaN/GaN Heterostructures", 1999 Materials Research Society Fall Meeting, Boston, MA

57. H. M. Ng, D. Doppalapudi, T. D. Moustakas, N. G. Weimann, and L. F. Eastman. The role of dislocation scattering in n-type GaN films. *Appl. Phys. Lett.* **73**, 821–3 (1998).
58. N. G. Weimann, L. F. Eastman, D. Doppalapudi, H. M. Ng, and T. D. Moustakas. Scattering of electrons at threading dislocations in GaN. *J. Appl. Phys.* **83**, 3656–9 (1998).
59. J. W. Hsu, M. J. Manfra, D. V. Lang, K. W. Baldwin, L. N. Pfeiffer, and R. J. Molnar. Surface morphology and electronic properties of dislocations in AlGaN/GaN heterostructures. *J. Electron. Mater.* **30**, 110–4 (2001)
60. G. Koley, and M. G. Spencer. Scanning Kelvin probe microscopy characterization of dislocations in III-nitrides grown by metalorganic chemical vapor deposition. *Appl. Phys. Lett.* **78**, 2873–5 (2001).
61. B. S. Simpkins, D. M. Schaadt, E. T. Yu, and R. J. Molnar. Scanning Kelvin probe microscopy of surface electronic structure in GaN grown by hydride vapor phase epitaxy. *J. Appl. Phys.* **91**, 9924–9 (2002).
62. J. W. P. Hsu, M. J. Manfra, D. V. Lang, S. Richter, S. N. G. Chu, A. M. Sergent, R. N. Kleiman, L. N. Pfeiffer, and R. J. Molnar. Inhomogeneous spatial distribution of reverse bias leakage in GaN Schottky diodes. *Appl. Phys. Lett.* **78**, 1685–7 (2001).
63. E. J. Miller, D. M. Schaadt, E. T. Yu, C. Poblenz, C. Elsass, and J. S. Speck. Reduction of reverse-bias leakage current in Schottky diodes on GaN grown by molecular-beam epitaxy using surface modification with an atomic force microscope. *J. Appl. Phys.* **91**, 9821–6 (2002).
64. E. J. Miller, D. M. Schaadt, E. T. Yu, X. L. Sun, L. J. Brillson, P. Waltereit, and J. S. Speck. Origin and microscopic mechanism for suppression of leakage currents in Schottky contacts to GaN grown by molecular-beam epitaxy. *J. Appl. Phys.* **94**, 7611–5 (2003).
65. C. C. Shi, P. M. Asbeck, and E. T. Yu. Piezoelectric polarization associated with dislocations in wurtzite GaN. Appl. Phys. Lett. 74, 573–5 (1999).
66. J. P. Hirth and J. Lothe. *Theory of Dislocations* (McGraw-Hill, New York, 1968).
67. F. A. Ponce, D. P. Bour, W. Gotz, and P. J. Wright. Spatial distribution of the luminescence in GaN thin films. *Appl. Phys. Lett.* **68**, 57–9 (1996).
68. E. J. Tarsa, B. Heying, X. H. Wu, P. Fini, S. P. Denbaars, and J. S. Speck. Homoepitaxial growth of GaN under Ga-stable and N-stable conditions by plasma-assisted molecular-beam epitaxy. *J. Appl. Phys.* **82**, 5472–9 (1997).
69. A. F. Wright and U. Grossner. The effect of doping and growth stoichiometry on the core structure of a threading edge dislocation in GaN. *Appl. Phys. Lett.* **73**, 2751–3 (1998).
70. J. Elsner, R. Jones, M. I. Heggie, P. K. Sitch, M. Haugk, T. Frauenheim, S. Oberg, and P. R. Briddon. Deep acceptors trapped at threading-edge dislocations in GaN. *Phys. Rev. B* **58**, 12571–4 (1998).
71. K. Leung, A. F. Wright, and E. B. Stechel. Charge accumulation at a threading edge dislocation in gallium nitride. *Appl. Phys. Lett.* **74**, 2495–7 (1999).
72. S. J. Rosner, E. C. Carr, M. J. Ludowise, G. G. Girolami, and H. I. Erikson. Correlation of cathodoluminescence inhomogeneity with microstructural defects in epitaxial GaN grown by metalorganic chemical-vapor deposition. *Appl. Phys. Lett.* **70**, 420–2 (1997).
73. C. Youtsey, L. T. Romano, and I. Adesida. Gallium nitride whiskers formed by selective photoenhanced wet etching of dislocations. *Appl. Phys. Lett.* **73**, 797–9 (1998).
74. P. M. Bridger, Z. Z. Bandic, E. C. Piquette, and T. C. McGill. Correlation between the surface defect distribution and minority carrier transport properties in GaN. *Appl. Phys. Lett.* **73**, 3438–40 (1998).
75. B. Heying, E. J. Tarsa, C. R. Elsass, P. Fini, S. P. Denbaars, and J. S. Speck. Dislocation mediated surface morphology of GaN. *J. Appl. Phys.* **85**, 6470–6 (1999).
76. D. M. Schaadt, E. J. Miller, E. T. Yu, and J. M. Redwing. Lateral variations in threshold voltage of an $Al_xGa_{1-x}N/GaN$ heterostructure field-effect transistor. *Appl. Phys. Lett.* **78**, 88–90 (2001).
77. K. V. Smith, E. T. Yu, J. M. Redwing, and K. S. Boutros. Scanning capacitance microscopy of AlGaN/GaN heterostructure field-effect transistor epitaxial layer structures. *Appl. Phys. Lett.* **75**, 2250–2 (1999).

78. K. V. Smith, X. Z. Dang, E. T. Yu, and J. M. Redwing. Charging effects in AlGaN/GaN heterostructures probed using scanning capacitance microscopy. *J. Vac. Sci. Technol. B* **18**, 2304–8 (2000).
79. K. V. Smith, E. T. Yu, C. Elsass, B. Heying, and J. S. Speck. Localized variations in electronic structure of AlGaN/GaN heterostructures grown by molecular-beam epitaxy. *Appl. Phys. Lett.* **79**, 2749–51 (2001).
80. D. M. Schaadt, E. J. Miller, E. T. Yu, and J. M. Redwing, "Quantitative analysis of nanoscale electronic properties in an $Al_xGa_{1-x}N/GaN$ heterostructure field-effect transistor structure," *J. Vac. Sci. Technol. B* **19**, 1671–4 (2001).
81. D. M. Schaadt and E. T. Yu. Scanning capacitance spectroscopy of an AlGaN/GaN heterostructure field-effect transistor: analysis of probe tip effects. *J. Vac. Sci. Technol. B* **20**, 1671–6 (2002).
82. E. T. Yu, P. M. Asbeck, S. S. Lau, and G. J. Sullivan. Piezoelectric Effects in AlGaN/GaN Heterostructure Field-Effect Transistors. *Electrochemical Society Proceedings* **98–2**, 468–78 (1998).
83. X. Zhou, E. T. Yu, D. Florescu, J. C. Ramer, D. S. Lee, and E. A. Armour. Observation of subsurface monolayer thickness fluctuations in InGaN/GaN quantum wells by scanning capacitance microscopy and spectroscopy. *Appl. Phys. Lett.* **85**, 407–9 (2004).
84. X. Zhou, E. T. Yu, D. I. Florescu, J. C. Ramer, D. S. Lee, S. M. Ting, and E. A. Armour. Observation of In concentration variations in InGaN/GaN quantum-well heterostructures by scanning capacitance microscopy. *Appl. Phys. Lett.* **86**, 202113–1–3 (2005).

Polarization in Wide Bandgap Semiconductors and their Characterization by Scanning Probe Microscopy

Goutam Koley, MVS Chandrashekhar, Chistopher I. Thomas, Michael G. Spencer

1 Introduction

In this chapter, we discuss recent developments in the characterization of polarization doped wide bandgap semiconductor (WBS) heterojunctions with an emphasis on characterization by scanning probe microscopy (SPM). Wide bandgap semiconductors, such as SiC and GaN, have a wide variety of applications in electronic and optoelectronic devices [1–6]. The operation of devices made from these materials is dominated by the polarization properties of III-N and SiC semiconductor crystals. For example, polarization doped 2 dimensional electron gas (2DEG) channels in undoped GaN/AlGaN and SiC polytypic heterojunctions have been fabricated with current densites in excess of 1A/mm. Other effects, such as surface charge instability stemming from polarization, offer interesting potentials and challenges. While some of these effects may be studied using more conventional techniques such as current-voltage (IV) and capacitance-voltage (CV) techniques (described in sections 2 and 3), others require probes capable of investigation on the nanometer scale. Transmission electron microscopy (TEM) can be used to investigate such polarization-induced phenomena with atomic resolution, but involves destructive and laborious sample preparation.

SPM techniques offer powerful, non-destructive tools to effectively and conveniently probe electronic materials and devices. Since the discovery of scanning tunneling microscopy (STM) in 1981 by Binnig and Rohrer [7], the field of SPM has advanced to encompass a large number of variations including, (i) STM, (ii) Atomic force microscopy (AFM), (iii) Electrostatic force microscopy (EFM), (iv) Scanning capacitance microscopy (SCM), (v) Scanning conduction microscopy (SCM), (vi) Magnetic force microscopy (MFM), (vii) Piezoelectric force microscopy (PFM), and (vii) Near field scanning optical microscopy (NSOM). Of the above modes, the first five modes have been extensively used for semiconductor characterization [8–15]. For WBS characterization, especially III-N, AFM, EFM and SCM modes are the most important due to the strong polarization of III-N and SiC. In the latter half of this chapter, we focus on the characterization of III-N by EFM operated in feedback mode, also known as scanning Kelvin probe microscopy (SKPM). SKPM is attractive because it provides quantitative information that is straightforward to interpret.

To fully understand the significance of SKPM, the basic physical properties of WBS and their heterostructures need to be understood. Consequently, this chapter has been divided into three parts, (i) polarization in SiC and III-N (ii) SiC and III-N heterojunctions, and (iii) the applications of AFM and SKPM to WBS materials and heterostructures.

1.1 Polarization in III-N

Although discussed in greater detail elsewhere in this book, we briefly review the polarization properties of III-N in this chapter. III-N are unique among the III-V semiconductors because nitrogen is the smallest and most electronegative of the group-V elements. The metal-nitrogen covalent bond has a greater degree of ionicity than other III-V covalent bonds. Since the wurtzite III-N do not have inversion symmetry along the *c*-axis, *the strong ionicity of the metal-nitrogen bond results in a large macroscopic polarization along the [0001] crystal direction [16,17].* Strong polarization is also present along the [0001] direction in SiC, and to a lesser extent in the [111] direction of zincblende GaAs and InP. *Since this polarization effect occurs in the equilibrium lattice of the III-N at zero strain, it is termed* ***spontaneous polarization*** P_s [16,17]. *The degree of non-ideality of the crystal lattice* governs the strength of the spontaneous polarization.

In the GaN hexagonal close packed (hcp) structure for example, each of the Ga (N) atoms is bonded to four N (Ga) atoms in a tetrahedral arrangement. (see Fig. 1).

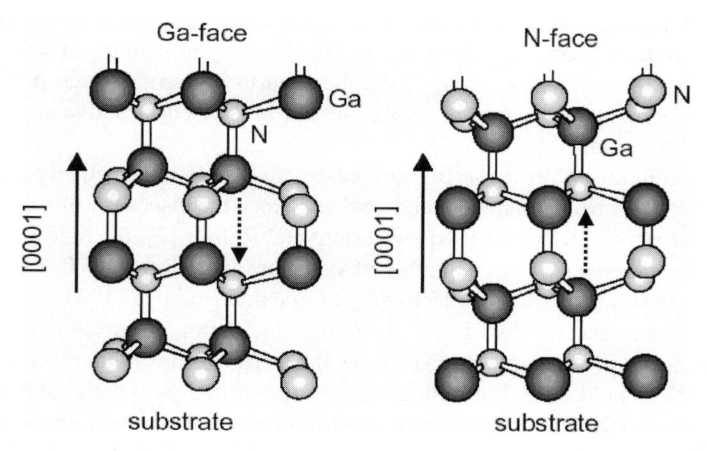

Fig. 1 Atomic arrangement in Ga-face and N-face GaN crystals. The dotted arrow shows the direction of the spontaneous polarization.

Table 1 Effects of lattice non-ideality on the strength of spontaneous polarization in III-V nitrides

Material	AlN	GaN	InN
c_0/a_0 [a]	1.6010	1.6259	1.6116
$P_{SP}(\mathrm{C/m^2})$ [a]	−0.081	−0.029	−0.032

[a] Ref. [6]

It can be shown that the P_s of an ideal hcp non-centrosymmetric crystal is zero, if

$$c_0/a_0 = \sqrt{8/3}$$

As the c_0/a_0 ratio decreases, the three back bonds are at a wider angle from the c-axis. As a result, the compensation polarization decreases giving rise to strong macroscopic P_s. Table 1 shows the c_0/a_0 ratio and the P_s for GaN, InN and AlN. As the lattice non-ideality increases (c_0/a_0 ratio decreases away from $\sqrt{8/3}$ of the ideal hcp structure), the value of P_s increases from GaN to InN to AlN.

If the c_0/a_0 ratio of the III-N lattice is changed, the polarization of the crystal is affected. One way to change the c_0/a_0 ratio is through strain, such as in pseudomorphic epitaxial growth. During or after pseudomorphic epitaxy, the material expands or contracts to match the in-plane lattice constants of the underlying layer. Consequently, the c_0/a_0 ratio changes which in turn changes the polarization strength. *The additional polarization in strained III-N layers is termed* ***piezoelectric polarization*** P_z [16, 17]. For example, if the crystal is under biaxial compressive stress, the in-plane lattice constant a_0 decreases and the vertical lattice constant c_0 increases, increasing the c_0/a_0 ratio toward the ideal lattice value. This decreases the total polarization of the crystal, since P_z and P_s act in opposite directions. On the other hand, if the c-plane is under tensile stress, the in-plane lattice constant increases and the vertical lattice constant decreases, lowering the c_0/a_0 ratio away from the idea value. This increases the overall polarization, since the P_z and the P_s now act in the same direction.

Calculation of interface bound charge and two-dimensional electron gas (2 DEG) density requires values for both P_s and P_z. The P_s in a given material is constant, while the P_z is a function of strain and can be calculated from [18]

$$P_z = 2\frac{a - a_0}{a_0}\left(e_{31} - e_{33}\frac{C_{13}}{C_{33}}\right), \tag{1}$$

where, a_0 is the equilibrium lattice constant, a is the actual lattice constant, e_{31} and e_{33} are the piezoelectric coefficients, and C_{13} and C_{33} are the elastic constants. For easy reference, the piezoelectric constants for GaN and AlN are given in Table 2.

In wurtzite III-N, e_{31} is always negative while e_{33}, C_{13} and C_{33} are always positive (see Table 2), therefore $(e_{31} - e_{33}C_{13}/C_{33})$ will always be a negative number. As a consequence, P_z in III-N is always negative for layers under tensile stress $(a > a_0)$ and positive for layers under compressive stress $(a < a_0)$. Noting that P_s in III-N

Table 2 Piezoelectric constants of wurtzite AlN and GaN used in the calculations

Material	e_{31} (C/m^2)	e_{33} (C/m^2)	C_{13} (Gpa)	C_{33} (Gpa)
GaN	−0.49[a]	0.73[a]	103[b]	405[b]
AlN	−0.60[a]	1.46[a]	108[b]	373[b]

[a] Ref. [16]
[b] Ref. [19]

Table 3 Basic properties of the major SiC polytypes

	Bandgap (eV)	Crystal Structure	P_s (C/m^2)	Hexagonality (%)
2H*	3.3	Wurtzite	0.04	100
4H	3.2	Wurtzite	0.02	50
6H	3	Wurtzite	0.01	33.33
15R	2.9	Rhombohedral	–	40
3C	2.3	Cubic	0	0

* Work on 2H SiC is limited due it's metastablity [23]

is always negative (Table 1), for layers under tensile stress, P_s and P_z are parallel to each other, and in those under compressive stress the two polarizations are anti-parallel.

1.2 Polarization in Silicon Carbide

Semiconductor heterojunctions are usually formed by compositional changes such as $GaAs/Al_xGa_{1-x}As$, $Si/Si_{1-x}Ge_x$, and $GaN/Al_xGa_{1-x}N$. Semiconductor heterojunction can also be formed using different crystal configurations (polytypes) of the same material. The different polytypes are realized through an abrupt change in stacking sequence [20]. Silicon Carbide (SiC) exhibits over 200 polytypes [21] and is capable of forming heteropolytype junctions. While there are other materials that show this tendency (for example ZnS, GaN, and CdS) SiC polytypes have markedly different electronic properties. For ZnS, GaN, and CdS the variation in bandgap energy (E_g) with polytype is < 0.3 eV, whereas for SiC E_g can vary by as much as 1 eV. Table 3 shows the E_g of the major polytypes as a function of the polytype hexagonality. The covalent bonds in SiC are strongly ionic (similar to III-N) due, in part, to the small size of the C atom. Thus, SiC wurtzite and rhombohedaral polytypes can demonstrate spontaneous polarization. Qteish et al. [22] calculated the direction and magnitude of the P_s in purely hexagonal 2H-SiC. For 2H-SiC, the magnitude of the spontaneous polarization P_s was predicted to be $4.32 \times 10^{-2} C/m^2$, which is higher than in GaN. The P_s vector was found to point from the carbon to silicon atom i.e. in the [000-1] direction. In order to estimate the P_s in the major SiC polytypes the 2H-SiC value is multiplied by the hexagonality parameter. Table 3 shows that the adjusted spontaneous polarization values for 4H and 6H-SiC (extrapolated linearly from the degree of hexagonality) compare favorably with those of GaN.

The E_g of 3C-SiC with respect to its hexagonal counterparts, makes it an ideal choice for the formation of SiC polytype heterojunctions. 3C-SiC/4H-SiC and 3C-SiC/6H-SiC heterojunctions, with conduction band offsets of 0.99 eV and 0.7 eV respectively, are of the most interest. The valence band offset ΔE_v can usually be neglected as it is small ~0.1 eV [22]. Additionally, 3C is the thermodynamically favorable form of SiC [24] and readily forms on a hexagonal substrate.

The different SiC polytypes are mutually lattice matched to within 0.1% [24] in the *a*-plane over a wide range of temperatures, which allows the realization of internally epitaxial layers with excellent interfaces. Therefore as practical matter the P_z can be neglected. The direction of polarization in hexagonal SiC predicts the formation of a 2DEG on the carbon terminated surface (C-face), while a 2 dimensional hole gas (2DHG) is predicted for the silicon terminated surface (or Si-face).

2 III-N and SiC Heterostructures

A majority of compound semiconductor electronic and optoelectronic devices take advantage of the physics and electronics possible only with heterostructures. For WBS, the formation of heterostructures is more interesting due to their polarization properties.

2.1 III-N Based Heterostructures

The most studied nitride based heterostructure for transistor applications is AlGaN/GaN. This structure is grown on SiC or sapphire as shown in Fig. 2. Typically, the

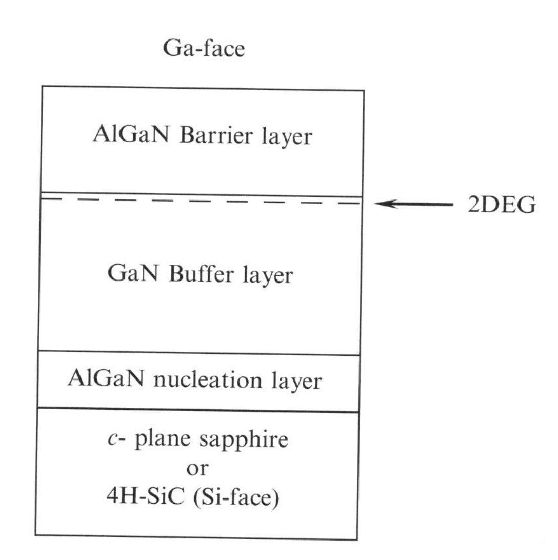

Fig. 2 Layer structures of nitride heterostructures grown by MOCVD.

GaN buffer layer is ~2 μm thick. The lattice stress is relaxed by this layer. The AlGaN barrier layer is pseudomorphic and assumes the in-plane lattice constants of the underlying GaN layer. The Al alloy composition and thickness typically varies from 25–35 % and ~20–30 nm respectively. Since GaN is a non-centrosymmetric crystal, there are two possible growth faces, the Ga-face and the N-face.

Normally, growth by metalorganic chemical vapor deposition (MOCVD) results in Ga face growth, while molecular beam epitaxy (MBE) produces either Ga-face or N-face materials. Ga-face films are the most common. GaN epitaxy improved dramatically in the 1980's, [25] making it possible to produce epitaxial layers of sufficient quality to fabricate electronic and optoelectronic devices. A detailed discussion can be found in Refs. [17], and [26–31].

Certain properties of III-N and SiC thin films and their heterostructures have set them apart from most other common semiconductors:

(1) The presence of polarization, which results in the creation of 2DEG in the absence of any intentional doping.
(2) For III-N, the absence a lattice and thermally matched substrate which causes high defects and dislocations densities.

The polarization of III-N and SiC gives rise to 2DEGs which do not originate from donor atoms. The 2DEGs form at interfaces, compensating the high polarization induced fixed sheet charge. The absence of parent donors is advantageous since it can significantly reduce the gate leakage current in high electron mobility transistors HEMTs under reverse bias. However, this poses another challenge. Since the overall semiconductor is charge neutral, the negative charge of the 2DEG must be compensated by positive charges, which are most likely present at the surface. These positive charges can be perturbed easily, making the 2DEG at the interface unstable.

Absence of native substrates has significant implications for III-N device technology. Lattice and thermal mismatch with non-native substrates (sapphire and SiC), causes high dislocations densities ($\sim 10^{10}\,\mathrm{cm}^{-2}$) in the III-N epitaxial films [32–34]. These dislocations have several deleterious effects on electronic and optoelectronic properties, including:

(i) Reduction of the 2DEG densities by surface states (induced by the dislocations) which trap electrons [30]
(ii) Reduction in electron mobility due to scattering by charged dislocations [35, 36]
(iii) Increase in HEMT gate leakage current under reverse bias, due to dislocation conduction [37]
(iv) Reduction in radiative recombination efficiency resulting in reduced light output in LEDs [34, 38].
(v) Degradation of device performance [39, 40].

From the above discussion if follows that non-destructive characterization of "as grown" III-N thin films, especially with respect to surface and interface electronic properties, is important. SKPM measurements measurements have proved valuable in characterizing surface and interface properties.

2.2 SiC Based Heteropolytype Structures

SiC it typically grown on a native substrate. 3C is the most thermodynamically stable polytype, and can readily be grown on hexagonal substrates to form a cubic/hexagonal stacking sequence. It is thermodynamically more difficult to grow the reverse heterostucture: hexagonal SiC on 3C. However, there have been reports of 6H grown unintentionally on 3C, either by stacking fault formation [41] or through carefully prepared MBE-grown templates [42]. Polytype conversions by annealing and transitions at high sublimation growth temperatures have also been reported, leading to alternative routes for preparation of hexagonal SiC on 3C [23]. Polytype control by these techniques is difficult and not suitable for systematic heterojunction investigation or for fabrication of transistors. Thus, inverted heterostructures (3C/hexagonal) are the most realizable device structures. Despite growth constraints, the 3C/6H or 3C/4H polytypes are attractive, by virtue of large conduction band offsets (CBO) and large induced P_s (see Table 4). Lattice-matched SiC HEMT's with performance comparable to the AlGaN/GaN HEMTs are possible.

There is a large body of work on polytype controlled epitaxy using standard growth techniques such as CVD, MBE, and sublimation [42–46]. The majority of these studies involve homoepitaxy of hexagonal SiC. There are a few reports of the growth of SiC heteropolytype structures using MBE. For a thorough discussion, see [42]. There are reports of heteropolytype junction growth using sublimation epitaxy and CVD [42–46].

SiC CVD is performed on Si-face substrates [42]. The surface free energy on the Si-face is higher than on the C face [42], leading to much more controlled nucleation and growth. A thorough review of Si-face heterojunction growth is given in ref. [45]. However, the 2DEG is produced on the C-face. Thus, a better understanding of the parameters controlling polytypes on the C-face is required. There have only been a few reports of hetero-epitaxy of 3C on the C-face [47].

Control of polytype uniformity during SiC growth involves the following parameters:

(i) Surface supersaturation providing over-potential for growth/deposition. Superstaturation is a function of temperature, growth rate and process parameters
(ii) Growth temperature which affects both surface supersation and surface diffusion length

Table 4 Heteropolytype junction parameters. Conduction band offset (CBO), spontaneous polarization charge (P_{SP}) and a-plane lattice mismatch are given.

Heterojunction	CBO (eV)	P_{SP} (C/m^2)[a]	P_{SP} direction	Lattice Mismatch (%)[b]
3C/4H-SiC	0.99	2.16×10^{-2}	[000-1]	0.08
3C/6H-SiC	0.7	1.44×10^{-2}	[000-1]	0.05

[a] Ref. [22]
[b] Ref. [24]

(iii) Substrate quality and surface preparation
(iv) Vicinality which controls step density and azimuthal misorientation. Vicinality also affects supersaturation and surface morphology

Homo-epitaxy of hexagonal SiC occurs through step-flow growth. The conditons for step flow growth are described in [48]. Step flow growth is a function of substrate miscut, C/Si ratio [42], SiC surface polarity [24], surface preparation [43] and temperature. Hetero-epitaxy of 3C SiC on hexagonal SiC substrates occurs via island nucleation on terraces. Island nucleation is favored by on on-axis substrates and reduced atom migration lengths (l_{sm}), low temperatures, and high supersaturation ratios. Despite the lower surface free energy on the C-face, l_{sm} values are much longer than on the Si-face, which has shorter migration length by virtue of more favorable desorption [48,49]. To promote 3C nucleation on the C-face and suppress hexagonal polytype growth, temperatures below 1400°C are desirable [44,48]. The low C-face free energy means that lower growth rates are required to prevent the onset of 3D growth and promote layer-by-layer (as opposed to step-flow) growth for optimal material quality. By varying the C/Si ratio ($R_{Si/C}$), the surface morphology can be controlled. Scanning electron micrographs of 3C-SiC grown on 4H substrates for varying $R_{Si/C}$ are shown in Figures 3 and 4, which show how morphology changes as the growth proceeds from Si to C rich. At low $R_{Si/C}$, excess silicon pools on the surface and serves as nucleation sites for the growth of amorphous SiC (Figure 3a) An optimal condition is shown in Figure 3b. The optimized growth displays island morphology, believed to originate from the twinned nature of the 3C polytype, giving rise to double positioning boundaries (DPBs) [42,45]. Further, changes in $R_{Si/C}$ results in columnar growth as shown in Figure 4.

One of the major challenges in the investigation of SiC heterostructures is the identification of polytype. This is particularly difficult in heterojunctions as a thin layer of 3C SiC must be differentiated from a thick hexagonal substrate. SiC heteropolytype structures may be studied using various techniques, some of which are briefly discussed below.

a)

b)
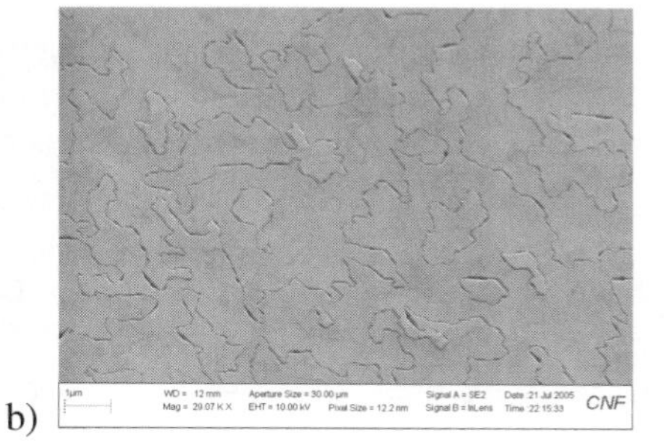

Fig. 3 Scanning electron micrographs documenting change of morphology with inlet $R_{C/Si}$ ratio on on-axis <0001> C-face 4H SiC substrates at 1400° C and a growth rate of ~300 nm/hr. a) C/Si = 0.8, b) C/Si = 1 (1 μm scale bar). The optimized morphology shows island nuclei believed to originate from double positioning boundaries (DPB's)

Fig. 4 Scanning electron micrographs documenting change of morphology with inlet $R_{C/Si}$ ratio on on-axis <0001> C-face 4H SiC substrates at 1400°C and a growth rate of $\sim$300 nm/hr. C/Si = 1.5 (100 nm scale bar)

Photoluminescence (PL): As SiC is an indirect gap semiconductor, it does not exhibit significant bandgap luminescence at room temperature, as excitons cannot be bound strongly at this temperature. To identify polytype bandgap through luminescence, the samples must be cooled to low temperatures <77 K to stimulate exciton formation and subsequent luminescence. Furthermore, at low temperatures, phonons are frozen out and do not significantly broaden luminescence peaks. Lasers with photon energy greater than the bandgap energy e.g. 325 nm He/Cd are used as PL excitation sources. In a related technique, electroluminescence, luminescence is stimulated with an electron beam of energy $\sim$1-20 keV.

Interestingly, there are certain features in the luminescence spectrum that increase in intensity with temperature. These are related to impurity levels that are activated with temperature and become optically active [42, 46].

Bai et.al [41] studied 3C stacking faults in 4H SiC, and measured Stark shifts in the 3C SiC excitonic levels. These Stark shifts were attributed to spontaneous polarization induced electric fields in the quantum well. No piezoelectric component was assumed because good polytype lattice match leads to unstrained layers. Through Schrödinger equation modeling of the 3C/4H quantum well, they attributed the measured Stark shift to a spontaneous polarization induced electric field of 1.3 MV/cm, which is about half that predicted by Qteish et al [22]. Bai's measured value was, however, consistent with the value extracted by Chandrashekhar et al. [52] from studies of a 3C/4H heterostructure (treated in greater detail in 1.3.2). The discrepancy between Qteish's [22] predictions and measurements may be attributed to interface charges, surface charges etc., although polarization field screening was systematically discounted by Bai et. al [41]. This discrepancy bears further investigation, and must be resolved to understand the nature of the 3C/4H SiC quantum well.

Electron diffraction techniques: 3C on hexagonal SiC substrates may be identified through electron diffraction techniques such as electron backscattered diffraction (EBSD), Kikuchi electron diffraction and X-ray photoelectron diffraction. These processes are induced by relatively low-energy electrons, typically $< 10\,\mathrm{keV}$ (corresponding to a penetration dept of $< 0.5\,\mu\mathrm{m}$ in SiC), making them very surface sensitive. Cubic crystals typically exhibit three-fold symmetry in the observed diffraction patterns, while hexagonal crystals display six-fold symmetry, by which the polytypes may be distinguished. For a further discussion of these techniques, see Fissel [42].

Raman spectroscopy: Raman spectroscopy involves mapping the phonon modes in a crystal. These are different for hexagonal and cubic crystals. The modes for 3C and hexagonal polytypes of SiC have been mapped [53, 54]. Because 3C/hexagonal Raman lines overlap, it is difficult to distinguish 3C from hexagonal regions.

3 Interface and Surface Charge in SiC and III-N Heterojunctions

Free charge at the AlGaN/GaN or SiC heteropolytype interfaces are created without any dopant impurities. Never the less, since the structures are charge neutral, the 2DEG must be compensated by positive charges. The origin of this 2DEG is discussed through evaluation of the 2DEG and piezoelectric bound charge density based on their relationship with the surface barrier potential and semiconductor parameters.

3.1 Charges at the Interface and Surface

3.1.1 III-N Heterostructures

III-N epitaxial layers are associated with a polarization P, which consists of spontaneous and piezoelectric components. A polarization gradient exists at a III-N heterostructure interface, which gives rise to a bound charge density σ_{int}. The bound charge is given as,

$$\begin{aligned}\sigma_{\mathrm{int}} &= P_{tot,layer1} - P_{tot,layer2} \\ &= (P_{SP} + P_{PE})_{layer1} - (P_{SP} + P_{PE})_{layer2}\end{aligned} \tag{2}$$

where the $\mathrm{P_{SP}}$ and $\mathrm{P_{PE}}$ denote spontaneous and piezoelectric polarizations respectively. B*ound* charges induced by a change in polarization of the two layers will attract compensating *mobile* charges at the interface. Positive bound charges induce a negative mobile sheet charge, and vice versa. Note that since the GaN buffer layer (Fig. 2) is relaxed, it has only spontaneous polarization, whereas pseudomorphic

AlGaN layer assumes the GaN in-plane lattice constant, and has both spontaneous and piezoelectric polarization. The bound charge at the AlGaN/GaN interface is then given by [16]

$$\begin{aligned}\sigma_{\text{int}} &= (P_{SP})_{GaN} - (P_{SP} + P_{PE})_{AlGaN} \\ &= P_{SP}(0) - P_{SP}(x) - 2\frac{a(0) - a(x)}{a(x)}\left(e_{31}(x) - e_{33}(x)\frac{C_{13}(x)}{C_{33}(x)}\right)\end{aligned} \tag{3}$$

In Fig. 5 the polarization induced bound charge is plotted against Al barrier layer alloy composition. The charge density induced by the each of the spontaneous and piezoelectric polarization components is also shown for comparison. We can see both polarization sources contribute to the sheet charge over the entire range of Al alloy composition. A closer inspection reveals that spontaneous polarization is more important for lower Al composition, contributing to 60.3 % of the total sheet charge at $x = 0.1$. At higher Al-contents, its contribution slowly decreases to 57.5% at $x = 0.35$, and 50.7% at $x = 1.0$.

Understanding polarization induced bound charge, we can calculate the magnitude of the interfacial 2DEG. For this calculation of the 2DEG, the surface barrier height is required. If the surface is bare, then knowing the surface barrier height is not easy and specialized measurements have to be made, e.g. by Kelvin probe technique, as will be discussed later. However, for nitride HFETs, gate metals form schottky barriers to the AlGaN surface, the height of which can be measured by capacitance-voltage, photo-emission spectroscopy, etc. Figure 6 shows the conduction band profile normal to the AlGaN/GaN interface together with sheet charge densities at the interface and surface.

The electrostatics of heterostructure quantum wells are modified by quantum size effects predicting carrier concentration and electric field distribution by virtue of strong carrier confinement. The classical Poisson equation must thus be solved

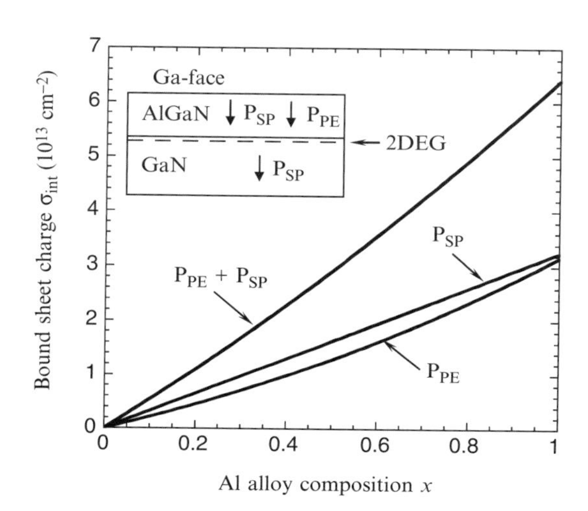

Fig. 5 Variation of the bound charge at the AlGaN/GaN interface (Ga-face material) with Al alloy composition of the AlGaN barrier layer.

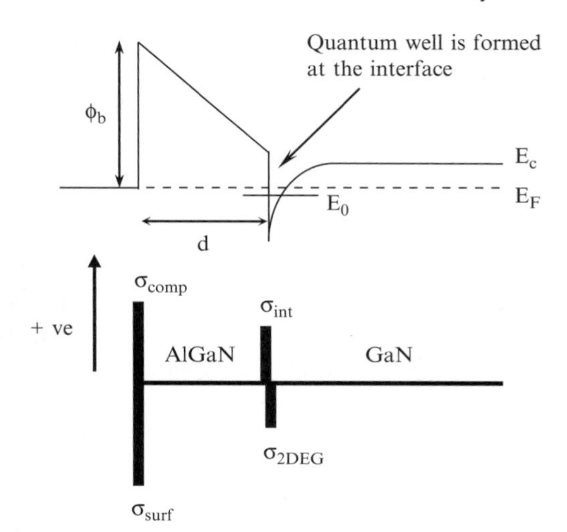

Fig. 6 Conduction band diagram of AlGaN/GaN heterostructure showing the 2DEG. Sheet charge densities at the surface and interface are also indicated for convenience. Note that there exists a charge neutrality condition across the AlGaN/GaN interface.

self-consistently with the Schrödinger equation, using iterative numerical methods (as closed form solutions are difficult to obtain). For the general case of non-constant transverse effective mass, the Schrödinger equation for electrons in the z (transverse) direction can be expressed as

$$\left[-\frac{\hbar^2}{2}\frac{d}{dz}\left(\frac{1}{m_{||}^*(z)}\frac{d}{dz}\right)+V(z)\right]\psi_i(z)=E_i\psi_i(z) \tag{4}$$

where E_i and $\psi_i(z)$ are the energy eigenvalue and wavefunction of the ith subband, $m_{||}^*(z)$, the effective mass in the z-direction, and V(z) is the electron potential energy, given by the conduction band edge,

$$V(z)=-q\phi(z)+\Delta E_C(z) \tag{5}$$

where $\phi(z)$ is the electrostatic potential, and $\Delta E_C(z)$ is a step function to describe the heterojunction conduction-band discontinuity. $\phi(z)$ can be calculated using the Poisson equation in the presence of polarization,

$$\frac{d}{dz}\left[-\varepsilon(z)\frac{d\phi(z)}{dz}+P(z)\right]=q(p(z)-n(z)) \tag{6}$$

where undoped heterostructures are assumed. $\varepsilon(z)$ is the static dielectric constant, which is $9.5\varepsilon_0$ for GaN and $9.7\varepsilon_0$ for SiC, P(z) is the sum of spontaneous and piezoelectric polarization, and p(z) and n(z) are the free hole and electron concentrations, respectively. In SiC, the piezoelectric contribution to polarization may be ignored because of the near perfect lattice matching of the various polytypes [24] and the low values of thermal coefficient of expansion. The electron concentration can be calculated semi-classically using

$$n(z) = \int_0^\infty \frac{D(V(z))}{1 + 2\exp(V_F - V(z))} dV \tag{7}$$

where degenerate Fermi statistics are used. D(V(z)) is the density of states in the conduction band. V_F is the Fermi level energy, and is typically taken to be the reference.

To obtain a self-consistent solution, an estimated potential profile $V(z)$ is used to solve the Schrödinger equation (4), to obtain the electron distribution in the conduction band normalized by equation (7). This electron distribution is then fed back into the Poisson equation (6) to obtain the refined potential iteratively until convergence. Other numerical methods may be used, with the aim of simplifying computation [50]. To obtain the 2DEG sheet charge density, the total charge distribution in the well can be integrated.

Although accurate calculations of electron distribution in structures can be obtained through numerical simulations using the coupled Schrödinger-Poisson equation described above, a simple semi-classical electrostatic analysis can be used to calculate the 2DEG sheet density. Taking the example of AlGaN/GaN heterostructure, and assuming charge neutrality to hold between the sheet charge densities at the surface and the interface [see Fig. 6], leads to an analytical expression for the 2DEG sheet charge density (n_s) as [16]

$$n_s = \frac{\sigma_{\text{int}}}{q} - \left(\frac{\varepsilon_0 \varepsilon(x)}{dq^2}\right)(q\phi_b(x) + E_F(x) - \Delta E_c(x)), \tag{8}$$

where σ_{int} is the sheet charge (Fig. 5), q is the electronic charge, ε_0 is the permittivity of free space, ε, d, and x are the relative dielectric constant, thickness, and barrier layer Al-mole fraction, ϕ_b is the Schottky barrier height, E_F is the Fermi energy level at the hetero-interface with respect to the GaN conduction band edge, and ΔE_c is the conduction band offset at the AlGaN/GaN interface. Equation 8 can be used to calculate the sheet charge density from the parameters used in Ref. [16].

From equation (8), the 2DEG density can be calculated as a function of Al alloy composition (for a fixed AlGaN layer thickness) or as a function of AlGaN layer thickness (for a fixed Al composition). Calculated free-electron sheet density versus aluminum mole fraction is shown in Fig. 7. Two different AlGaN barrier layer thicknesses (100 Å and 300 Å) are plotted to illustrate the effect of barrier thickness variation. We find that decreasing AlGaN barrier thickness reduces the 2DEG density by increased Schottky barrier depletion. Also shown for reference, in Fig. 7, is the fixed polarization charge plotted in a logarithmic scale. We notice from Fig. 7 that increasing AlGaN barrier thickness causes the 2DEG densities to approach the fixed polarization induced charge densities, as the effect of surface barrier related depletion is mitigated. It should be noted that the surface barrier height controls the charge dipole across the AlGaN barrier, which in turn controls the 2DEG density, and any change in surface barrier is reflected as a change in the 2DEG sheet density. This is why SKPM, which measures bare surface barrier height directly with nanoscale resolution, is so important for nitride heterostructure characterization.

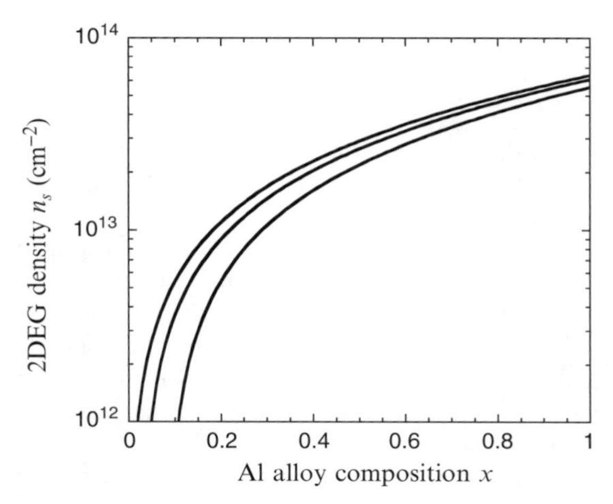

Fig. 7 Variation of the 2DEG density with Al alloy composition for 10 nm and 30 nm thick AlGaN barrier layers. The bound sheet charge density as a function of Al composition is also shown for reference.

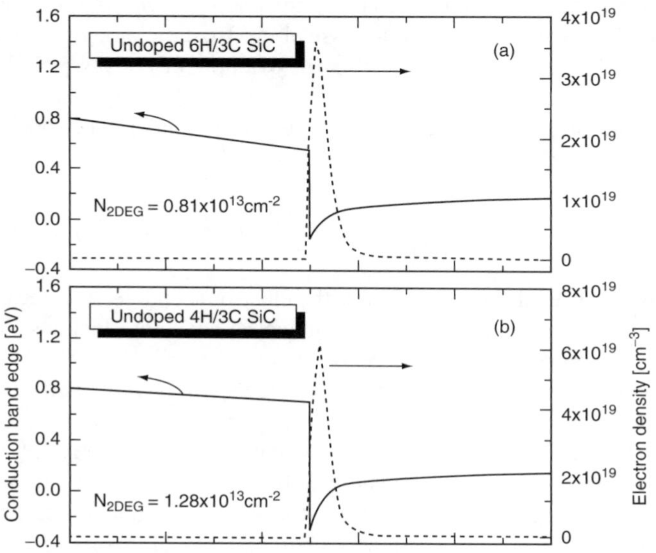

Fig. 8 Predicted electron concentration in a hexagonal/cubic SiC heteropolytype quantum well. The Schottky barrier in this case is on the hexagonal side, whereas as grown heteropolytypic structures would have a Schottky barrier on the cubic side. Courtesy Polyakov and Schwierz [50].

3.1.2 SiC Heteropolytype Structures

Carbon face: As discussed in sections 1.2 and 2.2 2DEG formation is favored on the (0001) C-face of 4H/6H SiC. Spontaneous polarization surface induces free electron mirror charge in the quantum well formed in the non-polar 3C SiC. The large conduction band offset between 3C and 4H-SiC (0.99 eV)/6H-SiC (0.7 eV), allows the interfacial quantum well to accommodate large polarization induced charge densities. Figure 8 shows the band diagram and predicted normal free carrier profile in these quantum wells [50]. Figure 9a) [51] shows the temperature dependence of van der Pauw geometry Hall mobility for a 2DEG in a 3C/4H SiC heteropolytype junction grown at 1400C. For comparison, a repeat polytype (homoepitaxy) sample grown at similar conditions on 6H SiC is shown [51]. The sheet charge density in the 3C/4H-SiC sample was $\sim 3 \times 10^{13}\,\text{cm}^{-3}$, and did not vary by more than ~5% over the temperature range of the measurement, whereas the charge in the repeat polytype sample froze out at low temperatures. Such persistent charge and conductivity at low temperatures imply the presence of a degenerate electron gas. Furthermore, the relatively high mobility ($\sim 300\,\text{cm}^2/\text{Vs}$) precludes pure hopping conduction (and hence degenerate doping). The discrepancy in sheet charge with theory was ascribed to degenerate carrier conduction in parallel with the 2DEG conduction. Mobility at low temperatures was limited by ionized impurity scattering, while that at high temperatures was limited by phonon scattering. The high temperature mobility in the repeat polytype sample $\sim T^{-2}$ whereas the 2DEG sample displayed mobility $\sim T^{-1}$, as expected for a 2DEG, in contrast with bulk conduction. It is expected that further

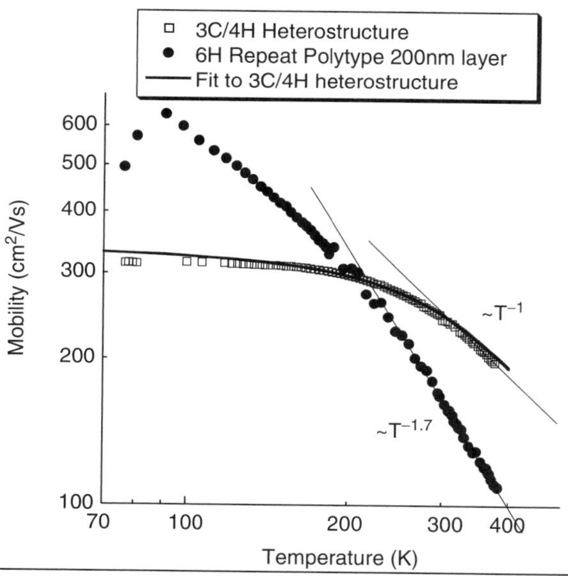

Bandgap (eV)	2.3
Effective mass m_n (3C)	$0.41\,m_0$
n_{s2D} (cm^{-2})	1.3×10^{13}
Ionized impurity concentration (cm^{-3})	1.7×10^{19}
Static dielectric constant	9.72
Optical dielectric constant	6.52
Degenerate mobility μ_{3D}(cm^2/Vs)	20
Longitudinal elastic constant (GPa)	500
Optical phonon energy (meV)	102.8
Acoustic deformation potential (eV)	11

a)

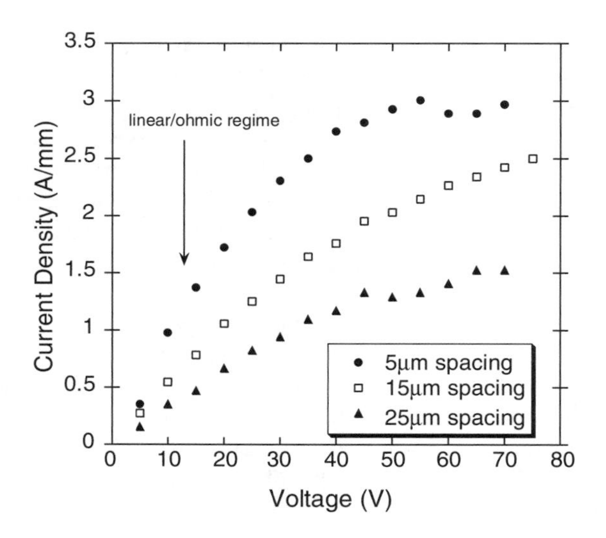

b)

Fig. 9 a) Temperature dependence of Hall mobility for 3C/4H heteropolytype junction. A 6H repeat polytype sample is shown for comparison. Relevant parameters for 3C SiC are given. The electron sheet concentration is $3 \times 10^{13}\,\mathrm{cm}^{-3}$. b)Pulsed IV characteristic of mesa isolated 2 terminal TLM patterns on 3C/4H SiC heteropolytype structure for pad to pad spacings varying from $5\,\mu\mathrm{m}$ to $25\,\mu\mathrm{m}$. The corresponding sheet resistance is $642\,\Omega/\mathrm{sq}$. The electron saturation velocity was estimated to be $\sim 6 \times 10^6\,\mathrm{cm/s}$.

optimization of growth conditions will decrease unintentional impurity incorporation and improve material quality to achieve mobilities $>1000\,\mathrm{cm}^2/\mathrm{Vs}$, comparable to GaN.

Figure 9 b) shows the pulsed I-V transmission line model (TLM) characteristics of the 3C/4H SiC heteropolytype junction fabricated on mesa-isolated regions. Stable saturation currents as high as 3A/mm were seen, comparable with GaN/AlGaN. Currents as high as 6A/mm were observed in $2\,\mu\mathrm{m}$ wide channels, but were unstable, possibly due to heating at the unoptimized tunneling contacts. Currents at voltages $>40\,\mathrm{V}$ showed some instability, possibly due to the contacts or the unpassivated surface, which is important in determining the carrier density and hence current. The electron saturation velocity, v_s, corresponding to 3A/mm is $6\times 10^6\,\mathrm{cm/s}$, $\sim 1/3$ of the ideal SiC value of $2 \times 10^7\,\mathrm{cm/s}$. These high currents and saturation velocities suggest the utility of SiC heteropolytypes in microwave applications.

Silicon Face: The $< 0001 >$ Si–face of 4H/6H SiC favors the formation of two dimensional hole gases. The spontaneous polarization leaves a fixed negative charge on the surface of the hexagonal SiC, implying a mirror free-hole charge in the nonpolar 3C. Considering the small valence-band offset $\sim 0.1\,\mathrm{eV}$ [22], compared to the large predicted polarization sheet charge ($\sim 1.5 \times 10^{13}\,\mathrm{cm}^{-2}$), the confinement of the mirror free-hole charge appears uncertain. However, the large sheet charge also induces a large polarization field, leading to high band banding, forcing an almost

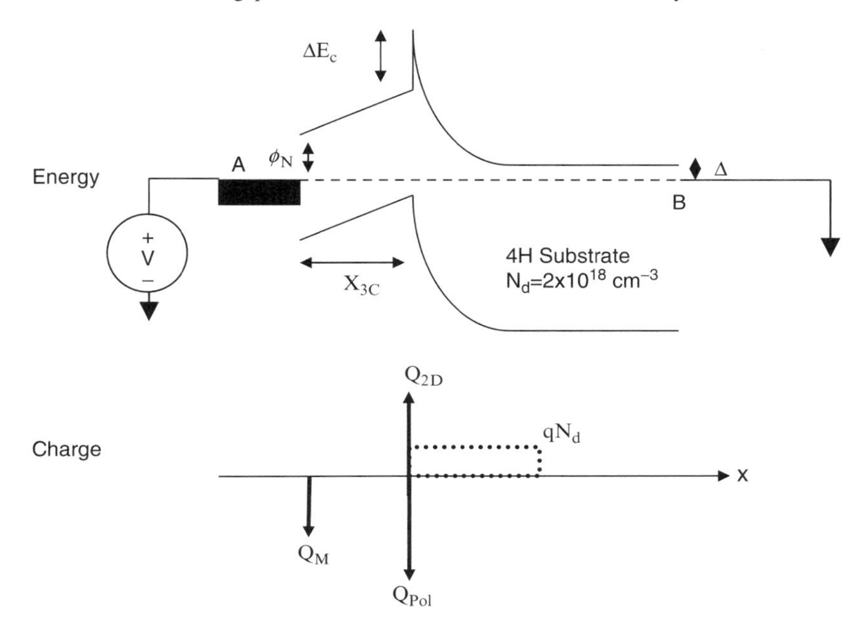

Fig. 10 Energy band diagram and charge balance for 3C/4H heteropolytype junction grown on the Si-face of n^+ 4H SiC. A metal Schottky barrier of height ϕ_N is shown on the left.

purely-polarization induced quantum-well in the valence band. Figure 10 illustrates this situation for 3C-SiC grown on a n^+ 4H SiC substrate [52]. The valence band offset will be ignored for the remainder of this section.
Invoking Gauss' law at the hetero-interface,

$$F_2 = F_1 + \frac{Q_{Interface}}{\varepsilon} = \frac{Q_M}{\varepsilon} + \frac{[Q_{Pol} + Q_{2D}]}{\varepsilon} = -\frac{qN_d x_d}{\varepsilon} \tag{9a}$$

$$Q_M = -qN_d x_d - [Q_{Pol} + Q_{2D}], \tag{9b}$$

where x_d is the depletion width in the substrate, Q_{pol} the fixed polarization charge, Q_{2D} the hole gas free charge, Q_M the charge on the Schottky metal, N_d the doping concentration in the substrate.

Taking an energy loop from A to B, starting and ending at the Fermi level,

$$\frac{\phi_N}{q} + \frac{Q_M}{\varepsilon} X_{3C} + \frac{\Delta E_C}{q} - \frac{qN_d x_d^2}{\varepsilon} - \frac{\Delta}{q} = 0 \tag{10}$$

As the applied voltage is very positive, the hole-gas is completely depleted (i.e. $Q_{2D} = 0$), we can replace ϕ_N with $\phi_N - V$ and solve (9) and (10) simultaneously,

$$x_d + X_{3C} = \frac{X_{3C} + \sqrt{X_{3C}^2 + 4\frac{C - V - \frac{[Q_{Pol}]X_{3C}}{\varepsilon}}{\frac{qN_d}{\varepsilon}}}}{2} \tag{11}$$

where $C = (\phi_N + \Delta E_C - \Delta)/q \simeq 1.5\,V$, assuming a Schottky barrier height of 0.5 eV. This thickness corresponds to the measured capacitance in the extreme positive voltage region (see Figure 11). Furthermore, one can predict the turn-on voltage V_T of this heterodiode by setting $x_d = 0$, at which point all band-bending in the substrate is relieved and electrons can flow unimpeded into the 3C region.

$$V = C - \frac{[Q_{pol}]X_{3C}}{\varepsilon} = V_T \tag{12}$$

Figure 11 shows measured C-V and I-V characteristics of this heterojunction. The turn-on voltage and forward bias capacitance are measured, and can be inserted into

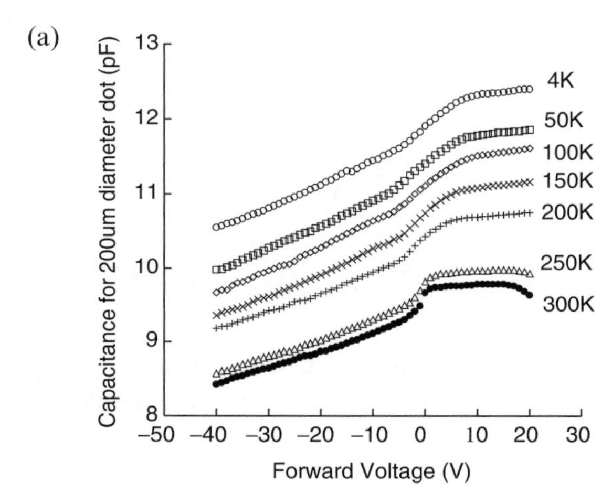

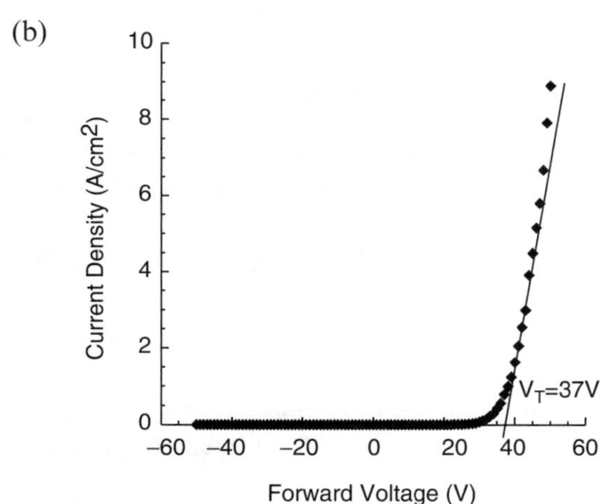

Fig. 11 (a) 1 MHz C-V characteristic, and (b) IV characteristic of 3C/4H heteropolytype junctions grown on n^+ SiC [52].

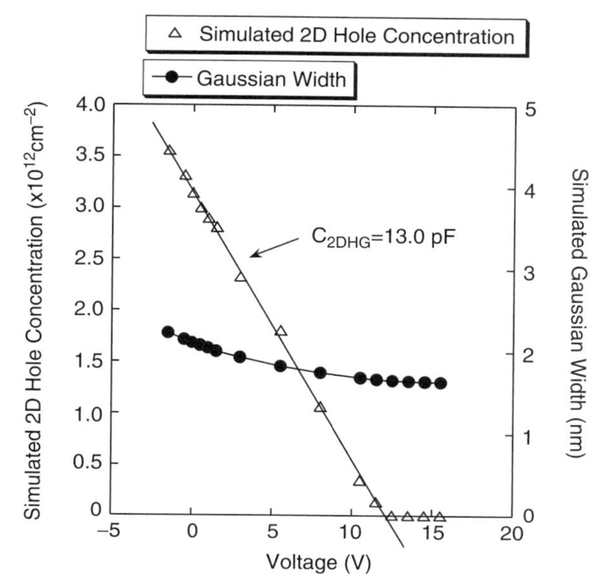

Fig. 12 Simulated 2DHG concentration and confinement as a function of applied voltage for the situation illustrated in Figure 10. The 2DHG full width half maximum is <3 nm over the entire range simulated [52].

equations (11) and (12). Solving these equations simultaneously yields the fixed polarization charge Q_{pol} and the thickness of the 3C epitaxial layer X_{3C}. The polarization charge density extracted using this method, $9.7 \times 10^{12}\,cm^{-2}$, agrees well with that extracted from polarization induced Stark shift measurements in photoluminescence studies of stacking faults in SiC epitaxial layers [41].

It is interesting to note that the 2DHG does not show up explicitly in the C-V characteristic because of the high frequency, but is inferred, in analogy with a MOS capacitor. To predict the 2DHG density, the measured polarization charge inserted into a Schrödinger-Poisson solver yields the charge density and confinement of the 2DHG as a function of applied voltage (Figure 12). Thus, the 2DHG is strongly confined by the polarization alone (a zero valence band offset was assumed to simplify the calculation).

Realization of SiC HEMT's: The lattice-matched SiC hetero-polytype junction promises HEMT's which allow high current for fast switching microwave transistors by virtue of high sheet carrier densities and high carrier mobility. Without a heterojunction, high carrier concentration can only be achieved through doping, which severely degrades carrier mobility in concentrations above $10^{17}\,cm^{-3}$. To realize HEMT's, conduction in the (0001) plane must be employed. This is dependent on material and unintentional impurity concentration (section 2.2).

Because 3C SiC must be grown on on-axis substrates, it is susceptible to heavy twinning. Twin boundaries may affect leakage from a Schottky gate metal. In addition, the usual issues of ohmic contact formation and optimization of gate metallization must be addressed.

Despite remaining technological challenges, polarization doped SiC channels appear suitable for high current devices. C-face DEG's displays significant current capability, while the Si-face 2DHG offers an alternative to p-channel MOSFETs in SiC, for high hole mobility devices.

3.2 Surface States and Their Significance

Surface states and charge instability: At the free surface of a semiconductor, unsaturated bonds are present, which can give rise to electronic states in the forbidden gap. These states can trap electrons or holes, and affect the semiconductor device properties. III-N, devices are very sensitive to surface states and their charge capture, because a charge-neutrality condition exists between the surface and interface charges as discussed in section 1.

The charge neutrality condition implies that any change in surface charge density will be mirrored by an equal change in mobile electron density. For example, if surface states capture electrons, and the net surface charge becomes more negative, then the 2DEG density will decrease. This phenomenon has been observed widely in AlGaN/GaN heterostructrue field effect transistors (HFETs), and will be discussed in greater detail in section 4.

On the other hand, if AlGaN/GaN heterostructures are exposed to above bandgap energy light, then the photo-generated excess electrons increase the 2DEG density. The excess charge density decreases to its thermodynamic equilibrium value following a stretched exponential decay, with time constant up to several days.

Sensors based on charge instability: The instability of the 2DEG at the heterostructure interface, which is a major problem for HFETs, can be exploited to develop sensors for molecules that cause the surface charge to change. Since the III-N layers are polarized, it can be expected that polar molecules would affect the surface charge more than non-polar molecules, and can be easily sensed. This was indeed observed by researchers, when they used a gateless AlGaN/GaN FET structure to perform the sensing experiments. It was observed that polar molecules such as water, ethanol, methanol, and acetone, in contact with the exposed surface, changed the surface 2DEG, and hence the drain current. However, non-polar molecules such as gycerol had no effect on the drain current [58].

Surface states as origin of the 2DEG: Considering the charge balance across the AlGaN barrier layer in an AlGaN/GaN heterostructure, we observe that there are four major components in the overall charge balance equation: (a) polarization charge at the AlGaN surface (air/AlGaN interface) σ_{surf}, (b) polarization charge at the AlGaN/GaN interface σ_{int}, (c) 2DEG charge at the AlGaN/GaN interface σ_{2DEG}, and finally, (d) surface charges σ_{comp} (refer to Fig. 6). As will be shown later,

σ_{surf} is much larger than σ_{int}. Therefore, the compensating surface charge (which must be positive) needs to balance the difference in the polarization charges, and the 2DEG.

By natural extension, surface donor states contributing to the positive compensating charge have been considered to be the origin of the 2DEG [59]. In other words, it was proposed that the 2DEG originates from donor states at the surface that are ionized if they lie above the Fermi level at the surface. *A validation of the surface donor theory, as well as an estimate of the surface donor density, can thus be directly obtained by measuring the 2DEG density and the surface barrier height (2DEG density and surface barrier height are related for a fixed Al alloy composition, by equation 3) in a series of samples where the surface barrier as well as the 2DEG varies.* These measurements were performed, and the surface donor density for AlGaN/GaN heterostructures with a fixed Al alloy composition calculated, thus validating the surface donor model [59].

Surface states passivation: To prevent HEMT degradation by charge trapping at surface states, means to passivate the surface has been studied. Usually, surface passivation is achieved by deposition of a dielectric layer (usually silicon nitride or silicon dioxide) on the free surface. This surface passivation can significantly improve the performance of III-N HFET devices [2]. There are many opinions regarding the physical mechanism by which surface passivation improves device performance. It is possible that the surface pasivation (i) reduces the density of active surface states [60], and/or (ii) reduces the negative charge density at the surface, causing the 2DEG to increase, and surface barrier to decrease [61].

It is not clear if either of these mechanisms is responsible for the improvements caused by surface passivation, or whether the actual mechanism varies with crystal quality, device fabrication process, or stoichiometry of the passivation layer. There is some evidence that non-stoichiometric and slightly conducting silicon nitride (SiN_x) layers improve device performance more than perfectly stochiometric insulating layers. From Kelvin probe measurements of surface barrier height it was observed that the surface barrier reduces after silicon nitride passivation, while the charge instability also reduces significantly. Data and discussion presented below.

4 SPM Characterization of Heterostructures

With the understanding of the basic III-N and SiC properties, we are now ready to focus on actual measurements and their implications for polarization doped heterostructure device technology. In this section, we will first discuss the basics of Kelvin probe microscopy, and then its applications to the characterization of III-N and SiC thin films and heterostructures.

4.1 Basics of Kelvin Probe Microscopy

The total force on the probe tip (Fig. 13) (at a distance z from the sample) is given by the expression [62–65],

$$F_{tot} = F_{capa} + F_{coul} = \frac{1}{2}\frac{\partial C}{\partial z}(V_{app} - V_{con})^2 + \frac{Q_s Q_t}{4\pi\varepsilon_0\varepsilon_r z^2} \tag{13}$$

where Q_t is the charge on the tip, Q_s is the charge on the surface, V_{con} is the contact potential difference [qV_{con} is equal to the difference in work functions of the tip and the sample $(\phi_{tip} - \phi_{mat})$], and ε_r is the dielectric constant of the medium between the probe-tip and the sample. The first term in the RHS arises due to capacitive interaction, and the second term due to coulombic interaction between the tip and the sample. Now, we put $V_{app} = V_{dc} + V_{ac}\sin(\omega t)$, and substitute Q_t by the sum of image charge on tip, $-Q_s$ and $Q_e = CV_{app}$. The total force, F_{tot}, is then obtained after simplification of equation (13) as

$$\begin{aligned} F_{tot} =& \frac{1}{4}\frac{\partial C}{\partial z}V_{ac}^2 + \frac{1}{2}\frac{\partial C}{\partial z}(V_{dc} - V_{con})^2 - \frac{Q_s(Q_s - CV_{dc})}{4\pi\varepsilon_0\varepsilon_r z^2} \\ &+ \left[\frac{\partial C}{\partial z}(V_{dc} - V_{con}) + \frac{CQ_s}{4\pi\varepsilon_0\varepsilon_r z^2}\right]V_{ac}\sin(\omega t) \\ &- \frac{1}{4}\frac{\partial C}{\partial z}V_{ac}^2\cos(2\omega t) \\ =& F_{const} + F_{\omega} + F_{2\omega} \end{aligned} \tag{14}$$

Thus the total force is a sum of three components: a constant term, F_{const}, one varying with frequency ω, F_{ω}, and another varying with frequency $2\omega, F_{2\omega}$. From equation (14), F_{ω} is given by

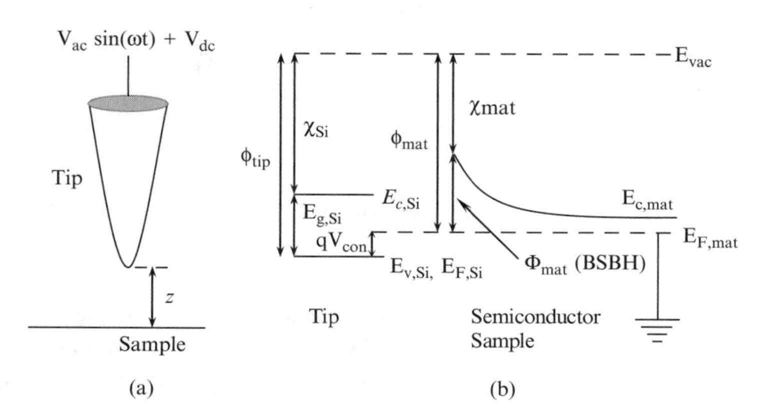

Fig. 13 a) Schematic diagram illustrating the tip-sample configuration and the electrical biases applied to the tip and the sample during measurement. b) Schematic band diagram showing the Fermi levels and the vacuum level for the tip and a semiconductor sample.

$$F_{\omega} = \left[\frac{\partial C}{\partial z}(V_{dc} - V_{con}) + \frac{CQ_s}{4\pi\varepsilon_0\varepsilon_r z^2} \right] V_{ac} \sin(\omega t), \tag{15}$$

where V_{dc} is the dc feedback voltage generated by the feedback loop with respect to the sample bias (ground potential). In feedback mode (Kelvin probe mode), V_{dc} applied to the probe tip nullifies F_{ω}. V_{dc} is given by simplification of equation (15) as

$$V_{dc} = V_{con} - \frac{CQ_s}{4\pi\varepsilon_0\varepsilon_r z^2 \left(\frac{\partial C}{\partial z}\right)} \tag{16}$$

When operated in feedback mode at a distance sufficiently far from the surface, it is possible to neglect the Columbic term, and the dc feedback voltage V_{dc} becomes exactly equal to the contact potential V_{con}. The work function is then given by

$$\phi_{mat} = \phi_{tip} - qV_{dc} \tag{17}$$

For semiconductors, often the bare surface barrier height (BSBH), Φ_{mat}, of the sample [see Fig. 13 (b)] is an important parameter to measure. From equation (17) and Fig. 13(b), we obtain

$$\Phi_{mat} = \phi_{tip} - \chi_{mat} - qV_{dc}, \tag{18}$$

where χ_{mat} is the electron affinity of the semiconductor sample. *From equation (18) changes in Φ_{mat} exactly equal changes in feedback voltage V_{dc}.* This is important for many practical applications (i.e. for comparing samples with similar electron affinity) where often the exact value of χ_{mat} is not known.

4.2 Characterization of Charge Instability

Below we discuss surface charge instability effects in AlGaN/GaN heterostructures. We specifically discuss two different techniques for studying the charge instability: (i) by exposing the heterostructure to above bandgap energy UV illumination, and (ii) by injecting electrons on to the heterostructure surface by a reverse biased schottky contact (usually the gate of a transistor).

Investigation of charge instability caused by UV laser exposure: When exposed to UV light (with energy higher than the bandgap) an AlGaN/GaN heterostructure, electron-hole pairs are generated, and then separated by the high built-in field in the AlGaN layer. It has been observed that the holes reach the surface (surmounting the valence band discontinuity), and the electrons collect at the heterostructure interface thereby increasing the 2DEG density [66]. Thus, the charge dipole across the AlGaN barrier layer and the surface barrier decrease. The reduction in surface barrier above bandgap energy illumination is not uncommon, and has often been seen in semiconductors.

However, there are two aspects in which the surface barrier reduction in AlGaN/GaN heterostructure is different: (i) the time constant of the recovery process after UV illumination, and (ii) spatial confinement of the local photogenerated carriers. The first point is significant, as it provides an opportunity to model the recovery process, where BSBH is used as a fitting parameter, and hence can be independently estimated with a reasonable accuracy. The second point is counter-intuitive, but indicates deep hole traps on the AlGaN surface. Both of these points are discussed in detail below.

III-N samples (Ga-face, grown on sapphire) were studied by UV illumination using a He-Cd laser (325 nm). Initial measurements were carried out on a set of three samples: sample #1 consisting of 16.5 nm barrier layer with 35% Al composition on a GaN buffer layer grown by metalorganic chemical vapor deposition (MOCVD), sample #2 consisting of GaN(2.5 nm)/AlGaN(6 nm)/GaN layers grown by MBE with 40% Al composition in the barrier layer, and sample #3 consisting of a simple GaN epilayer grown by MOCVD [66]. Figure 14 shows the relative variation of electronic surface potential (same sign as the surface barrier height in eV) for sample #1, 2 and 3 as the UV laser is switched on and off. A large decrease in surface potential is observed for all the samples as the laser is switched on, however, for the heterostructure samples (#1 and #2), the rise transient of the surface potential, after the laser is switched off, is much slower as compared to the GaN sample (sample #3).

These observations are explained as follows. When samples are illuminated, electron-hole pairs are generated in the GaN buffer layer (for 35% Al content the barrier layer will be transparent to the 325 nm light). The photogenerated holes

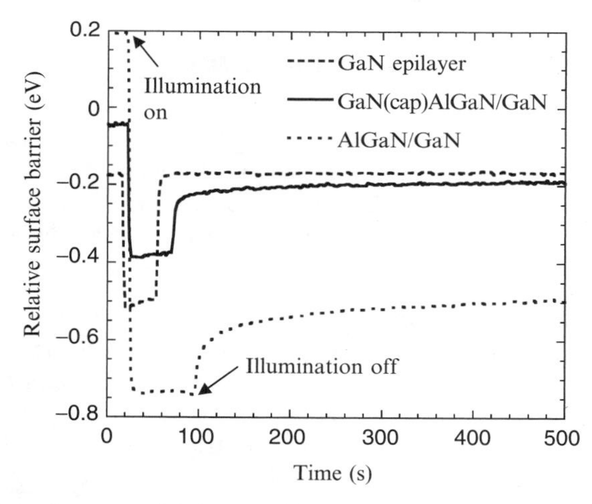

Fig. 14 Relative surface potential (electronic) transients for AlGaN/GaN (sample #1) and GaN(cap)/AlGaN/GaN (sample #2) heterostructures, and GaN epilayer (sample #3) induced by switching the UV laser ON and OFF. No significant rise time constant is observed for sample #3 in contrast with the HFET samples.

are swept towards the surface (assisted by the built-in electric field in the barrier layer) and electrons accumulate at the AlGaN/GaN interface. Thus, the net charge dipole across the barrier layer is reduced, which decreases the surface barrier height. As the barrier reduces from the equilibrium value, electrons at the interface also move toward the surface and recombine with holes allowing a steady state to be achieved under illumination. After illumination, the surface potential changes slowly by recombination of excess holes trapped at the surface and thermionically emitted electrons from the quantum well [66]. Movement of holes toward the surface is confirmed by increased reverse bias gate current under UV exposure. On the other hand, the increase in 2DEG density was confirmed by an increase in current flowing between two ohmic contacts, and decay (with time constant similar to surface potential increase) after illumination [67].

From Fig. 14, is is observed that the GaN sample does not show any significant decay time constant whereas the AlGaN/GaN heterostructure (sample #1) and GaN capped heterostructure (sample #2, grown by MBE) show a significant transients (such a long transient was also shown by MOCVD grown GaN capped heterostructure). Since samples with similar top layer (sample #2 and 3) show different responses whereas those with dissimilar top layer (sample #1 and 2) shows similar responses, we conclude that response is governed by the barrier potential and not the nature of the exposed surface. The rising voltage transient is a stretched exponential indicating a self-limiting process. This is expected since the barrier increases as electrons move toward the surface and neutralize holes, reducing further flow of electrons. To fit the experimental data a simple model is proposed based on the thermionic emission of electrons from the quantum well [66]. The thermionic emission current per unit area, J, is given by

$$J = \frac{d}{dt}(q\sigma_{net}) = A^*T^2 \exp\left(-\phi_B/kT\right), \tag{19}$$

and the surface barrier height ϕ_B as

$$\phi_B = \Delta E_C + q^2\sigma_{net}d/\varepsilon\varepsilon_0 - E_F, \tag{20}$$

where A^* is the effective Richardson's constant, T is the absolute temperature, ΔE_C is the conduction band offset, σ_{net} is the net charge at the interface, d and ϖ are the thickness and permittivity of the AlGaN layer, E_F is the Fermi level, and q is the magnitude of the electronic charge. The net polarization charge at the interface was assumed to be $1.46 \times 10^{13}\,\text{cm}^{-2}$ [59]. The value of A^* was taken as $26.4\,\text{A/cm}^2/\text{K}^2$ (assuming electrom effective mass to be isotropic and $0.22\,\text{m}_0$). ΔE_C and E_F were calculated from standard formulas [18]. From equations (19) and (20) the shape of the rising transient can be obtained for an initial value of ϕ_B (immediately after illumination) as a fitting parameter [see Fig. 15 (a)]. Figure 15 (b) shows the theoretical rise transient fitted to the experimental data. The initial value of ϕ_B used in calculations is $\sim$0.71 eV. From ϕ_B and the change in surface potential under illumination, $\Delta\phi$ [see Figs. 14 and 15(a)], the surface barrier (before illumination) for AlGaN surface can be estimated to be $\sim$1.65 eV at 35% Al alloy composition in

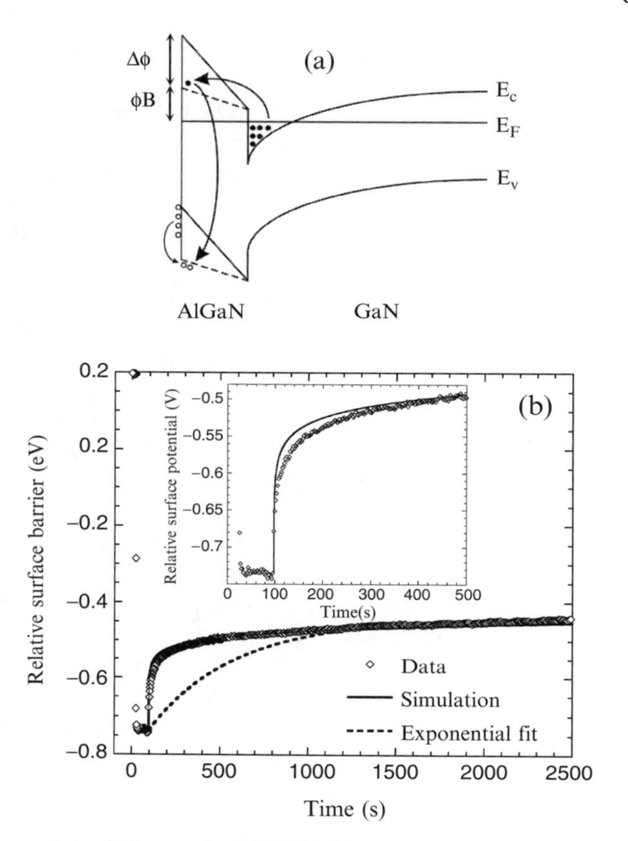

Fig. 15 (a) Schematic band diagram for AlGaN/GaN sample (sample #1) explaining the observed UV laser induced transients. (b) Theoretically calculated rise transient fitted to the experimental data, using initial value of surface potential (just after turning off laser), ϕ_B [see (a)], as a fitting parameter. An exponential fit with a time constant of 500s is also shown (dotted line) to highlight the non-exponential nature of the rise transient. Inset shows a magnified view of the rise transient and the theoretical fit for clarity.

the barrier layer. A dc offset is also added to the calculation of surface potential to match the observed surface potential (to facilitate comparison with the observed trends). As seen in Fig. 15(b) the theoretical curve fits well with experimental data. In contrast, a purely exponential rising curve does not satisfactorily match the experimental trend. This clearly points out the strong non-exponential nature of the rise transient.

Based on the method described above, the variation of BSBH of AlGaN/GaN heterostructures as a function of Al alloy composition and thickness of the barrier layer can be obtained. The surface potential recovery transients are slower for higher Al composition in the barrier layer, since the electrons have to overcome a higher barrier to recombine, because of a higher conduction band offset. A significant switch-on transient is also observed (as the laser is switched on) since the

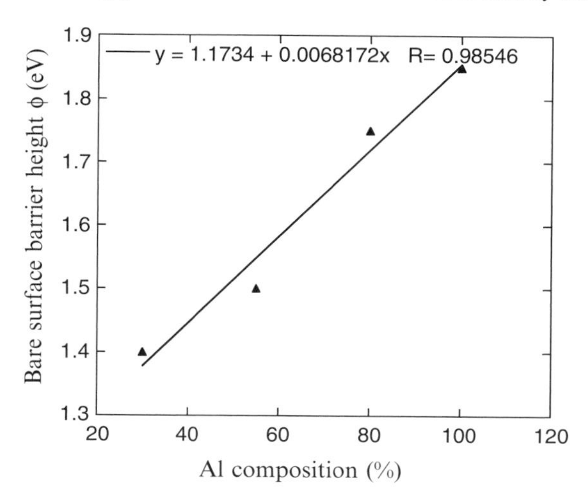

Fig. 16 Variation of the BSBH with Al alloy composition of the barrier layer.

valence band offset, preventing the movement of holes toward the surface, increases with Al alloy composition.

In Fig. 16 the variation of the BSBH with Al alloy composition is shown. The BSBH values were calculated based on the simulations described in the preceding section. The BSBH increases with increasing Al composition with a slope of 0.68. Compared to the schottky barrier height slope (versus Al alloy composition) of 1.3 [18], this slope is much smaller. We note that the BSBH of GaN obtained from extrapolation at zero Al content, is found to be 1.17 eV, which is rather large compared to the generally accepted BSBH value of 0.8 eV. More accurate BSBH values would be obtained with larger number of data points.

Studies were also conducted regarding the variation of BSBH with the thickness of the AlGaN barrier layer. Figure 17 shows the variation of the surface barrier with AlGaN layer thickness for 35% Al composition (AlGaN/GaN structure shown in Fig. 2). We see that the surface barrier increases from 1.15 eV to $\sim$1.7 eV as the barrier layer thickness increase from 50 Å to 440 Å. Such a trend is expected if surface donors are the origin of electrons (more discussion on the origin of 2DEG later), and 2DEG increases with barrier thickness (Fig. 17). As the thickness of the barrier layer increases, the surface Fermi energy level falls with respect to the conduction band edge. This in turn increases the surface barrier height, emptying more donor states and increasing the 2DEG density. It should noted, however, that the surface barrier height does not continue to increase indefinitely, and saturates around 200–250 Å barrier thickness. It is probable that beyond this thickness, the surface state density increases sharply, or the barrier layer becomes partially relaxed.

It is possible to vary the surface barrier height by locally illuminating the heterostructure surface. Exposed regions have lowered surface barriers, while the surface barrier of the unexposed regions remain unchanged. A GaN (25–30 Å thick)

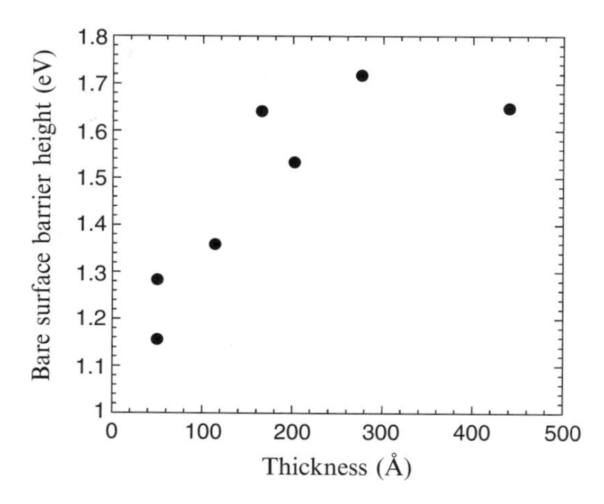

Fig. 17 Variation of the BSBH with AlGaN barrier layer thickness for AlGaN/GaN heterostructures grown by MOCVD.

capped AlGaN (300 Å, 35% Al)/GaN heterostructure sample grown on SiC substrate was exposed through a mask having squares with sides of 1(9), 2(7), 5(5), 10(3), and 20(2) μm. The numbers in the bracket indicate the number of each size of squares in a particular bordered area. The imprinted surface barrier image on is shown in Fig. 18. The patterns, once imprinted, were found to persist for a few days to a few weeks. The possibility of such a localized variation of surface barrier is rather surprising and counter-intuitive. This is because we would expect the photo-generated holes reaching the surface to be able to move fairly quickly, and not remain confined in the exposed regions. *The fact that they remain confined in the exposed regions only (or at least move very slowly), points to the significant density of fairly deep hole-traps at the surface* [67]. More studies are required to determine the position and density of such traps in the forbidden gap. The lateral spread of the patterns, is ~1–2 μm on every side from the exposed region [see Fig. 18(a)]. This is on the order of diffusion length of GaN, of ~1 μm [68], [69].

The idea of surface barrier patterning is indeed very interesting, and important information about the surface mobility of holes, and hole trap density and energy can possibly be obtained from these studies. We should also mention here that the surface barrier can be patterned with high-energy electrons (50–100 KeV) as well, using e-beam lithography [67]. The areas exposed to the e-beam have lower barrier. This can be explained by noting that highly energetic electrons would create electron-hole pairs in the material, which would then separate, as discussed before, to lower the surface barrier.

Investigation of charge instability by electron injection: The problem of charge instability in III-N heterostructures has also been studied by negative charge injection on the exposed AlGaN surface in AlGaN/GaN HFETs. The negative charges were injected from the gate of the HFETs under appropriate biasing conditions.

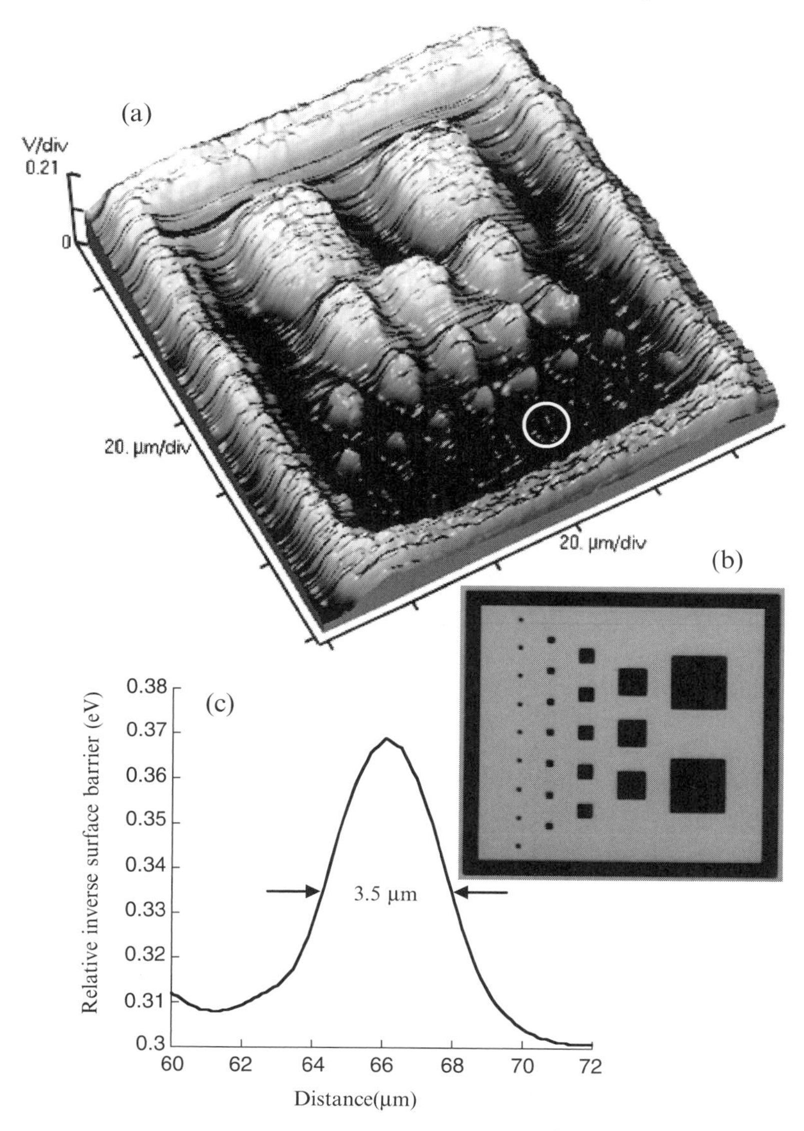

Fig. 18 (a) Surface barrier patterns imprinted on a GaN (25–30 Å thick) capped AlGaN (300 Å, 35% Al)/GaN heterostructure grown on SiC substrate. The quartz mask had square patterns of 1, 2, 5, 10, and 20 μm as shown in (b). (c) Line scan through the peak shown by the white circle in (a).

This study of negative charge (electron) accumulation on the surface is exactly opposite to the situations discussed above (accumulation of holes at the surface after UV illumination), and provides complementary information. The significance of the electron accumulation study is very broad, since it is linked to the important

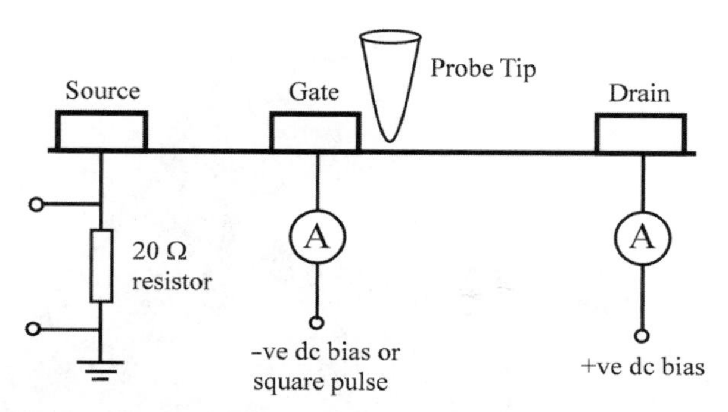

Fig. 19 Schematic diagram of the measurement set up. The probe tip is positioned on the exposed surface for surface potential measurements. A 20 Ω resistor is placed in series with the source to measure the drain current and surface potential simultaneously with respect to time.

phenomenon of current slump or current collapse observed in III-N heterostructures [70–72].

The nitride heterostructures used in this study consist of typical layer structures of AlGaN barrier layer (30–35% Al, 250–300 Å)/GaN buffer layer (1–2 μm)/ nucleation layer/substrate grown along the *c*-axis by MOCVD [23]. The transistors were fabricated with sub-micron rectangular and mushroom gates with both T-gate and U-gate layout geometries. The rectangular gates are typically 0.6–0.8 μm long, while the mushroom gates have a footprint of ∼0.3 μm with ∼0.2–0.3 μm overhangs on each side. Typical source-gate and gate-drain spacings are 1 μm and 2 μm respectively. To measure the surface potential changes caused by electron accumulation, surface potential measurements were conducted in the region between the gate and the drain of HFET devices. A schematic diagram for biasing and current measurements is shown in Fig. 19. To investigate the link between surface accumulation of charges and the drain current reduction, simultaneous measurements of surface potential and drain current were carried out. For measurement drain current, a 20 Ω resistor is connected in series with the source as shown in Fig. 19. The voltage across this resistor was recorded by the computer (simultaneously with the surface potential signal), from which the drain current was calculated. To inject electrons to the surface states, the devices were stressed continuously for two minutes under a gate bias of −12 V and a drain bias of 20 V. To measure the drain current transient after stress, a low drain bias of 1 V was applied, while the gate bias was kept at 0 V. The effects of bias stresses under UV illumination were also studied for which a He-Cd laser operating at 325 nm was used.

Figure 20 shows the surface potential and the drain current transient measured on the same unpassivated device (100 μm wide with a T-gate layout geometry) as before, after the device was subjected to the bias stress mentioned above. As seen from comparison of the two transient responses, both have very similar recovery transients and both the transients flatten out after similar time interval (after ∼200 s). For surface potential measurements the tip was positioned ∼0.4–0.5 μm from the gate footprint toward the drain side. A zero scan size was chosen for measurement

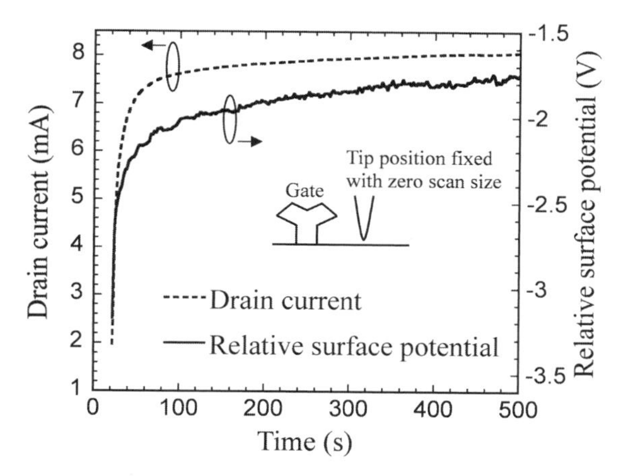

Fig. 20 Simultaneously measured drain current and surface potential transients (keeping the probe tip fixed at a certain position with zero scan size as shown in the inset) after the device underwent stress.

(see inset of Fig. 20). For drain current measurements 1 V was applied to the drain while the gate was biased to 0 V. The real value of surface potential would thus be further lowered by a few tenths of a volt depending on the electrostatic potential of the position where the probe tip was placed. It is to be noted that the accuracy of the surface potential measured is estimated to be around 70–80 % due to tip convolution, and more importantly, the effect of the cantilever [73]. From Fig. 20, we observe that the lowest surface potential is $\sim -3.2\,\text{V}$, and so the corrected potential taking into account the worst case cantilever effect and the effect of applied drain bias ($\sim$0.6 V) is $-5.2\,\text{V}$ (equal to $-3.2\,\text{V}/0.7 - 0.6\,\text{V}$). Due to the difference in work functions of the tip and the sample surface, the unstressed surface potential is $\sim -1\,\text{V}$ (see unstressed surface potential in Fig. 21 near gate), and so the change in surface potential due to stress can be roughly estimated to be $-4.2\,\text{V}$, which can almost shut off the channel ($V_{th} = -5.5\,\text{V}$) as has been observed. However, one should be careful about such comparisons since the drain current is dependent on the overall potential profile at the surface, and not just the potential of a particular position.

To determine the actual profile of the surface potential after stress, Kelvin probe measurements were made in scanning mode over a distance of 1 μm from the edge of the gate. The device scanned was an unpassivated rectangular gate device with a larger access region for measurements. It was 150 μm wide with a T-gate layout geometry, and fabricated on AlGaN/GaN layers grown on a sapphire substrate. After stress, the spatial variation of surface potential over a distance of 1 μm was recorded at different intervals of time starting from t = 0 min (drain and gate voltages were 0 V during measurements), and is shown in Fig. 21. As can be seen from the scans, most of the surface potential changes occur close to the gate. Furthermore, the surface potential initially decreases with distance and reaches a minimum at $\sim$0.3 μm from the edge of the gate before starting to increase smoothly again and

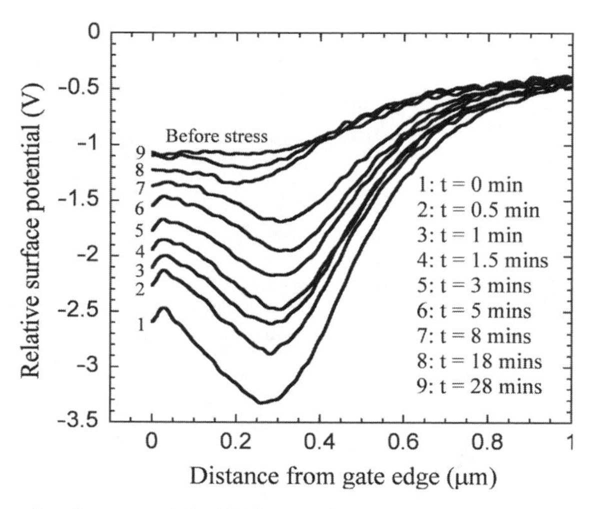

Fig. 21 Variation of surface potential with distance from the edge of the gate after the device was subjected to stress. Scans were taken at different intervals of time after stress to understand the time evolution of the surface potential profile.

finally becoming negligible ∼1 μm away. For regions closer to the gate, the tip will be positioned farther away from the surface due to the height of the gate metallization (∼500 nm) and conical nature of the tip (half angle ∼12°). While that would reduce the accuracy of the measurements to a certain degree, it is not the sole reason for surface potential variation near the gate (in other words for the occurrence of surface potential minimum away from the gate). This is seen from the line scans at later time (i.e. at $t = 8$ minutes) when the surface potential becomes almost flat near the gate thus indicating that the measured potential is almost independent of the distance of the tip from the surface. Measurements at different intervals of time indicate approximately the same position for the surface potential minimum. Over the entire region of the scan the surface potential rises slowly but quite regularly with time, and the surface potential profile almost reaches the profile before stress after ∼28 minutes.

In another experiment, simultaneous variation of surface potential and drain current under UV illumination was recorded on an HFET device (100 μm wide, rectangular gate with T-gate layout geometry, on sapphire) after it was stressed. While both the surface potential and the drain current transients were being measured after stress, the laser was switched on at $t = 96$ s and switched off at $t = 172$ s. The resulting surface potential and drain current transients are shown in Fig. 22. After the laser was switched on, the surface potential changed dramatically by ∼1.7 V while the drain current changed by ∼0.9 mA. After the laser was switched off the surface potential reduced by ∼0.2 V while the drain current reduced by ∼0.45 mA. Since, after stressing, the drain current and the surface potential usually reaches a relatively steady value (increasing at a very slow rate) around $t = 100$ s, we can estimate the final effect of the laser illumination as a net increase in surface potential

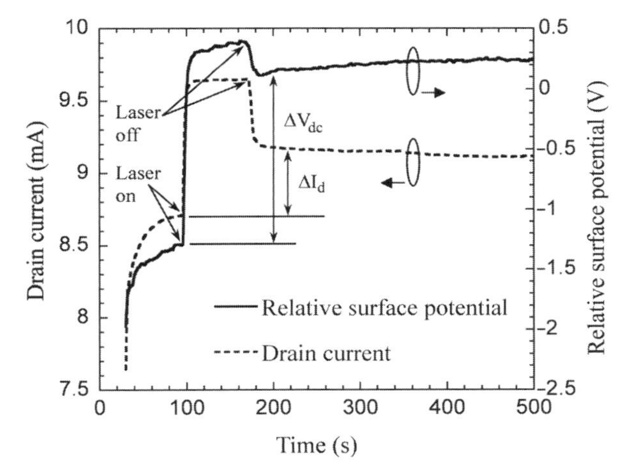

Fig. 22 Simultaneously measured drain current and surface potential transients after type II stress with the laser switched on and off. The change in surface potential (after laser was switched off), ΔV_{dc}, resulted in a large change in drain current, ΔI_d. The tip was positioned ~0.2–0.3 μm from the gate during measurements.

(ΔV_{dc}) and drain current (ΔI_d) by ~1.5 V and ~0.45 mA, respectively. Thus, it can be concluded that the laser illumination resulted in an increase in 2DEG density, and helped the device to quickly regain the highest possible drain current.

From the experiments described above it is clear that after stress, the drain current of the device is reduced, and slowly recovers after stress is removed. The reduction in drain current is not observed if the device is illuminated with UV light, and very much suppressed if the device is passivated (more discussion about passivation effects later on). Since SiN_x passivation is not expected to modify the properties of the underlying layers, we infer that the surface is responsible for causing the reduction in drain current. This is also supported by surface potential measurements, which indicate negative potential at the surface after stressing while both the drain and the gate biases are zero. Such a negative potential measured on the surface near the gate would indicate presence of negative charges or electrons at the surface (near the gate). Since there was no drain current flowing during the stress, the possibility of highly energetic electrons from the channel reaching the surface, can be safely ruled out. *Therefore, it can be concluded that the accumulation of negative charges near the gate is due to electrons tunneling from the gate under high bias stress.* Figure 23 (a) shows the schematic band diagram under drain and gate bias stress. The high field near the gate helps the electrons to tunnel into the surface states. Measurements of surface potential variation in a real device were made to verify the existence of high field near the gate. Figure 23 (b) shows the variation of surface potential [inverted to make it have the same sign as electronic energy for comparison with Fig. 23 (a)] from the gate toward the drain. The gate and the drain biases were −5 and +5 V, respectively. As seen from Fig. 23 (b), most of the voltage difference of 10 V appears near the gate within ~0.2 μm from the edge. Thus,

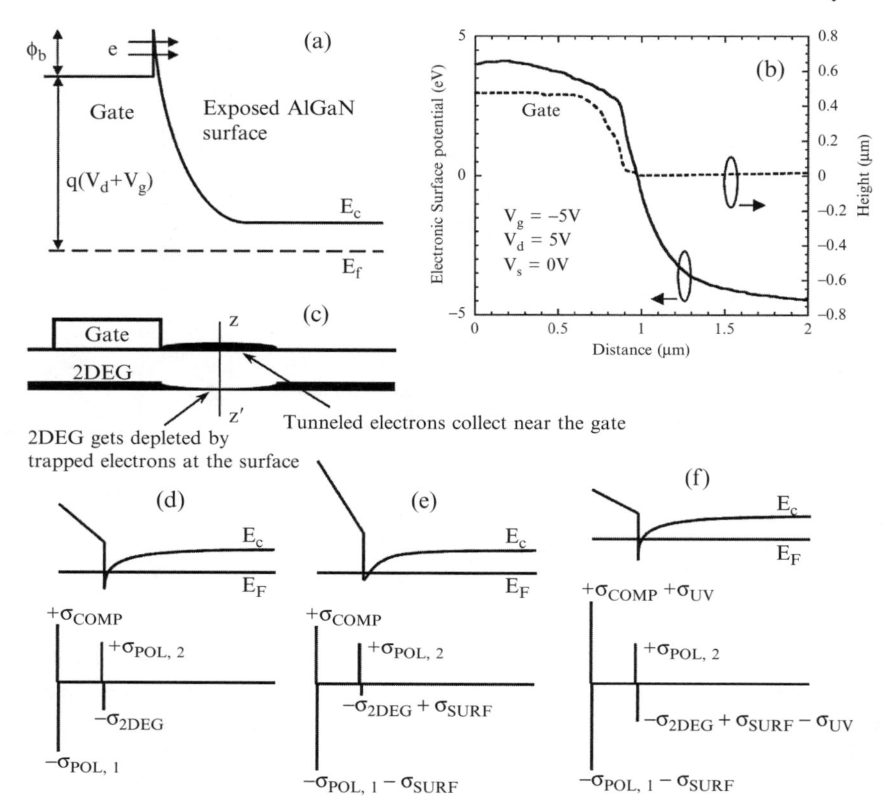

Fig. 23 (a) Schematic band diagram explaining the tunneling of electrons from the gate to the surface when a device is stressed by applying biases $-V_g$ and V_a to the gate and the drain, respectively. (b) Actual measurement of the surface potential variation (inverted for easier comparison with conduction band profile) from the gate toward the drain. (c) Tunneled electrons trapped near the gate reduce the 2DEG by raising the surface barrier. Band diagrams along the zz′ line are shown (d) before stress, (e) after stress and (f) after UV illumination of the stressed device. The reference Fermi level for measurements, as well as the sheet charge distributions at the surface and interface, are also shown for the respective cases.

it is demonstrated that there exists a very high field near the gate. Aided by this high field, electrons tunnel from the gate and get trapped by the surface states near the gate [see Fig. 23(c)]. This explains our observation of large changes in surface potential near the gate due to stress (Fig. 21). Due to high field near the gate it is possible that the electrons quickly move away from near the gate and preferentially accumulate at a distance farther (for the device measured, ∼0.3 μm farther) from the edge of the gate (see Fig. 21). Figures 23 (d) and (e) show the schematic band diagram and the sheet charge densities at the surface and the interface of the device before and after stress, respectively, along the zz′ axis shown in Fig. 23(c). It is seen that the accumulation of the electrons at the surface after stress causes the separation between the surface conduction band and the reference Fermi level (the Fermi level at the AlGaN/GaN interface in equilibrium with the Fermi levels of the ohmic contacts; not the surface quasi-Fermi level) to increase, which reduces the 2DEG [see Fig. 23(c)]. *Now, this region of electron accumulation can act as a virtual gate and reduce the channel current.*

4.3 Surface States Characterization and Passivation

Throughout this chapter we have mentioned that surface states are extremely important from the perspective of III-N technology. However, it is quite difficult to accurately measure the density of active surface states, which is further complicated by the variations in growth techniques, surface preparation, and other processing steps. Consequently, there have been attempts to passivate the surface to avoid the detrimental effects of surface states altogether. In this section, we will discuss measurements carried out to estimate the density of surface states that might be responsible for 2DEG formation at the interface (based on the surface donor theory). The reduction in surface and interface charge instability due to surface passivation with SiN_x will also be discussed.

Surface states and the origin of 2DEG: Interesting observations were made from the comparison of BSBH and 2DEG density variation with AlGaN thickness for a series of Ga-face AlGaN/GaN heterostructure samples grown on *c*-plane sapphire by MOCVD (structure in Fig. 2) [74]. The AlGaN barrier layer thickness was varied from 50–440 Å, while the Al alloy composition was kept constant at ~35% (Al composition measured by X-Ray diffraction). The calculated BSBH values as well as the 2DEG density are plotted against the AlGaN barrier thickness in Fig. 24.

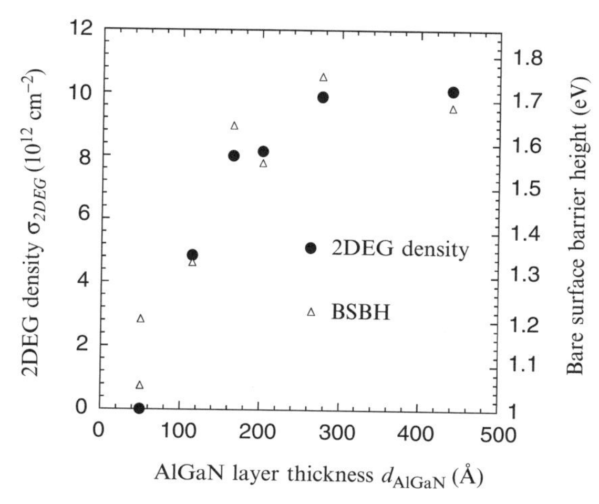

Fig. 24 2DEG density and BSBH for AlGaN/GaN heterostructures plotted against thickness of the AlGaN barrier layer (with 35% Al alloy composition).

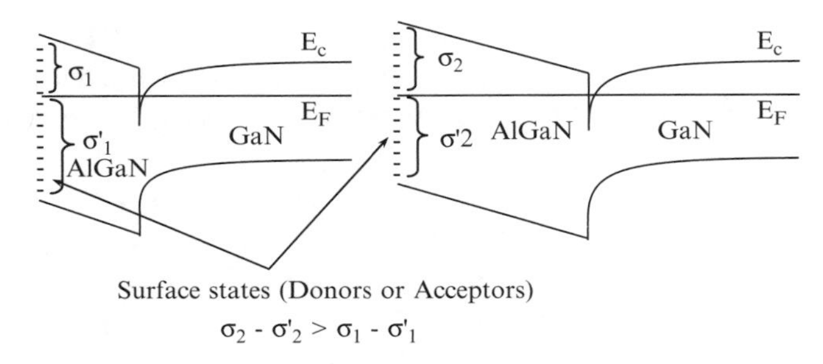

Fig. 25 Schematic band diagram showing the surface donor states as the origin of the 2DEG density at the interface. As the surface barrier increases, the positive charge at the surface (this is separate from the negative polarization charges at the surface), due to ionization of the surface donors (and de-occupation of the acceptor states, if any) increases, increasing the 2DEG. σ_1 and σ_2 are the ionized donor charge densities, and σ_1' and σ_2' are the occupied acceptor charge densities.

We find that for lower AlGaN thickness the surface barrier increases very regularly, but after ~200 Å the surface barrier becomes almost constant at ~1.7 eV. A remarkable similarity is observed between this trend and the 2DEG density variation with AlGaN thickness, which also starts saturating after ~200 Å. Such a similarity can be explained quite appropriately if we assume surface donors as the origin of electrons as illustrated by the band diagrams in Fig. 25. At equilibrium, the Fermi level at the surface is below some of the donor energy states, and those ionized states contribute the electrons that give rise to the 2DEG at the interface. As the thickness of the barrier layer increases, the surface Fermi level is lowered more and more with respect to the conduction band, increasing the surface barrier, causing more and more donor states to be emptied, thereby increasing the 2DEG density.

Mathematically, the 2DEG density is related to the surface donor states (where for generalization, acceptor states have also been assumed to be present) as

$$\sigma_{2DEG} = \int_{E_v}^{E_c} N_D(E)\left[1 - f(E)\right] dE - \int_{E_v}^{E_c} N_A(E) f(E) dE - \sigma_{surf} + \sigma_{in} \tag{21}$$

where σ_{2DEG}, σ_{surf}, and σ_{in} are the electron density (magnitude) at the interface, the polarization induced bound charge (magnitude) at the surface, and at the interface, respectively. N_D and N_A are the donor and acceptor type trap densities ($cm^{-2}\,eV^{-1}$) at the surface, and E_c and E_v are the conduction and valence band edges, respectively. All traps are assumed to be present within the bandgap. It should be noted here that since $\sigma_{surf} >> \sigma_{in}$ (the polarization gradients are much different at the AlGaN/GaN and AlGaN/air interface), so the ionized donors not only provide for the 2DEG, but also compensate for $(\sigma_{surf} - \sigma_{in})$, and any charged acceptors as well. However, here we are concerned only with those ionized donors that contribute electrons to form the 2DEG. The experimental observations in this study support the surface donor model previously proposed based on 2DEG measurements on AlGaN/GaN heterostructures [59]. However, this model takes into account a distribution of donor states throughout the bandgap, rather than single-energy donor states as proposed in the previous models [59]. The model presented here is also supported by other experimental observations, since the BSBH varies with AlGaN thickness, while in the previous models proposing single-energy donor states, the BSBH has to remain almost constant once the Fermi level reaches the energy level of the donors (and 2DEG starts forming). Since the variation of the 2DEG and the BSBH have almost the same functional form versus barrier thickness, so the 2DEG will vary almost linearly with the BSBH as shown in Fig. 26. The surface donor density (of those contributing to the formation of the 2DEG) is therefore constant between BSBH of ~1.0 and 1.8 eV, and given by the slope of the least square fit line as $1.58 \times 10^{13}\,\text{cm}^{-2}\,\text{eV}^{-1}$. For comparison, the surface donor density measured by Hasegawa et al. [75] (using schottky diode I-V characteristics and drain current transients in AlGaN/GaN heterostructure field effect transistors) in an earlier study, has also been presented in Fig. 26.

Effects of surface passivation: To mitigate performance degradation of devices, the exposed surface of III-N hetrostructures are passivated with dielectric layers. SiN_x or SiO_2 are commonly used. It has been observed that after passivation, the surface potential decreases significantly (several tenths of eV) [67], and the 2DEG density increases [67,70]. There is also evidence that the charge instability, caused by external perturbation, is significantly reduced after passivation. AlGaN/GaN heterostructures were passivated by depositing a 200 nm layer of SiN_x by plasma enhanced CVD at 300°C.

The reduction in charge instability was observed for AlGaN/GaN HFETs. Drain current and surface potential transients measured on devices before and after

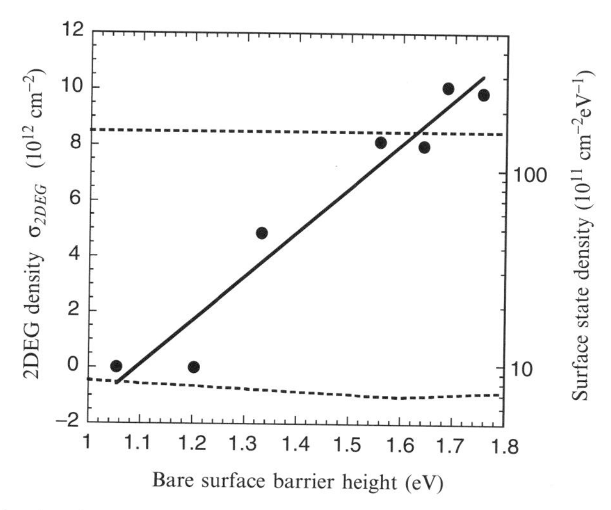

Fig. 26 2DEG density plotted against BSBH for AlGaN (35% Al)/GaN heterostructures with varying AlGaN thickness. The slope of the least square fit line ($y = -17.268 + 15.824x$) gives an approximate value of the density of surface donor states. Here y is in cm^{-2}, and x is in eV. Comparison of the surface donor density from the present work, and the surface state density from Ref. [75], as a function of energy from the conduction band, is also shown (dotted lines).

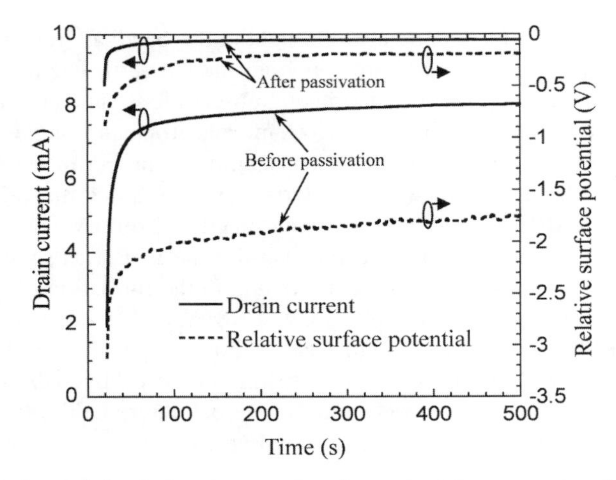

Fig. 27 Comparison of the drain current and surface potential recovery transients for similar and consecutive devices before and after passivation. Before measurement the devices underwent stress.

passivation and are shown in Fig. 27. Before measurement, these devices underwent stress for 2 minutes as discussed in the previous section. As can be seen from Fig. 27, both the drain current and the surface potential transient magnitudes are reduced significantly after passivation. The microwave output power for these devices also increased significantly from $\sim$2–2.5 W/mm to $\sim$3–3.5 W/mm at 10 GHz. Similar measurements of surface potential and drain current transients on another passivated device (with 7–7.5 W/mm output power density as mentioned above) after stressing resulted in no significant transients. *These results support the correlation between the surface charge instability and microwave power performance of AlGaN/GaN HFETs and has been discussed in detail in Ref. [71, 72].*

5 Summary

In this chapter, we have discussed the polarization properties of wide bandgap semiconductors such as SiC and III-N. III-N are typically grown on non-native substrates, giving rise to material defects, dislocations and charge instability on unpassivated surfaces.

Characterization of III-N material and devices has been discussed, with an emphasis on SKPM. A major objective of these characterizations is to unravel the causes behind the interface charge instability commonly observed in III-N heterostructures. Due to wide ranging application, the field of III-N technology has expanded at a rapid rate in recent years, and still remains a very exciting and important field of research. Due to the unique material properties of III-N, their characterization using traditional methods does not always yield sufficient information, especially with regards to their electronic properties. The SKPM technique, also relatively new, has been useful in this respect. SKPM has yielded several interesting results regarding III-N heterostructures, which have been described in detail in the chapter, and are summarized below:

(i) Bare surface barrier height of AlGaN/GaN heterostructures
(ii) Electronic properties of dislocations
(iii) Charge instability in AlGaN/GaN heterostructures
(iv) Surface state characterization and validation of the surface donor model
(v) Direct measurement of surface accumulation of charge caused by device stressing, and its correlation with microwave output power
(vi) Effects of passivation on heterostructure materials and HFET devices

SiC heteropolytype structures were discussed with an emphasis on polarization-doped electronic channels. This lattice matched system holds promise owing to the possibility of realizing high current microwave transistors, operating in fields akin to GaN/AlGaN HEMTs. The superior thermal conductivity of SiC should make SiC polarization-doped transistors very realiable. The growth and characterization of the material properties of these heterostructures were discussed briefly. The electronic properties of the polarization doped channels were demonstrated. Indeed, the similarity of the AlGaN/GaN system to the SiC heteropolytype system suggests that SKPM and other scanning probe techniques will reveal the nature of the surface and interface of these heterostructures, which may shed light on the influence of DPB's and other material issues, some of which were discussed in this chapter.

Many of the polarization induced nanoscale phenomena discussed in this chapter are of great scientific and practical interest. The GaN/AlGaN 2DEG, for example has successfully been used for high power microwave devices and is currently being commercialized. A thorough understanding of polarization induced phenomena relies on the ability to investigate nanoscale dimensions. Scanning probe techniques are ideally suited to this have been applied to the characterization of a wide variety of phenomena in III-N, as well as other semiconductors. The popularity of these techniques is also due to the quantitative and repeatable measurement data that can be obtained readily and easily interpreted. SKPM, performed in conjunction with conventional AFM, yields very important information regarding the nanometer scale electronic properties of materials and devices [76]. The power of SPM makes the study of polarization-dependent phenomena in semiconductors a very rich area for research and development in this nanoscale world.

References

1. U. K. Mishra, P. Parikh, and Yi-Feng Yu, *Proc. IEEE* **90**, 1022 (2002).
2. L. F. Eastman, V. Tilak, J. Smart, B. M. Green, E. M. Chumbes, R. Dimitrov, H. Kim, O. Ambacher, N. Weimann, T. Prunty, M. Murphy, W. J. Schaff, and J. R. Shealy, *IEEE Trans. Electron Devices* **48**, 479 (2001).
3. S. Nakamura and G. Fasol, *The blue laser diode*, Springer-Verlag, Berlin, 1997.

4. M. Razeghi, *Proc. IEEE* **90**, 1006 (2002).
5. Raghunathan, R. Alok, D. Baliga, B.J., *IEEE Elec. Dev. Lett.*, **16**, 226 (1995)
6. Shenoy, J.N. Cooper, J.A., Jr Melloch, M.R., *IEEE Elec. Dev. Lett*, **18,** 93 (1997)
7. G. Binnig, H. Rohrer, Ch. Gerber, and E. Weibel, *Appl. Phys. Lett.* **40**, 178 (1982).
8. E. T. Yu, S. L. Zuo, W. G. Bi, C. W. Tu, A. A. Allerman, and R. M. Biefeld, *J. Vac. Sci. Technol. A* **17**, 2246 (1999).
9. K. Shiojima, J. M. Woodal, C. J. Eiting, P. A. Grudowski, and R. D. Dupuis, *J. Vac. Sci. Technol. B* **17**, 2030 (1999).
10. S. Keller, G. Parish, P. T. Fini, S. Heikman, C. H. Chen, N. Zhang, S. P. DenBaars, U. K. Mishra, and Y. F. Yu, *J. Appl. Phys.* **86**, 5850 (1999).
11. M. Tanimoto and O. Vatel, *J. Vac. Sci. Technol. B* **14**, 1547 (1996).
12. F. Robin, H. Jacobs, O. Homan, A. Stemmer, and W. Bächtold, *Appl. Phys. Lett.* **76**, 2907 (2000).
13. G. Koley and M. G. Spencer, *Appl. Phys. Lett.* **78**, 2873 (2001).
14. P. J. Hansen, Y. E. Strausser, A. N. Erickson, E. J. Tarsa, P. Kozodoy, E. G. Brazel, J. P. Ibbetson, U. Mishra, V. Narayanamurti, S. P. DenBaars, and J. S. Speck, *Appl. Phys. Lett.* **72**, 2247 (1998).
15. J. W. P. Hsu, M. J. Manfra, R. J. Molnar, B. Heying, J. S. Speck, *Appl. Phys. Lett.* **81**, 79 (2002).
16. F. Bernardini, V. Fiorentini, and D. Vanderbilt, *Phys. Rev. B*, **56**, 1002 (1997).
17. O. Ambacher, *J. Phys. D: Appl. Phys.* **31**, 2653 (1998).
18. O. Ambacher, J. Smart, J. R. Shealy, N. G. Weimann, K. Chu, M. Murphy, W. J. Schaff, L. F. Eastman, R. Dimitrov, L. Wittmer, M. Stutzmann, W. Rieger, and J. Hilsenbeck, *J. Appl. Phys.* **85**, 3222 (1999)
19. A. F. Wright, *J. Appl. Phys.* **82**, 2833 (1997).
20. U. Starke, J. Schardt, J. Bernhardt, M. Franke, and K. Heinz, *Phys. Rev. Lett.*, **82**, 2107 (1999)
21. C. Cheng, R. J. Needs and V Heine, *J. Phys. C: Solid State Phys.*, **21,** 1049 (1988)
22. A. Qteish, Volker Heine, and R. J. Needs, *Phys. Rev. B*, **45**, 6534 (1992)
23. J.A. Powell, Herbert A. Will, *J. Appl. Phys.*, **43**, 1400 (1972)
24. A. Fissel, *J. Crystal Growth*, **212**, 438 (2000)
25. H. Amano, N. Sawaki, I. Akasaki, and Y. Toyoda, *Appl. Phys. Lett.* **48**, 353 (1986).
26. S. C. Jain, M. Willander, J. Narayan, and R. V. Overstraeten, *J. Appl. Phys.* **87**, 965 (2000).
27. R. Dimitrov, M. Murphy, J. Smart, W. Schaff, J. R. Shealy, L. F. Eastman, O. Ambacher and M. Stutzmann, *J. of Appl. Phys.* **87**, 3375 (2000).
28. B. Heying, E. J. Tarsa, C. R. Elsass, P. Fini, S. P. Denbaars, and J. S. Speck, *J. Appl. Phys.* **85**, 6470 (1999).
29. M. J. Murphy, K. Chu, H. Wu, W. Yeo, W. J. Schaff, O. Ambacher, J. Smart, J. R. Shealy, L. F. Eastman, and T. J. Eustis, *J. Vac. Sci. Technol. B* **17**, 1252 (1999).
30. O. Ambacher, B. Foutz, J. Smart, J. R. Shealy, N. G. Weimann, K. Chu, M. Murphy, A. J. Sierakowski, W. J. Schaff, L. F. Eastman, R. Dimitrov, A. Mitchell, and M. Stutzmann, *J. Appl. Phys.* **87**, 334 (2000).
31. V. Ramachandran, R. M. Feenstra, W. L. Sarney, L. Salamanca-Riba, J. E. Northrup, L. T. Romano, and D. W. Greve, *Appl. Phys. Lett.* **75**, 808 (1999).
32. S. D. Lester, F. A. Ponce, M. G. Craford, and D. A. Steigerwald, Appl. Phys. Lett. **66**, 1249 (1995).
33. F. A. Ponce, MRS Bull. **22**, 51 (1997).
34. H. Amano, S. Kamiyama, I. Akasaki, Proc. IEEE **90**, 1015 (2002).
35. Nils G. Weimann, L. F. Eastman, D. Doppalapudi, H. M. Ng, and T. D. Moustakas, *Appl. Phys. Lett.* **83**, 3656 (1998).
36. Debdeep Jena, Arthur C. Gossard, and Umesh K. Mishra, *Appl. Phys. Lett.* **76**, 1707 (2000).
37. J. W. P. Hsu, M. J. Manfra, D. V. Lang, S. Richter, S. N. G. Chu, A. M. Sergent, R. N. Cleiman, L. N. Pfeiffer, and R. J. Molnar, *Appl. Phys. Lett.* **78**, 1685 (2001).
38. X. G. Qiu, Y. Segawa, Q. K. Xue, Q. Z. Xue, and T. Sakurai, *Appl. Phys. Lett.* **77**, 1316 (2000).
39. S. Nakamura, *Semicond. Sci. Technol.* **14**, R27 (1999).

40. E. Munoz, E. Monroy, J. L. Pau, F. Calle, F. Omnes, and P. Gibart, *J. Phys. Condens. Mater* **13**, 7115 (2001).
41. S. Bai, R. P. Devaty, W. J. Choyke, U. Kaiser, M. F. MacMillan, *Appl. Phys. Lett*, **83**, 3171 (2003)
42. A. Fissel, *Physics Reports*, **379**, 149 (2003)
43. Hiroyuki Matsunami, Tsunenobu Kimoto, *Mat. Sci. Eng.* **R20**, 125 (1997)
44. Glass, R. C.; Henshall, D.; Tsvetkov, V. F.; Carter, C. H., Jr., *Phys. Stat. Sol. (B), Appl. Res.*, **202**, 149 (1997)
45. Philip G. Neudeck, J. Anthony Powell, and Andrew J. Trunek, Xianrong R. Huang, and Michael Dudley, *Mat. Sci. For.* **389–293**, 311 (2002)
46. A. A. Lebedev, G. N. Mosina, I. P. Nikitina, N. S. Savkina, L. M. Sorokin and A. S. Tregubova, *Tech. Phys. Lett.*, **27**, 1052 (2006)
47. H.S. Kong, J.T. Glass, R.F. Davis, *Appl. Phys. Lett.*, **49**, 1074 (1986)
48. Tsunenobu Kimoto and Hiroyuki Matsunami, *J. Appl. Phys.*, **78**, 3132 (1995)
49. David J. Larkin, Philip G. Neudeck, J. Anthony Powell, and Lawrence G. Matus, *Appl. Phys. Lett.*, **65**, 1659 (1994)
50. V. M. Polyakov and F. Schwierz, *J. Appl. Phys.*, **98**, 023709 (2005)
51. MVS Chandrashekhar, C.I. Thomas, J. Lu, M.G. Spencer, *Appl. Phys. Lett* **91,** 102727 (2007)
52. MVS Chandrashekhar, C.I. Thomas, J. Lu, M.G. Spencer, *Appl. Phys. Lett* **90,** 173509 (2007)
53. Nakashima, S.; Harima, H. *Physica Stat. Sol. (A), Appl. Res*, **162**, 39 (1997)
54. J. C. Burton, L. Sun, M. Pophristic, S. J. Lukacs, and F. H. Longa, Z. C. Feng, I. T. Ferguson, *J. Appl. Phys.*, **84**, 6268 (1998).
55. H. Kroemer, Wu-Yi Chien, J. S. Harris, Jr. and D. D. Edwall, *Appl. Phys. Lett.*, **36**, 295 (1980).
56. Sghaier, N. Bluet, J.-M. Souifi, A. Guillot, G. Morvan, E. Brylinski, C., *IEEE Trans. Elec. Dev.*, **50**, 297 (2003).
57. Hideharu Matsuura, Masahiko Komeda, Sou Kagamihara, Hirofumi Iwata, Ryohei Ishihara, Tetsuo Hatakeyama, Takatoshi Watanabe, Kazutoshi Kojima, Takashi Shinohe, Kazuo Arai, *J. Appl. Phys.*, **96,** 2708 (2004).
58. R. Neuberger, G. Muller, O. Ambacher, and M. Stutzmann, *Phys. Stat. Sol.* **185**, 85 (2001).
59. J. P. Ibbetson, P. T. Fini, K. D. Ness, S. P. Denbaars, J. S. Speck, and U. K. Mishra, *Appl. Phys. Lett.* **77**, 250 (2000).
60. A. Vertiatchik, L. F. Eastman, W. J. Schaff, and T. Prunty, *Electron. Letts.* **38**, 388 (2002).
61. X. Z. Dang, E. T. Yu, E. J. Piner, and B. T. McDermott, J. Appl. Phys. **90**, 1357 (2001).
62. B. D. Terris, J. E. Stern, D. Rugar, and H. J. Mamin, *Phys. Rev. Lett.* **63**, 2669 (1989).
63. R. M. Nyffenegger, R. M. Penner, and R. Schierle, *Appl. Phys. Lett.* **71**, 1878 (1997).
64. G. Koley and M. G. Spencer, *J. Appl. Phys.* **90**, 337 (2001).
65. P. M. Bridger, Z. Z. Bandif, E. C. Piquette, and T. C. McGill, *Appl. Phys. Lett.* **74**, 3522 (1999).
66. G. Koley, H. Cha, C. I. Thomas, and M. G. Spencer, Appl. Phys. Lett. 81, 2282 (2002).
67. G, Koley, H. Cha, J. Hwang, W. J. Schaff, L. F. Eastman, and M. G. Spencer, *J. Appl. Phys.* **96**, 4253 (2004).
68. Z. Z. Bandić, P. M. Bridger, E. C. Piquette, and T. C. McGill, *Solid-State Electron* **44**, 221 (2000).
69. L. Chernyak, A. Osinsky, and A. Schulte, *Solid-State Electron* **45**, 1687 (2001).
70. B. M. Green, K. K. Chu, E. M. Chumbes, J. A. Smart, J. R. Shealy, and L. F. Eastman, *IEEE Electron Device Lett.* **21**, 268 (2000).
71. G. Koley, V. Tilak, L. F. Eastman, and M. G. Spencer, *IEEE Trans. Electron Devices* **50**, 886 (2003).
72. R. Vetury, N. Q. Zhang, S. Keller, and U. K. Mishra, *IEEE Trans. Electron Devices* **48**, 560 (2001).
73. G. Koley, M. G. Spencer, and H. R. Bhangale, *Appl. Phys. Lett.* **79**, 545 (2001).
74. G. Koley and M. G. Spencer, *Appl. Phys. Lett.* **86,** 042107 (2005).
75. H. Hasegawa, T. Inagaki, S. Ootomo, and T. Hashizume, *J. Vac. Sci. Technol. B* **21**, 1844 (2003).
76. G. Koley and M. G. Spencer, *Encyclop. NanoSci. Nanotech.* **4**, 327 (2003).

Functionally Graded Polar Heterostuctures: New Materials for Multifunctional Devices

Debdeep Jena, S. Pamir Alpay, and Joseph V. Mantese

1 Introduction

Mixing materials of different compositions is an ancient art. As early as 3000 B.C., metallic alloys - brass & bronze were used for sculpture work. Over the last century, major strides were made in the art of crystal growth of metals, dielectrics, and semiconductors. An alloy offers an opportunity to exploit physical, electrical, and optical properties of materials which are either intermediate, or absent in its constituent materials. This has been the driving force behind the study and discovery of new generations of alloys. With the advent of epitaxial growth techniques such as Molecular Beam Epitaxy (MBE) and Metal-Organic Chemical Vapor Deposition (MOCVD), such hybrid materials can now be engineered at the atomic scale.

Polar semiconductors of the wurtzite crystal structure possess spontaneous and piezoelectric polarization, the most relevant examples being the III-V Nitride semiconductors GaN, AlN, and InN. The electric dipoles in each unit cell of such materials are 'frozen' once the crystal is grown, and can contribute to the formation of bound as well as mobile charges. Closely related to such polar semiconductors are ferroelectric crystals, in which the electric dipoles can be flipped with external electric fields. Some ferroelectric crystals have electronic bandgaps comparable to wide-bandgap semiconductors, and are only recently being investigated in that light; they are referred to as 'ferroelectric semiconductors'. Finally, ferromagnetic crystals are analogous in that they possess a magnetic moment (spin) in each unit cell. Controlled doping of semiconductors with magnetic atoms allows one to marry the properties of semiconductors and ferromagnets in a new class of materials called Dilute Magnetic Semiconductors (DMS), the alloy $Ga_{1-x}Mn_xAs$ being the prime example. Thus, by the technique of alloying, the distinction between the seemingly different classes of materials - semiconductors, ferroelectrics, and ferromagnets is slowly becoming blurry. The same alloy material can exhibit the functionality of different material classes, and hence such materials are labeled 'multifunctional'.

In this chapter, the science and the applications of a new class of compositionally 'graded' alloy materials is presented. A rich range of physical phenomena emerge with the compositional grading of crystals that possess electronic or magnetic polarization. The chapter covers recent findings in compositionally graded polar semiconductors, ferroelectrics, and ferromagnets. The second section of this chapter will concentrate on compositionally graded III-V Nitride semiconductor heterostructures, and their device applications. The third section will generalize the results achieved in the semiconductors to allied phenomena in ferroelectrics and ferromagnets. The final section will outline some unsolved problems in this field and will offer the authors' perspective for future research in this exciting area.

2 Graded Polar Nitride Semiconductor Heterostructures

2.1 Polarization in Nitrides: A Tutorial

Polarization is an important property of the III-V nitride semiconductors. Owing to the large ionicity of the Ga-N bond, the material possesses a large piezoelectric polarization component. In addition, the stable phase of the III-V nitride semiconductor family is the wurtzite structure with a non-ideal c/a ratio. The uniaxial nature of the crystal coupled with this non-ideality causes a large spontaneous polarization, which is absent in zinc-blende polar crystals. This fact has been described in detail in other chapters of this book.

Figure 1 shows the energy gap against lattice constants of III-V semiconductors. The stable phase of the III-V nitrides is wurtzite, though the zinc-blende version has also been grown. The direct energy gaps of GaN (3.4eV), AlN (6.2eV) and InN ($\sim$ 0.7eV) cover a wider spectrum than the zinc-blende III-V semiconductors[1]; indeed, they cover the *entire* visible spectrum and *beyond*. The lattice constants and direct energy gaps are listed in Table 1. The table also lists mechanical constants (c_{ij}), piezoelectric moduli (e_{ij}) and the magnitude of spontaneous polarization in the III-V nitrides; in comparison, zinc-blende structure semiconductors have *much smaller* piezoelectric constants and *no* spontaneous polarization. This subsection is intended to be a primer for the phenomena of electronic polarization in semiconductors and therefore, orders of magnitudes, and design rules for *using* polarization in nitride heterostructures are emphasized. For a rigorous treatment of polarization, please refer to the chapters by Boguslawski & Bernholc, and Abbacher and Cimalla in this book.

The bonds in *all* III-V and II-VI compound semiconductors are polar owing to the difference in the ionicity of the constituent atoms[2]. Symmetry properties of the crystal structure dictate the presence/absence of spontaneous polarization. The cubic zinc-blende structure symmetry forbids spontaneous polarization, whereas

[1] The II-VI semiconductor ZnO has attracted increased attention owing to electronic and structural properties very similar to GaN.

[2] The chemical bonds of IV-IV materials such as SiC and SiGe are also polar.

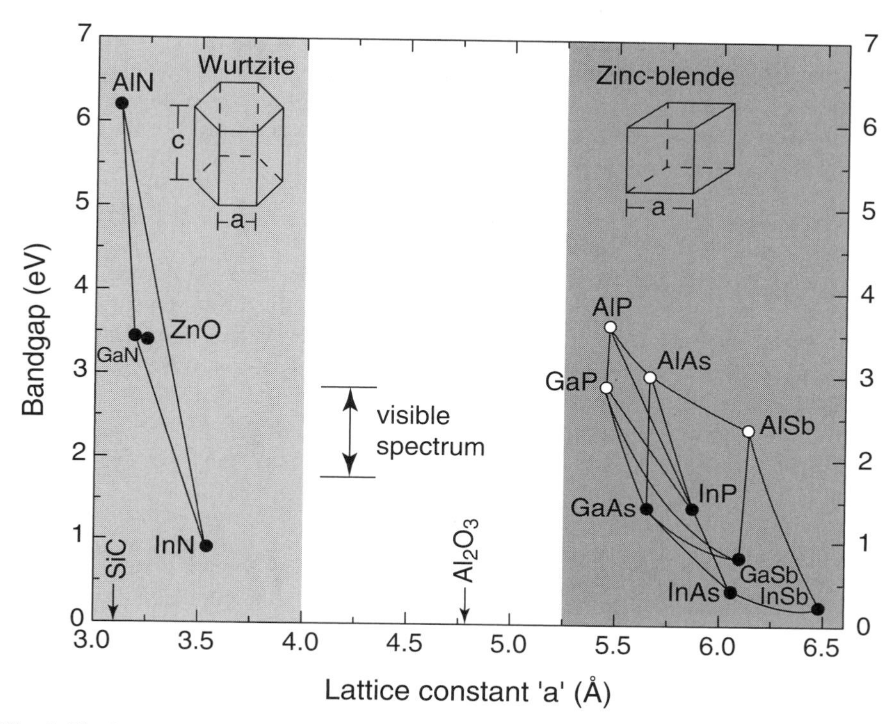

Fig. 1 The bandgap-lattice constant plot. All bandgaps plotted are the direct gaps; indirect gap semiconductors are shown by open circles. Wurtzite crystals are characterized (see insets) by two lattice constants (a,c) of which the a-lattice constant is used for the figure. The a-lattice constants of the common substrates for growth of III-V Nitrides - SiC and Al_2O_3 (sapphire) are indicated by arrows.

Table 1 Comparative material properties [1]

Material	Lattice constant $a_0(c_0)$Å	Direct Gap (eV)	P_{sp} (C/m^2)	e_{33} (C/m^2)	e_{31} (C/m^2)	c_{13} (GPa)	c_{33} (GPa)
GaN	3.189 (5.185)	3.4	−0.029	0.73	−0.49	103	405
InN	3.544 (5.718)	0.7	−0.032	0.97	−0.57	92	224
AlN	3.111 (4.978)	6.2	−0.081	1.46	−0.60	108	473
ZnO	3.249 (5.205)	3.4	−0.057	0.89	−0.51		
GaAs	5.653	1.4	0	−0.12	+0.06		
InAs	6.058	0.4	0	−0.03	+0.01		
InP	5.870	1.4	0	+0.04	−0.02		
GaSb	6.096	0.8	0	−0.12	+0.06		
InSb	6.479	0.2	0	−0.06	+0.03		
AlAs	5.661	3.0	0	−0.01	+0.01		
AlP	5.467	5.5	0	+0.04	−0.02		
AlSb	6.136	2.4	0	−0.04	+0.02		
GaP	5.451	2.9	0	−0.07	+0.03		

non-ideal wurtzite structures allow it[3]. Both wurtzite and zinc-blende structures exhibit piezoelectric polarization.

The piezoelectric polarization of the crystal, defined in terms of the moduli e_{ijk} and d_{ijk}, relate the piezoelectric field along the ith direction to the strain (ϵ_{jk}) and the stress (σ_{jk}) by the relation

$$P_{pz,i} = e_{ijk}\epsilon_{jk} = d_{ijk}\sigma_{jk}, \tag{1}$$

where the Levi-Civita convention of summation over repeated indices is employed. Since the moduli are symmetric in the indices j,k, the third rank-tensors e_{ijk}, d_{ijk} contract to a simpler 3×6 matrix form e_{ij}, d_{ij} [2]. For the wurtzite crystal, symmetry arguments further reduce the piezoelectric strain-moduli to just three independent constants e_{13}, e_{33}, e_{15}; the matrix is written as

$$e = \begin{pmatrix} 0 & 0 & 0 & 0 & e_{14} & 0 \\ 0 & 0 & 0 & e_{15} & 0 & 0 \\ e_{31} & e_{31} & e_{33} & 0 & 0 & 0 \end{pmatrix}. \tag{2}$$

Stress and strain coefficients are related in a crystal by the relation $e_{jk} = c_{ij}d_{ik}$ where the elastic coefficients c_{ij} form a tensor that may be conveniently put in a matrix notation

$$\begin{pmatrix} c_{11} & c_{12} & c_{13} & 0 & 0 & 0 \\ c_{12} & c_{11} & c_{13} & 0 & 0 & 0 \\ c_{13} & c_{13} & c_{11} & 0 & 0 & 0 \\ 0 & 0 & 0 & c_{44} & 0 & 0 \\ 0 & 0 & 0 & 0 & c_{44} & 0 \\ 0 & 0 & 0 & 0 & 0 & \frac{1}{2}(c_{11}-c_{12}) \end{pmatrix} \tag{3}$$

for the wurtzite structure. For analysis of polarization in III-V nitrides, the two elastic coefficients c_{13}, c_{33} necessary have been listed in Table 1. The piezoelectric moduli e_{13}, e_{33} are listed for all semiconductors in the same table. In Figure 2, we plot the two piezoelectric moduli of all the semiconductors listed in the table. It is clear from the plot that the III-V nitride semiconductors have piezoelectric coefficients that are an *order of magnitude* larger than the other III-V semiconductors. Thus, much higher polarization charges are expected for strained wurtzite III-V nitride semiconductors than for III-V semiconductors in the zinc-blende family.

From now on, attention is directed to polarization along the $[0001]$ direction of the wurtzite crystal since most III-V nitride crystals are grown epitaxially along this direction. Also, since most of the material presented in this chapter will deal with growth on GaN substrates, strain will be referenced to the relaxed lattice of GaN.

The piezoelectric polarization along the $[0001]$ direction ($i = 3$) may be written simply as

$$P_3 = e_{33}\epsilon_3 + e_{31}(\epsilon_1 + \epsilon_2), \tag{4}$$

[3] Spontaneous polarization is forbidden by symmetry in ideal wurtzite crystals, where the ratio $c/a = \sqrt{8/3}$. See chapters the chapters by Ambacher Cimalla, and Butte & Grandjean in this book.

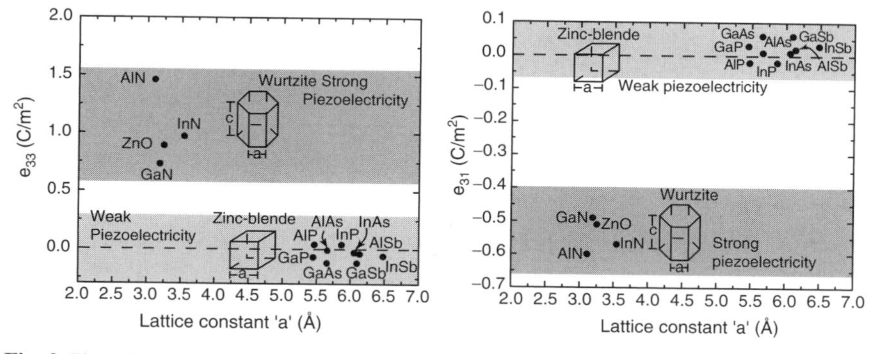

Fig. 2 Piezoelectric constants e_{33}, e_{31} of various semiconductors. The III-V Nitrides of wurtzite structure exhibit much large piezoelectric polarization than the zinc-blende family.

and the strain components ϵ_3, ϵ_1 are related by the elastic coefficients by the relation $\epsilon_3 = -2\epsilon_1(c_{13}/c_{33})$. Using this, one obtains the piezoelectric polarization

$$P_{pz,[0001]} = 2(e_{31} - e_{33}\frac{c_{13}}{c_{33}})\epsilon_1, \tag{5}$$

where $\epsilon_1 = (a - a_{GaN})/a_{GaN}$ is the in-plane strain. The piezoelectric polarization for alloys are typically obtained by a linear interpolation of the moduli of the binary constituents [3] (Vegard's law). For non-linear polarization coefficients, refer to the chapter by Ambacher and Cimalla. It is worthwhile to note that the piezoelectric polarization along the $[111]$ direction in the zinc-blende III-V semiconductors has been studied in some detail [4]. However, the low piezoelectric constants of the material system makes polarization an unattractive tool for band-engineering in such materials.

In addition to the large piezoelectric polarization, the III-V nitride semiconductors also exhibit strong spontaneous polarization in the wurtzite phase. The magnitude of spontaneous polarization in the III-V nitrides is shown in Figure 3 (and listed in Table 1). Also shown in the figure for comparison is the spontaneous polarization in two ferroelectric materials that are strongly insulating. The magnitude of spontaneous polarization in the ferroelectrics is an order of magnitude larger than the III-V nitrides.

Figure 4 illustrates the microscopic picture of polarization in a slab of strain-free GaN. Each unit cell can be thought to contain a charge dipole that is formed due to the spatial separation of the centroids of the negative charge due to the electron clouds and the positive charge due to the nuclei. The dipoles in every layer of unit cells neutralize each other in the bulk of the semiconductor, but form sheet charges[4] $\pm\sigma_\pi$ on the surfaces. A free Ga-face surface develops a negative sheet charge, and a positive sheet charge forms on the N-face surface. The net electric field in the bulk of the semiconductor is therefore not zero, but $F_\pi = \sigma_\pi/\epsilon$ by Gauss's law. Note that

[4] The subscript π indicates that the quantity is due to polarization.

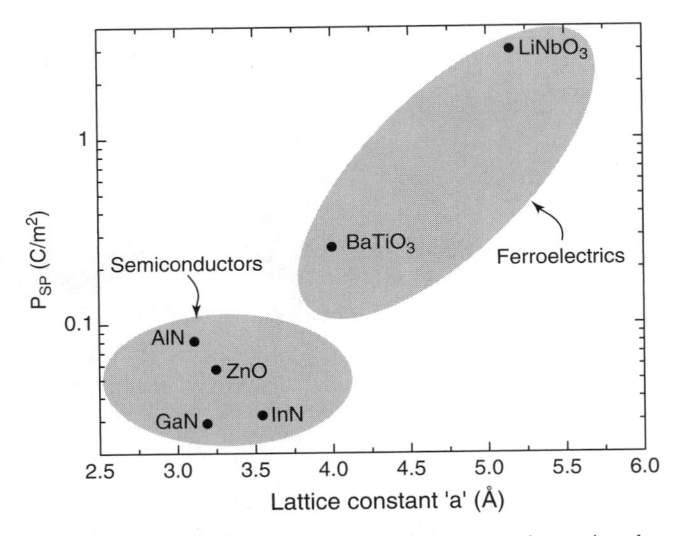

Fig. 3 Spontaneous polarization constants of ferroelectrics and wurtzite semiconductors.

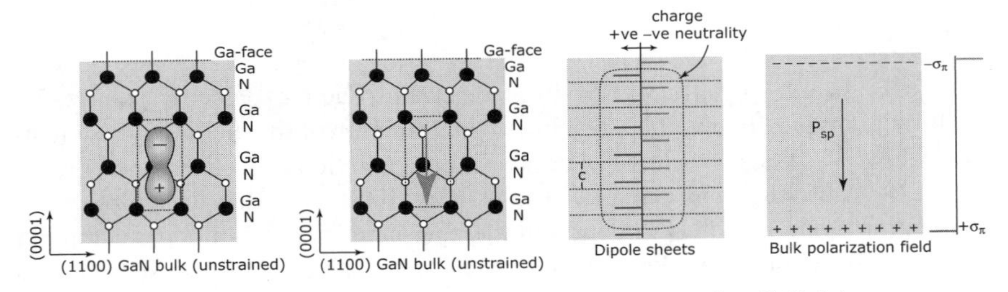

Fig. 4 Microscopic picture of spontaneous polarization in a freestanding GaN slab.

this is the field if the bound sheet charge dipole that forms on the free surfaces is not neutralized by the flow of mobile charges.

To obtain an estimate for the magnitude of spontaneous polarization, the surface-polarization charge density for GaN is $P_{sp}/e \approx 1.8 \times 10^{13}$/cm^2, and the same charge is 5×10^{13}/cm^2 for AlN. These charges originate due to minute displacements of the electron clouds and the positively charged nuclei of the constituent atoms of the semiconductor crystal, and are large enough to affect the electrical properties of the material drastically at surfaces and interfaces. Since the atomic sheet density in semiconductors is $\sim 10^{15}$/cm^2, roughly one out of 100 atoms can be pictured to be contributing to the polarization charge. In comparison, every unit cell in a ferroelectric material contributes to the polarization charge.

An estimate of the electric field arising from the polarization sheet charges gives $F_\pi \approx 1 - 10$ MV/cm. Such large fields cause a major redistribution of charges in the system. That is the underlying theme of this chapter (and this book). What is interesting is that these large fields are frozen into the system by means of the crystal structure and they do have striking effects on the band diagrams, charge

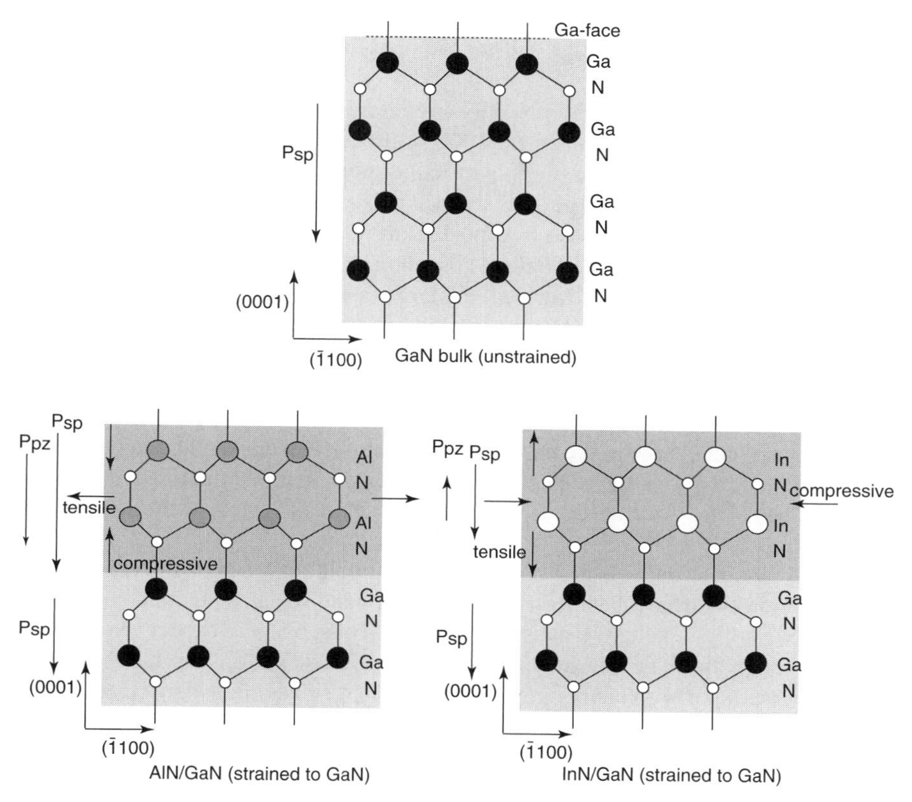

Fig. 5 Crystal structures and polarization fields in GaN, AlN on GaN and InN on GaN. GaN is relaxed, and AlN and InN are coherently strained leading to a piezoelectric component of the polarization.

accumulation and depletion, and in general, the steady-state picture is much different from the flat-band situations one is accustomed to in traditional heterostructure device physics.

Figure 5 depicts the situation at heterojunctions of GaN with AlN and InN. Traditionally the III-V nitrides are grown Ga-face up (i.e., if one cuts a plane containing vertical bonds, the surface is Ga atoms) along the [0001] direction of the wurtzite structure. Our studies will be confined to strain-relaxed GaN, and coherently strained alloys grown on it. In such cases, the spontaneous polarization field in GaN points in the direction depicted. The lower left figure depicts a layer of AlN grown pseudomorphically on GaN. AlN has a smaller in-plane lattice constant than GaN, and thus will have an in-plane tensile strain and a compressive strain in the [0001] direction, and $\epsilon_1 = (a_{AlN} - a_{GaN})/a_{GaN} < 0$. This causes a piezoelectric polarization field in the AlN layer to add to the spontaneous polarization field. The bound polarization sheet charge density at the GaN-AlN interface is thus given by the difference

$$\frac{\sigma_\pi}{e} = \frac{1}{e}(\mathbf{P}_{AlN} - \mathbf{P}_{GaN}) \cdot \mathbf{n} = 6.4 \times 10^{13}/cm^2. \quad (6)$$

For the case of InN grown on GaN, InN is compressively strained in the plane of growth and has a tensile strain in the [0001] direction, and thus the piezoelectric component of polarization points opposite to the spontaneous polarization direction. The spontaneous polarization of InN is very close to that of GaN, and the polarization sheet charge at the interface is almost entirely piezoelectric. The small lattice mismatch between AlN and GaN allows psuedomorphic layers of AlN thinner than $\sim$5 nm to be grown on GaN. However, the large lattice mismatch between InN and GaN does not allow for thin InN pseudomorphic layers.

In a compositionally uniform, unstrained bulk polar material that possesses spontaneous polarization, the total polarization $\mathbf{P}$ is constant in space and the volume-density of polarization charge vanishes (mathematically, $\nabla \cdot \mathbf{P} = 0$ but $\mathbf{P} \neq 0$). Due to the absence of net charge in the bulk, the macroscopic electric field has a magnitude $\mathbf{F}_\pi = -\mathbf{P}/\epsilon$, where ϵ is the dielectric constant of the semiconductor. This electric field can be related to the surface charges $\sigma_\pi = \mathbf{P} \cdot \mathbf{n}$ of a slab of the material by Gauss' law.

Metals cannot sustain electric fields inside their bulk; they screen the field by flow of conduction electrons. A dielectric on the other hand sustains the field completely till the field is high enough to cause breakdown. Thus, one can expect the response of a semiconductor to be intermediate between these cases. Consider the material to be a semiconductor with energy gap E_g. Let $F_\pi = \sigma_\pi/\epsilon$ denote the unscreened electric field created due to the spontaneous polarization-induced surface charges σ_π. The valence band can supply electrons if required to satisfy electrostatic boundary conditions. Therefore, the energy gap E_g sets a limit on the thickness of the slab beyond which the polarization-dipole would be neutralized by the flow of electrons from the valence band; this critical thickness is given by $d_{cr} = E_g/F_\pi$. This phenomenon can be called the 'closing of the gap' due to polarization. However, if there are surface states in the gap, the critical thickness would be reached at a smaller total thickness of the layer. The situation is depicted in Figure 6. The flow of charges creates a neutralizing dipole to the surface polarization dipole, and in the process flattens the bands. The polarization-induced neutralizing dipole charges should be mobile, and respond to lateral electric fields for a perfect crystal. However, if there are deep electronic (surface) states in the bandgap, these polarization-induced carriers can be bound (or localized) by strong Coulombic forces, and in such a case will have far less contribution to electrical conductivity. They can be extracted out of the traps into the bands by either optical and thermal excitation, and can contribute to a photo-excited or thermally-excited hopping conductivity. A careful experimental study of the charges formed entirely due to polarization at free *surfaces* of III-V nitride crystals has not been performed till date.

For Ga-face GaN, the critical thickness for the closing of the gap is $d_{cr} \approx 10$ nm. Most experimental GaN samples are typically thicker, and thus the polarization dipole is neutralized. If GaN is grown on another material (typically sapphire or SiC), the neutralization condition still holds, though the critical thickness would change. The interface charges should be strongly localized at the defect states in the

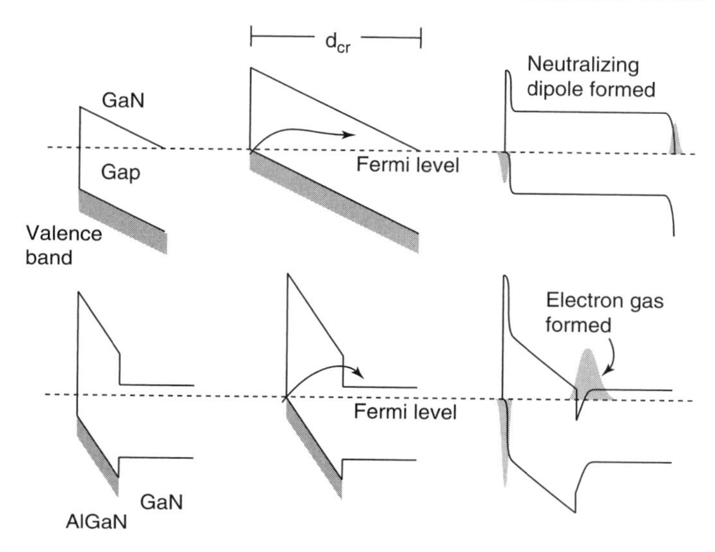

Fig. 6 Closing of the energy gap due to polarization assisted band-bending in GaN and AlGaN/GaN structures.

nucleation layer of real GaN crystals. If not, they appear as a degenerate conductive layer at the growth interface. This is harmful for device applications, since a conductive polarization-induced sheet of charge will be a path of leakage for lateral current flow. Such degenerate carrier layers have been reported in GaN grown on sapphire and have prompted the 'two-layer' model [5] for bulk conductivity. Hsu et.al. have used scanning-probe microscopy techniques to investigate the properties of this degenerate layer [6]. Look et. al. have studied the transport properties of the degenerate layer at the interface [5] by Hall effect, and it has also been observed by capacitance-voltage profiling of free carriers. However, the carrier concentrations observed have been much higher than that demanded by the spontaneous polarization alone. Various explanations including diffusion of shallow dopants into the GaN have been suggested. However, the recent development of Fe-doped semi-insulating GaN [7] by introduction of extra states in the gap is effective in rendering the degenerate electron gas immobile, quite possibly because the states are highly localized in space.

For extremely thin epitaxial layers, the situation is quite different. Large electric fields can be sustained over short distances without fear of the polarization-induced closing of the gap, thus causing large band-bending. In Figure 6, the process for a thin peudomorphic AlGaN layer grown on GaN is depicted. The spontaneous and piezoelectric polarization add in the AlGaN layer, and there exists a small critical thickness beyond which the polarization dipole at the AlGaN surface and the AlGaN/GaN interface is neutralized by the formation of a two-dimensional electron gas at the AlGaN/GaN interface. The existence of surface states in the AlGaN energy gap lowers the critical length of AlGaN; this is analyzed by Ibbetson et al. [8].

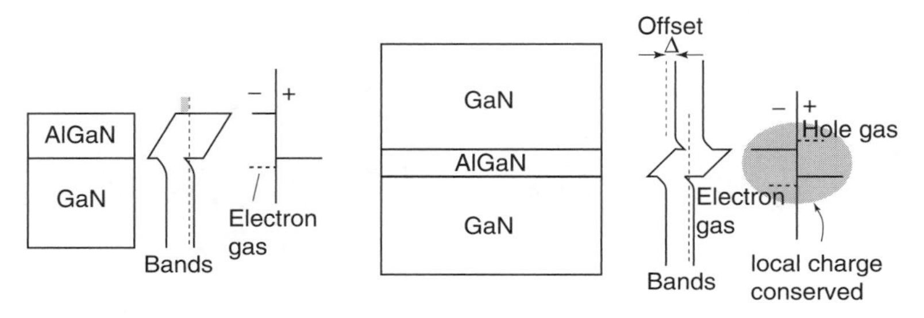

Fig. 7 Band diagrams for AlGaN/GaN and GaN/AlGaN/GaN heterostructures showing polarization induced two-dimensional electron and hole gases and effective band offsets.

The discontinuity of slopes of the band diagram across the heterojunction is equal to the fixed polarization sheet charge at the interface. Unlike the polarization-induced electron gas in bulk GaN, the polarization induced two-dimensional electron gas (2DEG) at the AlGaN/GaN interface exhibits a high mobility owing to the high purity of the interface that is achieved using epitaxial growth techniques. Thus, this constitutes an attractive technique for devices, analogous to modulation doping in AlGaAs/GaAs heterostructures. Here, the doping is achieved from surface states, the transfer of electrons facilitated by the large electric field in the AlGaN layer. The field in the AlGaN layer decreases as the AlGaN layer thickness is increased, and approaches zero as the charge in the 2DEG equals the fixed polarization sheet charge at the interface. Thus the charge available at the 2DEG can be modulated by changing the AlGaN thickness (or the Al-composition of the AlGaN layer), and does not require the introduction of shallow donors. This has obvious advantages for the transport properties of the 2DEG, as described in other chapters of this book.

Figure 7 depicts an AlGaN/GaN heterostructure and another structure with a thick GaN cap layer grown on AlGaN. For the second structure, the polarization dipole has to be neutralized *locally* to prevent the closing of the gap and maintain charge and energy equilibrium (i.e., satisfy Poisson and Schrodinger equations self-consistently). The outcome is a mobile 2DEG at the lower interface, and a hole accumulation layer at the top interface. The hole gas at the top interface could be quantized, depending on the strength of the electric field at that interface[5]. The coupling between the electron and hole gases and their density can be tuned by changing the thickness of the AlGaN layer. Note that there exists a critical length for the two carrier gases to form; this critical distance is crudely estimated to be the thickness of AlGaN that will close the gap: $d_{cr} \approx E_G^{AlGaN}/F_\pi$ where F_π is the electric field in the AlGaN layer, and depends on the alloy composition of the AlGaN barrier. As long as the AlGaN thickness is lower than the critical length, polarization will introduce an *effective* 'staggered' band-offset $\Delta E_c = \Delta E_v = \Delta_\pi$ in the band diagram

[5] Note that electrostatics requires positive charges to form at the top interface to compensate the negative charges in the 2DEG. In the presence of a large density of donor-like trap states, a positive space charge can do the same job as a hole gas. However, the volume-density of these traps required would be unreasonably high, since the 2DEG densities are close to $10^{13}/cm^2$.

far from the AlGaN layer. The magnitude of this offset is given by $\Delta_\pi = F_\pi \times d_{AlGaN}$ which interestingly is *tunable* for constant AlGaN composition by tuning the AlGaN *thickness*. If the AlGaN layer is thin enough, it will be transparent to the flow of carriers across it by tunneling, and the situation can be said to be a polarization-induced band offset. Such an idea was investigated by Keller et al. [9] by introducing an extremely thin AlN layer in GaN. The effective band offset Δ_π decreases as the AlGaN thickness increases and the electron-hole gas pair is formed, and goes to zero as the electron-hole gas density approaches the polarization dipole. We note here that the hole gas in such a structure is yet to be experimentally observed.

Finally, instead of a sharp heterojunction, interesting phenomena occur if the junction is graded. A compositionally graded polar material has a non-vanishing divergence of polarization, since polarization is dependent on the material composition which changes spatially. For example, polarization of AlGaN is larger than GaN, and keeps increasing as the Al-composition increases. The microscopic picture is depicted in Figure 8, which can be understood in the same fashion as Figure 4. The strength of the dipole changes with the Al composition, which breaks the *local*

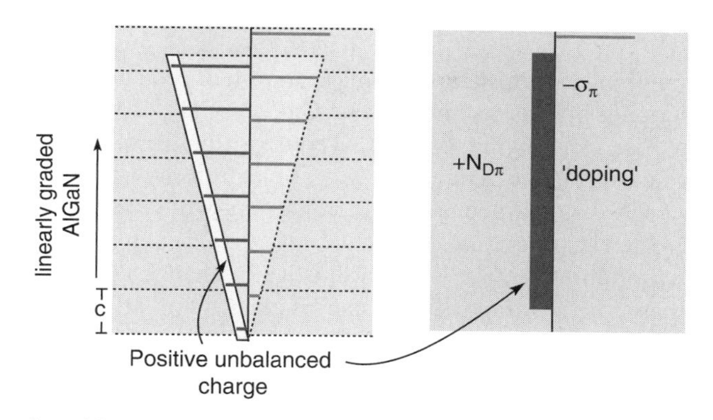

Fig. 8 Illustration of formation of net bulk charge in a layer with polarization gradient.

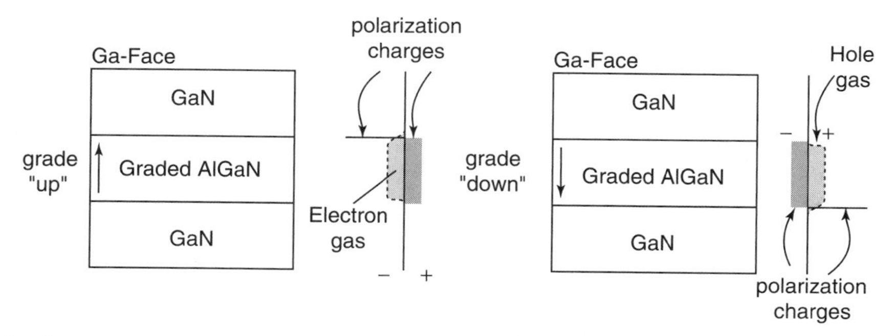

Fig. 9 Charge distributions in compositionally graded AlGaN on GaN showing the formation of three-dimensional electron and hole gases.

charge neutrality in the bulk. A bound volume charge density is obtained as a result, satisfying the *global* charge neutrality requirement $eN_{D\pi}t = \sigma_{\pi}$, where $N_{D\pi}$ is the volume charge density in cm^{-3}, and t is the thickness of the graded layer.

Compositional grading of a polar semiconductor therefore results in a bound polarization 'bulk' charge density given by $_{D\pi} = \nabla \cdot \mathbf{P}$. In Figure 9, such a situation is depicted[6]. If a graded AlGaN layer is grown on Ga-face GaN with increasing Al-composition towards the Ga-face, it results in a net positive polarization charge. The charge profile is determined by the grading scheme; a linear grading results in an approximately constant polarization charge (the electrostatics will be described later). This polarization charge is fixed (bound), and does not contribute to electrical conductivity directly. However, the electric field that is produced due to this charge is large enough to attract carriers from available sources of charge (the surface, remote dopants, etc) and a three-dimensional mobile electron slab will be formed. This technique of polarization 'bulk-doping' does not require the introduction of shallow donors or acceptors. The density of carriers generated in this manner will have no temperature dependence since they are not thermally activated from the donor/acceptor sites, and should exhibit high mobilities due to the removal of ionized impurity scattering. This is indeed observed in practice, and is presented later in this chapter.

Another important manifestation of polarization-induced bulk doping is the possibility of p-doping by this technique. For Ga-face GaN, if one compositionally grades 'down', i.e., from higher to lower Al compositions towards the Ga-face, it is possible to get mobile holes in the same way as electrons by grading up. The situation for N-face GaN is complementary - grading 'up' will create holes and grading 'down' will create electrons.

Thus, polarization effects are seen to drastically change the band-diagrams of III-V nitride based heterostructures and can be expected to affect many traditional design techniques of electronic and optical devices. We now turn our attention to those effects.

The III-V nitride semiconductors burst into the limelight with the advent of light emitting diodes (LED) and the blue laser. The effect of polarization in optical devices is particularly striking, and it offers a direct probe to the polarization field magnitudes.

In Figure 10 a typical AlGaN/GaN (or GaN/InGaN) multiple quantum-well (MQW) band-diagram is depicted. This structure forms the active region of most optoelectronic devices based on GaN. The lattice is matched to GaN, i.e., AlGaN or InGaN are coherently strained. As a result, the unscreened polarization fields cause a 'sawtooth' type of band diagram for the well and the barrier regions. The injection of free carriers into the structures causes a partial screening of the polarization field by a spatial separation of the electron and hole gases. Thus, the transition energy is red-shifted, and the oscillator-strength is reduced due to a lower electron-hole wavefunction overlap. This effect has been observed [10, 11] and remains an area of active research. In addition, the large strain in the structure causes shifts in the band-

[6] The source of the mobile charges is not shown in this figure.

edge energies, which cause further variation from the ideal, flat-band situations with fixed band edge energy in the AlGaAs/GaAs optical device structures. In general, for efficient light emitters, electronic polarization reduces the efficiency of interband optical transitions. Epitaxial growth of the nitride semiconductors in directions that have either lower polarization, or no polarization at all is being actively pursued for increasing the efficiency of light emission. The effect of polarization fields in optical devices is described in more detail in the chapter by Buttë and Grandjean in this book.

Polarization in the III-V nitrides is used for generating a 2-Dimenasional Electron Gas (2DEG) at AlGaN/GaN interfaces, which serves as the channel for the high-electron mobility transistor (HEMT). The 2DEG densities in such channels are epitaxially tunable over a large range $10^{12} - 3 \times 10^{13}/\text{cm}^2$ and typical low-field mobilities achieved at room temperature are in the range of 1500 $\text{cm}^2/\text{V}\cdot\text{s}$. We note in passing that AlGaN/GaN HEMTs out-perform the traditional AlGaAs/GaAs pHEMTs in output power due to high breakdown voltages owing to the wide bandgap, high saturation velocities, and large 2DEG densities with reasonably high mobility [12].

A problem in AlGaN/GaN HEMTs is the drop of gain at high frequencies due to the coupling of the 2DEG carriers to charges in the surface states. The surface charges are slow in their response to electric field variations, and thus the 2DEG channel of the HEMT, being an image of the surface states, does not respond to a fast electric field sweep. This problem has been attacked by passivating the surface (with a suitable dielectric like SiN or SiO_2), though it is not currently understood why it works. Another approach to this problem is by the introduction of 'surrogate' dopant layers near the surface, which take over the role of the surface states for the channel. This surrogate dopant layer, which can either be n- or p-type, was shown to work first by Jimenez et al. [13], and is actively being pursued as a possible solution to RF-dispersion at an epi-level.

It is interesting to note that the technique of achieving a polarization-induced 2DEG for the channel of a field-effect transistor was proposed much before the

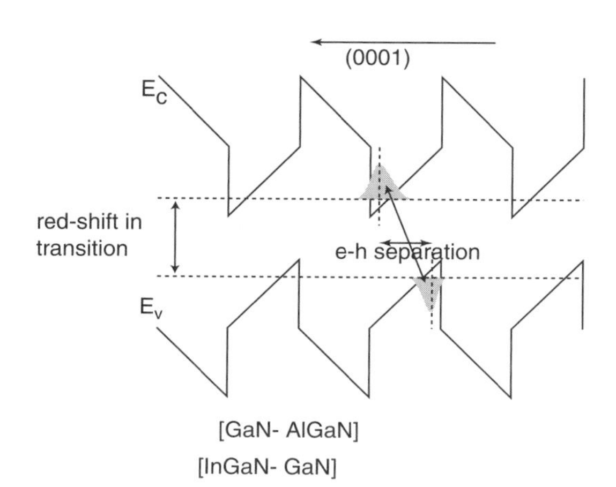

Fig. 10 Polarization induced red-shift and reduction of oscillator strength in optical devices.

advent of the III-V nitrides. Kuech et al. [14], building on the work by Shanabrook and Mailhot et al. [15], had proposed that the piezoelectric effect in the [111] direction of zinc-blende structures of InGaAs/AlInAs/InGaAs would be able to create a mobile 2DEG which would serve as a channel of the transistor. However, such a transistor was never reported in the zinc-blende material system. Almost *all* present field effect transistors of the GaN material system use the polarization-induced 2DEG as the channel.

Polarization-induced three-dimensional electron slabs offer an attractive alternative to highly doped channels. The high-conductivity channel required in traditional field effect transistors (MESFET, JFET) requires a high doping, with an associated drop in carrier mobility due to ionized impurity scattering. Ionized impurity scattering is greatly reduced in polarization-doped channels, as will be demonstrated in the polarization-doped junction FET described in later in this chapter.

While HEMTs, MESFETs and JFETs are essentially unipolar devices, the heterojunction bipolar transistor (HBT) employs both majority and minority carriers. The factor preventing the usage of the many advantages offered by GaN/AlGaN materials for HBTs is the p-type conductivity of GaN. The activation energy of holes due to the most common acceptor Mg is rather large ($E_A \sim 160$ meV). This results in highly resistive p-layers due to low efficiency of ionization of the holes into the valence band at room temperature. This problem has been attacked by using polarization as a tool by Kozodoy et al. [16]. Polarization-induced band-bending helps the generation of more holes from the deep acceptors by reducing the activation barrier. The density of such hole gases were reported to be temperature-independent, and they exhibited improved mobility. However, vertical transport is still a problem due to the valence-band barriers formed by the thick AlGaN layers. The recently acquired ability of growing coherently strained thin AlN layers [9] that are transparent to vertical flow of carriers makes it possible to further extend this idea. Asbeck et al. [17] have proposed that one can use a polarization-doped base layer instead of Mg-acceptor doping for the n-p-n HBT (described in the chapter by Yu and Asbeck in this book).

To conclude this tutorial, the phenomenon of electronic polarization in the III-V nitride semiconductors is a fascinating new tool for engineering semiconductor device structures. One can use polarization to induce free carriers (doping), and facilitate activation of deep carriers. One can create tunable heterojunction band offsets using polarization. One can convert a normal metal-AlGaN-GaN Schottky junction to a tunnelling ohmic contact *without* doping by tuning the AlGaN barrier thickness [18]. One can achieve below-gap optical transitions, and generate very high density, high mobility 2DEGs at heterojunctions. It can be safely predicted that this list of applications of polarization in III-V nitride semiconductors will expand in the future.

In this tutorial, it was described how a graded alloy of two materials of different polarizations can be used to achieve effective bulk doping. The rest of this section is devoted to experimental proof of this new form of doping. Slabs of high-mobility carriers are achieved in graded AlGaN by employing this form of doping. The samples are characterized structurally and electrically, and carrier charge profile and transport are studied. Comparison of transport properties of the three-dimensional electron slabs (3DES) with donor-doped carriers and 2DEG carriers highlight many features of the nature of carrier transport. Polarization-doped carriers exhibit better conductivity than their donor-doped counterparts; this is an attractive property for many applications. Magnetotransport analysis is performed on the 3DES and clearly resolved Shubnikov de-Haas oscillations are observed. The effective mass of electrons, quantum scattering time, and the alloy scattering potential are extracted from the magnetoresistance study. The polarization-induced 3DES is then integrated into a prototype transistor, the PolFET.

2.2 Electrostatics and Dipole-Engineering

Doping in semiconductors has been a much researched topic. The traditional shallow 'hydrogenic' doping technique is very well understood and gainfully employed. A good understanding of the role of ionized dopant atoms on carrier scattering in semiconductors led to the concept of modulation doping, which improved low temperature carrier mobilities in quantum-confined structures by many orders of magnitude [19].

Discontinuity of polarization across a sharp AlGaN/GaN heterojunction $\Delta P_{hj} = P_{tot}^{AlGaN}(x) - P_{Sp}^{GaN}$ results in a fixed polarization sheet charge of density $\sigma_\pi = \Delta P_{hj}/e$ at the interface. Compositional grading of the AlGaN/GaN heterojunction over a distance should spread the positive polarization sheet charge into a *bulk* 3D polarization background charge. The charge profile is given by the divergence of the polarization field, which changes only along the growth [0001] direction ($N_{D\pi}(z) = \nabla \cdot P = \partial P(z)/\partial z$). This fixed charge profile will depend on the nature of the grading; a linear grade results in an approximately uniform profile given by $N_{D\pi}(z) = [P(z_0) - P(0)]/z_0$. Here $P(z_0)$ is the polarization (spontaneous+piezoelectric) of $Al_xGa_{1-x}N$ at the local Al composition at $z = z_0$.

This fixed background charge attracts free carriers from remote donor states to satisfy Poisson's equation and charge neutrality. The end result of the charge rearrangements makes the polarization bulk charge act as a local donor with zero activation energy. This is illustrated schematically in Figure 11. Instead of thinking in terms of traditional ionized dopants and mobile carriers (which are *monopoles*), it is easier to think of the polarization *dipole* and the resultant neutralizing dipole formed. In Figure 11, the polarization dipole is shown in the top frame - it has a negative bound sheet charge on the surface, and an equal and opposite positive bound volume charge in the bulk. The neutralizing dipole has a high mobility bulk 3DES, and a positive sheet charge at the surface, which is typically not mobile. The 'neutralization' is not complete typically, resulting in a depletion region and a net electric field near the surface. In most Nitride heterostructures, the design criteria have to account for dipoles, and therefore, it is referred to here as 'dipole engineeering'.

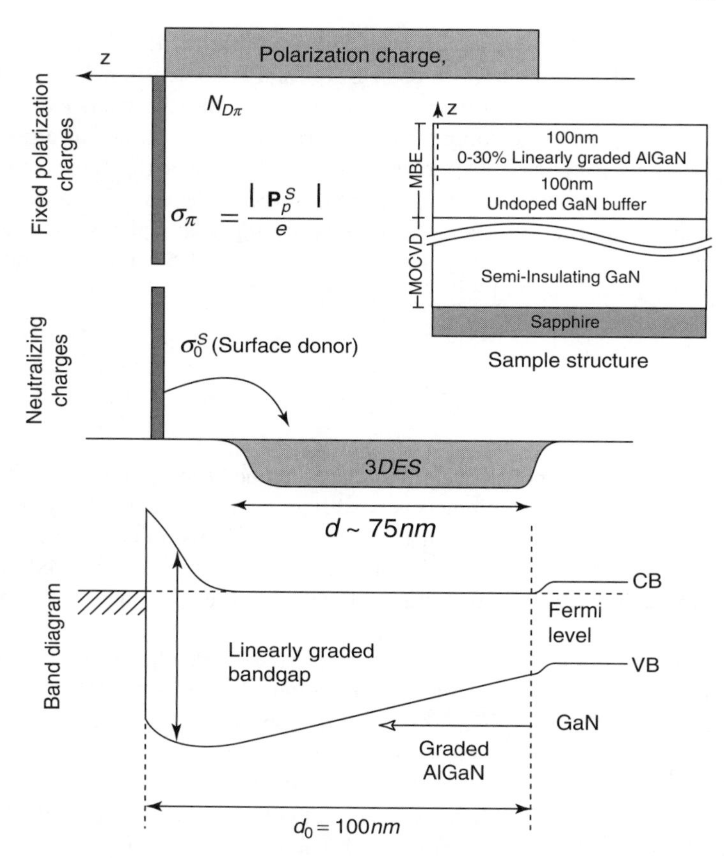

Fig. 11 Schematic of charge control showing polarization charges and formation of the 3DES. The band diagram shows depletion of the 3DES from the surface potential. Also shown is the epitaxial layer structure that generates the 3DES.

The mobile three-dimensional electron slab (3DES) thus formed should be usable just as impurity-doped carriers. However, removal of ionized impurity scattering should result in higher mobilities. Such polarization-induced electron slabs should in principle be similar to the modulation doped three-dimensional electron slabs in wide parabolically graded quantum wells in the AlGaAs/GaAs system [20, 21]. The mobile 3DES should not freeze out at low temperatures (as shallow donor-doped bulk carriers do), and should exhibit high mobilities at low temperatures.

If the grading is performed in the [0001] direction with[7] the position (z-) dependent composition $x(z)$, the polarization-doping achieved is given by

$$N_{D\pi} = \frac{\partial P[x(z)]}{\partial z}. \tag{7}$$

[7] Note that here x denotes the Al composition of AlGaN, and z represents the spatial co-ordinate in the growth direction.

The polarization (in cm^{-2}) of $Al_xGa_{1-x}N$ coherently strained on GaN is [3]

$$P[x] = [\underbrace{2x + 1.1875x^2}_{piezo} + \underbrace{3.25x}_{spontaneous}] \times 10^{13}. \tag{8}$$

The grading scheme employed will determine the doping profile. For a linear grading from GaN to $Al_{x_0}Ga_{1-x_0}N$ over a length z_0, $x(z) = x_0(z/z_0)$, and the doping thus achieved in the graded region (expressed in cm^{-3}) is

$$N_{D\pi}(x_0, z_0) = \frac{x_0}{z_0}(5.25 + 2.375x_0\frac{z}{z_0}) \times 10^{21}, \tag{9}$$

where z_0 is expressed in Å.

The interesting features of this form of doping are the control parameters. The doping density can be controlled by changing the alloy composition x_0, or the *thickness* z_0 of the graded layer. If the graded-layer thickness z_0 is smaller than the thermal de-Broglie wavelength ($\lambda_{dB} = h/\sqrt{2m^\star k_BT} \approx 170$Å) for non-degenerate carriers (or the Fermi-wavelength for degenerate carriers, $\lambda_F \approx 30$ nm for $n_{3d} = 10^{18}$ cm^{-3}), the free-electron gas is *quantized* in the z-direction and one would have a quasi-2DEG. However for wide slabs of graded regions ($z_0 \gg \lambda_{dB}, \lambda_F$), the electron gas is three-dimensional.

For a nominal thickness of 1000Å, the effective 3D doping density is shown in 12(a) as a function of the linear grading. The doping profile is very linear with the

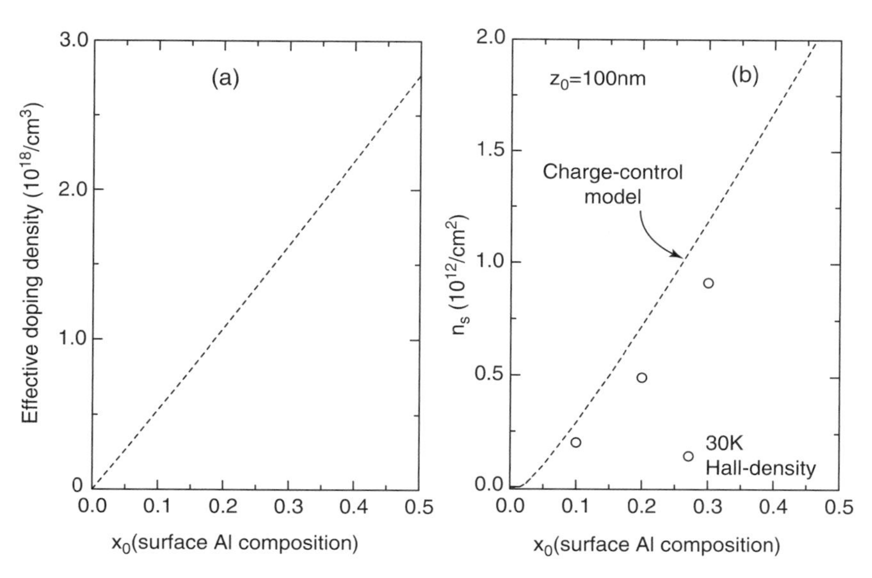

Fig. 12 (a): The calculated 3-D and sheet densities of free carriers achievable by polarization-doping. The experimentally measured Hall sheet-densities are shown in Fig(b). The thickness of the graded region is 1000 Å, and the doping density is shown as a function of the surface composition of the linearly graded region.

alloy composition - the small parabolic dependence on the surface alloy composition in Equation 9 can thus be neglected in charge-control analysis. Doing so, a surface pinning at $\Phi_S(x_0) = (1 + x_0)$ eV from the conduction band edge causes a depletion of the free carriers to a depth

$$z_{depl} = \sqrt{\frac{2\epsilon\Phi_S(x)}{eN_{D\pi}(x_0, z_0)}}. \tag{10}$$

After this depletion, the thickness of the conducting layer is $z_0 - z_{depl}$, and the total integrated sheet carrier density can be obtained from either a self-consistent solution of the Poisson equation, or estimated to be roughly equal to $N_{D\pi} \times (z_0 - z_{depl})$. The resultant *sheet* charge density that can be measured is calculated self-consistently and shown in 12(b), along with three values measured in this work. The small disagreement with theory may be attributed to the simplified model and perhaps the values of the polarization coefficients themselves. However, the principle of polarization bulk-doping is proved with this result. A description of the experiment that leads to the results in Figure 12 is now discussed.

2.3 Epitaxial Growth and Structural Properties

To verify the idea of polarization bulk-doping, five samples were grown by molecular beam epitaxy (MBE). High-resistivity semi-insulating (SI) GaN on sapphire grown by metal-organic chemical vapor deposition (MOCVD) [7] was used as the template for the growths. For all five samples, a 100nm buffer layer of undoped (Ga-face) GaN was grown by MBE, followed by a different cap layer for each. The relevant layer structures for the five samples are shown schematically in Figure 13. The top 100nm of sample 1 is bulk shallow donor doped with Si (activation energy $E_D = 20$ meV, and concentration $N_D = 10^{18}\text{cm}^{-3}$). Samples 2, 3 and 4 are linearly graded AlGaN/GaN structures for studying polarization bulk doping; they are

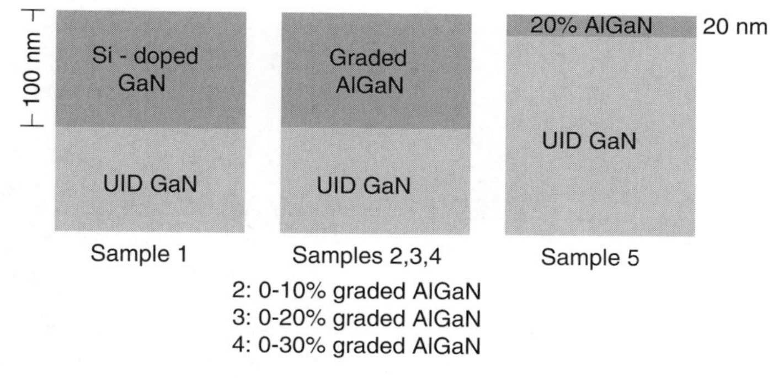

Fig. 13 The five sample structures chosen for study of polarization bulk-doping.

graded from GaN to 10%, 20% and 30% AlGaN respectively over $z_0 = 100$ nm. Sample 5 is a 20 nm $Al_{0.2}Ga_{0.8}N/GaN$ which houses a conventional 2DEG at the heterojunction. Samples 1 and 5 are control samples. A description of the growth process will now be given.

The growth regime phase diagram pioneered by Heying et al. [22] sets useful guidelines in choosing the correct growth regime for an application. The growth rate is limited by the N_2 flow rate through a EPI Unibulb Nitrogen-plasma source using a ultra-pure (99.9995% purity) source, which in turn is further purified by an inert-gas purifier. The plasma source creates the active nitrogen species that incorporates during the growth process. Gallium and aluminum are supplied from conventional effusion cell sources. Typical plasma conversion efficiencies are extremely low, and the growth rate is limited by the rate of Nitrogen flow. The Ga flux is thus used to control the material properties. As Heying et al. [22] have demonstrated, the Ga-rich regime of growth where excess Ga rides the surface during growth results in good transport characteristics, owing to good crystalline quality. Since our chief interest lies in the study of charge profiles and transport characteristics for electronic device applications, the metal(Ga)-rich regime is well suited for this purpose. This regime is adopted throughout for the MBE-grown samples.

A crucial ingredient in the growth of graded AlGaN layers is the precise control of the Al flux so that the aluminum composition changes in a controlled fashion with depth. The dependence of Al composition in the alloy on the aluminum flux was mapped out with X-Ray diffraction analysis. The dependence of flux on the cell temperature is Arrhenius-type owing to the activation energy for evaporating Al from the effusion cell. The alloy composition was related to the cell temperature, for which a direct computer control is available. A computer program was then used for changing the Al-cell temperature quasi-continuously for achieving the grading desired. Since the cell temperature has a finite settling time, the discrete nature of changes in cell temperature is smeared out, helping the alloy composition to be more homogenous.

Thus the $Al_xGa_{1-x}N$ alloy formed is essentially an *analog* alloy. By an analog alloy, it is implied that certain Ga atom sites are occupied by Al atoms in a way that on at length scales of the lattice constant, the metal site occupation is random, but on the scale of 10s of nm or more, the statistical average composition of Al and Ga atoms is in the ratio $x : (1-x)$.

The growth temperature is $T_{Gr} = 720 - 750$ Celsius, and the growth pressure (determined by the N_2 carrier gas flow) is 10^{-5} torr. All samples typically exhibit small Ga-droplets at the surface for the Ga-rich growth regime. The Ga flux used is in the range of $(1 - 1.2) \times 10^{-6}$ torr to maintain the Ga-rich regime.

Triple-crystal X-Ray diffraction data around the GaN (0002) peak of samples 1-4 is shown in Figure 14. The data points match very well with the theoretical solid curves calculated using dynamic-diffraction theory [23], reflecting the high degree of control of Al composition and growth rate in MBE. Atomic force microscopy (AFM) of the sample surfaces revealed step-flow growth and pseudomorphically strained graded AlGaN surfaces without relaxation.

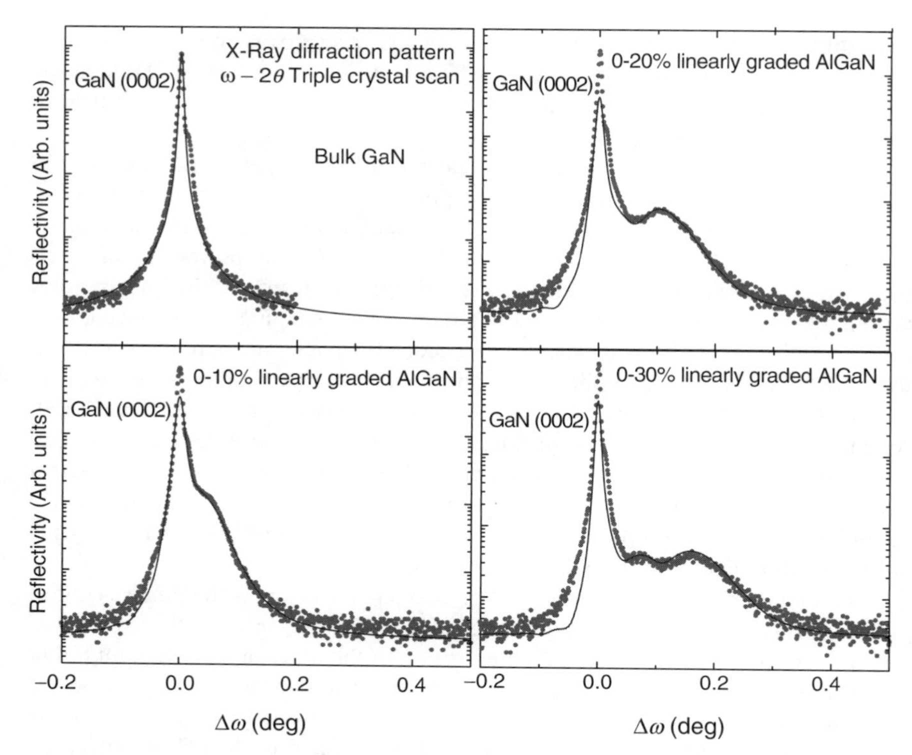

Fig. 14 X-Ray characterization of Samples 1-4 : The solid lines are calculated and the dots are measured.

Secondary Ion Mass Spectrometry (SIMS) was performed on an extra sample specifically grown for that purpose. The SIMS sample had a 20 nm GaN cap to prevent errors in SIMS profiling near the surface. Figure 15 shows the aluminum composition profile in the graded AlGaN layer. As can be seen from the profile, the linearity of Al composition in the graded layer is very accurately controlled. SIMS also revealed background oxygen concentration in the MBE GaN layer to be identical to the underlying MOCVD layer accompanied with a small increase in the AlGaN layers. Any background oxygen (which acts as a shallow donor in (Al)GaN) may provide a small amount of thermally activated carriers which can be frozen out at low temperatures.

2.4 Electronic Properties

Samples were also grown by metal-organic chemical vapor deposition (MOCVD) and were identical (surface, electrically and structurally) to the MBE grown samples, proving the robustness of the technique of polarization doping. To verify the

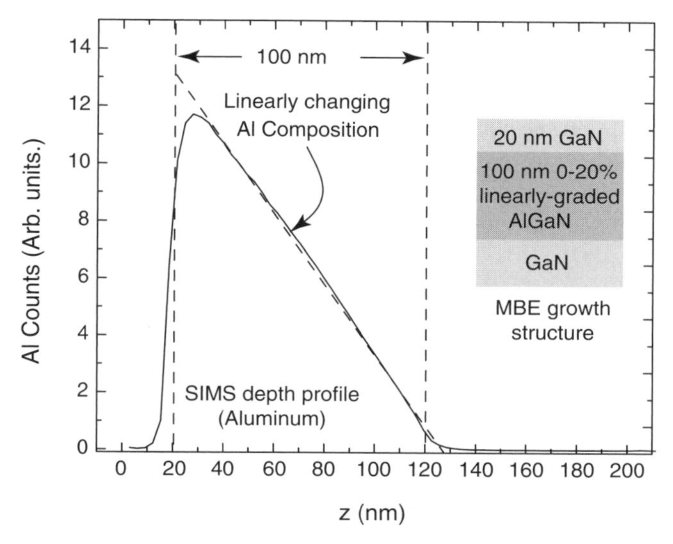

Fig. 15 SIMS profile for Aluminum composition of graded layer.

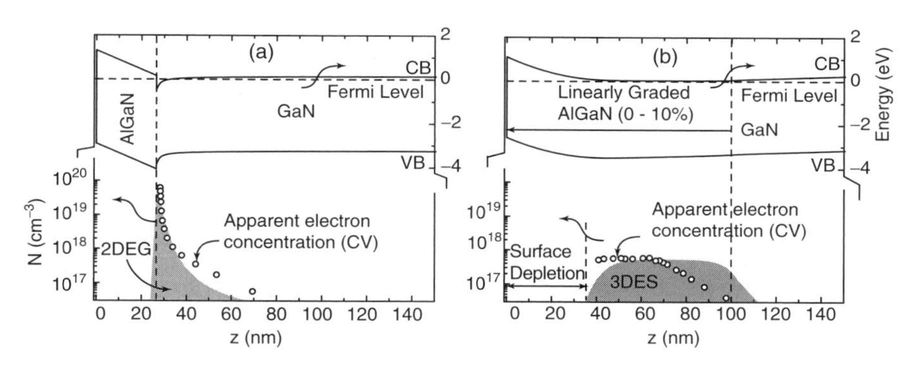

Fig. 16 Capacitance-voltage charge profile of a 2DEG and a 3DES.

charge distribution of the 3DES, two samples were grown separately by MOCVD for capacitance-voltage ($C - V$) profiling. The first sample is a 26 nm $Al_{0.3}Ga_{0.7}N$/GaN heterostructure with a 2DEG at the heterojunction, and the second is a graded AlGaN structure, where the AlGaN is graded from 0-10% over a thickness of 100 nm (similar to MBE Sample 2). Poisson and Schrödinger equations were solved self-consistently [24] to get the band diagrams and the charge profiles for the two situations. Figure 16 shows the calculated band diagram and the calculated real (zero gate bias) charge profiles for both structures in shaded gray. Polarization coefficients from [25] were used to simulate the fixed charges. Also shown in the figure are the apparent charge profiles (circles) extracted from a raw $C - V$ measurement. The apparent carrier profiles in 16(a) and (b) prove that the 2DEG at the heterojunction has *indeed* been spread out to form a 3DES as a result of the grading. The surface barrier height-induced band bending causes a partial depletion of the 3DES.

Thus, the characterization techniques of X-Ray, SIMS, AFM, and C-V profiling strongly support the claim of polarization bulk-doping occurring in the graded AlGaN layers. The transport properties are the most important since they determine the usefulness of the technique as compared to conventional doping techniques, and is the topic of the next section.

2.5 Transport Properties of Polarization-induced 3D Electron Slabs

Temperature-dependent ($T = 20 - 300$K) Hall measurements were performed on all the five MBE grown samples. Table 2 shows room-temperature and 30K Hall measurement data for all five samples. The table includes the free carrier density in bulk GaN and polarization-induced 3DES and 2DEG densities calculated by solving Schrödinger and Poisson equations self consistently for samples 2-5. The room temperature sheet conductivity $\sigma = qn\mu$ is also shown.

Temperature-dependent carrier densities and mobilities for samples 1, 4, and 5 are plotted in Figure 17 for comparison. Carriers in the 0-30% graded AlGaN sample mimic the transport characteristics of modulation doped 2DEGs and 3DESs characterized by a lack of activation energy, leading to a temperature independent carrier density. Carriers in the bulk donor-doped sample show the characteristic freeze-out associated with the hydrogenic shallow donor nature of Si in bulk GaN. A fit to theoretical dopant activation yielded an activation energy $E_D = 20$ meV with a doping density (fixed by the Si flux in MBE) $N_D = 10^{18}\text{cm}^{-3}$. The activation energy of Si closely matches that reported by Gotz et al [26]. 2DEG carrier mobilities (Sample 5) are higher than the shallow donor doped and polarization doped carriers both at room temperature and low temperatures.

Of special interest to device engineers is the room-temperature mobility, and especially the conductivity $\sigma = en\mu$. From Table 2, the room-temperature charge-mobility product of the polarization doped 3DES (Sample 4) is more than *double* of that of the comparable donor-doped sample (Sample 1). Furthermore, the trend

Table 2 Samples for Polarization doping experiment

Sample	Theory	n_s (cm^{-2})		μ (cm^2/V·s)		300K Conductivity
		30K	300K	30K	300K	($10^{-4}\Omega^{-1}$)
1	-	$7.3\cdot10^{11}$	$7.0\cdot10^{12}$	139	329	2.3
2	$2.5\cdot10^{12}$	$2.0\cdot10^{12}$	$1.7\cdot10^{12}$	1441	386	0.7
3	$5.8\cdot10^{12}$	$4.9\cdot10^{12}$	$7.8\cdot10^{12}$	2556	598	4.7
4	$9.0\cdot10^{12}$	$9.1\cdot10^{12}$	$8.9\cdot10^{12}$	2605	715	6.4
5	$7.7\cdot10^{12}$	$7.7\cdot10^{12}$	$7.8\cdot10^{12}$	5644	1206	9.4

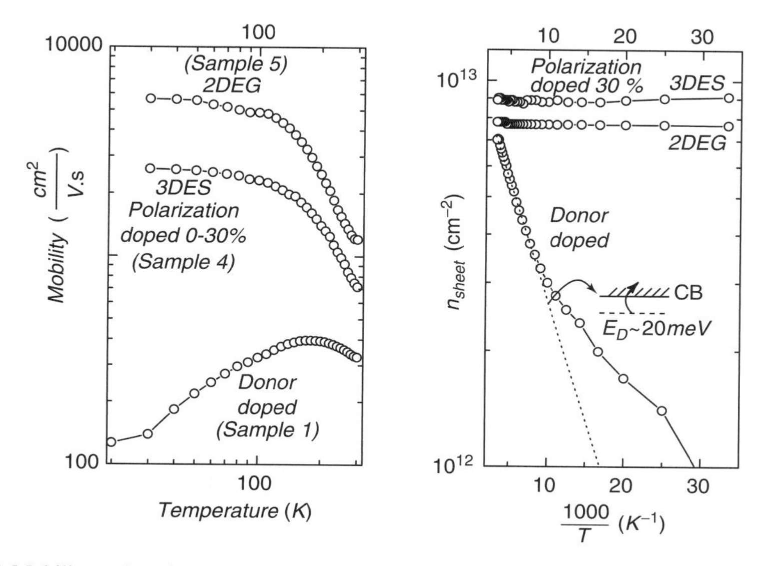

Fig. 17 Mobility and carrier concentrations of Samples 1, 4 and 5. The polarization-doped carriers have a temperature dependence of mobility that is intermediate between the donor-doped carriers and two-dimensional electron gas. The carrier concentration for the polarization-doped structure does not change with temperature.

with increasing alloy composition suggests that the conductivity *increases* with increasing carrier density (achieved by either grading to higher aluminum composition for the same thickness, or decreasing the thickness for same grading composition). This trend is very useful for the design of high conductivity layers required in many device structures.

The point of interest here is the more than an order of magnitude improvement of carrier mobility at low temperatures for the polarization doped 3DESs over comparable donor doped samples. In donor-doped GaN, thermally activated carriers freeze out at low temperatures, leading to electrons with lower energy and less effective screening. This causes severe ionized impurity scattering, lowering the mobility. However, the removal of ionized impurity scattering in the polarization-doped structures, aided by the *complete lack* of carrier freezeout at low temperatures results in much improved mobilities. Alloy disorder scattering is expected to be a strong candidate for limiting low-temperature mobility since the 3DES is completely housed in a linearly graded disordered alloy potential. This motivates the study of scattering mechanisms for such 3DES structures.

To investigate the effect of grading over different thicknesses, a series of samples were grown graded from GaN $\rightarrow$ $Al_{0.3}Ga_{0.7}N$ over thicknesses of 30, 50, and 100 nm. The evolution of the carrier distribution from a sharply peaked $\delta-$function like sheet charge in a 2DEG to bulk-like uniform density 3D slabs (3DES) density is shown in Figure 18. The band diagrams and the carrier concentrations were calculated self-consistently by solving Poisson's equation. The integrated sheet

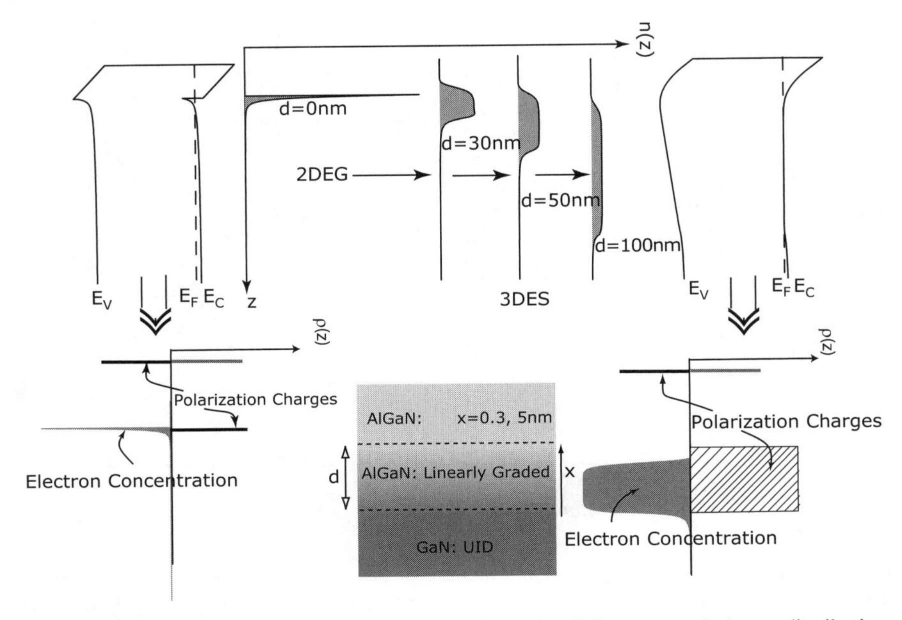

Fig. 18 Sample structures and calculated self-consistent band diagrams and charge distributions in polarization-doped graded AlGaN layers.

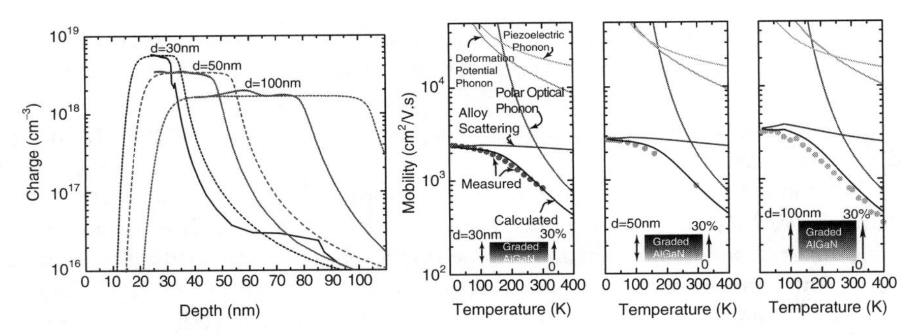

Fig. 19 (Left: Experimental carrier density profiles in graded AlGaN layers measured by C-V, and (Right): Theoretical and experimental temperature dependent electron mobility data, showing individual scattering mechanisms.

density $n_s = \int n(z)dz$ is the same for all structures. Figure 19 (left) shows the experimentally measured carrier density profiles in the three graded AlGaN samples. The carrier density and the spatial extent of the mobile electrons are evidently tuned by polarization-induced doping. The three other plots in Figure 19 show the experimentally measured temperature dependent carrier mobility in the polarization-doped 3DES, along with theoretical calculations (see [27], and [28]).

The total temperature-dependent mobility for the three graded AlGaN samples is calculated considering all important scattering mechanisms. Figure 19 shows the theoretical fits with the individual scattering mechanisms to the experimentally measured mobility. At high temperatures close to 300K, polar optical phonon scattering expectedly plays a major role in limiting the mobility. However, over the *entire* range of temperatures, alloy-disorder scattering is rather strong.

Alloy scattering limited electron mobility for an arbitrary degeneracy of three-dimensional carriers is given by

$$\mu_{alloy}(x) = \frac{2e\hbar}{3\pi m^{\star} V_0^2 \Omega(x) x(1-x)} \frac{k_B T}{n_{3d}} \ln(1+e^{\zeta}), \tag{11}$$

where x is the alloy composition, $\zeta = E_F/k_B T$, and V_0 is the alloy-scattering potential. Since the 3DES is in a *graded* alloy where x changes linearly the growth direction ($x[z] = x_0(z/z_0)$), the mobility is found by a spatial average ($\langle\mu\rangle^{-1} = \int dz \mu^{-1}(x[z]) / \int dz$). The alloy scattering potential needed to obtain a fit to the experimental data is $V_0 = 1.8$eV, which is smaller than the conduction band offset between AlN and GaN ($\Delta E_c = 2.1$eV). Other scattering mechanisms are found to be weak. Ionized impurity scattering is much weaker than traditional bulk-doped structures, and is frozen out at low temperatures. Thus, this constitutes a *measurement* of the alloy-scattering potential for AlGaN. It is a relatively clean measurement, since all other scattering mechanisms are effectively suppressed at low temperatures. Alloy scattering is *isotropic* since the alloy scattering potential is of a short-range nature. The angular dependence of the dominant scattering mechanism can be probed if one has access to the quantum scattering time. This is possible by measuring magnetoresistance oscillations, which the subject of the next subsection.

The 3DES generated by distributed polarization undergoes the traditional scattering mechanisms that are known to limit electron mobility - optical and acoustic phonon scattering, alloy scattering, ionized impurity scattering, and charged dislocation scattering. However, for graded polar semiconductors, a new scattering mechanism comes into play. This scattering mechanism originates from the coupling of alloy disorder and the dependence of polarization on the local alloy composition. The 'polarization disorder' is depicted schematically in figure 20.

Alloy disorder scattering originates from the random atomic potential variations in the crystal lattice of an analog alloy. For example, if one considers AlGaN, some Ga atom sites are occupied by Al atoms, and the local electronic potential is different from that in a perfect periodic potential. These fluctuations lead to alloy disorder scattering. However, in a graded polar alloy, the replacement of a Ga atom by an Al atom also changes the local electric dipole moment, since AlN has a larger polarization than GaN. This is shown schematically in Figure 20 for a polar alloy $A_xB_{1-x}C$, which is an alloy of binary materials AC and BC. Let's say AC is more polar than BC (AC could be AlN and BC could be GaN). Then, the electric dipole moment in a unit cell of AC will be larger than that in a cell of BC. If one considers the random spatial variation of the dipole moments, the electron will scatter from the non-periodic electronic potential resulting from them. Such a form of scattering has

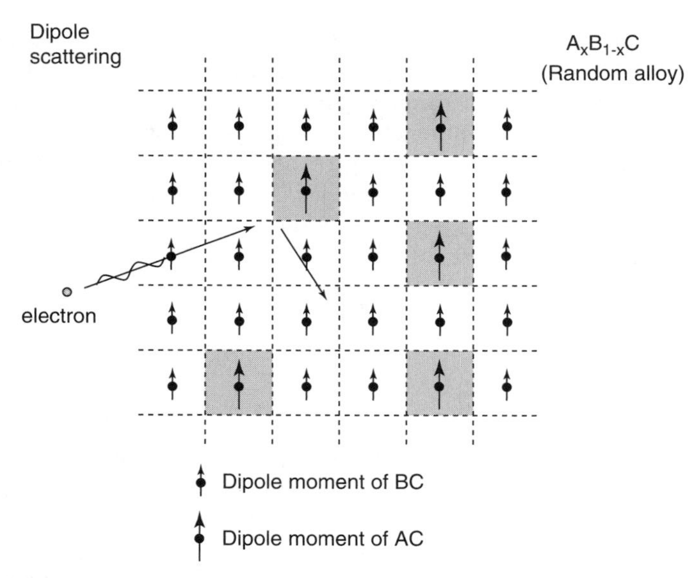

Fig. 20 Dipole scattering due to coupling of alloy disorder with electronic polarization.

been studied recently [29] for polarization-induced 2DEGs, and extended to 3DES in semiconductors and ferroelectric alloys in [30]. In addition, due to the coupling of interface roughness and the strong polarization, there is another form of dipole scattering that occurs at sharp Al(Ga)N/GaN heterojunctions [31].

The dipole-scattering limited electron mobility for non-degenerate carrier distributions was derived in [30] to be

$$\mu_{dipole} \approx \frac{8}{3}\left(\frac{e\hbar}{m^{\star}k_BT}\right)\left(\frac{N_C}{n_0}\right)\left(\frac{a_0}{d_0}\right)^2, \tag{12}$$

where N_C is the conduction band effective density of states, n_0 the mobile electron concentration, a_0 is the effective Bohr radius in the semiconductor, and $e \times d_0$ is the dipole moment due to polarization non-periodicity. The mobility reduces as the *square* of the dipole moment, and therefore this form of scattering is especially severe in alloys of strongly polar crystals. The most severe effect is expected to be in ferroelectrics, and such scattering starts affecting high Al-composition AlGaN alloy layers.

Though dipole scattering is not the dominant scattering mechanism at room temperature for most AlGaN alloys, this form of scattering is expected to be severe in ferroelectric alloys, especially those that have bandgaps in the semiconducting region. Ferroelectric alloys such as $BaSrTiO_3$ (BST) can be expected to suffer from dipole scattering. This situation is depicted in Figure 21, where the effect of dipole scattering is compared for weakly and strongly polar semiconductors, and ferroelectrics. The comparison is made for room-temperature mobilities at three

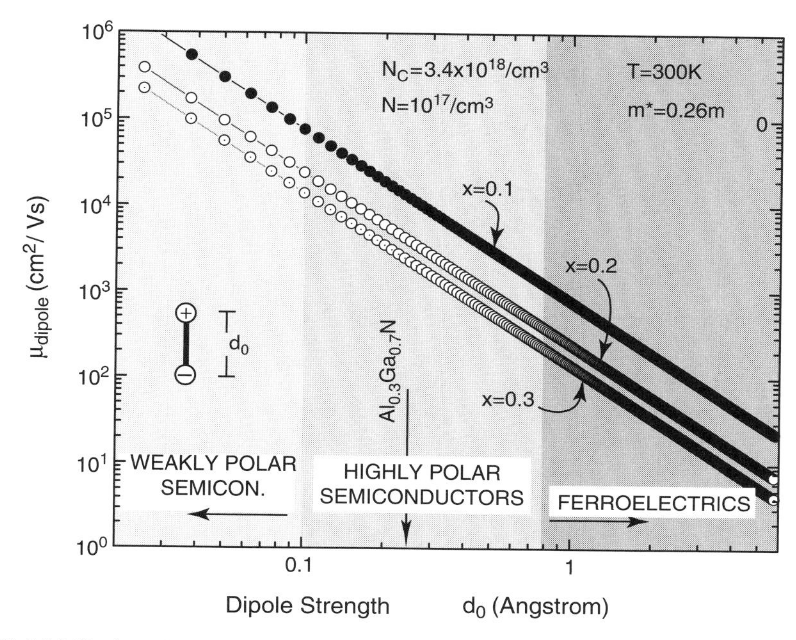

Fig. 21 Mobility in various classes of alloys due to dipole scattering.

alloy compositions ($x = 0.1, 0.2$, & 0.3), plotted against the magnitude of polarization which is captured by the length of a dipole in every unit cell. It can be seen that ferroelectric alloys will have mobilities in the range of ~ 10 cm^2/V.s. The dipole length in 30% AlGaN is shown in the figure. This new form of scattering will have to be considered in the future, as highly polar semiconducting and ferroelectric alloys layers find more widespread usage in microwave (and other) devices.

2.6 Quantum Magnetotransport Properties

Magnetic quantum effects on the carrier transport properties at low temperatures provide a valuable probe for determining various properties of a semiconductor [32, 33]. The carrier effective mass, quantum scattering time, and dominating scattering processes can be extracted from quantum magnetoresistance oscillations.

In the presence of a quantizing magnetic field, the unperturbed 3-dimensional density of states (DOS) $g_{3d}(\varepsilon)$ undergoes Landau quantization to quasi 1-dimensional DOS, and acquires an oscillatory component. 22 is a schematic of this process. Dingle [34] showed that inclusion of collisional broadening removes the divergence at the bottom of each 1-dimensional subband and damps the DOS oscillation amplitudes exponentially in $1/B$.

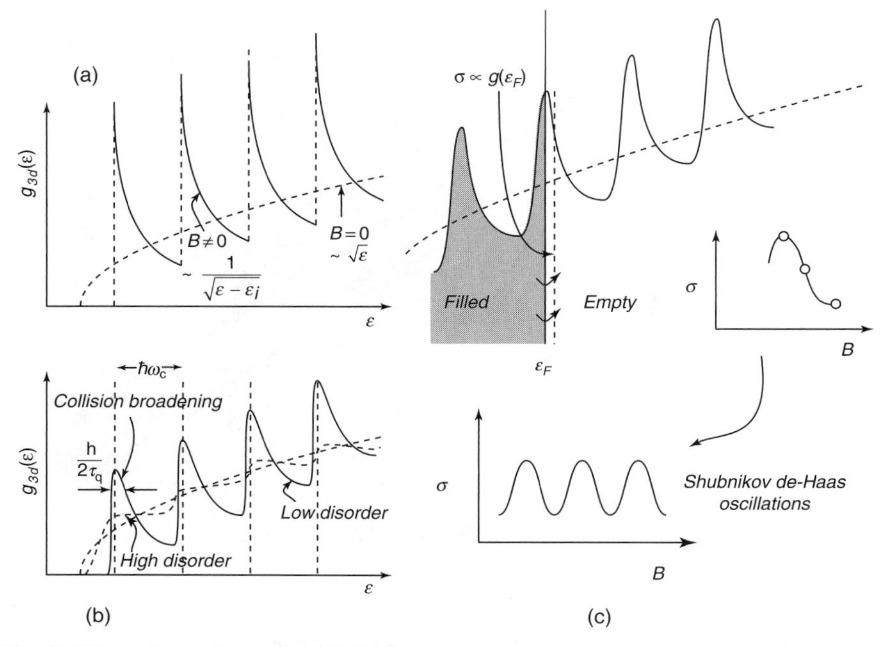

Fig. 22 Illustration of the origin of Shubnikov de-Haas magnetoresistance oscillation.

As is well known from the theory of the magnetic quantum effects, this oscillation of the density of states manifests in oscillations of both the diamagnetic susceptibility (manifesting in the de-Haas Van Alphen effect) and transport coefficients (manifesting in the Shubnikov de-Haas or SdH oscillations). In particular, the transverse ($\mathbf{B} \perp \mathbf{E}$) magnetoresistance R_{xx} exhibits oscillations in $1/B$. From Figure 22, this process can be pictured as an oscillation of the DOS available for carriers at the Fermi-level to scattering as the magnetic field changes. Kubo [35] derived the transverse magnetoresistance at high magnetic fields using the density-matrix approach to solve the transport problem.

The expression for R_{xx} can be decomposed into a background part and an oscillatory contribution [32] $R_{xx} = R_{xx}^{Back} + \Delta R_{xx}^{osc}$. The background term is attributed to sample inhomogeneities and disorder. The amplitude (A) of the oscillatory component can be cast in a form [33] that is simple to use and captures the physical processes reflected in the measured magnetoresistance -

$$\Delta R_{xx}^{osc} \propto A = \underbrace{\frac{\chi}{\sinh\chi}}_{D_t(T)} \times \underbrace{\exp(-\frac{2\pi\Gamma}{\hbar\omega_C})}_{D_c(B)} \times (\frac{\hbar\omega_c}{2\varepsilon_F})^{1/2} \cos(\frac{2\pi\varepsilon_F}{\hbar\omega_C} - \delta), \quad (13)$$

where ΔR_{xx}^{osc} is the oscillating part of the magnetoresistance with the background removed, and $\omega_c = eB/m^{\star}$ is the cyclotron frequency. $\chi = 2\pi^2 k_B T/\hbar\omega_c$ is a temperature dependent dimensionless parameter and Γ is the collisional broadening energy due to quantum scattering events. δ is a phase factor that is unimportant for our study. The terms $D_t(T)$ and $D_c(B)$ in equation 13 are the temperature and collision damping terms respectively; it is easily seen that in the absence of damping of the oscillations due to temperature ($\lim_{T\to 0} D_t(T) = 1$) and in the absence of damping due to collisions ($\lim_{\Gamma\to 0} D_c(B) = 1$), the magnetoresistance would exhibit a weakly modulated ($\sim B^{1/2}$) cosine oscillations in $1/B$. In fact, the two damping terms $D_t(T), D_c(B)$ are used as probes to tune the temperature and magnetic field *independently* to extract the effective mass and the quantum scattering time. The period of the oscillatory cosine term yields the carrier density of the 3DES since the period is linked to n_{3d}. R_{xy}, the Hall resistance is linear with B, and should show plateaus at the minima of R_{xx} when a *small* number of Landau levels are filled.

The sample chosen for magnetotransport studies has a polarization-doped 3DES in a 100 nm $0-30\%$ graded AlGaN layer (see [36]). For magnetotransport measurements on the 3DES, ohmic contacts were formed in a Van-der Pauw geometry. The sample was immersed in a ^3He low-temperature cryostat with a base temperature of 300 mK. Magnetic fields in the range $0\text{T} \leq B \leq 14\text{T}$ were applied. R_{xx} and R_{xy} was measured as in the geometry depicted in the inset in Figure 23 using a standard low-frequency lock-in technique.

Figure 23 shows the measured (raw) magnetoresistance R_{xx} as a function of temperature and magnetic field. The oscillations are seen to be damped with increase in temperature, and they become more pronounced at high magnetic fields. A clear background negative parabolic feature is seen. Negative parabolic magnetoresistance has been observed in many other systems and there exist differing opinions regarding its origin [37–40] including weak localization, hopping and electron-electron interactions. The negative magnetoresistance part is not discussed any

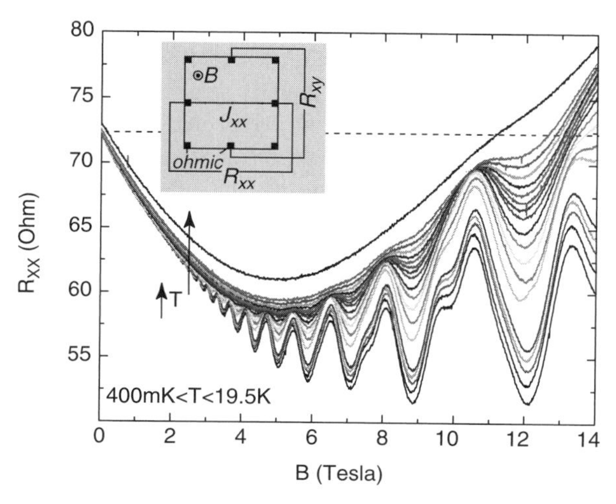

Fig. 23 The raw measured magnetoresistance oscillations. The inset is the device geometry. (Measurements courtesy Angela Link, WSI, Munich, Germany.)

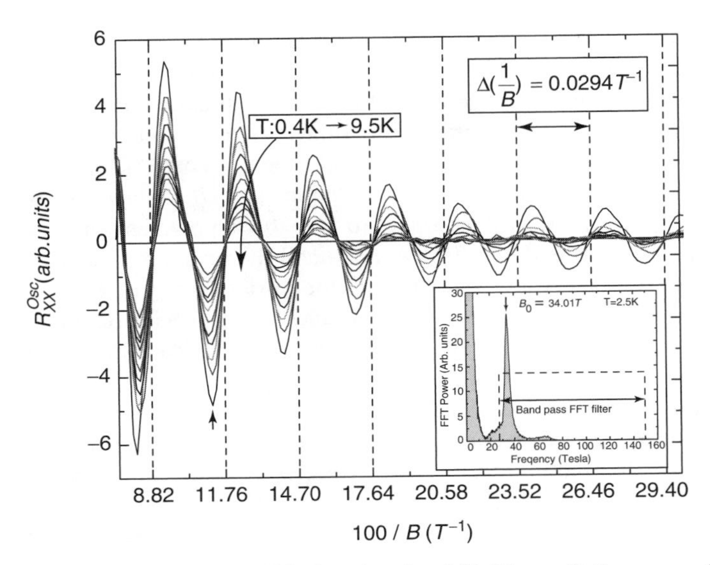

Fig. 24 The oscillatory component ΔR_{xx}^{osc} plotted against $1/B$. The oscillations are periodic with period $\Delta(1/B) = 0.0294T^{-1}$, and are damped with both increasing temperature (different curves), and increasing $1/B$. Also shown in the inset is a typical FFT power spectrum (at $T = 2.5K$) showing a peak at the fundamental period, and the band-pass window used to filter the oscillatory component ΔR_{xx}^{osc}.

further, since it would take us too far from the theme of this chapter. Nevertheless, such behavior has also been reported for AlGaN/GaN 2DEG structures [41,42], and is an interesting topic in its own right.

We first analyze the oscillatory component of the transverse magnetoresistance (Figure 24). This figure is plotted in the following manner: we first take the raw R_{xx} vs B data and interpolate it to create an equally spaced $N = 2^{15}$ size FFT window. We then find the FFT power spectrum. This is repeated for R_{xx} measured at different temperatures. A typical FFT power spectrum (at T = 2.5K) is shown in the inset of 24. There is a clearly resolved peak at the fundamental oscillation period $B_0 =$ 34.01T. A band pass filter [Δf_{pass} = 28 - 150 T] is then employed to remove the background, which removes the negative parabolic contribution. The resulting ΔR_{xx}^{osc} for various temperatures $0.4\text{K} < T < 9.5\text{K}$ is shown in the plot against $1/B$.

As is clear from the plot, the period of oscillations is $\Delta(\frac{1}{B}) = 0.0294T^{-1} = 1/B_0$. The oscillations are strongly damped with increasing $1/B$ as well as with increasing temperature, as predicted by the theory.

In Figure 25 the plot of the oscillatory transverse magnetoresistance ΔR_{xx}^{osc} and R_{xy} at T = 400mK plotted against the applied magnetic field is shown. The Hall mobility determined from the slope of the R_{xy} curve is $\mu_H \simeq 3000$ cm^2/V·s, which is higher than 77K low-field Hall mobility of $\mu_{77K} \simeq 2500$ cm^2/V·s. Also, assuming that the 3DES is spread over a thickness d, the sheet carrier density of the 3DES is calculated to be $n_{3d} \times d = 1/R_H e = B/eR_{xy} = 7.2 \times 10^{12}$cm^{-2}. This is consistent

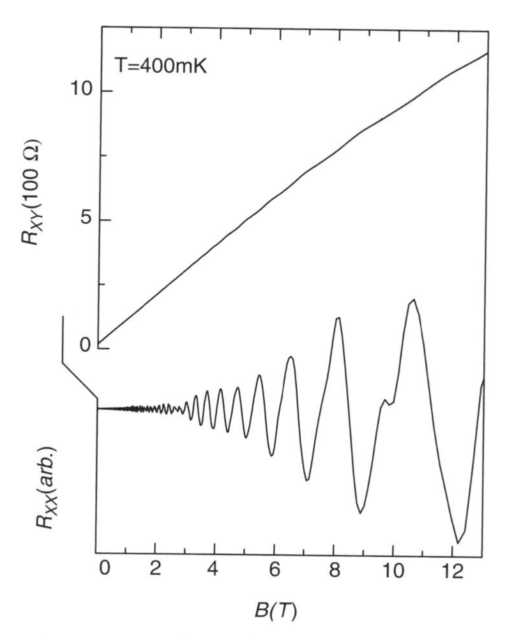

Fig. 25 Magnetotransport measurement data at $T = 400mK$. The figure also shown as an inset the geometry used for measuring R_{xx}, R_{xy}. The R_{xx} shown is the oscillatory component with the background removed.

with the 77K low-field Hall measured sheet density of $7.5 \times 10^{12}\text{cm}^{-2}$. The spread of the 3DES is calculated from a self-consistent Poisson-Schrodinger band calculation to be $d = 75$ nm due to 25 nm depletion of the 3DES from the surface potential. This depletion in the graded AlGaN layer was verified by capacitance-voltage profiling (see Figure 16). Thus, the Hall 3-dimensional carrier density is $n_{3d} \sim 10^{18}\text{cm}^{-3}$.

Carrier concentration from SdH oscillations

First, we observe from Equation 13 that the period $\Delta(\frac{1}{B})$ is linked to the carrier density of the 3DEG by the relation $\Delta(1/B) = e\hbar/m^{\star}\varepsilon_F = (2e/\hbar)(3\pi^2 n_{3d})^{-2/3}$. From the plot, the period $\Delta(1/B) = 0.0294\text{T}^{-1}$ yields a direct measurement of the 3-dimensional carrier concentration $n_{3d}^{SdH} = 1.1 \times 10^{18}\text{cm}^{-3}$. Thus, the carrier density measured from the quantum oscillations is close to the carrier density measured by classical Hall technique ($n_{3d} = 10^{18}cm^{-3}$).

Effective mass

The effective mass of carriers is determined by fitting (for the fitting procedure, see Sladek, [43]) the measured amplitude damping (Figure 24) with temperature at a fixed B to the temperature-damping term $D_t(T) = \chi/\sinh\chi$ of Equation 13. This is done in Figure 26(a). For the peak at $B = 8.9$T (arrow in 24) the effective mass is found to be $m^{\star} = 0.21 m_0$; we get the same effective mass for the amplitude

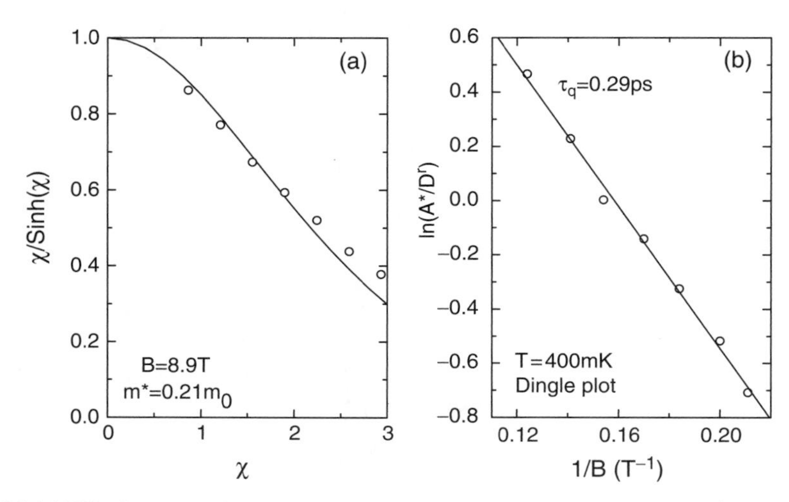

Fig. 26 (a) Effective mass plot at $B = 8.9T$ where the data (dots) are fit to the $\chi/\sinh\chi$ (line) damping term [43]. (b) Dingle plot for extraction of the quantum scattering time [34].

peaks at $B = 10.5T$. The band-edge electron effective-mass in pure GaN (AlN) is $m^\star_{GaN} = 0.20m_0(m^\star_{AlN} = 0.32m_0)$ [44]. From a linear interpolation for the 3DES experiencing an *average* Al-composition of $\langle x \rangle = 0.11$ we expect an effective mass of $0.21m_0$, which is in good agreement with the measured value. The value is also close to the effective mass measured for two-dimensional electron gases at AlGaN/GaN heterojunctions by the Shubnikov de-Haas method [45–48].

Scattering mechanisms

The next important parameter of the 3DES that is measured from the SdH oscillations is the collisional broadening energy Γ (due to Dingle). This term is a measure of the smearing of the delta-function discontinuities in the DOS due to quantum scattering events, and it appears as the imaginary part of the single-particle self energy function [32]. Collisional broadening energy is linked to the quantum scattering time τ_q and the Dingle temperature T_D by the relation $\Gamma = \hbar/2\tau_q = \pi k_B T_D$. This quantity is experimentally accessible from a controlled Landau damping of the oscillation amplitudes with $1/B$ at a fixed temperature; in other words, by tuning $D_c(B)$. Equation 13 can be cast in the form

$$\ln\left(\frac{\bar{A}}{(\frac{\hbar\omega_c}{2\varepsilon_F})^{\frac{1}{2}} D_t(T)}\right) = C - \left(\frac{\pi m^*}{e\tau_Q}\right)\frac{1}{B} \tag{14}$$

for extracting τ_q and the related quantities Γ, T_D. Here $\bar{A}$ are the extremum points of the damped oscillations, forming the exponentially decaying envelope. Equation 14 suggests that a plot of the natural log of the left side quantity against $1/B$ ('Dingle plot') should result in a straight line whose slope is $-\pi m^\star/e\tau_q$. Since we have already measured the effective mass $m^\star$, we can extract τ_q from the slope.

Figure 26(b) shows the Dingle plot for $T = 400$mK, yielding a $\tau_q = 0.29$ps. An averaging of the quantum scattering times over a range of low temperatures yields a value $\tau_q^{av} = 0.3$ps. The quantum scattering time does not show any discernible trend with temperature in this range. We calculate the corresponding level broadening $\Gamma = 1.1$ meV and Dingle temperature $T_D = 4$ K. The Landau level separation at $B = 10$T is $\hbar\omega_C = 5.8$ meV, sufficiently larger than both the measured collisional broadening of the Landau levels ($\Gamma = 1.1$ meV) and the thermal broadening $k_B T = 0.09$ meV at $T = 1K$, thus satisfying the conditions required for clear Shubnikov de-Haas oscillations.

The difference between quantum scattering time and the momentum scattering time is described in [49]. For isotropic scattering events with no angular preference, the quantum and momentum scattering times are the same $\tau_m/\tau_q = 1$. If the dominant scattering process has a strong angle dependence, the ratio is much larger than unity. This fact has been utilized to identify the dominant scattering mechanism in modulation-doped AlGaAs/GaAs two-dimensional electron gases [50].

The classical (or momentum) scattering time τ_m is directly measured from mobility via the Drude relation $\mu = e\tau_m/m^\star$. Low-temperature Hall mobility gives $\tau_m = 0.34$ps for the 3DES. Within limits of experimental error, $\tau_m/\tau_q \sim 1$. Thus, the ratio indicates that the dominant scattering mechanism at low temperatures is probably [51] of a short range (isotropic) nature.

Size-effect scattering [52] that occurs if the width of the 3DES is much less than the mean-free path of electrons is negligible since our 3DES has a mean free path $\lambda = \hbar k_F \mu/e \approx 60$ nm whereas the width of the 3DES is $d_0 \approx 75$ nm. The chief scattering mechanisms that can affect mobility are alloy disorder scattering, charged dislocation scattering (owing to a high density of dislocations $N_{disl} \sim 10^9 \text{cm}^{-2}$), and ionized impurity scattering from the remote donors at the surface states.

Hsu and Walukiewicz [51] show that remote ionized impurity scattering strongly favors small angle scattering, thus causing the ratio $\tau_c/\tau_q \gg 1$. Since $\tau_c/\tau_q \approx 1$ is observed for the 3DES, remote ionized impurity scattering is unimportant.

The ratio of classical to quantum scattering times due to charged dislocation scattering is calculated [53] to be

$$\frac{\tau_m}{\tau_q}|_{disl} = 1 + 2k_F^2\lambda_{TF}^2, \tag{15}$$

where $\lambda_{TF}^2 = 2\epsilon\varepsilon_F/3e^2 n_{3d}$ is the Thomas-Fermi screening length of the degenerate 3DES. The ratio for our 3DES is 2.3, which is larger than the observed value; thus, we exclude dislocation scattering to be the most important scattering mechanism.

So we converge on alloy scattering as the dominant scattering mechanism at low temperatures. The alloy scattering potential V_0 is of a short range nature, which makes the scattering process isotropic and $\tau_m/\tau_q \sim 1$, as observed. This is a confirmation of the conclusion arrived at previously in this chapter from a comparison of calculated and measured temperature-dependent mobility. Thus, it strengthens the claim of the clean measurement of the alloy scattering potential of AlGaN ($V_0 = 1.8$eV).

In summary, by exploiting the polarization charges in the AlGaN/GaN semiconductor system, it is possible to create a 3DES without intentional doping. The mobile carriers in the 3DES are degenerate at low temperatures and exhibit a high mobility. The polarization-doped layers have much better transport properties than comparable donor-doped layers, which makes them attractive for applications in various devices. The lack of carrier freezeout enables one to observe Shubnikov de-Haas oscillations in magnetotransport measurements of the 3DES, which reveal several important parameters. First, the temperature damping of oscillations reveals the effective mass of electrons to be very close to that in bulk GaN ($m^{\star} = 0.21m_0$). Next, the quantum scattering time of electrons in the 3DES is found from the Dingle plot to be $\tau_q = 0.3$ps. The ratio of the classical (momentum) scattering time to the quantum scattering time is found to be close to unity $\tau_m/\tau_q \approx 1$. The ratio suggests that short-range scattering dominates transport properties at low temperatures. This scattering mechanism is identified to be alloy scattering. This lets us extract another valuable parameter, the alloy scattering potential in $Al_xGa_{1-x}N$ to be $V_0 = 1.8$ eV.

Degenerate three-dimensional electron gases are an interesting playground for study of collective phenomena such as spin-density waves, Wigner crystallization, and integral and fractional quantum-Hall effects in 3-dimensions (for a clear account of many predicted effects, see Halperin [54]). Polarization-doped electron slabs presented in this chapter provide an interesting addition to the few existing techniques (such as parabolic grading in AlGaAs/GaAs heterostructures) [21] for creating such electron populations, overcoming the thermal freezeout effects associated with *impurity-doped* semiconductors. The wide *tunability* of slab thickness and electron density offered by polarization-doping makes it an attractive system to study such effects.

2.7 Device Applications of Polarization-'Doped' Graded Nitride Layers

The doping technique developed in graded polar alloy semiconductors is attractive from the viewpoint of devices. The critical advantage it offers is the high conductivity of the graded channels compared to similar shallow donor-doped layers. This property provides the necessary motivation for incorporating the doping technique in a field-effect transistor structure. This subsection describes such a polarization-doped Field-Effect Transistor, or the PolFET.

The III-V nitride family of semiconductors possess properties that are very desirable for high power and high speed device applications [12]. Among the properties are the high saturation velocities $v_{sat} \approx 2 \times 10^7$cm/s and the wide bandgap that leads to large breakdown fields $E_{Br} \approx 5 \times 10^6$ V/cm, making them attractive for many device applications. The use of GaN in field-effect transistors has been largely limited to the high-electron mobility transistor (HEMT). Field effect transistors (FETs) such as the metal-semiconductor FET (MESFET), the metal-insulator FET (MISFET) or the junction-FET (JFET) have not gained enough popularity. Nevertheless, there have been attempts to fabricate of such devices for tackling

Table 3 GaN-based MESFET & JFET results

Year	Group	Growth	L_G μm	I_{DS}^{max} mA/mm	g_m^{max} mS/mm	N_D 10^{17} cm^{-3}	μ_{300K} cm^2/V·s
1993	Khan et. al. [56]	MOCVD	4.0	170	23	1	350
1994	Binari et. al. [57]	MOCVD	1.4	$\approx$330	20	2.7	400
1996	Zolper et. al. [58]	MOCVD	1.7	$\approx$33	7	–	–
1999	Egawa et. al. [59]	MOCVD	2.0	281	33	1.1	585
2000	Zhang et. al. [60]	MOCVD	0.8	270	48	24	270
2001	Gaska et. al. [55]	MOCVD	1.5	300	70	15	100
2002	Jimenez et. al. [13]	MBE	0.7	600	60	10	$\approx$200
2004	Rajan et. al. [61]	MOCVD	0.7	430	67	10	700

some of the long-standing problems plaguing HEMTs (Gaska et al., [55]) such as linearity, aging effects, and RF-dispersion.

Table 3 shows a timeline of results in GaN-based MESFET or JFET technology. From the first GaN-based MESFET reported by Khan et al. [56], there have been steady improvements of the device characteristics such as maximum drain-source currents and the device transconductance. A point that is of importance in this chapter is that *all* results till this work have used a donor-doped layer as the channel material of the device. The mobility of electrons such doped channels is typically low for high doping densities (see Table 3), and has been one of the main reasons impending the popularity of GaN-based MESFETs and JFETs.

One might ask why fabricate MESFETs or JFETs when one can use the sharp AlGaN/GaN heterojunction 2DEGs for high quality HEMTs? One possible advantage that MESFETs or PolFETs can offer is related to a reduction of the hot-phonon induced saturation velocity reduction in HEMTs (see [62–64]). Due to the very high volume density of carriers in a typical AlGaN/GaN HEMT, a very large density of polar optical phonons accumulate at the heterojunction. Due to the extremely strong electron-phonon Fröhlich interaction, the rate of emission of Longitudinal Optical (LO) phonons is $\sim$10 fs, but their decay rate into acoustic and Transverse Optical (TO) modes takes as long as $\sim$1 ps. Therefore, there is a large buildup of non-equilibrium LO phonons in the 2DEG channel, which scatter the electrons and lower their saturation velocity. This places a limit on the high-speed performance of the HEMT. However, if the volume-density of carriers is lowered, the saturation velocity is observed to increase (see [27,65]). For achieving low volume-densities of carriers (for high saturation velocities), and simultaneously high values of low-field electron mobilities, the polarization-doped 3D electron slabs might offer an elegant solution for future high-speed nitride-based transistor technologies.

2.7.1 The PolFET

Donor-doping of GaN at densities of $N_D \approx 10^{18}$ cm^{-3} leads to a mobility in the range of 200$-$300 cm^2/V·s owing to combined ionized impurity and optical phonon scattering effects. This was observed in the control sample (Sample 1) described earlier, and has indeed been observed by many groups. Scattering by ionized impurities is virtually eliminated in polarization-induced two-dimensional electron gases (2DEGs) that form at AlGaN/GaN heterojunctions, thus improving the room-temperature mobility to typical values of ($\mu_{2DEG} \approx 1500$ cm^2/V·s).

As opposed to a conventional n-doped channel [13,60] or an ion-implanted channel [58] in earlier GaN-based FETs, the novel technique of polarization bulk-doping is employed. The polarization-doped layer forms a three-dimensional electron slab (3DES) with a improved carrier mobility. For an effective doping of $N_D \approx 10^{18}$ cm^{-3}, the 3DES exhibits a room-temperature mobility of $\mu_{3DES} \approx 700$ cm^2/V·s as compared to a mobility of $\mu_{imp} \approx 300$ cm^2/V·s for shallow donor (Si) doped GaN. The mobility in the graded AlGaN layers is limited by a combination of alloy scattering and polar optical phonon scattering at 300K, and reached values ranging from 700 – 900 cm^2/V.s.

Figure 27 shows the layout, layer structure, and the band diagram of the device. The structure consists of a 2μm Fe-doped semi-insulating (SI) GaN grown on a c-plane sapphire substrate by metal-organic chemical vapor deposition (MOCVD). Using this SI GaN as a template, we grow the device layers by molecular beam epitaxy (MBE) under Ga-rich conditions. The MBE grown epitaxial layers consist of an unintentionally doped (UID) GaN spacer of 100 nm followed by a 100 nm graded AlGaN layer, and capped by a 50 nm Mg-doped p-GaN layer. The nominal Mg doping in the p-capped layer is $N_A \approx 10^{19}$ cm^{-3}. The polarization-doping is $N_D \approx 10^{18}$ cm^{-3}. The transport properties of such graded AlGaN layers was studied by Hall measurements and revealed a room temperature mobility of $\mu \approx 700$ cm^2/V·s.

The device structure was obtained by deposition of Pd/Au/Ni ohmic contact on the p-type layer for the gate contact. The access regions between the gate and drain and gate and source were etched away by reactive ion etch (RIE). Ti/Al/Ni/Au layers were deposited to form the source and the drain ohmic contacts to the polarization-doped 3DES. Different devices on the die were isolated by a 300 nm deep mesa-etch. The devices fabricated had a T-shaped layout with 0.7μm gate length, 0.7μm gate-source spacing, and 1.5μm gate-drain spacing, and a gate width of 150μm. The final device structure is shown in 27 with the schematic band-diagram and the channel. The gate ohmic to the p-GaN lets the p-n junction act as a gate to modulate the 3DES spread in the graded AlGaN channel, thus enabling modulation of the source-drain current.

Figure 28 shows the DC current-voltage (IV) characteristics of the PolFET. The gate-drain p-n junction had a turn-on voltage of 1.9 V. This enables us to measure drain-source currents starting at a gate voltage of $V_G = 1$ V and stepping down the gate voltage in steps of $\Delta V_G = -1$ V. The DC IV curve was measured in this manner. The device exhibits a maximum DC drain-source current of $I_{DS}^{max} = 430$ mA/mm. Channel pinch-off is obtained at a gate voltage $V_P = -6$ V.

Figure 29 shows the transconductance ($g_m - V_{GS}$) plot for the PolFET. A maximum transconductance of $g_m^{max} = 67$ mS/mm was achieved at a gate-source voltage of $V_{GS} = -4$ V. The transconductance drops at higher gate voltages - the cause of this roll-off is similar to the roll-off of transconductance observed in AlGaN/GaN HEMTs [12]. The contact-resistance of the ohmics to the polarization-doped

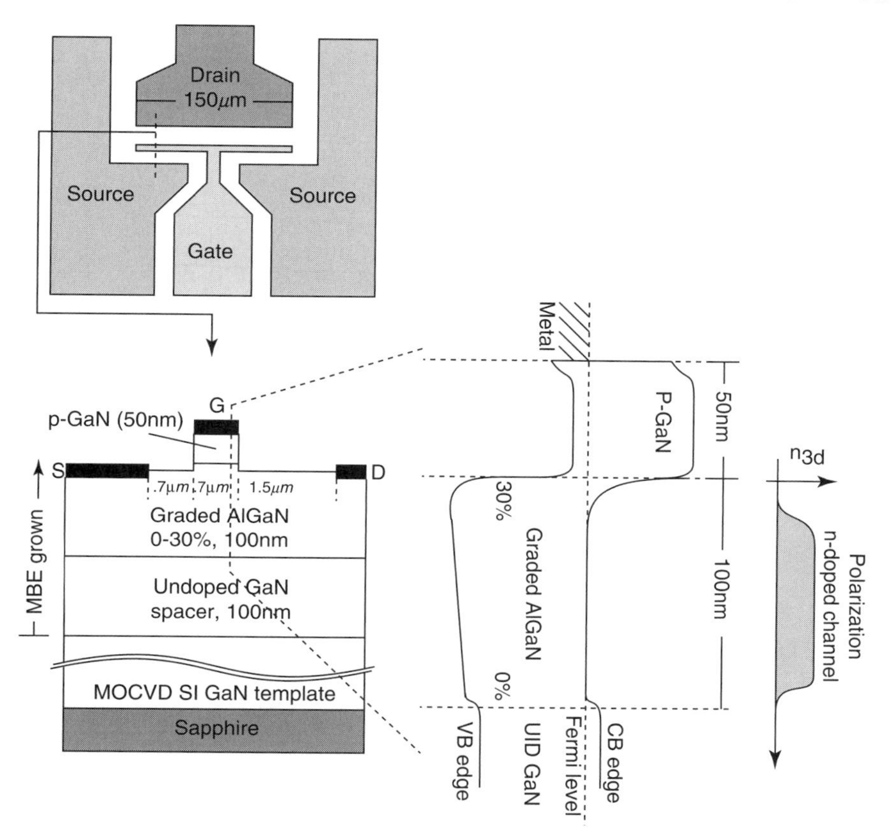

Fig. 27 Polarization doped channel Junction Field-Effect Transistor (PolFET). The top view of the device, the layer structure, and the band diagram are shown. The device was fabricated by Dario Buttari and Ana Jimenez.

channel was in the range of $R_c \approx 0.3 \pm 0.5\Omega$mm. The value is low and comparable to the contact resistance of ohmics of 2DEG-channels in HEMTs, which is an encouraging sign. Frequency dependent measurements yielded a $f_T = 8.45$ GHz, and $f_{max} = 10.9$ GHz for this prototype device. The gate-drain breakdown voltage was measured to be $V_{BD} \approx 60$ V by a two-terminal measurement where the source was left floating and the breakdown cutoff was a current of 1 mA/mm.

The first results on the PolFET demonstrated that a polarization-doped channel is a viable alternative to achieving high-conductivity channels in GaN-based structures and in polar semiconductors in general. The channels have several advantages, and from Table 3, it is evident that the PolFET is among the best devices in this family. The existing technology for ohmic contacts in HEMTs works well for these channels, which is an advantage. The doping control by changing thickness and/or composition is very beneficial, and can be suitably exploited. This device has been improved upon and much better performance has recently been achieved in PolFET technology ([61]).

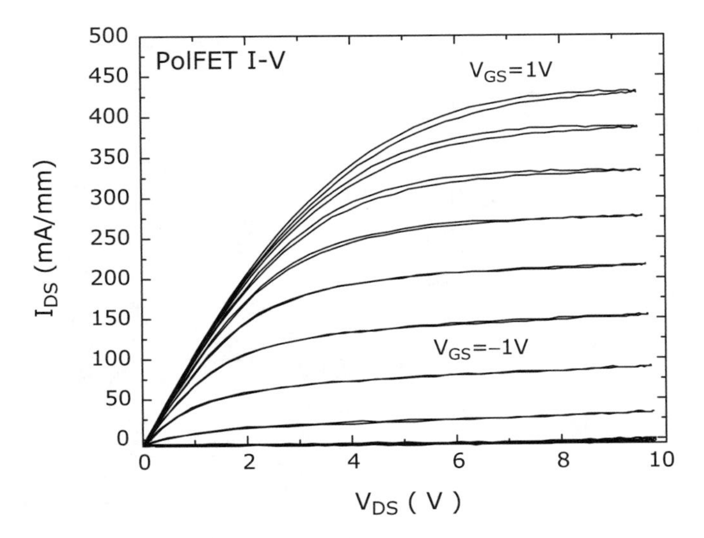

Fig. 28 Current-Voltage characteristic of the PolFET.

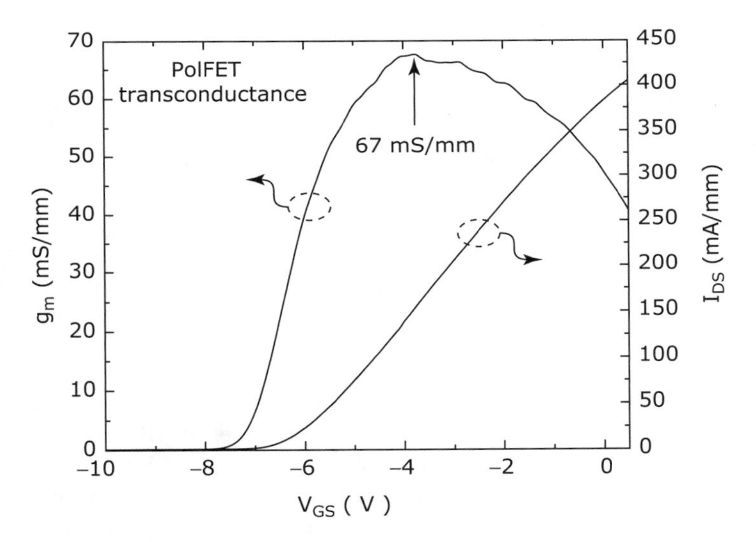

Fig. 29 Transconductance of the PolFET.

2.7.2 Other Device Applications

Besides serving as a channel material for FETs, the technique of polarization bulk-doping may be put to several other uses such as for regrown ohmic-contacts. The additional band-discontinuity achieved at a regrown polarization-doped Al-GaN ohmic contact will serve as an efficient hot-electron launcher from the source into the FET channel, reducing transit times. The flexibility of polarization doping by grading (by controlling alloy composition and/or graded layer thickness independently) is an added attraction. An interesting extension is the possibility of achieving polarization doped p-type carriers with higher mobilities by grading down from AlGaN for Ga-face III-V nitrides (or grading up from GaN to AlGaN in N-face III-V nitrides). In such polarization doped 3D hole slabs (3DHS), one might need to supply holes through remote acceptors. This might solve the problems associated with the high activation energy of the commonly used acceptor (Mg) for GaN. Demonstration of the n-type 3DES present the first step towards realizing the proposed enhancement of base conductivity in AlGaN/GaN heterojunction bipolar transistors by exploiting the strong electronic polarization properties of the III-V nitride semiconductors [17]. An important effect on device design would be the requirement of compensation doping in graded III-V (Al)GaN layers for removing unwanted mobile carriers that will necessarily result from polarization doping. The most important application of polarization-induced doping might well be in Ultra Violet (UV) and deep-UV optical devices for high-density data storage and readout. Such deep-UV devices require high Al-composition AlGaN layers, which are rather difficult to dope with traditional donors and acceptors.

This concludes the section on graded polar semiconductor heterostructures. In the next section, the concept of polarization grading is extended to ferroelectric and ferromagnetic materials, and the universal physical phenomena underlying their properties is presented.

3 Universal Physics of Functionally Graded Ferroelectric and Ferromagnetic Alloys

3.1 Order Parameters in Ferroic (Ferroelectric, Ferromagnetic, & Ferroelastic) Materials

Ferroics such as ferroelectrics, ferro- or ferrimagnets, and ferroelastics are high energy-density materials that store, convert, and release energy (electrical, magnetic, and mechanical) in a well-controlled manner, making them highly useful as sensors, actuators, and non-volatile memory elements. Their physical properties are sensitive to changes in external conditions such as temperature, pressure, electric, and magnetic fields.

Ferroics are principally distinguished by four main characteristics -

- First, their property-specific order parameter(s) (e.g., polarization, magnetization, or self-strain, for ferroelectrics, ferromagnets, and ferroelastics, respectively) spontaneously assume non-zero values below a threshold temperature T_C even in the absence of an applied stimulus. The emergence of such an order parameter may be due to a displacive transformation, (electrical or magnetic) ordering, or a combination of both. This transition is accompanied by a structural phase change from a highly symmetric crystalline structure, to one in which there

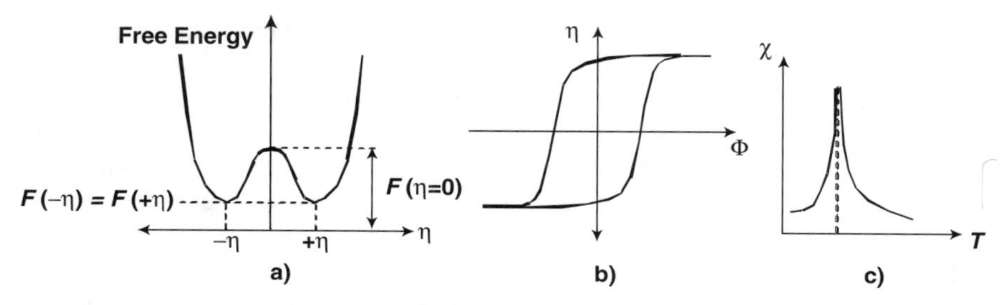

Fig. 30 a) Double-well potential for a typical ferroic material below T_C showing equilibrium states. The order parameter η corresponds to polarization, magnetization, and self-strain of a ferroelectric, ferromagnetic, and ferroelastic material, respectively. b) Stimulus-response behavior of a typical ferroic. Φ refers to electric field, magnetic field, and stress. c) Ferroic susceptibility versus temperature curve displaying anomaly near the phase transformation temperature. χ corresponds to dielectric susceptibility, magnetic susceptibility, and elastic compliance in a ferroelectric, ferromagnetic, and ferroelastic material, respectively.

is an asymmetry (spontaneous symmetry breaking). Such behavior is a direct result of the nature of the underlying internal potentials, which are characterized by double well minima as shown in Figure 30a. Here, two possible response states are shown for the three most common ferroic systems.

- Secondly, ferroics exhibit hysteresis in their stimulus-response behavior: e.g., polarization vs. applied electric field, magnetization vs. applied magnetic field, and strain vs. applied stress. Figure 30b shows the response of these three material systems with generalized hysteresis graphs for any ferroic system. The hysteretic characteristics of the stimulus-response behavior of ferroic systems are due to domain phenomena.
- Another distinguishing characteristic of ferroic materials is their large and nonlinear generalized susceptibilities. For example, the dielectric response, the magnetic susceptibility, and the elastic moduli in ferroelectric, ferromagnetic, and ferroelastic materials, respectively, show a λ-type critical behavior near the phase transformation temperature T_C (Figure 30c).
- Finally, the ferroic phase transformation can be induced via an external field conjugate to the order parameter.

Ferroelectrics, such as $BaTiO_3$ or $PbTiO_3$, can sense changes in temperature, electric field, and the stress state [66]. They can actuate due to their piezoelectric properties. New generations of ferroelectric IR detectors for night-vision applications are based on the strong variation of the spontaneous polarization with temperature near the phase transformation temperature T_C. Ferro- and ferrimagnetic materials (for example, Fe- and Ni-alloys or ferrites such as $CoFe_2O_4$) can be employed for the very same purposes where the sensing and actuating is accomplished by magnetic stimulation/response [67]. The non-zero spontaneous polarization or magnetization in these materials even in the absence of an external stimulus such as an applied electric or magnetic field can be utilized for applications such as non-volatile Random Access Memories (NVRAM). In ferromagnetic materials, the spin of the electron is employed instead of its charge to create a remarkable new generation of spintronic devices which may turn out to be smaller, more versatile and more robust than those currently making up silicon chips and circuit elements. Many technologically important materials and minerals undergo a ferroelastic transformation [68]. The classical example in minerals is lead phosphate, $Pb_3(PO_4)_2$. In metals, this phase transformation is called a martensitic transformation and is the reason of shape-memory and pseudoelastic deformation phenomena in certain alloys. The high-T_C superconductor $YBa_2Cu_3O_{7-x}$ undergoes a tetragonal-to-orthorhombic transformation at 970 K and associated with this is a 2% self-strain. $SrTiO_3$ has a cubic perovskite structure which transforms to a tetragonal lattice via the rotation of the oxygen tetrahedral below 100 K. In fact, both ferroelectric and ferromagnetic transformations are improper ferroelastic phase transformations with a self-strain (or lattice deformation) that spontaneously emerges below T_C.

To discuss the physics of ferroic materials in general, we will employ the Landau theory of phase transformations. It was first introduced in 1930s to describe a complex problem in solid-state phase transformations-ordering phenomena in metallic alloys [69]. The order-disorder transformation involves a change in the crystal symmetry. The material transforms from a high-symmetry disordered phase to a low-symmetry ordered phase. The broken symmetry in the crystal upon ordering can be characterized by an order parameter. Landau has shown that the Helmholz free energy of an order-disorder transformation can be expressed very simply as a polynomial expansion of the order parameter that describes the degree of order (or disorder). Landau theory has been used quite successfully to describe phase transformations in a variety of materials systems, including (but not limited to) ferromagnetic, martensitic (ferroelastic), and ferroelectric transitions with appropriate order parameters. For example, proper ferroelectric, ferromagnetic, and ferroelastic phase transformations can be described via the Landau potential with the polarization P_i, magnetization M_i, or the self-strain as the order parameter, respectively. The obvious advantage of the Landau phenomenology is that a variety of physical properties can be explained *without referring to all the degrees of freedom* in complex materials systems such as ferroelectrics and ferromagnets where phenomena are of subatomic origin.

Let us consider the expansion of the free energy of a ferroic phase transformation of a single-domain system with three order parameters η_i such that it is a harmonic function of the order parameters (and thus does not contain odd powers):

$$\begin{aligned} G(\eta_i) = \int_V dV[\alpha_{ij}\eta_i\eta_j + \beta_{ijkl}\cdot\eta_i\eta_j\eta_k\eta_l + A_{ijkl}(\nabla_i\eta_j\cdot\nabla_k\eta_l) + \ldots \\ + \delta_{ijk}x_{ij}\eta_k) + \frac{1}{2}q_{ijkl}x_{ij}\eta_k\eta_l - (\frac{1}{2}\Phi_i^D - \Phi_i)\eta_i + F_{el,T}(C_{ijkl})], \end{aligned} \tag{16}$$

where α_{ij}, β_{ijkl}, and A_{ijkl} are the free energy expansion coefficients, δ_{ijk} and q_{ijkl} are the bilinear and linear-quadratic coupling coefficients between the order parameter and the strain x_{ij}, C_{ijkl} are the elastic coefficients, Φ_i is an externally applied electrical or magnetic field, and Φ_i^D is the internal depolarization or demagnetization field for ferroelectric of ferromagnetic materials systems, respectively. In the

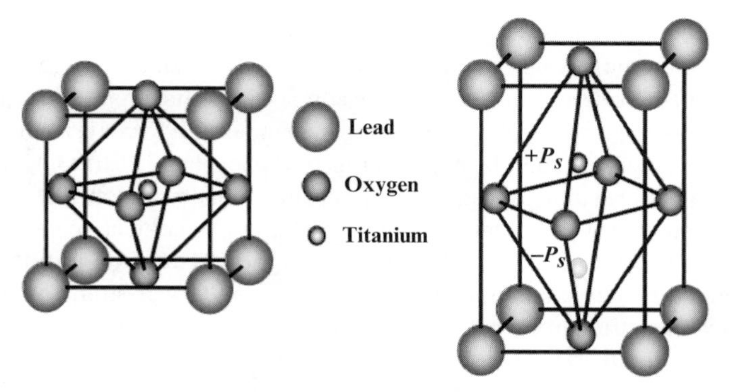

Fig. 31 The perovskite crystal structure of PbTiO3 (a) above, (b) below the ferroelectric phase transformation temperature. The tetragonality of the ferroelectric phase is highly exaggerated.

above relation, $F_{el,T}$ is the total elastic energy which has two components:

$$F_{el,T}(C_{ijkl}) = F_{el} + \frac{1}{2}\sigma_{ij}xkl, \tag{17}$$

where F_{el} is the internal strain energy due to variations in the self-strain and the last term is the elastic energy of an applied stress field, σ_{ij}. Proper ferroelectric, ferromagnetic, and ferroelastic phase transformations can be described via the above relation with the polarization P_i, magnetization M_i, or the self-strain e as the order parameter, respectively.

As a simple example, let us concentrate on a ferroelectric transformation where upon cooling the paraelectric cubic $m3m$ phase transforms to a ferroelectric tetragonal $4mm$ phase. Figure 31a shows the prototypical perovskite structure of $PbTiO_3$ (and $BaTiO_3$) above the transformation temperature T_C where the TiO_6 octahedra are linked in a regular cubic array forming the high-symmetry $m3m$ prototype for many ferroelectric forms. The small 6-fold coordinated site in the center of the octahedron is filled by a small, highly-charged Ti^{+4} and the larger 12-fold coordinated "interstitial" site between octahedral carries a larger Pb^{+2}. The oxygen ions sit at the face-centers of the cubic cells. The spontaneous polarization in $PbTiO_3$ arises from the spontaneous non-centrosymmetric displacement of Ti^{+4} and O^{-2} ions relative to Pb^{+2} ions. Figure 31b shows an example of such a shift in the tetragonal phase, resulting a net polarization along the c-axis of the tetragonal unit cell. As can be seen from Figure 31b, the spontaneous polarization and the accompanying lattice distortion can be along the positive or the negative z-axis, or the (001) direction.

For stress-free single-crystals, the (Helmholtz) free energy density can be described via a Landau-Ginzburg-Devonshire (LGD) potential that should be constructed in such a way that the existence of these two equivalent ferroelectric ground states are taken into account. This can be accomplished by expanding the free energy (per unit volume) in terms of the polarization only with respect to even powers, such that:

$$F_L(P,T) = F_0 + \frac{1}{2}\alpha P^2 + \frac{1}{4}\beta P^4 + \frac{1}{6}\gamma P^6, \tag{18}$$

where F_0 is the energy in the paraelectric state, P is the polarization (which is the order parameter of the phase transformation), and α, β, and γ are the expansion (or the dielectric stiffness) coefficients. α is a temperature dependent coefficient and its dependency is given by the Curie-Weiss law:

$$\alpha = \frac{T - T_C}{\epsilon_0 C} \tag{19}$$

where T_C and C are the Curie-Weiss temperature and constant respectively, and ϵ_0 is the permittivity of free space. The other two expansion coefficients are usually taken to be independent of temperature.

The Landau potential should exhibit only one minima at $P = 0$ above T_C corresponding to the centrosymmetric cubic non-polar phase. Below T_C, the free energy curve should have two minima corresponding to two identical ferroelectric ground states but with opposite orientation of the polarization direction along the "easy" (001) orientation (P_S and -P_S). The schematic free energy potentials at various critical temperatures are illustrated in Figure 32 for a 2nd-order and Figure 33 for a 1st-order ferroelectric phase transformation. It can be shown that if $\beta < 0$, the phase transformation from the paraelectric state is of 1st-order (i.e., discontinuity in P_S, and lattice parameters at T_C and thermal hysteresis in the same parameters around T_C) and it is of 2nd-order if $\beta > 0$ (i.e., gradual variation in P_S, and lattice parameters below T_C with no thermal hysteresis).

The spontaneous polarization P_S in the tetragonal phase can be obtained from the condition for thermodynamic equilibrium $\partial F_L / \partial P = 0$ such that:

$$P_S^2(T) = \frac{-\beta + (\beta^2 - 4\alpha\gamma)^{1/2}}{2\gamma} \tag{20}$$

The spontaneous polarization is due to the non-centrosymmetric displacements of the ions in the crystal and results in a tetragonal distortion. The structural

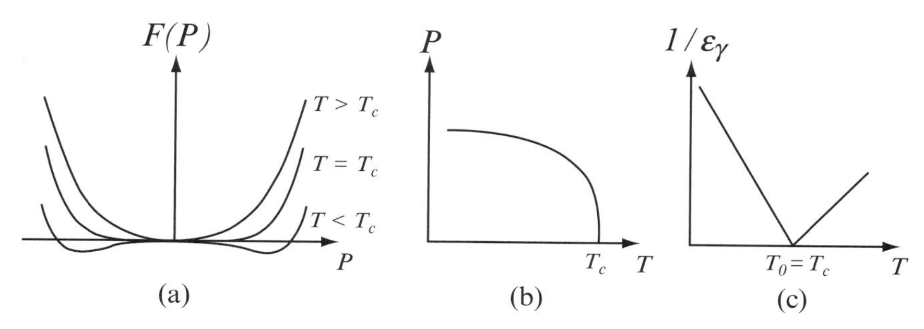

Fig. 32 Second-order transition: (a) Free Energy versus polarization, (b) spontaneous polarization versus temperature, (c) reciprocal susceptibility versus temperature.

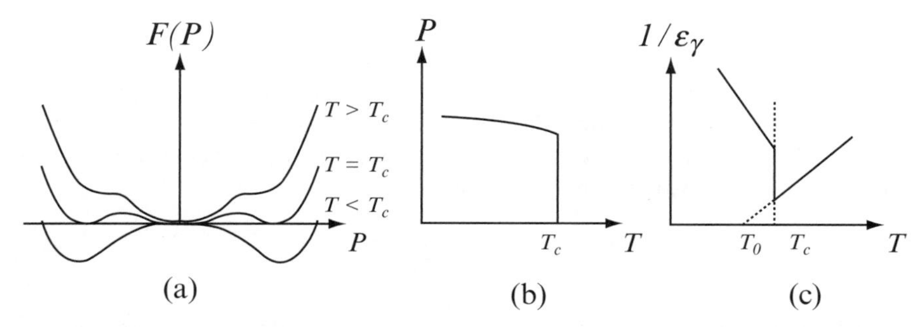

Fig. 33 First-order transition: (a) Free Energy versus polarization, (b) spontaneous polarization versus temperature, (c) reciprocal susceptibility versus temperature.

component of the phase transformation can be described by the self-strains. It should be noted that the self-strain is not actually a strain *per se* but rather a geometric description of the lattice distortion. The components of the self-strain tensor are defined as:

$$x_1^0 = x_2^0 = \frac{a - a_0}{a_0} = Q_{12}P_S^2, x_3^0 = \frac{c - a_0}{a_0} = Q_{11}P_S^2, \tag{21}$$

where a and c are the lattice parameters in the tetragonal ferroelectric state and a_0 is the lattice parameter of the cubic phase. Q_{ij} is a 4th-rank tensor (in the contracted notation) that couples the polarization with the self-strain, called the electrostrictive coefficients.

The effect of applied stresses and electric fields can be incorporated into the Landau potential as well. The energy of elastic stresses in a pseudo-cubic crystal is given by:

$$F_\sigma = \frac{1}{2}S_{11}(\sigma_1^2 + \sigma_2^2 + \sigma_3^2) + S_{12}(\sigma_1\sigma_2 + \sigma_1\sigma_3 + \sigma_2\sigma_3) + \frac{1}{2}S_{44}(\sigma_4^2 + \sigma_5^2 + \sigma_6^2), \tag{22}$$

where S_{ij} are the elastic compliances at constant polarization and σ_i are the components of the applied elastic stress tensor. It should be kept in mind that the applied stress is coupled with the polarization due to the electrostrictive effect and the energy of this coupling can be expressed by:

$$F_C = x_1^0 = x_1^0(\sigma_1 + \sigma_2) + x_3^0\sigma_3 = Q_{12}P^2(\sigma_1 + \sigma_2) + Q_{11}\sigma_3P^2. \tag{23}$$

The energy contribution of an applied electric field is given by:

$$F_E = E \cdot P, \tag{24}$$

where E is the electric field along the easy axis. Taking into account these contributions, the total energy (per unit volume) is described by the Gibbs free energy:

$$G(P, T, \sigma_i, E) = F_L - F_\sigma - F_C - F_E. \tag{25}$$

The electrical and electromechanical properties of a ferroelectric can then be obtained by taking the appropriate derivatives of the above free energy functional. For example, the dielectric susceptibility along the easy axis under zero stress is given by $(\partial^2 G/\partial P^2)^{-1}$ such that:

$$\chi_3 = (\alpha + 3\beta P^2 + 5\gamma P^4)^{-1}, \tag{26}$$

and the relative dielectric constant along the easy axis is defined as:

$$\epsilon_r \epsilon_0 = 1 + \chi_3, \rightarrow \epsilon_r \approx \chi_3/\epsilon_0 \tag{27}$$

since $\chi_3 \gg 1$.

The (converse) piezoelectric coefficient that describes the strain due to an applied electric field along the (001) direction is defined through the equation, $d_{33} = \chi_3(\partial^2 G/\partial P \partial \sigma_3)^{-1}$, such that:

$$d_{33} = \epsilon_0 \chi_3 Q_{11} P. \tag{28}$$

The pyroelectric response that describes the variation in the spontaneous and induced polarization (or the dielectric displacement) with temperature is given by:

$$p = \frac{\partial P_S}{\partial T} + E\frac{\partial \epsilon_3}{\partial T}, \tag{29}$$

where the last term describes the temperature dependence of polarization induced by an applied (external) electric field.

We note that the Landau potential can be somewhat more complex for the low temperature phases of $BaTiO_3$ where the ionic displacement (or correspondingly the polarization) can be along the (110) directions in the orthorhombic phase or along the (111) directions in the rhombohedral phase. For these polymorphs, the expansion of the free energy has to be carried out for multiple order parameters; two for the orthorhombic and three for the rhombohedral phase.

The Landau formalism is based on the assumption that the polarization of the system is homogenous and there are no (local) variations in the order parameter. This is valid at relatively low temperatures where fluctuations of the order parameter are small. However, thermal vibrations close to the phase transformation temperature T_C may result in strong inhomogeneities in the polarization, especially in the short range. As shown by Ginzburg, the energy interaction between these regions with different polarizations can be approximated by:

$$F_G = \frac{1}{2}A(\frac{dP}{dz})^2 \approx \frac{1}{2}\delta^2|\alpha|(\frac{dP}{dz})^2 \tag{30}$$

for fluctuations in one dimension, where δ is a characteristic length, of the order of the linear dimensions of the region size.

The revised Landau potential that incorporates these polarization inhomogeneities is thus given by:

$$F_L(P,T) = F_0 + \frac{1}{2}\alpha(T)P^2 + \frac{1}{4}\beta P^4 + \frac{1}{6}\gamma P^6 + \frac{1}{2}A(T)(\frac{dP}{dz})^2. \tag{31}$$

The above relation can also be obtained on the basis of considerations associated with the possibility of expanding the thermodynamic potential into a series in powers not only of P but of the derivatives of P as well. The parameter A is always (taken to be) positive, thus the gradient term in the above relation acts as a restoring force that serves to damp out the spatial variations in P.

One of the shortcomings of a generalized Landau theory is that the hysteresis behavior observed experimentally can be obtained with one significant difference. The theoretically predicted value of the coercive field may be two- to three-orders of magnitude larger than the experimentally observed values. This is because of the fact that switching in the LGD treatment is due to thermodynamic instability of the polarization with respect to an applied field in the reverse direction rather than the nucleation and growth of electrical domains.

3.2 Functionally Graded Electrets and Magnets

3.2.1 Ferroelectric-Paraelectric (Dielectric) Bilayers

Polarization-graded ferroelectrics with smooth composition, temperature, or stress gradients can be viewed as bilayer structures in the limit of ever-increasing number of bilayer couples. This simple structure will serve as a guide to describe the physics of more complicated graded structures. We are particularly interested in how the individual layers are coupled through electrostatic, mechanical, and electromechanical interactions. The theoretical model employed will be basic Landau phenomenology and Maxwell relations.

Consider two uncoupled, unconstrained FE layers with equal lateral dimensions, shown in Figure 34a. In its most general form, the energy density of layers 1 and layer 2 in their uncoupled, unconstrained state can be expressed as:

$$\begin{aligned} F_1 &= F_{0,1} + \tfrac{1}{2}aP_1^2 + \tfrac{1}{4}bP_1^4 + \tfrac{1}{6}cP_1^6 \\ F_2 &= F_{0,2} + \tfrac{1}{2}dP_2^2 + \tfrac{1}{4}eP_2^4 + \tfrac{1}{6}fP_2^6, \end{aligned} \tag{32}$$

where $F_{0,i}$ is the energy of layer i in its high-temperature PE state, P_i are the polarizations of layers 1 and 2, and $a,b,c,d,e,$ and f are Landau coefficients. a and d are temperature dependent with their temperature dependency given by the Curie-Weiss law, i.e., $a = (T - T_{C,1})/\epsilon_0 C_1$ and $d = (T - T_{C,2})/\epsilon_0 C_2$, where ϵ_0 is the permittivity of free space, $T_{C,i}$ and C_i are the Curie-Weiss temperature and constant of layer i. The other coefficients for both materials are assumed to be temperature independent.

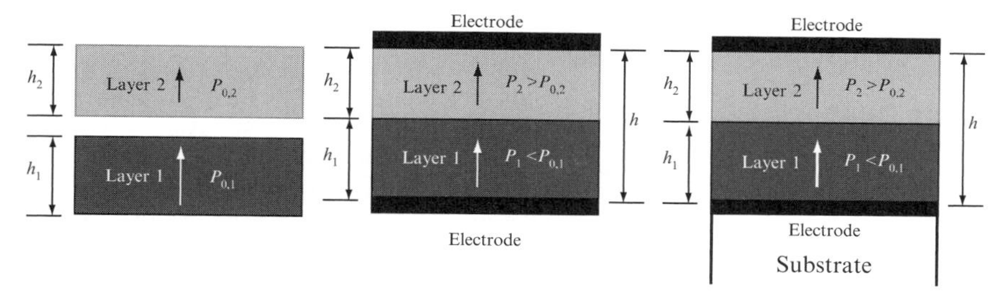

Fig. 34 a) Two freestanding FE layers. The initial polarization in layers 1 and 2 are $P_{0,1}$ and $P_{0,2}$, respectively. b) A bilayer constructed by joining the layers in a), sandwiched between metallic top and bottom electrodes, c) a heteroepitaxial bilayer made up of the bilayer in b) on a thick substrate. Due to interlayer coupling, $P_1 < P_{0,1}$ and $P_2 > P_{0,2}$.

The spontaneous polarization for each layer ($P_{0,1}$ and $P_{0,2}$ for layers 1 and 2, respectively) is given by the condition of thermodynamic equilibrium, $\partial F_i / \partial P_i = 0$.

Suppose layer 1 with thickness h_1 is joined with layer 2 with thickness h_2 and the bilayer is sandwiched between metallic electrodes [Figure 34b)]. A FE superlattice consisting of sets of identical bilayers with the same short circuit conditions can be treated analogously. The relative volume fraction of layer 2 is h_2/h. We will assume that both h_1 and h_2 are much larger than the characteristic correlation lengths of each layer. When these layers are coupled, the actual polarization of each layer is expected to be different from its "decoupled" value due to the electrical interaction between the layers. The internal electric fields $E_{D,1}$ and $E_{D,2}$ in layer 1 and layer 2 due to the polarization mismatch establish new polarization states, i.e., P_1 and P_2 in layer 1 and 2, respectively, see Figure 34b. The internal electric fields $\mathbf{E_D}$ are related to the difference in polarization of each layer and can be determined through the Maxwell relations $\nabla \times \mathbf{E_D} = 0$ and $\nabla \cdot \mathbf{E_D} = (1/\epsilon_0)(\rho_f - \nabla \cdot \mathbf{P})$ where ρ_f is the free charge density. For perfectly insulating bilayers, the internal fields in each layer are given by [70]

$$E_{D,1} = -\frac{1}{\epsilon_0}(P_1 - < P >) = \frac{\alpha}{\epsilon_0}(P_2 - P_1), \tag{33}$$

and

$$E_{D,2} = -\frac{1}{\epsilon_0}(P_2 - < P >) = \frac{1-\alpha}{\epsilon_0}(P_1 - P_2), \tag{34}$$

where $< P > = (1-\alpha)P_1 + \alpha P_2$ is the average polarization.

It is clear that $E_{D,2} > 0$ enhances the polarization of layer 2 whereas $E_{D,1}$ attempts to decrease the polarization of layer 1 since it lies anti-parallel to the polarization vector ($E_{D,1} < 0$). Therefore, in equilibrium, it is expected that $P_1 < P_{0,1}$ and $P_2 > P_{0,2}$ so as to decrease the initial polarization difference. The total free energy functional incorporating the potential energies of the internal fields $E_{D,1}$ and $E_{D,2}$ is given by:

$$F_\Sigma = (1-\alpha)[F_1(P_1) - EP_1 - \frac{1}{2}\xi E_{D,1}P_1] + \alpha[F_2(P_2) - EP_2 - \frac{1}{2}\xi E_{D,2}P_2] + F_S/h$$
$$= (1-\alpha)[F_1(P_1) - EP_1] + \alpha(1-\alpha)[F_2(P_2) - EP_2] + \frac{1}{2}\alpha(1-\alpha)\frac{\xi}{\epsilon_0}(P_1 - P_2)^2 + F_S/h. \tag{35}$$

where E is an applied electrical field parallel to the polarization and F_S is the energy of the interfaces between the layers. We assume that the layers are relatively thick compared to the correlation length of ferroelectricity which is of the order of 1 nm [71]. Therefore, we can neglect the interface energy F_S/h even for thin bilayers with thickness of about 100 nm. This simplification does not affect polarization and the stress state within the individual layers. The polarization is given by the continuity of the normal component of the electrical displacement across the interfaces in each layer [72]. The polarizations beyond the interface area are constant in each layer and do not depend on distribution of polarization near the interface. Similarly, due to the condition of mechanical compatibility across the interfaces, the internal stresses arising from the misfit between the films and the substrate are homogeneously distributed throughout the volume of the individual layers if these layers are free of dislocations and other defects [73]. In equation 35 we have introduced a coefficient ξ which is essentially a measure of the free charge density with respect to the bound charge at the interlayer interface such that $\xi = 1 - \rho_f/\rho_b$, where ρ_b is the bound charge density. The two limiting values, $\xi = 1$ and $\xi = 0$ correspond to perfect insulating and semiconducting FE bilayers, respectively. The latter condition implies that there are sufficient free charges with high mobility to compensate for the internal fields due to the polarization mismatch. Thus, for $\xi = 0$ (and for $\xi < 0$), there is no electrostatic contribution to the total free energy due to the internal electrical field.

The equilibrium polarization of each layer is given by the simultaneous solution of the equations of state $\partial F_\Sigma/\partial P_1 = 0$ and $\partial F_\Sigma/\partial P_2 = 0$:

$$\frac{dF_1}{dP_1} = E + \xi\frac{\alpha}{\epsilon_0}(P_2 - P_1),$$
$$\frac{dF_2}{dP_2} = E + \xi\frac{1-\alpha}{\epsilon_0}(P_1 - P_1). \tag{36}$$

In the absence of an external electric field ($E = 0$), the thermodynamic potential for each layer can be extracted from the total free energy as:

$$FF_1(P_1) = F_{0,1} + \frac{1}{2}aP_1^2 + \frac{1}{4}bP_1^4 + \frac{1}{6}cP_1^6 - \frac{1}{2}\xi E_{D,1}P_1,$$
$$F_2(P_2) = F_{0,2} + \frac{1}{2}dP_2^2 + \frac{1}{4}eP_2^4 + \frac{1}{6}fP_2^6 - \frac{1}{2}\xi E_{D,1}P_1. \tag{37}$$

The analysis can be extended to a heteroepitaxial bilayer grown on a thick cubic substrate [Figure 34c] by incorporating the elastic energy of the internal stresses that results in renormalized Landau coefficients [74]:

$$
\begin{aligned}
a' &= a - x_1 \frac{4Q_{12,1}}{S_{11,1}+S_{12,1}}, \\
b' &= b + \frac{4Q_{12,1}^2}{S_{11,1}+S_{12,1}}, \\
d' &= d - x_2 \frac{4Q_{12,2}}{S_{11,2}+S_{12,2}}, \\
e' &= e + \frac{4Q_{12,2}^2}{S_{11,2}+S_{12,2}}.
\end{aligned} \tag{38}
$$

where $S_{ij,i}$ and $Q_{ij,i}$ are the elastic compliances at constant polarization and electrostrictive coefficients of material i, respectively. $x_i = (a_S - a_i)/a_S$ are the (polarization-free) misfit strains of layer i with respect to substrate, where a_i are the unconstrained equivalent cubic cell constants of layer i and a_S is the lattice parameter of the substrate. For a pseudomorphic bilayer with $h < h_\rho$ where h_ρ is the critical thickness for misfit dislocations, these misfit strains are not independent and the relation between them is given by $x_2 = 1 - [a_2(1-x_1)/a_1]$.

In order to quantify the interlayer coupling effect explicitly, we can rearrange equation 35 and normalize the coefficients,

$$
\begin{aligned}
F_\Sigma &= (1-\alpha)[F_1(P_1) + \frac{1}{2}\frac{\xi}{\epsilon_0}\alpha P_1^2] + \alpha[F_2(P_2) + \frac{1}{2}\frac{\xi}{\epsilon_0}(1-\alpha)P_2^2] + F_{el} - JP_1P_2 \\
&= (1-\alpha)[F_{0,1} + \frac{1}{2}\bar{a}P_1^2 + \frac{1}{4}\bar{b}P_1^4 + \frac{1}{6}\bar{c}P_1^6] + \alpha[F_{0,1} + \frac{1}{2}\bar{d}P_1^2 + \frac{1}{4}\bar{e}P_1^4 \\
&\quad + \frac{1}{6}fP_1^6] + F_{el} - JP_1P_2,
\end{aligned} \tag{39}
$$

where the normalized coefficients are given by:

$$
\bar{a} = \begin{cases} a + \frac{\xi}{\epsilon_0}\alpha & unconstrained \\ a' + \frac{\xi}{\epsilon_0}\alpha & heteroepitaxial \end{cases}, \tag{40}
$$

$$
\bar{d} = \begin{cases} d + \frac{\xi}{\epsilon_0}(1-\alpha) & unconstrained \\ d' + \frac{\xi}{\epsilon_0}(1-\alpha) & heteroepitaxial \end{cases}, \tag{41}
$$

and $\bar{b} = b$, $\bar{e} = e$ for the unconstrained bilayer and $\bar{b} = b'$, $\bar{e} = e'$ for the heteroepitaxial bilayer. F_{el} is the elastic energy of the polarization-free misfit:

$$
F_{el} = \frac{x_1^2}{S_{11,1}+S_{12,1}} + \frac{x_2^2}{S_{11,2}+S_{12,2}}, \tag{42}
$$

which is absent for the unconstrained bilayer. The last term in equation 39 is the interlayer coupling and J is the coupling coefficient given by:

$$
J = \alpha(1-\alpha)\frac{\xi}{\epsilon_0}. \tag{43}
$$

The relation for the coupling coefficient indicates that the coupling between layers can be controlled by either the fraction of each layer or by modifying the free carrier content quantified by coefficient ξ. For example, if there are sufficient free charge carriers within the material that are mobile, ξ is zero and the bilayer is decoupled and behaves as if it is in a series connection (i.e., with an electrode layer between the ferroelectric layers). We note that the interlayer interactions also modify first Landau coefficients via equations 40 and 41 and hence may alter phase transformation characteristics.

Using the equilibrium polarization of each layer given by $\partial F_{\Sigma}/\partial P_1 = 0$ and $\partial F_{\Sigma}/\partial P_2 = 0$, the (small-signal) average dielectric response of a perfectly insulating bilayer (ξ=1) can be determined as:

$$< \epsilon_R > \simeq \frac{1}{\epsilon_0}\frac{d < P >}{dE} = \frac{1}{\epsilon_0}[(1-\alpha)\frac{\delta P_1}{E} + \alpha\frac{\delta P_2}{E}], \tag{44}$$

where $< P > = (1-\alpha)P_1 + \alpha P_2$ is the average polarization, $\delta P_i = P_i(E) - P_i(E=0)$ as $E \rightarrow 0$, and P_i satisfying equations 36.

It is clear that the electrostatic energy of the internal fields in equation 37 serves to introduce a symmetry-breaking element in the otherwise symmetric Landau potentials having even powers of the polarization. Consider the example of an equi-fraction ($\alpha = 1/2$), stress-free $BaTiO_3$-$Ba_{0.9}Sr_{0.1}TiO_3$ (BT-BST 90/10) bilayer. In Figure 35, we plot the normalized free energy as a function of the net polarization of each layer. The "degree" of symmetry-breaking strikingly varies with the variations in the density of free charges in the bilayer. For instance, for $\xi = 0$ [Figure 35a], the two layers are electrically screened from one another and can act essentially independent and are unrelated entities. This is due to the complete compensation of the electrical field resulting from bound charges at the interlayer interface through free charges. An equivalent construction of this condition would be two ferroelectric layers with a metallic electrode between them, i.e., two dielectrics with a series connection. Both layers display typical symmetric double well potentials, whose minima correspond to two energetically identical FE ground states. The equilibrium polarizations of the layers assume their values in single-crystal form. Due to the free charges, a relative large polarization difference ΔP (for BT-BST90/10 bilayer, $\Delta P \approx 0.056 C/m^2$) can be maintained between the two layers.

The electric coupling between layers due to the polarization mismatch is evident for non-zero values of ξ where this coupling is enhanced with increasing ξ, i.e., a reduction in the free charges. Figures 35 (b)-(d) show that the otherwise symmetric double wells of BT and BST90/10 are skewed towards one FE equilibrium state with $P > 0$. The other FE ground state with $P < 0$ becomes *metastable*. The strongest value of electrical coupling between the layers corresponds to a bilayer made up of two completely insulating FEs, i.e., no free charges, $\xi = 1$. This results in only one stable FE ground state in both layers and a small polarization difference between two layers ($\Delta P = 0.037$ C/m^2). The other FE ground state becomes *unstable* due to the electrical interaction between the layers. This shows that the initial "uncoupled" polarization gradient in insulating graded FEs becomes smoother due to the

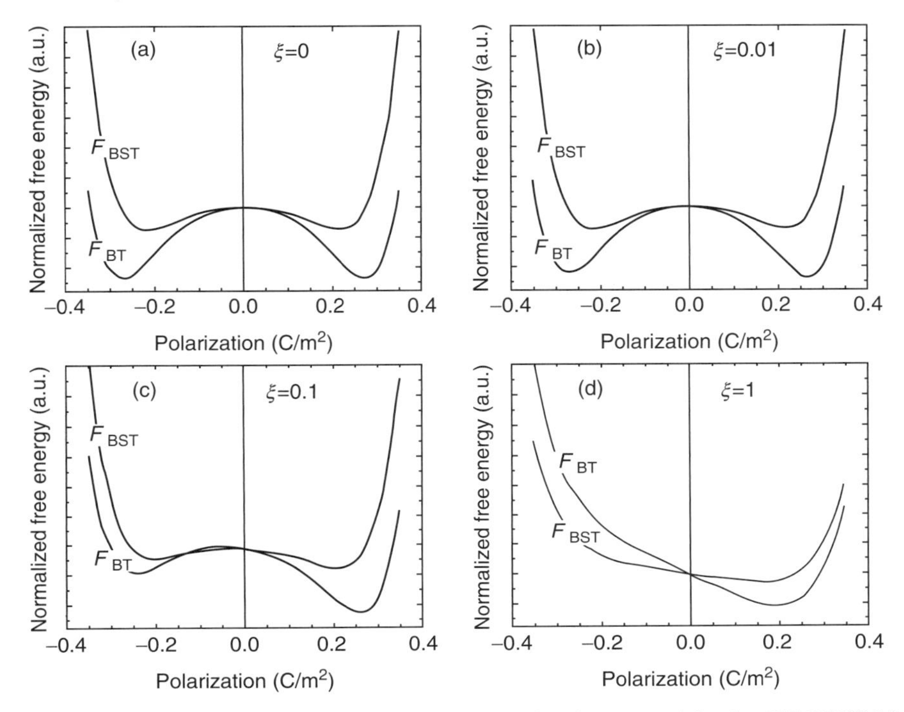

Fig. 35 Free energy potentials as a function of polarization for an equal-fraction BT BST90/10 bilayer: (a) $\xi = 0$, (b) $\xi = 0.01$, (c) $\xi = 0.1$, and (d) $\xi = 1$.

electrostatic interactions between the layers thus resulting in a smaller polarization difference that seeks to minimize the internal electric field. The above results are consistent with both the *ab initio* and density functional theory calculations [75, 76] wherein they also concluded that compositional variations result in a broken inversion symmetry that likewise leads to asymmetric thermodynamic potentials. We show that whenever the free charge density is less than the bound charges density, an internal potential arises from the compositional inhomogeneity always resulting in asymmetric potentials.

We consider a heteroepitaxial (001) ST BT bilayer on a thick (001) ST substrate such that $x_{ST} = 0\%$ in Eq. (8) at RT. The strain in the BT layer is $x_{BT} = 2.28\%$. The equilibrium polarization of a BT film at this strain level is 0.38 C/m^2. The spontaneous polarization in the BT layer in the ST BT bilayer for $E = 0$ decreases from this value with increasing volume fraction of the ST layer until it completely vanishes at a critical relative thickness of $\alpha_{ST} = 0.66$ (Figure 36a). Figure 36b plots the difference between polarizations in the coupled BT and ST layers $\Delta P = P_{BT} - P_{ST}$. ΔP is small (typically less than 1% of the polarization of BT) indicating that the induced polarization in the ST layer almost equals to polarization in BT layer. As the relative fraction of the ST layer increases, there is a commensurate rise in the depoling field in the BT layer as well as a drop in the internal field in the ST layer that induces polarization. Eventually, a critical relative thickness is reached at

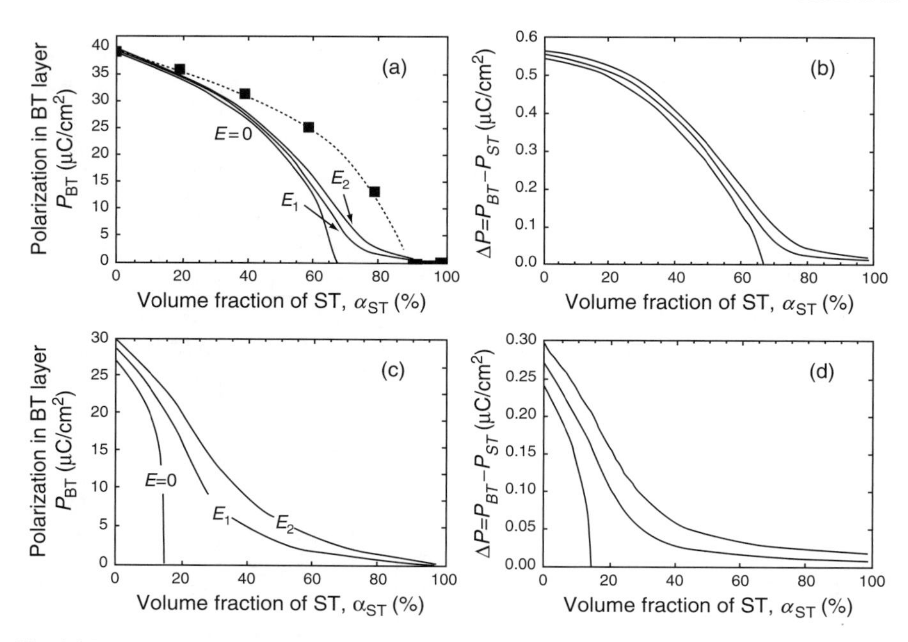

Fig. 36 Dependence on ST fraction, α_{ST}, of polarization in BT layer under different external fields (E=0, E_1=100 kV/cm, E_2=200 kV/cm) and the polarization difference, $\Delta P = P_{BT} - P_{ST}$: (a) and (b) for constrained BT ST bilayer; x_{BT}= 2.28% and x_{ST} = 0% (solid squares: first-principles results from Ref. [77]); (c) and (d) for unconstrained BT ST bilayer.

$\alpha_{ST} = 0.66$ which corresponds to $P_1 = P_2 = 0$, the only solutions of Eqs. (5) for $\alpha_{ST} \geq 0.66$. The equilibrium polarization in BT layer and the polarization difference between two layers for a completely relaxed and unconstrained system has similar behavior as shown in Figure 36c and d. Without the biaxial internal stress, the polarization in the BT is $\sim 0.27\text{C/m}^2$ for $\alpha_{ST} \approx 0$ and disappears at a critical relative thickness $\alpha_{ST} = 0.14$.

The dielectric responses of the heteroepitaxial and unconstrained BT ST bilayers are presented in Figure 37 that display an anomaly at the critical relative thicknesses. The analogy with temperature dependence of polarization and dielectric response as well as with smearing of the dielectric anomaly under an applied electric field is obvious (Figure 36).

Comparison of results of our scale independent analysis in bilayers with results of first principal calculations for superlattices allows one to conclude that the effects of electrostatic interactions considered above should be observed in thin films as well. The polarization of superlattices with period equaling five atomic planes [77] shown in Figure 37a (solid squares) demonstrates a similar dependence on the layer fraction as a macroscopic ferroelectric-paraelectric bilayer. A critical fraction of $\sim$0.9 can be expected on the basis of extrapolation of microscopic data to the zero polarization (dashed line in Figure 36a). It is clear that increasing the period of the superlattices should decrease the deviation between the results of macroscopic and microscopic analysis.

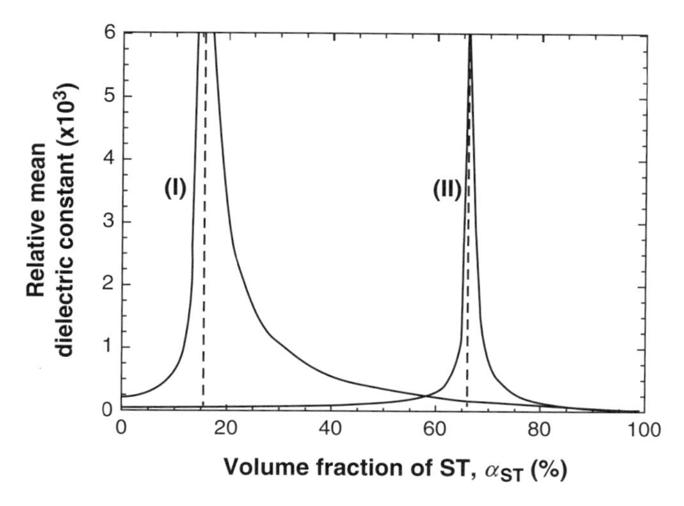

Fig. 37 Relative mean dielectric constant as a function of volume faction, α_{ST}: (I) for unconstrained BT ST bilayer; (II) for strained heteroepitaxial BT ST bilayer, x_{BT}= 2.28% and x_{ST} = 0%.

3.2.2 Theoretical Analysis of Polarization-Graded Ferroelectrics

Building upon the bilayer model, in this section we will analyze polarization-graded ferroelectrics. The potential of these more complicated structures follows from the bilayer analysis where the coupling between layers is established through internal electrical, mechanical, and electromechanical interactions. We will use the same analogy in the next section to establish the relations in compositionally graded ferromagnetic materials,

Consider a perovskite ferroelectric oxide such as $BaTiO_3$, which exhibits a cubic-tetragonal ferroelectric phase transformation. Systematic variations in the polarization can be achieved (and have been achieved experimentally, see [78–80]) in a number of ways including a variation in the composition of the material, impressing temperature gradients across the structure, or by imposing non-uniform external stress fields, as illustrated in Figure 38. For analysis, we assume a ferroelectric of thickness L sandwiched between two metallic electrodes with the easy axis of polarization along the z-axis such that $\mathbf{P} = [0,0,P(z)]$. The ferroelectric is assumed to be homogeneous along the $x-$ and $y-$ directions, reducing the problem to only one dimension. Taking into account the depoling effect and the energy of the internal stresses due to variations in the self-strain (or lattice parameter variations), the Gibbs free energy (per unit area) is given by:

$$G=\int_0^L dz[\frac{1}{2}\alpha P(z)^2+\frac{1}{4}\beta P(z)^4+\frac{1}{6}\gamma P(z)^6+\frac{1}{2}D(\frac{dP(z)}{dz})^2-\frac{1}{2}\xi E_D(z)\cdot P(z)+F_{el}(z)] \tag{45}$$

The above relation contains terms that were discussed previously: the Landau expansion in terms of polarization, the Ginzburg gradient term to account for the

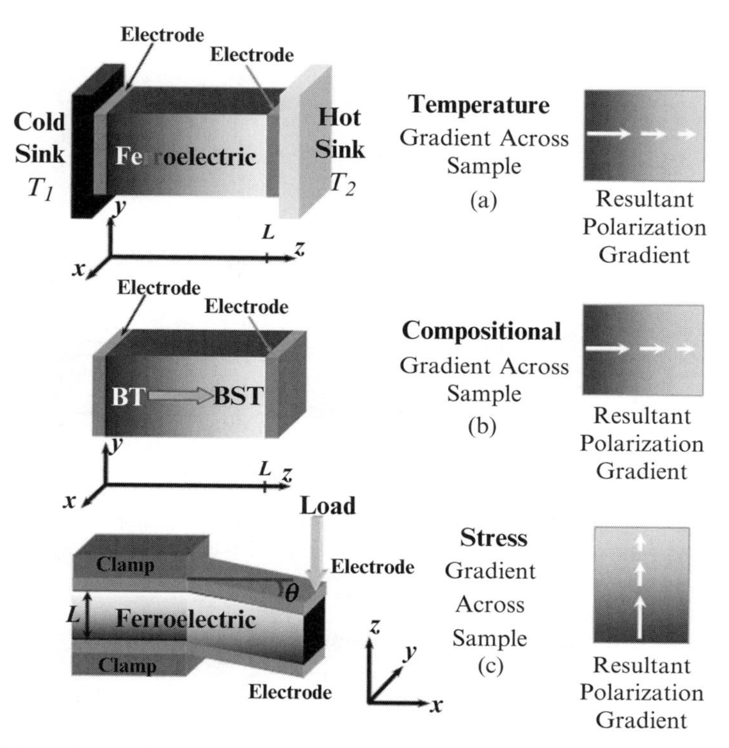

Fig. 38 Polarization-graded ferroelectrics may be achieved through: (a) temperature, (b) stress, and (c) compositional gradients.

inhomogeneous state of polarization, and the built-in internal electric field. Similar to this electrostatic field, there is a built-in, position dependent stress field within a ferroelectric with polarization variations. This is due the electrostrictive coupling between the polarization and the self-strain. There exists a biaxial stress state with equal orthogonal components in the xy-plane of each layer and the corresponding mechanical boundary conditions are given by $\sigma_1 = \sigma_2$, and $\sigma_3 = \sigma_4 = \sigma_5 = \sigma_6 = 0$, where σ_i are the components of the internal stress tensor. Thus, the total internal elastic energy due to the self-strain gradient for each "layer" can then be expressed as:

$$F_{el} = \frac{1}{2}(\sigma_1 x_1 + \sigma_2 x_2) \tag{46}$$

where $x_1 = x_2$, and $\sigma_1 = \sigma_2 = \sigma$ are the internal stresses and strains, respectively, in the xy-plane. We note that although the out-of-plane strain $x_3 \neq 0$, it will not have an effect on the strain energy since the material is unconstrained in the z-direction and hence $\sigma_3 = 0$. The in-plane strain is determined by the condition that both the average internal stress and the average momentum of the internal stress should be zero, such that

$$F_{el} = \bar{C}[Q_{12}[P(z)^2 - <P>^2] + (z - \frac{L}{2})\kappa]^2 \tag{47}$$

where Q_{12} is the electrostrictive coefficient in a direction normal to the polarization vector, κ is the radius of curvature, and $\bar{C}$ is an effective elastic constant given by:

$$\bar{C} = C_{11} + C_{12} - 2\frac{C_{12}^2}{C_{11}}, \tag{48}$$

and C_{ij} are the elastic moduli at constant polarization.

We will assume that the contribution of the internal electrostatic field is small. This can be achieved through very smooth gradients, by charge compensation, or by the formation of electrical domains as shown theoretically by Roytburd *et al.* [81]. Hence, the minimization of the free energy with respect to the polarization in the absence of an external electric field yields the Euler-Lagrange equation:

$$D\frac{d^2P}{dz^2} = AP + BP^3 + \gamma P^5 \tag{49}$$

with renormalized coefficients:

$$\begin{aligned} A &= \alpha + 4\bar{C}Q_{12}[(z - \frac{L}{2})\kappa - Q_{12} < P >^2], \\ B &= \beta + 4\bar{C}Q_{12}^2. \end{aligned} \tag{50}$$

For compositionally graded ferroelectrics, A, B, and D are functions of the composition, and therefore are location dependent [i.e., $A(z), B(z)$, and $D(z)$]. For temperature-graded ferroelectrics, A and D are a function of the temperature and are location dependent. Assuming that steady-state heat transfer is established (i.e., the temperature across the ferroelectric bar in Figure 38 is a linear function of the position), the normalized coefficient A in equation 50 becomes:

$$A(z) = \frac{L(T_1 - T_0) + z(T_2 - T_1)}{L\epsilon_0 C} + 4\bar{C}Q_{12}[(z - \frac{L}{2})\kappa - Q_{12} < P >^2]. \tag{51}$$

A strain-graded ferroelectric can be analyzed in terms of a simple cantilever beam setup. A bending force is applied along the $z-$direction resulting in a systematic variation along the $z-$direction for the normal strain (compressive or tensile). The resulting stress conditions in the cantilever beam are $\sigma_1 \neq 0$ and $\sigma_2 = \sigma_3 = 0$. For a small bending angle, the coupling between the polarization and the applied bending force can be neglected yielding the Euler-Lagrange equation:

$$D\frac{d^2P}{dz^2} = (\alpha - 2Q_{12}C^\star x_1)P + \beta P^3 + \gamma P^5, \tag{52}$$

where

$$C^\star = C_{11} - \frac{2C_{12}^2}{C_{11} + C_{12}} \tag{53}$$

is an effective modulus for bending and $x)1(z)$ is the position dependent normal strain due to the external bending force. The normal strain $x_1(z)$ is related to the bending angle θ and the length of the bent beam S through:

$$x_1(z) = (z - \frac{L}{2})\frac{\theta}{S}. \tag{54}$$

Basic electrostatic theory shows that an inhomogeneous distribution of the polarization is associated with a bound charge:

$$\rho_v = -\nabla_i \cdot P_j = -\frac{dP(z)}{dz}, \tag{55}$$

where ρ_v is the volume density of the bound charge. This bound charge generates a built-in electrical field and the resulting built-in potential V_{int} is given by:

$$V_{int} = -\frac{1}{C_F L}\int_0^L z\rho_v(z)dz = \frac{1}{C_F L}\int_0^L z(\frac{dP(z)}{dz})dz, \tag{56}$$

where C_F is the capacitance of the graded ferroelectric. Then, the charge offset due to this built-in potential is:

$$\Delta Q = C_Q V_{int} = \frac{k}{L}\int_0^L z(\frac{dP(z)}{dz})dz, \tag{57}$$

where C_Q is the capacitance of the load capacitor in a Sawyer-Tower circuit and $k = C_Q/C_F$.

Using the boundary conditions $dP/dz = 0$ at $z = 0$ and $z = L$ corresponding to complete charge compensation at the ferroelectric/electrode interfaces, we plot the polarization profile normalized with respected to the average polarization $< P >$ for the three cases in Figure 39. For the analysis we have chosen $BaTiO_3$

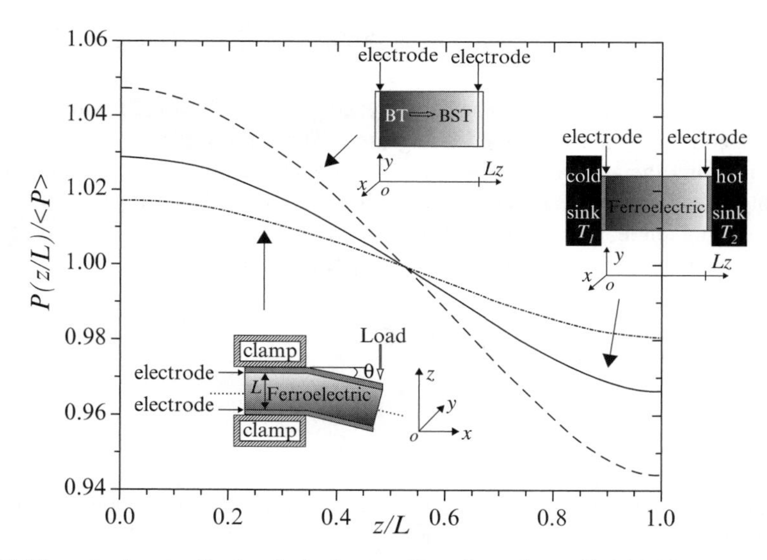

Fig. 39 Theoretical normalized polarization profiles along the z direction for temperature-, composition-, and strain-graded systems.

as the prototypical system for the temperature and strain graded ferroelectrics, and $Ba_xSr_{1-x}TiO_3$ for the analysis of a compositionally graded ferroelectric system with $0 < x < 1$.

The normalized polarization decreases monotonically across the structure with a variation of the polarization along z−direction predicted in all three graded systems. The polarization gradient diminishes close to the surfaces because of the boundary conditions. For the temperature graded $BaTiO_3$ system, the hot end and cold end are chosen to be at T_C and room temperature (RT=25°C) respectively. If the hot end temperature is higher than T_C, a paraelectric region with no spontaneous polarization will form at this end; a feature important for analyzing the charge offset behavior with respect to the temperature as well. Within the ferroelectric region, the polarization profile exhibits similar behavior as the one shown in Figure 39. The end point compositions for compositionally graded systems are $BaTiO_3$ (at $z = 0$) and $Ba_{0.75}Sr_{0.25}TiO_3$ (at $z = L$) and a linear relationship between the composition and the position is assumed. It is worth mentioning that the magnitude and the direction of the polarization gradient depend upon the temperature, composition and strain gradients. For example, the direction of the polarization gradient will be reversed if the positions of the hot and cold heat sinks are exchanged. In the strain graded $BaTiO_3$ system, the deflection angle θ of the level arm is taken as 1.5°. A more abrupt polarization gradient should be expected with increasing θ.

In addition to the polarization hysteresis offset, compositionally graded ferroelectrics have peculiar dielectric properties. The dielectric response of compositionally graded ferroelectrics as a function of temperature exhibits characteristics of a diffuse phase transformation, which is inherently linked, with the distribution of the phase transformation temperature resulting from the composition gradient across the ferroelectric. Figure 40 shows the variation in the average dielectric response as

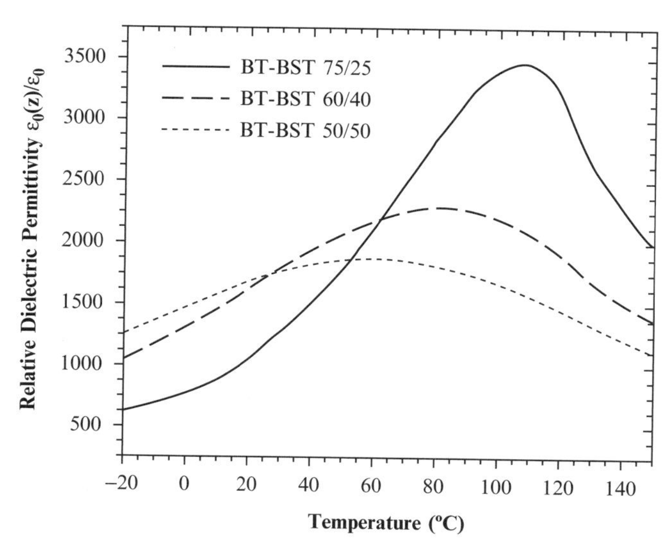

Fig. 40 The relative average dielectric permittivity as a function of T for a compositionally graded BT

a function of temperature for three compositionally graded $Ba_xSr_{1-x}TiO_3$ systems with one end fixed at $BaTiO_3$. In comparison to a sharp peak of the dielectric permittivity at T_C for bulk homogenous ferroelectrics, a diffused dielectric response with the temperature can be expected for compositionally graded ferroelectrics. The maximum in the dielectric permittivity is broadened. The extent of this broadening depends on the composition gradient. A steeper composition gradient can give rise to a broader maximum. This behavior has been documented experimentally in compositionally graded $Ba_xSr_{1-x}TiO_3$ ferroelectric thin films where a more pronounced broad plateau region of the permittivity with the variation of the temperature was observed for $Ba_{0.5}Sr_{0.5}TiO_3$-$BaTiO_3$ graded thin film compared to $Ba_{0.75}Sr_{0.25}TiO_3$-$BaTiO_3$ graded film [82].

For polarization-graded ferroelectrics, the hysteresis loop is shifted "up" or "down" along the polarization axis, resulting in a charge offset. Hence, an effective (or pseudo-) pyroelectric coefficient p_{eff} can been defined as:

$$p_{eff} = \frac{d(\Delta Q)}{dT} = \frac{k}{L}\frac{d}{dT}\int_0^L z(\frac{dP}{dz})dz. \tag{58}$$

Figure 41 shows the theoretically calculated temperature dependence of the charge offset and the effective pyroelectric coefficient for compositionally graded $Ba_xSr_{1-x}TiO_3$. The end compositions are chosen to be $BaTiO_3$ and $Ba_{0.7}Sr_{0.3}TiO_3$ respectively. A maximum effective pyroelectric coefficient is expected to occur around 15°C, corresponding to the onset of paraelectricity at the $Ba_{0.7}Sr_{0.3}TiO_3$ end.

In addition to the unique dielectric and pyroelectric properties of polarization graded structures, a static bending can be expected in these materials, forming a vertical displacement. This is due to the built-in strain gradient that arises from the

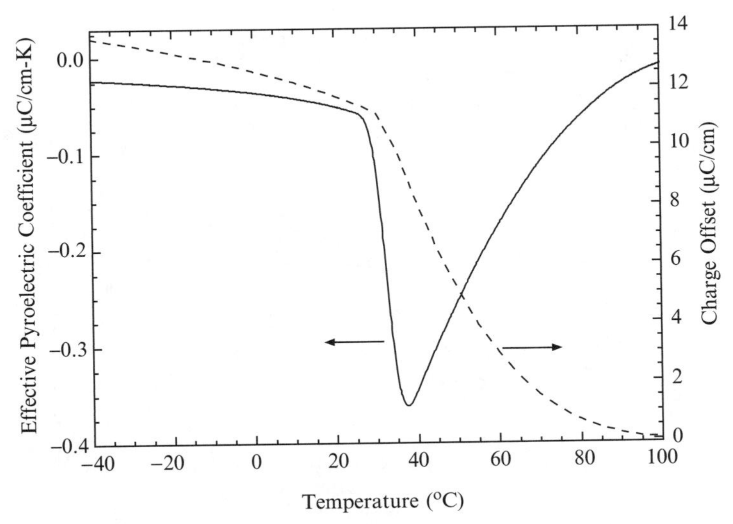

Fig. 41 Effective pyroelectric coefficient (solid line) and charge offset per unit area (dashed line) as a function of temperature in graded BT BST 75/25.

grading of the composition and concomitantly, the spontaneous self-strain. As an example, we consider two compositionally graded systems: BT-BST with $0 < x < 0.25$ and $PbTiO_3$-$Pb_{1-x}Zr_xTiO_3$ (PT-PZT) with $0 < x < 0.6$. A graded FE plate with a thickness much smaller than its lateral dimensions will form a dome shape similar to due to the built-in strain gradient. The static displacement of this dome-shaped structure can be determined from earlier equations, and the condtions of mechanical equilibrium as [83]:

$$d(E) = \frac{1}{\kappa} - \sqrt{(\frac{1}{\kappa})^2 - (\sin\frac{L}{2})^2}, \tag{59}$$

where κ and L are the curvature and the lateral dimension of the plate, respectively.

The results show a large non-linear strain along the plate thickness due to the built-in stress. For a plate with dimensions of 25.4mm×25.4mm×0.5mm graded along the plate thickness, the dome that forms may have a height as much as ~101% and ~18% of the plate thickness for PT-PZT 60/40 and BT-BST 75/25, respectively. Graded BT-BST 75/25 has a lower spontaneous displacement due to an overall a smaller spontaneous polarization and self-strain compared to graded PT-PZT 60/40 [84]. This effect is comparable to RAINBOW™ (Reduced and Internally Biased Oxide Wafer) [85–88] and THUNDER™ (Thin Unimorph Driver) [89, 90] ceramics of similar physical dimensions [85]. It should be noted that the dome shape in graded ferroelectrics is a result of the spatial variation of the self-strain while the dome shape in THUNDER™ devices is due to the thermal expansion coefficient difference between layers.

For actuator applications, the relative displacement under the external electric field is of primary interest. We can define this dynamic response as the "unblocked" displacement, ΔD, relative to the static curvature resulting from the compositionally induced strain gradient as:

$$\Delta D(E) = \frac{d(E) - d(E=0)}{d(E=0)} \times 100\%. \tag{60}$$

Fig. 42 shows ΔD for a 25.4mm×25.4mm×0.5mm graded PT-PZT 60/40, PT-PZT 80/20, and BT-BST 75/25 plates, respectively. As the field increases, the relative displacement can reach as high as 23% at 50kV/cm for PT-PZT 60/40 though it is lower in PT-PZT 80/20 and BT-BST 75/25 due to the smaller polarization gradient. It is important to point out the applied electric field opposes the spontaneous polarization within the graded structure. As the opposing electric field is increased, it will eventually suppress and reverse the polarization and the bending direction. In the case of the electric field along the polarization, the displacement will decrease with increasing electric field, i.e. negative ΔD. The reason for this is that the end with weaker polarization is more sensitive to the electric field, for example PZT end and BST end, which has higher susceptibility and is thus much easier to be polarized in the presence of the electric field. Therefore, the polarization gradient will be washed out and the resulting built-in strain will diminish.

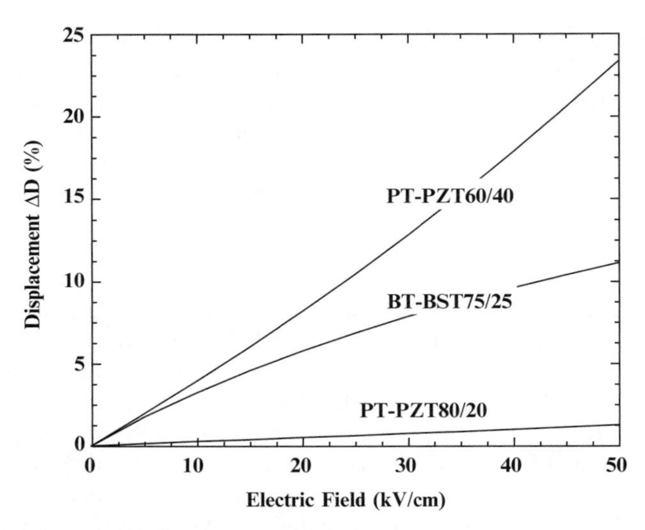

Fig. 42 Displacement relative to the static deformation caused by the compositionally induced strain gradient, ΔD, as a function of external electric field for PT-PZT 60/40, PT-PZT 80/20 and BT-BST 75/25.

3.3 *Functionally Graded Ferromagnets*

The concepts developed for polarization graded ferroelectrics can be immediately expanded to include all ferroic systems. The unique pproperties of graded ferroelectrics arise from an internal "built-in" electric field as was shown in the previous sections. To summarize, for electrically insulating polarization-graded ferroelectrics (where the dielectric displacement, **D**, obeys the relation $\nabla \cdot \mathbf{D} = 0$), a straightforward application of Maxwell's equations relates a spatial variation in polarization to a dielectric based internal electrostatic potential according to:

$$\nabla^2 V(\mathbf{r}) = \frac{1}{\epsilon_0} \nabla \cdot \mathbf{P}(\mathbf{r}) \tag{61}$$

where $\mathbf{P}(\mathbf{r})$ is the position dependent dielectric polarization (induced and spontaneous).

Intuitively, it follows that this approach should be extendable to other ferroic systems, most particularly to magnetization-graded ferromagnets, the subject of this report. From Maxwell's equations (where $\nabla \cdot \mathbf{B} = 0$, with **B** the magnetic induction), a gradient in magnetization, **M**, yields an internal *magnetostatic potential* according to the relation:

$$\nabla^2 \phi(\mathbf{r}) = \nabla \cdot \mathbf{M}(\mathbf{r}). \tag{62}$$

The magnetic scalar potential, $\phi(\mathbf{r})$, is defined through $-\nabla\phi(\mathbf{r}) = \mathbf{H_{int}}(\mathbf{r})$ where $\mathbf{H_{int}}$ is the internal magnetic field. The spatial invariance of $\mathbf{M}(\mathbf{r})$ arises from both the permeability and the spontaneous magnetization.

To fully understand the origin of this internal field bias we must go beyond Equation (35) and account for the interdependence of the magnetization and the internal strains that arise as a result of the compositional gradient. Here, we again employ the Landau theory of phase transformations as a guide.

Let us consider a compositionally graded ferromagnetic bar of length L. The grading is along the length L ($z-$direction) which coincides with the easy magnetization direction of a ferromagnet with $\mathbf{M}$=[0,0,$M(z)$]. If there are no external fields acting on the sample, the free energy functional is given by:

$$F[M(z)] = \int_0^L dz[\alpha(z)M(z)^2 + \beta(z)M(z)^4 + \frac{1}{2}A(\frac{dM(z)}{dz})^2 - \frac{1}{2}N_D M(z)^2 + F_{EL}[M(z), C_{ijkl}(z), \lambda_{ijkl}(z)]] \tag{63}$$

where $(1/2)N_D M(z)^2 = H_D M(z)$ is the magnetostatic energy due to the demagnetization, H_D is the demagnetizing field, and N_D is a demagnetization factor that depends only on the geometry of the sample, C_{ijkl} are elastic moduli at constant magnetization, and λ_{ijkl} are the magnetostrictive coefficients. The demagnetization is negligible if the length along which the magnetization varies is much larger than the lateral dimensions. For such samples, the free energy relation can be organized via blending in the elastic energy into the first two terms of the Landau expansion via renormalized coefficients $\bar{\alpha}$ and $\bar{\beta}$ yielding [68, 91]:

$$F[M(z)] = \int_0^L dz[\bar{\alpha}[C_{ijkl}, \lambda_{ijkl}]M(z)^2 + \bar{\beta}[C_{ijkl}, \lambda_{ijkl}]M(z)^4 + \frac{1}{2}A(\frac{dM(z)}{dz})^2 + \bar{F}_{el}[C_{ijkl}(z)]], \tag{64}$$

where $\bar{F}_{el}$ is the magnetization-free elastic energy.

The equilibrium magnetization profile is then given by the equation of state, $\partial F/\partial M = 0$, in the form of a Euler-Lagrange equation:

$$A\frac{d^2M(z)}{dz^2} = 2\bar{\alpha}M(z) + 4\bar{\beta}M(z)^3. \tag{65}$$

These relations clearly show that the physics of graded ferromagnets can be described in a fashion quite similar to graded ferroelectrics. A direct implication of this is that we should expect unique properties from graded ferromagnets as well. Obviously, these are a shift in the magnetization hysteresis loops, a pseudo piezo- and pyromagnetic response based on the magnetization offset, and diffuse magnetic susceptibilities that depend on the "strength" of the grading.

This brings us to an end of our discussion of functionally graded ferroelectric and ferromagnetic alloy materials. The connection between these materials and the graded polar semiconductor heterostructures can be gauged by the similarity of electrostatic Maxwell's equations required to understand them. The major differences that distinguish the physics of these three classes of materials are related to their respective conductivities, and the structural phase transitions that characterize traditional ferroelectrics, but are absent in semiconductors and ferromagnets. However, as alluded to in the introduction to this chapter, the differences become blurry in multifunctional materials that possess more than one of the three characteristics: semiconducting, ferroelectric, or ferromagnetic. For such materials, one has to start using the universal physics outlined in this subsection.

4 Summary and Challenges

In summary, a rich range of physical phenomena were outlined for graded polar heterostructures composed of semiconductors, ferroelectrics, or ferromagnets. Graded polarization doping was demonstrated in III-V nitride semiconductors, and concrete examples showed the technological relevance of this doping technique. Similarly, it was shown that graded ferroelectrics and ferromagnets show features that are not possible otherwise. The physics of materials that bridge the properties of these three classes of materials is yet to be completely developed. Quite possibly, our understanding of multifunctional materials will develop with progress in epitaxial growth capabilities, and availability of high-quality defect free layer structures.

Ferroelectric-semiconductor hybrid layered structures are already being investigated; the chapter by Wu, Singh, & Singh in this book describes various device applications of such structures. Recently, 2DEGs have been observed at perovskite heterostructures very similar to those in traditional semiconductor heterostructures [92]. The presence of doping and vacancies in perovskite structures has been observed to lead to blue emission [93, 94]. These phenomena are traditionally studied by semiconductor physicists and device engineers. There are indications that with the improvement of epitaxial growth technologies, a rich range of physical phenomena will be uncovered in perovskite structures (of which the ferroelectrics are the most widely studied). By using lessons learnt from traditional semiconductor devices, and by exploiting the richness of properties of perovskites, a wide range of new devices are possible. Some possible applications are in sensors, microwave devices, and optoelectronic devices. The major challenges are the capability to grow structurally and chemically perfect multifunctional materials, and to develop technologies to probe their new physical properties.

References

1. E. archive New Semiconductor Materials Characteristics and Properties *http://www.ioffe.rssi.ru/SVA/NSM/*.
2. P. Yu and M. Cardona, *Fundamentals of Semiconductors, Physics and Materials Properties*. Berlin: Springer Verlag, 1st ed., 1996.

3. O. Ambacher, B. Foutz, J. Smart, J. R. Shealy, N. G. Weimann, K. Chu, M. Murphy, A. J. Sierakowski, W. J. Schaff, L. F. Eastman, R. Dimitrov, A. Mitchell, and M. Stutzmann, "Two-dimensional electron gases induced by spontaneous and piezoelectric polarization in undoped and doped AlGaN/GaN heterostructures," *J. Appl. Phys.*, vol. 87, p. 334, 2000.
4. C. Mailhiot and D. L. Smith, "Electronic structure of [001]- and [111]-growth-axis semiconductor superlattices," *Phys. Rev. B*, vol. 35, p. 1242, 1987.
5. D. C. Look and R. J. Molnar *Appl. Phys. Lett.*, vol. 70, p. 3377, 1997.
6. J. W. P. Hsu, D. V. Lang, S. Richter, R. N. Kleiman, A. M. Sergent, and R. J. Molnar *Appl. Phys. Lett.*, vol. 77, p. 2673, 2000.
7. S. Heikman, S. Keller, S. P. DenBaars, and U. K. Mishra *Appl. Phys. Lett.*, vol. 81, p. 439, 2002.
8. J. P. Ibbetson, P. T. Fini, K. D. Ness, S. P. DenBaars, J. S. Speck, and U. K. Mishra, "Polarization effects, surface states, and the source of electrons in AlGaN/GaN heterostructure field effect transistors," *Appl. Phys. Lett.*, vol. 77, p. 250, 2000.
9. S. Keller, S. Heikman, L. Shen, I. P. Smorchkova, S. P. DenBaars, and U. K. Mishra *Appl. Phys. Lett.*, vol. 80, p. 4387, 2002.
10. P. Waltereit, O. Brandt, A. Trampert, H. T. Grahn, J. Menniger, M. Ramsteiner, M. Reiche, and K. Ploog, "Nitride semiconductors free of electrostatic fields for efficient white light-emitting diodes," *Nature*, vol. 406, p. 865, 2000.
11. R. Cingolani, A. Botchkarev, H. Tang, H. Morkoç, G. Traetta, G. Coli, M. Lomascolo, A. Di Carlo, F. Della Sala, and P. Lugli *Phys. Rev. B*, vol. 61, p. 2711, 2000.
12. U. K. Mishra, P. Parikh, and Y. F. Wu, "AlGaN/GaN HEMTS: An overview of device operation and applications," *Proceedings of the IEEE.*, vol. 90, p. 1022, 2002.
13. A. Jimenez, D. Buttari, D. Jena, R. Coffie, S. Heikman, N. Zhang, L. Shen, E. Calleja, E. Munoz, J. Speck, and U. K. Mishra *IEEE Elect. Dev. Lett.*, vol. 23, p. 306, 2002.
14. T. F. Kuech, R. T. Collins, D. L. Smith, and C. Mailhiot, "Field-effect transistor structure based on strain-induced polarization charges," *J. Appl. Phys.*, vol. 67, p. 2650, 1990.
15. E. S. Snow, B. V. Shanabrook, and D. Gammon *Appl. Phys. Lett.*, vol. 56, p. 758, 1990.
16. P. Kozodoy, I. P. Smorchkova, M. Hansen, H. Xing, S. P. DenBaars, U. K. Mishra, A. W. Saxler, R. Perrin, and W. C. Mitchel *J. Appl. Phys.*, vol. 75, p. 2444, 1999.
17. P. M. Asbeck, E. T. Yu, S. S. Lau, W. Sun, X. Dang, and C. Shi, "Enhancement of base conductivity via the piezoelectric effect in AlGaN/GaN HBTs," *Solid-State Electron.*, vol. 44, p. 211, 2000.
18. M. Singh, Y. Zhang, J. Singh, and U. K. Mishra *Appl. Phys. Lett.*, vol. 77, p. 1867, 2000.
19. L. Pfeiffer, K. W. West, H. L. Stormer, and K. W. Baldwin *Appl. Phys. Lett.*, vol. 55, p. 1888, 1989.
20. M. Shayegan, T. Sajoto, M. Santos, and C. Silvestre *Appl. Phys. Lett.*, vol. 53, p. 791, 1988.
21. A. C. Gossard, M. Sundaram, and P. F. Hopkins, *Epitaxial Microstructures, Semiconductors and Semimetals, vol 40.* San Diego: Academic Press, 1st ed., 1994.
22. B. Heying, R. Averbeck, L. F. Chen, E. Haus, H. Riechert, and J. S. Speck *J. Appl. Phys.*, vol. 88, p. 1855, 2000.
23. O. Brandt, P. Waltereit, and K. Ploog *J. Phys. D: Appl. Phys.*, vol. 35, p. 577, 2002.
24. G. L. Snider *1DPoisson*, http://www.nd.edu/g̃snider/.
25. F. Bernardini, V. Fiorentini, and D. Vanderbilt, "Spontaneous polarization and piezoelectric constants of III-V nitrides," *Phys. Rev. B*, vol. 56, p. R10 024, 1997.
26. W. G. Götz, N. M. Johnson, C. Chen, H. Liu, C. Kuo, and W. Imler *Appl. Phy. Lett.*, vol. 68, p. 3144, 1996.
27. J. Simon, K. Wang, H. Xing, D. Jena, and S. Rajan, "Carrier transport and confinement in polarization-induced 3D electron slabs: Importance of alloy scattering in AlGaN," *Appl. Phys. Lett.*, vol. 88, p. 042109, 2006.
28. S. Rajan, S. DenBaars, U. K. Mishra, H. Xing, and D. Jena, "Electron mobility in graded AlGaN alloys," *Appl. Phys. Lett.*, vol. 88, p. 042103, 2006.
29. D. Jena, A. C. Gossard, and U. K. Mishra, "Dipole scattering in polarization induced III-V Nitride two-dimensional electron gases," *J. Appl. Phys.*, vol. 88, p. 4734, 2000.

30. W. Zhao and D. Jena, "Dipole scattering in Highly Polar Semiconductor Alloys," *J. Appl. Phys.*, vol. 96, p. 2095, 2004.
31. D. N. Quang, N. H. Tung, N. V. Tuoc, N. V. Minh, and P. N. Phong, "Roughness-induced piezoelectric charges in wurtzite group-III-nitride heterostructures," *Phys. Rev. B*, vol. 72, p. 115337, 2005.
32. L. M. Roth and P. M. Argyres *Semiconductors and Semimetals*, vol. 1, p. 159, 1966.
33. C. Hamaguchi *Basic Semiconductor Physics*, p. 280, 2001.
34. R. B. Dingle *Proc. Roy. Soc.*, vol. A211, p. 517, 1952.
35. R. Kubo, H. Hasegawa, and N. Hashitsume *J. Phys. Soc. Japan*, vol. 14, p. 56, 1959.
36. D. Jena, S. Heikman, J. S. Speck, A. C. Gossard, U. K. Mishra, A. Link, and O. Ambacher, "Magnetotransport properties of a polarization-doped three-dimensional electron slab," *Phys. Rev. B*, vol. 67, p. 153306, 2003.
37. G. Bauer and H. Kahlert, "Low-Temperature Non-Ohmic Galvanomagnetic Effects in Degenerate n-type InAs," *Phys. Rev. B*, vol. 5, p. 566, 1972.
38. Y. Katayama and S. Tanaka *Phys. Rev.*, vol. 153, p. 873, 1967.
39. M. R. Boon *Phys. Rev. B*, vol. 7, p. 761, 1973.
40. B. L. Altshuler, D. Khmelnitzkii, I. A. Larkin, and P. A. Lee *Phys. Rev. B*, vol. 22, p. 5142, 1980.
41. T. Wang, Y. Ohno, M. Lachab, D. Nakagawa, T. Shirahama, S. Sakai, and H. Ohno *Appl. Phys. Lett.*, vol. 74, p. 3531, 1995.
42. A. F. Brana, C. Diaz-Paniagua, F. Batallan, J. A. Garrido, E. Munoz, and F. Omnes *J. Appl. Phys.*, vol. 88, p. 932, 2000.
43. R. J. Sladek *Phys. Rev.*, vol. 110, p. 817, 1958.
44. I. Vurgaftman, J. R. Meyer, and L. R. Ram-Mohan *J. Appl. Phys.*, vol. 89, p. 8815, 2001.
45. S. Elhamri, R. S. Newrock, D. B. Mast, M. Ahoujja, W. C. Mitchel, J. M. Redwing, M. A. Tischler, and J. S. Flynn *Phys. Rev. B*, vol. 57, p. 1374, 1998.
46. W. Knap, S. Contreras, H. Alause, C. Skierbiszewski, J. Camassel, M. Dyakonov, J. L. Robert, J. Yang, Q. Chen, M. A. Khan, M. L. Sadowski, S. Huant, F. H. Yang, M. Goian, J. Leotin, and M. S. Shur *Appl. Phys. Lett.*, vol. 70, p. 2123, 1997.
47. A. Saxler, P. Debray, R. Perrin, S. Elhamri, W. C. Mitchel, C. R. Elsass, I. P. Smorchkova, B. Heying, E. Haus, P. Fini, J. P. Ibbetson, S. Keller, P. M. Petroff, S. P. DenBaars, U. K. Mishra, and J. S. Speck *J. Appl. Phys.*, vol. 87, p. 369, 2000.
48. Z. W. Zheng, B. Shen, R. Zhang, Y. S. Gui, C. P. Jiang, Z. X. Ma, G. Z. Zheng, S. L. Gou, Y. Shi, P. Han, Y. D. Zheng, T. Someya, and Y. Arakawa *Phys. Rev. B*, vol. 62, p. R7739, 2000.
49. D. Jena and U. K. Mishra, "Quantum and classical scattering times due to charged dislocations in an impure electron gas," *Phys. Rev. B*, vol. 66, p. 241307, 2002.
50. J. P. Harrang, R. J. Higgins, R. K. Goodall, P. R. Ray, M. Laviron, and P. Delescluse *Phys. Rev. B*, vol. 32, p. 8126, 1985.
51. L. Hsu and W. Walukiewicz *Appl. Phys. Lett.*, vol. 80, p. 2508, 2002.
52. W. Walukiewicz, P. F. Hopkins, M. Sundaram, and A. C. Gossard *Phys. Rev. B.*, vol. 44, p. 10909, 1991.
53. D. Jena and U. K. Mishra, "Quantum and classical scattering times due to dislocations in an impure electron gas," *Phys. Rev. B*, vol. 66, p. 241307(Rapids), 2002.
54. B. I. Halperin, "Possible States of a Three-Dimensional Electron Gas in a Strong Magnetic Field," *Jpn. J. Appl. Phys.*, vol. 26, p. (suppl.3), 1987.
55. R. Gaska, M. S. Shur, X. Hu, J. W. Yang, A. Tarakji, G. Simin, A. Khan, J. Deng, T. Werner, S. Rumyantsev, and N. Pala *Appl. Phys. Lett.*, vol. 78, p. 769, 2001.
56. M. A. Khan, A. R. Bhattarai, J. N. Kuznia, and D. T. Olson *Appl. Phys. Lett.*, vol. 63, p. 1214, 1993.
57. S. C. Binari, L. B. Rowland, W. Kruppa, G. Kelner, K. Doverspike, and D. K. Gatskill *Electron. Lett.*, vol. 30, p. 1248, 1994.
58. J. C. Zolper, R. J. Shul, A. G. Baca, R. G. Wilson, S. J. Pearton, and R. A. Stall *Appl. Phys. Lett.*, vol. 68, p. 2273, 1996.
59. T. Egawa, K. Nakamura, H. Ishikawa, T. Jimbo, and M. Umeno *Jpn. J. Appl. Phys. Part 1*, vol. 38, p. 2630, 1999.

60. L. Zhang, L. F. Ester, A. G. Baca, R. J. Shul, P. C. Chang, C. G. Willinson, U. K. Mishra, S. P. DenBaars, and J. C. Zolper, "Epitaxially-grown GaN junction field effect transistors," *IEEE Trans. El. Dev.*, vol. 47, p. 507, 2000.
61. S. Rajan, H. Xing, S. DenBaars, U. K. Mishra, and D. Jena, "AlGaN/GaN polFETs for microwave power applications," *Appl. Phys. Lett.*, vol. 84, p. 1591, 2004.
62. Matulionis, A., "High-Field transport in III-V Nitride FETs - a Hot Phonon Bottleneck," *Hot Carriers in Semiconductors (Conference), Chicago*, p. (In press), 2005.
63. K. Wang, J. Simon, N. Goel, and D. Jena, "Optical study of hot-electron transport in GaN: Signatures of the hot-phonon effect," *Appl. Phys. Lett.*, vol. 88, p. 022103, 2006.
64. C. H. Oxley and M. J. Uren, "Measurement of Unity Gain Cutoff Frequency and Saturation Velocity of a GaN HEMT Transistor," *IEEE Trans. Electron. Dev.*, vol. 52, no. 2, p. 165, 2005.
65. Liberis, J. and Ramons, M. and Kiprijanovic, O. and Matulionis, A. and Goel, N. and Simon, J. and Wang, K. and Xing, H. and Jena, D., "Hot-phonons in Si-doped GaN," *Appl. Phys. Lett.*, vol. 89, p. 202117, 2006.
66. E. Fatuzzo and W. J. Merz, *Ferroelectricity*. New York: John Wiley and Sons, Inc., 1967.
67. J. Smit and H. P. J. Wijn, *Ferrite*. New York: John Wiley and Sons, Inc., 1959.
68. E. Salje, *Phase Transitions in Ferroelastic and Co-elastic Crystals*. Cambridge: Cambridge Univeristy Press, 1990.
69. L. Landau and E. Lifshitz, *Statistical Physics*. Oxford: Pergamon Press, 1980.
70. R. Kretschmer and K. Binder, "Surface effects on phase transitions in ferroelectrics and dipole magnets," *Physical Review B*, vol. 20, no. 3, pp. 1065–1075, 1979.
71. B. Strukov and A. Levanyuk, *Ferroelectric Phenomena in Crystals*. Berlin: Spring-Verlag, 1998.
72. L. D. Landau and E. M. Lifshitz, *Electrodynamics of Continuous Media*, vol. 8. Butterworth-Heinemann: Elsevier, 2nd ed., 1984.
73. L. D. Landau and E. M. Lifshitz, *Theory of Elasticity*, vol. 7. Butterworth-Heinemann: Elsevier, 2nd ed., 1984.
74. N. A. Pertsev, A. G. Zembilgotov, and A. K. Tagantsev, "Effect of mechanical boundary conditions on phase diagrams of epitaxial ferroelectric thin films," *Physical Review Letters*, vol. 80, no. 9, pp. 1988–1991, 1998.
75. N. Sai, B. Meyer, and D. Vanderbilt, "Compositional inversion symmetry breaking in ferroelectric perovskites," *Physical Review Letters*, vol. 84, pp. 5636–5639, 2000.
76. N. Sai, K. M. Rabe, and D. Vanderbilt, "Theory of structural response to macroscopic electric fields in ferroelectric systems," *Physical Review B*, vol. 66, pp. 104108–104125, 2002.
77. J. B. Neaton and K. M. Rabe, "Thoery of polarization enhancement in epitaxial batio3/srtio3 superlattices," *Applied Physics Letters*, vol. 82, no. 10, pp. 1586–1588, 2003.
78. J. Mantese, N. Schubring, A. L. Micheli, M. Mohammed, R. Naik, and G. W. Auner, "Slater model applied to polarization graded ferroelectrics," *Applied Physics Letters*, vol. 71, no. 14, pp. 2047–2049, 1997.
79. J. Mantese, N. Schubring, A. L. Micheli, M. Thompson, R. Naik, G. W. Auner, I. B. Misirlioglu, and S. P. Alpay, "Stress induced polarization-graded ferroelectrics," *Applied Physics Letters*, vol. 81, p. 1068, 2002.
80. W. Fellberg, J. Mantese, N. Schubring, and A. L. Micheli, "Origin of the "up", "down" hysteresis offsets observed from polarization-graded ferroelectric materials," *Applied Physics Letters*, vol. 78, no. 4, pp. 524–526, 2001.
81. A. L. Roytburd and J. Slutsker, "Thermodynamics of polydomain ferroelectric bilayers and graded multilayers," *Applied Physics Letters*, vol. 89, no. 4, p. 042907, 2006.
82. R. Slowak, S. Hoffmann, R. Liedtke, and R. Waser, "Functional Graded High-K $(Ba_{1-x}Sr_x)TiO_3$ Thin Films for Capacitor Structures with Low Temperature Coeffcient," *Integrated Ferroelectrics*, vol. 24, p. 169, 1999.
83. L. B. Freund, "Some elementary connections between curvature and mismatch strain in compositionally graded thin films," *Journal of the Mechanics and Physics of Solids*, vol. 44, no. 5, pp. 723–736, 1996.
84. "The average spontaneous polarization P_S and the in-plane self-strain are approximately 0.68 $Coul/m^2$ and 1 percent for PT and 0.23 $Coul/m^2$ and 0.1 percent for BT."

85. G. H. Haertling, "Rainbow ceramic-a new type of ultra-high-displacement actuator," *American Ceramic Society Bulletin*, vol. 73, no. 1, p. 93, 1994.
86. G. H. Haertling, "Method for making monolithic prestressed ceramic devices," 1995.
87. W. D. Nothwang, M. W. Cole, and R. W. Schwartz, "Stressed-biased actuators: Fatigue and lifetime," *Integrated Ferroelectrics*, vol. 71, pp. 249–255, 2005.
88. R. W. Schwartz, L. E. Cross, and Q. M. Wang, "Estimation of the effective d(31) coefficients of the piezoelectric layer in rainbow actuators," *Journal of the American Ceramic Society*, vol. 84, no. 11, pp. 2563–2569, 2001.
89. K. M. Mossi, G. V. Selby, and R. G. Bryant, "Thin-layer composite unimorph ferroelectric driver and sensor properties," *Materials Letters*, vol. 35, no. 1-2, pp. 39–49, 1998.
90. K. M. Mossi, R. G. Bryant, and P. Mane, "Piezoelectric composites as bender actuators," *Integrated Ferroelectrics*, vol. 71, pp. 221–232, 2005.
91. Z.-G. Ban, S. P. Alpay, and J. Mantese, "Fundamentals of graded ferroic materials and devices," *Physical Review B*, vol. 67, p. 184104, 2003.
92. A. Ohtomo and H. Y. Hwang, "A high-mobility electron gas at the laalo3/srtio3 heterointerface," *Nature*, vol. 427, pp. 423–426, 2004.
93. J. Mannhart and D. G. Schlom, "Semiconductor physics: The value of seeing nothing," *Nature*, vol. 430, pp. 620–621, 2004.
94. H. Y. Hwang, "Perovskites: Oxygen vacancies shine blue," *Nature Materials*, vol. 4, pp. 803–804, 2005.

Polarization in GaN Based Heterostructures and Heterojunction Field Effect Transistors (HFETs)

Hadis Morkoç and Jacob Leach

Introduction

The need for computers to handle large volumes of data for high speed computing, real time signal processing, telecommunication, imaging, low noise and high frequency amplification, and high power compact amplifiers has generated an unequalled interest in advancing the speed of electronic devices and circuits. RF, microwave, and millimeter wave systems for telecommunication and many other traditional uses require devices with ever increasing performance in terms of noise figure and gain at frequencies exceeding 100 GHz, and also most pertinently to the subject matter of this chapter – increased power. All of these driving forces have resulted in intense activity of new device concepts as well as heterostructures based on new semiconductors.

In semiconductors, excluding Si, forms other than the ubiquitous MOSFET are used for high performance FETs due to the lack of gate quality dielectrics. Until the advent of Modulation Doped FETs (MODFETs), this device was a MEtal Semiconductor FET (MESFET). The speed of the device depends on carrier transit time under the gate from the source side to the drain side, and the delays inherent in the devices such as those caused by capacitors and resistors. In scaling MESFETs for short channel and fast devices, one encounters limits in terms of doping requirements and the proximity of the gate with respect to the conducting channel. In MODFETs, these limits are alleviated considerably. The semiconductor form that was first used for MODFETs was GaAs-based. While exploring the properties of quantum wells and superlattices, it was discovered that when the large bandgap AlGaAs is doped with a donor impurity near the junction it forms with the adjacent small bandgap GaAs, the electrons donated by donors in the wider bandgap material diffuse to the lower energy conduction band of GaAs where they are confined due to the heterointerfacial potential barrier. The electron gas which forms the conduction channel thus is formed in equilibrium by doping the large bandgap material, and in the process does not get affected, except remotely, by the donors and possesses nearly impurity scattering free transport.

Let us now consider briefly the principles of operation of MODFETs. In polar semiconductors such as the GaN system grown along the [0001] direction, polarization induced free carriers either due to strain or compositional variation or both can be used for channel. Although this does not occur on non-polar surfaces, for technological reasons, the polar surface is the only one used for FETs. Because there is no intentional doping involved in some cases, the term heterojuction FETs (HFETs) is used to describe the devices termed MODFET in the GaAs system. For an in depth treatment of MODFETs the reader is advised to refer to a two volume text on the topic by *Morkoç et al* [1] and references therein, and a review article appearing in the June 1986 issue of IEEE Proceedings by *Drummond et al.* [2]

In MODFETs, the source and drain ohmic contacts are made directly to the two-dimensional electron gas (2DEG) while the gate electrode between these two terminals modulates the current. A schematic cross-sectional diagram of a pseudomorphic InGaAs channel MODFET on GaAs or InP substrate is shown in Figure 1. Large low field mobilities, relatively large maximum electron velocities, and large electron concentrations provided by many of the compound semiconductors are ideal for high performance FETs. Proximity of the gate to the conduction channel in MODFETs, as well as the confinement provided by the field profile in the substrate side of the channel, reduces if not eliminates the enormous increase observed in the output conductance as the gate length is scaled down. In addition, the change in the threshold voltage of this all-epitaxial structure as the gate length is reduced, unlike in conventional Metal Semiconductor FETs (MESFETs), is minimal. These last two properties have served well in scalability of these devices so much so that deep submicron ($\leq 0.2\,\mu m$) small-scale digital circuits have already been fabricated successfully. These same proximity and carrier confinement effects are also responsible for the gains obtained at or over 100 GHz, considered impossible only a few

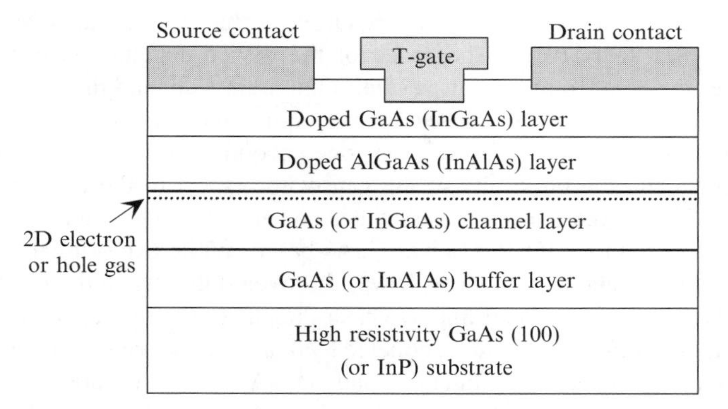

Fig. 1 Cross-sectional view of a pseudomorphic, used loosely for coherently strained, channel MODFET on GaAs or InP substrates. On GaAs substrates, the mole fraction of InAs of coherent channel is limited to below 35%. On InP, on the other hand, a range of 53 to 80%, later extended to 100%, has been explored.

years back. The confinement potential induced by the channel charge on the substrate side alone is not sufficient, as it diminishes near the drain side, particularly in deep saturation.

To circumvent this problem, a semiconductor with a larger bandgap than that forming the channel can be incorporated just below the channel. In GaAs channels, AlGaAs (and in InGaAs channels InAlAs) can serve this function well. Alternatively, the channel itself can be made out of a smaller bandgap material as in the case of pseudomorphic InGaAs channel devices imbedded in an otherwise GaAs/AlGaAs structure. The InAs mole fraction in the pseudomorphic device is limited to a maximum of about 35% to avoid the deleterious crystalline defects that otherwise result and deteriorate the device performance quite substantially. Even though the AlGaAs/InGaAs based pseudomorphic MODFET became the dominant compound semiconductor FET in the marketplace with its low noise, high gain, and reasonably high power handling capability, high RF power came at the expense of power combining schemes at a substantial cost. This is where GaN based HFETs came into the picture.

GaN's large bandgap, large dielectric breakdown field, fortuitously good electron transport properties [3–5] (an electron mobility possibly in excess of 2,000 $cm^2\, V^{-1}s^{-1}$ and a predicted peak velocity approaching $3 \times 10^7\, cms^{-1}$ at room temperature, although the velocity deduced from device current or speed is much lower and is a topic of raging debate), and good thermal conductivity are trademarks of high power/temperature electronic devices [6]. To give a flavor, *Sheppard et al.* [7] have reported that 0.45 μm gate, high power modulation doped FETs (MODFETs) on SiC substrates exhibited a power density of 6.8 W/mm in a 125 μm-wide device and a total power of 4 W (with a power density of 2 W/mm) at 10 GHz. Other groups have also reported on the superior power performance of GaN-based MODFETs on SiC and sapphire substrates with respect to competing materials, particularly at X band and higher frequencies, [8–11] keeping in mind that record levels belong to structures grown on SiC substrates. This is due to high thermal conductivity of SiC and also enhanced quality of GaN heterostructures on SiC. For starters, GaN grown on SiC does not undergo an in-plane rotation around the c-axis and is thus void of the associated disorder.

GaN/AlGaN MODFETs prepared at HRL laboratories by MBE on SiC substrates have exhibited a total power level of 6.3 W at 10 GHz from a 1-mm wide device. A saturated power density of 6.6 W/mm with a power added efficiency of 35% at 20 GHz also resulted [12]. Credited to the supreme thermal conductivity of SiC, the power level in these devices at the time was not thermally limited as the power density extrapolated from a 0.1-mm device is 6.5 W. When four of these devices are power combined in a single stage amplifier, an output power of 22.9 W with a power added efficiency of 37% was obtained at 9 GHz [13]. Equally impressive is the noise figure of 1.0 and 1.75 dB which was obtained at 10 and 20 GHz, respectively, [14] and which improved to 0.85 dB at 10 GHz with an associated gain of 11 dB. In terms of the linearity a $0.15\,\mu m \times 100\,\mu m^2$ device yields an output third-order intercept point (OIP3) of 23 dBm at $V_{ds} = 3\,V$, and $V_{gs} = -5\,V$, where a noise figure (NF) of 1.0 and 1.75 dB was obtained at 10 and 20 GHz, respectively [15]. The drain breakdown voltages in these quarter micron gate devices are about 60 V, which are in part responsible for such a record performance [16].

With further optimization, standard devices have a power output up to 9.8 W/mm at 8 GHz [17] with gradually increasing power densities and total power levels [18]. Polyimide passivated devices with 0.23 μm gate lengths yielded a peak power density of 7.65 W/mm at 18 GHz [19]. These data suggest that polyimide can be an effective passivation film for reducing surface states. With field plates, which spreads the field between the gate and drain more uniformly (which has also been applied to SiC-based devices [20]), thereby allowing larger drain breakdown voltages, [21, 22] power densities up to 30 W/mm have been reported [23]. As will be detailed soon, CW power levels over 200 W at C band have been achieved already. On the cutoff frequency front, short channel devices exhibited with $f_T = 101\,\text{GHz}$ and $f_{max} = 155\,\text{GHz}$ [24, 25].

Applications of high power GaN-based HFETs include amplifiers operative at high power levels, high temperatures and in unfriendly environments. Examples include radar, missiles, satellites, as well as in low-cost compact amplifiers for wireless base stations. A good deal of these applications is currently met by the pseudomorphic HFETs developed earlier [26].

As mentioned, AlGaN/GaN heterostructures have been the subject of many investigations because of their potential for use in high-temperature, high-power devices [6, 27–29]. Due to the large band discontinuities and polarization-induced screening charge, interface bound two dimensional electron gas (2DEG) concentrations exceed $10^{13}\,\text{cm}^{-2}$. While polarization effects cause a redistribution of weakly bound and free charges, unlike implications made in the literature they cannot directly produce free electrons to form a 2DEG [30–33]. GaN-based systems, and issues dealing with heterointerfaces must include a discussion of polarization. Polarization induces a field, which in turn affects the interface charge through screening in that mobile carriers move to where the fixed polarization charge with opposite polarity is. Because nitrides are large bandgap materials, they tend to be n-type and the hole concentration is extremely low. Consequently, the mobile carriers are normally electrons donated by intentional donors, donor-like defects, or contacts.

In this chapter, operational principles of HFETs, with particular attention to the polarization issues pertinent to HFETs, will be discussed. This will be followed by analytical expressions describing HFET operation, results of HFET simulations, technology, and finally performance of nitride-based HFET devices in terms of DC and RF.

1 Heterojunction Field Effect Transistors (HFETs)

With its reduced impurity scattering and unique gate capacitance-voltage characteristics, the HFET has become the dominant high frequency device. Among the HFET's most attractive attributes are close proximity of the mobile charge to the gate electrode and high drain efficiency. As in the case of emitters, the GaN-based

HFETs have quickly demonstrated record power levels at high frequencies with very respectable noise performance and large drain breakdown voltages.

In HFETs, the carriers that form the channel in the smaller bandgap material are donated by the larger bandgap material, ohmic contacts or both. Because the mobile carriers and their parent donors are spatially separated, short-range ion scattering is nearly eliminated, which leads to mobilities that are characteristic of nearly pure semiconductors. A Schottky barrier is then used to modulate the mobile charge that in turn causes a change in the drain current. Because of this heterolayer construction, the gate can be placed very close to the conducting channel, resulting in large transconductances [1]. Figure 2 illustrates a schematic representation of a GaN/AlGaN HFET heterostructure in which the donors in the wider bandgap AlGaN or the polarization-induced free carriers when grown on polar surfaces provide the carriers. In an HFET device under bias, the carriers can also be provided by the source contact.

As mentioned above, in heterojunctions containing donors and acceptors, and shallow defects, the associated free carriers within the Fermi statistics diffuse to the semiconductor with the smaller bandgap where they are confined, due to potential barriers, to potential minima. The resulting charge separation due to free carriers causes an internal electric field, *screening field*. In addition, an electric field can also be induced by the application of an external voltage such as done through the use of Schottky barriers, metal-oxide semiconductor structures, and p-n junctions. Spontaneous and strain-induced piezoelectric polarization can influence the final status of the interfacial free-charge density in these heterostructures. Any shallow defects (induced fields would change the ionization ratio), free carriers, and surface contacts must be included for a complete treatment.

The AlGaN barrier [34] or on the AlGaN surface [35] have been suggested as the source of electrons. Positive surface charge concept has also been suggested to account for the experimental observations in the form of dependence of the 2DEG density on the thickness and/or alloy composition of the AlGaN barrier [36, 37]. Typically, the AlGaN barrier is grown on a relatively thick GaN layer to form the basis for the 2DEG system in a Ga-polarity sample. The inherent lattice mismatch causes a biaxial tensile strain, and the thermal mismatch causes a biaxial compressive strain in the growth plane. The resultant strain induces a macroscopic electric field in the polar material. In addition, due to the particular crystal structure of the wurtzite lattice, a spontaneous polarization field is also found in both materials even in the absence of strain. In most current heterostructures, where the growth takes place along the (0001) direction, both the spontaneous and induced polarizations are directed opposite to the growth direction. The effect of polarization field on the position of the band edges has been calculated by several groups [30, 34, 38–40]. Polarization induced fields in Ga-polarity samples, just as any negative gate voltage induced field in FETs, increase the conduction band edge in AlGaN barrier with distance from the interface. In the presence of free, weakly bound and surface charges, the internal polarization field is screened by a redistribution of these charges. The surface states may be in the form of donor-like states, which donate their electrons to the lowest unoccupied energy states at the AlGaN/GaN interface.

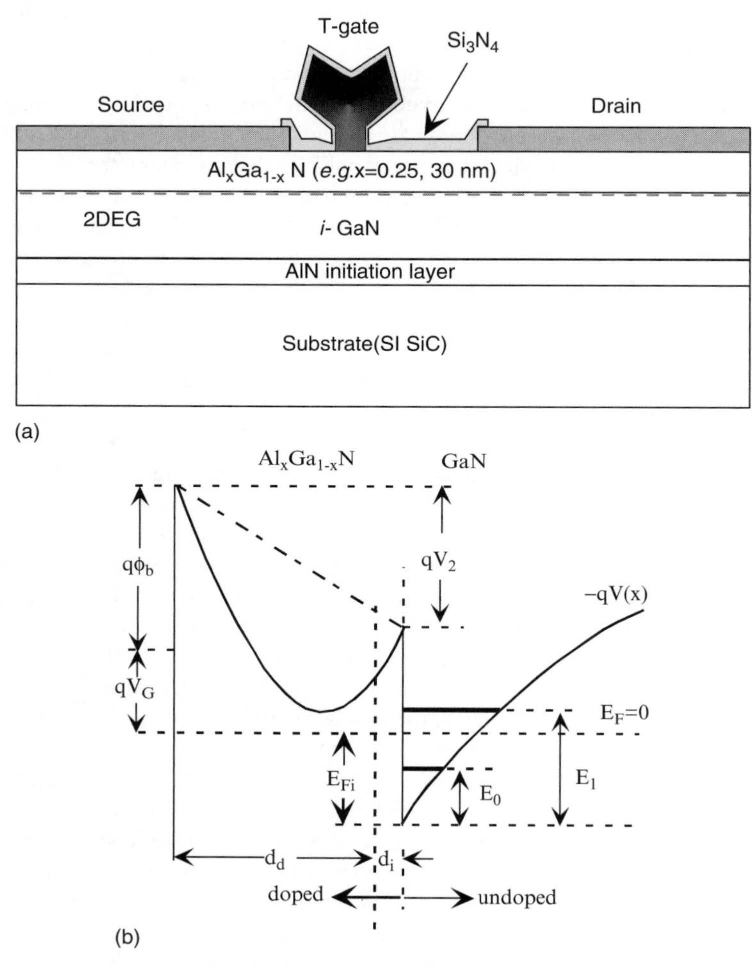

Fig. 2 (a) Schematic representation of an AlGaN/GaN modulation doped field effect transistor with a T-gate HFET. (b) Schematic band structure of an AlGaN/GaN modulation doped heterostructure (some undoped relying on instead the polarization charge only) in which the free carriers are provided to the GaN layer by the dopant impurities placed in the larger bandgap AlGaN barrier layer or by polarization charge. The band bending in the barrier for the doped barrier case is represented by a solid line whereas that for the literally undoped barrier is shown with a broken line.

The holy grail of GaN/AlGaN heterostructures is the debate on the origin of the carriers which end up at the interface. The observed dependence of the 2DEG density on the thickness and composition of the AlGaN barrier has been linked to surface donor states, the binding energy of which is roughly equal to the Schottky barrier height in n-type GaN [35]. This may point to the same surface states being responsible for a possible and weak Fermi level pinning, which is not that notable on the surface of GaN. In addition, pinning of the surface Fermi level in n-type GaN requires electrons to be transferred from bulk donors to surface acceptor states, whereas an excess of surface donors is required to form 2DEG in an AlGaN/GaN heterostructure. This inconsistency can be resolved; however, by assuming that the surface defects are amphoteric (i.e. they can act as either acceptors or donors depending on the circumstances) [41]. The polarization fields present in nitride heterostructures are strong enough to shift the Fermi energy at the surface of the AlGaN barrier below the charge state level of the surface defects for Ga-face growth. This causes the surface defects on the barrier to transform from being acceptor-like to donor-like surface defects, which can provide the electrons for the 2DEG at the AlGaN/GaN interface [41]. Hsu and Walukiewicz [33, 42] elaborated on the surface donor like defect that is likely to form at growth temperatures and its manifestation as a source of carriers confined at the underlying interface between the AlGaN top layer and GaN below it. The model calculations appeared to be somewhat insensitive to parameters such as donor formation energy and surface Fermi level. One sensitive parameter is the strength of the polarization field, which will be discussed below.

1.1 Polarization Issues as Pertained to HFETs

Group III-V nitride semiconductors exhibit highly pronounced polarization effects. Semiconductor nitrides lack center of inversion symmetry and exhibit piezoelectric effects [43] when strained along <0001>. Piezoelectric coefficients in nitrides are almost an order of magnitude larger than in many of the traditional group III-V semiconductors [43–47]. The strain induced piezoelectric and spontaneous polarization charges have profound effects on device structures. The piezoelectric effect has two components. One is due to lattice mismatch (misfit) strain while the other is due to thermal strain caused by the thermal expansion-coefficient difference between the substrate and the epitaxial layers. The low symmetry in nitrides, specifically the lack of center of inversion symmetry present in zincblende and wurtzite structures, may be interpreted as some sort of non-ideality, which is not the case. Non-vanishing spontaneous polarization is allowed in an ideal wurtzite structure [50, 51] This spontaneous polarization is noteworthy particularly when heterointerfaces between two nitride semiconductors with varying electronegativity are involved. This manifests itself as a polarization charge at hetero-interfaces. Spontaneous polarization was only understood fully not too long ago by *King-Smith* and *Vanderbilt*, [50] and *Resta* et al. [51].

In heterojunction devices such as HFETs where strain and heterointerfaces are present, the polarization charge is present and is inextricably connected to free carriers, which are indeed present. As such, polarization charge affects device operation in all nitride-based devices, particularly HFETs, and thus must be taken into consideration in device design unless non-polar surfaces such as the a-plane are used. The quality of films on non-polar surfaces has not kept pace with those on polar basal plane which make the topic of discussion quite relevant. As mentioned above, polarization charge arises from two sources: piezoelectric (PE) effects and the difference in spontaneous polarization between AlGaN and GaN, even in the absence of strain. These charges exist in all compound semiconductors to varying degrees unless self-cancelled by the symmetry of the particular orientation under consideration such as the non-polar surfaces/interfaces.

In relative terms, spontaneous polarization is larger than the piezoelectric polarization in AlGaN/GaN-based structures. In the case of InGaN/GaN structures, spontaneous polarization is relatively small but not as small as the earlier predictions called for, but still noteworthy, as spontaneous polarizations in GaN and InN are not as different from one another. However, the strain-induced piezoelectric polarization can be sizable. If and when defect-associated relaxation occurs, reducing the strain in the films, the strength of the piezoelectric polarization is lowered. Spontaneous polarization and piezoelectric polarization affect the band diagram of heterostructures. The effects are very large and can easily obscure the engineered designs.

Polarization is dependent on the polarity of the crystal, namely whether the bonds along the c-direction are from cation sites to anion sites, or visa versa. The convention is that the [0001] axis points from the face of the N plane to the Ga plane, and marks the positive z-direction. In other words, when the bonds along the c-direction (single bonds) are from cation (Ga) to anion (N) atoms, the polarity is said to be the Ga polarity, and the direction of the bonds from Ga to N along the c-direction marks the [0001] direction which is generally taken to be the +z-direction. By a similar argument, when the bonds along the c-direction (single bonds) are from anion (N) to cation (Ga) atoms, the polarity is said to be the N polarity, and the direction of the bonds from N to Ga along the c-direction marks the direction which is generally taken to be the -z-direction. To shed further light, the Ga polarity means that if one were to cut the perfect solid along the c-plane where one breaks only a single bond, one would end up with a Ga-terminated surface. A schematic representation of the spontaneous polarization in a model GaN/AlN/GaN wurtzitic crystal is shown in Figure 3.

The spontaneous polarization $\boldsymbol{P}_{spont}$ (also commonly referred to as $\boldsymbol{P}_o$) in a solid has not always been well defined, although much better understanding of it has been emerging. Only those differences in **P** between two phases which can be linked by an adiabatic transformation that maintains the insulating nature of the system throughout are well defined. For example, one phase can be considered unstrained and the other strained. *Vanderbilt* proved that the polarization difference ΔP between the wurtzite and zincblende phases could be calculated by considering an interface between the two phases and by defining $\boldsymbol{P}_{spont}$ to be zero in the zincblende phase. In short, by calculating the integral of a quantum mechanical Berry phase along a line in the Brillouin Zone (BZ) from one end to the other in the bulk wurtzite symmetry leads to polarization **P** with respect to that in zincblende (which is zero by definition because zincblende is cubic and cannot have a spontaneous polarization in an infinite bulk periodic crystal). The Berry phase actually represents an overlap integral between the periodic part of the Bloch function at **k** and a neighboring k-point, $\mathbf{k}'$. *Zorroddu* et al. [48] and *Bernardini* et al. [43] showed that the charges accumulating at each interface in a self-consistent calculation can be obtained from the $\Delta\boldsymbol{P}$ of the two bulk layers forming the heterointerface. The relation between the charge and

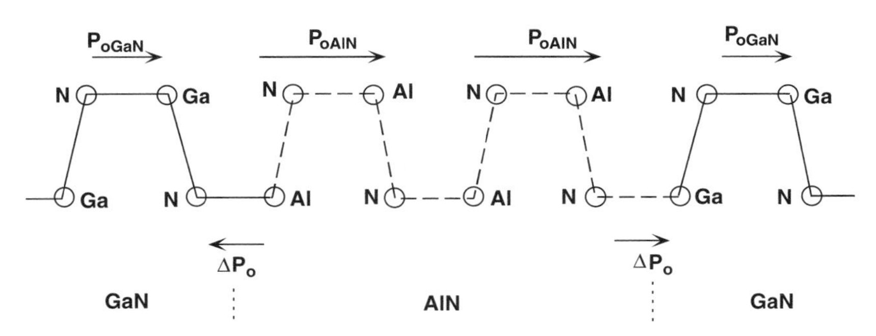

Fig. 3 Schematic depicting the convention used for determining the polarity and crystalline direction in wurtzitic nitride films. The diagram shows the case for a Ga-polarity film with its characteristic bonds parallel to the c-axis (horizontal in the figure) going from the cation (Ga or Al) to the anion (N). The spontaneous polarization components P_{OGa} and P_{OAlN} for a periodic GaN/AlN structure are also indicated with that for AlN having a larger magnitude. The spontaneous polarization is negative and thus points in the $[000\bar{1}]$ direction. Caution must be exercised here as there is no long-range polarization field, just that it is limited to the interface. The polarization in AlN is larger in magnitude than in GaN. There exists a difference in polarization at the interface, ΔP_O pointing in the $[000\bar{1}]$ direction for both GaN/AlN interfaces.

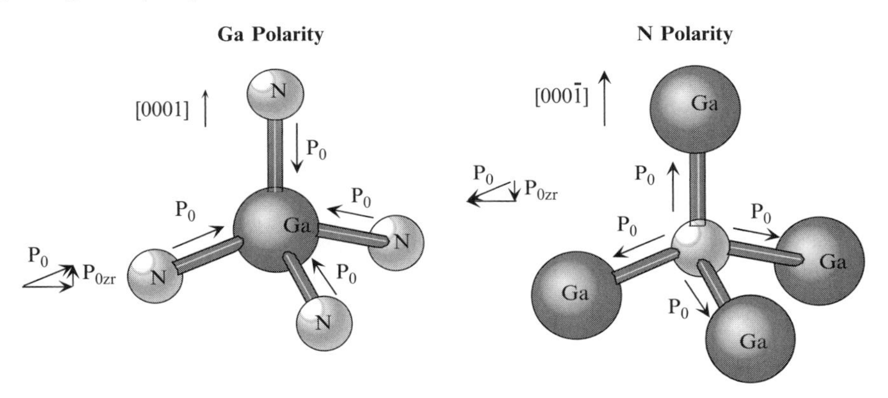

Fig. 4 Ball and stick configuration of an ideal GaN tetrahedron with proper c/a ratio and internal parameter u for both Ga and N-polarity in a relaxed state.

P follows basically from Gauss's law. The bound charge density $\rho_b = -\nabla.P$. This means that across an abrupt interface with P_1 on one side and P_2 on the other side: one gets $P_2 - P_1 = \sum_i \rho_s$ (surface charge density at the interface with the appropriate signs).

Even though it is overly simplistic, a graphical picture of polarization due to strain (piezo component) and heterointerfaces (spontaneous component), the latter is in the case of different ionicity, can be obtained which is helpful. Shown in Figure 4 is a ball and stick diagram of a tetrahedral bond between Ga and N in Ga and N-polarity configurations showing the polarization vector due to the electron cloud being closer to the N atoms. Actually, the cumulative polarization due to the triply bonded atoms is along the direction of the single bond. The in-plane and vertical components of polarization due to pairs of atoms cancel one another if the tetrahedron is ideal.

However, when a Ga-polarity film is under homogeneous in-plane tensile strain, the cumulative z-component, [0001] direction, of the polarization associated with the triple bonds decreases causing a net polarization which would be along the $[000\bar{1}]$ direction, as shown Figure 5. In a nitrogen polarity film, the same occurs except the net polarization would be in the opposite, [0001], direction. When an in-plane and homogeneous compressive strain is present, the net polarization would be in the [0001] direction in the Ga-polarity case and $[000\bar{1}]$ direction in the N-polarity case, as shown in Figure 6.

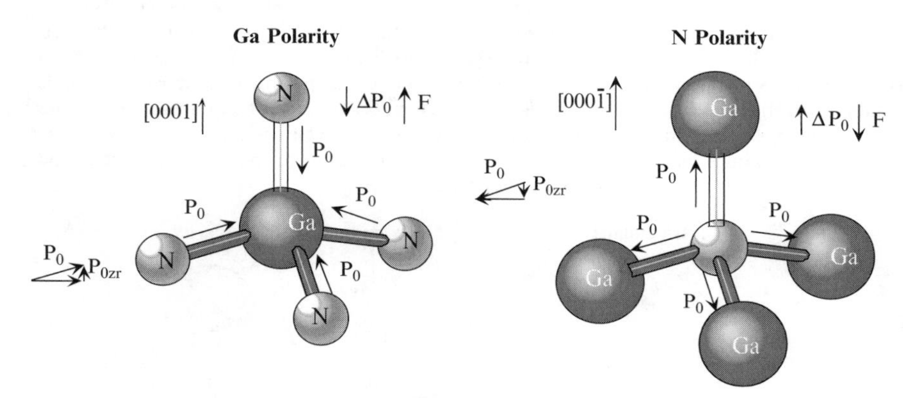

Fig. 5 Ball and stick configuration of a GaN tetrahedron for both Ga and N polarity with a homogeneous in-plane tensile strain showing a net polarization in the [0001] direction for Ga-polarity and $[000\bar{1}]$ polarization for N-polarity.

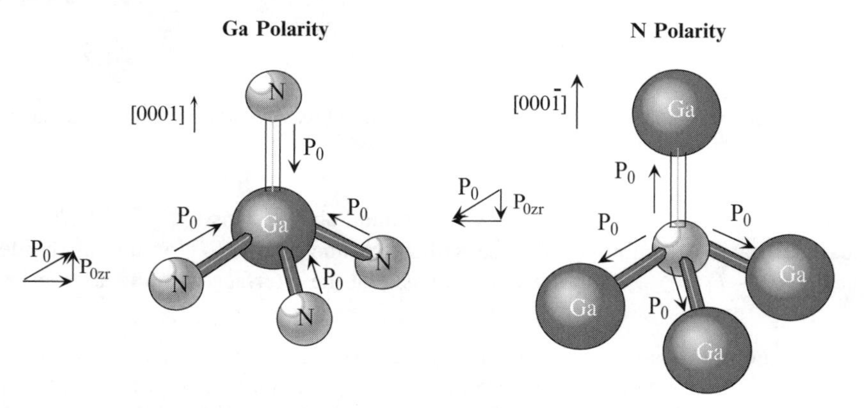

Fig. 6 Ball and stick configuration of a GaN tetrahedron for both Ga and N polarity with a homogeneous in-plane compressive strain showing a net polarization in the $[000\bar{1}]$ direction for Ga-polarity and [0001] polarization for N-polarity.

The same graphical argument can be used to attain a mental image of spontaneous polarization at heterointerfaces as well. For example if we were to construct two tetrahedra, one representing a GaN bilayer and another on top of it representing an AlN bilayer, the top N atom shown in Figure 4 for the Ga-polarity configuration would make triple bonds with it. Because AlN has a higher electronegativity difference than GaN, the net component of the polarization vector in the [0001] direction in triply bonded N with Al is larger in amplitude than in the GaN-tetrahedron, there would be a net interfacial polarization in the $[000\bar{1}]$ direction even without strain. In short, the source of piezoelectric polarization is strain in an electronegative binary. That for spontaneous polarization is the change in electronegativity across an interface such as the AlN and GaN interface.

Substrates upon which nitride films are grown lack the wurtzitic symmetry of nitrides. Consequently, the polarity of the films may not be uniform. In this type of inversion domain, the wrong type of bonds, e.g. Ga-Ga and N-N, are formed at the boundary projected on $(11\bar{2}0)$ plane. Inversion domains combined with any strain in nitride-based films lead to flipping piezoelectric (PE) fields with untold adverse effects on the characterization of nitride films in general and the polarization effect in particular, and on the exploitation of nitride semiconductors for devices. Such flipping fields would also cause much increased scattering of carriers as they traverse in the c-plane. Having made the case, it should be mentioned that if proper measures are taken, the Ga-polarity films grown by OMVPE and to a lesser extent by MBE are nearly or completely inversion domain boundary free even on sapphire substrates. However, in the case of MBE, unless optimum AlN buffer layers are employed or N-polarity films are grown by incorporation of GaN initiation layers, inversion domain boundaries do occur, sometimes in high concentrations.

The magnitude of the polarization charge, converted to number of electrons can be in the mid $10^{13}cm^{-2}$ level for AlN/GaN heterointerfaces, which is huge by any standard. For comparison, the interface charge in the GaAs/AlGaAs system used for HFETs is less than 10% of this figure. An excellent review of the polarization effects can be found in reference [52].

The Schrödinger's equation and Poisson equation can be used self-consistently in order to study the channel formation and current flow mechanisms in GaN-based HFETs [53, 54]. Several approaches have been used to define the system Hamiltonian used in the Schrödinger equation, namely effective masses [55], **k.p** expansion [56] and tight-binding expansion [57–60]. The use of sophisticated models such as k.p or tight-binding is justified, perhaps even made necessary, by the complex wurtzite band structure, particularly for determining the valence band states. Thus, calculations of optical processes involving band to band transitions must consider the details of the band structure beyond the simple effective mass approximation (EMA) [61, 62]. However, when only the conduction band processes are of interest, EMA is still a very accurate means of determining the properties of interest. In fact, nitride based semiconductors in the wurtzite structure possess a conduction band with a Γ minimum, which can be described reasonably well within such an approximation. Within the effective mass theory Schrödinger's equation takes the form [55, 63, 64]:

$$-\frac{h^2}{2}\frac{d}{dz}\left(\frac{1}{m_z^*}\frac{d}{dz}\right)\psi + eV(z)\psi = E\psi \tag{1}$$

where $m^*(z)$ is the (position dependent) effective mass, *V(z) is* the electrostatic potential, ψ *is* the electron wavefunction, and E is the energy level.

Or in terms of the Hamiltonian, the Schrödinger's equation can be expressed as

$$H = H_0 + eF_S z = -\frac{\hbar^2}{2m_z^*}\frac{\partial^2}{\partial z^2} + eF_S z \tag{2}$$

Non-parabolicity may induce deviations from the simple parabolic band model. However this will not substantially change the results that are presented here. For a triangular potential barrier which can approximately represent the potential distribution at the heterointerface with a 2D charge being present, we can use the following boundary conditions:

$$V(z) = \begin{cases} eF_S z & z > 0 \\ \infty & z \leq 0 \end{cases} \text{with } F_S = \frac{eN_D W}{\varepsilon} \tag{3}$$

where F_S is the electric field and z is the distance along the growth direction, normal to the interface, being 0 at the interface.

The solution of the wave-function is given by [65, 66]

$$\psi(z) = Ai\left[\frac{2m_z^* eF_S}{\hbar^2}\left(z - \frac{E_i}{eF_S}\right)\right] \tag{4}$$

with $Ai(\cdot)$ being the Airy function given by

$$Ai(u) = \frac{1}{\pi}\int_0^\infty cos(\left(\frac{1}{3}t^3 + ut\right) dt \tag{5}$$

The eigenvalues for energy are:

$$E_i \approx \left(\frac{\hbar^2}{2m_z^*}\right)^{1/3}\left[\frac{3\pi eF_S}{2}\left(i + \frac{3}{4}\right)\right]^{2/3} \tag{6}$$

with the $(i + 3/4)$ replaced with 0.7587, 1.7540, and 2.7575 for the first three lowest, sub-bands, respectively, for the exact eigenvalues. The parameter i takes values of $i = 0, 1, 2, 3$ with 0 representing the ground state and the rest the excited states. The average value for z (where the 2DEG can be approximated to be) for one sub-band occupation can be found from:

$$< z >= \int \psi^* z \psi dz \approx \frac{2E_i}{3eF_S} \tag{7}$$

A variational method introduced by Fang and Howard [67] can be used as well in which case $\psi(z) = Az\exp[-az]$, where A is a normalization constant and a is a variational parameter, as a wave function of ground state. From the normalization condition $\int \psi^*(z)\psi(z)dz = 1$ the normalization constant A can be found as $A = 2a^{3/2}$.

The expectation value for the total energy is given by

$$\begin{aligned}\langle E\rangle &= \int \psi^*(z)H\psi(z)dz \\ &= \int \psi*(z)\left(-\frac{\hbar^2}{2m_z^*}\frac{\partial^2}{\partial z^2} + eF_S z\right)\psi(z)dz \\ &= \frac{\hbar^2 a^2}{2m_z^*} + \frac{3}{2}\frac{eF_S}{a}\end{aligned} \tag{8}$$

The variational parameter can be found from $\frac{d}{da}\langle E\rangle = 0$, i.e. $2\hbar^2 a^3 - 3m_z^* eF_S = 0$. Then wave function of the ground state is

$$\psi(z) = 2(a)^{3/2} z\exp[-az] \text{ with } a = \left(\frac{3}{2}\frac{m_z^* eF_S}{\hbar^2}\right)^{1/3}. \tag{9}$$

The average position of electrons in the ground state is

$$\langle z\rangle = \int \psi*(z)z\psi(z)dz = \frac{3}{2a} = \frac{3}{2\left(\frac{3}{2}\frac{m_z^* eF_S}{\hbar^2}\right)^{1/3}} = \frac{3\hbar^{2/3}}{\left[12m_z^* eF_s\right]^{1/3}} \tag{10}$$

Assuming that only the ground state is occupied at an $Al_{0.2}Ga_{0.6}N/In_{0.1}Ga_{0.9}N$ interface with $-0.0055C/m^2$ interfacial charge, the electric field is $F_S = -5.7068 \times 10^7\,V/m$, the variational parameter $a = 5.9819 \times 10^8\,\mathrm{m}^{-1}$, and the average position of electrons is $\langle z\rangle = 2.5\,nm$ from the interface. Here effective masses used for GaN and InN are 0.2 and 0.11, respectively, and a linear interpolation is used to find the effective mass of $In_{0.1}Ga_{0.9}N$.

In the nitride semiconductors with wurtzite phase, spontaneous and piezoelectric polarization effects are present, [68] which necessitate that the Poisson equation be solved for the displacement field, $D(z)$

$$\frac{d}{dz}D(z) = \frac{d}{dz}\left(-\varepsilon(z)\frac{d}{dz}V(z) + P(z)\right) = q\left(p(z) - n(z) + N_D^+ - N_A^-\right) \tag{11}$$

where ε is the dielectric constant, P the total polarization, $n(p)$ the electron (holes) charge concentration and $N_D^+(N_A^-)$ the ionized donor(acceptor) density.

In the self-consistent procedure, potential V is obtained using Equation 11 and initial guess of the mobile charge concentration, and then inserted into the Schrödinger's equation, Equation 1 which is solved to get the energy levels and wavefunctions of the systems. The new electron charge density is obtained by applying Fermi statistics as follows:

$$n_{2D}(z) = \frac{m(z)\,k_B T}{\pi \hbar^2} \sum_i |\varphi_i(z)|^2 \, ln\left[1 + e^{\frac{E_F - E_i}{k_B T}}\right] \tag{12}$$

where E_F is the Fermi level, E_i the energy of the i-th quantized level, T the temperature and k_B the Boltzmann constant. The calculated density is than plugged into Poisson equation (Equation 11) and the iteration repeated until convergence is achieved. Convergence of the self-consistent algorithm can be improved adopting special relaxation techniques. Here a first order expansion of the model reported in reference [69] was used.

Spontaneous polarization $P(\mathrm{C/m^2}), Z^*$ Born, or the transverse component of the charge tensor Z_3^T, piezoelectric constants $(\mathrm{C/m^2})$, elastic constants (GPa), and the ratio $R = -2C_{31}/C_{33}$ of wurtzitic nitrides, as obtained in the LDA and GGA approximations [48] are tabulated in Table 1. Also tabulated following the elastic constants e_{33} and e_{31} is $\varepsilon_{31}^{(p)}$ which is the applicable piezoelectric constant in the context of experiments dealing with current flow across the sample. The constant e_{31} is relevant to systems in depolarizing fields such as nitride nanostructures [80].

Table 1 Elastic constants and spontaneous polarization charge in nitride semiconductors. The data in bold letters are associated with density functional theory (DFT) in the generalized gradient approximation (GGA) which is more accurate than others reported prior. Moreover, the resultant predictions are in relatively better agreement with experimental data as well as the bowing parameters observed in polarization charge in alloys. References [43,49], and [70]

	AlN	GaN	InN
e_{33}^* [C/m^2]	1.46	0.73	0.97
LDA	1.8	0.86	1.09
GGA	**1.5**	**0.67**	**0.81**
e_{31} [C/m^2]	−0.60	−0.49	−0.57
LDA	−0.64	−0.44	−0.52
GGA	**−0.53**	**−0.34**	**−0.41**
e_{31}^p LDA	−0.74	−0.47	−0.56
GGA	**−0.62**	**−0.37**	**−0.45**
C_{33} (Gpa) GGA	377	354	205
C_{31} (Gpa) GGA	94	68	70
P$_0$[C/m^2]	−0.081	−0.029	−0.032
LDA	−0.10	−0.032	**−0.041**
GGA	**−0.090**	**−0.034**	−0.042
P$_0$[C/m^2], ideal wurtzite structrure	−0.032	−0.018	−0.017
$R = -C_{31}/C_{33}$ LDA	−0.578	−0.40	−0.755
GGA	**−0.499**	**−0.384**	**−0.783**
$[e_{31} - (C_{13}/C_{33})e_{33}]$ in [C/m^2]	**−0.86**	**−0.68**	**−0.90**

e_{31} and e_{33} are piezoelectric constants
C/m^2 is Coulomb per square meter
C_{31} and C_{33} are elastic stiffness coefficient or elastic constants.
e_{31}^p This is referred as the proper piezoelectric constant

Values given in Table 1 are taken from a series of publications by Bernardini, Fiorentini, and Vanderbilt. The data in bold are those reported in an earlier publication [43] and the remaining data points are taken from a later publication [49]. Following the initial reports of piezoelectric and spontaneous polarization, [43] the authors returned to the topic [48] as the values of the initial parameters were not consistent with other reports [71]. Bernardini et al. [70] re-analyzed the polarization as obtained using the Berry phase method within two different **density-functional theory** (DFT) exchange-correlation schemes. Specifically, the authors used the Vienna ab initio simulation package (VASP) and the pseudopotentials provided therewith, as in reference [71]. The newer calculations were carried out using both the **generalized gradient-corrected local density-approximation** (GGA) to density-functional theory in the Perdew-Wang PW91 version, and the **local-density approximation** (LDA) in the Ceperley-Alder-Perdew-Zunger form. Ultrasoft potentials were used treating Ga and In d electrons as valence at a conservative cut-off of 325 eV. Finally, the reciprocal-space summation was done on a (888) Monkhorst-Pack mesh. The results of the refined GGA calculations in terms of spontaneous polarization, piezoelectric and elastic constants so calculated are tabulated in Table 1 along with earlier calculations. The two sets of data, the earlier and refined are within 10% agreement. The results of more refined LDA calculations are also provided. For deference to earlier work, the piezoelectric constants, albeit an incomplete list dealing with e_{14}, for GaN were estimated theoretically for cubic GaN [72] and used in early investigation of piezoelectric effects in GaN, [73] and deduced from the mobility data [74] which is indirect particularly in samples containing many scatterers.

The data in bold letters is recommended.

1.1.1 Piezoelectric Polarization

In a polarizable medium, the displacement vector can be expressed in terms of two components due to the dielectric nature of the medium and the polarizability nature of the medium as [75]

$$\overrightarrow{D} = \varepsilon \overrightarrow{E} + 4\pi \overrightarrow{P} \text{ in cgs and } \overrightarrow{D} = \varepsilon \overrightarrow{E} + \overrightarrow{P} \text{ in mks units} \tag{13}$$

where $\overrightarrow{E}$ and $\overrightarrow{P}$ represent the electric-field and polarization vectors. Considering only the piezoelectric component, the piezoelectric polarization vector is given by: [76]

$$\overrightarrow{P}_{PE} = \overleftrightarrow{e}\vec{\varepsilon} \tag{14}$$

where $\overleftrightarrow{e}$ and $\vec{\varepsilon}$ are the piezoelectric and the stress tensors. In order to gain a quantitative understanding of the piezoelectric polarization, the piezoelectric tensor which is defined as the derivative of the polarization with respect to strain must be considered.

In hexagonal symmetry, electric polarization is related to strain through the electric piezoelectric tensor as: [77]

$$\begin{pmatrix} P_x \\ P_y \\ P_z \end{pmatrix} = \begin{pmatrix} 0 & 0 & 0 & 0 & e_{15} & 0 \\ 0 & 0 & 0 & e_{24} & 0 & 0 \\ e_{31} & e_{31} & e_{33} & 0 & 0 & 0 \end{pmatrix} \begin{pmatrix} \varepsilon_{xx} \\ \varepsilon_{yy} \\ \varepsilon_{zz} \\ \varepsilon_{yx} \\ \varepsilon_{zx} \\ \varepsilon_{xz} \end{pmatrix} \quad \text{Note that } e_{24} = e_{15} \text{ for hexagonal symmetry} \tag{15}$$

where P_i, e_{ij}, and ε_{ij} represent the electric polarization, electric piezoelectric coefficient, and strain, respectively.

Before proceeding, let us make some general comments about strain. Strain-stress relationship or Hooke's law can be used to describe the deformation of a crystal ε_{kl}, due to external or internal forces or stresses σ_{ij},

$$\sigma_{ij} = \sum_{k,l} C_{ijkl}\varepsilon_{kl} \tag{16}$$

where C_{ijkl} is the fourth-ranked elastic tensor and represents the elastic stiffness coefficients in different directions in the crystal, which due to the C_{6v} symmetry can be reduced to a 6×6 matrix using the Voigt notation: $xx \to 1, yy \to 2, zz \to 3, yz, zy \to 4, zx, xz \to 5, xy, yx \to 6$. The elements of the elastic tensor can be rewritten as $C_{ijkl} = C_{mn}$ with $i, j, k, l = x, y, z$ and $m, n = 1, \ldots, 6$. With this notation, Hooke's law can be reduced to

$$\sigma_i = \sum_j C_{ij}\varepsilon_j \tag{17}$$

Or as treated in reference [78]:

$$\begin{vmatrix} \sigma_{xx} \\ \sigma_{yy} \\ \sigma_{zz} \\ \sigma_{xy} \\ \sigma_{yz} \\ \sigma_{zx} \end{vmatrix} = \begin{vmatrix} C_{11} & C_{12} & C_{13} & 0 & 0 & 0 \\ C_{12} & C_{22} & C_{13} & 0 & 0 & 0 \\ C_{13} & C_{13} & C_{33} & 0 & 0 & 0 \\ 0 & 0 & 0 & C_{44} & 0 & 0 \\ 0 & 0 & 0 & 0 & C_{55} & 0 \\ 0 & 0 & 0 & 0 & 0 & C_{66} \end{vmatrix} \begin{vmatrix} \varepsilon_{xx} \\ \varepsilon_{yy} \\ \varepsilon_{zz} \\ \varepsilon_{xy} \\ \varepsilon_{yz} \\ \varepsilon_{zx} \end{vmatrix} \tag{18}$$

with $C_{66} = \frac{C_{11}-C_{12}}{2}$. If the crystal is strained in the (0001) plane, and allowed to expand and constrict in the [0001] direction, the $\sigma_{zz} = \sigma_{xy} = \sigma_{yz} = \sigma_{zx} = 0$, $\sigma_{xx} \neq 0$ and $\sigma_{yy} \neq 0$, and the strain tensor has only three non-vanishing terms, namely,

$$\varepsilon_{xx} = \varepsilon_{yy} = \frac{a - a_0}{a_0}, \varepsilon_{zz} = \frac{c - c_0}{c_0} = -\frac{C_{13}}{C_{33}}(\varepsilon_{xx} + \varepsilon_{yy}) \tag{19}$$

Where a and a_0, c and c_0, represent the in-plane and out of plane lattice constants of the epitaxial layer and the relaxed buffer (substrate), respectively. The above assumes that the in-plane strain in x and y directions is identical, namely $\varepsilon_{xx} = \varepsilon_{yy}$. When the crystal is uniaxially strained in the (0001) c-plane and free to expand and constrict in all other directions, σ_{zz} is the only non-vanishing stress term, and the strain tensor is reduced to

$$\begin{vmatrix}\varepsilon_{yy}\\ \varepsilon_{zz}\end{vmatrix} = \frac{1}{C_{13}^2 - C_{11}C_{33}} \begin{vmatrix} C_{12}C_{33} - C_{13}^2 \\ C_{11}C_{13} - C_{12}C_{13} \end{vmatrix} \varepsilon_{xx} \quad (20)$$

The nomenclature used in the literature for parameters surrounding strain, and piezoelectric and elastic constants vary in that e.g. e_{xx} and e_{11} are interchangeably used. The two-way transformation of $x \equiv 1$, $y \equiv 2$, $z \equiv 3$ can be used to convert from one nomenclature to the other. Likewise P_{PE}, P^{PE}, P_3^{PE}, P_3^{pz}, and P^{pz} are commonly used in literature interchangeably to depict piezoelectric polarization. If the subscript 3 is also used as in P_3^{PE}, it specifically indicates that for biaxial *in-plane* strain, the only non-vanishing polarization is along the c-axis. Even in cases where the subscript is not employed, the underlying assumption is that the polarization is along the c-direction, as growth of nitride semiconductor structures is performed predominantly on the basal plane. In the same vein, P_0, P^{sp} and P_{sp} indicate spontaneous polarization along the c-axis. ΔP_{sp} and ΔP_0 both represent differential spontaneous polarization at a heterointerface.

The components of the piezoelectric polarization tensor given by Equation 15 can be expressed in terms of a summation, using P_i^{pz} instead of P_{PE}, as

$$P_i^{pz} = \sum_j e_{ij}\varepsilon_j \text{ with } i = 1,2,3 \text{ and } j = 1, \,..,6 \quad (21)$$

where P_i^{pz} is the i-th component of the piezoelectric polarization.

The wurtzite symmetry reduces the number of independent components of the elastic tensor, e, to three, namely e_{15}, e_{31} and e_{33}. The third independent component of the piezoelectric tensor, e_{15}, is related to the polarization induced by a shear strain which is not applicable to the epitaxial growth schemes employed. The index 3 corresponds to the direction of the c-axis. It is clear that the piezoelectric properties of the Wz structures are somewhat more complicated. If we restrict ourselves to structures with growth along the [0001] direction or **z**-direction, or along the c-axis, only the e_{31} and e_{33} components need to be considered. The piezoelectric polarization in the [0001] direction can be obtained by setting $i = 3$. The electric polarization component in the c-direction, which is designated by z- in the above nomenclature, is given by

$$P_z = e_{31}\varepsilon_{xx} + e_{31}\varepsilon_{xx} + e_{33}\varepsilon_{zz} = 2e_{31}\varepsilon_{xx} + e_{33}\varepsilon_{zz} \quad (22)$$

For isotropic basal plane strain, the strain components $\varepsilon_{xx} \equiv \varepsilon_{\perp}/2$, and thus Equation 22 can be written as

$$P_z \equiv P_3^{pe} = e_{31}\varepsilon_{\perp} + e_{33}\varepsilon_{zz} \quad (23)$$

In hexagonal symmetry, strain in the z-direction can be expressed in terms of the basal plane strain $\varepsilon_\perp$ through the use of Poission's ratio which is expressed in terms of the elastic coefficients C_{ij} as $\varepsilon_{zz} = -2(C_{13}/C_{33})\varepsilon_{xx} = -(C_{13}/C_{33})\varepsilon_\perp$. In the case of externally applied pressure in addition to mismatch strain, the out of plane strain can be related to the in-plane strain through $\varepsilon_{zz} = -\left[(p + 2C_{13}\varepsilon_{xx})\right]/C_{33}$, where p is the magnitude of compressive pressure (in the same unit as the elastic coefficients). In terms of the nomenclature again, it should also be noted that $\varepsilon_1 \equiv \varepsilon_{11} \equiv \varepsilon_{xx}$ and $\varepsilon_3 \equiv \varepsilon_{33} \equiv \varepsilon_{zz}$ in other notations used in the literature (and also in this text)

$$P_z = \left(e_{31} - e_{33}\frac{C_{13}}{C_{33}}\right)\varepsilon_\perp \tag{24}$$

Piezoelectric polarization is also described in terms of piezoelectric moduli in the literature which is treated here for completeness. In terms of piezoelectric moduli, d_{ij}, which are related to the piezoelectric constants by

$$e_{ij} = \sum_k d_{ik}C_{kj} \text{ with } i = 1,2,3 \text{ and } j = 1,\ldots,6, \text{ and } k = 1,\ldots,6, \tag{25}$$

Using Equation 25 in Equation 21 and strain-stress relationship (stress is equal to the product of elastic constant and strain in a tensor form), the piezoelectric polarization can be expressed in terms of piezoelectric moduli as

$$P_i^{pz} = \sum_j d_{ij}\sigma_j \text{ with } i = 1,2,3 \text{ and } j = 1,\ldots,6 \tag{26}$$

Symmetry considerations lead to $d_{31} = d_{32}$, $d_{15} = d_{24}$, and all other components $d_{ij} = 0$, and thus Equation 26 reduces to a set of three equations:

$$P_1^{pz} = \frac{1}{2}d_{15}\sigma_5, P_2^{pz} = \frac{1}{2}d_{15}\sigma_4, \text{ and } P_3^{pz} = d_{31}(\sigma_1 + \sigma_2) + d_{33}\sigma_3 \tag{27}$$

For biaxial strain, which is the case with epitaxial layers, additional conditions are imposed in that $\sigma_1 = \sigma_2$, $\sigma_3 = 0$. Moreover, the shear stresses are negligible which leads to $\sigma_4 = \sigma_5 = 0$. Consequently, in cases primarily applicable to epitaxial layers grown along the c-direction, the piezoelectric polarization is left with only one non-vanishing component, which is in the growth direction and is given by, using Equation 27,

$$P_3^{pz} = 2d_{31}\sigma_1 \tag{28}$$

Utilizing stress-strain relationship

$$\sigma_1 = \varepsilon_1\left(C_{11} + C_{12} - 2\frac{C_{13}^2}{C_{33}}\right) \tag{29}$$

we obtain

$$P_3^{pz} = 2d_{31}\varepsilon_1 \left(C_{11} + C_{12} - 2\frac{C_{13}^2}{C_{33}} \right) \tag{30}$$

where P_3^{pz} represents the piezoelectric polarization along the c-direction and similar to Equation 24 which expresses the same in terms of piezoelectric constants as opposed to piezoelectric moduli. For hexagonal crystals the relations between piezoelectric constants and piezoelectric moduli expressed in Equation 25 can be reduced to

$$\begin{aligned} e_{31} &= e_{32} = C_{11}d_{31} + C_{12}d_{32} + C_{13}d_{33} = (C_{11} + C_{12})d_{31} + C_{13}d_{33} \\ e_{33} &= 2C_{13}d_{31} + C_{33}d_{33} \\ e_{15} &= e_{24} = -C_{44}d_{15} \\ e_{ij} &= 0 \text{ for all other components.} \end{aligned} \tag{31}$$

We have so far focused on lattice mismatch induced strain. However, the thermal expansion coefficients of the layers used to compose many of the heterostructures are different which upon cooling from growth temperature could lead to thermal induced strain. A larger effect in this vein, however, is that caused by differences in the thermal expansion coefficient between the substrate and epitaxial stack used. In that case, the piezo-component would have two parts, namely the lattice mismatch (lm) or misfit strain and the thermal strain (ts) leading to $P^{pe} = P^{lm} + P^{ts}$. Another issue that must be considered is that the electric field induced due to strain (piezoelectric field) in adjacent layers of a heterostructure comprised of A and B (i.e. A for AlGaN, B for GaN and A, B for AlGaN/GaN) is

$$E_{A,B} = E_{A,B}^{sp} + E_{A,B}^{pe} \tag{32}$$

If for example A is composed of a ternary, then the linear interpolation for both *sp* and *pe* polarizations can be used to a first extent.

Another relevant but not discussed nearly as much topic is that the properties that cause the piezoelectric polarization can also lead to pyroelectric effects. Such phenomena can be rather important in nitride-based devices, as the junction temperature is high by the nature of the applications such as lasers and high-power amplifiers. Consequently, the thermally induced electric field, pyroelectric effect, would most likely be present [79] with consequences similar to those ascribed to polarization effects.

If one considers distortion of the u parameter as well, the piezoelectric polarization can be expanded as

$$\partial P_3 = \frac{\partial P_3}{\partial a}(a - a_0) + \frac{\partial P_3}{\partial c}(c - c_0) + \frac{\partial P_3}{\partial u}(u - u_0) \tag{33}$$

The internal u parameter is defined as the average value of the projection of the vector connecting a nitrogen atom with its first neighbor in the $(000\bar{1})$ direction

Table 2 Structural parameters of wurtzitic AlN, GaN and InN reported in reference [43], and updated with density functional theory (DFT) in the generalized gradient approximation (GGA) approximation in reference [48]. It should be mentioned that GGA produced data for structural as well as polarization related parameters, see Table 1, are in better agreement with refined experimental data

	a_0(Å)			c_0/a_0			u_0		
	Ref [43]	GGA [48]	Exp.	Ref [43]	GGA [48]	Exp.	Ref [43]	GGA [48]	Exp.
AlN	5.814	3.1095	3.1106	1.6190	1.6060	1.6008	0.380	0.3798	0.3821
GaN	6.040	3.1986	3.1890	1.6336	1.632	1.6263	0.376	0.3762	0.377
InN	6.660	3.5848	3.538	1.6270	1.6180	1.6119	0.377	0.377	

along this same direction. The three parameters, a, c, and u are not independent of each other. If the partial derivatives in Equation 33 are known, following reference [43], one can write for the two piezoelectric constants:

$$e_{33} = c_0 \frac{\partial P_3}{\partial c} + \frac{4qc_0}{\sqrt{3}a_0^2} Z^* \frac{\partial u}{\partial c} \tag{34}$$

$$e_{31} = \frac{a_0}{2} \frac{\partial P_3}{\partial a} + \frac{2q}{\sqrt{3}a_0} Z^* \frac{\partial u}{\partial a} \tag{35}$$

where

$$Z^* = \frac{\sqrt{3}a_0^2}{4q} \frac{\partial P_3}{\partial u} = Z_3^T \tag{36}$$

is the axial component of the Born, or the transverse component of the charge tensor Z_3^T. Structural parameters, which are useful in treating the polarization issue in nitride semiconductors, are tabulated in Table 2.

In Equation 34, it is implicit that the vector connecting the cation with the anion has a modulus u_c associated with the internal cell parameter, and points in the direction of the c axis. The first term in Equation 34 and Equation 35 signifies the term called the clamped-ion term, and represents the effect of the strain on the electronic structure. The second term represents the effect of *internal* strain on the polarization. The derivatives of u with respect to c and a in Equation 34 and Equation 35 are related to the strain derivatives of u through $c_0 du/dc = du/d\varepsilon_3$ and $a_0 du/da = 2du/d\varepsilon_1$.

A comprehensive table including experimental and calculated values of elastic compliance, elastic constants and piezoelectric constants as well as the Poisson number of wurtzite binary group III nitrides at room temperature is shown in Table 3.

In addition to binaries, the nitride heterojunction systems utilize ternary and to a lesser extent quaternary alloys as well. Knowing the piezoelectric parameters of the end binary points is generally sufficient, to a first order, to discern parameters for more complex alloys. For example, in the case of $Al_xGa_{1-x}N$, the piezoelectric

Table 3 Experimental and predicted elastic compliance, elastic constants and piezoelectric constants as well as the Poisson number of wurtzite binary group III nitrides at room temperature (theory – references [43], [48], and [83], experiment from [85]. For an expanded list of elastic constants, see reference [84]

Parameter	GaN			AlN			InN	
	Theory ref. [83]	Theory Ref: [43, 48]	experiment	Theory ref. [83]	Theory Ref: [43, 48]	experiment	Theory ref. [83]	Theory Ref. [43, 48]
C_{11} (GPa)	367		370	396		410	223	
C_{12} (GPa)	135		145	137		140	115	
C_{13} (GPa)	103	68	110	108	94	100	92	70
C_{33} (GPa)	405	354	390	373	377	390	224	205
C_{44} (GPa)	95		90	116		120	48	
Expt of above all from [89]								
$\nu(0001)$	0.52	0.38	0.56	0.58	0.50	0.51	0.82	0.68
$S_{11}(10^{-12}\,\mathrm{Nm}^{-2})$	3.267		3.326	2.993		2.854	6.535	
$S_{12}(10^{-12}\,\mathrm{Nm}^{-2})$	−1.043		−1.118	−0.868		−0.849	−2.724	
$S_{13}(10^{-12}\,\mathrm{Nm}^{-2})$	−0.566		−0.623	−0.615		−0.514	−1.565	
$S_{33}(10^{-12}\,\mathrm{Nm}^{-2})$	2.757		2.915	3.037		2.828	5.750	
$S_{44}(10^{-12}\,\mathrm{Nm}^{-2})$	10.53		11.11	8.621		8.333	20.83	
$e_{31}(\mathrm{Cm}^{-2})$		−0.34			−0.53	−0.58 [86]		0.41
$e_{33}(\mathrm{Cm}^{-2})$		0.67			1.50	1.55 [86]		0.81
$e_{24} = e_{15}(\mathrm{Cm}^{-2})$	−0.22 [87]		−0.30 [88]			−0.48 [86]		
$d_{31}(10^{-12}\,\mathrm{Cm}^{-2}\,\mathrm{Pa})$		−1.253			−2.298	−2.65		−3.147
$d_{33}(10^{-12}\,\mathrm{Cm}^{-2}\,\mathrm{Pa})$		2.291			5.352	5.53		6.201
$d_{15}(10^{-12}\,\mathrm{Cm}^{-2}\,\mathrm{Pa})$		−1.579			−2.069	−4.08		−2.292

polarization vector expression, using linear interpolation within the framework of Vegard's law, can be described as: [76]

$$\mathbf{P}^{\mathrm{pe}} = \left[x\overleftrightarrow{e}_{AlN} + (1-x)\overleftrightarrow{e}_{GaN}\right]\vec{\varepsilon}(x) \tag{37}$$

The same argument can be extended to piezoelectric polarization in quaternary alloys such as $\mathrm{Al_xIn_yGa_{1-x-y}N}$ in a similar fashion as

$$\vec{P}^{pe} = \left[x\overleftrightarrow{e}_{AlN} + (1-x)\overleftrightarrow{e}_{InN} + (1-x-y)\overleftrightarrow{e}_{GaN}\right]\vec{\varepsilon}(x) \tag{38}$$

The linear interpolation is very convenient and does give reasonably accurate values. However, as will be discussed next, while the Vegard's law applies to the alloys, the polarization charge itself is not a linear function of composition [80, 81].

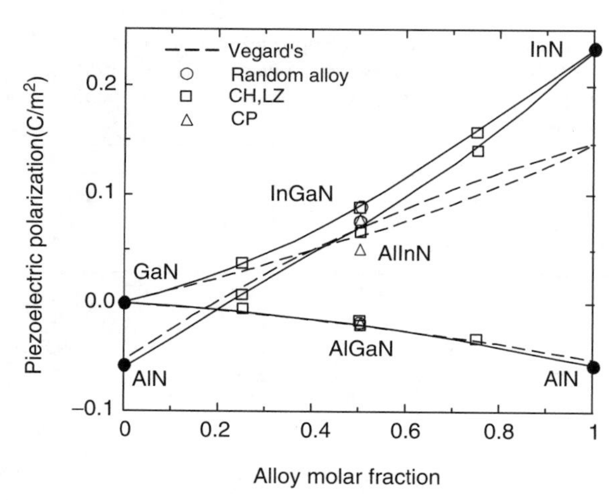

Fig. 7 Piezoelectric component of the macroscopic polarization in ternary nitride alloys epitaxially strained on a relaxed GaN layer (template). Open symbols represent the directly calculated values for random alloy (circles), CH-like and LZ-like (squares), and CP-like (triangles) structures, respectively. Dashed lines represent the prediction of linear piezoelectricity, while the solid lines are the prediction of Equation 37 using the nonlinear bulk polarization as shown in Figure 8. Courtesy of F. Bernardini and V. Fiorentini.

The model heterostructure considered by Bernardini and Fiorentini [82] is a coherently strained alloy grown on a relaxed binary buffer layer (bulk for this purpose) in which the in-plane lattice parameter $a_{\text{alloy}} = a_{\text{GaN}}$. The piezoelectric component is the difference between the total polarization to be obtained and the spontaneous polarization. Shown in Figure 7 is the piezoelectric polarization as a function of the alloy composition. Symbols represent the calculated polarizations for AlGaN, InGaN, and InAlN alloys as a function of compositions.

The piezoelectric constants e can be calculated for the equilibrium state of the binary, AN, and as such they do not depend on strain. The dashed lines in Figure 7 represent the piezoelectric term as computed from the above relations using the piezoelectric constants computed for the binaries [43]. The Vegard's law of Equation 37 when combined with $P_{AlN}^{pe} = e_{33}\varepsilon_3 + 2e_{31}\varepsilon_1$ clearly fails to reproduce the calculated polarization, and misses the strong non-linearity of the piezoelectric term evident in Figure 7. This is due to a valid nonlinearity of the bulk piezoelectricity of the binary constituents which is of nonstructural origin. It should be stated that bowing due to the microscopic structure of the alloys is negligible. The argument forwarded by Bernardini and Fiorentini [86] is that they calculate the piezoelectric polarization as a function of the basal strain for AlN, GaN, and InN while optimizing all structural parameters. The results depicted in Figure 8 clearly indicate that the piezoelectric polarization of the binaries is an appreciably nonlinear function of the lattice parameter a, which is related basal strain. Because all lattice parameters closely follow Vegard's law, the nonlinearity cannot be related to deviations from linearity in the structure. Bernardini and Fiorentini [82] substitute the *nonlinear* piezoelectric polarization computed for the binaries into the Vegard

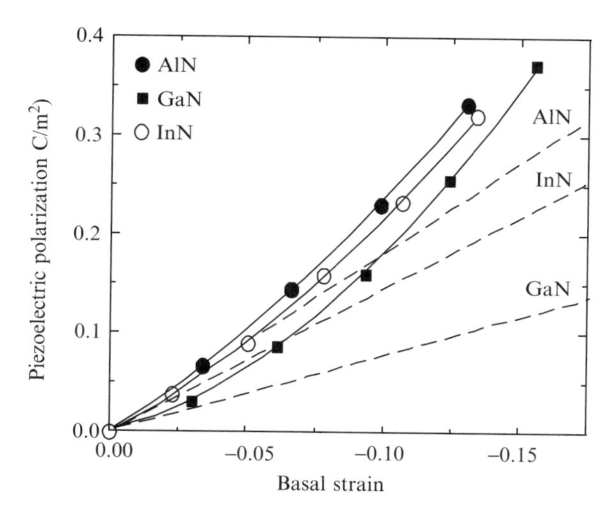

Fig. 8 Piezoelectric polarization in binary nitrides as a function of basal strain (symbols and solid lines) compared to linear piezoelectricity prediction (dashed lines). The *c* and *u* lattice parameters are optimized for each strain. Courtesy of F. Bernardini and V. Fiorentini.

interpolation, Equation 37. In doing so, they obtain excellent agreement with the polarization calculated directly for the alloys as shown with solid lines in Figure 7. This led them to conclude that the nonlinearity in bulk piezoelectricity dominates over any effects related to disorder, structure, bond length and angle alternation, etc. Importantly, they concluded that Vegard's law still holds in calculating the piezoelectric polarization of the III-V nitrides alloys, provided that the nonlinearity of the bulk piezoelectric of the constituents is accounted for. Serendipitously, this means that the piezoelectric polarization of any nitride alloy at any strain can be found by noting the value for x (the composition), followed by calculating the basal plane strain, $\varepsilon(x)$ from Vegard's law and P_{pe} from Equation 38 using the nonlinear piezoelectric polarization of the binaries (Figure 8). This approach is of paramount value in the modeling of nitride heterostructures, especially those with high In content, and AlInN alloys.

We should be cognizant of the fact that not only do extended defects in general cause local strain and therefore inhomogeneous strain but they also attract and often trap impurities, point defects, and free charge. Moreover, the strain component affects the charge through the piezoelectric component. As such, the effect of dislocations on piezoelectric polarization should be included to fully consider the true nature of III-Nitride epilayers.

1.1.2 Spontaneous Polarization

Spontaneous polarization calculated for the binary nitride semiconductors were tabulated in Table 1. For ternary and quaternary alloys the simplest approach is to use a linear combination of the binary end points taking into account that the mole fraction can be used under the auspices of the Vegard's law. However, this linear interpolation falls short of agreeing with the experimental variation of spontaneous polarization with respect to the mole fraction. Consequently, non-linear models have been developed which are discussed after the linear interpolation is applied for simplicity. The linear interpolation for spontaneous polarization in quaternary alloys such as $Al_xIn_yGa_{1-x-y}N$ can be expressed as

$$\overline{P}^{sp}(x,y) = x\overline{P}^{sp}_{AlN} + y\overline{P}^{sp}_{InN} + (1-x-y)\overline{P}^{sp}_{GaN} \tag{39}$$

The ternary cases can be obtained by simply setting either x or y to zero. This is again predicated on the assumption that polarization charge obeys Vegard's law, as shown below. Using the GGA calculation results for the spontaneous polarization and linear interpolation, as in Equation 39, for ternaries one gets:

$$\begin{aligned} \mathbf{P}^{sp}_{Al_xGa_{1-x}N} &= -0.09\mathrm{x} - 0.034(1-\mathrm{x}) \\ \mathbf{P}^{sp}_{In_xGa_{1-x}N} &= -0.042\mathrm{x} - 0.034(1-\mathrm{x}) \\ \mathbf{P}^{sp}_{Al_xIn_{1-x}N} &= -0.09\mathrm{x} - 0.042(1-\mathrm{x}) \end{aligned} \tag{40}$$

In reality, the actual polarization can deviate considerably from the linear approximation. The polarization in an alloy can be described in a generic measurable quantity to a first approximation by a parabolic model involving a *bowing* parameter, similar to that used for the bandgap of alloys.

The spontaneous polarization for a ternary $P_{sp}(A_xB_{1-x}N)$ with A and B representing the metal components and N representing nitrogen is given by: [80, 81, 84]

$$P^{sp}_{A_xB_{1-x}N} = xP^{sp}_{AN} + (1-x)P^{sp}_{BN} - b_{AB}x(1-x) \tag{41}$$

P^{sp}_{AN} and P^{sp}_{BN} are the spontaneous polarization terms for the end binaries forming the alloy. The bowing parameter is by definition

$$b_{AB} = 2P_{AN} + 2P_{BN} - 4P_{A_{0.5}B_{0.5}N} \tag{42}$$

which requires only the knowledge of the polarization of the ternary alloy at the mid point, i.e. molar fraction $x = 0.5$. Knowledge of the bowing parameter from Equation 42 would lead to determination of the spontaneous polarization at any composition. For $Al_xGa_{1-x}N$ Equation 41 and Equation 42 take the form

$$P^{sp}_{Al_xGa_{1-x}N} = xP^{sp}_{AlN} + (1-x)P^{sp}_{GaN} - b_{Al_xGa_{1-x}N}x(1-x) \tag{43}$$

with

$$b_{Al_xGa_{1-x}N} = 2P_{AlN} + 2P_{GaN} - 4P_{Al_{0.5}Ga_{0.5}N} \tag{44}$$

The first two terms in Equation 43 are the usual linear interpolation terms between the binary constituents. The third term, quadratic, represents the nonlinearity. Higher-order terms are neglected because their contribution is estimated to be less than 10%. Using the numerical GGA values in Table 1 for the spontaneous polarization in AlN and GaN and the bowing parameter for random alloy AlGaN given in references [80] and [81] leads to

$$P^{sp}_{Al_xGa_{1-x}N} = -0.09x - 0.034(1-x) + 0.0191x(1-x)$$

and

$$P^{sp}_{In_xGa_{1-x}N} = -0.042x - 0.034(1-x) + 0.0378x(1-x)$$

and

$$P^{sp}_{Al_xIn_{1-x}N} = -0.090x - 0.042(1-x) + 0.0709x(1-x) \tag{45}$$

To understand the physical origin of the spontaneous polarization bowing, Bernardini and Fiorentini [82] decomposed the spontaneous polarization into three distinct components based on their genesis, namely the internal structural and bond alternation, volume deformation, and disorder. The internal structural and bond alternation (strain) can be caused by varying cation-anion bond lengths. The volume deformation can be due to compression or dilation of the bulk binaries from their original equilibrium lattice constants to the alloy values. The disorder effect is due to the random distribution of the chemical elements on the cation sites. Bernardini and Fiorentini [82] showed that in ordered alloys the structural contribution is dominant, that the volume deformation accounts for one-third of the bowing found in random alloys, and that the effect of disorder appears insignificant in terms of its effect on the bowing of spontaneous polarization.

1.2 Analytical Description of HFETs

In order to qualitatively demonstrate the effect of charge stored at the heterointerface on mobility and carrier velocity, we present in the following an analytical description of the operation of HFET. For the sake of simplicity, we avoid consideration of quantitative structural and device analyses, and present a model that accounts primarily for the basic and important features of HFETs, reference [2]. The model, which is developed for the GaAs/AlGaAs system and does not account for polarization charge is based on the concept that the amount of charge which is depleted from the barrier donor layer is accumulated at the interface, while the Fermi level is kept constant across the hetero-interface. Specific issues arising from the use of GaN particularly the polarization issues will later be added to the model. The electron sheet charge with no external gate bias (or hole charge in the case of p-channel HFET which is very unlikely in the case of GaN) provided by the donors in the barrier layer may be given by

$$n_{s0} = \left[\frac{2\varepsilon_2 N_d}{q}\left(\Delta E_c - E_{f2} - E_{fi}\right) + N_d^2 d_i^2\right]^{1/2} - N_d d_i \tag{46}$$

where E_{f2} is the separation between the conduction band in the barrier layer and the Fermi level, N_d is the donor concentration in the barrier layer, ε_2 is the dielectric constant of the barrier layer, q is the electronic charge, ΔE_c is the conduction band discontinuity, E_{fi} is the Fermi level with respect to the conduction band edge in the channel layer, and d_i is the thickness of the undoped layer in the barrier layer at the heterointerface. A graphical description of the aforementioned parameters as well as the band edge profile for an AlGaN/GaN heterostructure with gate bias on the surface is shown in Figure 2.

The electron charge stored at the heterointerface is given by

$$n_s = \frac{\rho}{\beta} \ln \left[\left\{ 1 + e^{\beta(E_{fi}-E_0)} \right\} \left\{ 1 + e^{\beta(E_{fi}-E_1)} \right\} \right] \tag{47}$$

where $\beta = q/kT$, and $E_0 = \gamma_0 n_s^{2/3}$ and $E_1 = \gamma_1 n_s^{2/3}$ are the positions of the first and the second quantum (potential) states at the interface. These states correspond to a triangular well formed by the interfacial stored charge. The energy reference is the bottom of the conduction band edge in GaAs. It is assumed here that these lowest energy states are the only ones that are either filled or partially filled. The constants γ_0 and γ_1, which are dependent on the effective mass of the channel material used, and the density of states $\rho(\rho = q\, m^*/\pi h^2$, where m^* is the effective mass of electrons, and h is the Planck's constant), are derived on the basis that the quantum well may reasonably be triangular in shape. Depending on the value of the applied voltage, the gate on the surface of the barrier layer depletes some or all of the stored charge at the interface. Thus only a simultaneous solution of Equation 46 and Equation 47 can result in the determination of the Fermi level provided the interface sheet charge concentration is known. The determination of the sheet charge concentration can similarly be carried out from the same equations if the Fermi level is known. With a gate voltage present, Equation 46, which depicts the equilibrium situation, must be replaced by

$$n_s = \frac{\varepsilon_2}{qd} \left[V_g - \left(\phi_b - V_{p2} + \frac{1}{q} E_{fi} - \frac{1}{q} \Delta E_c \right) \right] \tag{48}$$

where ϕ_b is the Schottky barrier height of the gate metal deposited on the barrier layer, V_g is the applied gate to source bias voltage, d_d is the thickness of the doped barrier layer, $d = d_d + d_i$ and $V_{p2} = qN_d d_d^2/2\varepsilon_2$, and ε_2 represent the dielectric constants in the barrier (*AlGaN*). Similar treatments which have their roots in the models developed for AlGaAs/GaAs MODFETs, but modified to take the polarization charge into account can be found in a book chapter by Karmalkar et al. [89] and a paper by Rashmi et al. [90]. In addition a MESFET model which is somewhat applicable to parts of this treatment can be found in reference [91].

The interface charge concentration in presence of a gate bias may be expressed by

$$n_s = \frac{\varepsilon_2}{q(d + \Delta d)} \left\{ V_g - V_{off} \right\} \tag{49}$$

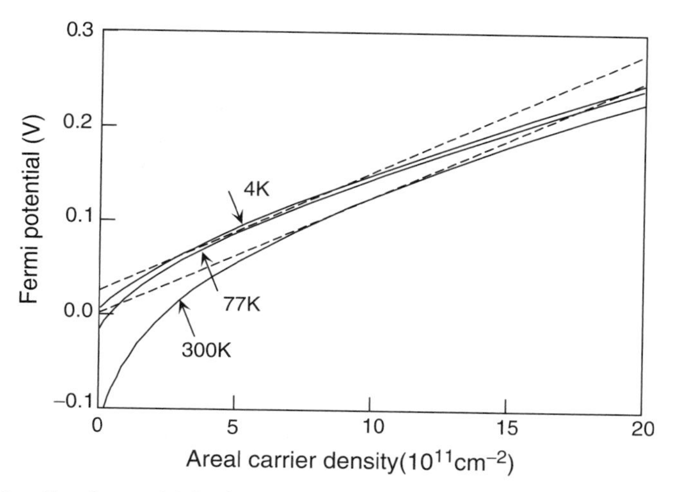

Fig. 9 Interface Fermi potential (E_{Fi}) vs. the sheet carrier concentration for a GaAs/AlGaAs heterointerface with 2DEG shown for a pictorial view of the parameters used in derivations.

where $\Delta d = \varepsilon_2 a_F/q$ and for the case without polarization-induced interfacial charge

$$V_{off} = \phi_b - \frac{1}{q}\Delta E_c - V_{p2} + \frac{1}{q}\Delta E_{F0} \tag{50}$$

Δd is typically about 2–4 nm for GaN and 8 nm for GaAs, and represents in pictorial terms the location of the peak density of the 2DEG from the heterointerface. In addition, the terms ΔE_{F0} and a_F are determined from the extrapolations. For example, ΔE_{F0}, which is a temperature dependent quantity, is the residual value of the Fermi level for zero interface sheet density obtained by extrapolating linearly from the linear region of the curve, as shown by dashed lines in Figure 9 for two temperatures, 77 K and 300 K for a GaAs/AlGaAs heterointerface.

Under ideal conditions V_{off} would represent the threshold voltage. However, experimentally the current and through it the free sheet charge in the channel is plotted against the gate voltage and the extrapolation of the linear region to zero charge would result in the threshold voltage. Due to substantial sub-threshold voltage, the gate voltage leading to nearly zero current flow in the channel would be smaller (in absolute value it is larger for an n-channel FET).

The parameter a_F represents the slope of the curve, which is reasonably linear for a wide range of sheet charge except near the vanishing values, relating the Fermi level to the sheet charge. Utilization of this displaced linear approximation leads to the interface Fermi level to be expressed as

$$E_{Fii} = \Delta E_{F0}(T) + a_F n_s \tag{51}$$

For example, for the GaAs/AlGaAs system, $a_F \approx 0.125 \times 10^{-16}\,\mathrm{V\,m^{-2}}$, and $\Delta E_{F0} \approx 0$ at 300 °K and 0.025 meV at T $\leq$ 77 °K. Similar figures can be obtained for the GaN/AlGaN system neglecting, for the time being, the effect of the polarization charge at the interface.

In a field-effect transistor, the drain bias produces a lateral field. For long channel devices and/or for very small drain biases it is generally assumed that the channel voltage, which varies along the channel between the source and the drain and finally reaches a value equal to the drain voltage, is added to the gate potential. When it is done, Equation 49 becomes a function of the distance, x, along the channel:

$$n_s = \frac{\varepsilon_2}{q(d+\Delta d)}\{V_g - V_{off} - V(x)\} \tag{52}$$

with $V(x)$ as the channel potential.

For a GaN-based modulation doped FET including the variety dependent only on the polarization induced charge, the terms Δd and ΔE_{F0} will be neglected in which case one can relate the sheet carrier concentration to the gate voltage above threshold as

$$n_s = \frac{\varepsilon_2}{qd}\{V_g - V_{off}\} \text{ or the interface charge } Q_s = \frac{\varepsilon_2}{d}(V_g - V_{off}) \tag{53}$$

and with interfacial polarization charge of n_p and again neglecting the effect of the Δd and ΔE_{F0} on the threshold voltage, but including the charge due to bulk doping, $N_B W_d$ (donor level and depletion layer thickness in the channel layer) one gets

$$V_{off} = \phi_b - \frac{1}{q}\Delta E_c - V_{p2} - \frac{qd}{\varepsilon_2}(n_p + N_B W_d) \tag{54}$$

It is assumed that the doped layer thickness is very small compared to the AlGaN barrier because Equation 54 lumps the bulk charge as interfacial charge of equal amount without considering its distributed nature and assumes its distance to the gate metal to be the same as the thickness of the barrier layer.

For a long channel MESFET, the case is similar in MODFETs in terms of the expressions that follow. The current voltage relationships can be found in many elementary device texts. Suffice it to say that the drain current in an n-type long channel device before saturation, where constant mobility can be assumed, is given by

$$I_D = G_0 V_p \left[\frac{V_D}{V_P} + \frac{2}{3}\left(\frac{-V_G}{V_P}\right)^{3/2} - \frac{2}{3}\left(\frac{V_D - V_G}{V_P}\right)^{3/2}\right] \tag{55}$$

where G_0, V_p represent the full channel conductance (assuming no depletion at all), and total gate voltage inclusive of the gate built-in voltage, required to pinch off the channel, respectively. They are given by

$$G_0 = \frac{qN_D \mu a Z}{L} \quad \text{and} \quad V_P^2 = \frac{qN_D}{2\varepsilon}a^2 \tag{56}$$

where N_D is the channel doping level assumed to be equal to the electron concentration (should be the electron concentration when these two parameters deviate substantially), Z is the width of the gate, L is the gate length, a is the total channel thickness, and ε is the dielectric constant of the channel.

For an n-channel device, the applied gate voltage $V_G < 0$. In the current saturation regime, as defined by pinch-off due to the reverse biased gate-drain junction in which case $V_D - V_g = V_P$, the drain current is given by

$$I_{DS} = G_0 V_p \left[\frac{V_G}{V_P} + \frac{2}{3} \left(\frac{-V_G}{V_P} \right)^{3/2} + \frac{1}{3} \right] \tag{57}$$

Equation 55 is applicable below saturation and Equation 57 is applicable in saturation.

Returning to modulation doped FETs, for small values of $V(z)$, it may be assumed that the constant mobility regime is inapplicable and that

$$I_d = q n_s \mu Z \frac{dV(x)}{dx} = \frac{\mu Z \varepsilon_2}{d} \left\{ V_G - V_{off} - V(x) \right\} \frac{dV(x)}{dx} \tag{58}$$

where μ is the charge carrier mobility and Z is the width of the gate. By integrating Equation 58 from the source to the drain while keeping in mind that the drain current remains constant throughout the channel and $V(x=0) = 0; V(x=L) = V_{DS}$, one obtains

$$I_{DS} = \frac{\mu Z}{L} \frac{\varepsilon_2}{d} \left[(V_G - V_{off}) V_{DS} - \frac{V_{DS}^2}{2} \right] = \beta_d \left[V_{Geff} V_{DS} - V_{DS}^2/2 \right] \tag{59}$$

where $V_{Geff} = (V_G - V_{off})$, and V_{DS} is the drain-source voltage, $\beta_d = \frac{\mu Z \varepsilon_{21}}{dL}$, and L is the intrinsic channel length or in practical terms the gate length. The current reaches saturation when the drain voltage is increased to the point where the field in the channel exceeds its critical value thereby causing the velocity to saturate.

Under these circumstances and utilizing Equation 52, the drain current may be calculated following the steps

$$I_{DS} = q Z v_{sat} n_s = \frac{\varepsilon_2 Z v_{sat}}{d} \left\{ V_G - V_{off} - V_{DSS} \right\} = \beta_d V_0 \left[V_{Geff} - V_{DSS} \right] \tag{60}$$

where V_{DSS} is the saturation drain voltage, I_{DS} is the saturation current, $V_0 = v_{sat} L/\mu$, and v_{sat} is the saturation velocity. For GaN devices Δd can be neglected because the maximum of the 2DEG is only about 4 nm from the interface (which is smaller than the thickness of the barrier layer, typically greater than 20 nm).

Equating Equation 59 and Equation 60 at the saturation point, one can solve for drain saturation voltage as

$$V_{DSS} = V_G + V_0 - \left(V_0^2 + V_{Geff}^2 \right)^{1/2} \tag{61}$$

which when substituted back into Equation 60 yields

$$I_{DS} = \beta_d V_0^2 \left[\left\{ 1 + \left(\frac{V_{Geff}}{V_0} \right)^2 \right\}^{1/2} - 1 \right]. \tag{62}$$

The treatment presented above is known as the two-piece model, implying that an abrupt transition takes place from the constant mobility regime to the constant velocity regime. A more accurate picture is one in which this transition is smoother allowing the use of a phenomenological velocity-field relationship for a more accurate description of the HFET operation. The simplest of all these pictures is one that neglects the peak in the velocity-field curve and assumes Si-like velocity-field characteristics. One such characteristic may be expressed by

$$v = \frac{\mu F(x)}{1 + \mu F(x)/v_{sat}} = \frac{\mu_0 F(x)}{1 + \frac{F(x)}{F_C}} \tag{63}$$

where $F(x)$ represents the electric field in the channel, and is equal to $(F(x) = -dV(x)/dx), \mu$ is low field mobility and $F_C = v_{sat}/\mu$ is the electric field at the saturation point. It may be noted that this field is not constant throughout the channel. Recognizing that

$$I_D = Q_{total}(x) Z v(x) \tag{64}$$

where $Q_{total} = \frac{\varepsilon_2}{d}(V_G - V_{off} - V(x))$ we get for the drain current I_D:

$$I_D = \frac{Z \varepsilon_2 v(x)}{d} \left[V_G - V_{off} - V(x) \right] \tag{65}$$

Using the expression for $v(x)$ given in Equation 63 above, Equation 65 may be simplified as:

$$I_D = \frac{Z \varepsilon_2 \mu v_{sat}}{d} \left[\frac{dV(x)/dx}{v_{sat} + \mu dV(x)/dx} \{ V_G - V_{off} - V(x) \} \right] \tag{66}$$

where $v_{sat} = \mu F_c, F_c$ being the field where the velocity assumes its saturation value. By integrating Equation 66 from the source end $(x = 0)$ of the channel to the drain end of the channel $(x = L)$, while keeping in mind that the drain current must be constant throughout the channel, one may obtain an expression for the drain current in the framework of a procedure elucidated by Lehovec and Zuleeg [92]. One thus obtains

$$\begin{aligned} I_D &= \frac{Z \varepsilon_2 \mu v_{sat}}{d} \left[\frac{V_{Geff} V_D - V_D^2/2}{v_{sat} L + \mu V_D} \right] \\ &= \frac{1}{1 + \mu V_D / v_{sat} L} \left\{ \frac{\mu Z}{L} \frac{\varepsilon_2}{d} \left[(V_G - V_{off}) V_D - \frac{V_D^2}{2} \right] \right\} \end{aligned} \tag{67}$$

Note that when the saturation velocity v_{sat} approaches infinity, Equation 67 reduces to Equation 59 which is valid for the constant mobility case, and corroborates with the gradual channel approximation which is valid for long channel HFETs. Following the procedure of Lehovec and Zuleeg [92], using Equation 66 and Equation 67, and assuming velocity saturation, the drain saturation current I_{DS} may be determined as

$$I_{DSS} = \frac{2\left(V_G - V_{off}\right)^2 Z\varepsilon\mu}{L(d+\Delta d)}\left[1+\{1+\xi_d\}^{1/2}\right]^{-2} \tag{68}$$

where

$$\xi_d = \frac{2\mu\left(V_g - V_{off}\right)}{v_{sat}L}. \tag{69}$$

Note that when $V_0 \gg (V_g - V_{off})$ drain current saturation occurs due to pinch-off, but not due to velocity saturation as is expected for the long channel devices.

The transconductance is an important parameter in HFETs and is defined by

$$g_m = \left(\frac{\partial I_d}{\partial V_g}\right)_{V_d=const} \tag{70}$$

For the saturation regime, the transconductance may be expressed by

$$g_m^{sat} = \left(\frac{\partial I_{dss}}{\partial V_g}\right)_{V_d=const} = \frac{2Z\varepsilon\mu\left(V_g - V_{off}\right)}{(d+\Delta d)L}\left\{(1+\xi_d)+\left((1+\xi_d)^{1/2}\right)\right\}^{-1} \tag{71}$$

The maximum transconductance is obtained when the sheet charge density is fully undepleted under the gate, which leads to

$$g_m^{max} = \frac{Zq\mu n_s}{L}\left[1+\left\{\frac{q\mu n_s(d+\Delta d)}{\varepsilon v_{sat}L}\right\}^2\right]^{-1/2} \tag{72}$$

For very short gate lengths, which occur in essentially all modern HFETs, the second term in the bracket dominates, and Equation 72 reduces to

$$g_m^{max} \approx \frac{Z\varepsilon v_{sat}}{d+\Delta d} \tag{73}$$

The measured transconductance is actually smaller than that given by Equation 73 in that the source resistance, which will be defined shortly, acts as a negative feedback. Taking the circuit effects into account, the measured extrinsic transconductance may be given by

$$\left(g_m^{max}\right)_{ext} = \frac{g_m^{max}}{1+R_s g_m^{max}}. \tag{74}$$

1.2.1 Examples for GaN and InGaN Channel HFETs

Let us now consider two cases, one with *GaN* and the other with *InGaN* channel FET with an *AlGaN* barrier layer. If the GaN channel is grown on a thick GaN buffer layer it can be assumed relaxed. Therefore no piezoelectric charge would be induced by GaN. If the InGaN channel is made thin enough for it to be coherently strained, it would be under compressive strain and there would be a piezoelectric polarization. The AlGaN layer on the GaN channel layer would be under tensile strain, also leading to piezoelectric polarization. For simplicity let us assume that in the case of InGaN channel, both AlGaN and InGaN layers assume the in-plane lattice constant of GaN. This example is depicted in Figure 10 which also shows the compositional gradient induced spontaneous polarization.

Briefly, tensile strained AlGaN and compressively strained InGaN would have piezoelectric polarizations as indicated in Figure 10. In addition, all three layers would have spontaneous polarization at each compositional gradient, again as shown in Figure 10 where the length of the arrows represents the relative values. We should point out that the spontaneous polarization is negatively larger in InGaN than GaN which means that InGaN on GaN interface would attract holes which are not assumed present in the system and therefore neglected. However, the polarization at the AlGaN/GaN interface and also that at the AlGaN/InGaN interface would attract electrons.

Armed with the above derivations, the output IV characteristics for an AlGaN/InGaN HFET and AlGaN/GaN HFET (in this example to maintain generality the

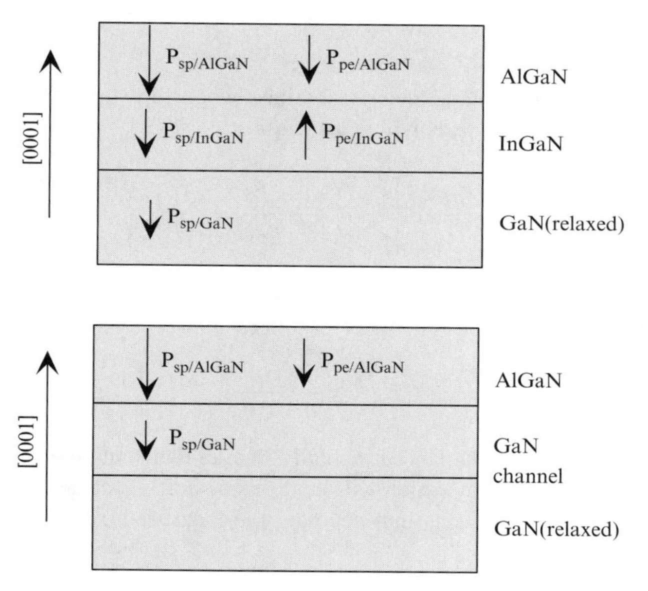

Fig. 10 Schematic representation of an AlGaN/GaN and an AlGaN/InGaN channel structure indicating the piezo and spontaneous polarization under the assumption that both InGaN and AlGaN assume the in-plane lattice constant of GaN in which case InGaN would be under compressive and AlGaN would be under tensile strain.

AlGaN layer is doped similar to the AlGaAs/GaAs modulation doped case) can be calculated. A model calculation can be perfomed for a layer structure consisting of a GaN buffer which is relatively thick and relaxed, InGaN channel which is 20 nm thick, and the top layer of AlGaN, part of which next to the interface is undoped. To make matters somewhat simpler, one can neglect the effect of strain on the band structure while using the appropriate bowing parameters for bandgap calculations. To again make the calculations simpler the linear interpolation for determining the spontaneous and piezoelectric polarization charge can be employed. The nonlinear approach with stated bowing parameters can be used for a more accurate determination. For the model calculations let use assume that the mole fraction in $Al_xGa_{1-x}N$ is 20% (x = 0.2), the doping level in $Al_xGa_{1-x}N$ is $10^{18}\,\text{cm}^{-3}(N_D)$, and the thickness of the doped $Al_xGa_{1-x}N$ is 20 nm. The thickness of the undoped $Al_xGa_{1-x}N$ is 2 nm. In the case of the $In_yGa_{1-y}N$ channel, the composition $y = 0.10$ and its thickness is 20 nm. The gate length is 1 micron and centered in a 3 micron channel. The gate width is $Z = 100$ microns, and the Schottky built-in voltage is 1 V. Let us also assume a saturated velocity of $1 \times 10^7\,\text{cm/s}$ and to make the picture even simpler, let us also assume that the access resistances are zero.

As in the case of GaAs/AlGaAs MODFET, the voltage required for depleting the doped part of the $Al_{0.2}Ga_{0.8}N$ can be calculated as

$$V_p = \frac{qN_D}{2\varepsilon_2}d_d^2 = \frac{(1.6\times10^{-19}C)(1\times10^{24}m^{-3})}{2(7.5438\times10^{-11}F/m)}(20\times10^{-9}m)^2 = 0.4242\,V$$

Using the values reproduced from Table 1 and Table 2, one can calculate the polarization induced charge in $Al_{0.2}Ga_{0.8}N/In_{0.1}Ga_{0.9}N/GaN$ and $Al_{0.2}Ga_{0.8}N/GaN$ systems. Let us now consider the case of the $Al_{0.2}Ga_{0.8}N/GaN$ system first.

1.2.2 The Case of $\mathbf{Al_{0.2}Ga_{0.8}N/GaN}$

With the aid of Equation 50 and assuming that $\Delta E_{F0} = 0$, $V_{off} = 0.3098\,V$ and the charge at the interface due to dopants in the barrier is given by

$$Q = \frac{7.5438\times10^{-11}F/m}{20\times10^{-9}m}(V_G - 0.3098V),$$

which for $V_G = 0$ (no applied voltage to the gate)

is $Q = -0.0012C/m^2$ corresponding to $n_{so} = \frac{Q_s}{q} = 7.5\times10^{15}electrons/m^2$.

Spontaneous polarization charge:

$$\begin{aligned} P^{SP}_{Al_xGa_{1-x}N} &= -0.09x - 0.034(1-x), \text{ that for } x = 0.2 \text{ is} \\ P^{SP}_{Al_{0.2}Ga_{0.8}N} &= -0.0452C/m^2 \\ P^{SP}_{GaN} &= -0.034C/m^2 \text{ and} \end{aligned}$$

The net charge at the AlGaN/GaN interface due to spontaneous polarization is

$$\Delta P^{SP} = -0.0452 - (-0.034) = -0.0112\,C/m^2$$

The piezoelectric polarization is only contributed from the AlGaN layer:

$$P^{PE}_{AlGaN} = 2\left(\frac{a_{GaN} - a_{AlGaN}}{a_{AlGaN}}\right)\left(e_{31} - e_{33}\frac{C_{13}}{C_{33}}\right)$$
$$= 0.01112(-0.548) = -0.0061\,\mathrm{C/m^2}$$

The total charge at the interface due to polarization is a sum of piezo and spontaneous charges:

$$P_{total} = P^{PE}_{AlGaN} + \left(P^{SP}_{Al_{0.2}Ga_{0.8}N} - P^{SP}_{GaN}\right) = P^{PE}_{AlGaN} + \Delta^{SP} = -0.00173\,C/m^2$$

which in terms of the electron concentration is $n_p = (Q_p)/q = 1.08 \times 10^{16}$ $electrons/m^2$ with no applied gate voltage.

The total charge at the interface is the sum of the doping induced and that polarization induced charge:

$$Q_{total} = Q + Q^{SP} + Q^{PE} = -0.0185\,C/m^2$$

The total charge is also equal to $Q_{total} = CV_{off} = \frac{\varepsilon_2}{d}V_{off}$ which gives rise to a V_{off} value of $V^{l}_{off} = -4.905\,V$

The current-voltage or the output characteristics of this HFET can be calculated with the aid of Equation 67 which is repeated here for convenience:

$$I_D = \frac{1}{1 + \mu V_D / v_s L}\left\{\frac{\mu Z}{L}\frac{\varepsilon_2}{d}\left[(V_G + |V_{off}|)V_D - \frac{V_D^2}{2}\right]\right\}$$

where, $Z = 100\,\mu\mathrm{m}$, $L = 1\,\mu\mathrm{m}$, $v_s = 1 \times 10^7\,\mathrm{cm/s}$, $d = 20\,\mathrm{nm}$, mobility for GaN is $1000\,\mathrm{cm^2/Vs}$ ε_2 is the dielectric constant for $Al_{0.2}Ga_{0.8}N$ layer $\varepsilon_2 = \varepsilon_{AlGaN} = (8.9 - 1.9x)\varepsilon_0 = 8.52\varepsilon_0$

The calculated output characteristic for the $Al_{0.2}Ga_{0.8}N$/GaN HFET under consideration is shown in Figure 11.

1.2.3 The Case of $\mathbf{Al_{0.2}Ga_{0.8}N/In_{0.1}Ga_{0.9}N}$

In a similar fashion to that which we just undertook for the AlGaN/GaN, the pertinent parameters for an $Al_{0.2}Ga_{0.8}N/In_{0.1}Ga_{0.9}N$ MODFET can be calculated. The voltage required to deplete the dopant charge in $Al_{0.2}Ga_{0.8}N$ (the same as in the case of $Al_{0.2}Ga_{0.8}N$/GaN MODFET)

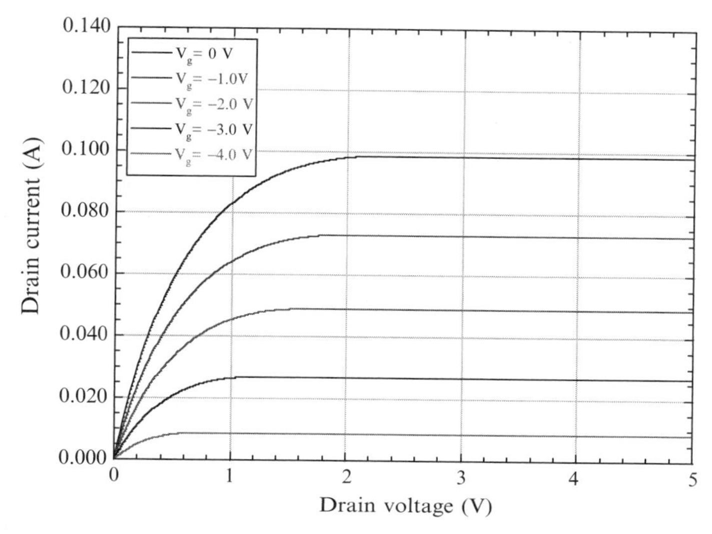

Fig. 11 The calculated output characteristic for the $Al_{0.2}Ga_{0.8}N$/GaN MODFET with parameters described in the text with low field mobility of $1000\,\mathrm{cm^2/Vs}$ and a saturation velocity of 10^7 cm/s, ($V_{off} = -4.905$ V).

$$V_p = \frac{qN_D}{2\varepsilon_2}d_d^2 = \frac{(1.6\times10^{-19}C)(1\times10^{24}m^{-3})}{2(7.5438\times10^{-11}F/m)}(20\times10^{-9}m)^2 = 0.4242\,V.$$

With the aid of Equation 50 and assuming that $\Delta E_{F0} = 0$, $V_{off} = 0.038\,V$ The charge at the interface due to dopants in the in the $Al_{0.2}Ga_{0.8}N$ barrier is given by

$$Q_{doping} = \frac{7.5438\times10^{-11}F/m}{20\times10^{-9}m}(V_G - 0.038\,V)$$

which for $V_G = 0$ (no applied voltage at the gate) becomes $Q_{doping} = -1.43\times10^{-4}C/m^2$.

Spontaneous polarization charge for $P^{SP}_{Al_{0.2}Ga_{0.8}N} = -0.0452\,C/m^2$ and for $P^{SP}_{In_{0.1}Ga_{0.9}N} = -0.0348\,C/m^2$ leading to a net spontaneous polarization charge at the $Al_{0.2}Ga_{0.8}N/In_{0.1}Ga_{0.9}N$ interface of $Q^{SP} = \Delta P^{SP} = -0.0104\,C/m^2$.

Assuming that AlGaN assumes the lattice constant of relaxed InGaN, the PE charge is $P^{PE}_{AlGaN} = -0.020\,C/m^2$. In the case of AlGaN and InGaN in-plane lattice constants being the same as that for GaN, the charge is $Q^{PE} = P^{PE}_{AlGaN} + P^{PE}_{InGaN} = 0.0054\,C/m^2$. In the case where InGaN is relaxed, $Q^{PE} = P^{PE}_{AlGaN} + P^{PE}_{InGaN} = -0.0085\,C/m^2$.

Neglecting the spontaneous polarization between the $In_{0.1}Ga_{0.9}N$ channel and the GaN buffer, the total charge at the interface is the sum of calculated charges thus far:

$$Q_{total} = Q_{doping} + Q^{SP} + Q^{PE} = -0.0051\,C/m^2.$$

Utilization of $Q_{total} = \frac{\varepsilon_{AlGaN}}{d}(V_G - V_C(x) - V_{off}^{total})$ leads to $V_{off}^{total} = -1.352V$.

The electron mobility of $In_{0.1}Ga_{0.9}N$ is determined by relating it to GaN using the effective masses in the following form.

$$\mu_{In_{0.1}Ga_{0.9}N} = \frac{m_e^*(GaN)}{m_e^*(In_{0.1}Ga_{0.9}N)}\mu(GaN) = \frac{0.2m_0}{0.191m_0}0.1m^2/Vs = 0.1047m^2/Vs$$

assuming that the relaxation time in both semiconductors is the same.

The current-voltage, or the output characteristics of this MODFET is can be calculated with the aid of Equation 67 which is repeated here for convenience:

$$I_D = \frac{1}{1+\mu V_D/v_s L}\left\{\frac{\mu Z}{L}\frac{\varepsilon_2}{d}\left[(V_G + |V_{off}|)V_D - \frac{V_D^2}{2}\right]\right\}$$

where again, $Z = 100\,\mu\text{m}$, $L = 1\,\mu\text{m}$, $v_s = 1 \times 10^7\,\text{cm/s}$, $d = 20\,\text{nm}$, mobility for GaN is $1000\,\text{cm}^2/\text{Vs}$.

ε_2 is the dielectric constant for $Al_{0.2}Ga_{0.8}N$ layer

$$\varepsilon_2(x) = (8.9 - 1.9x)\varepsilon_0 = 7.5438 \times 10^{-11}F/m$$

The calculated output characteristic for the $Al_{0.2}Ga_{0.8}N/In_{0.1}Ga_{0.9}N$ MODFET under consideration is shown in Figure 12.

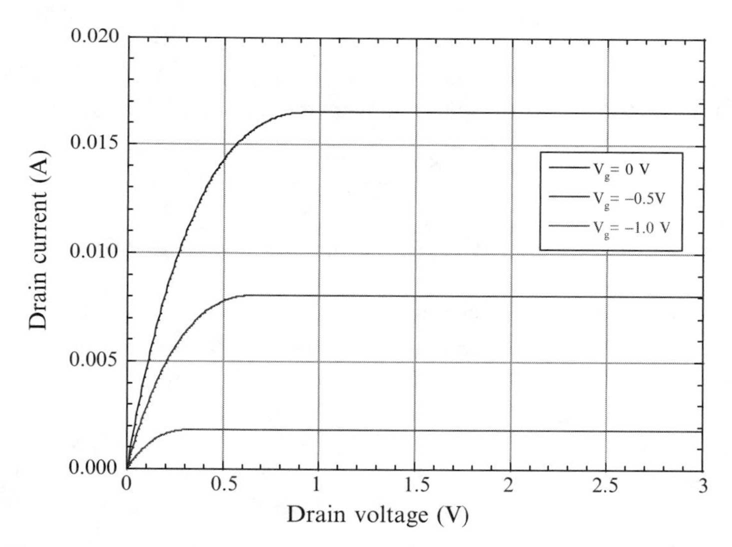

Fig. 12 The calculated output characteristic for the $Al_{0.2}Ga_{0.8}N/In_{0.1}Ga_{0.9}N$ HFET with parameters described in the text. The effect of spontaneous polarization between InGaN and the GaN buffer is neglected as that contribution is small relative to the strain induced piezoelectric polarization, ($V_{off} = -1.352\,V$).

1.3 Numerical Modeling of Sheet Charge and Current

Having treated GaN and InGaN channel HFETs analytically, in the following, numerical results of the electron density and distribution as well as the band edge profile for both normal and inverted modulation doped FETs including the polarization charge are presented. The presentation follows that reported by Sacconi et al. [53]. The range of parameter values and structural design has been chosen to not only represent the most commonly used structures, but also those that would demonstrate how the important properties would change with structural design changes. This also applies to N-polarity samples discussed here not because they are technologically important, but because of the need for a full understanding of how polarization charge affects the parameters of importance for FETs.

Let us consider two HFET structures, namely a single heterojunction AlGaN/GaN Normal Modulation Doped FET (NMODFET) where the AlGaN donor layer is grown on top of the GaN channel layer, and an "inverted" GaN/AlGaN/GaN MODFET (IMODFET) where the channel layer is grown on top of the AlGaN donor layer. It should be noted that for the inverted n-channel device to function requires N-polarity samples, which are not as yet competitive with Ga-polarity samples in devices. The NMODFET structure consists (from the gate to the substrate) of a 150 Å n-doped ($n = 10^{18}\,cm^{-3}$) AlGaN, 50 Å unintentionally doped AlGaN layer and a thick GaN buffer. The IMODFET consists (from the gate to the substrate) of a 300 Å unintentionally doped GaN layer, 50 Å unintentionally doped AlGaN, 150 Å n-doped ($n = 10^{18}\,cm^{-3}$) AlGaN, 300 Å unintentionally doped AlGaN layer, and a thick GaN layer.

A residual doping of $10^{17}\,cm^{-3}$ in both GaN and AlGaN layers is assumed, a figure which would be updated downward as the deposition technologies are improved. A Schottky barrier height (ϕ_B) of 1.1 eV for the metal-GaN interface and a $\phi_B = 1.2\,eV$ for the metal-AlGaN interface is used. Calculations have been performed for $Al_xGa_{1-x}N$ with Al concentrations of $x = 0.1, 0.2, 0.3, 0.4$. Both $[0001]$ and $[000\bar{1}]$ growth directions are considered. In the simulations, an effective mass of 0.19, which is somewhat smaller than the commonly used $0.22m_0$, for electrons and 1.8 for holes, which is the midrange among the reported values, in both GaN and AlGaN layers have been used. The band gaps and band discontinuities of the AlGaN layers used are tabulated in Table 4.

Table 4 Band Gap and Conduction band discontinuities used for $Al_xGa_{1-x}N$ and $Al_xGa_{1-x}N/GaN$ heterointerface, respectively

x (Al)	E_G [eV]	ΔE_C [eV]
0.1	3.62	0.17
0.2	3.85	0.33
0.3	4.09	0.51
0.4	4.35	0.69

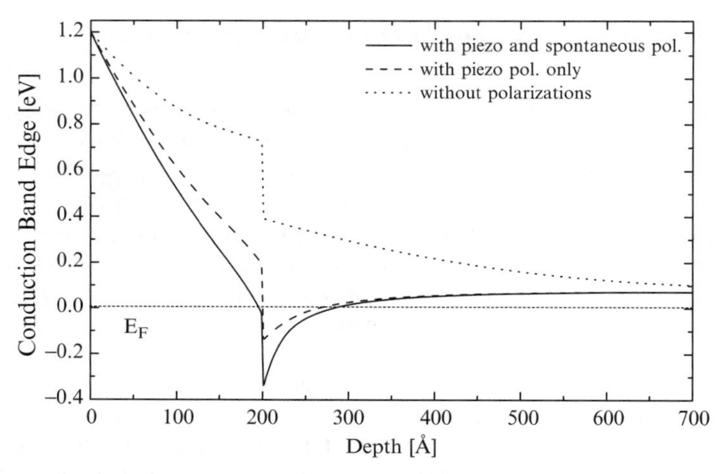

Fig. 13 Calculated conduction band edge for the N-MODFET structure grown in the [0001] direction for $V_G = 0$ with and without polarization fields.

The conduction band edge profile for the NMODFET grown in the [0001] direction is depicted in Figure 13 for the cases *i*) with both spontaneous and piezoelectric polarization fields, *ii*) without considering the polarization fields, *iii*) with only the piezoelectric polarization fields. The difference in piezoelectric and spontaneous polarization between AlGaN and GaN layer manifests itself as a fixed 2D-charge density at the interface between the two materials. For the [0001] growth direction considered in Figure 13 the difference in polarization between the two materials induces a positive charge ($\sigma = +1.12 \times 10^{13}\,\mathrm{cm}^{-2}$) at the $Al_{0.2}Ga_{0.8}N/GaN$ interface. Free electrons are then attracted by this positive bound polarization charge, tending to accumulate at the interface and thus forming a conductive channel. Moreover, the high electric field due to the interface charge favors the build up of a large channel density and of a strong channel confinement. Within the AlGaN layer, the strong electric field compensates the space charge contribution coming from the ionized donors. Consequently, it prevents the appearance of the parasitic channel that would otherwise form in the doped AlGaN layer [56, 93].

The comparison reported in Figure 13 between the three cases with different contribution of the polarization fields shows the importance of considering both spontaneous and piezoelectric polarizations in GaN based device modeling. In fact, by neglecting the spontaneous polarization, as was done recently, [94–97] the channel electron density is underestimated [56]. Clearly the sign of the polarization charge is crucial. For the same N-MODFET structure grown in the $[000\bar{1}]$ direction, the resulting polarization charge would be negative (with the same magnitude) and electrons would be repelled from the channel as shown in Figure 14.

The distribution of the free electron charge in the channel is shown in Figure 15 for several values of the Al concentration of the AlGaN layer. Increasing the Al content induces a larger polarization charge at the GaN/AlGaN interface and consequently a higher channel electron concentration.

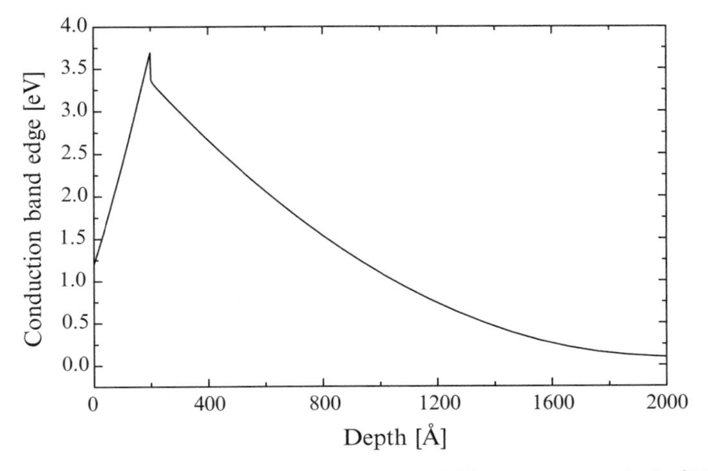

Fig. 14 Calculated conduction band edge for the N-MODFET structure grown in the $[000\bar{1}]$ direction for $V_G = 0$.

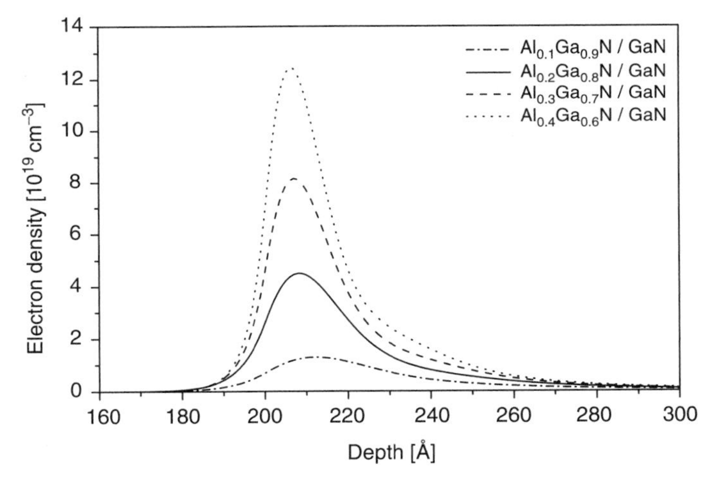

Fig. 15 Electron density distribution in the channel of the [0001] grown N-MODFET for $V_G = 0$ for several Al contents of the $Al_xGa_{-1-x}N$ layer.

The calculations we have shown so far are obtained by considering only the polarization charge at the AlGaN/GaN interface. In reality, however, polarization charges that form at the metal-AlGaN and at the end of the GaN buffer region should be accounted for. The metal-AlGaN charge is completely screened by the charges induced on the metal surface and can therefore be neglected. On the other hand, the charges at the end of the buffer region may induce large deviation with respect to the situation depicted above. Oberhuber *et al.* [39] have considered a -$\sigma/2$ charge at the interface between the GaN and a nucleation region. The exact amount of such charge depends, however, on the morphology of the heterojunction and may differ from the theoretical value $\sigma = \Delta P/q$. The situation is less critical if the bottom interface is far away from the main AlGaN/GaN heterojunction. In this case, the polarization charge that arises can be completely screened by the residual doping in the GaN buffer layer. On the contrary, if such interface is close to the AlGaN/GaN heterojunction, the polarization charge can completely deplete the channel. In the simulations considered here a thick GaN substrate is assumed. Thus, the effect of the polarization charge at the end of the GaN substrate is completely screened.

The band edge profile and electron densities for the I-MODFET grown in the $[000\bar{1}]$ direction are shown in Figure 16 and Figure 17 respectively. A comparison of the conduction band edges with and without polarization charges is also plotted. As for the N-MODFET, the presence of the fixed and positive polarization charge at the GaN/AlGaN interface induces the formation of a channel not present in absence of the polarization charge. For the I-MODFET a $-\sigma$ polarization charge is also present at the end of AlGaN region (i.e. at the AlGaN/GaN interface). Similar to the [0001] grown N-MODFET, a larger Al content of the AlGaN layer induces a larger polarization charge at the GaN/AlGaN interface and consequently an increasing of electron concentration in the channel. Naturally, for [0001] orientation, the interface-charge forms below the AlGaN layer which is not what is desired for an I-MODFET. What is desired is the formation of the electron sheet layer on top of the AlGaN layer, which is possible when the $[000\bar{1}]$ orientation is employed. The structure in its present shape, i.e., the [0001] polarity would show FET performance providing that the AlGaN layer is completely depleted but with small transconductance. If the AlGaN is not depleted, then the device would function as a MESFET dominated by transport in the AlGaN layer unless the gate potential is large enough to deplete the AlGaN layer. To eliminate the formation of an interface

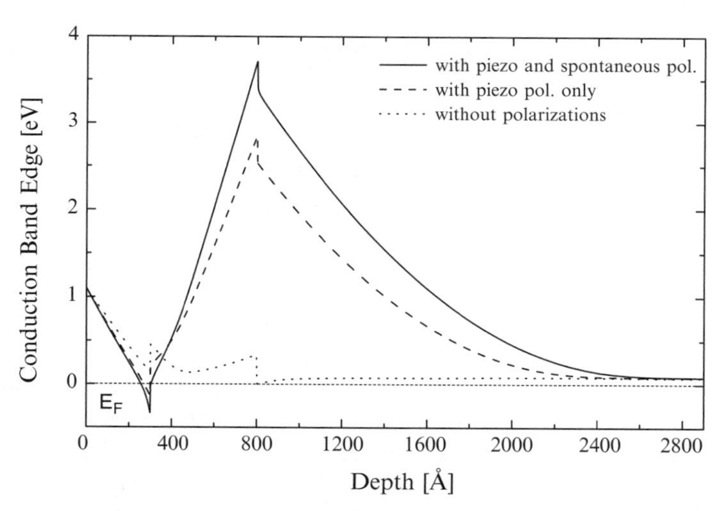

Fig. 16 Conduction band edge for the I-MODFET structure grown in the $[000\bar{1}]$ direction for $V_G = 0$ with and without polarization fields.

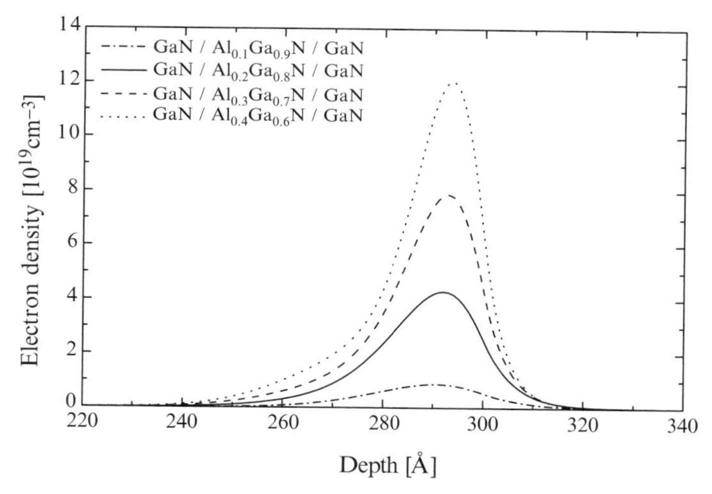

Fig. 17 Electron density distribution in the channel of the $[000\bar{1}]$ grown I-MODFET for $V_G = 0$ for several Al contents of the $Al_xGa_{-1-x}N$ layer.

electron charge at the bottom of the AlGaN layer, the bottom heterointerface should be graded substantially. In that case, the [0001] polarity would cause the band diagram to accumulate holes at the top interface if they are present. That top interface would accumulate electrons in the $[000\bar{1}]$ polarity.

The channel charge density is therefore controlled by two factors: *i*) the Gate bias as in traditional N-MODFET device, ii) the Al content of the AlGaN layer, which tailors the polarization field. Charge control in nitride based devices can be achieved by adjusting two independent parameters and thus offers a wide degree of flexibility with respect to traditional devices. This can be seen from the sheet charge concentration in the channel as obtained by integrating the electron density distribution along the z-direction. Considering the explicit dependence of the sheet charge density on the gate voltage V_G, we have:

$$n_s(V_G) = \int n(V_G, z)dz \tag{75}$$

Figure 18 and Figure 19 show the sheet electron density in the channel as a function of gate bias for several Al contents in the AlGaN layer for both [0001] grown N-MODFET and $[000\bar{1}]$ grown I-MODFET, respectively. Naturally, the channel electron density increases for larger Al contents of the AlGaN layer. We note also that the density is higher for the N-MODFET with respect to the I-MODFET because the particulars relating to the band bending on the top interface of the I-MODFET. As mentioned earlier, the I-MODFET structure is intended for use with $[000\bar{1}]$ for investigative purposes only as the body of work in the AlGaAs/GaAs system showed the N-MODFET to be desired device structure.

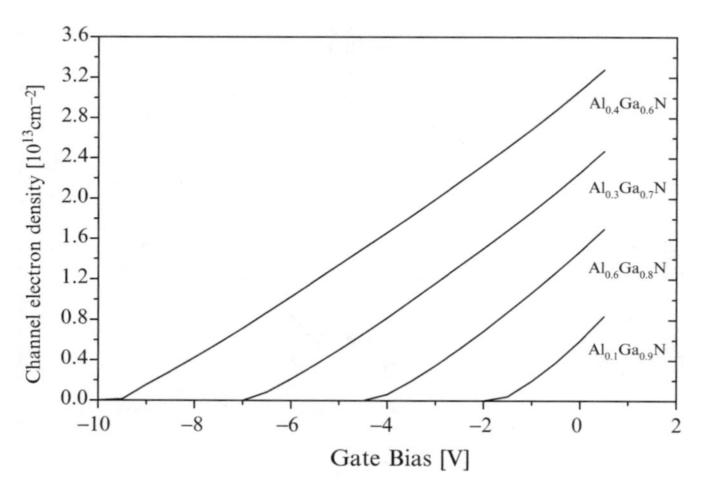

Fig. 18 Channel electron density as a function V_G (for several AlN mole fractions) for the N-MODFET grown along the [0001] direction.

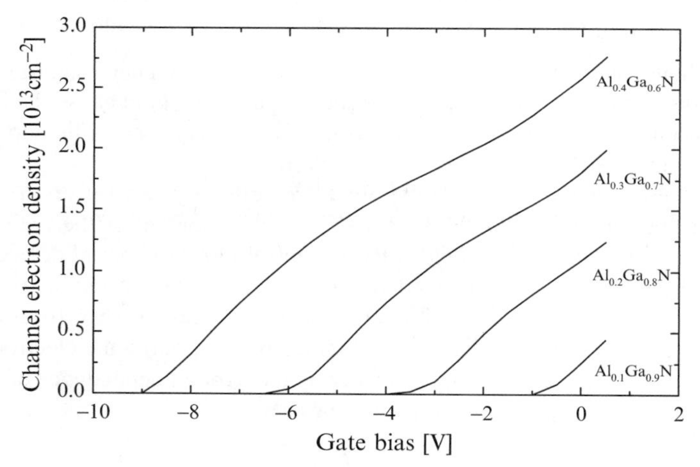

Fig. 19 Channel electron density as a function V_G (for several AlN mole fractions) for the I-MODFET grown along the $[000\bar{1}]$ direction.

Finally, the equilibrium static charge in undoped MODFET structures and current voltage characteristics of the same have been calculated, in a similar fashion as that reported above. The simulated structure is a normal MODFET with Ga-polarity layers wherein the AlGaN barrier is situated on top of the GaN channel layer [53]. Figure 20 (a) and (b) display the conduction band edge and electron density profile, respectively, for a Normal MODFET with the $Al_xGa_{1-x}N$ barrier thickness of 20 nm and Al content x = 0.1, 0.2, 0.3, and no doping in neither $Al_xGa_{1-x}N$ nor GaN. Compared to the intentionally or unintentionally doped cases (shown in Figure 13 and Figure 15), the conduction band well at the heterointerface becomes

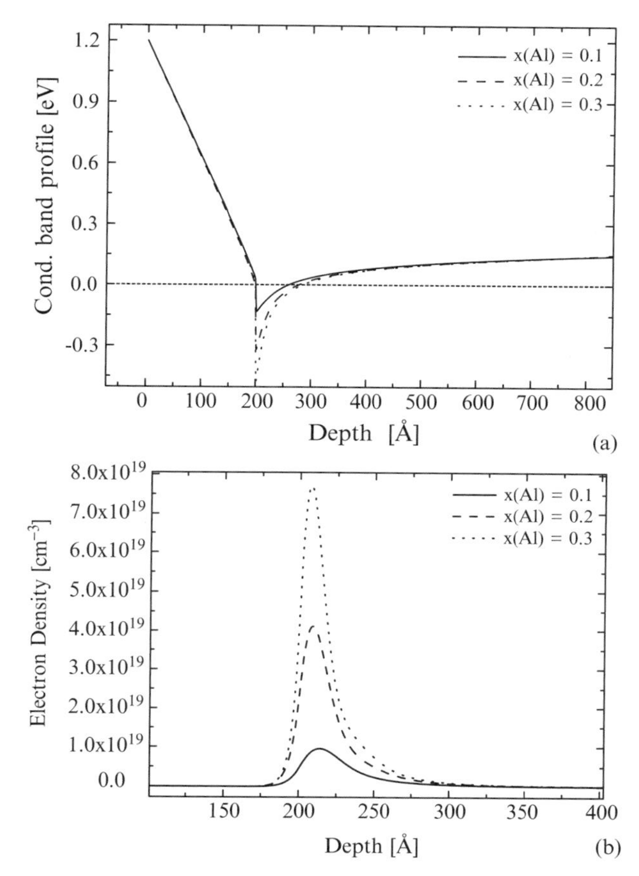

Fig. 20 (a) conduction band edge; (b) electron density profile for the for a Normal MODFET with the $Al_xGa_{1-x}N$ barrier thickness of 20 nm and Al content x $= 0.1, 0.2, 0.3$, and no doping neither in $Al_xGa_{1-x}N$ nor GaN, $V_G = 0$

a little shallower and the electron density reduces slightly. Figure 21 (a) and (b) display the conduction band profile and electron density distribution, respectively, for an NMODFET with $Al_xGa_{1-x}N$ barrier thickness of 30 nm and the same set of Al mole fractions as above, and no doping neither in AlGaN nor GaN. A thicker $Al_xGa_{1-x}N$ layer, up to a point, induces a larger electron density at heterointerface because of larger band bending in AlGaN.

Sacconi et al [53] have implemented a quasi-2D [98–100] model for the calculation of the current-voltage characteristics of the nitride MODFETs. This model makes use of the exact value of the sheet charge density in a MODFET device channel, obtained from the self-consistent Schrödinger-Poisson solution presented above.

We assume a FET model shown in Figure 22 where the x-axis is along the channel and the z-axis is along the growth direction. The model also considers the

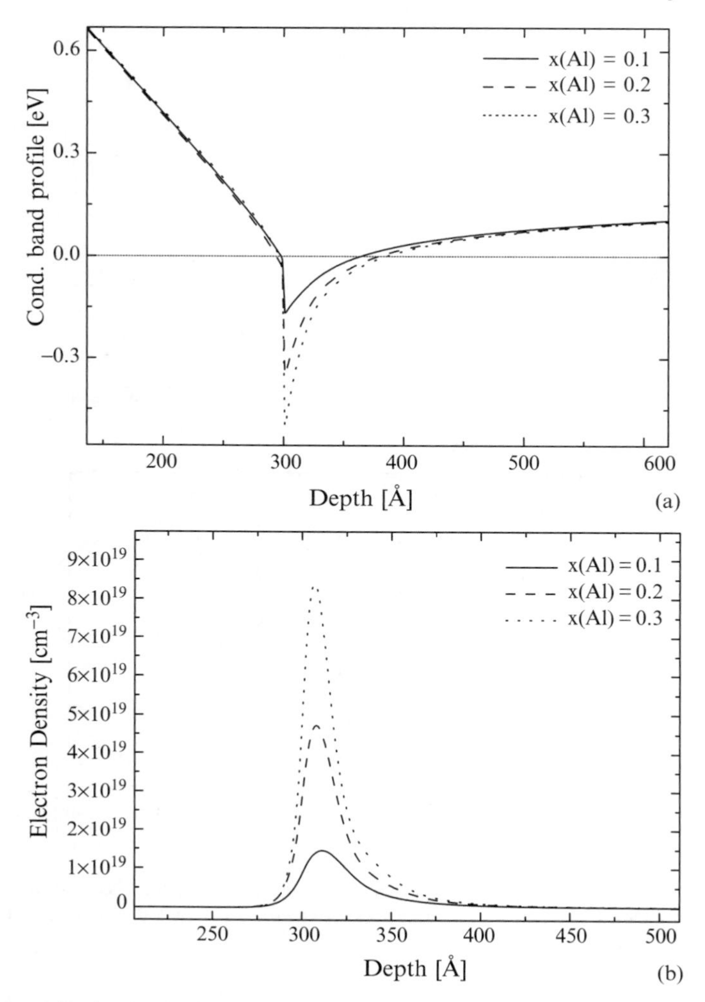

Fig. 21 (a) conduction band edge; (b) electron density profile for a Normal MODFET with the $Al_xGa_{1-x}N$ barrier thickness of 30 nm and Al content x = 0.1, 0.2, 0.3, and no doping neither in $Al_xGa_{1-x}N$ nor GaN, $V_G = 0$

presence of a drain (R_D) and source (R_S) resistance. When a drain bias (V_D) is applied, the potential along the channel may be considered as varying gradually from the source bias (V_S) to V_D. In this situation it is still possible to calculate the sheet charge density n_s at every section grid, provided that one considers the $V(x)$ potential (on the top surface) for each point of the channel. Because for n-channel devices, V_D is positive and V_S is zero, $V(x)$ contributes to the channel depletion and the sheet charge density n_s for the generic x section of the FET will be therefore:

$$n_s(x) = n_s\left(V_G - V\left(x\right)\right) \tag{76}$$

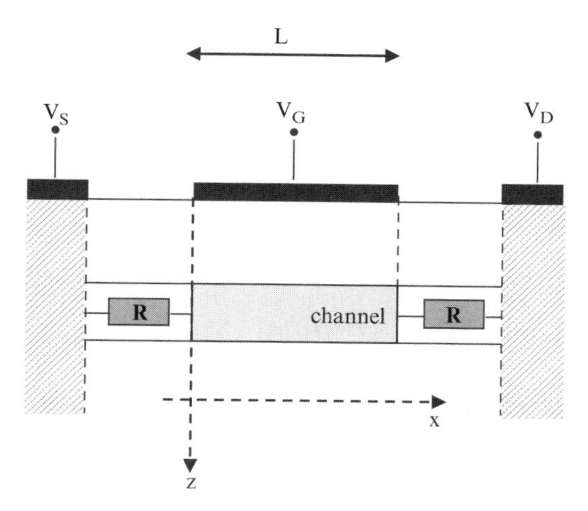

Fig. 22 Schematic representation of the quasi-2D FET model used.

By neglecting diffusion contributions, the source-to-drain current I_{DS} is given by:

$$I_{DS} = -qZv(x)\,n_s(x) \tag{77}$$

where Z is the gate width and $v(x)$ the electron mean velocity, supposed independent from transverse coordinate.

The dependence of the drift velocity on the longitudinal electric field is empirically given by Equation 63.

Parasitic components are included explicitly through the drain and source resistances (R_D, R_S)

$$\begin{aligned} V_S^e &= V_S + I_{DS}R_S \\ V_D^e &= V_D - I_{DS}R_D \end{aligned} \tag{78}$$

where V_D^e and V_S^e represent the effective bias boundaries of the gate region on the drain and source sides, respectively. For a certain value of I_{DS}, we can calculate the V_D by solving Equation 77. The explicit equation for the current is:

$$I_{DS} = qZ\frac{\mu_0 F(x)}{1 + \frac{F(x)}{F_C}}n(V_G - V(x)) \tag{79}$$

The numerical solution is based on the discretization of this expression into N sections, each with amplitude h, so that $Nh = L$, where L is the gate length. Given the $(i-1)^{\text{th}}$ section potential, the ith potential $V_i = V_{i-1} + F_i h$ where F_i is the ith section electric field. We have then the N relations:

$$I_{DS} = qZ\frac{\mu_0 F_i}{1 + \frac{F_i}{F_C}}n(V_G - V_{i-1} - F_i h) \tag{80}$$

Because the $(i-1)^{\text{th}}$ section potential is known from the previous step, this is a non-linear equation in the unknown F_i. Solving iteratively for all the N sections, one obtains the value of the drain voltage V_D consistent with the assumed current.

Repeating this procedure for a suitable range of values of I_{DS}, one obtains the set of corresponding values of V_{DS} and thus the MODFET I-V characteristics, which are elaborated on below.

1.4 Calculated I-V Characteristics

In this section we discuss the simulated I-V characteristic of the normal and inverted MODFET, obtained for a gate length of $L = 0.3\,\mu\text{m}$. We have chosen a drain and source contact resistivity of about $1\,\Omega\text{mm}$, which is consistent with experimentally measured values on these types of devices [100]. We use a saturation velocity of $2.5 \times 10^7\,\text{cm/s}$, (reference [4]), while for the low field mobility we choose a value of $\mu_0 = 1100\,\text{cm}^2/\text{Vs}$, slighter higher than the GaN bulk value, according to the experimental and theoretical results for similar devices [4, 56, 101–103].

In Figure 23 we show the I_{DS} vs. V_{DS} for the [0001] polarity MODFET for several gate (V_{GS}) voltages. The results are presented for both $x = 0.2$ (Figure 23 (a)) and $x = 0.4$ (Figure 23 (b)), Al concentration of the top layer. For $x = 0.2$, the MODFET reaches pinch-off for a bias voltage of $V_{GS} = -4.4\,\text{V}$ while for $x = 0.4$ the pinch-off is reached at $V_{GS} = -9.5\,\text{V}$. On the other hand, the saturation drain current for $x = 0.2$ is $I_{DS} = 2.4\,\text{A/mm}$ at $V_{GS} = 0$ and it increases up to 5.76 A/mm for an Al content of $x = 0.4$. Thus, the current flowing in the devices depends strongly on the Al contents of the top layer. This is essentially due to the increasing of the channel electron density induced by the increasing of the polarization charge going from an Al content of 0.2 up to 0.4. This peculiarity of the MODFET should be considered in the design of these devices because fluctuations of the alloy composition of the top layer may induce large variations with respect to nominal electrical values of the device. It should also be pointed out that the gate leakage would determine the extent of gate voltage that can be applied to the gate. For a gate bias of 9.5 V and AlGaN layer thickness of 20 nm, the vertical field under the gate near the source can reach 4.75 MV/cm. This means that MODFETs utilizing large mole fractions of Al may require thin AlGaN layers or recessed gates to keep the gate voltage smaller.

A similar situation is obtained for the I-MODFET with the $GaN/AlGaN/GaN$ structure grown in the $[000\bar{1}]$ direction, meaning with N-polarity. The calculated I_{DS} vs. V_{DS} characteristics are reported in Figure 24 (a) and (b) for $x = 0.2$ and $x = 0.4$ Al composition of the $AlGaN$ layer, respectively. Also in this case the pinch-off bias depends critically on the Al composition and varies from $-3.9\,\text{V}$ for $x = 0.2$ up to -9.0 for $x = 0.4$. Saturation currents are lower for the I-MODFET at $x = 0.2$ with respect to the equivalent MODFET structure. Such difference, however, is negligible for the case with $x = 0.4$.

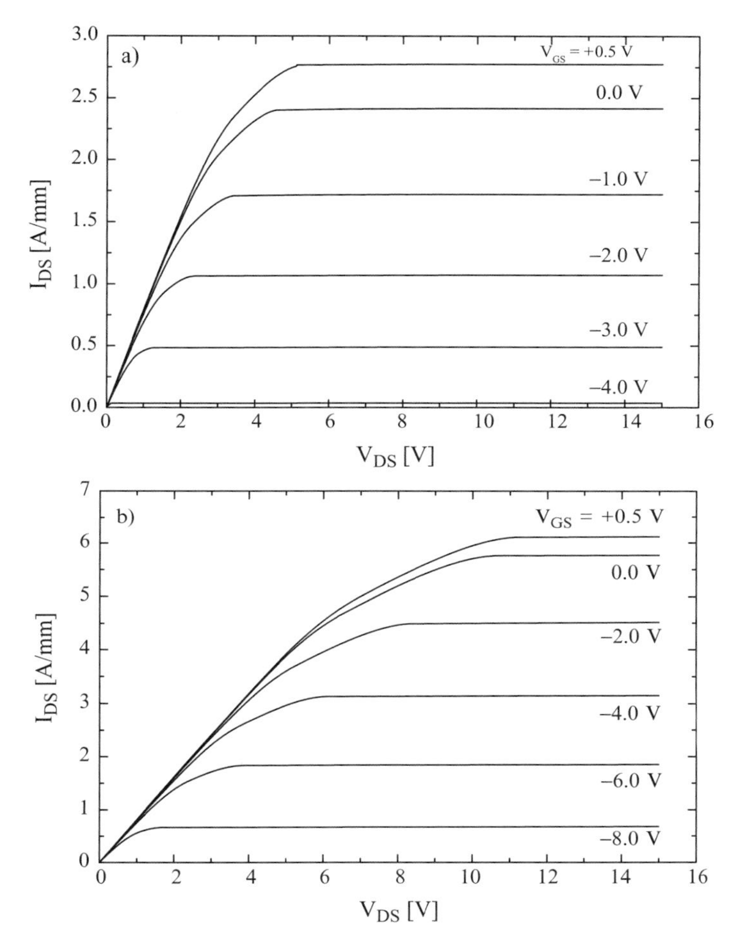

Fig. 23 I_{DS} vs. V_{DS} for the [0001] polarity Normal MODFET for several gate (V_{GS}) voltages and for $x = 0.2$. (*b*) I_{DS} vs. V_{DS} for the [0001] polarity Normal MODFET for several gate (V_{GS}) voltages and for $x = 0.4$.

1.5 Experimental Considerations

The transit time under the gate of a submicron HFET is on the order of a few picoseconds. In view of this, the charging time of the input (C_{gs}) and the feedback (C_{dg}) capacitance through the input resistance R_{in} in the equivalent circuit, shown in Figure 25, determines the speed of response. Generally two parameters, the current gain cut-off frequency and maximum oscillation frequency, are figures of merit to gauge the expected high frequency performance of a HFET. Among them, the current gain cut-off frequency is defined as the frequency at which the current gain goes to unity.

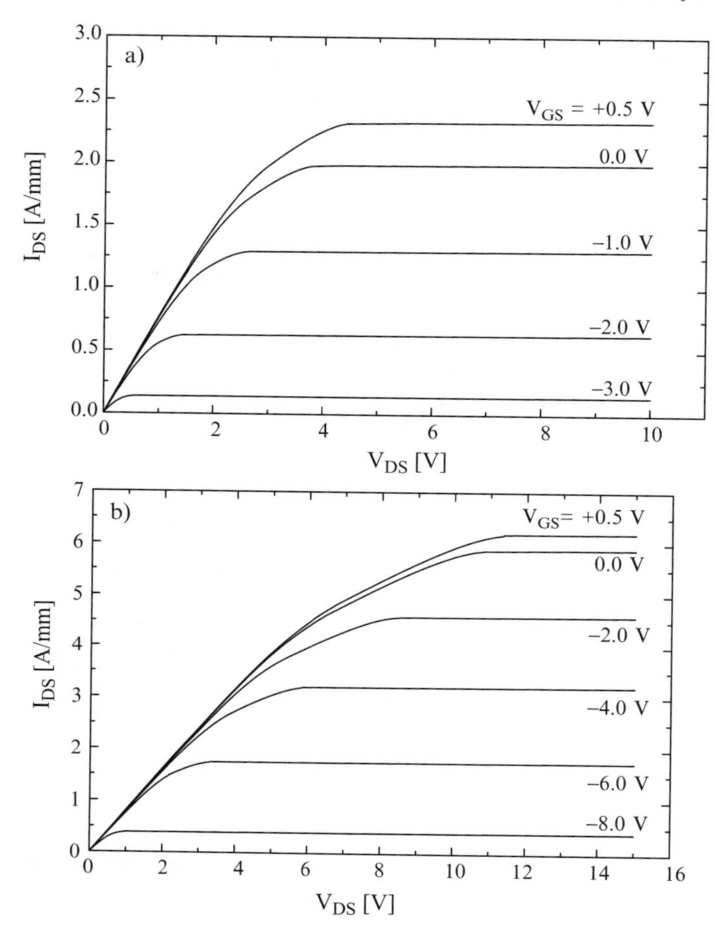

Fig. 24 (a). I_{DS} vs. V_{DS} for the $[000\bar{1}]$ polarity Inverted MODFET for several gate (V_{GS}) voltages and for $x = 0.2$. (b). I_{DS} vs. V_{DS} for the $[000\bar{1}]$ polarity Inverted MODFET for several gate (V_{GS}) voltages and for $x = 0.4$.

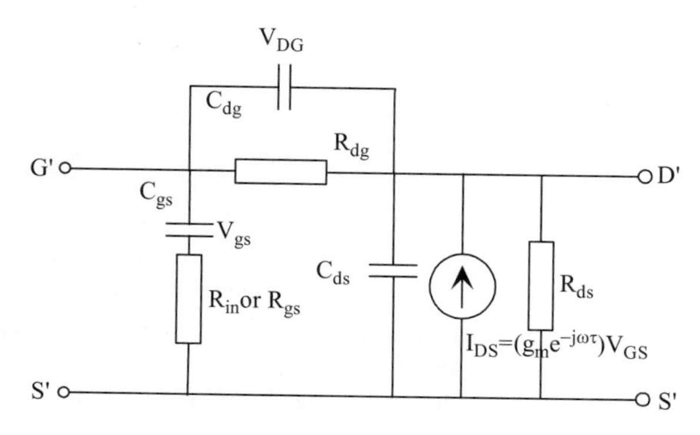

Fig. 25 High frequency equivalent circuit of an intrinsic FET.

$$f_T = \frac{g_m}{2\pi(C_{gs}+C_{gd})} \approx \frac{v_{sat}}{2\pi L} \tag{81}$$

where C_{gs} and C_{gd} are the gate-source capacitance, and gate drain feedback capacitance, respectively. The feedback capacitance (C_{gd}) is smaller than the gate source capacitance (C_{gs}) and is typically neglected which when utilized leads to

$$f_T = \frac{g_m}{2\pi C_{gs}} = \frac{v_{sat}}{2\pi L} \tag{82}$$

From Equation 82 it may be noted that the higher the saturation velocity and the smaller the gate length, the higher the value of f_T.

The maximum oscillation frequency, defined as the frequency at which the power gain goes to unity may be given by: [104]

$$f_{\max} = \frac{f_T}{2\left(r_1 + f_T \tau_3\right)^{1/2}}. \tag{83}$$

If R_s is the series resistance, C_{dg} is the drain-gate capacitance, and G_d is the differential drain conductance, then the parameter r_1 is

$$r_1 = (R_g + R_i + R_s)\, G_d \tag{84}$$

and the feedback time constant τ_3 is

$$\tau_3 = 2\pi R_g C_{dg}. \tag{85}$$

If we consider yet more of the extrinsic elements, the f_{max} term can be expressed as: [89]

$$f_{max} = \frac{f_T}{2\left\{1 + [R_s + R_g]\, G_d + 2\left(C_{dg}/C_{gs}\right)\left[\left(C_{dg}/C_{gs}\right) + g_m\left(R_s + G_d^{-1}\right)\right]\right\}^{1/2}} \tag{86}$$

The parameters have their usual meanings in that R_s and R_g represent the source and gate extrinsic resistances. The terms C_{dg} and C_{gs} represent the drain to gate feedback capacitance and gate to source capacitance, respectively. The device can be measured in the scattering, or the s-parameter configuration for short, which then can be converted to two terminal y-parameters. With years of experience and knowledge of intrinsic and extrinsic device parameters as well as the feedback components, an equivalent circuit such as that shown in Figure 25 can be developed. By matching the two terminal y-parameters calculated from the s-parameters and those of equivalent circuit of Figure 25, one can determine the values of the elements shown in the same Figure 25. It is however, a very cumbersome task to rely on this matching only. Independent determination of as many parameters as possible increases the chances that the parameters extracted are unique. To this end, source and drain and to some extent extrinsic gate resistances can be determined from the dc characteristics. In addition, the inductive parameters can be calculated based on the specifics of the bonding wires or the ribbon. Having those parameters fixed, one can then get reasonable values for the rest of the circuit parameters through a fitting. With those circuit parameters in hand one can predict the device performance at high frequencies as well as designing input and output matching circuits for amplifiers. For small signal devices the current sources I_{dg} and I_{gs} can be neglected.

Changing the band discontinuity (if the barrier is doped) changes the interface sheet concentration and, in addition, the polarization charge due to strain and compositional grading. This will lead to a change in the interface sheet carrier density (Equation 48 with Equation 54 to include the polarization charge), and thus the current level (Equation 58, Equation 59, and Equation 60). Theoretically, larger sheet carrier concentrations, and higher mobilities and velocities lead to larger transconductances (Equation 73), larger current (Equation 81) and power gains (Equation 83) at high frequencies. Better carrier confinement at the heterointerface also assists in confining the carriers in the channel, which reduces the output conductance G_d. Thus, an examination of Equation 83, Equation 84, and Equation 85 indicates that, if larger f_{max} is desired, it can be achieved by reducing r_1. Unlike the GaAs system, the strain induced polarization due to the compressively strained InGaN and the spontaneous as well as the strain induced polarization due to tensile strained AlGaN are against each other and smaller interface charge thus results. However, InGaN as a channel may have benefits in terms of reduced current lag and reduced hot phonon effects.

Traditional small signal considerations have to be augmented by large signal specific issues. The main parameter facing a power device is the maximum power level that can be obtained and the associated gain. In many applications, the noise figure of the device must also be considered. In simple terms, if the device has large drain breakdown voltage, high gain at high frequencies, and high drain efficiency, the stage is set for a desirable device. Even in a well-designed semiconductor device with all the accolades, the thermal wall is a very formidable one. Thus, it is imperative that the effect of temperature on device performance is accounted for accurately. As in small signal modeling, the first step in power modeling is to establish the basic device geometrical factors that are needed to calculate the current-voltage characteristics. Once these are known, the output characteristics superimposed with the load line can be used to estimate the power level that can be obtained from the device provided that it is not limited by the input drive as shown in Figure 26. In Class-A operation, the maximum power that can be expected from the drain circuit of a device is given by

$$P_{\max} = \frac{I_{dson}(V_b - V_{knee})}{8} \tag{87}$$

where I_{dson} is the maximum drain current (this is the drain current with a small positive voltage on the gate electrode), V_b is the drain breakdown voltage, and V_{knee} is the knee voltage as shown in Figure 26. The allowable positive gate voltage ($\approx$1 V) will depend on the channel doping and the work function of the gate metal. The positive gate voltage is limited by the onset of forward Schottky-diode current. The DC load line shown in Figure 26 would be used in a Class-A RF amplifier with the maximum drain voltage $V_d = V_b/2$. The slope of the load line is $1/R_L$ where R_L is the value of

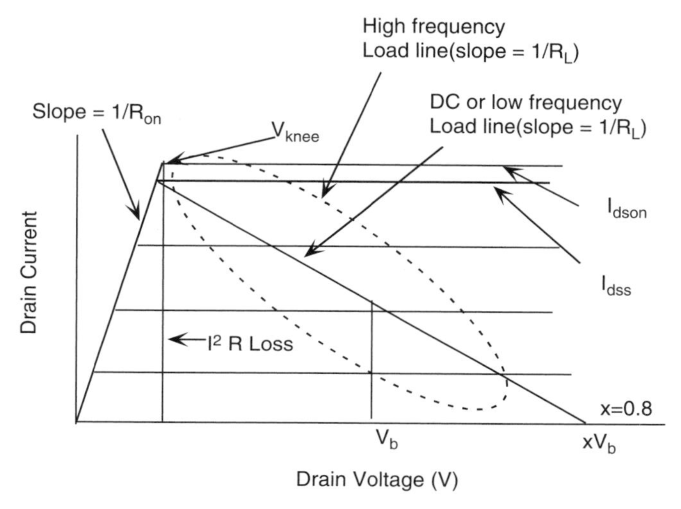

Fig. 26 Simplified schematic representation of I-V characteristics with a DC or low frequency load line and for high frequency load line, the latter shown as oval with broken lines. The divergence between the DC and high frequency RF load line is caused by inductive and capacitive elements of the transistor and matching elements with their associated phase dispersion and thus a nonlinear load line.

the load resistance at the output of the FET. What can be gleaned from Equation 87 is that V_b and I_{dson} must be made as large as possible. The utility of wide bandgap semiconductors such as GaN in this juncture is that the drain breakdown voltage is larger than that in conventional group III-V semiconductors.

In general, the drain can be swung up to voltages within 80% of the drain breakdown for a 20% margin of safety. It should be pointed out that the maximum drain current in nitride semiconductor-based HFETs is in the same ballpark as that of more conventional semiconductors. This implies that increased power handling capability is a direct result of large breakdown voltages and thermal conductivity and the fact that higher junction temperatures can be tolerated. Ability to increase drain bias increases the load resistance and makes it easier to match impedances, particularly in devices with large gate widths. In power devices, power dissipation within the device increases the junction temperature and alters the output characteristics. On the one hand, higher junction temperatures with respect to the case temperature would enhance the heat dissipation to the power of four of the temperature differential, but along with it comes reduced current and increased series resistances, which in turn increase the heat to be dissipated. Moreover, the thermal conductivity of the semiconductor decreases with increased temperature, exacerbating the situation. Consequently, the effect of junction temperature on the output characteristics must be taken into consideration. Temperature-dependent material parameters, if known, can be used to calculate the output characteristics with respect to temperature. However, a more pragmatic approach, particularly when the aforementioned parameters and/or models required are not available, can be taken in which one measures the

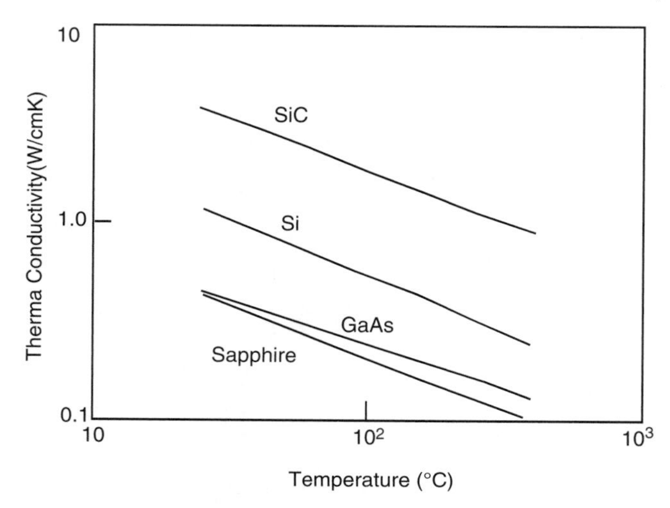

Fig. 27 Thermal conductivity vs. temperature for SiC, Si, GaAs, and sapphire. Reference [105].

output characteristics of the device under consideration as a function of temperature. The junction temperature is critically dependent on the substrate thermal conductivity that is available for various substrates including GaN. The functional dependence of thermal conductivity on temperature is

$$\chi(T) = \chi_{T_0}(T/T_0)^{-r} \tag{88}$$

where the power coefficient, r, is 0.559, 0.443, 0.524, and 0.544 for Si, GaAs, SiC, and sapphire, respectively, [105] and χ_{T_0} represents the room temperature (T_0) thermal conductivity. Thermal conductivity of sapphire, SiC, GaAs, and Si as a function of temperature is shown in Figure 27. In Figure 27, χ_{T_0} has also been appropriately reduced to account for the doping of the substrate material.

2 AlGaN/GaN HFET Performance

To reiterate, HFET's performance is due to the conduction channel that allows large sheet carrier concentrations to be maintained and its unique capacitance-voltage relationship [110]. Moreover, spatial separation of scattering centers (such as ionized donors) from the electrons leads to low field transport nearly void of ionized impurity scattering. As discussed in detail above, what is somewhat unique to GaN and its alloys as opposed to GaAs-based cousins is the spontaneous and the strain-induced piezoelectric polarization that causes redistribution of mobile and weakly bound charge and charge collected from metal contacts. In this section, the performance of mainly the AlGaN/GaN devices is discussed.

GaN-based devices are primarily geared toward power applications. However, there may be small signal applications as well requiring the robustness of GaN.

While these two types of devices share much in common there are notable differences. As the name suggests, the output power from a power device is the main output characteristic, as well, RF gain should be consistent in that either the maximum output current or the drain voltage should be large; the latter is easier to increase by choosing a semiconductor with large breakdown field such as GaN. The former requires large mobility/velocity and carrier concentration, which cannot be increased as readily as the breakdown field by choosing a semiconductor of relevance. High drain voltage operation also provides a large output resistance making it easier for impedance matching and power combining if necessary. Moreover, high power added efficiencies can be had this way in class A and AB operations where the gate swing actually extends beyond pinch-off. In small signal devices, parameters of primary importance have to do with gain and noise associated with the device. Small signal devices are mostly used for some aspects of communications, either on the transmitted side as medium power amplifiers or the receiving side as low noise receivers. In addition, up and coming collision avoidance systems (77 GHz), millimeter wave radar operative at the 94 GHz atmospheric window, and automatic toll collection types of applications are expected to make up a large part of market penetration.

For small signal devices, low noise figures and high operating frequencies demand very short gate lengths, about 0.15 μm and shrinking, operating at moderately small drain biases such as 5 V or less depending on the applications. The epitaxial structure and layout is more or less standard and is shown in Figure 28.

Because the gate definition requires costly electron beam lithography for small signal devices, power FETs do not necessarily use short gate lengths unless the frequency of operation and the gain dictate it. As the cost of lithography becomes a smaller fraction of the total cost, the demand on gain would pave the way for small gate lengths to be used for power devices as well. While packaging is a ubiquitous issue for all devices, it takes on a special meaning in power FETs as in addition to

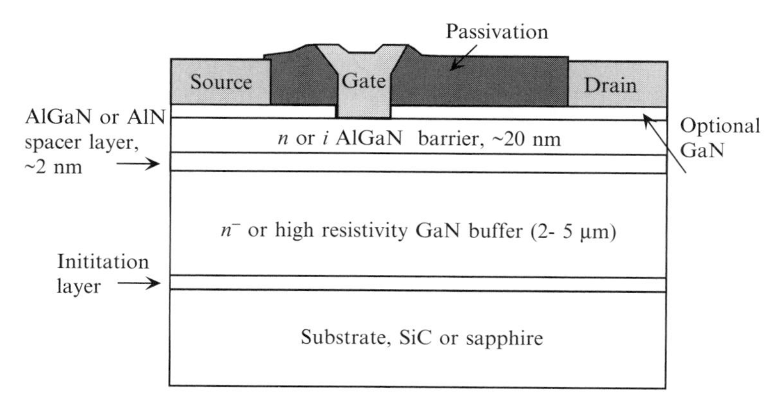

Fig. 28 A schematic cross sectional view of a small signal HFET utilizing GaN as channel and AlGaN as the barrier. The AlN interfacial barrier layer is helpful in reducing hot electron injection from the channel to the barrier layer.

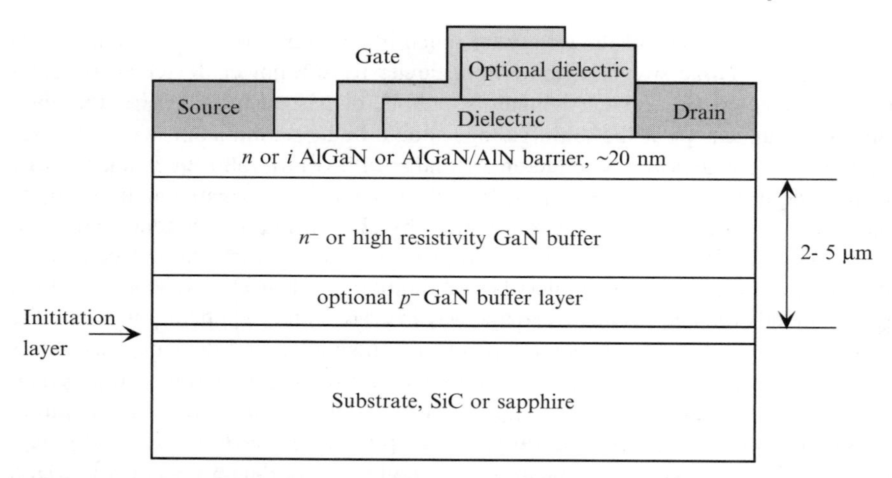

Fig. 29 A schematic cross-sectional view of a large signal HFET utilizing GaN as the channel and AlGaN as the barrier. Also the gate shown incorporates the field plate feature. The field plate can be constructed in such a way to allow independent bias for optimum field distribution and thus drain breakdown.

sealing the device and providing a simple way to transition it to the outside circuitry, the package must also dissipate large amounts of power.

A schematic of a power FET incorporating the features of field plates to spread the field between the gate and the drain and in the process increasing the drain breakdown voltage is shown in Figure 29. As mentioned above, and to be further shown in section 2.2., the drain breakdown voltage is the key for power performance. Increasing the gate-drain distance increases the breakdown voltage at the expense of increasing the drain resistance and lowering the output efficiency, the benefit outweighing the negative up to a point. Unfortunately, in FETs the electric field is nonuniform in the gate-drain region and the breakdown occurs near the gate. If this field somehow were to be spread more uniformly, large breakdown voltages (because the voltage is the integral of the field over distance) can be attained [20–23]. This is the concept behind the field plate design shown in Figure 29. In addition to a more complicated device fabrication procedure, the field plate increases the capacitance. However, in an optimized structure the overall benefit can be positive with up to five-fold increase in the drain breakdown voltage, [107] and a total power level of nearly 100 W [108].

The heat dissipation is a major problem however in GaN HFETs on sapphire substrates as the thermal conductivity of this substrate is about 0.3 W/cmK; it may even be somewhat lower. To make the matters worse, the thermal conductivity decreases rapidly as the temperature increases. Consequently, devices show a decreasing drain current (negative differential output conductance) as the drain bias is increased, and needless to say, the power performance is degraded. To overcome this, one must remove the sapphire substrate followed by mounting the structure on a substrate with better thermal conductivity, employ flip-chip mounting, or grow the structure on a substrate with better thermal conductivity. Among the substrates with better thermal conductivity are Si and in particular SiC. Layers on Si, however, are not of as high quality as one would like, which leaves SiC substrates, which are expensive and suffer from inferior surface – (characteristics or smoothness) due to the hardness of SiC. The early attempts to grow GaN layers by MBE on SiC were met with difficulty due to the surface damage roughness, though occasionally very high mobility could be obtained [109].

Two approaches can be employed to remove the surface damage. One is the mechanical chemical polish, which is very slow in coming, and the other is etching in H and Cl environments at very high temperatures such as 1500° C. The H cleaning process has been adopted as a standard procedure for MBE growth of GaN on SiC [110] with cleaning temperatures of about 1700° C. However, SiC substrates prepared by the sublimation method did not appear to survive this high temperature H etching process. Researchers have exploited in-situ H etching process [111] and HCl etching process [112] in the context of SiC epitaxy. Reports detailing these processes and their effects have appeared in the open literature already. The I-V characteristics of the particular device prepared in 1996 on SiC substrates are shown in Figure 30. The devices on SiC substrates did not exhibit the negative differential resistance, characteristic of sapphire substrates. Efforts soon to follow expended a good deal of effort on SiC substrates with performance initially comparable to

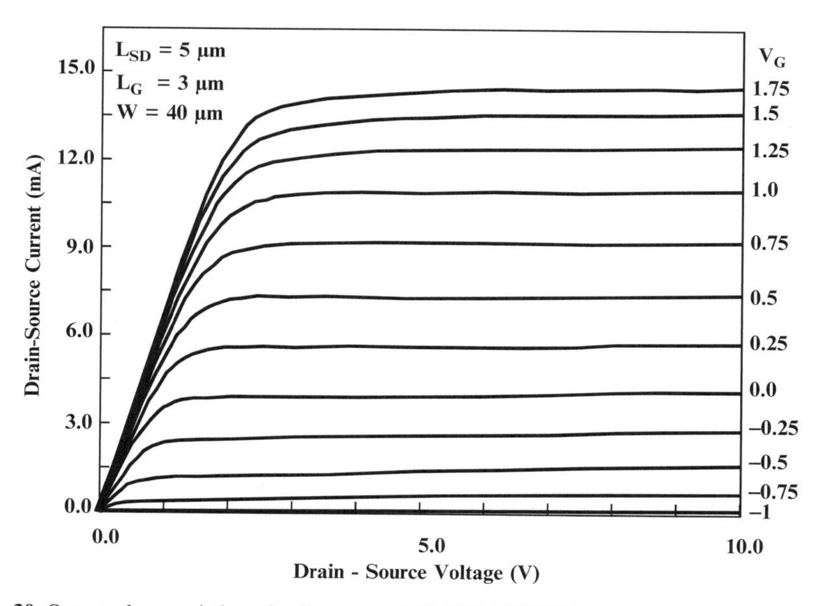

Fig. 30 Output characteristics of a 3μm gate AlGaN/GaN HFET grown on Leyl SiC substrate which is void of the output negative conductance. However, the Leyl substrates are highly conductive and not well suited for FETs due to RF shorting/loading. Nevertheless, experiments of this kind serve to prove the point that the negative output conductance observed in devices on sapphire are most likely of thermal origin.

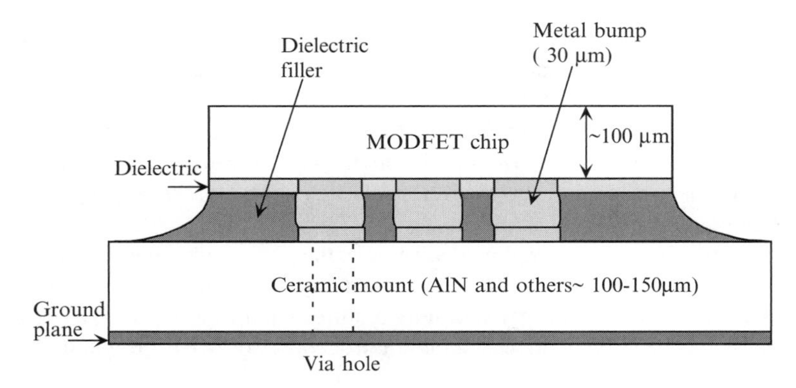

Fig. 31 Cross sectional view of a flip-chip mounted HFET device where the electrode spacing on the ceramic base match the gate, source and drain layout in the device. Metal bumps, typically made of In containing solder materials, are used for electrical connectivity between the device and the terminals on the ceramic base.

that on sapphire in terms of power output. Persistent effort in GaN HFETs on SiC substrates took these devices to their pinnacle with outstanding performance as discussed below. It should be pointed out that the pitch of gates for a power FET on a substrate with very good thermal conductivity can be made smaller than on a substrate with inferior thermal conductivity. Consequently, the chip size can be made much smaller in addition to other advantages.

Even with high thermal conductivity substrates, it may be necessary to flip-chip mount in case extreme power dissipation demands are put on the device, as shown in Figure 31. The flip-chip schemes allow a smaller physical distance between the heat sink and hottest part of the device. In addition to the power dissipation issue, the thermal expansion coefficient match between the semiconductor and the packaging material is critical. In small signal devices one can be satisfied with plastic encapsulants. However, in power devices, ceramic-based packages would be required. If the environmental conditions are severe, such as those required by military applications, hermetic metal packages might be required.

Electron mobility is a key parameter in the operation of n-channel FETs in that it affects the access resistances as well as the rate with which the carrier velocity increases with electric field. Ultimately, electron mobility is limited by the interaction of electrons with phonons, and in particular with optical phonons. This holds for bulk mobility as well as that in AlGaN/GaN modulation doped field effect transistors. The temperature dependence of mobility and the carrier concentration can be used to extract fundamental information regarding scattering mechanisms, [113, 114] which is discussed in great detail in reference [115]. As compared to the other III-V semiconductors such as GaAs, GaN possesses many unique material and physical properties, see reference [115]. However, even with improved material, the materials quality still remains as an obstacle for a thorough investigation of carrier transport. The earlier transport investigations had to cope with poor crystal quality and low carrier mobility, well below predictions [116–119].

We should point out that unintentionally doped GaN exhibits n-type conduction with a typical electron concentration of $\sim 10^{17}\,cm^{-3}$, with heavy compensation. Typical compensation ratios observed for OMVPE and MBE grown films are about 0.3, though a lower ratio of ~ 0.24 was reported for HVPE grown crystals [120–122]. Compensation reduces the electron mobility in GaN for a given electron concentration. Another point that should be kept in mind is that GaN layers are often grown on foreign substrates with very different properties. The degenerate layer at the interface (caused by extended defects and impurities), spontaneous polarization at heterointerfaces, and piezoelectric effects all should be considered. Experiments show that, even for thick GaN grown by HVPE, the degenerate interfacial layer has an important contribution to the Hall conductivity, especially at low temperatures where freeze-out occurs for the donors in bulk, leading to domination by the interfacial layer [123–125, 129]. In these cases, the measured data must be corrected to extract meaningful data [120]. The typical extended defect density of GaN grown by various techniques is $\sim 10^{9}\,cm^{-2}$, see reference [126]. In many cases, the dislocation and defect scattering may also limit the carrier mobility, especially at low temperatures [127, 128]. Finally, many material and physical parameters of GaN were not available for some of the previous simulations where those parameters were treated as adjustable parameters. Needless to say, reliable parameters are required in the calculation of the electron mobility and in the interpretation of experimental results accurately.

The room-temperature electron mobility values in bulk GaN grown with HVPE to a thickness of 60-μm was reported for GaN to be $950\,cm^2/Vs$ [129]. Freestanding GaN templates grown by HVPE exhibited room temperature electron mobilities approaching $1400\,cm^2/Vs$ [130]. Recent HVPE layers exhibit much higher mobilities on the surface of the layers, which approach that of freestanding templates. That reported for organo metallic vapor phase epitaxy (OMVPE) grown layers were in excess of $900\,cm^2/V.s$, reference [131], although the temperature dependence of mobility in this particular sample was rather unique. Early MBE layers exhibited mobilities as high as $580\,cm^2/Vs$ on SiC substrates, which at that time were not as commonly used as in recent times [132]. Typically, however, the MBE-grown films produce much lower mobility values of $100–300\,cm^2/Vs$ [133]. The lower mobilities have been attributed to both high dislocation densities [133–135] and elevated levels of point defects [136, 137].

Dislocations are considered by some to be an important scattering mechanism in films having dislocation densities above $1 \times 10^{8}\,cm^{-2}$ [133, 134]. We should keep in mind that these are preliminary attributes and more detailed experiments coupled with detailed analyses are needed to confirm the proposed models. Depending on the particulars of the growth and substrate preparation, GaN films grown by MBE typically have dislocation densities in the range of $5 \times 10^{9}–5 \times 10^{10}\,cm^{-2}$, reference [133]. With refined procedures, however, dislocation densities in the $8 \times 10^{8}–2 \times 10^{9}\,cm^{-2}$ can be obtained when grown directly on sapphire substrates with AlN or GaN buffer layers. Dislocation reduction, and other scattering centers that are inherently related to dislocations, is really the key to achieving high mobility GaN which goes to the heart of buffer layer and or early stages of growth. Based on the premise that the [002] X-ray diffraction is affected by screw dislocations and the (10–12) peak by edge dislocations and the fact that RF-nitrogen grown MBE layers produce excellent (0002) peaks (in the 40–120 arcsec range) while the (10–12) peaks are wider and weaker (in the 360 to 1200 arcsec range on sapphire) one can conclude that the majority of the dislocations in MBE layers are the propagating edge type.

The strength of MBE (that is, producing 2D-growth) does not bode well in dislocation reduction as the edge dislocations, which propagate along the c-axis, go right through the active portions of the sample. The details of this simplified picture is somewhat dependent on the particulars of the growth such as the group V/III ratio and growth temperature, details of which are discussed in reference [115]. Some sort of 3D growth at the early stages of the growth, as in the case of growth from vapor, followed by a smoothing layer would help reduce dislocation density. The more viable option is to use HVPE or OMVPE buffer layers for MBE growth. This approach in one effort led to record or near record bulk mobility of $1,150\,\mathrm{cm}^2/\mathrm{Vs}$ at room temperature and $53,500\,\mathrm{cm}^2/\mathrm{Vs}$ at 4.2 K in a 2DEG layer [138]. Further work on 2DEG led to increased mobilities [139]. A 2DEG sample prepared by MBE on an HVPE template having an areal carrier density of $2.35 \times 10^{12}\,\mathrm{cm}^{-2}$ exhibited a mobility of $75,000\,\mathrm{cm}^2/\mathrm{Vs}$, reference [140], which later was improved to $167,000\,\mathrm{cm}^2/\mathrm{Vs}$ in gated structures [141]. Ammonia MBE has also produced GaN with very high electron mobilities (about $60,100\,\mathrm{cm}^2/\mathrm{Vs}$ at 4 K) when grown on bulk GaN wafers which in turn were grown under high pressure and temperature conditions [142, 143]. It is clear that the buffer layers grown by the vapor phase epitaxy method helps eliminate a large portion of the problem faced by MBE, that is the poor quality of the buffer layer with its large edge dislocation content. The other long-standing obstacle for MBE, difficulties associated with preparation of sapphire and SiC substrates prior to growth, as in-situ high temperature treatment is not available, has been eliminated. In the case of sapphire, a high temperature anneal in O_2 environment produces atomically smooth and damage free surfaces [144]. In the case of SiC, some form of H_2 etching at elevated temperatures removes the surface damage caused by polishing [145]. Provided that the sample surface preparation prior to growth is done well, controlling the Ga/N ratio and substrate temperature in MBE growth causes the dislocation density across the homoepitaxial interface to remain more or less identical to that in the vapor phase grown template, allowing the other beneficial attributes of MBE to be brought to bear [146, 147].

2.1 Evolution of GaN FET Performance

Initial GaN HFETs utilized the background donors in the AlGaN layer, the density of which is not controllable, to say the least, and any other free and weakly bound electrons drawn to the interface. Congruent with the early stages of development and the defect-laden nature of the early GaN and $Al_xGa_{1-x}N$ layers, the HFETs exhibited very low transconductances (on the order of 20 mS/mm), and large on-resistances. In addition, they also exhibited a low-resistance state, which was relatively high to begin with, and a high-resistance state before and after the application of a high drain voltage (20 V). As in the case of GaAs/AlGaAs MODFETs, hot electron trapping in the larger bandgap material at the drain side of the gate is primarily responsible for the current collapse. The negative electron charge accumulated because of this trapping causes a significant depletion of the channel layer, more probably a pinch-off, leading to a drastic reduction of the channel conductance and the decrease of the drain current. This continues to be effective until the drain-source bias is substantially increased, leading to a space-charge injection and giving rise to an increased drain-source current.

With improvements in the materials quality available, the transconductance, current capacity, and drain breakdown voltage are all increased to the point that GaN-based HFETs are now strong contenders in the arena of high power devices/amplifiers, particularly at X band and higher frequencies. As is the case for FET device structures, improved and high resistivity buffer layers have once again played a pivotal role. For chronological purposes, a brief review of the latest class of HFETs with high transconductances and current levels is given.

A breakthrough regarding N-MODFETs based on GaN came in 1994 [148]. These devices with a gate length of 3 μm and gate width of 40 μm, exhibited transconductances of about 120 mS/mm with low on-resistances as they sported doped AlGaN donor layers and low resistance ohmic contacts. The I-V characteristics of an early N-MODFET device are shown in Figure 32. Shortly thereafter, devices with a gate length of 2 μm, gate width of 40 μm, and the drain-source separation of 4 μm exhibited drain currents of approximately 500 mA/mm and extrinsic transconductances of approximately $g_{em} = 185\,\mathrm{mS/mm}$, the output characteristics of which are shown in Figure 32 (b). The drain breakdown voltage for 1 μm gate-to-drain spacing was approximately 100 V, the exact value depending on the layer design and quality of the layered structure. With Hall effect measurements, the mobility and sheet carrier densities in the 2DEG were about $304\,\mathrm{cm^2/Vs}$ and $3.7 \times 10^{13}\,\mathrm{cm^{-2}}$, respectively, at room temperature.

What is unique to AlGaN/GaN HFETs as compared to their GaAs varieties is the polarization charge discussed earlier. Very high sheet carrier concentrations observed have been ascribed to the polarization induced charge. We have to recognize that ultimately, regardless of the source of the carriers, the strength of the electric field that can be accommodated by the semiconductor under the gate without excessive leakage will set an upper limit on the number of carriers that can be had at the interface. Use of multi-2DEG structures is one obvious method to increase the current capability of HFETs, and they have been employed. In those cases, the GaN layer is straddled by two doped AlGaN barriers that donate electrons to the channel, thus increasing the number of electrons available for current conduction. One can also consider doping the channel layers if need be, but with polarization induced charge present, this additional option does not seem to be needed. However, if and when the quality of layers of non-polar surfaces improve to be contenders, doped channels could be considered.

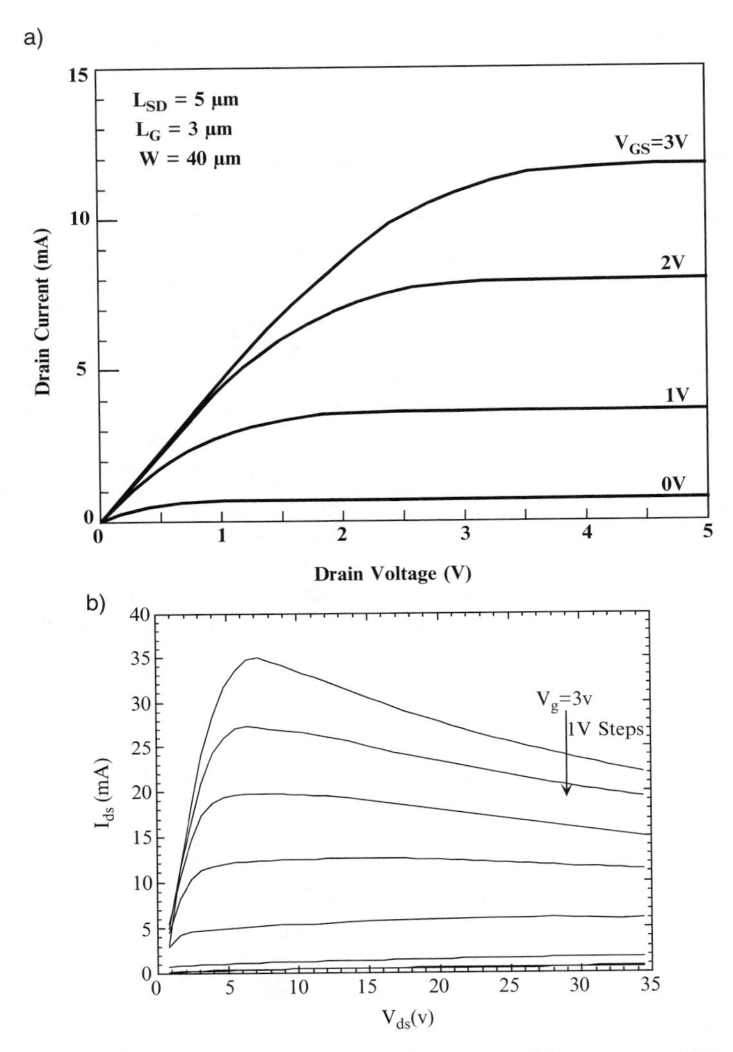

Fig. 32 (a). Output I-V characteristics of an early GaN HFET exhibiting respectable performance. The 3 µm gate device sported low resistance ohmic contacts and low leakage Schottky barriers. (b). Output I-V characteristics of a GaN HFET on sapphire with a 2 µm gate, which exhibits negative output conductance due to thermal effects associated with the relatively low thermal conductivity of sapphire.

The maximum drain saturation current I_{DS} corresponding to a drain-source voltage $V_{DS} = 7\,\text{V}$, and gate source voltage $V_{GS} = 3.5\,\text{V}$ in a double hetero-channel MODFET (DHCMODFET) is about 1100 mA/mm, which is important because in high power devices, the input is momentarily forward biased. The DHCMODFET has a room temperature extrinsic transconductance of $g_m = 270\,\text{mS/mm}$. The value of the total resistance R_T extracted from the linear region of the I-V curves is $4\,\Omega/\text{mm}$. Near pinch-off, the drain breakdown voltage for this device was about 80 V, indicating excellent power potential of the device. These measurements were made in a nitrogen-pressurized container to avoid possible oxidation of the contacts and probes. These devices maintain reasonable output characteristics at temperatures as high as 500° C with maximum drain current and extrinsic transconductance values of 380 mA/mm and 70 mS/mm, respectively. Cooling to room temperature restored the characteristics, which demonstrates the robustness of this material system and of the metallization employed. It should be repeated, however, that high power operation requires large drain breakdown voltages with the added benefit of having large output resistances, which ameliorates impedance matching.

MODFETs have progressed very quickly to a point where microwave performance has been established for a variety of devices with gate lengths as wide as 2 μm and as narrow as about 0.2 μm. To appreciate the development of this device, its microwave performance evolution is succinctly treated. A typical MODFET structure with a 2 μm gate length has been tested for small-signal S-parameters performed under bias conditions used for the power measurements (i.e., 15 V, −2.5 V, and 20 mA for the drain voltage, gate voltage, and drain current, respectively). The unity current gain cutoff frequency (f_T) and maximum frequency of oscillation (f_{max}) were 6 GHz and 11 GHz, respectively, at both 15 and 30 V bias. Values of f_T and f_{max} in excess of 50 GHz and 100 GHz have been reported for short channel (about 0.2 μm) devices, respectively. As touched upon earlier, devices on sapphire substrates suffer from the low thermal conductivity of sapphire and the devices exhibit negative differential resistance in the output characteristics. Remedies include better heat sinking by flip-chip mounting and, as touched upon above, the use of high resistivity 4H-SiC substrates, which provide good thermal conductivity but are hard to obtain.

Devices with submicron gate lengths, particularly with 0.25 μm or less, must utilize what is called the T-gate metallization for the gate to reduce gate resistance. A schematic view of one such device on SiC is shown in Figure 2 (a) along with an SEM image shown in Figure 33. Early GaN MODFET devices prepared on conducting 6H-SiC substrates exhibited output characteristics that lacked the negative resistance (i.e., they exhibited good heat sinking), see Figure 30 which bodes well for high power density and high power. There have been quite a few reports of MODFET power devices on high resistivity SiC (references [7, 10, 11]), and *p*-type SiC (reference [9]) substrates with phenomenal improvement in power handling capability notwithstanding the rapid progress on sapphire substrates. Increasingly, outstanding power levels became a reality with devices having near-half-micron or smaller gate lengths, a few examples of which are cited here. With 0.7-μm gate length devices on SiC substrates, where the gate-source spacing and gate-drain spacing were 0.5 and 0.8 μm, respectively, a total output power of 2.3 W in a device with a 1.28 mm gate periphery has been obtained [10]. The power gain at the 2.3 W output power point was 3.6 dB with power added efficiency (PAE) of 13.3% for a drain bias of 33 V. The current and power gain cutoff frequencies were 15 and 42 GHz, respectively. The contact resistance, though not the best,

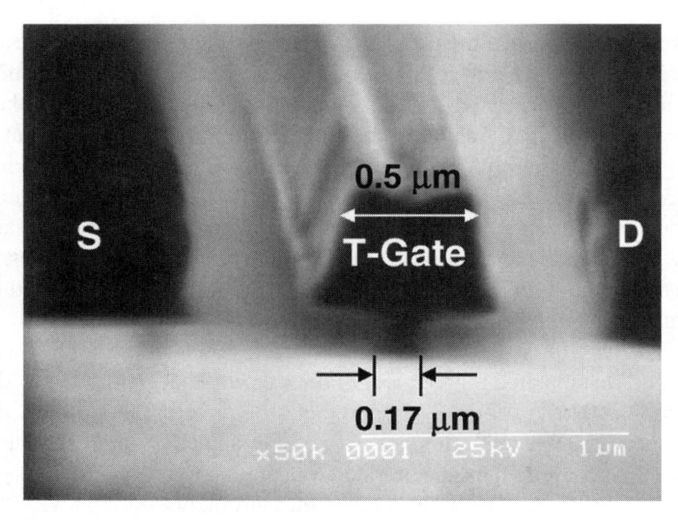

Fig. 33 An SEM cross-sectional image of a 0.17 μm foot print T-gate GaN based FET for reduced gate metal resistance. Courtesy of Primit Parikh, Cree Santa Barbara Technology Center.

was between 2.6 and 3.5 Ω mm. The maximum normalized transconductance was 270 mS/mm and the drain current was 293 mA/mm. For comparison, 0.7-μm gate-length $Al_{0.5}Ga_{0.5}N/GaN$ MODFETs on sapphire exhibited a current density of 1 A/mm, three-terminal breakdown voltages up to 200 V, and CW power densities of 2.84 and 2.57 W/mm at 8 and 10 GHz, respectively, representing a marked performance improvement for GaN-based FETs at the time. Because power and thermal issues are inextricable, we should point out that given the gate periphery, let alone the linear extrapolation from short gate width devices leads to a good deal of confusion as the pitch of the gate fingers used have a large effect. For comparison of various devices to be fair, the gate finger length and gate pitch must be the same. At the very least the chip size must be the same. Unfortunately, this has not been the case strictly speaking for GaN FETs when comparisons in the context of power densities have been made. Essentially, the total power per unit area of chip size would have to be used as power figure of merit.

Steady improvement in power performance has led to results at HRL laboratories with record-breaking performance [13]. The output I-V characteristics of a 250 nm gate length AlGaN/GaN HFET device of HRL on sapphire is shown in Figure 34. Typical DC characteristics include 600 mA/mm current performance and >60 V drain breakdown voltage. Small signal current and power gains as a function of frequency of a 0.17 μm × 150 μm gate AlGaN/GaN FET on SiC under bias conditions of 7 V drain bias and 390 mA/mm drain current are shown in Figure 35. Solid lines extrapolate the unilateral power gain and short circuit current gain out to 120 GHz (f_{max}) and 86 GHz(f_T).

In order to explore the impact of improved fabrication procedures, recessed gate [149, 150] AlGaN/GaN FETs have been fabricated by Moon et al., [149] using a

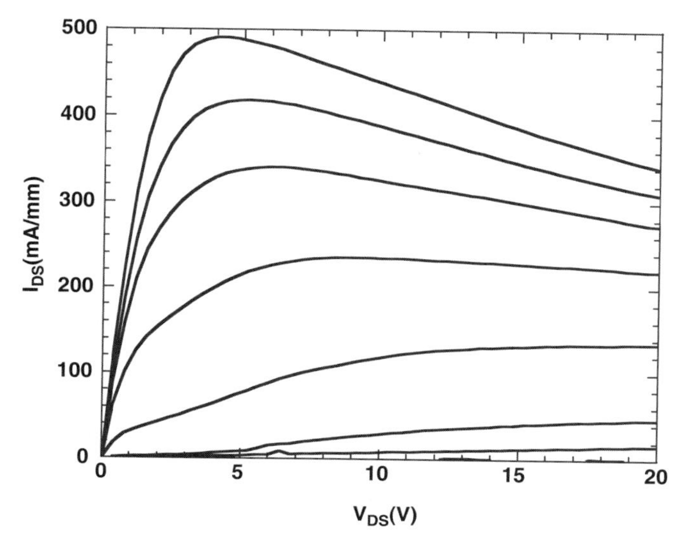

Fig. 34 The output I-V characteristics of a 0.25 μm gate AlGaN HFET on sapphire fabricated at HRL laboratories. Note the negative differential output conductance due to the poor thermal conductivity of sapphire substrate. Courtesy of Drs. N. X. Nguyen and C. Nguyen.

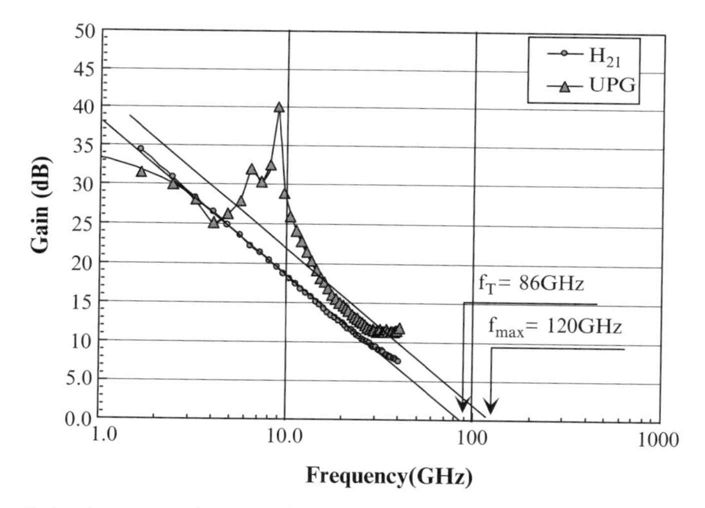

Fig. 35 Small signal current and power gains as a function of frequency of a 0.17 μm × 150 μm gate AlGaN/GaN FET on SiC under bias conditions of 7 V drain bias and 390 mA/mm drain current. Solid lines extrapolate the unilateral power gain and short circuit current gain out to 120 GHz and 86 GHz Courtesy of Primit Parikh, Cree Santa Barbara Technology Center.

low damage dry-etching method. The device threshold voltage was adjusted from −7 V to −2 V by changing the recess depth. A maximum drain current density of 1.5 A/mm was attained at a gate-source voltage of 4 V. For a 1 μm gate length, a peak intrinsic transconductance of 600 mS/mm was obtained. Small-signal S-parameter measurements showed f_{max} approaching 44 GHz with a gain of 13 dB at 8 GHz. At 8 GHz, a saturated power density of 4 W/mm was established, which is phenomenal for 1 μm gate length device. Long-term stability in terms of RF power operation was investigated at different RF input/output power levels. For example, at 65% of the saturated power level, no noticeable reduction in output power and gain was observed for over 60 hrs. Other recessed gate FETs on sapphire were characterized by 1.2 A/mm current levels for a gate length of 1.5 μm and stable operation for 500 h at 350° C was achieved [151]. In another effort [152] photo-electrochemical etching technique was used to recess the gate region before gate deposition. Although the FETs results were not satisfactory, mention is made here for this method of etching which can produce damage free surfaces.

In terms of RF power, 6.3 W of CW output power was obtained at 10 GHz from a 1 mm wide transistor device at HRL laboratories. More importantly, the power density remained nearly constant as the device size was scaled upward from a 0.1 mm width, where the device exhibits 6.5 W/mm, to 1.0 mm. These record-setting transistors of their time were epitaxially grown AlGaN/GaN heterostructures on semi-insulating SiC substrates by MBE. MBE growth also produces device characteristics with less than 5% standard deviation over the 2-inch diameter SiC substrate, a six-fold improvement over previously reported results. The HRL researchers have expanded their effort to include amplifiers with several cells and showed very good power scalability up to 2 mm of total gate periphery [13]. Using 250 nm gate devices, a CW output power of 22.9 W with an associated power added efficiency of 37% was measured for an amplifier at 9 GHz with 4 × 1 mm gate periphery devices. Furthermore, the same authors [13] also showed a CW power density of 4 W/mm at 20 GHz that represented the state of the art for any three terminal solid-state-device at this frequency of its the time.

In contrast to the earlier HRL devices whose DC characteristics were discussed above, the newer devices had a maximum drain current density exceeding 1.4 A/mm, and the peak transconductance at V_{ds} of 15 V was 250 mS/mm. The reverse bias gate to source breakdown voltage of the devices measured at 1 mA/mm of gate leakage current typically exceeded 80 V. Small-signal RF performance of HFETs were characterized in the 0.5 to 40.5 GHz range, and the cutoff frequencies accurately estimated, as discussed below. The best results are shown in Table 5 for 200 μm wide devices with various gate lengths.

Table 5 The current gain cutoff and maximum oscillation frequencies vs. gate length. Reference [13]

Gate length (nm)	Current gain cut off (f_T: GHz)	Maximum oscillation (f_{max}: GHz)
250	55	100
150	80	120
90	81	187
50	110	>140

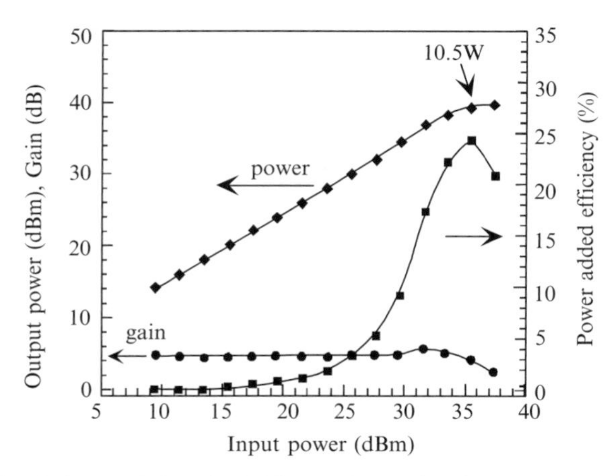

Fig. 36 Large signal characteristics of a 0.25 μm × 2 mm wide device at 10 GHz. The maximum CW output power of this device was 10.5 W. Reference [13].

Continuous wave power measurement of 0.1 mm, 1 mm, and 2 mm devices was performed at 10 GHz using a load-pull approach [13]. The gate length of the particular device subjected to this particular test was 250 nm. Maximum output power levels of 0.65 W, 6.3 W, and 10.5 W were measured for devices with 0.1 mm, 1 mm, and 2 mm total gate periphery, respectively, and scale nearly linearly. Shown in Figure 36 is the output power, power gain and power added efficiency versus input the power of a 2 mm periphery device with 250 nm gate length.

A much larger device with a total gate periphery of 24 mm fabricated at Cree Santa Barbara on a SiC substrate exhibited a total power output of 108 W at 2 GHz under a drain voltage bias of 52 V, corresponding to a power density of 4.5 W per mm. At the 2.3 dB compression point, the output power measured was 103 W. The top view of the packaged device along with the output power and gain vs. input power is shown in Figure 37 (a) and (b). The power devices fabricated at Cree Santa Barbara were also measured at a case temperature of 220° C in an effort to gauge the sensitivity of the device to environmental temperature. A power density of 7.2 W/mm with a PAE over 55% was obtained in a device with 0.5 mm gate periphery. The maximum power of 35 dBm (over 1 W) was measured at in input power of 23 dBm (over 100 mW) at a drain bias of 48 V. The device gain is obviously 12 dB.

Impressive power levels in C band have been obtained in many laboratories from GaN HFETs on SiC substrates. In one effort, a 24-mm wide power FET delivered 61 W and 156 W under CW and pulsed operating conditions, respectively at 4.0 GHz [153]. In this particular case, the internal matching circuit was attained in a package half-size as that for comparable power-level amplifiers in GaAs-based FETs. The developed GaN-FET amplifier with 24-mm gate periphery delivered a 61W output power with 10.2 dB linear gain and 42% power-added efficiency under CW operating conditions.

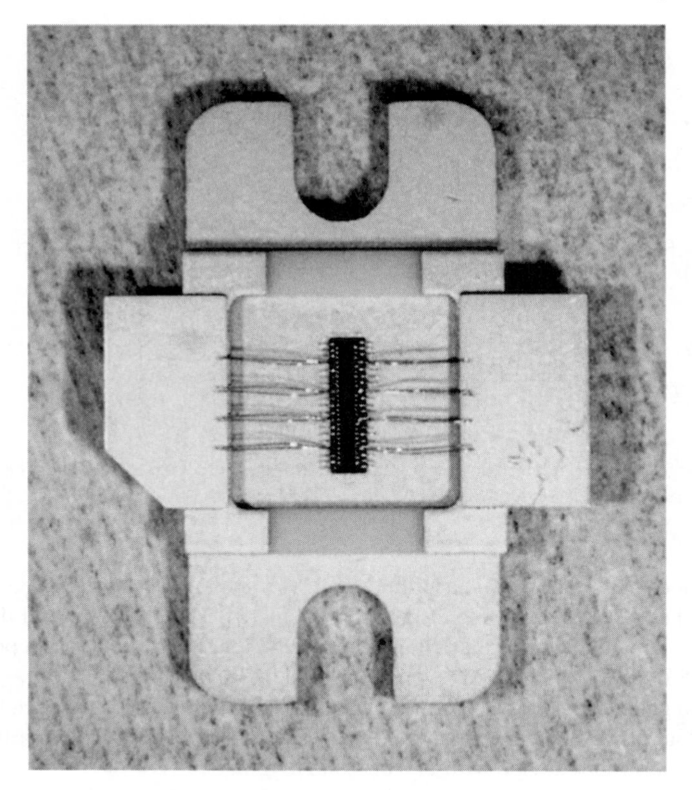

(a)

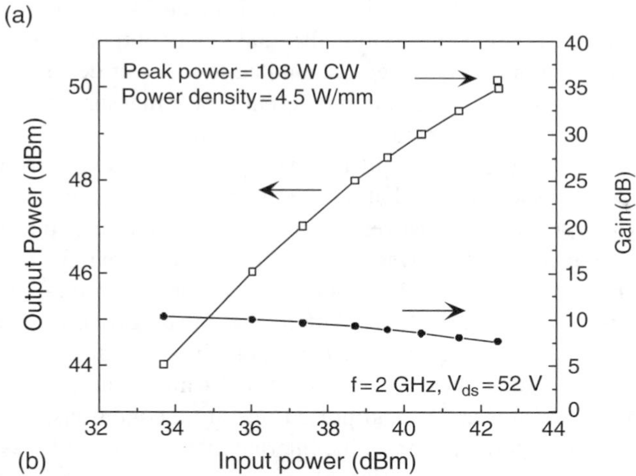

(b)

Fig. 37 (a) A photograph of a package device exhibiting 108 W of CW output power at 2 GHz. (b) Output power and gain vs. input power drive of the same device measured at 2 GHz at a drain bias of 52 V. Courtesy of Primit Parikh, Cree Santa Barbara Technology Center.

In another effort, recessed-gate FETs with a field plate a gate length of 1 μm exhibited an increased transconductance of 200 mS/mm without exhibiting drain current collapse due to the reduction of the peak electric field most likely [154]. Equivalent circuit analysis showed that the loss in gain owing to the additional gate feedback capacitance due to the FP electrode is more than compensated for as the FP allows one to increase the drain bias voltage to more than 30 V. A 48-mm-wide recessed-gate FP-FET biased at a drain voltage of 48 V exhibited a saturated output power of 197 W with a linear gain of 10.1 dB and a power-added efficiency of 67% at 2 GHz.

Yet in another effort with again recessed gate and field plate, it was noted the former processing step increased the transconductance, g_m, from 130 mS/mm to 220 mS/mm [155]. Moreover, the gate breakdown voltage (BV_{gd}) was improved from 160 V for the planar FP-FET to 200 V for the recessed FP-FET, resulting from one-order reduction in the reverse gate leakage-drain current, as well as suppressing the current collapse. At 2 GHz, a 32 mm-wide recessed FP-FET exhibited an output power of 149 W (4.7 W/mm) with 64% power-added efficiency and 8.7 dB linear gain with a drain bias of 47 V.

Similar efforts with recessed gates (by about 20 nm) and field plates led to high power levels in that a 48-mm-wide recessed FP FET exhibited an output power of 230 W (4.8 W/mm) with 67% power-added efficiency and 9.5-dB linear gain with a drain bias of 53 V at 2 GHz [156]. Similar to the above mentioned cases, the recessed-gate structure in an FP FET caused the transconductance to increase from 150 to 270 mS/mm, leading to an improvement in gain characteristics, and current collapse was minimized. Likewise, the field plate caused the gate breakdown voltage to improve from 160 V for the planar FP FET to 200 V for the recessed FP FET, resulting from one-order reduction in reverse gate leakage. Of importance to communication circuits, a third order intermodulation (OIP3) of −35 dBc (which required optimum power matching) was obtained with a PAE of 30% at 8-dB back off from saturated output power for a 4-mm-wide recessed FP FET, remaining nearly the same for drain voltages in the range of 25–48 V.

In terms of near-mm wave operation, measurements at higher frequencies were also made to determine the device response in terms of its power performance. Using a series of 150 nm × 200 μm gate devices, a CW output power density of 4 W/mm at 20 GHz was obtained [13]. The results of load-pull measurements at 20 GHz are shown in Figure 38 A device similar in dimensions, 200 nm × 200 μm, fabricated at Cree Santa Barbara Research Center and biased at 25 V and −4 V at the drain and gate, respectively, produced a maximum power level of 27 dBm for an input power of 20 dBm at 26 GHz corresponding to a peripheral power density of 5.1 W/mm, gain of 7 dB with a power added efficiency of about 25% at the maximum power level. When the drain voltage was reduced to 15 V from 25 V while keeping the gate voltage the same, the peripheral power density dropped to 3.2 W/mm.

As a prelude to power FETs in *Ka* band, Ionue et al. [157] explored AlGaN/GaN FETs on SI SiC with a gate width of 100 μm and a gate length of 0.09 μm, which exhibited a current gain cutoff frequency (f_T) of 81 GHz, a maximum frequency of oscillation (f_{max}) of 187 GHz, and a maximum stable gain of 10.5 dB at 30 GHz

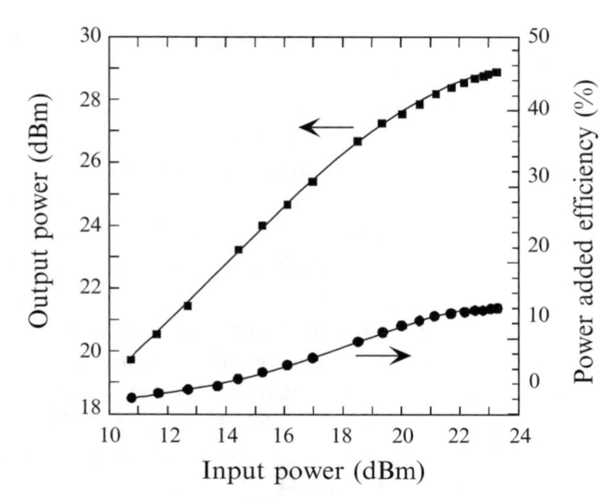

Fig. 38 Large signal characteristics of 0.15 µm × 200 µm GaN HFET at 20 GHz. The maximum CW power density obtained from this device was 4 W/mm. Reference [13].

(8.3 dB at 60 GHz). A point of important consequence is that channel electron velocities of 1.50–1.75 × 10^7 cm/s in a gate-length range of 0.25–0.09 µm were obtained from delay time analysis [158] in which the effect of source and drain resistances have been taken into account [159]. This raises the specter of hot phonon arguments, which have been identified as the cause of the deduced velocity falling below that predicted in GaN. With a 1.0 mm gate width and 0.25 µm gate length, a saturated power of 5.8 W with a linear gain of 9.2 dB and a power-added efficiency of 43.2%—has been achieved at 30 GHz using a single chip.

An HRL FET measured in terms of its S-parameters at −5.5 V and 12.5 V gate and drain bias voltages, respectively, and yielding a current gain cut-off and maximum power gain cut-off frequencies of about 48 and 100 GHz, respectively, exhibited minimum noise figure of 0.85 dB at 10 GHz with an associated gain of 11 dB as shown in Figure 39. The output I-V characteristics of the device, measured for its noise performance, are shown in Figure 34. A noise figure 0.75 dB with an associated gain of 12 dB at 10 GHz has been obtained at Cree Santa Barbara. This particular report is comparable to that by Moon et al. [160] in terms of the minimum noise at 10 GHz which approaches to that obtained in GaAs. Those authors reported low microwave noise performance from discrete AlGaN-GaN HFETs, with SD spacing of 1.5 µm and 0.15−µm gate length, at DC power dissipation levels comparable to that of GaAs-based low-noise FETs. At 1 V source-drain bias and DC power dissipation of 97 mW/mm, NF_{min} values of 0.75 dB at 10 GHz and 1.5 dB at 20 GHz were achieved, respectively.

Si substrates because of their availability, extremely high quality, relatively high thermal conductivity (nearly as good as GaN) and relatively low cost are always alluring no matter which semiconductor is being explored. The GaN family is no exception and the wurtzite polytype can be grown on the (111) plane of Si. In

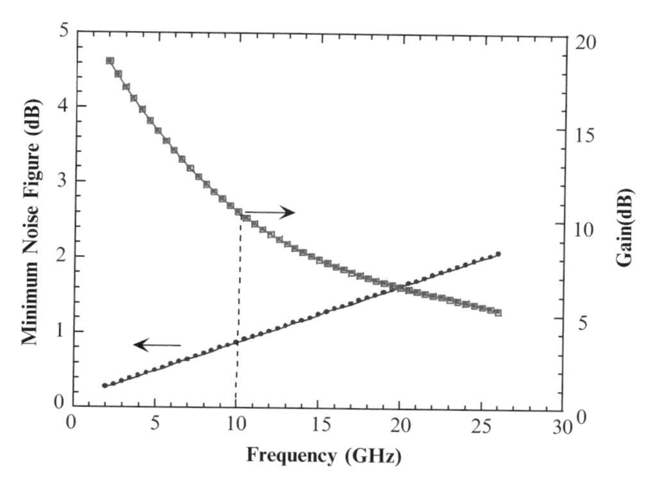

Fig. 39 The minimum noise figure and the associated gain of a 0.25 μm gate AlGaN HFET on sapphire fabricated at HRL laboratories. Courtesy of Drs. N. X. Nguyen and C. Nguyen.

addition to the lattice mismatch issue and a few other pertinent properties, such as the thermal expansion coefficient of Si being smaller than that of GaN (all nitrides for that matter), GaN grown on Si is left with residual tensile strain upon cooling due to this thermal mismatch. Here owing to tremendous progress made in GaN heterostructure FETs on Si substrates, a short review is presented.

In one investigation, the devices were reported to exhibit a saturation current of 0.91 A/mm, a peak extrinsic transconductance of 122 mS/mm, a unity current gain frequency of 12.5 GHz and $f_{max}/f_T = 0.83$, and a breakdown voltage at pinch-off of greater than 40 V, reference [161]. In another effort, [162] 0.3 μm gate length GaN heterostructure devices on 1Ω-cm p-Si(111) (which is too low for microwaves) exhibited static output characteristics with low output conductance and isolation approaching 80 V. Due to RF shunting by the substrates, devices showed a 25 GHz cutoff frequency, with near unity f_{max}/f_T ratio and 0.55 W/mm output power. Through backside processing, freestanding 0.4-mm HFET membranes with no thermal management were demonstrated and exhibited a significant improvement in their f_{max}/f_T ratio up to 2.5 at the cost of lower f_T and f_{max} along with an almost four-fold reduction of I_{dss}.

With steady improvement in materials quality, respectable $Al_{026}Ga_{074}N$–GaN heterojunction field-effect transistors fabricated in layers grown by OMVPE on high-resistivity 100-mm Si (111) substrates have been reported [163]. A maximum drain current density of ~1 A/mm, a three-terminal breakdown voltage of ~200 V, a cutoff frequency of 18 GHz, and maximum frequency of oscillation of 31 GHz, characterized the 2 × 50 μm × 0.7-μm gate-length devices. At a 50 V drain bias and at 2.14-GHz, a CW power density of 12 W/mm was attained with associated large-signal gain of 15.3 dB and a power-added efficiency of 52.7%. This is the highest power density ever reported from a GaN-based device grown on a silicon substrate,

Table 6 The RF performance of AlGaN/GaN FETs

Frequency, GHz	Power, W	Reference
30	5.8	[157]
4.2	156	[153]
2	197	[154]
10	50	[164]
2	149	[155]
2	230	[165]

and is competitive with the best results obtained from conventional device designs on any substrate.

The large-signal measurements were performed on-wafer using Focus Microwaves tuners. The data were collected for bias conditions of $V_{ds} = 50\,V$ and $V_{gs} = -1.1\,V$ at 25° C on thinned and back metallized wafers. No through-wafer vias were present. RF tuning for gain and output power was performed at an input power level of 13 dBm. At the point of optimum tuning, the source and load reflection coefficients at the device were $\Gamma_{source} = 0.77e^{j18^{\circ}}$ and $\Gamma_{load} = 0.85e^{j1^{\circ}}$. For a 50 V power sweep a 30.8 dBm saturated output power with associated large-signal gain of 15.3 dB and power-added efficiency (PAE) of 52.7% resulted, corresponding to a power density of 12 W/mm. Drain efficiency values of 70% were attained in class AB operation on larger devices when proper matching was possible to a reduced load impedance.

The device performance in terms of total power delivered is certain to improve with improved materials quality and processing along with improved matching circuitry and reduced microwave losses in the substrate material. However, a compilation of the power performance discussed is given in Figure 40 and tabulated in Table 6. It should be noted that the devices at lower frequency have much wider gate widths than those at higher frequencies.

2.2 Drain Voltage and Drain Breakdown Mechanisms

In power devices, the thermal limitation can never be eliminated completely as is the case in nitride devices, particularly when fabricated on sapphire substrates with a thermal conductivity of only approximately 0.3 W/cmK, which drops with increasing temperature. Inclusion of thermal limitations leads to results shown in Figure 41 for devices that compete in the high-power device arena [6, 105]. Because new device developments do in general compete with existing and alternative technologies, a brief account of competing technologies for the power arena is given below. The Si metal-semiconductor FET (MESFET) analytical curve, modeled for its simplicity, is slightly above the SiC analytical curve and indicates a maximum power density of 0.35 W/mm at $V_{dS} = 7\,V$ which is slightly lower than 0.39 W/mm. Note that

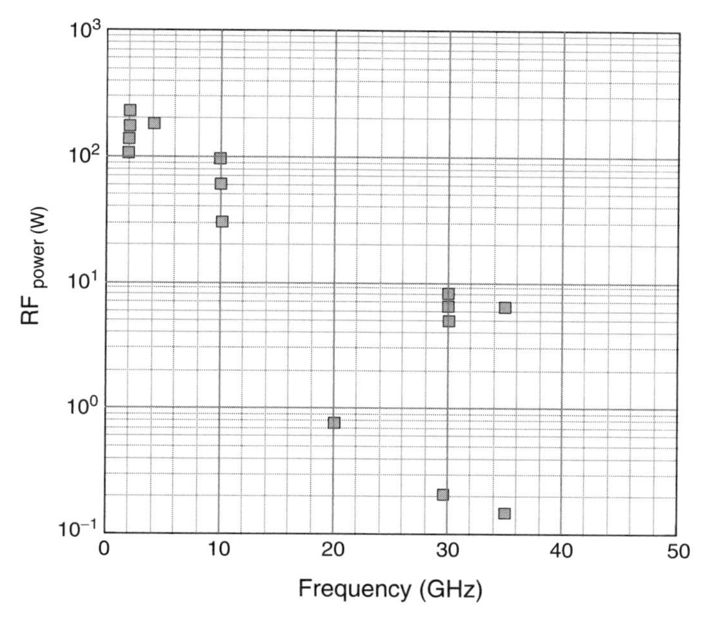

Fig. 40 Power vs. frequency performance of reported heterojunction field effect transistors utilizing AlGaN/GaN structures on Semi-insulating SiC substrates. Note that the devices at higher frequencies have much shorter gate widths the exact values of which are discussed above in the text.

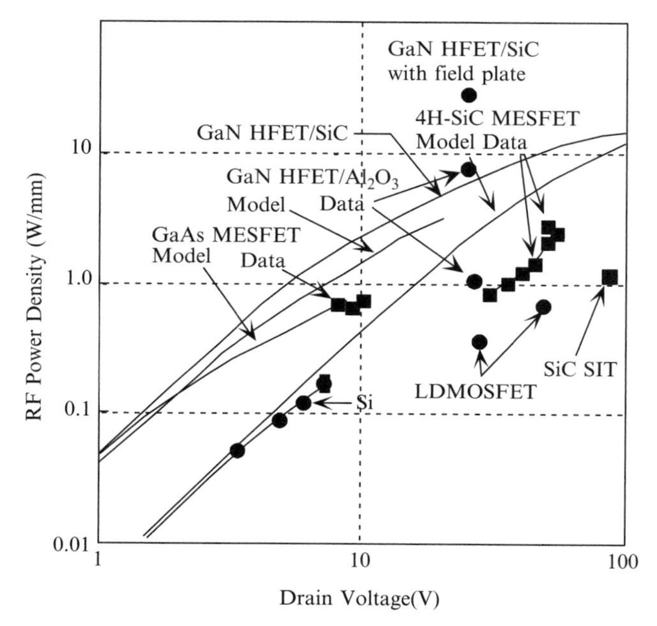

Fig. 41 Simulated and experimental RF power density data for Si, GaAs, SiC, and GaN FETs. The GaN data also include the field plate between the gate and drain to spread the electric field.

these are power densities only, and for high total power devices, thermal conductivity and drain breakdown voltage are dominating figures. Considering these together with device materials properties, GaN on SiC would have the best performance. Because Si RF MESFETs are unavailable, commercial Si RF metal-oxide semiconductor FET (MOSFET) results are used for comparison instead. At low voltages, the Si MOSFET data parallel the analytical curve suggesting the validity of the functional dependence of power density on drain voltage. Also shown are two higher power density data points 0.4 W/mm, $V_{dS} = 28$ V and 0.87 W/mm, $V_{dS} = 48$ V. These higher power densities were obtained with specially designed RF power MOSFETs that incorporate a lightly doped drain and field plates that significantly increase the breakdown voltage.

The GaAs analytical curve shows how well this device does in terms of the power density among all the devices at the lowest voltages primarily because of the higher electron mobility of GaAs. However, the low breakdown field limits the GaAs MESFET's drain voltage to about 8 V and power density to 0.63 W/mm including thermal effects. High-performance GaAs FET amplifiers with more complex device cross sections have achieved power densities as high as 1.4 W/mm at 18 V. However, typical commercially available GaAs MESFET power densities are below 1 W/mm. At 100 V, the SiC MESFET has calculated maximum power densities of 7.96 W/mm with thermal effects and 9.7 W/mm without thermal effects. The highest, at the time, demonstrated CW power density 3.3 W/mm ($V_{ds} = 50$ V) for a SiC MESFET [166] is also shown for comparison. Additional SiC data again illustrate the functional dependence of power density on drain voltage.

The GaN results of analytical models are highly dependent on the thermal conductivity of the substrate. With a sapphire substrate, the device is severely thermally limited to 2.24 W/mm at 30 V with a resulting channel temperature over 400° C. However with SiC substrates, [105] the analysis predicts that a GaN HFET could achieve 15.5 W/mm at 100 V while keeping the channel temperature at about 300 ° C. For higher junction temperatures and also with field plates, which spread the drain field, higher power densities have been obtained as indicated in the introductory material to this chapter. Moreover, it should be pointed out that the power density is an elusive figure in that it very much depends on the layout. For example if the gate fingers are spaced far apart the power density would be larger than if they are more closely spaced. In the end, maximum amount of power with the smallest amount of chip area is what counts and should be used as the figure of merit. However, chasing power density has become so pervasive and popular that only when these devices hit the marketplace would the reality set in. The key to further improvements lies with our ability to control the polarity of the films, to prepare inversion domain free material, and to reduce defects. If the past few years are any indication, substantial progress is in the wings.

The output power that can be extracted from the device is determined by the drain output characteristics, specifically the maximum current and maximum voltage that can be attained. To increase the output power, either the maximum drain current or the maximum drain voltage or both must be increased. Increased current reduces the output resistance whereas the increased voltage increases that resistance. The latter is easier to attain by using a semiconductor material with a large breakdown field and appropriate device designs, and therefore more conducive to impedance matching. If the gain and the input drive is sufficient and there are no anomalies, using class B-like operation, drain efficiencies in the range of 90% can be obtained as has been done in GaAs-based devices. In GaN there are anomalies, which are discussed briefly in Section 2.3., that prevents drain's efficiency from reaching anywhere near these figures.

Choosing a semiconductor with large a breakdown voltage provides an avenue with much better power prospects. For example going from GaAs to GaN, the breakdown voltage can be increased nearly ten fold, whereas the maximum currents would be comparable in both materials not considering the power density and heat dissipation issues. Therefore, the impetus for large drain voltage is tremendous, which is one of the reasons why GaN is such an attractive material for power amplifiers. It is therefore imperative we discuss briefly the issues involved in the maximum attainable drain voltage. Intrinsically, excluding the surface related issues and bulk bound contributions to premature current rise, the avalanche breakdown of the revere-biased gate-drain junction determines the upper voltage limit that can be applied even momentarily as that which occurs during a swing of the RF voltage. Essentially carriers introduced into a region of high field, such as the gate-drain region, multiply through avalanche multiplication causing a large current to flow, thus the term breakdown. It is a reversible process if the Joule heating is limited.

In the bulk, the critical field at which avalanche breakdown occurs scales approximately with the square of the energy band gap, or $E_{br} \propto E_g^2$, reference [167]. The breakdown voltage then scales with the fourth degree of the band gap [168]. At higher dopant concentrations the avalanche breakdown voltage is reduced, so the known concentration dependence $V_B \propto N_D^{-0.75}$ can be used, reference [168]. The final result for the avalanche break-down voltage for the asymmetric p^+-n junction, or Schottky contact-n junction, where n-type material has an energy band gap Eg and doping concentration N_D is: reference [169]

$$V_B = 23.6E_g^4(eV)\left(\frac{N_D\left(cm^{-3}\right)}{10^{16}}\right)^{-0.75} \quad (89)$$

FETs are planar device and surfaces play a large role as to the actual breakdown voltage; it would be different from that predicted by Equation 89. However, Equation 89 is very useful in judging the semiconductor and doping level dependence of the breakdown voltage. In addition, the issue of gate-drain junction breakdown is always marred with the gate leakage current whose origin is often time not well understood in an evolving technology such as GaN. Furthermore, a standard definition of breakdown voltage is also lacking. Already inconsistent temperature dependencies of Schottky diode breakdown voltages have been reported, some indicating that in at least some devices impact ionization process (avalanching) does not dominate the breakdown mechanism [170,171]. If so, the surface states would then be implicated. However, the field is yet in its infancy and not much is yet known about surface states in GaN except that passivation seems to have an effect.

Farahmand et al. [172] undertook a theoretical investigation of both DC and RF breakdown properties of GaN MESFETs of both wurtzite and zincblende polytypes. These calculations utilized a full-band ensemble Monte Carlo simulation with a numerical formulation of the impact ionization transition rate. The DC breakdown voltage in the wurtzite-phase GaN MESFET was found to be significantly larger than that in the zincblende phase because of relatively rapid electron heating in the zincblende polytype. The RF breakdown voltage was found to increase with increasing RF frequency. Holes were excluded from the model making the definition of breakdown voltage somewhat arbitrary. Farahmand et al. [172] defined the breakdown voltage as the drain–source voltage for which the drain current calculated with impact ionization is 3% higher than the drain current calculated without it. For RF breakdown, a large-signal RF bias was applied between the drain and source simulating on-state breakdown instead of the gate which is common in experiments to avoid computational complexity. The RF signal is assumed sinusoidal oscillating between V_{hi} and V_{lo} with an angular frequency, ω. In order to determine the RF breakdown the drain current was calculated with and without impact ionization.

The calculated drain currents, I_d, vs. drain–source voltage, V_{ds}, for gate–source voltage (inclusive of the Schottky barrier height), V_{gs}, varying from -0.1 to -5.1 V in 1 V steps are shown in Figure 42 for both the zincblende and wurtzite-phase MESFETs. The drain current for each bias was determined both with and without impact ionization. The solid and dashed lines in Figure 42 represent the calculated drain current in the presence and absence of impact ionization respectively. The relatively high output conductance of the devices is a consequence of the small gate length. The DC breakdown voltage is significantly higher in the wurtzite-phase device than in the zincblende-phase device for all gate biases. As an example, the breakdown voltage for $V_{gs} = -0.1$ V is 25 V for the zincblende variety and 45 V for the wurtzite variety.

The difference in energy gap between the wurtzite and zincblende phases of GaN is only about 5% and cannot account for the large difference in the breakdown voltage. To find the genesis one must consider that the impact ionization coefficients depend on the rate at which the carriers reach the threshold condition for impact ionization and the ionization transition rate at and above threshold [173]. The ionization transition rate as a function of energy in both zincblende and wurtzite GaN was calculated by Kolnik *et al.*, [174, 175] who determined that the transition rate in the zincblende phase is only slightly higher than that in the wurtzite phase in the entire range of energies considered. Thus, the marked difference between the calculated ionization coefficients is believed to be due primarily to the difference between the rates at which the carriers achieve the threshold for impact ionization.

Subsequent calculations [176, 177] of the ionization coefficients have shown that the electron distribution is much 'cooler' for the wurtzite phase than the zincblende phase of GaN for comparable electric field strengths which means that at a given field strength, fewer electrons reach threshold in the wurtzite polytype than in the zincblende polytype, leading to a lower ionization coefficient for wurtzite GaN. Moreover, the density of states within wurtzite GaN is greater than that within zincblende GaN which causes the electrons to be, on average, 'cooler' in the

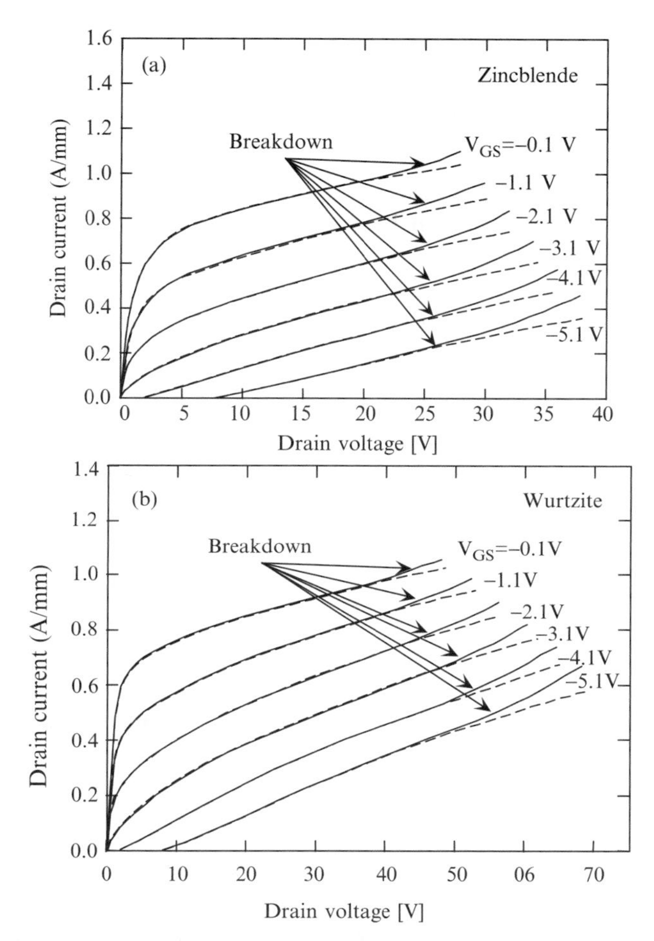

Fig. 42 Output characteristics for the zincblende polytype MESFET (a) and the wurtzite polytype MESFET (b). The gate–source voltage, V_{gs}, includes the Schottky barrier height. The drain currents have been calculated with impact ionization (solid lines) and without (dashed lines). The arrows show the breakdown locations on the I–V curves. The gate length is 0.1 μm, and the channel donor concentration is $3 \times 10^{17}\,cm^{-3}$. Courtesy of P. P. Ruden and Reference [172].

wurtzite phase than in the zincblende phase under the same bias conditions. The higher breakdown voltage for the wurtzite-polytype is therefore attributable to the difference in carrier temperature between the wurtzite and zincblende phases.

In terms of the experimental work, specifically in the study of Tan et al., [170] the gate leakage current was attributed to mainly surface states based on the observation that Schottky diodes of varying diameter fabricated in FET layers indicated the leakage current to be proportional to the periphery of the diodes, as opposed to the area. In addition to scaling with area or periphery, the temperature gate leakage current under dynamic bias meaning grounded source and reverse biased gate-drain, or static with source floating and reverse biased gate-drain terminals can be used to gain insight into the mechanisms involved. If the gate-drain bias is larger than the channel pinch off voltage (the gate voltage at which the channel is depleted for zero drain bias), the measurements are considered performed in a regime beyond pinch off.

The gate leakage current, I_g, beyond pinch-off has been measured as a function of the temperature using two-terminal measurements between the gate and drain (with the source floating) to eliminate the effect of the channel current [170]. The results showed the temperature dependence of I_g and the breakdown voltage from 20 to 200°C, to be a positive and a negative temperature coefficient, respectively. In other studies, the breakdown voltage exhibited a positive temperature coefficient as would be expected from impact ionization [171]. To ensure that true breakdown voltage dependence was observed, several devices were biased to destruction. An Arrhenius plot of the current, [170] beyond pinch-off, showed an activation energy of 0.21 eV over a range of $V_{DG} = 25–65$ V. The activation energy in this case has been reported to be very sensitive to surface preparation and any passivation, and exposure to light. These dependencies were interpreted as suggesting that perhaps a surface effect mechanism is responsible for the gate leakage in the devices explored by Tan et al. [170]. The model proposed is based on hopping along surface states, which beyond pinch-off and with increasing drain bias is fueled by electrons tunneling from the drain edge of the gate, where the field is highest, into surface states. These electrons tend to screen the depletion charge in the region between the gate and drain, and under equilibrium the electron transport rate is limited by their collection rate by the drain which relates to electron mobility. Illumination of the surface with light and increase in temperature enhances this conduction.

The drain breakdown voltage is somewhat elusive in that its determination is convoluted by leakage current. In fact, many practitioners define it as when the gate leakage reaches 1 mA/mm, which is determined by the amount of leakage that can be tolerated with any physics related to a breakdown mechanism. In an attempt to delineate the issues involved, Tan et al. [170] pushed some devices to destruction to ensure a correct interpretation of the breakdown voltage. In these devices, the temperature dependence of the breakdown voltage had a negative temperature coefficient, which suggested a breakdown mechanism other than that of impact ionization which is consistent [178] with some reports and contradicts other reports [171, 179]. Tan et al. [170] attributed their observations of negative temperature coefficient to a critical power density in the surface leakage current, beyond which surface heating exceeds cooling, leading to a positive temperature feedback, thermal runaway, and subsequent breakdown. As the ambient temperature is increased, the leakage current increases, causing a lower voltage threshold and hence the negative temperature dependence of the breakdown voltage. Figure 43 shows a three-terminal breakdown characteristic for different gate biases for a series of AlGaN/GaN HFETs each driven to catastrophic breakdown. The breakdown voltages are seen to reduce at high channel currents in response to the increase in device temperature due to the Joule heating.

An inspection of Figure 43 indicates that the breakdown voltage is dependent on the current flow through the devices assuming a value of 95 V at pinch-off

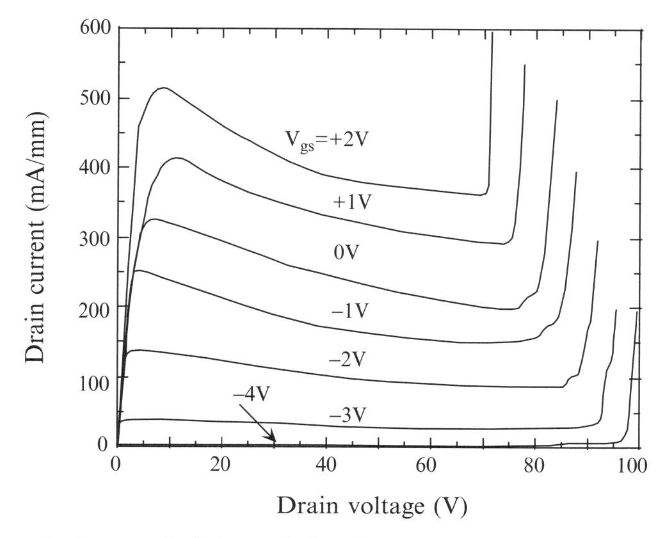

Fig. 43 Composite three-terminal characteristics of a series of 1 μm gate length devices, showing the breakdown voltage as a function of the drain current. Each gate value shown represents a different, but similar, device biased to destruction. Reference [170].

$(V_{gs} = -4\,V)$. This is approximately 30% less than the corresponding two-terminal gate–drain breakdown. The apparent reduction is attributed to an electrostatic interaction of the source seriously degrading gate–drain breakdown. This is supported by observations that in a 100 μm gate length HFET with a 2 μm gate–drain spacing there was no dependence of the breakdown voltage on whether two or three terminal configuration was used.

In an independent investigation, Nakao et al. [171] observed a positive temperature coefficient of the breakdown voltage which naturally is attributed to impact ionization. The apparent conflict among reports may be attributed to the status of device and the surface between the gate and drain. If the surface bound issues are eliminated, and the quality of the channel is improved, the impact ionization induced breakdown would be uncovered. In the devices investigated by Nakao et al., [171] the gate length (L_g) was varied from 1 to 10 μm, the gate-source and gate-drain spacing was 1.5 and 2 μm, respectively, and the gate width was 20 μm. The pinch off voltage of resulting HFETs was about −3 V. Nakao et al. [171] estimated the breakdown voltage, BV_{off} by measuring I_D-V_{DS} characteristics at off-state bias and the BV_{off} was defined as the V_{DG} at which the increase in I_{D} was 1 μA (0.05 mA/mm). The BV_{off} increased from 71 to 84 V as the temperature increased from 250 to 350 K indicating a positive temperature coefficient of 0.12 V/K, as shown in Figure 44. Nakao et al. [171] achieved this behavior in devices where an unspecified plasma treatment was applied to the gate region before metal deposition without which the breakdown voltage was lower as exemplified in Figure 45, which depicts the breakdown voltage dependence on the gate length for cases with and without such a plasma treatment for 2 μm gate drain gap.

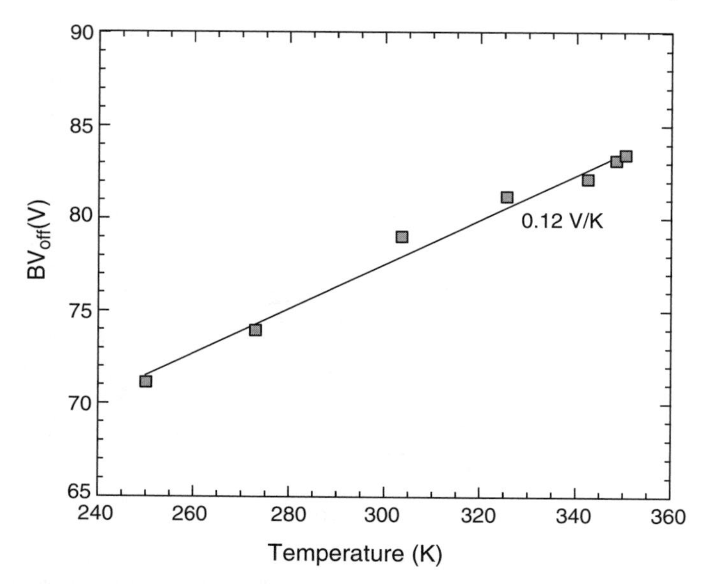

Fig. 44 Temperature dependence of the drain breakdown voltage BV_{off} in the off state, meaning that the device is in pinch-off and electron supply is not from the source. A positive temperature coefficient of 0.12 V/K was obtained. Reference [171].

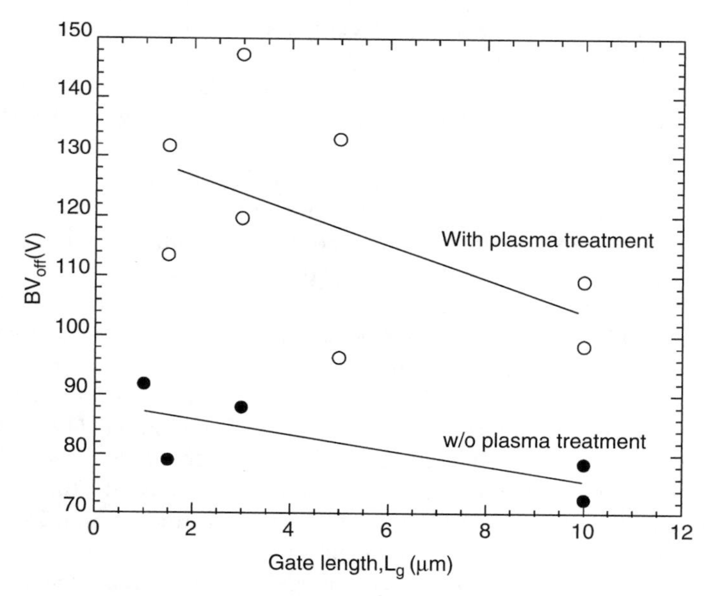

Fig. 45 The gate length, L_g dependence of BV_{off} for two types of processing steps. The Closed circles correspond to BV_{off} for the conventional process not employing a plasma surface treatment before the gate metal deposition. The open circles depict the case where a plasma treatment was incorporated in the process before gate deposition. Reference [171].

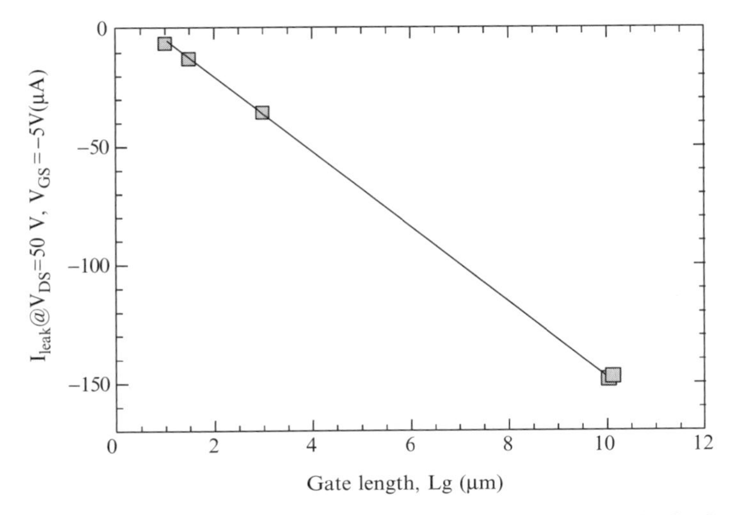

Fig. 46 The dependence of the gate leakage current, I_{leak} on the gate length with a fixed gate width indicative of leakage current being proportional to the area in FETs receiving a plasma treatment before gate deposition. Reference [171].

In devices with plasma treatment, the gate leakage current (I_{leak}) was reduced by two to three orders of magnitude (reference [180]). As one would expect, the BV_{off} values in the devices with small leakage current, I_{leak}, fabricated by plasma treatment were higher than those without plasma treatment. In addition, the leakage current scaled with the area of the device, not the periphery as is the case in the devices reported by Tan et al. [170]. This is depicted in Figure 46 for $V_{\text{DS}} = 50\,\text{V}$ and $V_{\text{GS}} = -5\,\text{V}$ as a function of L_{g}. The temperature dependence of I_{leak} measured up to 350°C was much smaller than that would be expected from an activated emission alone suggesting that leakage current has a good deal of tunneling component. This is quite inconsistent with Schottky barrier leakage current observed in bulk GaN layers where an exponential temperature as well as voltage dependence of the reverse current is observed. Moreover, the HFET devices fabricated on SiC substrates measured in author's laboratory confirmed the lack of strong dependence of the leakage current on temperature up about 300°C, but a strong dependence beyond this point up to the upper limit of measurement temperature of 500°C was observed. The important point is that the dependence is on area rather than the periphery, and the positive temperature coefficient of the breakdown voltage indicates the impact ionization to be the most likely cause of breakdown.

Wolff's [181] analytical expression for impact ionization coefficient $\alpha(E)$ is as follows:

$$\alpha(E) = \frac{qE}{E_i}\exp\left(\frac{-3E_{ph}E_i}{(qE\lambda)^2}\right) \tag{90}$$

where E is electric field, E_i is threshold energy for electron-hole pair generation, which can be assumed slightly larger than the bandgap of the semiconductor, E_{ph} is the optical phonon energy, and λ is the mean free path for optical phonon scattering assuming that this is the only scattering mechanism. Ascribing a length, l, within which the impact ionization occurs, further assuming that the field strength in that region is constant and given $E_{\max} = V_{DG}/l$, and I_{leak} is the primary electron current, the impact ionization current is given by

$$I_{ii} = \alpha E_{\max} I_{leak} l \tag{91}$$

Definition of the breakdown voltage, unless the breakdown is very sharp, requires some assumptions which are somewhat arbitrary and yet practiced widely. Generally, 1 mA for 1 mm of gate periphery is used although other criteria have been used. In the case of Nakao et al. [171] when the breakdown voltage is assumed to be the drain gate voltage V_{DG} at which the increase in drain current (ΔI_D) due to I_{ii} becomes a certain value (0.05 mA/mm in the present case), V_{Boff} is given by

$$I_{leak} \frac{E_i \Delta I_D}{qV_{Boff}} \exp\left[3E_{ph}E_i\left(\frac{1}{q\lambda V_{Boff}}\right)^2\right] \tag{92}$$

Equation 92 is plotted in Figure 47 with the following assumptions $E_i = 3.7\,\text{eV}$ [reference [182] and $E_{ph} = 92\,\text{meV}$ (a value of 42.3 meV has also been used in used in reference [115] to represent a convolution of various phonon processes that

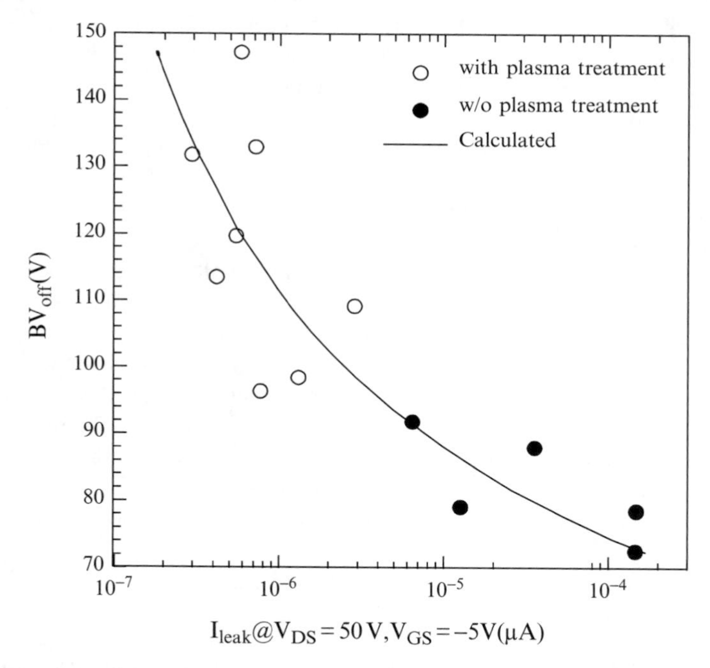

Fig. 47 The values of BV_{off} for devices plotted as a function of I_{leak}. The solid line is given by Equation 92. Reference [171].

may participate) [183]. Here $l/\lambda (= 220)$ is a fitting parameter. Assuming 3 nm for λ [reference [184]], 0.7 μm for l which is reasonable for a gate-drain spacing of 2 μm, Nakao et al. [171] calculated the breakdown voltage for a series of devices with measured leakage currents at low biases which used as the primary current, I_{leak}. The calculated and measured data are shown in Figure 47, reference [171]. Despite many fitting parameters in the form of assumptions which are not consistent across the board for GaN, the shape of the dependence of the breakdown voltage on the primary current indicates that at least in the devices of Nakao et al., [171] the impact ionization may be the primary breakdown mechanism. Considering all the results, one can surmise that there are many leakage paths and that if the surface state initiated leakage current is reduced sufficiently, and the quality of the GaN channel layer is sufficiently good, impact ionization process would be dominant.

2.3 Anomalies in GaN MESFETs and AlGaN/GaN HFETs

Field effect transistors in general and modulation doped field effect transistors in particular exhibit anomalies in their output *I–V* characteristics. Among the causes of these anomalies are channel carriers being trapped in the wide bandgap material and bulk, meaning the buffer layers. Additionally, unpassivated surface states, if present, could reduce the conducting charge, particularly between the gate and the drain region of the FETs [185]. These phenomena are depicted schematically in Figure 48. Some AlGaAs/GaAs HFETs, which are the predecessors of the current AlGaN/GaN HFETs, exhibited a lack of drain current for small drain biases at low temperatures which was then termed as "current collapse" [186]. This behavior was attributed to carrier injection from the channel to the AlGaAs barrier where the carriers are trapped at low temperatures. With below-gap light excitation, increasing temperature, and interchange of the source and drain terminals, the effect could be reversed. The GaAs buffer layer for the AlGaAs/GaAs-based device is of high quality so that its trapping effect is not dominant. The surface states in the AlGaAs/GaAs device were not deemed to have as profound an effect on the current voltage characteristics as in the GaN varieties. However, it is always a prudent approach to passivate the surface states regardless, as they affect the device operation with time.

In the GaN system, the surface states and/or defects play a much more important role on FET performance due in part to the relatively defective nature of the material and polarization fields as the layers are on polar surfaces and piezoelectric constants are relatively large as compared to GaAs. Anomalous characteristics, such as the so-called current collapse, kinks in the I–V characteristics, and long-term instabilities, have haunted the device from the time of early development. The current collapse, in the sense observed in the GaAs varieties, was observed in the very early stages of development in just a few cases and the term "collapse" has been very loosely used to describe all different types of anomalies. Binari et al. [185] reviewed the trapping behavior in GaN-based FETs around the time when the topic started to get some attention. To avoid confusion, the nomenclature adopted here is as follows:

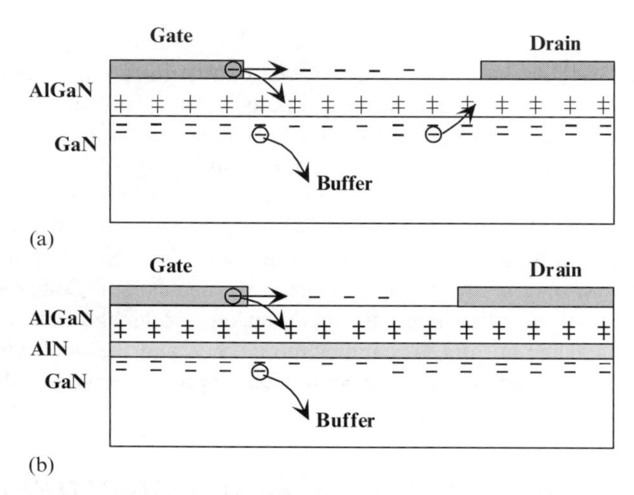

Fig. 48 a) Schematic representation of an AlGaN/GaN HFET indicating how the surface charge and charge injection into traps in the buffer layer and defects in AlGaN could serve in depleting conducting channel carriers. For simplicity, all the 2DEG is assumed to be due to donors placed in AlGaN. In addition, the regions from which carrier injection takes place are limited to the spacing between the gate and drain electrodes, the exact location is arbitrarily chosen. b) The same as in a) except with the insertion of an interfacial AlN layer which would help prevent injection of carriers from the channel to the barrier and or surface.

The current collapse is used to describe cases where the drain current actually goes down to zero or very small values and remains there for at least for some time. The anomalies in the IV characteristics manifesting themselves as unexpected slope change will be referred to as kinks. Any variation other than that expected from normal device operation, inclusive of unavoidable parasitics with frequency is referred to as dispersion. A form of this dispersion is the current lag in that the drain current does not reach what should be expected of the particular gate and drain biases based on the DC performance.

Preliminary investigations of these phenomena have been undertaken [187] with follow up studies of the current anomalies, mainly those caused by bulk traps in the buffer layers, [188] and transconductance dispersion with operation frequency [187, 189]. The characteristic time of the recharging process of the traps responsible for these anomalies in GaN ranges between nanoseconds and seconds. Generally, the trapping effects limit device performance in terms of noise even at relatively low frequencies [190–193]. However, in GaN-based FETs this takes on a different meaning in that dispersion in the output characteristic and transconductance with frequency limits the drain efficiency severely. An example is that the drain current lag prevents attainment of RF power congruent with the DC output characteristics of the device. Just to develop a sense of the extent of this effect, for a maximum drain voltage of 50 V (drain bias of 25 V) and maximum drain current of 1 A, one should normally get 6.25 W and 12.5 W in class A and class B operations, respectively, assuming an ideal case with zero saturation voltage, and no thermal limitations.

However, the observed values in the laboratory are in general substantially smaller. Simply, the drain current does not keep up with the gate bias voltage in response to high frequency large signal gate modulation [194].

Understanding the origin of the traps in GaN, particularly in the FET structures, their energy and physical location, and the physical mechanisms involved in trapping are of paramount importance for the optimization of device performance [195]. Owing to these devices being strong contenders in the marketplace for systems applications, these phenomena are getting a good deal of attention [196–204]. However, the reports are often contradictory, inconsistent, unclear, and simply put may be confusing to the uninitiated reader. As such it may not be clear as which if any report or any group of reports does go the heart of the physical phenomena. Some of these inconsistencies may be due to materials quality or the lack of it and its variation from sample to sample and of course from report to report, and to some extent the actual device fabrication technology employed.

The trapping mechanisms that are manifested as current lag are the so-called gate lag (lagging response of drain current after a pulse on the gate) and the drain lag (lagging response of the drain current after a pulse on the drain). In the work of reference [185], it was shown that like the conventional GaAs-based devices, gate lag measurements appear to be adversely affected by surface states while drain lag measurements appear to be advesly affected by traps within the buffer or barrier layers.

Reiterating, any negative charge below the channel and between the channel and the surface, including at the surface and in the underlying buffer layer, would lead to current lag. Current lag due to surface depletion in GaN MESFETs has been treated in some detail [205]. The carriers for barrier trapping could be provided by tunneling from the gate into the semiconductor and also from the channel into the barrier at high fields. The tunneling is assisted by the electric field perpendicular to the surface across the gate-to-barrier layer. Naturally, there are traps in the barrier, and the charge emission from the barrier traps is also affected by the field. Therefore, the influence of the field must be taken into account in characterizing the barrier traps whose characteristic times of the field-assisted emission may vary from hundreds of nanosecond to milliseconds [195]. Traps located in the buffer layer generally cause current kinks in the drain current, but one report attempts to link them to current collapse and drain lag [206]. The characteristic time of charge emission determined from the gate lag measurements extends to 10^3 s.

Charge temporarily trapped in the vicinity of the transistor channel can reduce the drain current level by as much as 90% [185]. Because the trap release time is very slow, the drain current remains affected and the drain current does not follow the gate voltage at high speed of modulation which can be hundreds of kHz. Because of a strong correlation of the effect with the semiconductor surface treatment in many devices from different laboratories, it has been convenient to conclude that at least some trapping centers are located on the surface [21]. In order for this model to be consistent with observations, it must also be assumed that the AlGaN surface contains large densities of the donor states [207]. The gate lag therefore has been associated with the ionized donor states located between gate and drain electrodes

[208]. Though reasonable, one should also consider the effect of the deposition of passivating layers on the underlying barrier properties, if for example a H-containing reactant such as SiH_4 or NH_3 are used, as in SiN_x passivation, H atoms available in a plasma-enhanced process could diffuse into the layer and form H-defect complexes, quenching the traps in the layer itself [209].

The temporal character of associated charge emission is typically a stretched exponent having a characteristic time in the range of seconds [208,210]. The stretched exponent dynamics being so complicated has made it nearly impractical for a quantitative analysis of these trapping centers. Trapped charge on the surface has been observed along with migration lengths up to 0.5–1 μm along the surface away from the gate contact with scanning Kelvin probe microscopy [210].

Filled surface states pinch the channel in the region between the gate and drain. This is schematically shown in Figure 49 where the load line and quiescent operating conditions for class A operation are shown. Also shown are the extremes of DC current as governed by the load line and RF current (shaded) which fails to follow the gate bias, with expected higher drain currents. The RF current can be determined

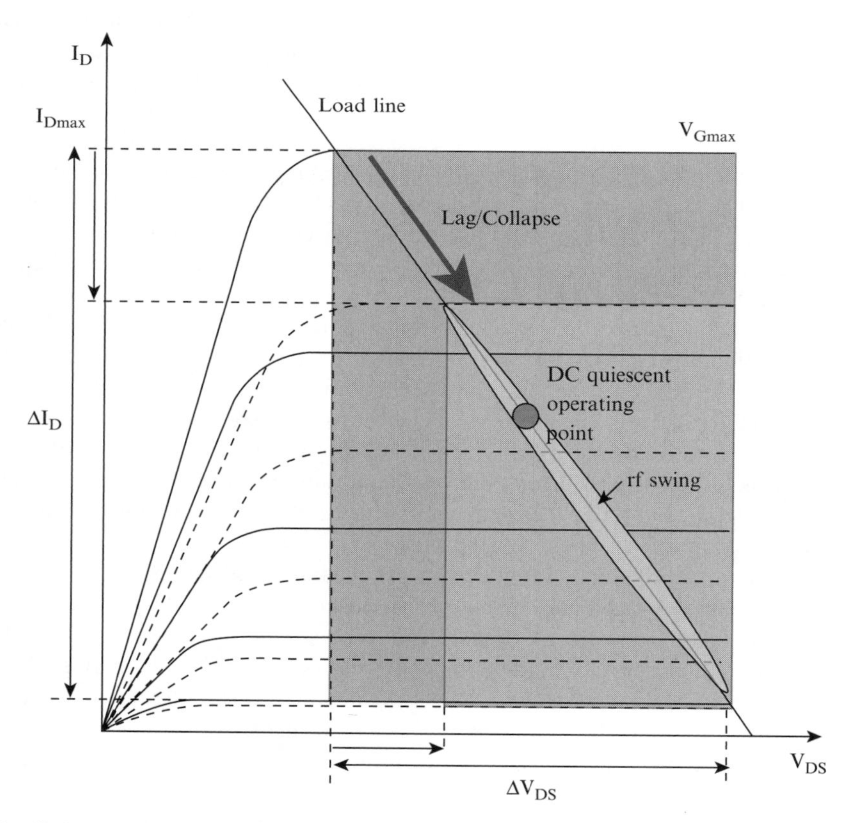

Fig. 49 Schematic representation of RF current lag superimposed on top of DC Drain (solid lines) and rf (Dashed lines) I–V characteristics with a DC load line.

by the use of the so-called load-pull tuning. It can also be measured under active loading conditions with the use of a high speed sampling scope for measuring the output RF voltage, in response to an RF input drive, where in the voltage measured can be converted to current knowing the load value.

Acknowledgements This work has been funded by a grant from the US Air Force Office of Scientific Research under the direction of Dr. G. L. Witt followed by Dr. K. Reinhardt. The authors would like to thank many of their colleagues and collaborators for their contribution to the research carried over a decade. Specifically, Prof. A. Di Carlo, Prof. F. Bernardini, Prof. V. Fiorentini, Prof. O. Ambacher, Prof. M. S. Shur, Prof. P. P. Ruden, Dr. P. Klein, Dr. S.C. Binari, Dr. O. Mitrofanov, Dr. M. Manfra, Dr. N. Nguyen, Dr. C. Nguyen, and Dr. Primit Parikh generously supplied figures from their work and in fact read portions of this text. HM would like to thank his present and past students for their contributions that got utilized in this chapter.

References

1. H. Morkoç, H. Ünlü, and G. Ji, "Fundamentals and Technology of MODFETs", Vols. I and II (Wiley and Sons, Wiley, Chichesters, West Sussex, UK (1991)
2. T.J. Drummond, W.T. Masselink and H. Morkoç, "Modulation Doped GaAs(Al,Ga)As Heterojunction Field Effect Transistors: MODFETs," Proc. of IEEE, Vol. 74(6), pp. 773–822, June 1986.
3. B. K. Ridley, "Exact electron momentum-relaxation times in GaN associated with scattering by polar-optical phonons", Journal of Applied Physics, vol. 84, no.7, 1, pp.4020–4021 Oct. (1998)
4. U. V. Bhapkar, M. S. Shur, "Monte Carlo calculation of velocity-field characteristics of wurtize GaN", J. Appl. Phys 82, 1649 (1997)
5. J. Kolnik, I.H. Oguzman, K.F. Brennan, R. Wang, P.P. Ruden, and Y. Wang, J. Appl. Phys. 78, pp. 1033–1038 (1995)
6. H. Morkoç, "Beyond SiC! III–V Nitride Based Heterostructures and Devices," in "SiC Materials and Devices," Y. S. Park, ed., Academic Press, Willardson and Beer Series, eds. Willardson and Weber, Vol. 52, Chapter 4, pp. 307–394, 1998.
7. S. T. Sheppard, K. Doverspike, W. L. Pribble, S. T. Allen, J. W. Palmour, L. T. Kehias and T. J. Jenkins, "High Power Microwave GaN/AlGaN HEMTs on Semi-insulating Silicon Carbide Substrates," IEEE Electron. Dev. Lett., 20(4), 161–163, (1999)
8. Y.-F. Wu, B. P. Keller, P. Fini, S. Keller, T. J. Jenkins, L. T. Kehias, S. P. DenBaars, U. K. Mishra, "High Al-Content AlGaN/GaN MODFET's for Ultrahigh Performance," IEEE Electron. Dev. Lett. 19(2), 50–53, 1998.
9. A. T. Ping, Q. Chen, J. W. Yang, M. A. Khan, and I. Adesida, "DC and Microwave Performance of High-Current AlGaN/GaN Heterostructure Field Effect Transistors Grown on p-Type SiC Substrates," IEEE Electron. Device Lett. 19(2), 54–56, 1998.
10. G. J. Sullivan, M. Y. Chen, J. A. Higgins, J. W. Yang, Q. Chen, R. L. Pierson, and B. T. McDermott, "High-power 10-GHz Operation of AlGaN HFET's on Insulating SiC," IEEE Electron. Dev. Lett. 19, 198–199, 1998.
11. S. Binari, J.M. Redwing, G. Kelner, and W. Kruppa, "AlGaN/GaN HEMTs Grown on SiC Substrates," Electron. Lett. 33(3), 242–243, 1997.
12. J.S. Moon, M. Micovic, P. Janke, P. Hashimoto, W-S. Wong, R.D. Widman, L. McCray, A. Kurdoghlian, and C. Nguyen, "GaN/AlGaN HEMTs operating at 20 GHz with continuous-wave power density $>6\,W/mm$", Electronics Letters, vol. 37, no. 8, pp. 528–530, 2 April 2001.

13. M. Micovic, A. Kurdoghlian, P. Janke, P. Hashimoto, D. W. S. Wong, J. S. Moon, L. McCray, and C. Nguyen, IEEE Trans. on ED, Vol. 48, No. 3, pp. 591–592, 2001.
14. J.S. Moon, M. Micovic, A. Kurdoghlian, R. Janke, P. Hashimoto, W-S. Wong, L . McCray, "Linearity of low microwave noise AlGaN/GaN HEMTs", Electronics Letters, vol. 38, no. 22, pp. 1358–1359, 24 Oct. 2002.
15. J.S. Moon, M. Micovic, A. Kurdoghlian, R. Janke, P. Hashimoto, W-S. Wong, L . McCray, "Linearity of low microwave noise AlGaN/GaN HEMTs", Electronics Letters, vol. 38, no. 22, pp. 1358–1359, 24 Oct. 2002.
16. N. Nguyen and C. Nguyen, private communication.
17. Y.-F. Wu, D. Kapolnek, J. Ibbetson, P. Parikh, B. Keller and U. K. Mishra, "Very high power density A1GaN/GaN HEMTs," IEEE Trans. on Electron Dev., Vol. 48, No. 3, pp. 586–590, March 2001.
18. V. Tilak, B. Green, V. Kapper, H. Kim, T. Prunty, J. Smart, J. Shealy, L. Eastman, IEEE Electron Device Lett., Vol. 22., pp. 504–506, Nov. 2001.
19. M. D. Hampson, S.-C. Shen, R. S. Schwindt, R. K. Price, U. Chowdhury, M. M. Wong, Ting G. Zhu, D. Yoo, R. D. Dupuis,and M. Feng, "Polyimide Passivated AlGaN–GaN HFETs With 7.65 W/mm at 18 GHz", IEEE Electron Device Letters, Vol. 25, No. 5, pp. 238–240, May 2004.
20. K. Chatty, S. Banerjee, T.P. Chow and R.J. Gutmann, Electron Device Letters, vol. 21, pp. 356–358, 2000.
21. Y. Ando, Y.Okamoto, H. Miyamoto, T. Nakayama, T. Ionue, and M. Kusuhara, "10-W/mm AlGaN-GaN HFET with field modulation plate", IEEE Electron Device Lett., Vol. 25, No. 5. pp. vol. 24, no. 5, pp. 289–291, May 2003.
22. N.-Q. Zhang, S. Keller, G. Parish, S. Heikman, S. P. DenBaars, and U. K. Mishra, "High breakdown GaN HEMTs with overlapping gate structure," IEEE Electron Device Lett., Vol. 21, pp. 421–423, Sept. 2000.
23. Y. F. Wu, A. Saxler, M. Moore, R. P. Smith, S. Sheppard, P.M. Chavarkar, T. Wisleder, U. K. Mishra, and P. Rarikh., "30-W/mm AlGaN/GaN HEMTs By Field Plate Optimization", IEEE Electron Device Letters, vol. 25, no. 3, pp. 117–119, March 2004.
24. W. Lu, J. Yang, M.A. Khan and I. Adesida, "A1GaN/GaN HEMTs on SiC with over 100 GHz fT and low microwave noise," IEEE Trans. on Electron Dev., Vol. 48, No. 3, pp. 581–585, March 2001.
25. Wu Lu, Jinwei Yang, M. Asif Khan, and Ilesanmi Adesida "AlGaN/GaN HEMTs on SiC with over 100 GHz fTand Low Microwave Noise", IEEE TED, VOL. 48, NO. 3, pp. 581–585, March 2001.
26. T. Henderson, M. I. Aksun, C.K. Peng, H. Morkoç, P.C. Chao, P.M. Smith, K.H.G. Duh and L.F. Lester, "Microwave Performance of a Quarter Micron Gate Low Noise Pseudomorphic InGaAs/AlGaAs Modulation Doped Field Effect Transistor," IEEE Electron. Dev. Lett., Vol. EDL-7, pp. 649–651, 1986.
27. Özgür Aktas, W. Kim, Z. Fan. S.N. Mohammad, A. Botchkarev, A. Salvador, B. Sverdlov, and H. Morkoç, Electron. Letts. Vol. 31, No. 16, pp. 1389–1390, (1995)
28. S. N. Mohammad, A. Salvador, and H. Morkoç, "Emerging GaN Based Devices," Proc. IEEE 83, 1306–1355, 1995.
29. S. C. Binari, J. M. Redwing, G. Kelner, and W. Kruppa, Electron. Lett. 33, 242 (1997)
30. Hadis Morkoç, Aldo Di Carlo and R. Cingolani, Solid State Electronics, Volume 46, Issue 2 pp. 157–202, (2002)
31. Hadis Morkoç, Aldo Di Carlo, and R. Cingolani, "GaN-Based Modulation Doped FETs" Low dimensional nitride semiconductors, edited by B. Gil, Oxford university Press, Oxford UK., pp. 341–414, 2002, ISBN 0 19 850974 X
32. Hadis Morkoç, Roberto Cingolani, and Bernard Gil, "Polarization Effects in Nitride Semiconductor Device Structures, and Performance of Modulation Doped Field Effect Transistors" Solid State Electronics, vol. 43, no. 10, pp. 1909–1927, Oct. 1999.
33. L. Hsu, W. Walukiewicz, "Effect of Polarization Fields on Transport Properties in AlGaN/GaN Heterostructures," J. Appl. Phys. 89, 1783 (2001)

34. L. Hsu and W. Walukiewicz, Appl. Phys. Lett. 73, 339 (1998)
35. I. P. Smorchkova, C. R. Elsass, J. P. Ibbetson, R. Vetury, B. Heying, P. Fini, E Haus, S. P. DenBaars, J. S. Speck, and U. K. Mishra, J. Appl. Phys. 86, 4520 (1999)
36. M. S. Shur, A. D. Bykhovsky, and R. Gaska, Solid-State Electron. 44, 205 (2000)
37. P. M. Asbeck, E. T. Yu, S. S. Lau, W. Sun, X. Dang, and C. Shi, Solid-State Electron. 44, 211 (2000)
38. M. S. Shur, A. D. Bykhovsky, and R. Gaska, Solid-State Electron. 44, 205 (2000)
39. R. Oberhuber, G. Zandler, and P. Vogl, Appl. Phys. Lett. 73, 818 (1998)
40. N. Maeda, T. Nishida, N. Kobayashi, and M. Tomizawa, Appl. Phys. Lett. 73, 1856 (1998)
41. W. Walukiewicz, Appl. Phys. Lett. 54, 2094 (1989)
42. L. Hsu, and W Walukiewicz, Phys. Rev. B56, 1520 (1997)
43. F. Bernardini, V. Fiorentini, and D. Vanderbilt, "Spontaneous Polarization and Piezoelectric Constants in III–V Nitrides," Phys. Rev. B, Vol. 56, R10024, 1997.
44. A. Bykhovski, B. Gelmont, M. Shur: J. Appl. Phys. 74, 6734 (1993)
45. D. Bykhovski, B. L. Gelmont, M.S. Shur: J. Appl. Phys. 81, 6332 (1997)
46. J. G. Gualtieri, J. A. Kosinski, A. Ballato: IEEE Trans. UFFC-41, 53 (1994)
47. G. D. O'Clock, M. T. Duffy: Appl. Phys. Let. 23, 55 (1973)
48. A. Zoroddu, F. Bernardini, P. Ruggerone, and V. Fiorentini, "First-principles prediction of structure, energetics, formation enthalpy, elastic constants, polarization, and piezoelectric constants of AlN, GaN, and InN: Comparison of local and gradient-corrected density-functional theory", Phys. Rev. B 64, 45208 (2001)
49. G. Burns, Solid State Physics (Academic Press, New York 1985), pp. 88–92.)
50. R.D. King-Smith, D. Vanderbilt, "Theory of polarization of crystalline solids", Phys. Rev. B Vol. 47, Issue 3. pp. 1651–1654 1651 15 January 1993.
51. R. Resta: Rev. Mod. Phys. 66, 899 (1994)
52. For a review see, R. Resta, "Macroscopic Polarization in Crystalline Dielectrics: the Geometric Phase Approach," Rev. Mod. Phys. 66, 899, (1994), and references therein.
53. Fabio Sacconi, Aldo Di Carlo, P. Lugli, and Hadis Morkoç, "Spontaneous and Piezoelectric polarization effects on the output characteristics of AlGaN/GaN heterojunction Modulation Doped FETs", IEEE Trans on Electron devices, special issue, Eds. U. K. Mishra and J. Zolper, IEEE Trans. on Electron Devices, Vol. TED-48, no. 3, pp. 450–457, (2001)
54. H. Morkoç, H. Ünlü, and G. Ji, Fundamentals and Technology of MODFETs, Vols. I and II (Wiley and Sons, Wiley, Chichesters, West Sussex, UK (1991)
55. G. Bastard, in "Wave Mechanics Applied to Semiconductor Heterostructures" , Edition de Physique, Paris , France, (1987).
56. R. Oberhuber, G. Zandler, and P. Vogl, Appl. Phys. Lett., 73, 818 (1998).
57. A. Di Carlo, S. Pescetelli, M. Paciotti, P. Lugli, and M. Graf, Solid State Comm. 98, 803 (1996)
58. A. Di Carlo, Phys. Stat. Solidi, 217, 703 (2000)
59. F. Della Sala, A. Di Carlo, P. Lugli, F. Bernardini, V. Fiorentini, R. Scholz, and J.M. Jancu, Appl. Phys. Lett., 74, 2002 (1999)
60. A. Di Carlo, F. Della Sala, P. Lugli, V. Fiorentini, F. Bernardini, Appl. Phys. Lett. 76, 3950 (2000)
61. R. Cingolani, A. Botchkarev, H. Tang, H. Morkoç, G. Coli', M. Lomascolo, A. Di Carlo, F. Della Sala, P. Lugli, Phys. Rev.B, 61, 2711 (2000)
62. A. Bonfiglio, M. Lomascolo, G. Traetta, R. Cingolani, A. Di Carlo, F. Della Sala, P. Lugli, A. Botchkarev, H. Morkoç, J. App. Phys, 87, 2289 (2000)
63. H. Morkoç, H. Ünlü, and G. Ji, Fundamentals and Technology of MODFETs, Vols. I and II (Wiley and Sons, Wiley, Chichesters, West Sussex, UK (1991)
64. P. Lugli, M. Paciotti, E. Calleja, E. Munoz, J.J. Sanchez-Rojas, F. Dessenne, R. Fauquembergue, J. L. Thobel, and G. Zandler: "HEMT Models and Simulations," in "Pseudomorphic HEMTs: Technology and Applications", Eds. R. Lee Ross, S. Swensson and P. Lugli, Kluwer Press, pp.141–163, Dordrecht, (1996)

65. M. Abramowitz and I. A. Stegun, eds. "Handbook of Mathematical Functions (National Bureau of Standards Applied Mathematic Series, NO. 55) (US Government Printing Office) 1964:
66. T. Ando, A. B. Fowler, and F. Stern, "Electronic properties of two-dimensional systems", Rev. Mod. Phys. Vol. 54, No. 2, pp. 437–672, April 1982.
67. F.F. Fang and W.E. Howard, "Negative field-effect mobility on (100) Si surfaces", Phys. Rev. Lett. Vol. 16(18), pp. 797–799, 1966.
68. V. Fiorentini, F. Bernardini, F. Della Sala, A. Di Carlo, and P. Lugli, Phys. Rev. B60, 8849 (1999)
69. A. Trellakis, A. T. Halick, A. Pacelli, and U. Ravaioli, J. Appl. Phys. 81, 7880 (1997)
70. Fabio Bernardini, Vincenzo Fiorentini, and David Vanderbilt, "Accurate calculation of polarization-related quantities in semiconductors", Phys. Rev. B, Vol. 63, 193201 (15 May 2001)
71. F. Bechstedt, U. Grossner, and J. Furthmüller, "Dynamics and polarization of group-III nitride lattices: A first-principles study", Phys. Rev. B 62, Issue 12, pp. 8003–8011, 15 September 2000.
72. M.A. Littlejohn, J. R. Hauser, T. H. Glisson: Appl. Phys. Lett. 26, 625 (1975)
73. A. Bykhovski, B. Gelmont, M. Shur: J. Appl. Phys. 77, 1616 (1995)
74. M. Shur, B. Gelmont, A. Khan: J. Electron. Mater. 25, 777 (1996)
75. See for example, C. Kittel, Introduction to Solid State Physics, 7th edition, Wiley, 1996; J. S. Blakemore, "Solid State Physics", 2nd Edition, Cambridge University Press, 1985; Walter Beam, "Electronics of Solids", McGraw Hill, 1965.
76. Vincenzo Fiorentini, Fabio Della Sala, Aldo Di Carlo, and Paolo Lugli, "Effects of macroscopic polarization in III-V nitride multiple quantum wells", Phys. Rev. B, Vol. 60, 8849–8858 (1999)
77. D. L. Rode, Phys. Rev., B2, 4036, (1970)
78. M. Suzuki, T. Uenoyama: Electronic and optical properties of GaN based quantum wells, in Group III-Nitride Semiconductor Compounds, Physics and Applications, ed. by B. Gil (Clarendon, Oxford 1998)
79. A.D. Bykhovski, V.V. Kaminski, M.S. Shur, Q.C. Chen, M. A. Khan: Appl. Phys. Lett. 69, 3254 (1996)
80. F. Bernardini and V. Fiorentini, "Nonlinear macroscopic polarization in III–V nitride alloys"Phys. Rev. B 64, 085207 (2001)
81. Fabio Bernardini and Vincenzo Fiorentini, "Erratum: Nonlinear macroscopic polarization in III-V nitride alloys, Phys. Rev. B vol. 64, 085207 (2001): Phys. Rev. B Vol. 65, 129903(E) (15 March 2002)
82. Fabio Bernardini and Vincenzo Fiorentini, "Nonlinear behavior of Spontaneous and Piezoelectric polarization", International Workshop on Physics of Light-Matter Coupling in Nitrides (PLMCN-1), September Sept 26–29, (2001) Rome Italy, phys. stat. sol.(a), Vol. 190, No. 1, 65–73, (2002)
83. A.F. Wright, "Elastic properties of zinc-blende and wurtzite AlN, GaN, and InN", J. Appl. Phys. Vol. Volume 82, Issue 6, pp. 2833–2839 September 15, 1997.
84. O Ambacher, J Majewski, C Miskys, A Link, M Hermann, M Eickhoff, M Stutzmann, F Bernardini, V Fiorentini, V Tilak, B Schaff and L F Eastman, "Pyroelectric properties of Al(In)GaN/GaN hetero- and quantum well structures", J. Phys.: Condens. Matter Vol. 14, pp. 3399–3434, (2002)
85. C. Deger, E. Born, H. Angerer, O. Ambacher, M. Stutzmann, J. Hornsteiner, E. Riha, and G. Fischerauer, "Sound velocity of AlxGa1 − xN thin films obtained by surface acoustic-wave measurements, Appl. Phys. Lett. Volume 72, Issue 19, pp. 2400–2402, May 11, 1998
86. K. Tsubouchi, K. Sugai, and N. Mikoshiba, in 1981 Ultrasonics Symposium, edited by B. R. McAvoy (IEEE, New York, 1981), Vol. 1, p. 375.
87. M. S. Shur, A. D. Bykhovski and R. Gaska Mater. Res. Soc. Int. J. Nitr. Semicond. Res. S 41 G16 (1999)
88. G. D. O'Clock, M. T. Duffy: Appl. Phys. Let. 23, 55 (1973)

89. For a review see, S. Karmalkar, M. S. Shur, and R. Gaska, "GaN based high electron mobility transistors" Chapter 3 in Wide Energy Bandgap Electronic Devices", Eds. F. Ren and J. Zolper. World Scientific, ISBN 981-238-246-1, 2003
90. Rashmi, Angu Agrawal, S. Sen, S. Haldar, and R. S. Gupta, "Analytical model for DC characteristics and small-signal parameters of AlGaN/GaN modulation-doped field-effect transistor for microwave circuit applications", Microwave and optical technology letters, Vol. 27, No. 6, pp. 413–419, December 20 2000
91. S. Bose, Adarsh, A. Kumar, Simrata, M. Gupta and R. S. Gupta, "A complete analytical model of GaN MESFET for microwave frequency applications", Microelectronics Journal Vol. 32, pp. 983–990, (2001)
92. Lehovec and R. Zuleeg, "Voltage-Current Characteristics of GaAs JFETs in the Hot Electron Range," Solid State Electron., Vol. 13, pp. 1415–1426, 1970.
93. K. Lee, M.S. Shur, T.J. Drummond and H. Morkoç, "Parasitic MESFET in (Al,Ga)As/GaAs Modulation Doped FETs and MODFET Characterization," IEEE Trans. Electron. Dev., Vol. ED-31, pp. 29–35, (1984)
94. E.T.Yu, G. J. Sullivan, P. M. Asbeck, C. D. Wang, D. Qiao, S. S. Lau, "Measurament of the piezoelectrically induced charge in GaN/AlGaN heterostructure field-effect transistors", Appl. Phys. Lett. 71, 2794 (1997)
95. P. Ramvall, Y. Aoyagi, A. Kuramata, P. Hacke, K. Horino "Influence of a piezoelectric field on the electron distribution in a double GaN/Al0.14Ga0.86N heterojuction", Appl. Phys. Lett. 74, 3866, (1999)
96. R. Gaska, J. W. Yang, A. Osinsky, A. D. Bykhovski, M. S. Shur, "Piezoeffect and gate current in AlGaN/GaN high electron mobility transistors", Appl. Phys. Lett. 71, 3673, (1997)
97. R. Gaska, J. W. Yang, A. Osinsky, A. D. Bykhovski, M. S. Shur, V. V. Kaminski, S. M. Soloviov, "The influence of the deformation on the two-dimensional electron gas density in GaN-AlGaN heterostructure", Appl. Phys. Lett. 72, 64, (1998)
98. B. Carnez, A. Cappy, A. Karzynski, E. Constant, and G. Salmer, "Modeling of a Submicrometer Gate Field Effect Transistor Including Effects of Nonstationary Electron Dynamics", J. Appl. Phys. 51, 784–790, (1980)
99. P. A. Sandborn, J. R. East, and G. I. Haddad. "Quasi-Two-Dimensional Modelling of GaAs MESFET's", IEEE Trans. Electron Dev., ED-34, 985–991 (1987)
100. C. M. Snowden and R. R. Pantoja. Quasi-Two-Dimensional modelling MESFET simulation for CAD", IEEE Trans. Electron Dev., ED-36, 1564–1573 (1989)
101. H. Morkoç: GaN-Based Modulation Doped FETs and UV detectors, Naval Research Reviews 51, 1 pp. 28–45 (1999)
102. M. J. Murphy, B. E. Foutz, K. Chu, H. Wu, W. Yeo, W. J. Schaff, O. Ambacher, L. F. Eastman, T. J. Eustis, R. Dmitrov, M. Stutzmann, and W. Riegerd, "Normal and inverted AlGaN/GaN based piezoelectric field effect transistors grown by plasma induced molecular beam epitaxy",. MRS Internet J. Nitride Semicond. Res. 4S1, G8.4 (1999)
103. Y.F. Yu, B. P. Keller, S. Keller, D. Kapolnek, P. Kozodoy, S. P. Denbaars, and U.K. Mishra, Appl. Phys. Lett. 69, 1438 (1996)
104. C. A. Liechti, "Microwave Field Effect Transistors", IEEE Trans. Microwave Theory Tech., Vol. MTT-24, pp. 279, 1976.
105. C. Weitzel, L. Pond, K. Moore, and M. Bhatnagar, "Effect of Device Temperature on RF FET Power Density," Proc. of Silicon Carbide, III-Nitrides and Related Materials, ICSI, August 1997, Stockholm, Sweden, Trans Publications, Ltd., Materials Science Forum, Vols. 264–268, pp. 969–972 (1998)
106. M. Moloney, F. Ponse, and H. Morkoç, "Gate Capacitance Voltage Characteristics of MODFETs: Its Effect on Transconductance," IEEE Trans. Electron. Dev. ED-32(9), 1675–1684, (1985)
107. S. Karmalkar, and U. K. Mishra, "Enahncement of breakdown voltage in AlGaN/GaN high electron mobility transistor using a field plate", IEEE trans. on Elec. Dev. Vol. TED-48. No. 8, pp. 1515–1521, August 2001.

108. Y. Okamoto, Y. Ando, K. Hataya, H Miyamoto, T. Nakayama, T. Inoue, and M. Kuzuhara,. "96 W AlGaN/GaN heterojunction FET with field-modulating plate", Electronics Letters, vol. 39, no. 20, pp. 1474–1475, 2 Oct. 2003.
109. M.E. Lin, S. Strite, A. Agarwal, A. Salvador, G.L. Zhou, N. Teraguchi, A. Rockett and H. Morkoç, "GaN Grown on Hydrogen Plasma Cleaned 6H-SiC Substrates," Appl. Phys. Letts., Vol. 62(7), pp. 702–704, (1993)
110. C. D. Lee, V. Ramachandran , A. Sagar, R. M. Feenstra, D. W. Greve, W. L. Sarney, L. Salamanca-Riba, D. C. Look, Bai Song Bai, W. J. Choyke, R. P. Devaty, "Properties of GaN epitaxial layers grown on 6H-SiC(0001) by plasma-assisted molecular beam epitaxy" TMS; IEEE. Journal of Electronic Materials, vol. 30, no. 3, pp. 162–9, March 2001.
111. J. A. Powell, D. J. Larkin and A. J. Trunek Use of Gaseous Etching for the Characterization of Structural Defects in Silicon Carbide Single Crystals. Silicon Carbide, III-Nitrides, and Related Materials. G. Pensl, H. Morkoç, B. Monemar and E. Janzen. Trans Tech Publications. 264–268: 421–424. (1998)
112. J. A. Powell, D. J. Larkin, P. G. Neudeck, J. W. Yang and P. Pirouz Investigation of Defects in Epitaxial 3C-SiC, 4H-SiC and 6H-SiC Films Grown on SiC Substrates. Silicon Carbide and Related Materials. M. G. Spencer, R. P. Devaty, J. A. Edmond et al. Bristol, IOP Publishing: 161–164. (1994)
113. D. L. Rode, Semiconductors and Semimetals, edited by R. K. Willardson and A. C. Beer (Academic, New York, 1975), Vol. 10, pp. 1–90.
114. K. Seeger, Semiconductor Physics, 2nd ed. (Springer, Berlin, 1982)
115. H. Morkoç "Handbook of Nitride Semiconductors and Devices, Vols.", Springer in press.
116. V. W. L. Chin, T. L. Tansley, and T. Osotchan, J. Appl. Phys. 75, 7365 (1994)
117. D. L. Rode and D. K. Gaskill, Appl. Phys. Lett. 66, 1972 (1995)
118. S. C. Jain, M. Willander, J. Narayan, R. Van Overstraeten, J. Appl. Phys. 87, 963 (2000)
119. B. K. Ridley, B. E. Foutz, and L. F. Eastman, Phys. Rev. B. Vol. 61, No. 24, pp. 1682–1689, 2000.
120. S. Dhar and S. Ghosh, J. Appl. Phys. 86, 2668 (1999)
121. V. W. L. Chin, T. L. Tansley, and T. Osotchan, J. Appl. Phys. 75, 7365 (1994)
122. D. C. Look, D. C. Reynolds, J. W. Hemsky, J. R. Sizelove, R. L. Jones, and R. J. Molnar, Phys. Rev. Lett. 79, 2273 (1997)
123. M. G. Cheong, K. S. Kim, C. S. Oh, N. W. Namgung, G. M. Yang, C. H. Hong, K. Y. Lim, E. K. Suh, K. S. Nahm, H. J. Lee, D. H. Lim, and A. Yoshikawa, Appl. Phys. Lett. 77, 2557 (2000)
124. W. Götz, L. T. Romano, J. Walker, N. M. Johnson, and R. J. Molnar, Appl. Phys. Lett. 72, 1214 (1998)
125. R. P. Joshi, Appl. Phys. Lett. 64, 223(1994)
126. P. Visconti, K. M. Jones, M. A. Reshchikov, R. Cingolani, H. Morkoç, and R. J. Molnar, Appl. Phys. Lett. 77, 3532 (2000)
127. H. M. Ng, D. Doppalapudi, T. D. Moustakas, N. G. Weimann, and L. F. Eastman, Appl. Phys. Lett. 73, 821 (1998)
128. Q. S. Zhu, and N. Sawaki, Appl. Phys. Lett. 76, 1594 (2000)
129. D. C. Look, D. C. Reynolds, J. W. Hemsky, J. R. Sizelove, R. L. Jones, and R. J. Molnar, Phys. Rev. Lett. 79, 2273 (1997)
130. D. Huang, F. Yun, P. Visconti, M. A. Reshchikov, D. Wang, H. Morkoç, D. L. Rode, L. A. Farina, Ç. Kurdak, K. T. Tsen, S. S. Park and K. Y. Lee, "Hall mobility and carrier concentration in GaN free-standing templates grown by hydride vapor phase epitaxy with high quality" Solid State Electronics, Vol. 45(5), pp. 711–715 (June 2001).
131. S. Nakamura, T. Mukai, and M. Senoh, J. Appl. Phys. 71, 5543 (1992)
132. M. E. Lin, B. Sverdlov, G. L. Zhou, and H. Morkoç, Appl. Phys. Lett. 62, 3479 (1993)
133. H. M. Ng, D. Doppalapudi, T. D. Moustakas, N. G. Weimann, and L. F. Eastman, Appl. Phys. Lett. 73, 821 (1998)
134. D. C. Look and J. R. Sizelove, Phys. Rev. Lett. 82, 1237 (1999)
135. N. G. Weimann, L. F. Eastman, D. Doppalapudi, H. M. Ng, and T. D. Moustakas, J. Appl. Phys. 83, 3656 (1998)

136. Z. Q. Fang, D. C. Look, W. Kim, Z. Fan, A. Botchkarev, and H. Morkoç, Appl. Phys. Lett. 72, 2277 (1998)
137. K. Wook, A. E. Botohkarev, H. Morkoç, Z. Q. Fang, D. C. Look, and D. J. Smith, J. Appl. Phys. 84, 6680 (1998)
138. B. Heying, I. Smorchkova, C. Poblenz, C. Elsass, P. Fini, and S. Den Baars, U. Mishra, and J. S. Speck, "Optimization of the surface morphologies and electron mobilities in GaN grown by plasma-assisted molecular beam epitaxy", Applied Physics Letters, Vol. 77, No. 18, pp. 2885–2887, 30 October 2000
139. M. J. Manfra, L. N. Pfeiffer, K. W. West, H. L. Stormer, K. W. Baldwin, J. W. P. Hsu, D. V. Lang, and R. J. Molnar, "High-mobility AlGaN/GaN heterostructures grown by molecular-beam epitaxy on GaN templates prepared by hydride vapor phase epitaxy" Appl. Phys. Letts. Vol. 77, Issue 18, pp. 2888–2890, October 30, 2000
140. M. J. Manfra, N. G. Weimann, J. W. P. Hsu, L. N. Pfeiffer, and K. W. West, S. Syed, H. L. Stormer, W. Pan, D. V. Lang, S. N. G. Chu, G. Kowach, and A. M. Sergent, J. Caissie, K. M. Molvar, L. J. Mahoney, and R. J. Molnar, "High mobility AlGaN/GaN heterostructures grown by plasma-assisted molecular beam epitaxy on semi-insulating GaN templates prepared by hydride vapor phase epitaxy", J. of Appl. Phys., Vol. 92, No. 1, pp. 338–345, 1 July 2002
141. M. J. Manfra, K. W. Baldwin, A. M. Sergent, K. W. West, R. J. Molnar and J. Caissie, "Electron mobility exceeding 160 000 cm2 /V s in AlGaN/GaN heterostructures grown by molecular-beam epitaxy", Appl. Phys. Lett. Vol. 85, No. 22, pp. 5394–5396, 29 November 2004.
142. E. Frayssinet, W. Knap, P. Lorenzini, N. Grandjean, J. Massies, C. Skierbiszewski, T. Suski, I. Grzegory, S. Porowski,G. Simin, X. Hu, and M. Asif Khan, M. S. Shur, R. Gaska, and D. Maude, Appl. Phys. Letts., Vol. 77. No. 16, pp. 2551–2553, (2000)
143. E. Frayssinet, W. Knap, P. Lorenzini, N. Grandjean, J. Massies, C. Skierbiszewski, T. Suski, I. Grzegory, S. Porowski, G. Simin, X. Hu, and M. Asif Khan, M. S. Shur, R. Gaska, and D. Maude, "High electron mobility in AlGaN/GaN heterostructures grown on bulk GaN substrates" Appl. Phys. Letts., Vol. 77. No. 16, pp. 2551–2553, (2000)
144. J. Cui, and A. Sun, M. Reshichkov, F. Yun, A. Baski, and H. Morkoç, "Preparation of Sapphire for High Quality III-Nitride Growth", MRS Internet Journal – The URL for the front page is http://nsr.mij.mrs.org/5/7/.
145. J. A Powell, D. J. Larkin and A. J. Trunek, "Use of Gaseous Etching for the Characterization of Structural Defects in Silicon Carbide Single Crystals," Silicon Carbide, III-Nitrides, and Related Materials. G. Pensl, H. Morkoç, B. Monemar and E. Janzen. Sweden, Trans Tech Publications. 264–268: 421–424, (1998).
146. E. J. Tarsa, B. Heying, X. H. Wu, P. Fini, S. P. DenBaars, and J. S. Speck, J. Appl. Phys. 82, 5472 (1997)
147. B. Heying, R. Averbeck, L. F. Chen, E. Haus, H. Riechert, and J. S. Speck, J. Appl. Phys. 88, 1855 (2000)
148. Ö. Aktas, W. Kim, Z. Fan, S.N. Mohammad, A. Botchkarev, A. Salvador, B. Sverdlov, and H. Morkoç, "High Transconductance-Normally-Off GaN MODFETs," Electron. Lett. 31(16), 1389–1390, (1995)
149. J.S. Moon, W-S, Wong, M. Micovic, M. Hu M, J. Duvall, M. Antcliffe, T. Hussain, P. Hashimoto, and L McCray, "High performance recessed gate AlGaN/GaN HEMTs", . Compound Semiconductors 2001. Proceedings of the Twenty-Eighth International Symposium on Compound Semiconductors. IOP Publishing. 2002, pp. 27–32.
150. T. Egawa, H. Ishikawa, M. Umeno, and T. Jimbo, "Recessed gate AlGaN/GaN modulation-doped field-effect transistors on sapphire", Appl. Phys. Lett. Vol. 76, No. 1, pp. 121–123, 3 January 2000
151. Takashi Egawa, Guang-Yuan Zhao, Hiroyasu Ishikawa, Masayoshi Umeno, E, and Takashi Jimbo, "Characterizations of Recessed Gate AlGaN/GaN HEMTs on Sapphire", IEEE Trans. on Electron. Dev. Vol. 48, No. 3, pp. 603–608 March 2001

152. Jong-Wook Kim, Jae-Seung Lee, Won-Sang Lee, Jin-Ho Shin Doo-Chan Jung, Moo-Whan Shin, Chang-Seok Kim, Jae-Eung Oh, Jung-Hee Lee, Sung-Ho Hahm, "Microwave performance of recessed gate Al0.2Ga0.8N/GaN HFETs fabricated using a photoelectrochemical etching technique" Materials Science and Engineering B Vol. 95, pp. 73–76 (2002)
153. Y. Okamoto, A. Wakejima, K. Matsunaga, Y. Ando, T. Nakayama, K. Kasahara, K. Ota, Y. Murase, K. Yamanoguchi, T. Inoue and H. Miyamoto, "C-band Single-Chip GaN-FET Power Amplifiers with 60-W Output Power", Microwave Symposium Digest, IEEE MTT-S International, pp. 491–494, 12–17 June 2005
154. Yasuhiro Okamoto, Yuji Ando, Tatsuo Nakayama, Koji Hataya, Hironobu Miyamoto, Takashi Inoue, Masanobu Senda, Koji Hirata, Masayoshi Kosaki, Naoki Shibata, and Masaaki Kuzuhara, "High-Power Recessed-Gate AlGaN–GaN HFET With a Field-Modulating Plate", IEEE Transactions on Electron Devices, V. 51, No. 12, p. 2217–2222, (2004)
155. Y. Okamoto, Y. Ando, K. Hataya;T. Nakayama, H. Miyamoto, T. Inoue, M. Senda, K. Hirata, M. Kosaki, N. Shibata, and M. Kuzuhara, "A 149W Recessed-Gate AlGaN/GaN FP-FET", Microwave Symposium Digest, 6–11 June 2004 IEEE MTT-S International, Vol. 3, pp. 1351–1354, 2004
156. Y. Okamoto, Y. Ando, K. Hataya, T. Nakayama, H. Miyamoto, T. Inoue, M. Senda, K. Hirata, M. Kosaki, N. Shibata, and M. Kuzuhara,,"Improved Power Performance for a Recessed-Gate AlGaN–GaN Heterojunction FET With a Field-Modulating Plate", IEEE Transactions on Microwave theory and techniques, Vol. 52, No. 11, pp. 2536–2540, (2004)
157. Takashi Inoue, Yuji Ando, Hironobu Miyamoto, Tatsuo Nakayama,Yasuhiro Okamoto,Kohji Hataya, and Masaaki Kuzuhara, "30-GHz-Band Over 5-W Power Performance of Short-Channel AlGaN/GaN Heterojunction FETs", IEEE Transactions on Microwave Theory and Techniques, Vol 53, No. 1, pp. 74–80 (2005)
158. N. Moll, M. R. Hueshen, and A. Fisher-Colbie, "Pulse-doped AlGaAs/InGaN's pseudomorphic MODFETs," IEEE Trans. ElectronDevices, vol. 35, no. 7, pp. 879–886, July1988.
159. P. J. Tasker and B. Hughes, "Importance of source and drain resistance to the maximum fT of millimeter-wave MODFETs," IEEE Electron Device Lett., vol. 10, no. 7, pp. 291–293, Jul. 1989.
160. J.S. Moon, M. Micovic, A. Kurdoghlian, P. Janke, P. Hashimoto, W-S Wong, L. McCray, and C. Nguyen. "Microwave noise performance of AlGaN-GaN HEMTs with small DC power dissipation", IEEE Electron Device Letters, vol.23, no.11, pp.637–639, Nov. 2002
161. P. Javorka, A. Alam, M. Wolter, A. Fox, M. Marso, M. Heuken, H. Lüth, and P. Kordoš, "AlGaN/GaN HEMTs on (111) Silicon Substrates", IEEE Electron Device Letters, Vol. 23, no. 1, pp. 4–6, January 2002
162. Eduardo M. Chumbes, A. T. Schremer, Joseph A. Smart, Y. Wang, Noel C. MacDonald, D. Hogue, James J. Komiak, Stephen J. Lichwalla, Robert E. Leoni, III, and James R. Shealy, "AlGaN/GaN High Electron Mobility Transistors on Si(111) Substrates", IEEE Trans. on Electron. Dev. Vol. 48, No. 3, pp. 420–425 March 2001
163. J. W. Johnson, E. L. Piner, A. Vescan, R. Therrien, P. Rajagopal,J. C. Roberts, J. D. Brown, S. Singhal, and K. J. Linthicum, "12 W/mm AlGaN–GaN HFETs on Silicon Substrates" IEEE Electron Device Letters, vol. 25, no. 7, pp. 459–461, July. 2004
164. Z. Yang, A. Koudymov, V. Adivarahan, J. Yang, G. Simin,and M. A. Khan, "High-Power Operation of III-N MOSHFET RF Switches" IEEE Microwave and wireless components letters, V. 15, No.12, pp. 850–852, 2005
165. Yasuhiro Okamoto,Yuji Ando, Koji Hataya, Tatsuo Nakayama, Hironobu Miyamoto, Takashi Inoue, Masanobu Senda, Koji Hirata, Masayoshi Kosaki, Naoki Shibata, and.Masaaki Kuzuhara, "Improved Power Performance for a Recessed-Gate AlGaN–GaN Heterojunction FET With a Field-Modulating Plate", IEEE Transactions on Microwave theory and techniques, Vol. 52, No. 11, pp. 2536–2540, 2004
166. K. E. Moore, C. E. Weitzel, K. J. Nordquist, L. L. Pond, III, J. W. Palmour, S. Allen, and C. H. Carter, Jr., IEEE Electron. Dev. Lett., 18(2), 69–70, (1997)
167. T. P Chow and R Tyagi, IEEE Trans. Electron Devices, Vol. 41, 1481 (1994)

168. S. M. Sze, "Physics of Semiconductor Devices", Wiley 2nd ed. 1982.
169. Z. Bandic, E. C. Piquette, P. M. Bridger, R. A. Beach, T. F. Kuech, and T. C. McGill, "Nitride based high power devices: Design and fabrication issues", Solid State Electronics, Vol. 42, No. 12, pp. 2289–2294, 1998.
170. W. S. Tan, P. A. Houston, P. J. Parbrook, D. A. Wood, G. Hill, and C. R. Whitehouse, "Gate leakage effects and breakdown voltage in metalorganic vapor phase epitaxy AlGaN/GaN heterostructure field-effect transistors", Applied Physics Letters, Vol. 80, No. 17, pp. 3207–3209, 29 April 2002
171. Takeshi Nakao, Yutaka Ohno, Shigeru Kishimoto, Koichi Maezawa, Takashi Mizutani, "Study on off-state breakdown in AlGaN/GaN HEMTs", phys. stat. sol. (c) 0, No. 7, 2335–2338 (2003)
172. Maziar Farahmand, Michael Weber, Louis Tirino, Kevin F. Brennan, and P. Paul Ruden, "Theoretical study of direct-current and radio-frequency breakdown in GaN wurtzite- and zinc-blende-phase MESFETs (metal–semiconductor field-effect transistors)", J. Phys.: Condens. Matter Vol. 13, pp. 10477–10486 (2001)
173. K. F. Brennan, "The Physics of Semiconductors with Applications to Optoelectronic Devices" (Cambridge: Cambridge University Press) p. 512 (1999)
174. J. Kolnik, I. H. Oguzman, K. F. Brennan, R. Wang and P. P. Ruden J. Appl. Phys. Vol. 79, p. 8838, (1996)
175. J. Kolnik, I. H, Oguzman, K. F. Brennan, J. Kolnik, R. Wang, and P. P. Ruden "Theoretical prediction of zinc blende phase GaN avalanche photodiode performance based on numerically calculated electron and hole impact ionization rate ratio", Mat Res Soc Symp Proc; Vol. 423, pp. 45–50, (1996)
176. J. Kolnik, I. H. Oguzman, K. F. Brennan, R. Wang and P. P. Ruden J. Appl. Phys. Vol. 81, p. 726, (1997)
177. I. H, Oguzman. E. Belloti, K. F. Brennan, J. Kolnik, R. Wang, and P. P. Ruden, "Theory of hole initiated impact ionization in bulk zincblende and wurtzite GaN", J Appl. Phys, Vol. 81, pp. 7827–7834, (1997)
178. A. P. Zhang, G. T. Dang, H. Cho, K. P. Lee, S. J. Pearton, J. I. Chyi, T. E. Nee, C. M. Lee, and C. C. Chuo, IEEE Trans. Electron Devices 48, 407 (2001)
179. V. A. Dmitriev, K. G. Irvine, and C. H. Carter, Jr., Appl. Phys. Lett. 68, 229 (1996)
180. S. Mizuno, Y. Ohno, S. Kishimoto, K, Maezawa, and T. Mizutani, "Large gate leakage current in AlGaN/GaN high electron mobility transistors". Jpn. Journal of Applied Physics Part 1-Regular Papers Short Notes & Review Papers, vol. 41, no. 8, pp. 5125–5126, Aug. 2002.
181. P. A. Wolff, Phys. Rev. 95 1415 (1954)
182. M. Reigrotzki, R. Redmer, N. Fitzer, S.M. Goodnick, M. Dur, W. Schattke, "Hole initiated impact ionization in wide band gap semiconductors", J. of Applied Physics, vol. 86, no. 8, pp. 4458–4463, 15 Oct. 1999.
183. B. K. Ridley, J. Phys.: Condens. Matter, Vol. 8, L511 (1996)
184. Y. Okuto and C. R. Crowell, Phys. Rev. B, Vol. 6, 3076 (1972)
185. S. C. Binari, K. Ikossi-Anastasiou, J. A. Roussos, W. Kruppa, D. Park, H. B. Dietrich, D. D. Koleske, A. E. Wickenden, and R. L. Henry, Special Issue of IEEE Electron Dev. Vol. 48, pp. 465–471, (2001)
186. R. Fischer, T.J. Drummond, J. Klem, W. Kopp, T. Henderson, D. Perrachione and H. Morkoç, "On the Collapse of Drain I-V Characteristics in Modulation Doped FETs at Cryogenic Temperatures," IEEE Trans. Electron. Dev., Vol. ED-31, pp. 1028–1032, 1984.
187. W. Kruppa, S. C. Binari, and K. Doverspike, "Low-frequency dispersion characteristics of GaN HFETs," Electronics Lett., vol. 31, pp. 1951–1952, 1995.
188. P. B. Klein, J. A. Freitas, Jr., S. C. Binari, and A. E. Wickenden, 'Observation of deep traps responsible for current collapse in GaN metal–semiconductor field-effect transistors,' Appl. Phys. Lett., 75, Issue 25, pp. 4016–4018 (1999)
189. E. Kohn, I. Daumiller, P. Schmid, N.X. Nguyen, and C.N. Nguyen, 'Large signal frequency dispersion of AlGaN/GaN HEMTs,' Electron. Lett., 35, 1022 (1999)
190. D.V. Kuksenkov, H. Temkin, R. Gaska, and J.W. Yang, 'Low-frequency noise in AlGaN/GaN heterostructure field effect transistors,' IEEE Electron Device Lett. 19, 222 (1998)

191. S.L. Rumyantsev, N. Pala, M.S. Shur, E. Borovitskaya, A.P. Dmitriev, M.E. Levinshtein, R. Gaska, M.A. Khan, J. Yang, X. Hu, and G. Simin, 'Generation-Recombination Noise in GaN/AlGaN Heterostructure Field Effect Transistors,' IEEE Trans Electron Dev., 48, 530 (2001)
192. P.H. Handel, "1/f Noise – an 'Infrared' Phenomenon", Phys. Rev. Letters Vol. 34, pp. 1492–1494 (1975)
193. P. H. Handel, "Nature of 1/f Phase Noise", Phys. Rev. Letters, Vol. 34, pp. 1495–1497 (1975)
194. C. Nguyen, N. X. Nguyen, and D. E. Grider, "Drain current compression in GaN MODFETs under large-signal modulation at microwave frequencies," Electronics. Lett., vol. 35, pp. 1380–1382, 1999.
195. Oleg Mitrofanov and Michael Manfra, "Mechanisms of gate lag in GaN/AlGaN/GaN high electron mobility transistors", Superlattices and Microstructures, Vol. 34, pp. 33–53, (2003)
196. S. Trassaert, B. Boudart, C. Gaquiere, D. Theron, Y. Crosnier, F. Huet, and M.A. Poisson, 'Trap effect studies in GaN MESFETs by pulsed measurements,' Elecronics Lett., 35, 1386 (1999)
197. S.L. Rumyantsev, M.S. Shur, R.Gaska, X. Hu, A.Khan, G. Simin, J. Yang, N. Zhang, S. DenBaars, and U.K. Mishra, 'Transient processes in AlGaN/GaN heterostructure field effect transistors,' Electron. Lett., 36, 757 (2000)
198. B.M. Green, K.K. Chu, E.M. Chumbes, J.A. Smart, J.R. Shealy, and L.F. Eastman, 'The effects of surface passivation on the microwave characteristics of undoped AlGaN/GaN HEMTs,' IEEE Electron Device Lett., 21, 268 (2000)
199. E.J. Miller, X.Z. Dang, H.H. Wieder, P.H. Asbeck, E.T. Yu, G.J. Sullivan, and J.M. Redwing, 'Trap characterization by gate-drain conductance and capacitance dispersion studies of an AlGaN/GaN heterostructure field-effect transistor,' J. Appl. Phys., 87, 8070 (2000)
200. A.V. Vertiatchikh, L.F. Eastman, W.J. Schaff, and T. Prunty, 'Effects of surface passivation of AlGaN/GaN heterostructure field effect transitor' Electron Lett. 38, 388 (2002)
201. S. Arulkumaran, T. Egawa, H. Ishikawa, and T. Jimbo, 'Comparative study of drain-current collapse in AlGaN/GaN high-electron mobility transistors on sapphire and semi-insulating SiC,' Appl. Phys. Lett., 81, 3073 (2002)
202. H. Marso, M. Wolter, P. Javorka, P. Kordoš, and H. Lüth, 'Investigation of buffer traps in an AlGaN/GaN/Si high electron mobility transistor by backgating current deep level transient spectroscopy', Appl. Phys. Lett., 82, 633 (2003)
203. A. Y. Polyakov, N. B. Smirnov, A. V. Govorkov, V. N. Danilin, T. A. Zhukova, B. Luo, F. Ren, B. P. Gila, A. H. Onstine, C. R. Abernathy, and S. J. Pearton, 'Deep traps in unpassivated and Sc2O3-passivated AlGaN/GaN high electron mobility transistors', Appl. Phys. Lett., 83, 2608 (2003)
204. P. B. Klein, "Photoionization spectroscopy in AlGaN/GaN high electron mobility transistors", J. Appl. Phys. 92, No. 9, pp. 5498–5502, November 1, 2002.
205. J. I. Izpura, "Drain current collapse in GaN metal–semiconductor field-effect Transistors due to surface band-bending effects", Semicond. Sci. Technol. Vol. 17, pp. 1293–1301 (2002)
206. P. B. Klein, S. C. Binari, K. Ikossi-Anastasiou, A. E. Wickenden, D.D. Koleske, R.L. Henry, and D.S. Katzer, 'Investigation of traps producing current collapse in AlGaN/GaN high electron mobility transistors,' Electron. Lett. 37, 661 (2001)
207. J.P. Ibbetson, P.T. Fini, K.D. Ness, S.P. DenBaars, J. S. Speck, and U. K. Mishra, 'Polarization effects, surface states, and the source of electrons in AlGaN/GaN heterostructure field effect transistors', Appl. Phys. Lett. 77, 250 (2000)
208. R. Vetury, Q. Zhang, S. Keller, and U.K. Mishra, 'The impact of surface states on the DC and RF Characteristics of AlGaN/GaN HFETs' IEEE Trans. Electron Devices 48, 560 (2001)
209. A. Hierro, S. A. Ringel, M. Hansen, J. S. Speck, U. K. Mishra, and S. P. DenBaars, Appl. Phys. Lett. 77, 1499 (2000)
210. G. Koley, V. Tilak, L.F. Eastman and M. Spencer, 'Slow transients observed in AlGaN/GaN HFETs: Effects of SiNx passivation and UV illumination', IEEE T Electron Dev., 50, 886 (2003)

Effects of Polarization in Optoelectronic Quantum Structures

Raphaël Butté and Nicolas Grandjean

1 Introduction

Although GaN epitaxial growth was already ongoing at the end of the 60's [1,2], the study of low dimensional nitride based heterostructures such as quantum wells (QWs) and quantum dots (QDs) started more than twenty years later. This is in contrast with what happened in the III-V arsenide system for which the growth of QWs began in the middle of the seventies [3,4]. The main reason for this delay is likely related to the poor crystal quality of GaN epilayers, which is a direct consequence of the lack of bulk GaN substrates. Meanwhile, the target of GaN technology was the achievement of *p*-type conductivity together with growth engineering aimed at improving the material quality. Once Akasaki and co-workers succeeded with *p*-type doping of GaN [5], the main objective of the III-V nitride research community – still very small at that time – was the fabrication of a blue light emitter. In 1993, Nakamura *et al.* first reported candela-class high brightness blue light-emitting diodes (LEDs) [6]. This was a true detonator in the semiconductor technology and the research activity in that field experienced an extraordinary boom. The laser diode (LD) was the next step [7], and since then, III-V nitride compounds are among the most studied semiconductors. The perspective of using GaN based white LEDs for high efficiency solid-state lighting makes them somehow the counterpart of silicon in electronics in the future optoelectronics industry. The point to be noticed is that blue LEDs and violet LDs employ InGaN/GaN QWs in their active region as they are efficient light emitters in this wavelength range. For example, it has been shown that a slight incorporation of indium in GaN QWs increases drastically the LED efficiency [8]. This can likely explain why most of the early studies dealing with nitride based QWs have been devoted to InGaN/GaN QWs and not to GaN/AlGaN ones. This situation has probably hindered fundamental studies and may be the reason why the experimental evidence of the built-in polarization field in nitride based QWs was only discovered a couple of years after the marketing of high brightness LEDs.

This Chapter will be organized as follows:

1- In the first part, we shall give an elementary description of the theory of spontaneous polarization and piezoelectric effects in III-V nitride semiconductor heterostructures. This section will first highlight the peculiarities of the wurtzite structure with respect to the zincblende one in terms of symmetry considerations. Then we will describe spontaneous polarization and piezoelectric effects in bulk III-V nitride materials. The link between stress and deformations in III-V nitride heterostructures will then be considered. It will be followed by a description of the optical properties of III-V nitride quantum heterostructures subject to built-in polarization fields e.g., the quantum confined Stark effect (QCSE) with a focus on the oscillator strength of optical transitions. Finally, a section will be devoted to the case of wurtzite QDs.
2- The second part of this Chapter will detail some specific experimental results essentially on GaN/AlGaN QWs which can be considered as a prototypical structure to probe polarization effects in III-V nitrides. It will first focus on experimental evidences of polarization fields in such heterostructures and the way to measure them. Then the optical properties of GaN/AlGaN QWs will be more specifically described before concentrating on that of GaN/AlN QDs. Finally, we will give a short overview of InGaN/GaN QW properties, in which the interplay between polarization and localization effects are quite complex.

2 Basic Elements of the Theory of Polarization in III-V Nitride Heterostructures

A pedestrian's guide to the theory of spontaneous polarization and piezoelectric effects

III-V nitride semiconductor compounds (GaN, AlN, InN and the related ternary or quaternary compounds) can crystallize either in the wurtzite (2-*H*) or in the zincblende (3*C*) structures. The stable phase, which will be the only one described afterwards, is the wurtzite phase which possesses the hexagonal symmetry. Such symmetry will induce several effects on this class of semiconductors:

- The hexagonal close-packed (*hcp*) structure exhibits a high (six fold) symmetry axis implying that III-V nitride compounds are anisotropic crystals. Consequently, their properties will differ strikingly from those of more conventional III-V semiconductors such as GaAs or AlAs which possess a cubic structure (zincblende). The wurtzite symmetry is such that the barycenters of positive and negative charges do not coincide inducing thereby a spontaneous polarization, *i.e.* a polarization at zero strain, oriented along the polar axis (here the [0001] direction) which will be shown to be very large. For this reason, III-V nitrides belong to the family of pyroelectric materials.
- In addition to spontaneous polarization effects, when the crystal lattice will be subjected to deformation, large piezoelectric effects will also come into play in direct correlation with the change of the crystal anisotropy.

2.1 The Wurtzite Structure

Crystals having the wurtzite structure belong to the space group – using Schoenflies notation – C_{6v}^4, the associated point group being C_{6v} [9]. The corresponding conventional unit cell will have a basis consisting of 2 X ions at (0 0 0) and (1/3 2/3 1/2) (where X = Ga, Al, In) and 2 nitrogen ions at $(0\,0\,u)$ and $(1/3\,2/3\,1/2+u)$ (Fig. 1).

Anions and cations form two interpenetrating hexagonal close-packed structures, shifted along the [0001] direction, *i.e.* along the so-called c-axis which is also known as the direction of stacking. The lattice parameter in the (0001) planes, the parameter a, and along the c-axis (parameter c) are thus distinct. The internal displacement parameter u is defined as the anion-cation bond length along the [0001] direction in units of c.

The room temperature values of a and c measured on GaN, AlN and InN are reported in Table 1 together with the parameters c/a and u.

In the ideal wurtzite structure, ions reside in perfect tetrahedral sites satisfying $c/a = \sqrt{8/3} = 1.633$ and $u = 0.375$. The negative value of $\Delta c/a$ measured on real crystals underlines the higher stability of the hexagonal phase with respect to the cubic structure. Besides, this parameter provides a measure of the ionic character of the bond. Thus it appears that the degree of ionicity is stronger for AlN than for GaN or InN.

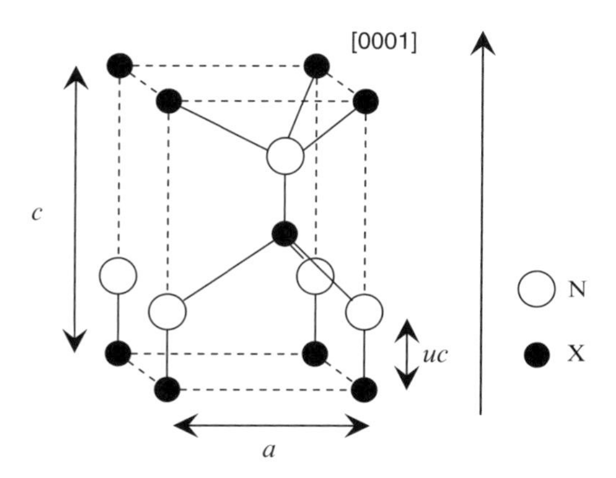

Fig. 1 Conventional wurtzite structure.

Table 1 Lattice parameters and internal displacement parameters of III-V nitride binary compounds in the wurtzite phase. $\Delta c/a$ reflects the departure of a given compound to the ideal hexagonal close-packed structure.

Parameter (T = 300 K)	GaN	AlN	InN
a(Å) [10]	3.189	3.112	3.545
c(Å) [10]	5.185	4.982	5.703
c/a	1.6259	1.6009	1.6087
$\Delta c/a$	−0.0071	−0.0321	−0.0243
u [11]	0.3769	0.3814	0.3787

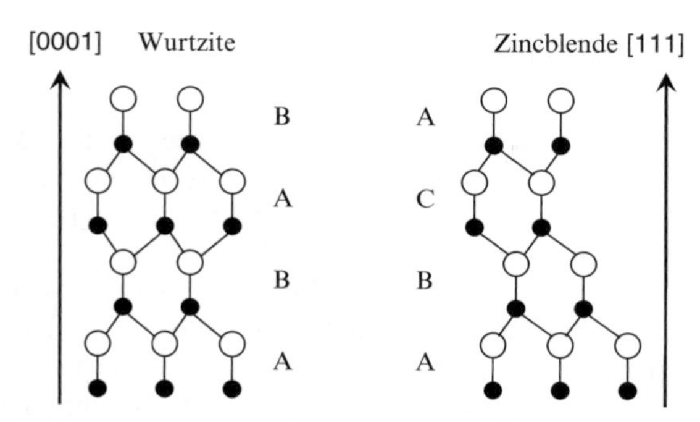

Fig. 2 Stacking sequences in wurtzite and zincblende structures.

The tetrahedral arrangement of the first ionic neighbors is identical for the two crystalline phases: wurtzite and zincblende. However these structures differ in the arrangement of the second ionic neighbors (Fig. 2). Indeed the (0001) planes of the wurtzite structure, which are equivalent to the (111) planes of the zincblende structure, are stacked following a sequence of the type ABABAB... differing from the sequence ABCABC... characterizing the cubic phase. The stacking sequence of the tetrahedra is thus said to be "staggered" in the zincblende structure and "eclipsed" in the wurtzite structure.

2.2 Strain and Internal Electric Field in III-Nitride Heterostructures

Semiconductor heterostructures, usually grown either by metalorganic vapor phase epitaxy (MOVPE) or by molecular beam epitaxy (MBE), are known to be highly sensitive to strain and deformations due to the different physical properties of the material constituents (lattice mismatch, thermal expansion coefficients, etc). This will be especially the case for heteroepitaxial structures [12]. The resulting deformations can alter significantly the electronic band structure through a variation of the crystal field [13].

As will be shown hereafter, in piezoelectric semiconductors these deformations create macroscopic polarization fields which will strongly modify the band structure through the QCSE [14]. The peculiar properties of III-V nitrides stem thus from their large piezoelectric constants and the spontaneous polarization which will induce huge electric fields (sometimes in excess of several MV/cm) even for unstrained quantum heterostructures [15, 16]. The combined effect of spontaneous and piezoelectric fields allowed explaining the lower transition photoluminescence (PL) energy of $In_xGa_{1-x}N$/GaN or GaN/$Al_yGa_{1-y}N$ QWs with respect to that of a flat band structure [17–19]. Consequently, in order to reach a precise understanding of the optical properties of III-nitride heterostructures, an accurate knowledge of their strain state is first required.

In this section we will introduce the formalism of the macroscopic elastic theory allowing describing the effect of stress in III-V nitride heterostructures. The separate contributions to the electric field induced both by spontaneous and piezoelectric polarizations, which cannot be neglected in quantum heterostructures made out of two different materials, will be described in detail. For this purpose, the elastic properties of III-V nitride compounds, which are closely linked to their piezoelectric properties, will also be described.

2.2.1 Determination of Polarization Fields

Besides stress effects induced by lattice-mismatch or thermal relaxation, which are common to crystalline heterostructures, III-V nitride structures are also characterized by electrostatic effects. This class of materials exhibits indeed polarization effects linked to their symmetry. We shall first start with a qualitative description of these piezoelectric properties, typical of bulk materials, before switching to the case of heterostructures.

Spontaneous Polarization and Piezoelectric Effects in Bulk Materials

(a) Spontaneous polarization

Certain crystals will exhibit a bulk spontaneous polarization even without external excitations. This effect is the so-called pyroelectric effect as a change in temperature can alter it [20]. From Neumann's principle, this physical property must be compatible with symmetry operations of the point group of the crystal under study. In particular, crystals characterized by an inversion center do not belong to the family of pyroelectric materials.

The wurtzite structure is the structure of highest symmetry compatible with the existence of spontaneous polarization. It is characterized by a high symmetry axis (c-axis) which is left unchanged through all transformations of the point group C_{6v} [9]. As a consequence the wurtzite structure will exhibit a spontaneous polarization along the (0001) axis. It is worth pointing out that nitrides distinguish themselves from other III-V materials in that they resemble more II-VI oxides and in some respects ferroelectric perovskites [15]. In Table 2, we report the spontaneous polarizations P_{sp} calculated by Bernardini *et al.* [11] from *ab initio* theory using the Berry-phase method [21–23] within the generalized gradient approximation (GGA). These values are relatively large for semiconductors as that of AlN is only 3–5 times smaller than in typical ferroelectric perovskites [24]. It is also seen that the larger

Table 2 Spontaneous polarizations for the III-V nitride binary compounds calculated within the GGA approach [11].

Materials	GaN	AlN	InN
P_{sp} (C/m^2)	−0.034	−0.090	−0.042

the deviation from the ideal wurtzite structure (*i.e.* the parameter u gets longer while c/a gets shorter), the larger the spontaneous polarization. This is especially the case of AlN which exhibits the highest ionic character among the three nitride binary compounds.

Besides the case of binary compounds, it is of high practical use to know the values of the spontaneous polarization in ternary or quaternary alloys. Bernardini and Fiorentini have investigated the variation of spontaneous polarization of ternary nitride alloys through *ab initio* calculations [25, 26]. By assuming that these alloys have a random microscopic structure they deduced that Vegard's law does not hold anymore. It was shown that the spontaneous polarization of random ternary nitride alloys, in C/m^2, can be expressed to second order in the composition parameter x as [26]:

$$\begin{aligned} P^{sp}_{Al_xGa_{1-x}N}(x) &= -0.090x - 0.034(1-x) + 0.019x(1-x), \\ P^{sp}_{In_xGa_{1-x}N}(x) &= -0.042x - 0.034(1-x) + 0.038x(1-x), \\ P^{sp}_{Al_{1-x}In_xN}(x) &= -0.090(1-x) - 0.042x + 0.071x(1-x). \end{aligned} \quad (1)$$

The first two terms in equations (1) correspond to the standard linear interpolation between the binary compounds, *i.e.* Vegard's law, while the last term is the bowing parameter describing the nonlinearity to quadratic order. Higher order terms are neglected since their contribution was estimated to be smaller than 10% in the worst case, namely that of AlInN alloys. This nonlinear change in spontaneous polarization *versus* lattice constant for the III-V nitrides is reported in Fig. 3 (solid lines). It is compared to Vegard interpolations (dotted lines). The peculiar case of $Al_xIn_yGa_{1-x-y}N$ alloys lattice matched to GaN is also shown as a dash-dotted line.

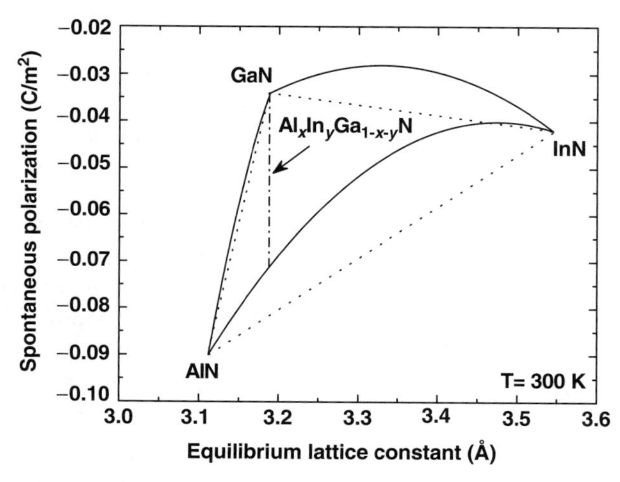

Fig. 3 Nonlinear spontaneous polarization *versus* lattice constant in $Al_xIn_yGa_{1-x-y}N$ alloys (solid lines). Dotted lines are Vegard interpolations linking III-V nitride binary compounds. The dash-dotted line shows the case of $Al_xIn_yGa_{1-x-y}N$ alloys lattice matched to GaN.

The reason for this nonlinearity has to be found in the different polarization response to changes in the lattice constant $a(x)$, *i.e.* changes in hydrostatic compression, in the ideal binary compounds. Thus the small bowing of AlGaN alloys stems from the reduced change in lattice parameter when going from GaN to AlN implying that the responses to hydrostatic pressure of both GaN and AlN are similar. The same does not hold anymore for AlInN alloys due to the significant change in lattice parameter when going from AlN ($a_{\mathrm{AlN}} \sim 3.11\,\text{Å}$) to InN ($a_{\mathrm{InN}} \sim 3.55\,\text{Å}$). The InGaN alloys constitute a somewhat intermediate situation though being closer to the case of AlInN alloys.

The general case of quaternary $\mathrm{Al}_x\mathrm{In}_y\mathrm{Ga}_{1-x-y}\mathrm{N}$ alloys can then be directly extrapolated from equations (1):

$$P^{sp}_{\mathrm{Al}_x\mathrm{In}_y\mathrm{Ga}_{1-x-y}\mathrm{N}}(x,y) = xP^{sp}_{\mathrm{AlN}} + yP^{sp}_{\mathrm{InN}} + (1-x-y)P^{\mathrm{sp}}_{\mathrm{GaN}} \tag{2}$$
$$+b_{\mathrm{AlGaN}}x(1-x) + b_{\mathrm{InGaN}}y(1-y) + xy\left[b_{\mathrm{AlInN}} - b_{\mathrm{AlGaN}} - b_{InGaN}\right]$$

where b_{AlGaN}, b_{InGaN} and b_{AlInN} are the bowing parameters of the III-V nitride ternary compounds. Numerically we get:

$$P^{sp}_{\mathrm{Al}_x\mathrm{In}_y\mathrm{Ga}_{1-x-y}\mathrm{N}}(x,y) = -0.090x - 0.042y - 0.034(1-x-y) \tag{3}$$
$$+0.019x(1-x) + 0.038y(1-y) + 0.014xy.$$

(b) Piezoelectric polarization

When exerting an external stress, the crystal will undergo a deformation inducing a modification of the whole polarization (Fig. 4). This phenomenon is known as the *direct* piezoelectric effect. The change in spontaneous polarization $\boldsymbol{P_{pz}} = \boldsymbol{P'} - \boldsymbol{P_{sp}}$ is commonly called piezoelectric polarization. The latter is more specifically induced by:

- the external deformation of the lattice unit which is described by a variation of the parameters a and c.

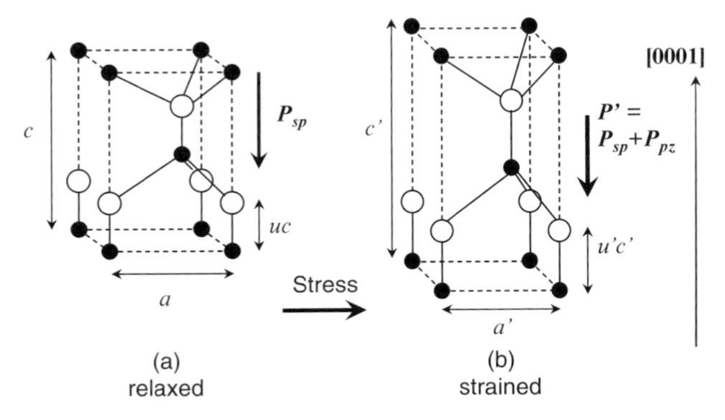

Fig. 4 Direct piezoelectric effect: external and internal deformations induce a variation of the spontaneous polarization $\boldsymbol{P_{sp}}$.

- changes in the anion-cation length described by a variation in the internal parameter u.

The geometry of the external stress being fixed, the internal deformation is determined by minimizing the energy of the deformed system, the latter being a function of electrostatic terms depending on the anion-cation bond length, *i.e.* it will depend on the internal parameter u.

An accurate description of the piezoelectric polarization for a generic ternary nitride $A_xB_{1-x}N$ alloy (A and B being either Ga, Al or In) is given in [25,26]. It was shown that contrary to the case of spontaneous polarization, Vegard's law holds. It is thus expressed as:

$$P^{pz}_{A_xB_{1-x}N}(x) = xP^{pz}_{AN}[\epsilon(x)] + (1-x)P^{pz}_{BN}[\epsilon(x)]. \quad (4)$$

This is due to the fact that the nonlinearity of the bulk polarization of piezoelectric origin, *i.e.* the piezopolarization, of the binary subcomponents as a function of basal strain ϵ is included. The corresponding strain dependent bulk piezopolarizations have been shown to express (in C/m^2) as [26]:

$$\begin{aligned} P^{pz}_{AlN} &= -1.808\epsilon + 5.624\epsilon^2 \text{ for } \epsilon < 0, \\ P^{pz}_{AlN} &= -1.808\epsilon - 7.888\epsilon^2 \text{ for } \epsilon > 0, \\ P^{pz}_{GaN} &= -0.918\epsilon + 9.541\epsilon^2, \\ P^{pz}_{InN} &= -1.373\epsilon + 7.559\epsilon^2 \end{aligned} \quad (5)$$

where $\epsilon(x) = [a_{sub} - a(x)]/a(x)$ is the basal strain of the alloy layer under study, $a(x)$ and a_{sub} being the lattice constants of the unstrained alloy at composition x and of the substrate, respectively.

The general assumption made in this Chapter regarding the heterostructures under consideration is generally that of pseudomorphic growth. Provided it is the case, basal strain ϵ can be directly calculated from the lattice constants which follow Vegard's law as a function of composition:

$$\begin{aligned} a_{Al_xGa_{1-x}N}(x) &= 3.189 - 0.077x, \\ a_{In_xGa_{1-x}N}(x) &= 3.189 + 0.356x, \\ a_{Al_{1-x}In_xN}(x) &= 3.112 + 0.433x. \end{aligned} \quad (6)$$

The change of the piezoelectric component of the macroscopic polarization in ternary alloys epitaxially strained on a GaN layer is shown in Fig. 5.

The general case of quaternary $Al_xIn_yGa_{1-x-y}N$ alloys can then be directly extrapolated from equations (4) and (5):

$$P^{pz}_{Al_xIn_yGa_{1-x-y}N}(x,y) = xP^{pz}_{AlN}[\epsilon(x,y)] + yP^{pz}_{InN}[\epsilon(x,y)] + (1-x-y)P^{pz}_{GaN}[\epsilon(x,y)] \quad (7)$$

where $\epsilon(x,y) = [a_{sub} - a(x,y)]/a(x,y)$ is the basal strain of the random quaternary alloy layer under study, $a(x,y)$ and a_{sub} being the lattice constants of the

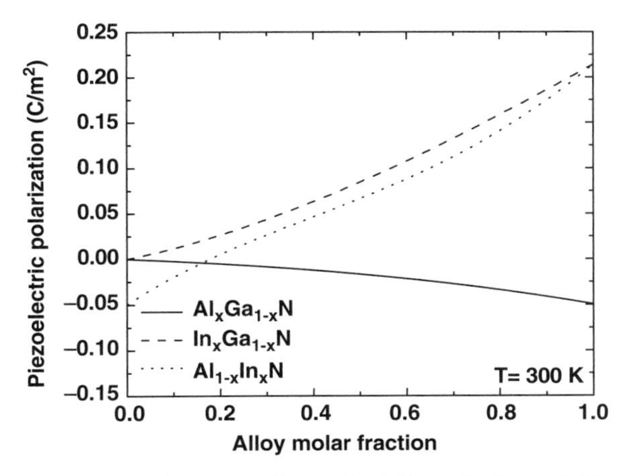

Fig. 5 Piezoelectric polarization in ternary alloys epitaxially strained on a GaN epilayer.

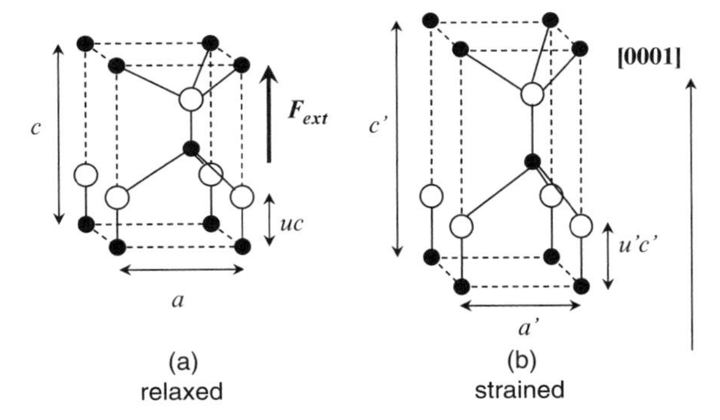

Fig. 6 Converse piezoelectric effect: an applied external electric field $\boldsymbol{F}_{ext}$ induces a deformation of the crystal unit.

unstrained alloy at composition x, y and of the substrate, respectively. $a(x,y)$ is obtained from (6):

$$a(x,y) = a_{\mathrm{Al}_x\mathrm{In}_y\mathrm{Ga}_{1-x-y}\mathrm{N}}(x,y) = 3.189 - 0.077x + 0.356y. \tag{8}$$

It is worth pointing out that the *converse* piezoelectric effect can also be observed when an applied electric field induces a deformation of the lattice (Fig. 6).

As will be described hereafter spontaneous and piezoelectric effects will play a major role in quantum heterostructures where strained thin layers, the well or dot materials, are pseudomorphically grown between higher energy barrier materials.

Stress and deformations in III-V nitride heterostructures

An accurate description of the stress state of quantum heterostructures made of piezoelectric materials requires taking into account electrostatic forces due to the electric fields present in the material. To do so, it is convenient to use the phenomenological formalism developed by Born and Huang [27] describing electromechanical coupling. In other words, we shall introduce fully coupled equations for internal electric field and strain in a III-V nitride QW or in a multiple (M) – QW system grown along the [0001] direction as described by Christmas *et al.* [28]. By "coupled equations", it is meant that the *direct* piezoelectric effect and the *converse* piezoelectric effect are fully taken into account.

The elements σ_{ij} of the stress tensor are a function of independent variables which are the elements ϵ_{kl} of the strain tensor and F_k of the electric field [27, 28]:

$$\sigma_{ij} = \sum_{k,l} C_{ijkl}\epsilon_{kl} - \sum_{k} e_{kij}F_k \quad \text{with} \quad i,j,k,l = x,y,z. \tag{9}$$

The first term of this expression can be identified to Hooke's law issued from elasticity theory. C_{ijkl} are the elastic stiffnesses of the materials which form a fourth-rank tensor. The second term results from the *converse* piezoelectric effect. The electric field is linked to strains through the piezoelectric constants e_{kij} which form a third-rank tensor.

Equation (9) can be simplified further by taking into account the hexagonal symmetry of III-V nitrides. Indeed tensors describing the intrinsic physical properties of such materials are left unchanged through symmetry operations of the point group C_{6v}. This way the number of independent elastic stiffnesses and piezoelectric constants is reduced to 5 and 3, respectively [20]. Consequently stress is given by:

$$\sigma_i = C_{ij}\epsilon_j - e_{ik}F_k \tag{10}$$

where the Voigt notation ($i = 1\ldots6$ and $j = 1\ldots6$) is used. In (10), the Einstein convention is used, *i.e.* repeated indices imply summation. In matrix notation we thus obtain:

$$\begin{pmatrix} \sigma_1 \\ \sigma_2 \\ \sigma_3 \\ \sigma_4 \\ \sigma_5 \\ \sigma_6 \end{pmatrix} = \begin{pmatrix} C_{11} & C_{12} & C_{13} & 0 & 0 & 0 \\ C_{12} & C_{11} & C_{13} & 0 & 0 & 0 \\ C_{13} & C_{13} & C_{33} & 0 & 0 & 0 \\ 0 & 0 & 0 & C_{44} & 0 & 0 \\ 0 & 0 & 0 & 0 & C_{44} & 0 \\ 0 & 0 & 0 & 0 & 0 & \frac{C_{11}-C_{12}}{2} \end{pmatrix} \cdot \begin{pmatrix} \epsilon_1 \\ \epsilon_2 \\ \epsilon_3 \\ \epsilon_4 \\ \epsilon_5 \\ \epsilon_6 \end{pmatrix} - \begin{pmatrix} 0 & 0 & e_{31} \\ 0 & 0 & e_{31} \\ 0 & 0 & e_{33} \\ 0 & e_{15} & 0 \\ e_{15} & 0 & 0 \\ 0 & 0 & 0 \end{pmatrix} \cdot \begin{pmatrix} F_1 \\ F_2 \\ F_3 \end{pmatrix}. \tag{11}$$

In a planar QW where the z axis is oriented along the growth direction, the boundary conditions are such that this heterostructure undergoes zero stress in the z direction, zero shear stresses and strains, and it has in plane symmetry of x and y directions. Consequently,

$$\begin{aligned} &\sigma_1 = \sigma_2, \\ &\sigma_3 = 0, \\ &\sigma_4 = \sigma_5 = \sigma_6 = 0, \\ &\epsilon_1 = \epsilon_2 = \epsilon, \\ &\epsilon_4 = \epsilon_5 = \epsilon_6 = 0. \end{aligned} \tag{12}$$

ϵ is the basal strain between the in plane lattice parameter of the well material and that of the barrier, which is assumed to be fully relaxed. From equations (10) to (12), we get:

$$\sigma_3 = 0 = 2C_{13}\epsilon + C_{33}\epsilon_3 - e_{33}F_3. \tag{13}$$

Then, the expression of the electrostatic displacement $\boldsymbol{D}$ writes:

$$D_m = \varepsilon_{mk}F_k + P_m, \tag{14}$$

where ε_{mk} is the dielectric tensor, which is diagonal, and P_m is the total or *built-in* polarization. In the QW, the total polarization is a function of both spontaneous and piezoelectric components:

$$\boldsymbol{P} = \boldsymbol{P_{sp}} + \boldsymbol{P_{pz}}, \tag{15}$$

where the piezoelectric polarization is given by:

$$P_m^{pz} = e_{mk}\epsilon_k. \tag{16}$$

By substituting equations (15) and (16) in equation (14), we obtain for the z component:

$$D_3 = \varepsilon_{33}F_3 + 2e_{31}\epsilon + e_{33}\epsilon_3 + P_3^{sp}. \tag{17}$$

As $\boldsymbol{D}, \boldsymbol{F}$ and $\boldsymbol{P_{sp}}$ are entirely oriented in the z direction, the other components of this system vanish so that subscripts can be dropped for the sake of simplicity. Electrical neutrality of the medium, supposed undoped, ensures that:

$$\nabla \cdot \boldsymbol{D} = 0 \tag{18}$$

so that $\boldsymbol{D}$ must be constant throughout all layers.

Now from equations (13) and (17), we can extract the expressions for the internal electric field F and the strain ϵ_3:

$$F_{coup} = \frac{2\epsilon(C_{13}e_{33} - C_{33}e_{31}) + C_{33}(D - P_{sp})}{C_{33}\varepsilon_{33} + e_{33}^2}, \tag{19}$$

$$\epsilon_{3,coup} = \frac{-2\epsilon(C_{13}\varepsilon_{33} + e_{31}e_{33}) + e_{33}(D - P_{sp})}{C_{33}\varepsilon_{33} + e_{33}^2}, \tag{20}$$

which are presently the coupled equations for the internal electric field and the [0001] strain in single III-V nitride QWs grown on a (0001) substrate (*i.e.* it applies essentially to the case of QWs grown on c-plane sapphire substrates and free-standing GaN substrates).

Note that quite often in the literature, analyses of QW heterostructures ignore the *converse* piezoelectric effect, which leads to the more common uncoupled model. This is justified *a posteriori* knowing that the fields determined with the uncoupled model are only a few percent higher ($\sim$2%) than when using the coupled model

[28]. In this latter case, the last term of equation (10) is neglected, which leads to:

$$F_{uncoup} = \frac{2\epsilon(C_{13}e_{33} - C_{33}e_{31}) + C_{33}(D - P_{sp})}{C_{33}\varepsilon_{33}}, \tag{21}$$

$$\epsilon_{3,uncoup} = -2\epsilon\frac{C_{13}}{C_{33}}. \tag{22}$$

In the case of single (S) QWs, the electrostatic displacement throughout the system is equal to the spontaneous polarization in the barrier except near the surface [29]:

$$D = P_{sp,B}. \tag{23}$$

$P_{sp,B}$ can then be substituted for D in equations (19)–(21). The field present in a SQW can thus be rewritten as:

$$F_{QW} = \frac{(P_{sp,B} - P_{sp,W})}{\varepsilon_W} - \frac{P_{pz,W}}{\varepsilon_W} = F_{sp,W} + F_{pz,W}, \tag{24}$$

where the subscripts W and B stand for QW and barrier, respectively.

If the barrier material is strained e.g., on a GaN or an AlN epilayer, the piezopolarization $P_{pz,B}$ from the barrier must be taken into account. Therefore equation (24) becomes:

$$F_{QW} = \frac{(P_{sp,B} - P_{sp,W})}{\varepsilon_W} + \frac{(P_{pz,B} - P_{pz,W})}{\varepsilon_W} = F_{sp,W} + F'_{pz,W}. \tag{25}$$

The built-in electric field calculated within the coupled model of the most common types of III-V nitride SQWs that can be encountered in the available literature and in commercially available devices are illustrated in Figs. 7 to 10. Calculations

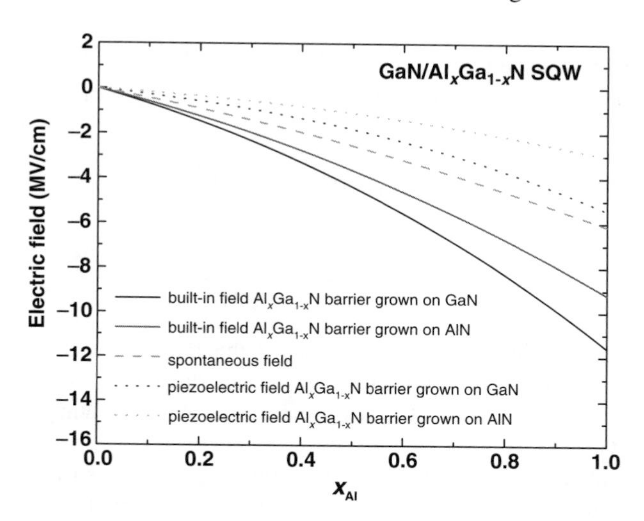

Fig. 7 Various components of the electric field in a single GaN/Al$_x$Ga$_{1-x}$N QW pseudomorphically grown either on a GaN or on an AlN epilayer *versus* Al molar fraction. Fields are in units of MV/cm with positive fields oriented along the [0001] direction.

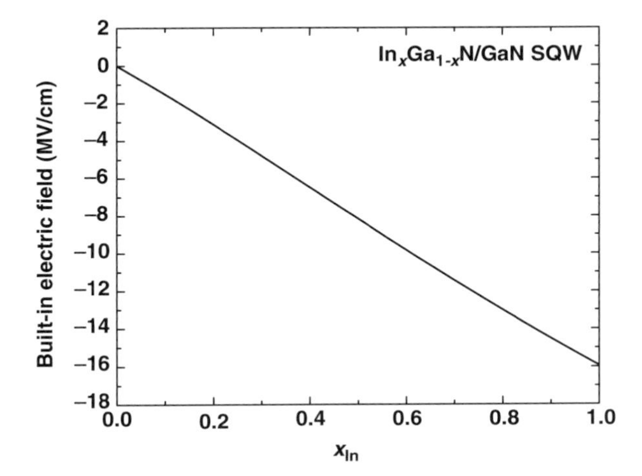

Fig. 8 Built-in electric field in a single $In_xGa_{1-x}N$/GaN QW pseudomorphically grown on a GaN epilayer *versus* In molar fraction. Fields are in units of MV/cm with positive fields oriented along the [0001] direction. As described in the text, for $In_xGa_{1-x}N$/GaN QWs the built-in electric field is mostly from piezoelectric origin.

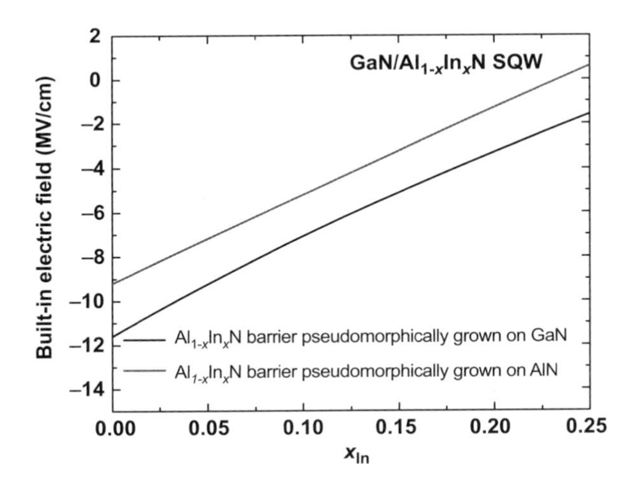

Fig. 9 Built-in electric field in a single GaN/$Al_{1-x}In_xN$ QW pseudomorphically grown either on a GaN or on an AlN epilayer *versus* In molar fraction. Fields are in units of MV/cm with positive fields oriented along the [0001] direction.

were performed taking into account the dependence of the spontaneous and the piezoelectric polarizations of III-V nitride alloys described in section 2.2.1.

When modeling multiple QWs (MQWs), the usual assumption is that the total potential drop should not exceed the bandgap [30]. A convenient approximation

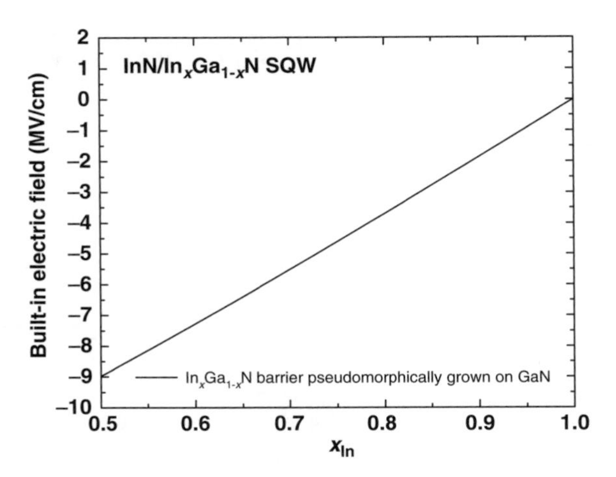

Fig. 10 Built-in electric field in a single $InN/In_xGa_{1-x}N$ QW pseudomorphically grown on a GaN epilayer *versus* In molar fraction. Fields are in units of MV/cm with positive fields oriented along the [0001] direction.

consists in applying the periodic boundary conditions [30]:

$$\sum_n l_n F_n = 0, \tag{26}$$

where the sum runs over all the layers, including the barrier layers and l_n is the thickness of layer n. In the framework of the uncoupled model, when equation (26) is combined with equation (14), the dielectric displacement throughout the sample, and consequently the total internal field in the j-th layer can be expressed as [30]:

$$F_{uncoup,j} = \frac{\sum_n l_n P_n/\varepsilon_n - P_j \sum_n l_n/\varepsilon_n}{\varepsilon_j \sum_n l_n/\varepsilon_n}. \tag{27}$$

As an illustration, we now show in Fig. 11 the calculated total internal polarization field *versus* Al and In molar fractions of an $Al_xIn_yGa_{1-x-y}N/GaN$ MQW system having active (l_W) and barrier (l_B) layers of the same thickness pseudomorphically grown on a GaN layer. Note that Fig. 11.a corresponds to the case where the GaN layers acts as the well material whereas Fig. 11.b corresponds to the case where the $Al_xIn_yGa_{1-x-y}N$ layers constitute the well material. In the specific case of identical barrier and well thicknesses, equation (27) then reduces to:

$$\begin{aligned} F_{uncoup,MQW} &= l_B \frac{(P_{sp,B} - P_{sp,W})}{(l_B\varepsilon_W + l_W\varepsilon_B)} + l_B \frac{(P_{pz,B} - P_{pz,W})}{(l_B\varepsilon_W + l_W\varepsilon_B)} \\ &= \frac{(P_{sp,B} - P_{sp,W}) + (P_{pz,B} - P_{pz,W})}{(\varepsilon_W + \varepsilon_B)}. \end{aligned} \tag{28}$$

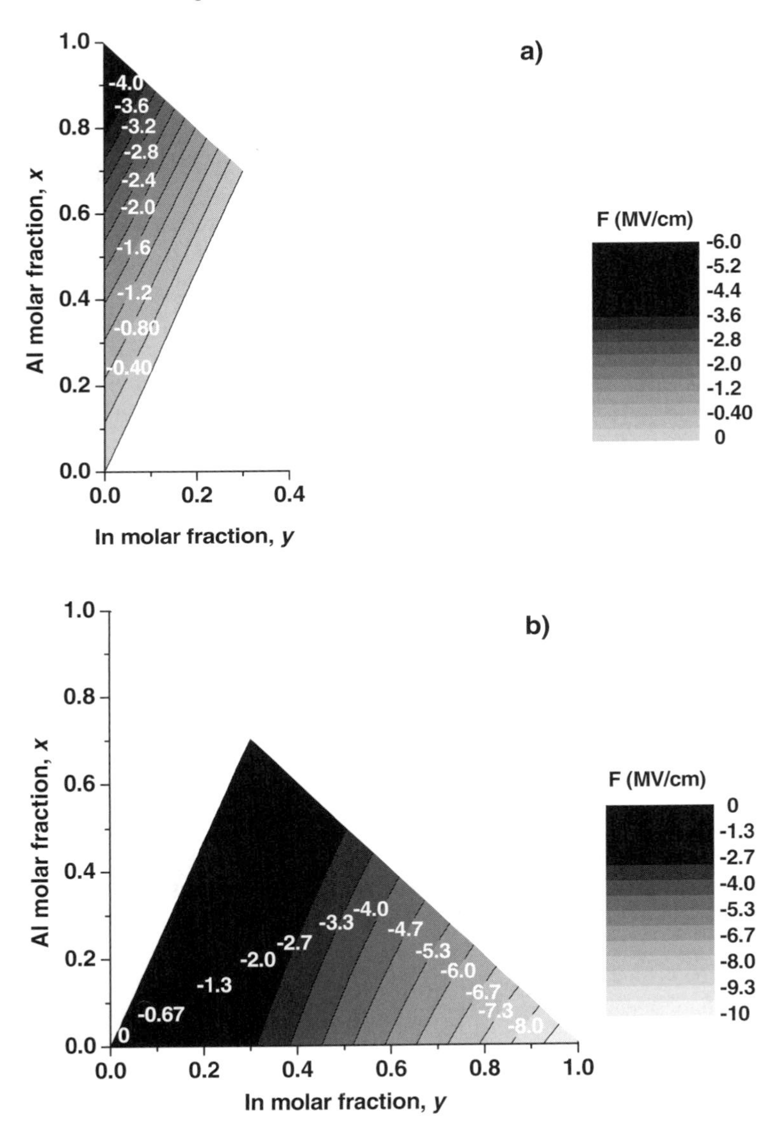

Fig. 11 Total internal electric field of an $Al_xIn_yGa_{1-x-y}N$/GaN MQW pseudomorphically grown on a GaN layer having active and barrier layers of the same thickness *versus* Al and In molar fractions. **a.** Case where the GaN layers acts as the well material. **b.** Case where the $Al_xIn_yGa_{1-x-y}N$ layers constitute the well material. Fields are in units of MV/cm with positive fields oriented along the [0001] direction.

From the previous analysis, it appears that polarization fields in III-V nitride heterostructures are not only the result of polarization differences but also of dielectric screening and geometrical factors [31, 32]. In this respect, the built-in electric field of the $Al_xIn_yGa_{1-x-y}N$/GaN MQW of Fig. 11 is about twice lower than in a single

QW structure made of the same couple of materials. In addition, it is worth pointing out that In and Al atoms have a different contribution to the built-in electric field. Indeed, it is seen that the higher the indium content, the larger the piezoelectric component. This relies for fairly obvious reasons on the large lattice mismatch between GaN and InN binary compounds which are also characterized by a small difference in spontaneous polarization. On the other hand, a rapid increase of the spontaneous component is observed with increasing Al content. In this latter case, this is due to the large difference existing between the spontaneous polarizations of AlN and GaN compounds. These two compounds having a reduced lattice mismatch compared to, e.g. GaN and InN, the contribution of the piezoelectric component is much less prominent.

Note that the coefficients e_{ij} are used to relate strain to the piezoelectric polarization (Eq. (16)). However, it is also possible making use of an alternative piezoelectric tensor expressed as a function of stress σ_j:

$$P_{pz,i} = d_{ij}\sigma_j, \tag{29}$$

where the d_{ij} are linked to e_{ij} through the relationship:

$$e_{ij} = d_{ik}C_{kj}. \tag{30}$$

This is of practical interest as d_{ij} coefficients are more often reported in the available literature than e_{ij} (Tab. 3).

Role of the internal parameter u

The formalism described previously does not make appear explicitly the stress-induced variation Δu of the internal parameter u. Note however that the latter is taken into account through the definition of the piezoelectric constants e_{ijk}, which

Table 3 Useful material parameters used to evaluate the internal electric field: elastic stiffnesses C_{ij}, relative dielectric constant ε_{33}, piezoelectric constants d_{ij}. Data are issued from the review paper of Vurgaftman and Meyer [10] except for the relative dielectric constants which are taken from Ref. [33].

Parameter	GaN	AlN	InN
C_{11} (GPa)	390	396	223
C_{12} (GPa)	145	137	115
C_{13} (GPa)	106	108	92
C_{33} (GPa)	398	373	224
C_{44} (GPa)	105	116	48
ε_{33}	10.28	10.31	14.61
d_{13} (pm/V)	−1.6	−2.1	−3.5
d_{33} (pm/V)	3.1	5.4	7.6
d_{15} (pm/V)	3.1	3.6	5.5

are essentially phenomenological parameters. The static equilibrium state of the strained crystal is thus a combination of both the external deformation of the lattice unit cell and the internal strain relaxation. The relation describing the minimization of the total energy of the system allows suppressing the internal parameter u of the formalism used.

To consider explicitly the internal deformation Δu, the contribution of the piezopolarization (16) to the total polarization (15) has to be rewritten as [15, 33]:

$$P_{pz} = 2e_{31}^{(0)} \cdot \epsilon_1 + e_{33}^{(0)} \cdot \epsilon_3 + \frac{eZ_B}{V} \cdot \Delta u, \quad (31)$$

where new piezoelectric coefficients, describing the response of the system to external deformations only, are introduced [15, 33]:

$$e_{31}^{(0)} = \left.\frac{\partial P_{pz}}{\partial \epsilon_1}\right|_u \quad \text{and} \quad e_{33}^{(0)} = \left.\frac{\partial P_{pz}}{\partial \epsilon_3}\right|_u . \quad (32)$$

The polarization resulting from the internal deformation Δu is expressed as a function of the electronic charge e, the Born effective charge Z_B associated to anion-cation pairs ($e.Z_B < 0$), the volume V of the unit cell. Δu being a function of external deformations ϵ_1 and ϵ_3:

$$\Delta u = 2\frac{\partial u}{\partial \epsilon_1} \cdot \epsilon_1 + \frac{\partial u}{\partial \epsilon_3} \cdot \epsilon_3, \quad (33)$$

we obtain by identifying (31) to (17), taking into account the various approximations introduced, the relationship between the piezoelectric coefficients $e_{ij}^{(0)}$ and e_{ij} [15]:

$$e_{31} = e_{31}^{(0)} + \frac{eZ_B}{V} \cdot \frac{\partial u}{\partial \epsilon_1} \quad \text{and} \quad e_{33} = e_{33}^{(0)} + \frac{eZ_B}{V} \cdot \frac{\partial u}{\partial \epsilon_3}. \quad (34)$$

The combination of equations (33) and (34) leads to:

$$\Delta u = \frac{V}{eZ_B} \cdot \left[2\left(e_{31} - e_{31}^{(0)}\right) \cdot \epsilon_1 + \left(e_{33} - e_{33}^{(0)}\right) \cdot \epsilon_3 \right]. \quad (35)$$

Calculated [11, 15] and measured [34] piezoelectric constants of III-V nitride binary compounds are reported in Table 4.

Table 4 Calculated [11, 15] and measured [34] piezoelectric constants of III-V nitride binary compounds.

Piezoelectric coefficients (C/m^2)	GaN			AlN			InN	
	[15]	[11]	[34]	[15]	[11]	[34]	[15]	[11]
$e_{31}^{(0)}$	0.45	–	–	0.36	–	–	0.45	–
e_{31}	−0.49	−0.34	−0.55	−0.60	−0.53	−0.60	−0.57	−0.41
$e_{33}^{(0)}$	−0.84	–	–	−0.47	–	–	−0.88	–
e_{33}	0.73	0.67	1.12	1.46	1.50	1.50	0.97	0.81

It has to be pointed out that the values of the piezoelectric coefficients of III-V nitride semiconductors are relatively high with values typically one order of magnitude larger than those of conventional III-V semiconductors (e.g., GaAs, InP) [11]. The other striking difference is the opposite sign of $e_{ij}^{(0)}$ and e_{ij}, which means that the contribution of internal deformations to the piezopolarization dominates over external deformations. This behavior is at the opposite of that typical of other III-V semiconductors. This is the signature of the larger ionicity of III-V nitrides making them closer in this respect to II-VI compounds. As a result, the internal deformation plays a significant role in establishing the equilibrium state of the deformed primitive unit cell.

2.3 *Effects of Polarization Fields on Optical Properties of III-V Nitride Quantum Heterostructures*

2.3.1 Case of Quantum Wells

Several approaches common to the field of quantum semiconductor heterostructures and in particular to the simpler case of planar QWs can be used to determine the energy of excitonic optical transitions. As described in the previous sections of this Chapter, one of the peculiarities of III-V nitride heterostructures grown along the c-axis resides in the presence of an electric field, oriented parallel to the growth axis, stemming from both spontaneous and piezoelectric origins, which will modify the shape of the electron and hole confinement potentials. Once a reasonable estimate of the built-in electric field is known, it is possible to solve the Schrödinger equation and thus to determine the wave functions and the position of the energy levels in the low density regime (*i.e.* the regime where optically or electrically injected free carriers do not screen the built-in electric field) [14, 35].

Thus in the framework of the envelope function formalism, the Schrödinger equation describing the relative motion of the exciton writes[1] [14]:

$$-\frac{\hbar^2}{2}\left[\frac{\partial}{\partial z_e}\left(\frac{1}{m_e^*}\frac{\partial}{\partial z_e}\right)+\frac{\partial}{\partial z_h}\left(\frac{1}{m_h^*}\frac{\partial}{\partial z_h}\right)\right]\phi_{\text{exc}}-\frac{\hbar^2}{2\mu_{\text{exc}}}\nabla_\rho^2\phi_{\text{exc}}$$
$$+\left[V_e(z_e)+V_h(z_h)\right]\phi_{\text{exc}}-\frac{e^2}{4\pi\varepsilon_0\varepsilon\sqrt{\rho^2+z^2}}\phi_{\text{exc}}=E_{\text{exc}}\phi_{\text{exc}}, \qquad (36)$$

where z is the space coordinate along the growth direction, m_e^* and m_h^* are the position dependent effective masses of the electron and hole, respectively, $V_e(z_e)$ and $V_h(z_h)$ are the confining potentials felt by electrons and holes, respectively, $\mu_{\text{exc}} = m_e^{-1} + m_h^{-1}$ is the reduced mass of the exciton, ρ is the relative in plane electron-hole distance, ϕ_{exc} and E_{exc} are the exciton wave function and exciton

[1] From the equation describing the motion of the center of mass of the exciton, it can be shown that the solution is identical to that of a plane wave having the kinetic energy of a free particle having the total mass of the exciton.

energy, respectively, ε_0 is the permittivity of vacuum and ε is the static dielectric constant. An efficient way to numerically solve the Schrödinger equation (36) consists in using either the three-point finite difference scheme or the matrix transfer method applied first to the case of a single particle (*i.e.* the situation without Coulomb interaction). A numerical approach based on a variational calculation can then be used to solve the full problem after decomposing ϕ_{exc} as the trial function:

$$\phi_{\text{exc}}(\rho, z_e, z_h) = f_e(z_e) f_h(z_h) \varphi_{\text{exc}}^{2\text{d}}(\rho), \tag{37}$$

where $f_e(z_e)$ and $f_h(z_h)$ are the electron and hole envelope wave functions, respectively, one of them having been determined in the single particle case, and $\varphi_{\text{exc}}^{2\text{d}}(\rho)$ is the ansatz $\varphi_{\text{exc}}^{2\text{d}}(\rho) = \frac{2}{a_B^* \sqrt{2\pi}} e^{-\frac{\rho}{a_B^*}}$ where a_B^* is the two dimensional exciton Bohr radius [14,36]. Such numerical methods are described in standard textbooks [35].

A more realistic description of the effects of polarization fields in quantum heterostructures requires a self-consistent approach in order to reproduce field screening by injected free carriers created either under optical or electrical pumping. Calculations are generally performed by solving self-consistently the Poisson's equation and the Schrödinger equation [31,37,38]. The symmetry of the problem allows reducing it to a one dimensional problem along the growth axis which is the polar axis. The 1D Poisson's equation thus writes:

$$\frac{d}{dz}\left[\varepsilon_0 \varepsilon \frac{d}{dz} \varphi_H(z)\right] = -e\left[N_D^+(z) - N_A^-(z) - n_{3D}(z) + p_{3D}(z) - n_{2D,i}(z)\right], \tag{38}$$

where $\varphi_H(z)$ is the Hartree potential of the electrostatic interaction due to mobile and immobile charges distributed in the system, $N_D^+(z)$ is the density of ionized donors, $N_A^-(z)$ is the density of ionized acceptors, $n_{3D}(z)$ and $p_{3D}(z)$ are the density of free electrons and free holes, respectively, $n_{2D,i}(z)$ is the two dimensional density of immobile polarization charges at the interface i (*i.e.* the difference of polarization to the right and to the left of the interface i). Once the Poisson's equation is solved, we get new confining potentials, the initial ones being corrected by the Hartree potential. The Schrödinger equation and the Poisson's equation are then iteratively solved until self-consistency is reached. The latter is achieved when successive iteration steps fall within a certain tolerance.

Note that the abovementioned analysis needs to be refined when aiming at describing effects involving higher lying energy levels within a given band. This is especially the case when an accurate description of intersubband transitions is required. An appropriate modeling should at least take into account the band nonparabolicity. This is usually done by considering an energy dependent effective mass. More subtle corrections can be obtained when taking into account the exchange interaction (Hartree-Fock approximation) or many-body corrections which become important at large carrier concentrations, namely the depolarization shift and the excitonic interaction [39].

The impact of the built-in electric field on the fundamental transition energy $E_{e_1-hh_1}$ in a single $In_{0.17}Ga_{0.83}N/GaN$ QW of 3 nanometers in the low density

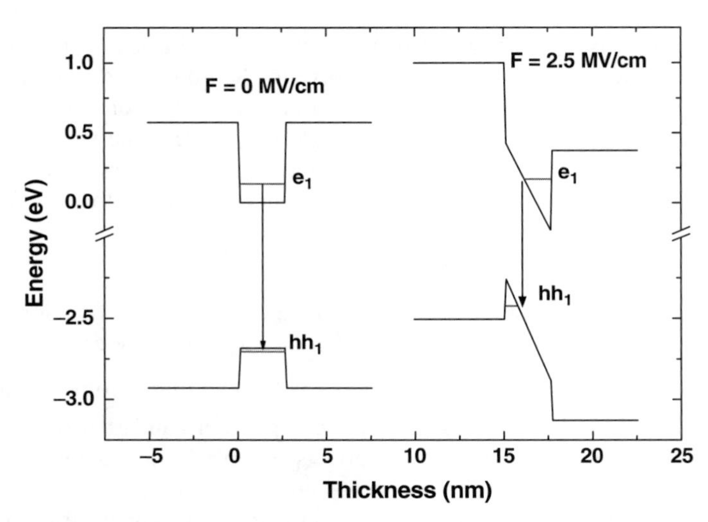

Fig. 12 Schematic of the band structure of a single $In_{0.17}Ga_{0.83}N$/GaN QW of 3 nanometers (∼12 monolayers (ML)) with and without built-in electric field.

regime, *i.e.* in the absence of field screening effects, is illustrated in Fig. 12. As a first approximation, the fundamental transition energy in such a heterostructure writes:

$$E_{e_1-hh_1} = E_{g,W} + e_1 + hh_1 - E_B - eF_W l_W \tag{39}$$

where $E_{g,W}$ is the bandgap of the well material,[2] e_1 and hh_1 are the confinement energies of electron and hole, respectively, E_B is the exciton binding energy, e is the electronic charge, F_W is the electric field in the QW, and l_W is the QW thickness. The last term of equation (39) describes the QCSE, which redshifts the transition energies in quantum heterostructures [14,17–19,40,41]. When the QCSE dominates over confinement effects (usually in QWs thinner than 2 nm, the effect of the built-in electric field is relatively weak [32]), the fundamental transition energy $E_{e_1-hh_1}$ of the quantum heterostructure of interest becomes smaller than the bandgap of the bulk material constituting the well. From the term $-eF_W l_W$, it appears that this redshift of the transition energy will be stronger in large wells experiencing large electric field values. As will be seen in the experimental section of this Chapter, it also explains why low energies can be reached in InGaN/GaN QWs with a rather low indium content.

Once electron and hole wave functions are known, it is possible to determine useful quantities such as the oscillator strength of an interband optical transition or its corresponding radiative lifetime. The former, at the Γ point of the Brillouin zone, writes [14]:

[2] It is worth pointing out that the bandgap of the well material to consider generally differs from that of a relaxed epilayer made of the same material. Thus the QW bandgap should be that taking into account the strain state of the layer. This is especially true for GaN QDs which are pseudomorphically strained on AlN layers where $E_{g,GaN} \sim 3.6\,\text{eV}$.

$$f_{osc} = \frac{2}{m_e \hbar \omega_{cv}} |d_{cv}|^2 \tag{40}$$

where $d_{cv} = \langle c |\boldsymbol{pe}| v\rangle$ is the dipolar matrix element with $|c\rangle$ and $|v\rangle$ the final and initial states of the optical transition of interest, $\boldsymbol{p}$ is the electron momentum operator and $\boldsymbol{e}$ is the unitary polarization vector, m_e is the free electron mass and $\hbar\omega_{cv}$ is the energy of the interband transition (usually the transition e_1-hh_1 in type I heterostructures, *i.e.* that between the first electron QW state (e_1) and the first heavy hole QW state (hh_1)). In a bulk crystal, $|c\rangle$ and $|v\rangle$ are the Bloch waves. In a QW, Bloch waves are modulated by envelope functions [14], so that in quantum systems, the following approximation is often made:

$$|d_{cv}|^2 = |\langle c|\boldsymbol{pe}\,|v\rangle|^2 \approx |\langle u_c|\boldsymbol{pe}\,|u_v\rangle|^2 \left| \int f_e(z) f_h(z) dz \right|^2 \tag{41}$$

where u_c and u_v are the periodic parts of the Bloch waves associated to the conduction band and the valence band, respectively, and $|\int f_e(z) f_h(z)dz|^2$ is the square modulus of the overlap integral between electron and hole wave functions. Finally we get:

$$f_{osc} = \frac{2}{m_e \hbar \omega_{cv}} |\langle u_c|\boldsymbol{pe}\,|u_v\rangle|^2 \left| \int f_e(z) f_h(z) dz \right|^2 . \tag{42}$$

It is thus seen that the interband oscillator strength, assimilated as a first approximation to that of excitons, is proportional to the square of the overlap integral between electron and hole envelope wave functions. Such a quantity can be easily computed when using the abovementioned methods.

The impact of the built-in electric field on the overlap between electron and hole envelope wave functions of the quantum well described in Fig. 12 is illustrated in Fig. 13. It is seen that concomitantly to the redshift of the fundamental transition energy $E_{e_1-hh_1}$; a spatial separation of electron and hole envelope wave functions

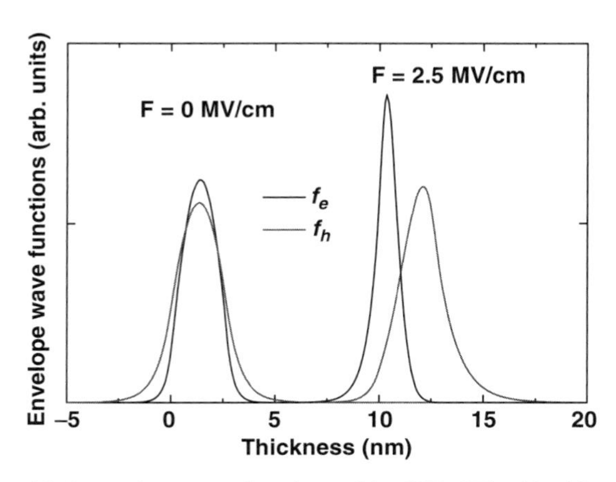

Fig. 13 Electron and hole envelope wave functions of the QW of Fig. 12 with and without built-in electric field.

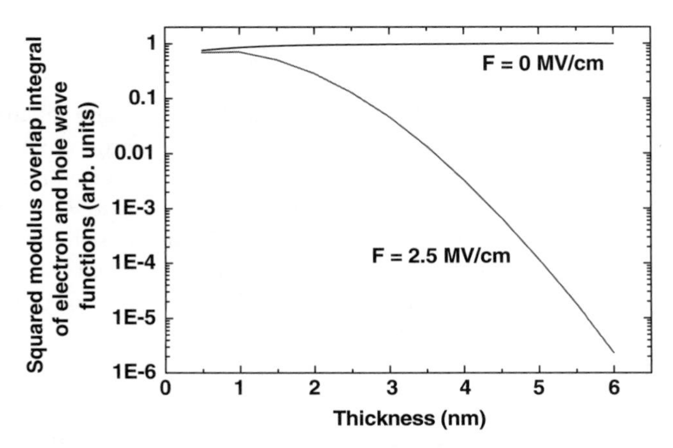

Fig. 14 Evolution of the squared modulus of the overlap integral between electron and hole envelope wave functions of a single $In_{0.17}Ga_{0.83}N$/GaN QW with increasing thickness when considering or not the impact of the built-in electric field.

occurs. This is due to the triangular shape of the confining potentials localizing the hole wave function on the substrate side and the electron wave function on the surface side, respectively. As a consequence, the stronger the QCSE, the lower the oscillator strength of the corresponding optical transition will be. As a further illustration of the QCSE, the evolution of the squared modulus of the overlap integral between electron and hole wave functions of a single $In_{0.17}Ga_{0.83}N$/GaN QW with increasing thickness in the low density regime is reported in Fig. 14.

The radiative lifetime τ_{rad} of the optical transition of interest is given, as a first approximation, by [42, 43]:

$$\tau_{rad} = \frac{2\pi\varepsilon_0 m_e c^3}{ne^2\omega_{cv}^2 f_{osc}} \tag{43}$$

where ε_0, m_e, e are the fundamental physical constants described previously, c is the speed of light in vacuum, n is the optical refractive index, ω_{cv} is the pulsation associated to the optical transition and f_{osc} is the corresponding excitonic oscillator strength given in equation (42). As can be seen from equation (43), a further consequence of the QCSE is an expected increase of the radiative lifetime. Indeed τ_{rad} will lengthen with increasing well width or QD height not only due to the decrease of f_{osc} but also due to the smaller transition energy. As a consequence, a nearly exponential increase of τ_{rad} can even be observed with increasing sizes for QWs/QDs above a certain width (height) [44–46].

Other effects arising from the QCSE are the occurrence of a Stokes shift and a decrease of the exciton binding energy, respectively. The Stokes shift relates here to the energy difference between the absorption edge and the bandedge PL energy of quantum heterostructures. Its precise origin is still an intense research subject in the field of III-V nitride quantum heterostructures, especially in the case of InGaN/GaN

QWs for which localizations effects come also into play. Nonetheless, Berkowicz *et al.* [44] and Lefebvre *et al.* [47] have unambiguously shown that for a given indium concentration, InGaN/GaN QWs exhibit a Stokes shift increasing with increasing well thickness (dot height), a behavior in agreement with that first observed more than two decades ago in GaAs quantum wells subject to an electric field applied perpendicular to the layers [40, 41]. The reason for such an effect has to be found in the progressive decrease in the oscillator strength of the fundamental transition energy with increasing thickness/height due to the decreasing overlap of electron and hole envelope wave functions in the triangular part of the confining potentials. On the other hand, higher energy levels which lie in the square part of the confining potentials are not so strongly affected by the built-in electric field, so that the corresponding electron and hole envelope wave functions exhibit a stronger absorption strength.

A decrease of the exciton binding energy with increasing well width (quantum dot height) is a further signature of the QCSE. This can be qualitatively understood by the decrease in the overlap of electron and hole envelope wave functions when increasing the well width, *i.e.* an increase of the two dimensional exciton Bohr radius [41, 48]. In other words, above a certain width the impact of the built-in electric field dominates over quantum confinement effects, which leads to a progressive separation of bound electron-hole pairs by the electric field *screening* the Coulomb interaction. As a consequence in wide III-V nitride quantum heterostructures, the exciton binding energy will become smaller than in their bulk counterpart leading to a disappearance of excitonic effects at a lower temperature/carrier density. Exciton binding energies have been calculated in GaN/ AlGaN QWs with an Al composition of $\sim$30%. It reaches a maximum of 55–65 meV for a well thickness of 5 MLs and rapidly decreases down to 10–15 meV for wide wells ($>$35 MLs) [49, 50].

Another effect of importance regarding the optoelectronic properties of III-V nitride heterostructures, though not directly related to the presence of a built-in electric field, resides in the reduced valence-band offset at GaN/InGaN heterojunctions [51, 52]. Indeed first-principles calculations predict a *natural* valence-band offset, *i.e.* the valence-band offset between unstrained materials, as small as 0.3 eV between unstrained InN and GaN [51]. The latter was shown to result from the *common-anion rule*, which does not hold anymore in the case of AlN/GaN heterojunctions for which a larger valence-band offset of 0.7 eV is calculated [51]. Van de Walle and Neugebauer have thus predicted an offset smaller than 0.1 eV for $In_{0.2}Ga_{0.8}N$/GaN QW heterostructures where the $In_{0.2}Ga_{0.8}N$ layer is compressively strained on GaN [51]. This aspect is of high practical interest since many III-V nitride-based optoelectronic devices are based on this type of QWs. Thus despite a large bandgap difference between III-V nitride binary compounds, holes will remain relatively weakly confined compared to electrons in heterostructures. The proximity of the valence-band continuum should thus act in a way such that thermal escape phenomena and coupling to higher valence band states should play a significant role in devices operating at room temperature and above.

Besides, growth related effects can also modify noticeably the emission properties of heterostructures. This is the case when e.g., surface segregation phenomena come into play. The resulting profile of the potential well will then depart from a structure having sharp interfaces as it usually leads to a triangular-like profile of the alloy composition. Such an effect can be reproduced by a phenomenological model accounting for the compositional profile of a segregated QW by means of a segregation coefficient R describing the exchange rate of a given atomic species between the nth monolayer and the $n+1^{\text{th}}$ monolayer [53]. In this model, the composition of the atomic species of interest in the nth monolayer is given by the formula [53]:

$$\begin{aligned} x_n &= x_0\left(1-R^n\right) && \text{in the quantum well with } 1 \leq n \leq N \\ \text{and } x_n &= x_0\left(1-R^N\right)R^{\mathrm{n-N}} && \text{in the barrier with } n > N \end{aligned} \tag{44}$$

where x_0 is the nominal composition of the segregating atomic species and N is the nominal thickness of the QW in MLs. This compositional change will then modify the built-in electric field in each ML. The electric field profile can be deduced by considering that the compositional difference between each layer of the QW will induce the presence of interfacial charges between each monolayer. The electric field of the nth monolayer, F_n, is thus the sum of these interfacial charges:

$$F_n = \frac{1}{2\varepsilon}\left(\sum_{i=1}^{n}(P_{i-1}-P_i) - \sum_{i=n+1}^{+\infty}(P_{i-1}-P_i)\right) \tag{45}$$

where ε is taken identical in the QW region and in the barriers.

As a consequence F_n writes:

$$F_n = (P_0 - P_n - P_n + P_\infty)/2\varepsilon \tag{46}$$

where P_0 and P_∞ correspond to the total polarizations of the barrier material on the substrate side and on the surface side, respectively. Provided that the barrier material is the same on each side of the well, we have $P_0 = P_\infty$, and thus equation (46) reduces to:

$$F_n = (P_0 - P_n)/\varepsilon. \tag{47}$$

As a consequence, the position of electron and hole energy levels as well as the overlap of their wave functions will be modified. For III-V nitrides, this situation is especially met in InGaN/GaN heterostructures where surface segregation of In occurs along the growth axis [54,55]. Such an effect will be further described in the experimental section of this Chapter.

2.3.2 Specific Case of Wurtzite Quantum Dots

So far we did consider only the case of planar QW heterostructures. However wurtzite QDs, essentially obtained *via* the Stranski-Krastanov (SK) growth mode, are receiving an increasing interest both on a fundamental point of view and an applied one [56–60]. Most of the studies focus on GaN/AlN QDs. This is a direct consequence of the reduced lattice-mismatch between AlN and GaN (2.4%), compared to the prototypical GaAs/InAs system (7.2%), making far less favorable the formation of AlGaN/GaN QDs. InGaN/AlGaN QDs are also studied [61–63] but work on this system remains more confidential.

The shape of wurtzite dots grown along the polar axis has been investigated through transmission electron microscopy, atomic force microscopy and reflection high energy electron diffraction (RHEED) measurements [64, 65]. These have shown that dots consist of a truncated hexagonal GaN pyramid defined by (0001) and (10–13) facets lying on a two-dimensional GaN wetting layer. Hexagonal GaN QDs are characterized by an angle of 30° for the inclined facets with respect to the c-plane with an aspect ratio between height and base diameter about 0.1–0.2 [64]. Several theoretical studies have been carried out in order to reproduce the electronic and optical properties of hexagonal GaN/AlN QDs [45,66–69]. All of them demonstrate that due to the built-in electric field, holes are localized in the wetting layer while electrons are pushed up to the top of the pyramid. This is in strong contrast to the situation occurring in self-assembled InAs/GaAs QDs where the hole is localized towards the top of the dot, above the electron [70, 71].

Interestingly enough, it has been shown that owing to large conduction- and valence-band offsets, energy levels in wurtzite GaN/AlN QDs can be determined analytically with a very satisfactory accuracy compared to experiment (in the low carrier density limit) assuming that electrons and holes are confined along the growth axis by a triangular well potential [69]. This is made possible since the confinement potential can be separated into a radial part and an axial part (*i.e.* along the c-axis). Besides, within this theoretical treatment the exact shape of the dots, namely a truncated hexagonal pyramid, can be assimilated to that of a truncated cone or even to a cylinder of radius r and height h for large dots or those with a small aspect ratio as it is usually the case for nitride QDs [68, 69]. Within this triangular well approximation, the confining axial potential $V(z)$ is linear for $z > 0$. The one-dimensional Schrödinger equation for the envelope wave function f_n within a band then writes [14, 72]:

$$-\frac{\hbar^2}{2m^*}\frac{d^2 f_n(z)}{dz^2} + eF_{QD} z f_n(z) = E_n f_n(z), \tag{48}$$

where F_{QD} is the built-in electric field within the dot, *i.e.* for $z > 0$. Equation (48) reduces to the well known Airy differential equation [73]:

$$\frac{d^2 f_n(\eta)}{d\eta^2} + \eta f_n(\eta) = 0 \tag{49}$$

when performing the change of variable:

$$z = \frac{1}{eF_{QD}} E_n - \left(\frac{\hbar^2}{2m^* eF_{QD}}\right)^{1/3} \eta. \tag{50}$$

Equation (49) can be solved exactly in terms of Airy function Ai(z) which leads to [67]:

$$f_n(z) = \mathcal{N} \times \exp\left(\sqrt{\beta (V_{BO} - E_n)} z\right), \quad z \leq 0, \tag{51a}$$

$$f_n(z) = \mathcal{N} \times \mathcal{C} \text{ Ai}\left[\left(\frac{\beta}{e^2 F_{QD}^2}\right)^{1/3} (eF_{QD} z - E_n)\right], z > 0, \tag{51b}$$

where

$$\mathcal{C} = \left(\text{Ai}\left[-\left(\frac{\beta}{e^2 F_{QD}^2}\right)^{1/3} E_n\right]\right)^{-1}, \tag{52}$$

$\beta = 2m^*/\hbar^2, V_{BO}$ is the GaN/AlN band offset, and $\mathcal{N}$ is a normalization constant. Note that equations (51a) and (51b) are solutions for the hole envelope wave functions. For the electron, the potential is inverted so that the substitution $z \rightarrow h - z$ has to be made. To obtain the axial confinement energies E_n, the boundary conditions have to be used which leads to the equation:

$$\text{Ai}\left(-\frac{\beta^{1/3} E_n}{(eF_{QD})^{2/3}}\right) = \frac{(eF_{QD})^{1/3}}{\beta^{1/6}\sqrt{V_{BO} - E_n}} \text{Ai}'\left(-\frac{\beta^{1/3} E_n}{(eF_{QD})^{2/3}}\right), \tag{53}$$

which has to be solved numerically. Nonetheless an analytical expression for the energies E_n can be obtained by considering that $V_{BO} \rightarrow \infty$, *i.e.* when applying the infinite barrier approximation which leads to:

$$E_n = -\frac{(eF_{QD})^{2/3}}{\beta^{1/3}} a_n, \tag{54}$$

where a_n is the $(n+1)^{\text{th}}$ zero of Ai(z). The latter can be determined to a very good approximation also for small values of n using the expression:

$$a_n \cong -\left[\frac{3\pi}{2}\left(n + \frac{3}{4}\right)\right]^{2/3}, \quad n = 0, 1, \ldots \tag{55}$$

so that equation (54) now writes:

$$E_n = \left(\frac{\hbar^2}{2m^*}\right)^{1/3} \left[\frac{3\pi e F_{QD}}{2}\left(n + \frac{3}{4}\right)\right]^{2/3}. \tag{56}$$

The fundamental transition energy in wurtzite GaN/AlN QDs grown along the polar c-axis then writes:

$$E_{fund,QD} = E_{g,QD} + E_0^e + E_0^h + E_{lat}^e + E_{lat}^h - E_{B,QD} - eF_{QD} h \tag{57}$$

where $E_{g,QD}$ is the bandgap of the QD material, E_0^e and E_0^h are the axial confinement energies of electron and hole, respectively, deduced from equation (56), E_{lat}^e and E_{lat}^h are the lateral confinement energies of electron and hole, respectively, and $E_{B,QD}$ is the exciton binding energy. E_{lat}^e and E_{lat}^h result from the in-plane field of about 1 MV/cm present in c-plane GaN/AlN QDs which provides an additional lateral confinement [69]. However, in the above analysis the impact of these lateral confinement energies E_{lat} can be neglected as their contribution is typically one to two orders of magnitude lower than E_n, which is also the case for $E_{B,QD}$ [69].

At this stage, it is worth pointing out that a rigorous description of c-plane GaN/AlN QDs implies considering a dependence of the built-in electric field F_{QD} different from that of a two-dimensional heterostructure. F_{QD} can be obtained from the analytical expression derived for the potential $\varphi(z)$ in the axial part of GaN/AlN dots by O'Reilly and coworkers in references [68, 69]:

$$\varphi(z) = JI_1 + \left(K + \frac{P_{spont,\mathrm{GaN}} - P_{spont,\mathrm{AlN}}}{4\pi\varepsilon\varepsilon_0}\right) I_2, \tag{58}$$

where:

$$J = \frac{-\epsilon_0 (1+\nu)(2e_{15} - e_{33} + e_{31})}{8\pi\varepsilon\varepsilon_0 (1-\nu)}, \tag{59}$$

$$K = \frac{\epsilon_0}{8\pi\varepsilon\varepsilon_0}\left(4e_{31} + 2e_{33} - \frac{1+\nu}{1-\nu}(2e_{15} + e_{31} + e_{33})\right), \tag{60}$$

with ϵ_0 the isotropic misfit strain and ν the Poisson's ratio, equal to $C_{12}/[2(C_{12} + C_{44})]$. I_1 and I_2 are analytical functions linear in z depending on the ratio $\zeta = h/r$:

$$I_1 \approx 4\pi\left(z - \frac{h}{2}\right)\left(-1 + \frac{2\zeta}{\sqrt{4+\zeta^2}} - \frac{\zeta^3}{(4+\zeta^2)^{3/2}}\right), \tag{61}$$

$$I_2 \approx 4\pi\left(z - \frac{h}{2}\right)\left(1 - \frac{\zeta}{\sqrt{4+\zeta^2}}\right). \tag{62}$$

F_{QD} will be simply given by the slope of $\varphi(z)$ around the point $z = h/2$. Note that the approximate expression of $\varphi(z)$ derived from equations (58) to (62) leads to very small errors ($<1\%$) for small values of ζ, a situation almost always fulfilled in GaN/AlN QDs. It can also be seen that the expression of $\varphi(z)$ involves the piezoelectric constant e_{15}, which does not play a role in the case of QWs, and for which many uncertainties still exist regarding its precise value.

Once transition energies and envelope wave functions are known, other useful quantities (radiative lifetime,...) can be derived in a way almost similar to that of QWs.

Finally we point out that another peculiarity of GaN/AlN QDs grown by plasma-MBE resides in the quasi-absence of intermixing [66, 74], that is interdiffusion of Al and Ga atoms, in strong contrast to InAs/GaAs QDs [70,71,75], which facilitates the previous theoretical analysis.

3 Experimental Manifestation of Polarization Fields in Group-III Nitride Based Nanostructures

3.1 Experimental Evidence of Polarization Fields

GaN/AlGaN QWs were first reported in 1990 [76]. At that time, the existence of a built-in polarization field was not mentioned. The QW transition energies measured by PL were slightly above the bandgap of bulk GaN, which was consistent with the expected low quantum confinement energies due to the large well thicknesses (10–30 nm) of the studied samples and heavy carrier effective masses of III-V nitrides. Hence the authors did not report any particular redshift compared to the GaN bandgap. Note that this is not surprising as screening of the internal electric field by free-carriers may occur for such wide wells. Actually, the observed blueshift of the QW luminescence with respect to the GaN buffer was probably related to strain effects induced by relaxed (or partially relaxed) AlGaN barriers. In 1995, MBE grown GaN QWs with a pretty good crystal quality were fabricated [77]. Their PL linewidth was less than 100 meV at room temperature. The samples were slightly Si-doped and the Al composition below 10%. As a consequence, the QCSE was still marginal as confirmed by the QW transition energy at 3.6 eV. At the same time, some theoretical papers dealt with the electronic properties of wurtzite GaN/AlGaN QWs to be used as active region in LDs but polarization effects were still ignored. One year later, the landmark paper of Bernardini, Fiorentini, and Vanderbilt was published [15]. In this publication, the authors first recalled that a strained wurtzite nitride based QW grown along the *c*-axis does experience a piezoelectric field because of the lattice-mismatch between the different binary compounds. Furthermore, they pointed out that nitride epilayers are also characterized by a spontaneous polarization field. In other words, a built-in electric field takes place in heterostructures even in the absence of strain. This was a rather new effect in the domain of semiconductor physics. Another important finding was that piezoelectric and spontaneous fields are very large and of the same order of magnitude, at least in the GaN/AlGaN system. For example, the value of the total polarization field calculated in [11] is as high as 10 MV/cm for the GaN/AlN material system. When considering GaN/AlGaN QWs with a typical Al composition ranging from 10 to 20%, this leads to an internal electric field of 1–2 MV/cm, which is almost one order of magnitude larger than those previously measured in InGaAs/GaAs QWs grown along the (111) direction. As a consequence, thick GaN based QWs should exhibit redshifted transition energies by several hundreds of meV due to a giant QCSE. Nitride based QWs should thus emit light at energies much lower than that of bulk GaN, provided that no screening of the internal electric field occurs.

Until 1998, there was no mention of QCSE in GaN/AlGaN QWs whereas a strong redshift of the luminescence energy of InGaN/GaN QWs was already reported [17, 78]. Then, Im *et al.* [18] studied high quality GaN/$Al_{0.15}Ga_{0.85}N$ QWs grown by MOVPE and reported sound evidence for QCSE: redshift of the

QW transition energy, PL emission below the bandgap of bulk GaN, and long carrier lifetime. The latter observation is one of the main fingerprints of QCSE as it is the direct consequence of the separation of electron and hole wave functions. Im and coworkers well accounted for the experimental data by considering a piezoelectric field of 350 kV/cm. In fact, the internal electric field they measured was resulting from both piezoelectric and spontaneous polarizations, even though it was not mentioned at that time. A few months later, Leroux *et al.* [19] first estimated the spontaneous polarization magnitude in III-V nitride compounds by carefully measuring the built-in electric field in high quality MBE grown GaN/AlGaN QWs. It is worth pointing out that from an experimental point of view it is difficult, nearly impossible, to discriminate spontaneous and piezoelectric polarizations because they both have the same physical origin.

In Fig. 15.a, the PL spectrum of a series of GaN/$Al_{0.17}Ga_{0.83}N$ QWs is reported. The well thickness varies from 1 nm to 8 nm. The corresponding PL peak energies are plotted in Fig. 15.b. It can be seen that the QW transition energy becomes lower than the GaN bandgap for a well thickness larger than 2 nm. Another observation is that it varies linearly for wide wells. This is of course one of the expected signatures of the QCSE: when the QW is thick enough, the quantum confinement energies are no longer dependent on the well width and the energy variation is only due to the Stark shift ($-eF_W l_W$). Thus the slope of the linear part of the PL energy provides a direct measurement of the internal electric field (0.78 MV/cm in the present case). Although it seems very easy to measure the electric field, an accurate determination is not straightforward. Actually, several parameters can lead to wrong values as discussed hereafter.

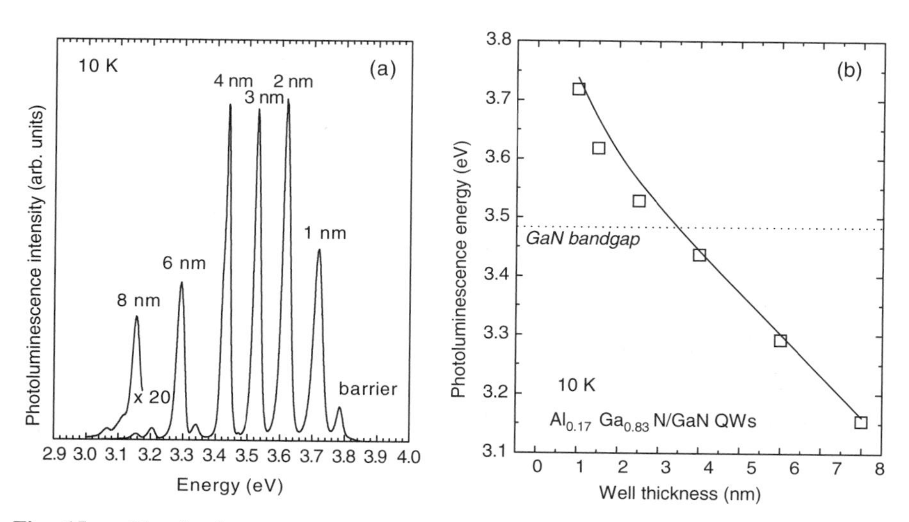

Fig. 15 a. Photoluminescence spectrum and **b.** transition energies of a series of GaN/$Al_{0.17}Ga_{0.83}N$ QWs.

3.2 Polarization Field Measurement

One has to keep in mind that PL data do not strictly correspond to e_1-hh_1 transitions. Indeed, luminescence is an extrinsic phenomenon which usually results from carriers localized in potential minima that are produced by impurities, deep levels, structural defects, alloy inhomogeneities, or QW interface roughness. In the latter case, well thickness and/or alloy fluctuations are the main factors responsible for carrier localization. For instance, let us consider a QW with significant thickness fluctuations: excitons are likely to be localized in the widest parts of the well, at least at low temperature. Consequently, if we ascribe the measured PL energy, coming from localized excitons, to free-exciton recombinations in a QW having an average well thickness deduced from transmission electron microscopy (TEM) or x-ray diffraction (XRD) experiments, we will overestimate the internal electric field. The localization energy, which corresponds to the energy difference between the emission lines of bound- and free-excitons can be estimated from temperature dependent experiments (Fig. 16.a). Let us consider that excitons are free at room temperature (note that this assumption is probably not valid for high Al content GaN/AlGaN QWs and for InGaN/GaN QWs), one can fit the data at high temperature by means of an empirical description such as the Varshni equation (Fig. 16.b). Then, the energy difference between the computed curve and the experimental data allows extracting the localization energy at low temperature [32]. In general, the localization energy does not affect dramatically the determination of the internal electric field when the Al content is low. This is no longer the case when the luminescence peak broadens, *i.e.* when the PL linewidth becomes larger than 50 meV. It is worth remembering

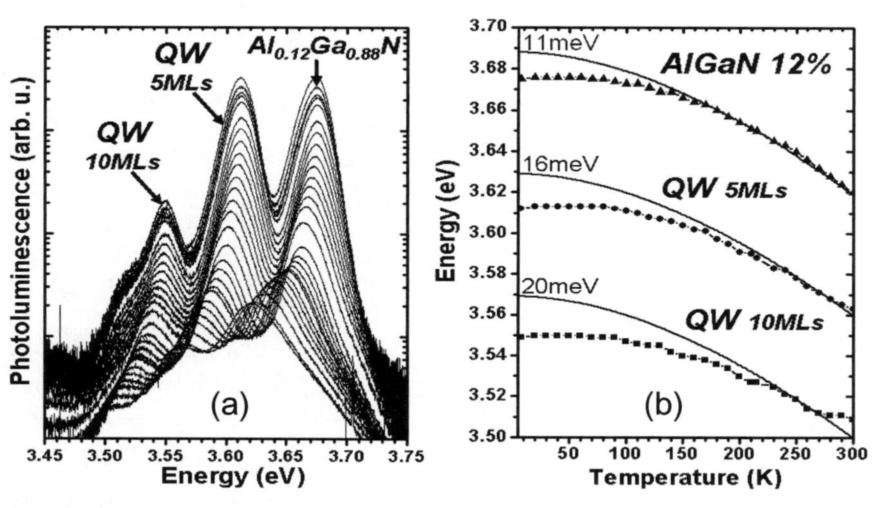

Fig. 16 a. Temperature dependent photoluminescence spectra of 5 ML and 10 ML thick GaN/$Al_{0.12}Ga_{0.88}N$ QWs and **b.** corresponding transition energies.

that PL excitation spectroscopy, reflectivity, or absorption give access to intrinsic transitions. Therefore, these spectroscopy tools are well suited to accurately determine the e_1-hh_1 transition energies (and upper excited levels) [44]. Conversely, these techniques are very sensitive to potential disorder leading to a critical broadening of the spectra. Furthermore, they cannot be used when the oscillator strength becomes low. Finally, PL experiments must be carefully carried out to avoid screening of the internal polarization field by photo-generated free-carriers [31, 79, 80]. Thus, very low power laser excitation densities must be used.

Another important factor that may influence the internal electric field is the crystal quality. The residual doping is a key factor influencing the magnitude of the internal polarization field. For *n*-type doping levels of a few $10^{18}\,\mathrm{cm}^{-3}$, the built-in electric field is partially screened [81]. Eventually, complete screening occurs at higher doping levels, typically in excess of $10^{19}\,\mathrm{cm}^{-3}$ [82]. The control of the residual doping is thus mandatory when aiming at extracting the intrinsic polarization field.

A fairly important structural parameter is the phenomenon of surface segregation. This effect has been previously reported in III-V arsenides [83–85], in particular in InGaAs alloys, but it is likely to occur in InGaN/GaN QWs as well [54]. It is worth noticing that this phenomenon is thermally activated and can then be strongly reduced by using low growth temperatures. The main consequence of surface segregation effects is to deeply modify the QW potential profile and therefore the transition energy as it produces a blueshift of the QW luminescence energy [86]. This blueshift would of course correspond to an apparent weaker internal electric field. However, the internal polarization field can still be determined even if surface segregation effects occur [86]. Actually, the QW transition energy is modified but its variation as a function of the well thickness remains unchanged, at least for wide wells. Finally, an important effect of surface segregation is the decrease of the oscillator strength of optical transitions because of the larger separation of electron and hole wave functions.

A last but tremendous difficulty when trying to determine the build-in electric field is obviously the accurate measurement of the well thickness. This can be done either *ex situ* by TEM and XRD experiments or *in situ* by monitoring RHEED intensity oscillations during the MBE growth process [87].

The existence of spontaneous polarization in group-III nitride based QWs is today well accepted. Combined with piezoelectric effect, it leads to giant internal electric fields of several MV/cm. The determination of the internal electric field has been mainly carried out *via* optical experiments [18, 19, 88–92]. Electron holography has also been proved to be successful [93]. The average value of the total polarization field for the GaN/AlN system is 5–7 MV/cm, which is slightly lower than theoretical predictions [11, 15]. Notice that a polarization field of 9–10 MV/cm was extracted from intersubband transition energies [39] and careful time-resolved PL experiments on GaN/AlN QDs [46]. Eventually, one has to mention that these values correspond to single heterostructures. Indeed, it has been shown both theoretically [30, 31] and experimentally [94] that the built-in electric field is reduced in MQWs compared to single QWs. Actually, the geometry of the sample is a key parameter. Thus, the presence of a 2D electron gas at the first AlGaN/GaN interface in a QW structure modifies the whole band bending and therefore the internal electric field [95].

3.3 Optical Properties of GaN/AlGaN Quantum Wells

Figure 17 shows the PL spectra of state of the art GaN/$Al_{0.22}Ga_{0.78}N$ QWs grown by MOVPE [96]. Two different growth temperatures have been investigated: standard growth conditions (with a high temperature (HT) of 1120°C) and low temperature (LT) growth conditions (930°C). The typical PL full-width at half-maximum (FWHM) is 31 meV for the LT growth conditions. The weak luminescence of the AlGaN barrier is visible on the high energy side. When the growth is performed at HT, both the QW and the barrier PL peaks broaden. Moreover, the PL intensity coming from the barrier increases probably because of photo-generated carrier trapping in alloy potential fluctuations.

What we learn from Fig. 17 is that the FWHM of the QW emission is mainly governed by the quality of the AlGaN barriers. This is exemplified in Fig. 18 where the micro-PL energy mappings ($50 \times 50\,\mu m^2$) of the two GaN/AlGaN QWs ($x = 0.22 - 0.24$) of Fig. 17 grown either at 1120°C (Fig. 18.a,c) or 930°C (Fig. 18.b,d) are compared. One can clearly see that the QW transition energy is correlated to the AlGaN barrier emission. Once the AlGaN barrier gets homogeneous, the QW luminescence energy peak remains constant leading to a sharp PL peak. This indicates

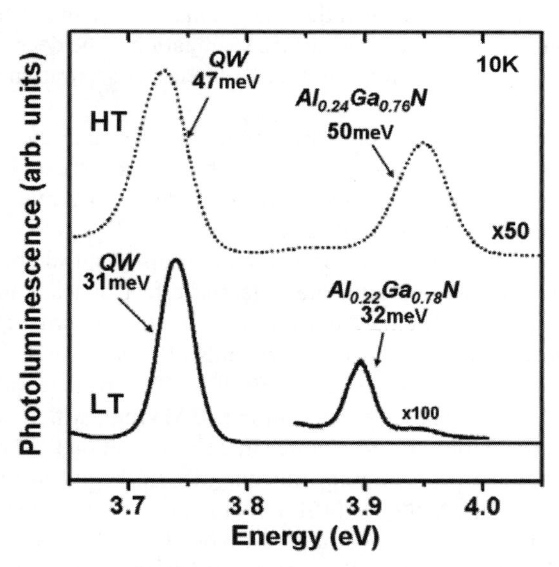

Fig. 17 Photoluminescence spectra of single GaN/AlGaN QWs grown at high and low temperatures.

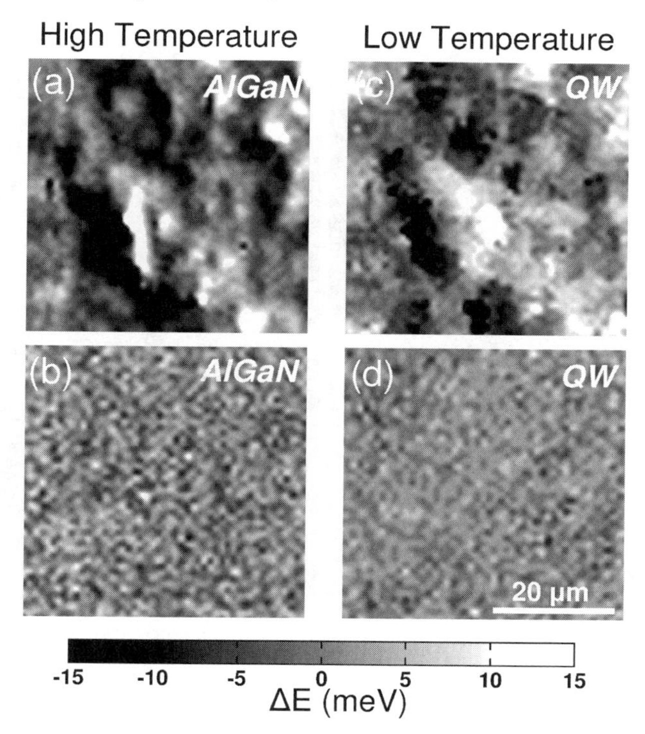

Fig. 18 Micro-photoluminescence mappings of single GaN/AlGaN QWs grown at low and high temperatures (same QWs as in Fig. 17). **a.** and **c.** correspond to the energy variation of the barrier and well PL peaks, respectively, when the growth is performed at HT. **b.** and **d.** correspond to the energy variation of the barrier and well PL peaks, respectively, when the growth is performed at LT.

that the QW transition energy is rather affected by AlGaN alloy disorder than by well thickness fluctuations.

At this stage of the discussion, one may wonder about the QW interface roughness, especially for low temperature growth conditions. Let us consider the GaN/AlGaN interface, *i.e.* the GaN surface morphology. Typical atomic force microscopy (AFM) images corresponding to GaN epilayers grown on *c*-plane sapphire substrate at high and low temperatures are displayed in Fig. 19. Although the overall surface morphology degrades at low temperature, it is always characterized by terraces indicating a step-flow growth mode, as previously observed in this temperature range by means of *in situ* measurements [97]. Thus, the surface morphology of GaN exhibits terraces whose length depends on the substrate/surface misorientation. The exciton Bohr radius being very small in nitride compounds, 3 nm in bulk GaN and probably less in QWs (at least for thin wells), excitons are marginally scattered by step-edges. The PL linewidth broadening will thus come mainly from composition fluctuation scattering due to alloy disorder in the barriers. This explains the PL mapping features in Fig. 18 showing that the QW energy is strongly correlated to the AlGaN barrier energy.

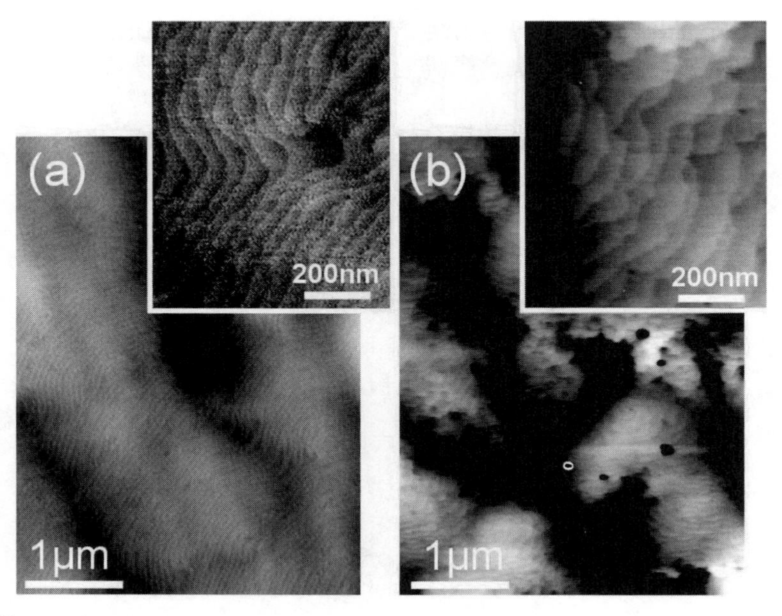

Fig. 19 Atomic force microscopy images of the GaN surface after growth carried out at **a.** high temperature and **b.** low temperature.

LT PL spectra (T = 8 K) of GaN/AlGaN QWs with a thickness of 10 MLs and an Al composition ranging between 5 and 30% are reported in Fig. 20. The first remark concerns the QW transition energy which is nearly independent of the Al composition in the AlGaN barrier. This peculiar feature appears only for a specific QW thickness of 10 MLs, as already reported [98,99]. This can be explained by the opposite variation of the quantum confinement energies and the QCSE with Al composition, 10 MLs corresponding to the perfect compensation of these two effects.

As shown in Fig. 20, the PL linewidth of MOVPE grown GaN/AlGaN QWs varies from 45 meV to 5 meV for Al contents between 30% and 5%. The value of 5 meV corresponds to the lowest value reported so far for nitride based QWs. Interestingly enough, when the excitation power becomes larger and larger a low energy band appears (Fig. 21.a). It is ascribed to biexciton recombinations as its integrated PL intensity varies quadratically with respect to the exciton integrated PL intensity (Fig. 21.b).

The presence of a large built-in electric field in GaN/AlGaN QWs has a profound impact on the dynamics of carrier recombinations and therefore on the QW efficiency. The radiative lifetime can be extracted from time-resolved PL experiments [18,44,79,80,100–102]. Thus radiative recombinations of excitons in 2.5 nm thick GaN/$Al_{0.07}Ga_{0.93}N$ MQWs exhibit an average decay time of 330 ps at 8 K [100]. Of course, the larger is the well (or the larger is the electric field), the longer is the radiative lifetime. This stems from the fact that, as a first approximation, the radiative lifetime is inversely proportional to the oscillator strength (cf. equation (43)), *i.e.* to

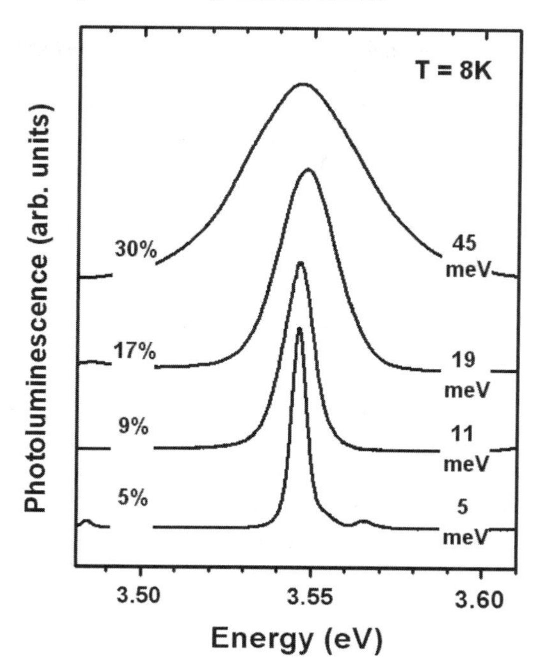

Fig. 20 Photoluminescence spectra of 10 ML thick single GaN/AlGaN QWs with different Al compositions in the barrier.

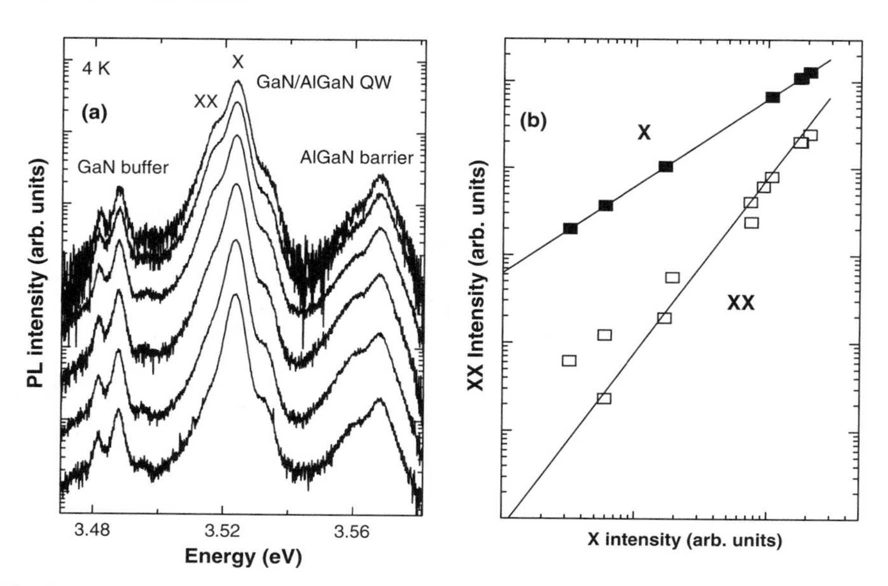

Fig. 21 **a.** Photoluminescence spectra of a single GaN/AlGaN QW ($x = 0.05$) for various excitation power densities **b.** Corresponding integrated PL intensity of exciton and biexciton lines, respectively.

the squared modulus of the overlap integral of electron and hole envelope wave functions. For $GaN/Al_{0.11}Ga_{0.89}N$ QWs, the 8 K PL decay time varies from 56 ps to 400 ps for well thicknesses varying from 3 MLs to 15 MLs [101]. These values can be well accounted for by considering an internal electric field of 450 kV/cm. The same behavior is observed for high Al compositions [89]. In the latter case, the lifetime at low temperature is of the order of a few ns for well thicknesses of 3–5 nm. A careful theoretical analysis of the decay time in GaN/AlGaN QWs has been performed [100]. These authors have shown that the PL emission is governed by charge accumulation and the loss of carriers from the QW ground state due to both radiative and nonradiative recombinations. The nonradiative lifetime at 8 K has been estimated to a few nanoseconds [79]. Temperature dependent measurements were carried out and a radiative lifetime of 2.4 ps in the low temperature limit for free-excitons was extracted for these QWs. This is much smaller than that found in GaAs based QWs confirming the larger oscillator strength of III-V nitrides with respect to III-As [100, 101]. The increase of the radiative lifetime for wide wells plays a key role on the QW efficiency. If one considers a constant nonradiative lifetime at a certain temperature, the increase of the quantum well thickness will automatically lead to a reduction of the oscillator strength and a larger sensitivity to nonradiative defects. As nitride heterostructures are generally affected by a huge density of threading dislocations ($>10^9\,cm^{-2}$ for heterostructures grown on c-plane sapphire substrates), thick GaN/AlGaN QWs are not well suited for opto-electronic applications. The evidence for carrier recombination on dislocations has been obtained by comparing the PL intensity evolution *versus* temperature for $GaN/Al_{0.10}Ga_{0.90}N$ QWs grown either on sapphire substrates (dislocation density of $5 \times 10^9\,cm^{-2}$) or on bulk GaN substrates (dislocation density less than $10^4\,cm^{-2}$) [102]. However, it has been found that the efficiency of high Al content GaN/AlGaN QWs, and *a fortiori* GaN/AlN QWs, is higher than that measured in low Al content QWs. This can be explained by invoking carrier localization induced by the QCSE. At first sight, this appears in contradiction with the previous discussion and rather counterintuitive. Actually, when the Al content is high, the QW transition energy varies very rapidly with the well thickness. For instance a 1 ML thickness variation in a GaN/AlN QW would induce in theory a redshift of at least 250 meV because of the QCSE. This means that well thickness fluctuations, which cannot be completely suppressed, provide efficient localization centers for carriers. As a consequence, the internal efficiency increases since the carrier diffusion toward dislocations is hindered, even at room temperature. In conclusion, GaN/AlN QWs, and to some extent high Al content QWs, can no longer be considered as true QWs but rather as QDs with in-plane localization of excitons due to well thickness fluctuations.

3.4 Optical Properties of GaN/AlN Quantum Dots

An attractive way to improve the internal quantum efficiency in highly defective materials has been proposed by Weisbuch and Nagle in 1990 [103] and experimentally demonstrated by Gérard *et al.* a few years later [104]. This approach relies on

QDs: once captured by QDs, carriers are strongly localized and can no longer diffuse toward nonradiative recombination centers. By analogy, it was then proposed that the high efficiency of InGaN based LEDs is the consequence of the formation of QD-like In-rich clusters induced by InGaN alloy phase segregation [105]. Even though this explanation is currently controversial [106–109], it has been the driving force for the study of self-assembled nitride based QDs. With this aim in mind, MBE is a well suited growth technique as it has been proved in the past to allow forming QDs by the so-called SK growth mode regime. This has been shown in the Ge/Si [110] and InAs/GaAs systems [111]. GaN QDs embedded either in an AlGaN [112] or in an AlN [57, 58, 113–116] matrix have been demonstrated. The island density depends on the growth technique. It can be larger than $10^{11}\,\mathrm{cm}^{-2}$ for MBE growth [57,58] and less than $10^{9}\,\mathrm{cm}^{-2}$ in the case of MOVPE [114, 116]. Typical GaN dots self-assembled on AlN by MOVPE are displayed in Fig. 22.a. Despite the huge density of dislocations usually present in AlN templates ($\sim 10^{10}\,\mathrm{cm}^{-2}$) and the large built-in electric field, an intense RT luminescence is observed, even in the visible range (Fig. 22.b). This well demonstrates the role of carrier localization to improve the efficiency in defective layers. The internal electric field has been first measured in GaN/AlN QDs grown by plasma-MBE [113]. A built-in electric field of 5.5 MV/cm was then extracted. Recently, careful time-resolved PL experiments have led to a value of 9 MV/cm [46]. This large polarization field is responsible for a huge redshift of the transition energy giving rise to light emission covering the whole visible range. Thus the PL energy can be continuously varied from blue to red by simply controlling the QD size [58]. In Fig. 22.b, a typical PL spectrum of GaN/AlN QDs is displayed. One can clearly see that the dots emit at energies much lower than the bulk GaN bandgap. The luminescence is broad and modulated by interference. It is worth recalling that such a large inhomogeneous broadening arises from the combined effects of dot height dispersion and QCSE. The emission of the wetting layer (WL) is located at high energy (>4 eV), which corresponds to a QW of 3–4 MLs. Such a WL is typical of the SK growth mode transition. The electronic

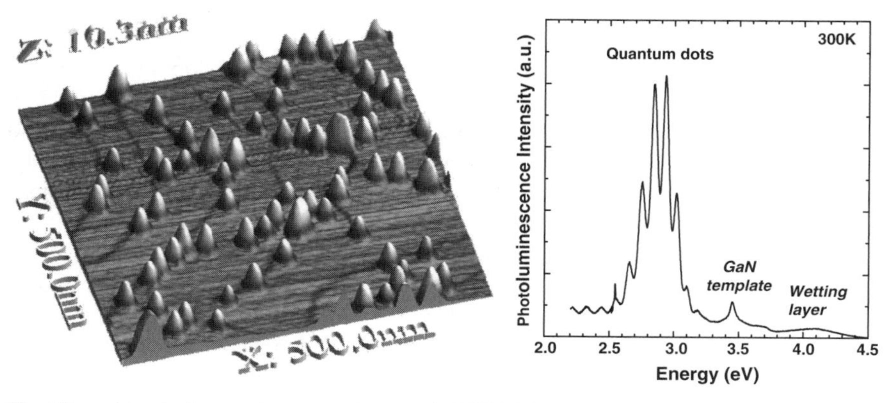

Fig. 22 a. Atomic force microscopy image of GaN/AlN quantum dots grown by MOVPE and **b.** corresponding PL spectrum measured at room temperature.

properties of GaN/AlN QDs have been extensively calculated [66, 67]. The effect of the internal electric field is similar to what happens in QWs primarily due to the low aspect ratio of the dots (0.15–0.2). Recently, single dot spectroscopy has been successfully carried out [117]. One of the main characteristics of small GaN QDs is the binding energy of the biexciton which is negative and large enough to allow single photon emission up to 200 K [60].

3.5 InGaN/GaN Quantum Wells: The Heart of Nitride Based Optoelectronic Devices

The existence of a large polarization field in GaN was pointed out in a few papers in the mid 90's [118, 119] but QCSE in InGaN/GaN QWs was unambiguously reported by Akasaki and coworkers in 1997 [78]. This might appear rather surprising as blue LEDs were already marketed a couple of years earlier [6]. Actually, one has to say that this is a common situation in the III-V nitride research area where device application is the main driving force, leaving fundamental physics issues a few steps behind the technology state of the art.

The QCSE in InGaN/GaN QWs was mainly experienced through the redshift of the QW transition energy as a function of well width [78]. Thus the polarization field was generally deduced from PL experiments. As already mentioned for GaN/AlGaN QWs, PL gives access to extrinsic transitions and carrier localization might affect the determination of the internal electric field. However, a reasonable estimate can be made if one considers constant localization energy whatever the well width. Also, it is important to point out that most of the samples studied in the literature were not single QWs but MQWs for which the internal electric field is reduced due to geometrical effects [30, 31]. In Fig. 23, electric field values extracted from the literature are reported as a function of the In composition. They correspond to the case of a single InGaN/GaN QW [44, 47, 120–125] and they have been corrected to account for the geometrical effect when required. It is interesting to notice that the dispersion is not so large once having considered the different parameters coming into play: the growth techniques, the different sample structures, the In composition and well thickness determination accuracy, the residual doping, and finally, the In composition fluctuations as well as the In surface segregation [54, 86, 126]. The reason for such a robust determination is quite simple: the electric field is so large that its related QCSE effect dominates over all the other factors. Eventually, one can remark in Fig. 23 that the data, except one, are slightly lower than those theoretically calculated [11, 15].

The existence of a large internal electric field is of course detrimental for the internal quantum efficiency because of the decrease of the oscillator strength of optical transitions. Actually, this could be limited by using a small well width, typically less than 2 nm, to warrant a sufficient overlap between electron and hole wave functions. The problem is then to achieve QWs emitting at long wavelengths. In principle, this might be possible by simply increasing the In content in the QWs. Unfortunately, the maximum In composition that can be reached while sustaining decent

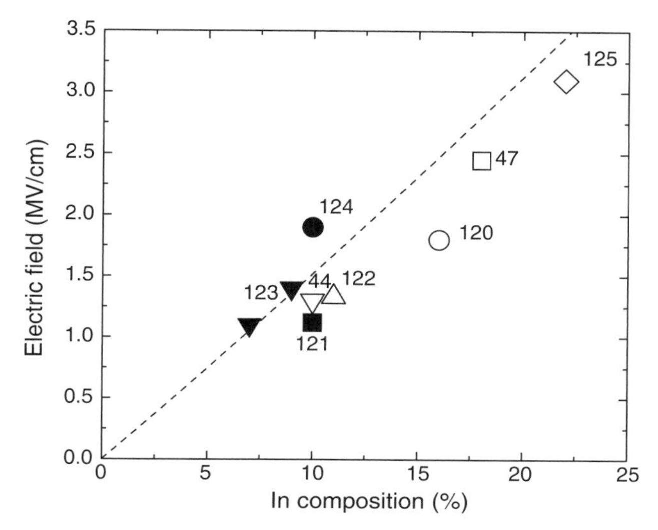

Fig. 23 Electric field values found in the literature for InGaN/GaN QWs having different In compositions. The dashed line corresponds to the theoretical value.

optoelectronic properties is typically 20–25% [127]. This is likely due to the large lattice mismatch between GaN and InN ($\sim$11%), which leads to a very small critical thickness before reaching plastic relaxation [128]. Combining thin well width and high In content allows achieving internal quantum efficiencies of $\sim$40% [129]. However, this limitation in the maximum In composition implies that light emission in the yellow-red range is not possible with thin QWs. This is why long wavelengths can only be reached by increasing the well thickness while keeping the In composition below 25% [127, 129]. The detrimental consequence is a strong collapse of the internal quantum efficiency because of the decrease of the oscillator strength [130].

As already mentioned, InGaN/GaN QWs are the heart of nitride based optoelectronic devices. One of the most puzzling features inherent to these heterostructures is their amazing capacity to emit light even though a huge density of dislocations is present. This has been explained by invoking carrier localization. Even if this explanation is widely admitted today, one must say that the question about the actual mechanism responsible for carrier localization is still an open debate. It is beyond the scope of this Chapter to thoroughly discuss and comment this topic but some key facts will be given hereafter.

Evidence for exciton localization and QD like behavior has been clearly demonstrated [108, 131–134]. Notice that the fingerprint of exciton localization is generally an S-shape behavior of the QW luminescence energy when plotted *versus* temperature [135]. However, the precise origin of the localization is not elucidated yet. The common picture until 2003 was to consider In rich clusters as being responsible for carrier trapping [105]. This analysis was based on cross-section TEM images showing strong In composition inhomogeneities. However, it was claimed later that the observation on TEM images of In rich clusters was an artefact resulting from e-beam irradiation [107]. On the other hand, Hangleiter and coworkers proposed in 2005 a novel explanation for the high-efficiency of InGaN/GaN QWs based on the self-screening of dislocations [106]. It relies on the formation of V-shaped defects surrounding threading dislocations in which the QW transition energy is larger than that on the basal planes acting thereby as a potential barrier for the diffusion of carriers toward dislocations. A similar mechanism was recently reported but without the need for V-shaped defects [109]. It was shown that the growth of InGaN/GaN QWs could lead to a built-in potential arising from a decrease in well thickness around the dislocations preventing the carriers to recombine on these nonradiative defects.

4 Conclusion

In this Chapter, it has been shown that to properly account for the optoelectronic properties of III-V nitride semiconductor heterostructures grown along the polar c-axis, a built-in electric field responsible for a large even a giant QCSE has to be considered. This built-in field has both spontaneous and piezoelectric origins. The former effect, namely the presence of an electric field in strain-free heterostructures, was a completely new phenomenon in the field of semiconductors when first highlighted [15]. The unambiguous experimental demonstration of such an electric field due to spontaneous polarization was first shown in GaN/AlGaN QWs [19]. The existence of such a field was further confirmed in strain-free heterostructures made of lattice-matched GaN/AlInN MQWs [16]. Besides, the large electric fields present in InGaN/GaN heterostructures, mainly from piezoelectric origin, allow obtaining light emitting devices operating in the visible even with limited indium contents [8]. However, in this latter case, this is achieved to the detriment of the internal quantum efficiency. Therefore the realization of efficient nitride-based light emitters in the green region of the visible spectrum grown along the polar axis remains a problem. The same limitation is encountered with InGaN laser diodes as the longest operating wavelength reported so far is not in the green range [136]. Two alternatives are currently explored to circumvent this problem: they consist in using either nonpolar (e.g. r-plane sapphire, free-standing m-plane GaN, or γ-$LiAlO_2$) or semipolar (e.g. {100} spinel, {110} spinel, {10-10} sapphire or (11–22) GaN) substrates for the growth in order to decrease or even to suppress completely the impact of polarization fields. These two approaches are still in their infancy as these research areas are currently rapidly evolving. Thus, the first demonstration of nonpolar LEDs dates back to 2000 [137] while nonpolar blue LEDs with an external quantum efficiency exceeding 1% were first reported at the beginning of 2006 [138]. Very recently, m-plane InGaN/GaN LEDs grown on m-plane GaN single crystals have reached an output power of 23.7 mW and an external quantum efficiency of 38.9% at a driving current of 20 mA [139]. One of the main advantages of such devices resides in the nearly wavelength-shift-free of the emission with increasing current densities due to the absence of screening effects [138, 139], a crucial aspect for demanding applications requiring a constant color rendering. Similarly, the concept of InGaN/GaN heterostructures grown along semipolar orientations was introduced a few years ago [140]. The novelty of this research area is exemplified by the fact that the exact orientation angle of semipolar planes is still debated [141]. Nonetheless, interesting perspectives for visible light emitters are present as testified by the recent demonstration of green InGaN/GaN LEDs grown on (11–22) GaN bulk substrates exhibiting an output power of 1.91 mW and an external quantum efficiency in excess of 4% at a driving current of 20 mA [142].

Finally, we point out that the presence of large built-in electric fields in polar nitride-based heterostructures might somehow reveal useful for some applications based on optical nonlinearities. More generally, the engineering of the internal electric field is likely to be fruitful for investigating novel physical phenomena and devices.

Acknowledgements The contribution of Drs. Amélie Dussaigne and Eric Feltin, two outstanding "growers" in their respective field, MBE and MOVPE, is gratefully acknowledged. This work has benefited from our daily interactions with our past and present colleagues from the Ecole Polytechnique Fédérale de Lausanne. N. Grandjean acknowledges his former co-workers, Drs. M. Leroux, B. Damilano and J. Massies, for their fruitful collaboration during his years of work at CRHEA-CNRS. We are also grateful to the Sandoz Family Foundation for its financial support and to Dr. B. Gil for a critical reading of the manuscript. Special thanks to Dr. P. Lefebvre for stimulating discussions all along this "nitride" story.

References

1. J. I. Pankove, J. E. Berkeyheiser, H. P. Maruska, and J. Wittke, Solid State Commun. **8**, 1051 (1970)
2. R. Dingle, D. D. Sell, S. E. Stokowski, and M. Ilegems, Phys. Rev. B **4**, 1211 (1971)
3. R. Dingle, W. Wiegmann, and C. H. Henry, Phys. Rev. Lett. **33**, 827 (1974)
4. R. C. Miller, D. A. Kleinman, W. A. Nordland, Jr., and A. C. Gossard, Phys. Rev. B **22**, 863 (1980)
5. H. Amano, M. Kito, K. Hiramatsu, and I. Akasaki, Jpn. J. Appl. Phys. **28**, L2112 (1989)
6. S. Nakamura, T. Mukai, and M. Senoh, Appl. Phys. Lett. **64**, 1687 (1994)
7. S. Nakamura, M. Senoh, S. Nagahama, N. Iwasa, T. Yamada, T. Matsushita, H. Kiyoku, and Y. Sugimoto, Jpn. J. Appl. Phys. **35**, L74 (1996)
8. S. Nakamura, Science **281**, 956 (1998)
9. G. Burns and A. M. Glazer: *Space Groups for Solid State Scientists*, 2nd edn. (Academic Press Inc., San Diego 1990)
10. I. Vurgaftman and J. R. Meyer, J. Appl. Phys. **94**, 3675 (2003)
11. F. Bernardini, V. Fiorentini, and D. Vanderbilt, Phys. Rev. B **63**, 193201 (2001)
12. M. A. Herman, W. Richter, and H. Sitter: *Epitaxy, Physical Principles and Technical Implementation* (Springer-Verlag, Berlin 2004)
13. P. Y. Yu and M. Cardona: *Fundamentals of Semiconductors, Physics and Materials Properties*, 2nd edn. (Springer-Verlag, Berlin 1999)
14. G. Bastard, *Wave Mechanics Applied to Semiconductor Heterostructures, Monographies de Physique* (les éditions de physique, Les Ulis 1988)
15. F. Bernardini, V. Fiorentini, and D. Vanderbilt, Phys. Rev. B **56**, R10024 (1997)
16. S. Nicolay, J.-F. Carlin, E. Feltin, R. Butté, M. Mosca, N. Grandjean, M. Ilegems, M. Tchernycheva, L. Nevou, and F. H. Julien, Appl. Phys. Lett. **87**, 111106 (2005)

17. S. Chichibu, T. Azuhata, T. Sota, and S. Nakamura, Appl. Phys. Lett. **69**, 4188 (1996)
18. J. S. Im, H. Kollmer, J. Off, A. Sohmer, F. Scholz, and A. Hangleiter, Phys. Rev. B **57**, R9435 (1998)
19. M. Leroux, N. Grandjean, M. Laügt, J. Massies, B. Gil, P. Lefebvre, and P. Bigenwald, Phys. Rev. B **58**, R13371 (1998)
20. J. F. Nye: *Physical Properties of Crystals, their Representation by Tensors and Matrices* (Oxford University Press, Oxford 1979)
21. R. D. King-Smith and D. Vanderbilt, Phys. Rev. B **47**, 1651 (1993)
22. D. Vanderbilt and R. D. King-Smith, Phys. Rev. B **48**, 4442 (1993)
23. R. Resta, Rev. Mod. Phys. **66**, 899 (1994)
24. R. Resta, M. Posternak, and A. Baldereschi, Phys. Rev. Lett. **70**, 1010 (1993)
25. F. Bernardini and V. Fiorentini, Phys. Rev. B **64**, 085207 (2001)
26. V. Fiorentini, F. Bernardini, and O. Ambacher, Appl. Phys. Lett. **80**, 1204 (2002)
27. M. Born and K. Huang: *Dynamical Theory of Crystal Lattices* (Oxford University Press, London 1968)
28. U. M. E. Christmas, A. D. Andreev, and D. A. Faux, J. Appl. Phys. **98**, 073522 (2005)
29. At the surface, the periodicity of the semiconductor crystal is broken. The resulting surface states, essentially dangling bonds and impurities adsorbed on the surface, create a continuum of localized states in the semiconductor bandgap and the corresponding surface charge is usually compensated by the creation of a macroscopic depletion layer. For a thick nitride layer, similarly the spontaneous polarization discontinuity across the surface must be compensated by a change in the electrostatic displacement. Consequently there will be a region of thickness l characterized by a two-dimensional electron gas whose surface density will equal the spontaneous polarization.
30. F. Bernardini and V. Fiorentini, Phys. Stat. Sol. (b) **216**, 391 (1999)
31. V. Fiorentini, F. Bernardini, F. Della Sella, A. di Carlo, and P. Lugli, Phys. Rev. B **60**, 8849 (1999)
32. N. Grandjean, B. Damilano, S. Dalmasso, M. Leroux, M. Laügt, and J. Massies, J. Appl. Phys. **86**, 3714 (1999)
33. F. Bernardini, V. Fiorentini, and D. Vanderbilt, Phys. Rev. Lett. **79**, 3958 (1997)
34. I. L. Guy, S. Muensit, and E. M. Goldys, Appl. Phys. Lett. **75**, 4133 (1999)
35. Paul Harrison: *Quantum Wells, Wires and Dots: Theoretical and Computational Physics of Semiconductor Nanostructures*, 2nd edn. (John Wiley & Sons Ltd, Chichester, New York 2005)
36. R. Zimmermann and D. Bimberg, Phys. Rev. B **47**, 15789 (1993)
37. J. M. Li, Y. W. Lü, D. B. Li, X. X. Han, Q. S. Zhu, X. L. Liu, and Z. G. Wang, J. Vac. Sci. Technol. B **22**, 2568 (2004)
38. S. Y. Lei, B. Shen, L. Cao, F. J. Xu, Z. J. Yang, K. Xu, and G. Y. Zhang, J. Appl. Phys. **99**, 074501 (2006)
39. M. Tchernycheva, L. Nevou, L. Doyennette, F. H. Julien, E. Warde, F. Guillot, E. Monroy, E. Bellet-Amalric, T. Remmele, and M. Albrecht, Phys. Rev. B **73**, 125347 (2006)
40. D. A. B. Miller, D. S. Chemla, T. C. Damen, A. C. Gossard, W. Wiegmann, T. H. Wood, and C. A. Burrus, Phys. Rev. Lett. **53**, 2173 (1984)
41. D. A. B. Miller, D. S. Chemla, T. C. Damen, A. C. Gossard, W. Wiegmann, T. H. Wood, and C. A. Burrus, Phys. Rev. B **32**, 1043 (1985)
42. D. L. Dexter, in *Solid State Physics*, edited by F. Seitz and D. Turnbull (Academic Press, New York 1958), Vol. 6, p. 353
43. J. Feldmann, G. Peter, E. O. Göbel, P. Dawson, K. Moore, C. Foxon, and R. J. Elliott, Phys. Rev. Lett. **59**, 2337 (1987)
44. E. Berkowicz, D. Gershoni, G. Bahir, E. Lakin, D. Shilo, E. Zolotoyabko, A. C. Abare, S. P. Denbaars, and L. A. Coldren, Phys. Rev. B **61**, 10994 (2000)
45. V. A. Fonoberov and A. A. Balandin, J. Appl. Phys. **94**, 7178 (2003)
46. T. Bretagnon, P. Lefebvre, P. Valvin, R. Bardoux, T. Guillet, T. Taliercio, B. Gil, N. Grandjean, F. Semond, B. Damilano, A. Dussaigne, and J. Massies, Phys. Rev. B **73**, 113304 (2006)

47. P. Lefebvre, A. Morel, M. Gallart, T. Taliercio, J. Allègre, B. Gil, H. Mathieu, B. Damilano, N. Grandjean, and J. Massies, Appl. Phys. Lett. **78**, 1252 (2001)
48. R. Cingolani, H. Botchkarev, H. Tang, H. Morkoç, G. Traetta, G. Coli, M. Lomascolo, A. di Carlo, F. Della Sella, and P. Lugli, Phys. Rev. B **61**, 2711 (2000)
49. P. Bigenwald, P. Lefebvre, T. Bretagnon, and B. Gil, Phys. Stat. Sol. (b) **216**, 371 (1999)
50. R. T. Senger and K. K. Bajaj, Phys. Rev. B **68**, 205314 (2003)
51. C. G. Van de Walle and J. Neugebauer, Appl. Phys. Lett. **70**, 2577 (1997)
52. C. G. Van de Walle and J. Neugebauer, Nature **423**, 626 (2003)
53. K. Muraki, S. Fukatsu, Y. Shiraki, and R. Ito, Appl. Phys. Lett. **61**, 557 (1992)
54. C. Kisielowski, Z. Liliental-Weber, and S. Nakamura, Jpn. J. Appl. Phys. Part 1 **36**, 6932 (1997)
55. N. Grandjean, J. Massies, S. Dalmasso, P. Vennéguès, L. Siozade, and L. Hirsch, Appl. Phys. Lett. **74**, 3616 (1999)
56. B. Daudin, F. Widmann, G. Feuillet, Y. Samson, M. Arlery, and J.-L. Rouvière, Phys. Rev. B **56**, R7069 (1997)
57. F. Widmann, J. Simon, B. Daudin, G. Feuillet, J.-L. Rouvière, N. T. Pelekanos, and G. Fishman, Phys. Rev. B **58**, R15989 (1998)
58. B. Damilano, N. Grandjean, F. Semond, J. Massies, and M. Leroux, Appl. Phys. Lett. **75**, 962 (1999)
59. K. Tachibana, T. Someya, Y. Arakawa, R. Werner, and A. Forchel, Appl. Phys. Lett. **75**, 2605 (1999)
60. S. Kako, C. Santori, K. Hoshino, S. Götzinger, Y. Yamamoto, and Y. Arakawa, Nat. Mater. **5**, 887 (2006)
61. H. Hirayama, S. Tanaka, P. Ramvall, and Y. Aoyagi, Appl. Phys. Lett. **72**, 1736 (1998)
62. K. Tachibana, T. Someya, and Y. Arakawa, Appl. Phys. Lett. **74**, 383 (1999)
63. B. Damilano, N. Grandjean, S. Dalmasso, and J. Massies, Appl. Phys. Lett. **75**, 3751 (1999)
64. F. Widmann, B. Daudin, G. Feuillet, Y. Samson, M. Arlery, and J.-L. Rouvière, MRS Internet J. Nitride Sem. Res. **2**, article *20* (1997)
65. B. Damilano, *PhD thesis*, University of Nice – Sophia Antipolis, France (2001)
66. A. D. Andreev and E. P. O'Reilly, Phys. Rev. B **62**, 15851 (2000)
67. V. Ranjan, G. Allan, C. Priester, and C. Delerue, Phys. Rev. B **68**, 115305 (2003)
68. D. P. Williams, A. D. Andreev, E. P. O'Reilly, and D. A. Faux, Phys. Rev. B **72**, 235318 (2005)
69. D. P. Williams, A. D. Andreev, and E. P. O'Reilly, Phys. Rev. B **73**, 241301(R) (2006)
70. P. W. Fry, I. E. Itskevich, D. J. Mowbray, M. S. Skolnick, J.J. Finley, J. A. Barker, E. P. O'Reilly, L. R. Wilson, I. A. Larkin, P. A. Maksym, M. Hopkinson, M. Al-Khafaji, J. P. R. David, A. G. Cullis, G. Hill, and J. C. Clark, Phys. Rev. Lett. **84**, 733 (2000)
71. D. M. Bruls, J. W. A. M. Vugs, P. M. Koenraad, H. W. M. Salemink, J. H. Wolter, M. Hopkinson, M. S. Skolnick, F. Long, and S. P. A. Gill, Appl. Phys. Lett. **81**, 1708 (2002)
72. C. Weisbuch and B. Vinter: *Quantum Semiconductor Structures, Fundamentals and Applications* (Academic Press, San Diego 1991)
73. *Handbook of Mathematical Functions with Formulas, Graphs, and Mathematical Tables*, edited by M. Abramowitz and I. A. Stegun (Dover Publications, Inc., New York 1965)
74. M. Arlery, J.-L. Rouvière, F. Widmann, B. Daudin, G. Feuillet, and H. Mariette, Appl. Phys. Lett. **74**, 3287 (1999)
75. N. Grandjean, J. Massies, and O. Tottereau, Phys. Rev. B **55**, 10189 (1997)
76. M. A. Khan, R. A. Skogman, J. M. Van Hove, S. Krishnankutty, and R. M. Kolbas, Appl. Phys. Lett. **56**, 1257 (1990)
77. A. Salvador, G. Liu, W. Kim, Ö. Aktas, A. Botchkarev, and H. Morkoç, Appl. Phys. Lett. **67**, 3322 (1995)
78. T. Takeuchi, S. Sota, M. Katsuragawa, M. Komori, H. Takeuchi, H. Amano, and I. Akasaki, Jpn. J. Appl. Phys. **36**, L382 (1997)
79. A. Reale, G. Massari, A. Di Carlo, P. Lugli, A. Vinattieri, D. Alderighi, M. Colocci, F. Semond, N. Grandjean, and J. Massies, J. Appl. Phys. **93**, 400 (2003)

80. P. Lefebvre, S. Kalliakos, T. Bretagnon, P. Valvin, T. Taliercio, B. Gil, N. Grandjean, and J. Massies, Phys. Rev. B **69**, 035307 (2004)
81. N. Grandjean, B. Damilano, S. Dalmasso, M. Leroux, M. Laügt, and J. Massies, Phys. Stat. Sol. (a) **176**, 219 (1999)
82. A. Bonfiglio, M. Lomascolo, G. Traetta, R. Cingolani, A. Di Carlo, F. Della Sala, P. Lugli, A. Botchkarev, and H. Morkoç, J. Appl. Phys. **87**, 2289 (2000)
83. F. Turco and J. Massies, Appl. Phys. Lett. **51**, 1989 (1987)
84. J. M. Moison, C. Guille, F. Houzay, F. Barthe, and M. Van Rompay, Phys. Rev. B **40**, 6149 (1989)
85. N. Grandjean, J. Massies, and M. Leroux, Phys. Rev. B **53**, 998 (1996)
86. A. Dussaigne, B. Damilano, N. Grandjean, and J. Massies, J. Cryst. Growth **251**, 471 (2003)
87. N. Grandjean and J. Massies, Appl. Phys. Lett. **72**, 1078 (1998)
88. R. Langer, J. Simon, V. Ortiz, N. T. Pelekanos, A. Barski, R. André, and M. Godlewski, Appl. Phys. Lett. **74**, 3827 (1999)
89. J. C. Harris, T. Someya, S. Kako, K. Hoshino, and Y. Arakawa, Appl. Phys. Lett. **77**, 1005 (2000)
90. A. Di Carlo, A. Reale, P. Lugli, G. Traetta, M. Lomascolo, A. Passaseo, R. Cingolani, A. Bonfiglio, M. Berti, E. Napolitani, M. Natali, S. K. Sinha, A. V. Drigo, A. Vinattieri, and M. Colocci, Phys. Rev. B **63**, 235305 (2001)
91. H. Tang, J. B. Webb, P. Sikora, S. Raymond, and J. A. Bardwell, J. Appl. Phys. **91**, 9685 (2002)
92. E. Kuokstis, W. H. Sun, C. Q. Chen, J. W. Yang, and M. Asif Khan, J. Appl. Phys. **97**, 103719 (2005)
93. J. Cai, F. A. Ponce, S. Tanaka, H. Omiya, and Y. Nakagawa, Phys. Stat. Sol. (a) **188**, 833 (2001)
94. M. Leroux, N. Grandjean, J. Massies, B. Gil, P. Lefebvre, and P. Bigenwald, Phys. Rev. B **60**, 1496 (1999)
95. J. L. Sánchez-Rojas, J. A. Garrido, and E. Muñoz, Phys. Rev. B **61**, 2773 (2000)
96. E. Feltin, D. Simeonov, J.-F. Carlin, R. Butté, and N. Grandjean, Appl. Phys. Lett. **90**, 021905 (2007)
97. M. V. Ramana Murty, P. Fini, G. B. Stephenson, C. Thompson, J. A. Eastman, A. Munkholm, O. Auciello, R Jothilingam, S. P. DenBaars, and J. S. Speck, Phys. Rev. B **62**, R10661 (2000)
98. N. Grandjean, J. Massies, and M. Leroux, Appl. Phys. Lett. **74**, 2361 (1999)
99. F. Natali, D. Byrne, M. Leroux, B. Damilano, F. Semond, A. Le Louarn, S. Vezian, N. Grandjean, and J. Massies, Phys. Rev. B **71**, 075311 (2005)
100. P. Lefebvre, J. Allègre, B. Gil, A. Kavokin, H. Mathieu, W. Kim, A. Salvador, A. Botchkarev, and H. Morkoç, Phys. Rev. B **57**, R9447 (1998)
101. P. Lefebvre, J. Allègre, B. Gil, H. Mathieu, N. Grandjean, M. Leroux, J. Massies, and P. Bigenwald, Phys. Rev. B **59**, 15363 (1999)
102. N. Grandjean, B. Damilano, J. Massies, G. Neu, M. Teissere, I. Grzegory, S. Porowski, M. Gallart, P. Lefebvre, B. Gil, and M. Albrecht, J. Appl. Phys. **88**, 183 (2000)
103. C. Weisbuch and J. Nagle, Science and Engineering of 1D and 0D Semiconductor Systems, NATO ASI Series, Plenum New-York, B214, 319 (1990)
104. J.-M. Gérard, O. Cabrol, and B. Sermage, Appl. Phys. Lett. **68**, 3123 (1996)
105. Y. Narukawa, Y. Kawakami, M. Funato, Shizuo Fujita, Shigeo Fujita, and S. Nakamura, Appl. Phys. Lett. **70**, 981 (1997)
106. A. Hangleiter, F. Hitzel, C. Netzel, D. Fuhrmann, U. Rossow, G. Ade, and P. Hinze, Phys. Rev. Lett. **95**, 127402 (2005)
107. T. M. Smeeton, M. J. Kappers, J. S. Barnard, M. E. Vickers, and C. J. Humphreys, Appl. Phys. Lett. **83**, 5419 (2003)
108. S. F. Chichibu, A. Uedono, T. Onuma, B. A. Haskell, A. Chakraborty, T. Koyama, P.T. Fini, S. Keller, S. P. Denbaars, J. S. Speck, U. K. Mishra, S. Nakamura, S. Yamaguchi, S. Kamiyama, H. Amano, I. Akasaki, H. J. Han, and T. Sota, Nat. Mater. **5**, 810 (2006)
109. S. Sonderegger, E. Feltin, M. Merano, A. Crottini, J.-F. Carlin, R. Sachot, B. Deveaud, N. Grandjean, and J. D. Ganière, Appl. Phys. Lett. **89**, 232109 (2006)

110. D. J. Eaglesham and M. Cerullo, Phys. Rev. Lett. **64**, 1943 (1990)
111. L. Goldstein, F. Glas, J.-Y. Marzin, M. N. Charasse, and G. Le Roux, Appl. Phys. Lett. **47**, 1099 (1985)
112. S. Tanaka, S. Iwai, and Y. Aoyagi, Appl. Phys. Lett. **69**, 4096 (1996)
113. F. Widmann, B. Daudin, G. Feuillet, Y. Samson, J.-L. Rouvière, and N. Pelekanos, J. Appl. Phys. **83**, 7618 (1998)
114. M. Miyamura, K. Tachibana, and Y. Arakawa, Appl. Phys. Lett. **80**, 3937 (2002)
115. K. Hoshino, S. Kako, and Y. Arakawa, Appl. Phys. Lett. **85**, 1262 (2004)
116. D. Simeonov, E. Feltin, J.-F. Carlin, R. Butté, M. Ilegems, and N. Grandjean, J. Appl. Phys. **99**, 083509 (2006)
117. S. Kako, K. Hoshino, S. Iwamoto, S. Ishida, and Y. Arakawa, Appl. Phys. Lett. **85**, 64 (2004)
118. A. Bykhovski, B. Gelmont, M. Shur, and M. A. Khan, J. Appl. Phys. **77**, 616 (1995)
119. A. D. Bykhovski, V. V. Kaminski, M. S. Shur, Q. C. Chen, and M. A. Khan, Appl. Phys. Lett. **68**, 818 (1996)
120. T. Takeuchi, C. Wetzel, S. Yamaguchi, H. Sakai, H. Amano, and I. Akasaki, Appl. Phys. Lett. **73**, 1691 (1998)
121. F. Chen, M. C. Cheung, P. M. Sweeney, W. D. Kirkey, M. Furis, and A. N. Cartwright, J. Appl. Phys. **93**, 4933 (2003)
122. H. Kollmer, Jin Seo Im, S. Heppel, J. Off, F. Scholz, and A. Hangleiter, Appl. Phys. Lett. **74**, 82 (1999)
123. F. Renner, P. Kiesel, G. H. Dohler, M. Kneissl, C. G. Van de Walle, and N. M. Johnson, Appl. Phys. Lett. **81**, 490 (2002)
124. I. H. Brown, I. A. Pope, P. M. Smowton, P. Blood, J. D. Thomson, W. W. Chow, D. P. Bour, and M. Kneissl, Appl. Phys. Lett. **86**, 131108 (2005)
125. A. Hangleiter, F. Hitzel, S. Lahmann, and U. Rossow, Appl. Phys. Lett. **83**, 1169 (2003)
126. N. Duxbury, U. Bangert, P. Dawson, E. J. Thrush, W. Van der Stricht, K. Jacobs, and I. Moerman, Appl. Phys. Lett. **76**, 1600 (2000)
127. M. Poschenrieder, F. Schulze, J. Blasing, A. Dadgar, A. Diez, J. Christen, and A. Krost, Appl. Phys. Lett. **81**, 1591 (2002)
128. M. J. Reed, N. A. El-Masry, C. A. Parker, J. C. Roberts, and S. M. Bedair, Appl. Phys. Lett. **77**, 4121 (2000)
129. D. Fuhrmann, C. Netzel, U. Rossow, A. Hangleiter, G. Ade, and P. Hinze, Appl. Phys. Lett. **88**, 071105 (2006)
130. S. Nakamura, Semicond. Sci. Technol. **14**, R27 (1999)
131. D. Hofstetter, J. Faist, and D. P. Bour, Appl. Phys. Lett. **76**, 1495 (2000)
132. I. L. Krestnikov, N. N. Ledentsov, A. Hoffmann, D. Bimberg. A. S. Sakharov, W. V. Lundin, A. F. Tsatsul'nikov, A. S. Usikov, Zh. I. Alferov, Yu. G. Musikhin, and D. Gerthsen, Phys. Rev. B **66**, 155310 (2002)
133. R. Seguin, S. Rodt, A. Strittmatter, L. Reißmann, T. Bartel, A. Hoffmann, D. Bimberg, E. Hahn, and D. Gerthsen, Appl. Phys. Lett. **84**, 4023 (2004)
134. H. Schömig, S. Halm, A. Forchel, G. Bacher, J. Off, and F. Scholz, Phys. Rev. Lett. **92**, 106802 (2004)
135. K. L. Teo, J. S. Colton, P. Y. Yua, E. R. Weber, M. F. Li, W. Liu, K. Uchida, H. Tokunaga, N. Akutsu, and K. Matsumoto, Appl. Phys. Lett. **73**, 1697 (1998)
136. S. Nagahama, T. Yanamoto, M. Sano, and T. Mukai, Phys. Stat. Sol. (a) **194**, 423 (2002)
137. P. Waltereit, O. Brandt, A. Trampert, H. T. Grahn, J. Menniger, M. Ramsteiner, M. Reiche, and K. H. Ploog, Nature **406**, 865 (2000)
138. A. Chakraborty, B. A. Haskell, H. Masui, S. Keller, J. S. Speck, S. P. DenBaars, S. Nakamura, and U. K. Mishra, Jpn. J. Appl. Phys. **45**, 739 (2006)
139. M. C. Schmidt, K.-C. Kim, H. Sato, N. Fellows, H. Masui, S. Nakamura, S. P. DenBaars, and J. S. Speck, Jpn. J. Appl. Phys. **46**, L126 (2007)
140. T. Takeuchi, H. Amano, and I. Akasaki, Jpn. J. Appl. Phys., Part 1, **39**, 413 (2000)
141. A. E. Romanov, T. J. Baker, S. Nakamura, and J. S. Speck, J. Appl. Phys. **100**, 023522 (2006)
142. M. Funato, M. Ueda, Y. Kawakami, Y. Narukawa, T. Kosugi, M. Takahashi, and T. Mukai, Jpn. J. Appl. Phys. **45**, L659 (2006)

Index

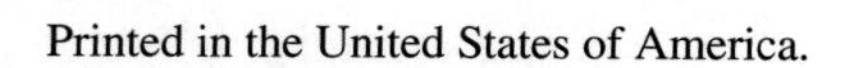

Printed in the United States of America.